Quantum Wave Mechanics

Quantum Wave Mechanics

Fourth Edition

Superata tellus sidera donat – Boethius, Consolatio Philosophia

"Overcome the earth and the stars shall be yours"

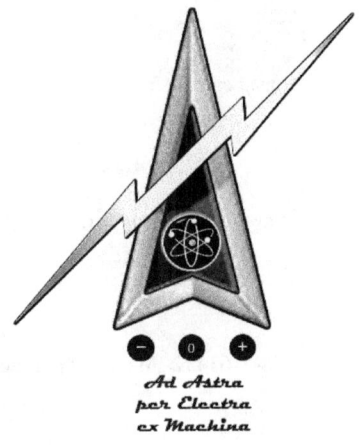

Larry J. Reed

To my parents who never knew the result of their great experiment

Copyright © 2019, 2020, 2022 by Larry J. Reed

All rights reserved.

No part of this publication may be reproduced, stored in a retrieval system, or transmitted in any form or by any means, electronic, mechanical, recording or otherwise, without the prior written permission of the author.

Printed on acid-free paper.

Library of Congress Control Number: 2018901065

ISBN: 978-1-63492-964-6   paperback

To order additional copies of this book, contact:

www.booklocker.com

CONTENTS

Preface ix

SECTION 1 – LIGHT

1. Photon model 1
2. Quantum vacuum 13
3. Electromagnetic 4-Potential 25
4. Soliton confinement 35
5. Electromagnetic field dimensions 37
6. Electromagnetic spectrum 45
7. Curvature and torsion 50
8. Polarization effects in a dielectric medium 53
9. Reflectors and dielectric lenses 60
10. Deflection of light 76
11. Origin of inertia 83
    11.1 Trapped waves in a standing wave resonator 83
    11.2 Confined light 83
    11.3 Contracted standing waves in motion 93
12. Electromagnetic wave propagation 95
13. Standing wave transformations 101
14. Phase-locked resonators with phase conjugate wave reflectors 111
15. Phase conjugate resonator experimental potential 116
16. Planck aether 132

SECTION 2 – ELECTRICITY

17. Electron model 148
18. Pair production and annihilation 208
19. Coulomb's law 225
    19.1 Electrostatics 225
    19.2 Generalization of Coulomb's law 226
20. Origin of the Electron fine structure constant α 228
    20.1 Background 228
    20.2 Electron charge-to-mass ratio 230
    20.3 Thomas precession 231
    20.4 Electron stability 235
21. Electric charge 239
    21.1 Dimensions of electric charge 239
    21.2 Electrical charge characteristics of elementary particles 247
    21.3 Relation of electric charge to topological charge 249
22. Complex numbers 258
23. Phasors 263
24. Quaternions 265

Contents

25. Spinors 267
26. Topological charge 269
27. Bound particle states 278
    27.1 Multiple electron and positron states 278
    27.2. Gluon fields 288
    27.3 Neutrinos 299
28. Lagrangian 305
29. Hamiltonian 307
30. Laplacian 309
    30.1 Wave function curvature 309
    30.2 Laplace equilibrium surfaces 310
    30.3 Laplace transforms 311
31. Tensors 315
32. Nonuniform force fields 318
    32.1 Dielectrophoretic force 318
    32.2 Van der Waals forces 321
    32.3 Magnetophoretic force 322
    32.4 Electromagnetic Magnus effect 323
    32.5 Ponderomotive (radiation pressure) force 323
    32.6 Traveling wave electromagnetic fields 323

SECTION 3 – GRAVITY

33. Gravitation 328
    33.1 Gravity of the matter 328
    33.2 Newton's law of gravitation 330
    33.3 Gravitational flux intensity 332
    33.4 Comparison of gravity and electricity 333
    33.5 Kepler's laws 345
    33.6 N-body gravitation 348
34. Gravitation as a harmonic phenomena 349
35. Gravitational frequency domain 357
36. Mass scaling 359
37. Generalized Newtonian gravitational law 361
38. Gravitational potential 363
39. Gravitational gamma 372
40. Newton's second law 374
41. Gravitational constant 378
    41.1 Newtonian gravitation constant G 378
    41.2 Gravitational G as a function of gamma 381
    41.3 Gravitational G as a function of permittivity 383
42. Newtonian gravitational force equation 388
    42.1 Gravitational force between electrons 388
    42.2 Relation between electrostatic force and gravitational force 394
    42.3 Gravitational force as a function of capacitance 395
    42.4 Force of gravity as a residual Coulomb force 396
    42.5 Imbalance force between dipoles 397
    42.6 Gravitational vector potential 399

Contents

43. Einstein field equation 401
44. Quantum gravity 407
    44.1   Introduction 407
    44.2   Quantum diagrams 408
45. Graviton model 413
46. Nonlinear gravitational field 419
47. Gravitational field of mass in motion 424
    47.1   Mass current 424
48. Wavefront (Moiré) interference patterns 426
    48.1   General 426
    48.2   Fresnel zones 426
49. Gravitational frequency redshift 445
    49.1   Photon frequency 445
    49.2   Gravitational time dilation 447
    49.3   Frequency shift differential 447
    49.4   Phase shift differential 448
50. Gravitational frequency spectrum 450
    50.1   Gravitational frequency range 450
    50.2   Fourier spectral analysis of Earth's gravitational spectrum 451
51. Coupled oscillators 457
    51.1   Oscillator synchronization 457
    51.2   Frequency arrhythmia 463
    51.3   Constant velocity (inertial frame) 463
    51.4   Constant acceleration (Rindler frame) 470
    51.5   Oscillator arrays 482
    51.6   Standing wave levitation and propulsion 483
52. Antigravity 502
    52.1   Alteration of gravitational potential 502
    52.2   Spectral energy density modulation 511
    52.3   Speculative design exercise 532
    52.4   Engineering the vacuum 592
53. Gravitation tonality 601
54. Visualization of dimensional relationships 623
    54.1   Ontological structure 623
    54.2   Dimensional conversions 624
    54.3   Graphical representations 663
    54.4   Creation of the universe 673

References 678

Index 685

Preface

This book attempts an explanation and geometrical description of a quantum field theory of light, electric charge, and gravity. Understanding the fundamental nature and interactions of such quantum fields is facilitated with knowledge of wave phenomena and physical properties of the vacuum that enable wave propagation. All light and matter is composed of quanta that share a fundamental characteristic in that they are composed of quanta that spin and with spin angular momentum of only certain discrete multiples of Planck's quantum constant h (integer spin bosons and half-interger spin fermions). Why does this manifest, highly localized, quantized spin wave effect occur and how does it result in concentrated energy in the form of matter? The estimated quantum mechanical energy content of the vacuum (10^{113} J/m^3) is vastly larger by ~122 orders of magnitude than the energy contained in the observable universe (10^{-9} J/m^3) composed of fermions and bosons. What accounts for this incredible mismatch in the minute fraction of energy in that we can directly perceive and experience (i.e., quanta with spin) and vacuum energy (fluctuations without spin) that which is inaccessible to observation? Just what are photons, electrons, and gravitons? How are they created from the seeming void of the vacuum, and how do they interact? Why is the speed limit of the universe set at a certain finite velocity of light? What exactly is electric charge? What is mass? How does mass interaction result in gravitational attraction? We seek an *explanation* for such phenomena, not just an ad hoc label description without visualization. Several interrelated themes are developed in terms of wave phenomena, energy density gradients, spin waves, and quantum effects in a physical vacuum. The subject matter and concepts discussed are necessarily speculative but are founded on known wave-mechanics principles. Major themes addressed include the following:

Light. A freely propagating photon wavetrain or light quanta in empty space is described as a helical traveling electromagnetic wave of quantized spin angular momentum moving at the velocity of light *semper et ubique*. Photons are classified as integer spin bosons. The physical vacuum as a polarizable medium enables wave propagation and appears ultimately to be quantized at the Planck scale. In the Winterberg Planck aether hypothesis, the vacuum is a Bose-Einstein condensate (BEC) superfluid composed of positive and negative Planck mass dipoles. Fundamental particles such as the photon and the electron are viewed as polarized quasi-particle wave excitations of much smaller Planck particles. Electromagnetic waves are conjectured to consist of spin density waves of Planck dipoles enabling formation of kink or antikink solitons. Similar to the exotic properties of supercooled $_3$He BEC liquid helium superfluid, spin waves are not tolerated by the vacuum but quickly become localized and isolated quantized vortices. A BEC condenstate represents a fifth state of matter in which particles collectively act in coherent waves oscillating in phase at the same frequency. The formation of electric charge q, magnetic vector potential **A**, electric field intensity **E**, and magnetic field intensity **H** in a vacuum devoid of matter may be understood in terms of the spatial and temporal density fluctuations and motion of Planck dipoles. The speed of light is a function of the Planck energy density of a polarizable vacuum characterized by the variable index of refraction K_{PV}.

Photons and electrons/positrons may be directly interconverted in high-energy processes of pair production and annihilation. Any viable model of the photon or electron must account for this interconvertability. Oscillation of electrons generates electromagnetic waves. Electrons can resonantly couple with electromagnetic waves. Photons and electrons can interact, for example, as plasmonic waves of free electrons in a metal surface, in secondary emission of electrons due to the photoelectric effect as in a photomultiplier, in an

Preface

ionized plasma as plasmons (quanta of electron waves), in a vacuum as in free-electron lasers, magnetrons, photomultipliers, etc., in photonic-excited condensed matter excitons, and in absorption of photons in photo-sensitive semiconductor P-N diodes or photon emission in laser diodes. Absorption of energy of a photon in a semiconductor can be transferred to an electron as potential energy. Photon emission occurs when the electron loses potential energy when electron-hole pairs recombine. P-N-P junctions of quantum dimensions can provide a storage medium for electrons as quantum mechanical standing wave traps. Photonic devices enable conversion of photons into an electron current and vice versa to generate an electrical signal or photo signal. Photon interaction with electrons can result in motion of matter as, for example, in particles suspended in an EM tractor beam by photophoretic forces in an optical trap.

In addition to frequency and spin, a photon traveling wave disturbance may be described in terms of curvature and torsion. The straight line motion of a photon in a gravity-free, zero-curvature vacuum reflects a balance in electric and magnetic energy. A change in torsion of a photon in an optically dense medium is associated with effective mass. The processes of electron/positron pair production and annihilation are described in terms of the geometry of a photon helicoid. During electron/positron pair creation, the increased curvature k and decreased torsion τ of a helical wave train due to Faraday rotation and Levi-Civita effects results in formation of two counter-rotating loops of opposite topological charge. Each loop contains two spinors corresponding to poloidal and toroidal rotation of a toroidal electron and positron spin wave. Electric charge is related to topological charge associated with precessional rotation and is quantized as a result of quantization of spin angular momentum described by Planck's constant $\hbar$ ($= 2\pi h$).

Mass. Mass is a fundamental, intrinsic property of matter attributed to the interaction of electromagnetic quantum fields, i.e., a wave interference effect. In the Einstein relation, mass $m = E/c^2$ where E is energy and c = the celerity of light. Energy is a measure of wavefront curvature. Mass is associated with retardation of energy flow and resultant time dilation. Wave energy packets are separated by nodes which obstruct energy flow restricting propagation. Mass corresponds to wave periodicity and is a measure of EM wave volumetric nodal density. Rest mass is observed only in fundamental particles with electric charge and is a ratio of charge to Compton angular frequency. Fundamental particles are viewed as standing wave resonant structures and not point-like objects. Travelling waves such as light and neutrinos acquire effective mass during propagation in regions of higher EM density. Standing waves acquire mass and inertia as a result of confinement of travelling waves as demonstrated in work by Jennison and Drinkwater. The self-referral dynamics of radiation trapped in a phase-locked cavity accounts for Newton's First Law of Motion, i.e., every object in a state of uniform motion tends to remain in motion unless an external force is applied to it. Hence, there is no need to attribute inertia to instantaneous interaction with the rest of the matter in the universe according to the Mach hypothesis. Mass and inertia are local phenomena. Mass may be understood as an interaction of electromagnetic fields resulting in accelerative wavefront curvature without recourse to the hypothetical, vaguely described Higgs field with unexplained mechanism for imparting mass to massless particles. The confinement of light, consisting of massless photons, in a fixed reference frame of a cavity resonator results in the creation of mass and inertia. At a sufficiently high energy level corresponding to the rest mass of the electron, the imbalance of electrostatic and magnetostatic fields results in topological confinement of a photon within a fixed volume of Compton radius R_C. Hence, fermions may be interpreted as spinning, phase-locked, topologically confined, standing wave resonant structures with electrons and positrons as the fundamental building blocks of matter.

Preface

In general, the motion of a push-pull phase-locked cavity resonator consists of an oscillatory sequenced series of accelerative jumps interspersed with coasting periods of constant velocity. In this respect, a cavity resonator is somewhat analogous to an inflated bouncing rubber ball alternately compressing and decompressing without internal dissipative losses. The rhythmic pulsation of a phase-locked resonator in motion generates longitudinal and transverse EM waves with frequency which varies with the cavity velocity. For matter (composed of resonant EM standing waves) in motion, the Lorentz contraction is interpreted as a physical wavelength compression due to variation in EM field energy density as measured by vacuum refractive index K_{PV}. A phase-locked resonator in motion exhibits an oscillatory, pulsing compression and expansion emitting dipole radiation transverse to the direction of motion. Interaction of these radiated waves with nearby electrons via the electromagnetic vector potential A^μ results in coupling of N number of electrons increasing their effective collective inertia as N^2.

The Lorentz transformations of motion in terms of velocity ratios compared with Ivanov-LaFreniere standing wave transformations in terms of standing wave ratios are shown to be equivalent. Ivanov and LaFreniere have shown that standing waves undergo wavelength (nodal) contraction in the direction of motion. An object in motion relative to a fixed observer undergoes a Lorentz contraction (wavelength compression) in the direction of motion and a Lorentz Doppler shift in frequency (reduction). The wavelength compression is a physical result of an increase in the vacuum energy density. Moving clocks which are made of standing matter waves undergo time dilation as a result. This is in keeping with de Broglie and Schrödinger's view that matter waves are real physical waves and not merely particle location probability amplitudes described in the Born interpretation. The EM wavelength contraction and frequency shift in a polarizable vacuum accounts for mass in motion and gravitational effects, including the energy change, deflection of light, gravitational frequency shift, and clock slowing. The speed of light c appears invariant in all inertial frames due to Lorentz contraction of the measurement apparatus and a concomitant Lorentz Doppler frequency shift. Spacetime remains Euclidean over scales comparable to wavelength. The apparent Lorentz space contraction and time dilation are the result of contraction of the nodal distance of the standing wave(s) which constitute the length of measurement. Time dilation is equivalent to a change in the size of the units of measurement which are undetectable to an observer as both the object and the comoving measurement apparatus undergo Lorentz transformation.

Fundamental particles of matter exhibit properties of standing EM waves trapped in a phase-locked resonator including Doppler frequency shifts in motion, inertia (resistance to motion) and de Broglie waves. Matter in motion relative to an observer exhibits de Broglie 'matter' waves as a modulated moving standing wave. The inverse effect of self-induced motion of matter may potentially be realized utilizing synthesized red- and blue-shifted Lorentz Doppler waves parametrically amplified in a phase conjugate phase-locked resonator. Energy of motion results from conversion of energy of the pump waves to the contracted moving standing wave formed from the signal wave and its counterpropagating phase conjugate wave within the resonator. Velocity of the resonator wave system is proportional to the wave phase difference while acceleration is proportional to the frequency difference. Synthesized matter waves would provide means for inertia modification and control as well as self-induced motion of matter. Such technology would enable EM wave-based propulsion without wheels, friction, reaction or expulsion mass. Inverse effects are not without precedent as, for example, inverse Doppler effect, inverse Sagnac effect, inverse Faraday effect, inverse Compton effect, inverse spin Hall effect, inverse Cherenkov effect, inverse Raman effect, inverse Cotton-Mouton effect, inverse Barnett effect (Einstein de Haas effect) and inverse piezoelectric effect, etc.

Preface

Electric charge. Traditionally, electric charge has been opaquely described as a separate dimension without geometrical description or explanation of its origin. In this book, a description of electric charge is detailed relating it to dimensions of mass, rotation rate and time which is interpreted as a rate of precession of closed loop standing waves and described by the fine structure constant. Spin momenta is associated with loop closure failure defects or dislocations in spacetime and resultant torsion stresses. The electron is described as a helical toroid standing wave formed from an energetic photon travelling wave with a full twist looped into a circle of a radius equal to the Compton wavelength. The photon helicoid may be envisioned as a twisted ribbon spinning around its longitudinal axis. The electron toroid geometry may be described in a twisted ribbon analogy as a spinning closed-loop Hopf strip – the simplest form of topological knot. The torus geometry is formed by a rotating charge path in the shape of a Hopf link with toroidal and poloidal components. The ½-spin characteristic of the electron arises as a result of a toroidal spin component of Compton frequency ω_C and a poloidal spin component of Zitterbewegung frequency equal to $2\omega_C$. The imbalance of the electrostatic and magnetostatic energy gives rise to the fine structure constant α. The charge-to-mass e/m ratio corresponds to a precession frequency equal to $\omega_{e/m}$. The whirl number is found equal to the inverse fine structure constant α^{-1}. Electric charge has mechanical dimensions of MLT^{-1} and represents an angular precession of ~1/137 ($\simeq 0.007$) radians/sec. Spin angular precession creates wave function interference obstructing energy flow resulting in electron mass. Charge represents an angular torsion deficit angle and is correlated with a mass deficit angle.

Gravity. All matter is composed of quantum oscillators emitting electromagnetic waves over the entire EM frequency spectrum. Gravity is viewed as a standing wave interaction between coupled oscillators interacting via EM traveling waves (e.g., bosons). Inertia and mass are the result of standing electromagnetic waves generated by an isolated oscillator within a phase-locked resonator (e.g., fermions). Electromagnetic resonant wave interactions in a polarizable vacuum (PV) model in Euclidean space exhibit geometrical spacetime curvature consonant with Einstein SR/GR. Acceleration of gravity is the result of a spectral energy density gradient and corresponds to the rate of change of rapidity. Inertial mass of matter in motion and gravitational mass of matter in a gravitational field are equivalent as both arise from acceleration into regions of increased EM flux energy density and nonlinear frequency dependent alterations in vacuum dielectric constant.

Gravitation in the Einstein General Theory of Relativity (GR) is ascribed to a curvature of space and time in an abstract mathematical representation. However, the GR theory metaphysical description of gravitation does not describe the physical mechanism for how matter induces curvature or how spacetime curvature influences motion of matter. It is argued that space and time are not physical objects but are merely the mathematical ordering of location of points in space and events in time. Time represents the flow of energy. In an optical theory of gravity, the deflection of light in a gravitational field is the result of variation in the vacuum refractive index K_{PV} which is a measure of the electromagnetic field energy density. Gravitation is equivalent to a dielectric gradient force in a polarizable vacuum as a result of local Fresnel zone variation in electromagnetic (EM) flux density and vacuum refractive index K_{PV}. The gravitational force $\mathbf{F}_g$ is proportional to the gradient of K_{PV}^2 ($\mathbf{F}_g \propto m\nabla K_{PV}^2$). Interference of electromagnetic waves from coupled oscillators produce Moiré patterns and Fresnel zones. EM wave front interference creates a Fresnel zone effect between coupled mass source oscillators concentrating the local flux density and increasing the electric permittivity gradient. Gravitons are illustrated as wave interference of counterpropagating phase conjugate photons reflected from Fresnel zone boundaries. Wavefront curvature provides an accelerative force indistinguishable from

Preface

gravity. The observed contraction of wavelength nodal distances is responsible for the perceived Lorentz spatial contraction and time dilation effects. The metric of curved spacetime corresponds to the wave front interference node metric. Hence, the gravitational field becomes quantized and spacetime remains Euclidean. The quantized gravitation field may be understood as purely an electromagnetic phenomena. As such, gravitational fields may be subject to modification by alteration of the local electromagnetic field density to check the propensity to fall or neutralize weight. Ability to effect at will modification of the gravitational spectral energy density gradient and concomitant neutralization of the local gravitational frequency shift differential will mark a significant technological achievement and prove a benchmark of human intellect.

A conundrum of modern physics is the apparent incompatibility between quantum mechanics and general relativity, each of which have had considerable success in describing aspects of the physical universe. It is asserted that the mathematical construct of spacetime curvature as represented in Einstein's GR applies not to spacetime itself but rather to wavefront curvature and nodal contraction of electromagnetic waves in spacetime. Based on investigations by Michael Faraday, the existence of electromagnetic waves theorized by James Clerk Maxwell was experimentally demonstrated by Heinrich Hertz, Nikola Tesla and others. The exact nature of just what is doing the 'waving' in electromagnetic waves has remained a mystery. What constitutes Faraday's invisible electric and magnetic field lines of force? What Planck scale vacuum elements are in contact to enable transmission of force? What accounts for the apparent tension and pressure? The vacuum is calculated to have enormous energy density and is characterized by quantum oscillators with zero point energy. What is the nature of such oscillators that support propagation of electromagnetic and gravitational wave disturbances? In this book, *Quantum Wave Mechanics*, electromagnetic fields and waves are conjectured to be disturbances in the physical vacuum composed of rotating quantized Planck dipoles consisting of coupled positive and negative Planck mass with net zero mass and angular momenta. Electric and magnetic field lines are manifestations of temporal and spatial spin alignments of groupings of adjacent Planck dipoles. The Planck vacuum is represented as an exceedingly dense energetic medium composed of Planck mass dipoles with characteristic Planck impedance and under symmetry breaking, easily induced to spin. Bosonic and fermionic fields correspond to resonant spin wave interactions between such dipoles. Bosons represent traveling waves while fermions represent standing wave structures both of which are electromagnetic. The spacetime metric is a mathematical overlay description of electromagnetic interactions characterizing the relative positioning of objects in space and ordering of events in time. The quantum vacuum may be understood as a foam of postive and negative curvature and mass. Planck masses are the underlying quintessence or "stuff" of the vacuum and are thought to arise spontaneously as a result of vacuum instability from a state of nothingness. Dark energy represents quantum wave function dissonance driving universe expansion.

Concept and inspiration is the aegis of design and invention. Knowledge and understanding of quantum wave mechanics effects in the physical vacuum is a *conditio sine qua non* for future theoretical development of funamental physics and may lead to new technological developments such as wave-based propulsion, enhanced energy conversion, vacuum engineering, programmable quantum dot nanostructures or artificial atoms and force field effects. The relation and interconversion of fundamental and derived dimensions of physical quantities and geometrical interpretations illustrated herein are intended to relate previous discoveries and provide new sources of insight and ideas as to the nature of physical reality and the universe in which we live.

Section 1 - Light

1. Photon Model

There is no greater mystery to me than that of light travelling through darkness. – Alexander Volkov

I happen to have discovered a direct relation between magnetism and light, also electricity and light, and the field it opens is so large and I think rich. – Michael Faraday

We can scarcely avoid the inference that light consists in the transverse undulations of the same medium which is the cause of electric and magnetic phenomena. – James Clerk Maxwell

For the rest of my life, I will reflect on what light is. – Albert Einstein

Light consists of photons, the quanta of electromagnetic fields. A freely-propagating photon in empty space (gravity-free, zero curvature vacuum) is represented as a helical, self-sustaining, traveling electromagnetic wavepacket of quantized spin angular momentum moving at the velocity of light. An electromagnetic (EM) wavefront consists of a multitude of in-phase photon torsional wave trains each of an ordered sequence of photon wave packets. The photon is categorized as a stable, massless boson having no electrical charge with spin angular momentum $s = \pm \hbar$. The photon is right-circularly polarized if $s_z = \hbar$ and left-circularly polarized if $s_z = -\hbar$. In the ansatz model described, the photon is posited to consist of a spin density wave disturbance in a quantum vacuum composed of positive and negative Planck mass dipoles. The spin field forms a kink or antikink soliton defined in terms of the spin direction. Like a soliton, a photon may be regarded as a local confinement of the energy of a wave field with particle-like behavior and may propagate without dissipation or change of form. The spin axis **s** is aligned with the direction of propagation vector **k** in either the forward or the backward direction depending on helicity. An individual photon may be either right or left circularly polarized represented as polar opposites on a Poincaré sphere. Linear polarization requires a superposition of an even number of photons of opposite spin. In a coherent wave, wave packets overlap and are in the same direction, in-phase and of the same frequency.

A conceptual helicoid model of the photon as a quantized wave packet of a single wavelength is illustrated in Fig. 1-1. A helicoid describes a minimal surface generated by a space curve which simultaneously rotates about an axis and translates parallel to that axis such that the ratio of the rotational velocity to the translational velocity is constant. A minimal surface has positive curvature in one principal direction and negative curvature in the orthogonal direction with zero mean curvature. See Fig. 1-2. Photon models consisting of transverse **EM** plane wavefronts composed of rotating orthogonal **EHV** dreibein (triad) have been proposed by Hunter et al [1] and Funaro [2, 3, 4] and expressed in tetrad formalism by Evans [5, 6]. The **EM** potential is defined by the dreibein A_μ^a where superscript a = 1, 2, 3 represents the tangent space and subscript μ the base manifold. The internal index **e** in O(3) electromagnetics, represents an orthonormal tangential space defined by unit vectors e^1, e^2, e^3 (complex circle basis) to the base manifold. The rotating **EHV** dreibein describes a helical geodesic propagation path (circular helix, constant curvature, constant torsion) which may be represented as a space curve in terms of the Frenet-Serret equations. A unit speed curve corresponds to arc length s. Curvature (departure from linearity) corresponds to the change in direction of the tangent vector **t** per unit arc length while the torsion (departure from planarity) corresponds to the change in direction of the binormal vector **b** (= **t** x **n**) which is orthonormal to the osculating plane defined by the curvature (radial) vector **n** and tangential vector **t**. The charge path trajectory describes a geodesic on a cylinder. Zero axial velocity (zero torsion) corresponds to a circle. Infinite velocity (infinite

1. Photon Model

torsion) corresponds to a line. In geometric terms, motion of an inextensible curve of constant torsion (= 1/r) and constant curvature k = ω_u associated with the sine-Gordon equation ($\omega_{uv} = 1/\rho^2 \sin\omega$) traces out a single-soliton Bianchi surface as it moves where at each instant will be an asymptotic line on the surface. The sine-Gordon equation in this representation includes one spatial and one temporal independent variable. The spatial motion of curves of constant torsion and curvature such as the Sine-Gordon curve are geometrically linked with soliton theory of nondispersive, solitary waves such as described by the Korteweg–de Vries (KdV) nonlinear wave equation.

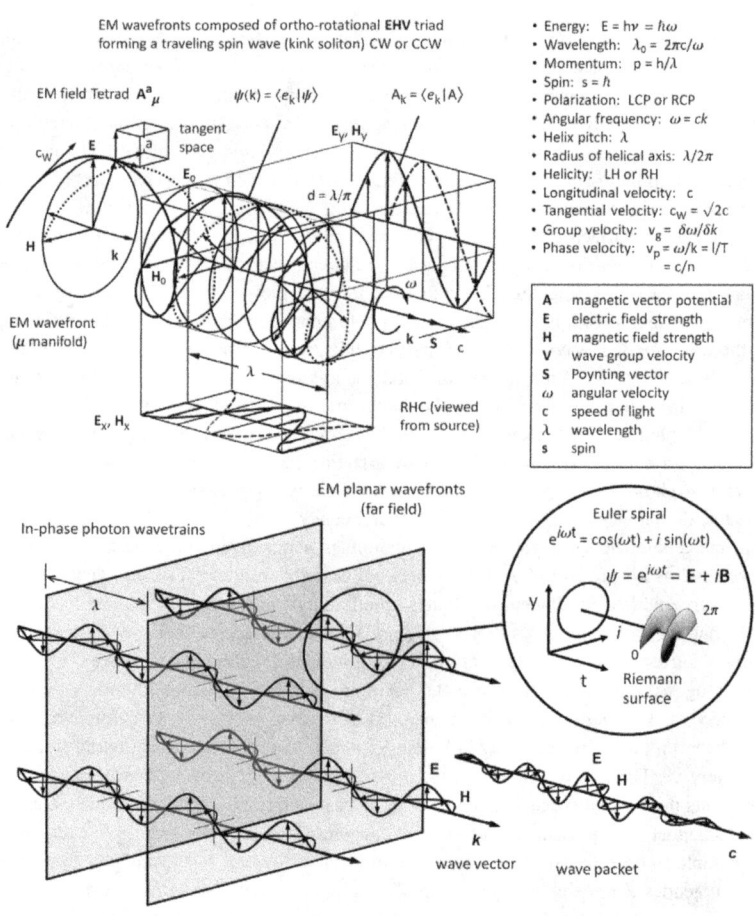

Fig. 1-1. Illustration of a monochromatic single-wave length photon represented as a torsional wave and a EM wavefront composed of photon wavetrains remote from source quantum harmonic oscillator. Photons are elementary excitations of the normal modes of the electromagnetic field with quantized energy $\hbar\omega_\lambda$ and represent quantization of Maxwell's equations. Energy flow is in the direction of the Poynting vector **S** (= **E** x **H**).

1. Photon Model

Photon as a Rankine Dual Vortex

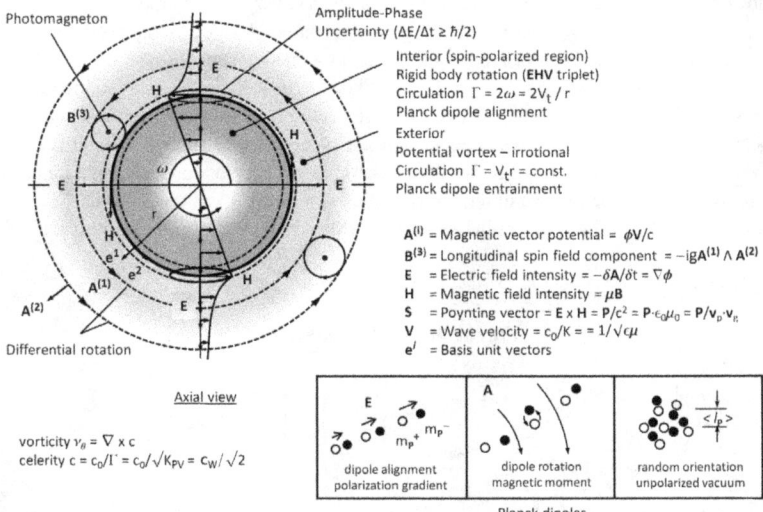

Photon Model Geometry in a Zero Curvature Vacuum

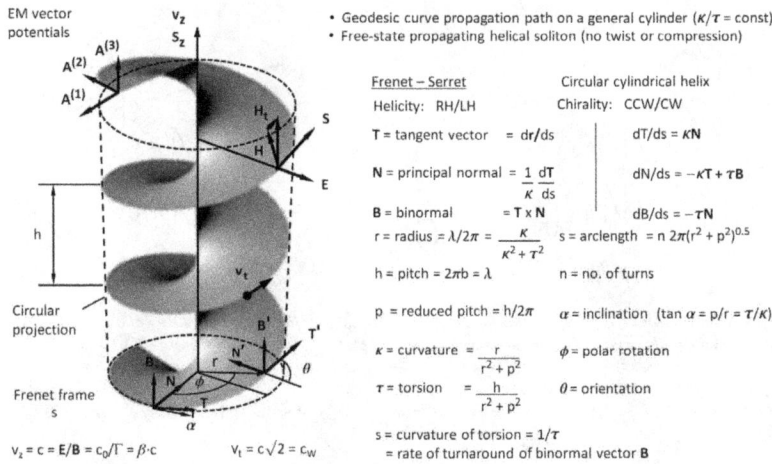

Fig. 1-2. Photon model detail illustrating electric and magnetic field component relationship to helical curvature and torsion. The motion of the Frenet frame along the unit speed curve is described by the Frenet-Serret formulas. The contravariant vector **T** is tangent to the parametized curve. Right-hand (RH) circular polarization is shown.

Ordinary photons do have spin, they have a notion of helicity so they spin around their direction of motion. – Roger Penrose

Electrons behave in exactly the same way as photons; they are both screwy. – Richard Feynman

3

1. Photon Model

A wave group consists of consecutive energy packets separated by nodes. Energy trapped between consecutive nodes cannot escape, hence, wave energy travels at the group velocity v_g (= dx/dt = dω/dk) in the direction of propagation denoted by the wave vector k. In a vacuum, the wave vector is given by $k = \omega/c$. In a material medium, the wave vector $k = \omega n/c = k_0/\sqrt{(1-\beta^2)}$ where n is the refractive index. An increase in the refractive index lowers the phase velocity. The refractive index of a zero curvature vacuum equals one. For a metamaterial with refractive index equal zero, the phase velocity and wavelength become infinite. An increase in the energy density of the medium acts to slow the group velocity compressing the wavetrain resulting in frequency up chirp. Conversely, a decrease in energy density of the medium results in frequency down chirp. This effect is illustrated in Fig. 1-3.

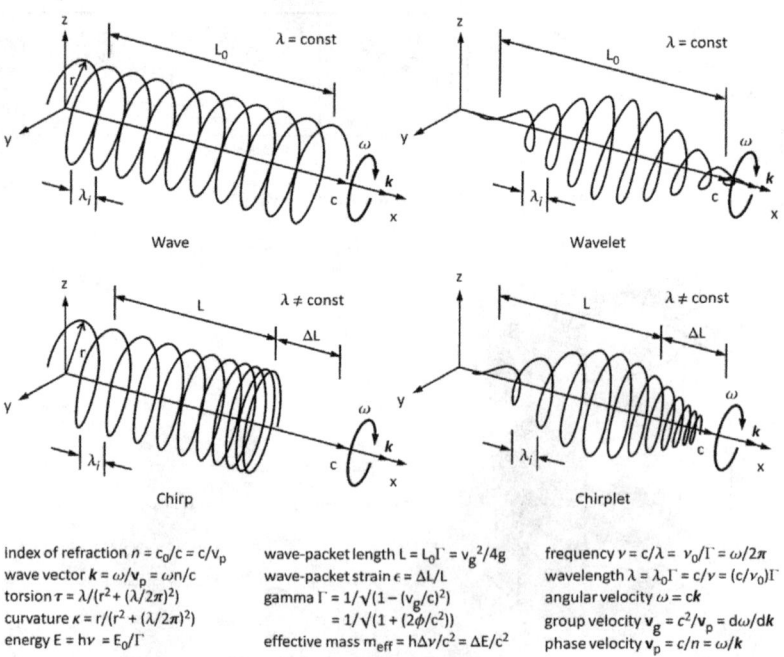

index of refraction $n = c_0/c = c/v_p$	wave-packet length $L = L_0\Gamma = v_g^2/4g$	frequency $\nu = c/\lambda = \nu_0/\Gamma = \omega/2\pi$
wave vector $k = \omega/v_p = \omega n/c$	wave-packet strain $\epsilon = \Delta L/L$	wavelength $\lambda = \lambda_0\Gamma = c/\nu = (c/\nu_0)\Gamma$
torsion $\tau = \lambda/(r^2 + (\lambda/2\pi)^2)$	gamma $\Gamma = 1/\sqrt{(1-(v_g/c)^2)}$	angular velocity $\omega = ck$
curvature $\kappa = r/(r^2 + (\lambda/2\pi)^2)$	$= 1/\sqrt{(1+(2\phi/c^2))}$	group velocity $v_g = c^2/v_p = d\omega/dk$
energy $E = h\nu = E_0/\Gamma$	effective mass $m_{eff} = h\Delta\nu/c^2 = \Delta E/c^2$	phase velocity $v_p = c/n = \omega/k$

Fig. 1-3. Wave pulse compression results in high frequency chirp. Increase in frequency is produced as a result of increase in refractive index n (= $c_0/c = c/v_p$) of the propagation medium which is a measure of EM energy density. Up chirp corresponds to an increase in frequency and torsion while down chirp corresponds to decrease in frequency and torsion. A complex-valued EM wavefunction ψ (= $\mathbf{E} + i\mathbf{B}$) $\mapsto e^{i\theta}\psi$ can be represented as a complex-valued wavelet time-frequency transform. Wavelet compression (i.e., change in scale factor α) corresponds to a chirplet transform commonly used in signal processing.

I therefore take the liberty of proposing...the name photon. – Gilbert N. Lewis

1. Photon Model

Near an oscillating dipole source in the near field evanescent region of the radiation pattern, the electric **E** and magnetic **H** fields are out-of-phase and the Poynting vector **S** (= **E** x **H**) has transverse components. This can result in unusual effects exploited in plasmonic devices, metamaterial magnetic mirrors and the like. The field intensities in the near field fall more steeply as $1/r^3$ than $1/r^2$ as in the far field region. Electromagnetic field propagation from an oscillating dipole antenna illustrating amplitude and phase change in electric **E**-field and magnetic **H**-field with distance from source generator is illustrated in Fig. 1-4. Wave impedance is reactive close to the antenna with transverse and longitudinal E-field components in the near-field. In the far-field, the transverse E-field and H-field are perpendicular to the direction of wave propagation in phase quadrature with the wavefront approximating a plane wave. The transition to a radiative field is marked by detachment of flux lines from the antenna and self-closure.

Emission of energy from a harmonic oscillator occurs during the first quarter period of oscillation and temporarily stored in the reactive field. Energy is radiated or reabsorbed by the dipole antenna during the next quarter period. The magnetic field appears initially to propagate at infinite velocity before slowing down to the speed of light within about a quarter wavelength. The electromagnetic field energy $E = \frac{1}{2}[\mathbf{E}^2 - c^2\mathbf{B}^2]$. In the near field, a dipole antenna field energy is primarily electric while a loop antenna field energy is largely magnetic. The average energy velocity of propagation equals the speed of light. The free-space impedance Z_0 (radiation resistance) of the vacuum to EM wave propagation is illustrated in Fig. 1-5 and depicts the mode transition from near- to far-field. Far-field impedance $Z_0 = 120\pi\ \Omega \simeq 376.7\ \Omega$. The Planck impedance Z_P is shown for comparison. In circa 350 BC, Aristotle declared the famous dictum *horror vacui* (Nature abhors a vacuum) based on the deduction that in a complete vacuum infinite speed would be possible because motion would encounter no resistance. Hence, if infinite speed was impossible, so to is a complete vacuum. The finite velocity of light in vacuo is a measure of impedance of wave energy flow ($c = 1/Z_P 4\pi\epsilon_0$). A complete vacuum devoid of a Planckian substructure as hypothesized would not support propagation of a photonic spin wave disturbance.

Light itself is a revelation. – James Turrell

If I pursue a beam of light with velocity c, I should observe such a beam of light as an electromagnetic field at rest though spatially oscillating. – Albert Einstein

Nothing is more practical than a good theory. – Ludwig Boltzmann

The ether undulates athwart the path of the wave's advance. – Robert Gascoyne-Cecil

But without a medium how can the propagation of light be explained? – Albert Michelson

The vacuum was characterized by Dirac as a state with an infinite number of zero energy and zero momentum quanta. Photon emission was considered as a transition from this vacuum state to a state of a single photon with finite momentum and energy; photon absorption consisted of the reversed transition. Schrödinger, on the other hand, in his formulation of quantum mechanics, conceived the idea of a "wave function" ψ as representing some kind of wave, and he interpreted the squared modulus $|\psi|^2$ as the density of electronic matter. For him, following de Broglie, waves were the fundamental entities. – Silvan S. Schweber

The same entity, light, was at once a wave and a particle. How could one possibly imagine its proper size and shape? To produce interference it must be spread out, but to bounce off electrons it must be minutely localized. This was a fundamental dilemma, and the stalemate in the wave-photon battle meant that it must remain an enigma to trouble the soul of every true physicist. It was intolerable that light should be two such contradictory things... – Banesh Hoffman

1. Photon Model

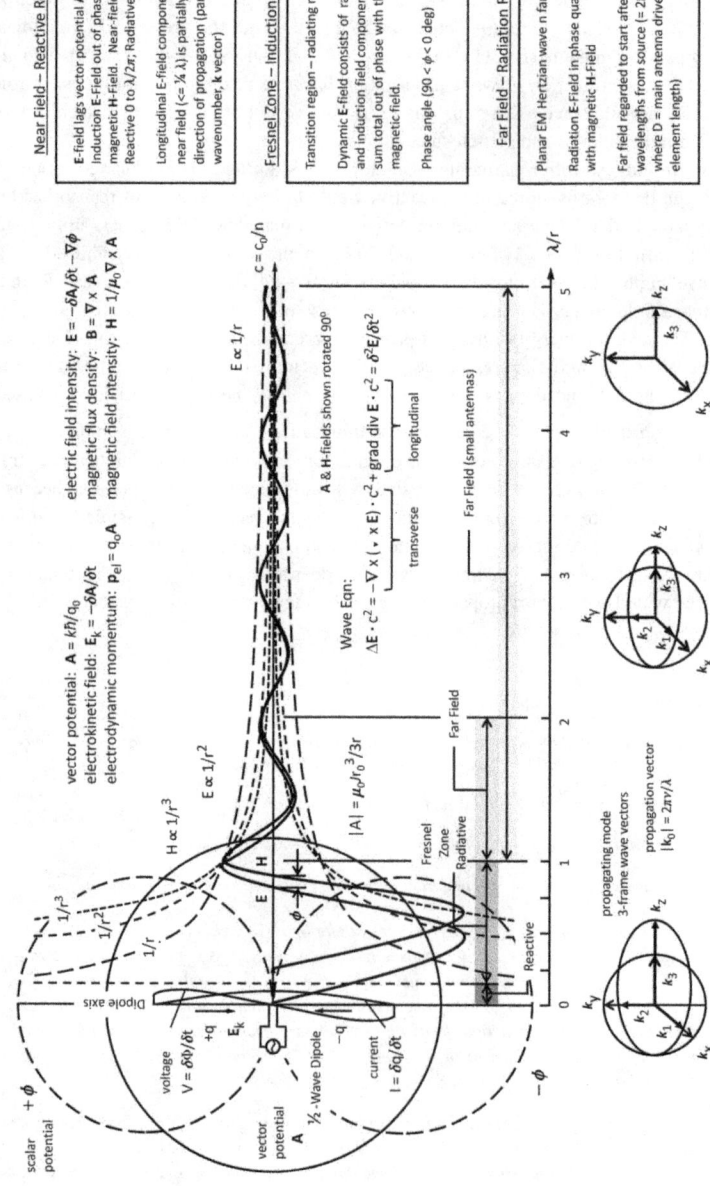

Fig. 1-4. Oscillating dipole electromagnetic wave propagation (½ wave dipole antenna at rest relative to an observer).

1. Photon Model

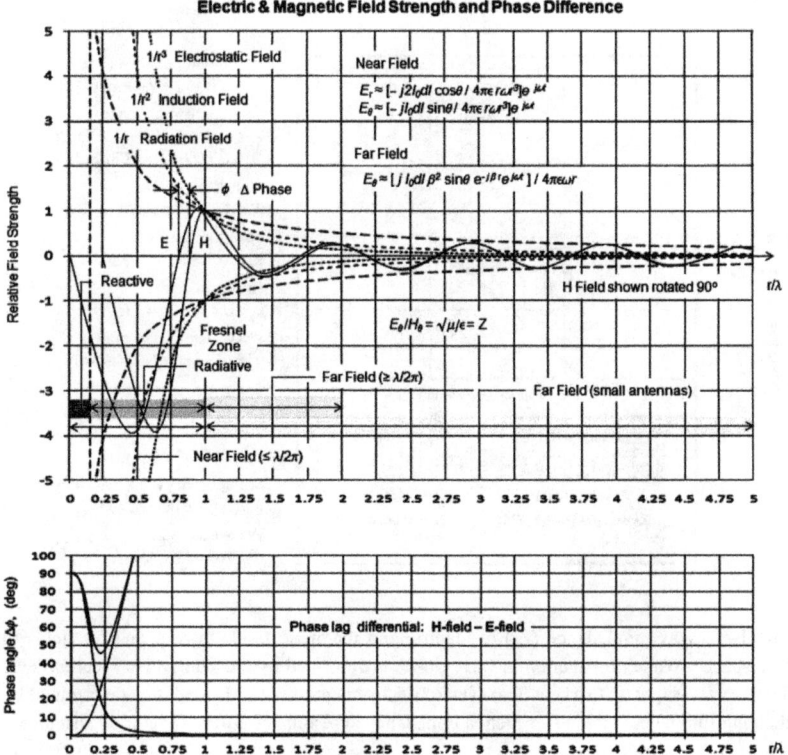

Fig. 1-5. **E** and **H** field phase lag is illustrated as a function of distance from the emitter. The phase difference falls to zero in the far field closed-loop region.

As shown in Fig. 1-6, the wave impedance ($Z_0 = |\mathbf{E}|/|\mathbf{H}|$) of electric and magnetic fields in the near field vary with frequency. The velocity of propagation becomes independent of frequency for a wave frequency much greater than the cutoff frequency ($\nu \gg \nu_c$). In the transverse electromagnetic mode (TEM), the free wavelength is much greater than the cutoff wavelength ($\lambda \gg \lambda_c$) and, in a dispersionless medium, the phase velocity equals the group velocity ($v_p = v_g$) for plane waves. In a waveguide at the cutoff frequency, the phase velocity is infinite and the group velocity is zero. The cutoff wavelength of the vacuum is taken as the Planck wavelength ($l_P = \sqrt{c^5/\hbar G/c^3} = 1.616 \times 10^{-35}$ m). For comparison, the impedance of an orbital electron in the 1st orbital of a hydrogen atom is given as $Z_0/\alpha = 51,649\ \Omega$ where α = the electron fine structure constant ($\alpha = Z_0/Z_e = v_e/c$) which reflects the ratio of the tangential speed of the electron velocity to the velocity of light. The Planck impedance is given by $Z_P = V_P/I_P = \hbar/\theta_P^2 = 1/4\pi\epsilon_0 c$ and is a quantized expression of the opposition to a change in flow of Planck charge q_P (= $e/\sqrt{\alpha}$). The Z_0/Z_P ratio (= 4π) implies a corresponding ratio of Planck mass tangential velocity to the velocity of light.

The speed of light is a constant of measure and not a constant of nature. – Konstantin Meyl

1. Photon Model

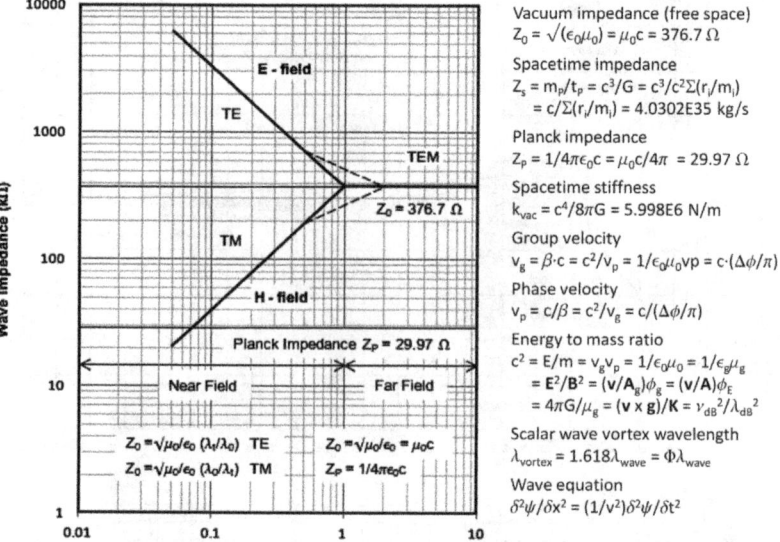

Fig. 1-6. Wave impedance (Z_0) of electric and magnetic fields in near and far fields in free-space. Wave propagation in waveguides is described by the Transverse Electric (TE), Transverse Magnetic (TM) or Transverse Electromagnetic (TEM) modes as determined by the longitudinal component. Antenna impedance at resonance equals radiation resistance.

Illustration of variation in wave vector (**k**) direction in relation to energy flow direction represented by the Poynting vector (**S**) in different types of media is shown in Fig. 1-7. In an ordinary wave, vectors **k** and **S** are aligned whereas in an extraordinary ray, the **k** and **S** vectors are not aligned. The group velocity v_g (the velocity of the wave group modulation envelope) is in the direction of propagation. The phase velocity v_p (the velocity of a point on the smaller constituent waves, e.g., zero-crossing or wave crest) is in the direction of momentum transfer. A travelling wave has a standing wave ratio SWR = 1 (no reflection) while a standing wave has an SWR > 1. In a material medium, the phase velocity v_p (= $1/\sqrt{\epsilon_0\mu_0}$) is a function of permittivity and permeability. Man-made metamaterials, such as split-ring resonators (SSRs), complimentary split-ring resonators (SCRRs), exhibit simultaneous negative permeability ϵ and permeability μ. Left-hand materials (LHMs) or negative index materials (NIMs) exhibit negative index of refraction resulting in opposed phase and group velocities. For LHMs, EM waves propagate towards the source, opposite to the direction of energy flow. A similar reverse effect occurs as well for Cherenkov radiation such as emitted when an electron passes through a dielectric medium of periodic structure comparable to the wavelength at a speed in excess of the phase velocity of light. Optical tractor beams make use of such effects to produce a negative radiation pressure pulling an illuminated object towards the source. With a zero-index material (ZRM), the phase velocity $v_p = \infty$ and the group velocity $v_g = 0$, such that no light propagates. A ZRM may be constructed, for example, by stacked layers of metamaterial with alternating positive and negative permittivity with effective zero permittivity resulting in phase velocity and wavelengths approaching infinity.

1. Photon Model

In a nondispersive medium, such as a vacuum below the Planck cutoff frequency ω_P, the group velocity equals the phase velocity ($v_g = v_p$). Non-dispersive waves (e.g., light, sound, shallow water tension waves, solitons) retain their envelope shape while their energy, momentum, and phase speed remains constant ($\omega = ck$). In a dispersive medium, waves of different wavelength travel with different speeds resulting in chromatic dispersion effects such as prismatic rainbows. In a bulk medium, chromatic dispersion results from a variation in refractive index with frequency. The phase velocity in a dispersive medium varies with frequency. For a normal dispersive medium, higher frequency components travel slower than low frequency components resulting in up-chirp. For a dispersive medium, such as water, the group velocity is greater than the phase velocity ($v_g > v_p$). Dispersive waves (e.g., deep water gravity waves, capillary waves) broaden as they propagate. The group velocity of the wave is usually the observed velocity and is the velocity at which energy is transmitted. Zero dispersion can occur when the material and waveguide dispersion cancel, an effect significant in fiber optics. Pulse broadening may also occur as a result of modal dispersion in waveguides.

For a light pulse propagating in an anomalous dispersive medium (ADM), higher frequency wave components travel faster than slower components producing down-chirp. The velocity of light is not necessarily *semper et ubique* a fixed upper limit. In a medium with anomalous dispersion, under some circumstances, the group velocity of a narrow-band pulse may exceed the velocity of light ($v_g > c$). This apparent contradiction with relativity theory is explained as causality is preserved as it is the signal front velocity v_f that is limited to the speed of light c and it is the signal wave front that conveys information, not necessarily the wave peak. Anomalous dispersion occurs when the frequency of incident light is approximately equal to the absorption resonance frequency of the medium. In a non-linear optical medium, an AC Kerr effect can give rise a self-induced phase modulation and frequency shift produced by change in the refractive medium by the electric field of the light wave. For a narrow-band pulse in a anomalous dispersive medium, the pulse peak becomes shifted towards the signal front as the pulse propagates. As a result, the peak of the pulse envelope described by the group velocity becomes larger than front velocity ($v_g > v_f$) which propagates at c. Under resonance conditions, the group velocity v_g may be positive or negative, subluminal or superluminal. Pulse propagation in non-dispersive, dispersive and anomalous dispersive medium are compared in Fig. 1-8.

Polarization of light, by convention, refers to the direction of the transverse vibration plane of the electric field. For unpolarized light, there is no preferred transverse direction. Polarized light may be linear, elliptical or circularly polarized depending on the rotation and amplitude of the electric field vector. A superposition of left- and right- hand circularly polarized photons can give rise to a light wave with linear polarization such that there is no net angular rotation. A circularly polarized beam exhibits a constant angular rotation as viewed along the axis of propagation with constant amplitude. A circularized polarized beam may be realized by out-of-phase superposition of two circularly polarized waves. If the amplitude varies with rotation, the beam is elliptically polarized. Polarization states of an electromagnetic wave may be described by a two-dimension complex (Jones) vector or represented on a Poincaré sphere such as shown in Fig. 1-9. For fully polarized light, the polarization state point lies on the surface; partially polarized states lie within the sphere.

All the fifty years of conscious brooding have brought me no closer to the question, "What are light quanta?" Of course today every rascal thinks he knows the answer, but he is deluding himself. – Albert Einstein

1. Photon Model

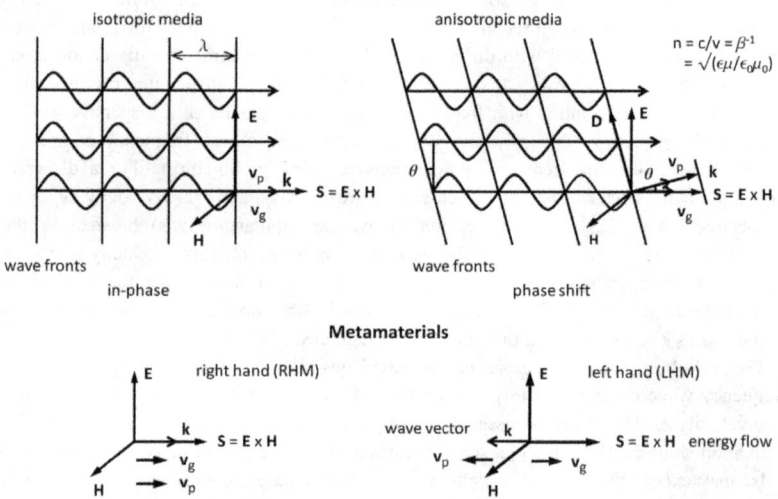

$\omega = 2\pi f = kv_p$	angular frequency
$k = 2\pi/\lambda$	wave propagation vector of phase wave
$v_p = \omega/k = \lambda f = \omega/nk$	phase velocity of a monochromatic wave
$v_g = d\omega/dk = c/n$	group velocity of packet of waves within freq. band $d\omega$ (e.g, envelope node velocity)

D	Electric flux density
E	Electric field intensity
H	Magnetic field intensity
k	Wave vector
S	Poynting vector

Fig. 1-7. Wave vector **k** defining the direction of planes of constant phase is, in general, not in the same direction as the energy flow vector **S** direction and depends on the type of media. Plane waves in isotropic media which are in-phase have wavefronts perpendicular to the direction of propagation. In anisotropic media, plane waves which are phase shifted have wavefronts that are inclined to the direction of propagation. Unlike right-hand metamaterials, left-hand metamaterials exhibit wave vectors opposite to the direction of energy flow. Chiral metamaterials allow for torsional deformation. A plane wave of definite wave vector **k** and polarization s has no localization in space or time and, hence, may be regarded as a one-photon state distributed over spacetime. A photon may be approximately localized in the form of a wave packet centered at a given location at a given time. A quantum mechanical n-photon Fock state $|n\rangle$ corresponds to a discrete number of field excitations $|k_1 s_1, k_2 s_2 ... k_n s_n\rangle$ where k_i is the wave vector and s_i the polarization for a given mode generated by repeated action of creation operators $\hat{a}^\dagger$ on the ground state $|\psi_0\rangle$.

I insist upon the view that 'all is waves' – Erwin Schrödinger

The Schrödinger equation came as a great relief, now we no longer had to learn the strange mathematics of matrices. – George Uhlenbeck

If I were there I would, as in the case of the Zeeman effect, plead for a formal dualistic theory; everything must be describable both in terms of the wave theory and in terms of light quanta. – Werner Heisenberg

There are only three basic actions needed to produce all of the phenomena associated with light and electricity. – R. P. Feynmann

1. Photon Model

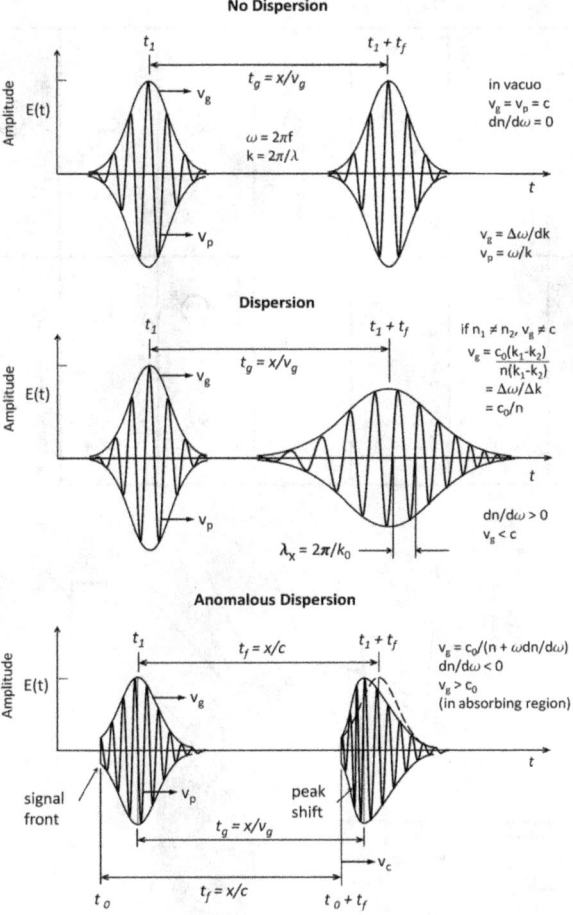

Fig. 1-8. Pulse propagation in a non-dispersive, dispersive and an anomalous dispersive medium. In the later case, the shifted pulse peak corresponds to superluminal group velocity ($v_g > c$). Causality is preserved as the front velocity of the propagated signal equals the speed of light. In waveguide dispersion, as in an optical fiber, light pulses travel faster in the core faster than in the cladding leading to pulse broadening and bit error(s).

The envelope (irradiance) propagates at v_g while the carrier wave propagates at v_p. For a single pulse, the group velocity is the velocity of the pulse amplitude. v_g equals v_p when $dn/d\omega = 0$ such as in a vacuum ($v_g = v_p = c$). In a dispersive medium where the refractive index n varies, $v_g < v_p$ and $v_g < c_0$. In regions near resonance where $dn/d\omega$ is positive, $v_g < v_p$. In anomalous dispersion, $dn/d\omega$ is negative ($v_g > c_0$) and absorption is strong. For a soliton pulse, dispersion and nonlinearity cancel out such that the pulse shape and spectrum are preserved during propagation. In a negative refractive index metamaterial, the phase wave vector is in the opposite direction to the group velocity and energy propagation.

1. Photon Model

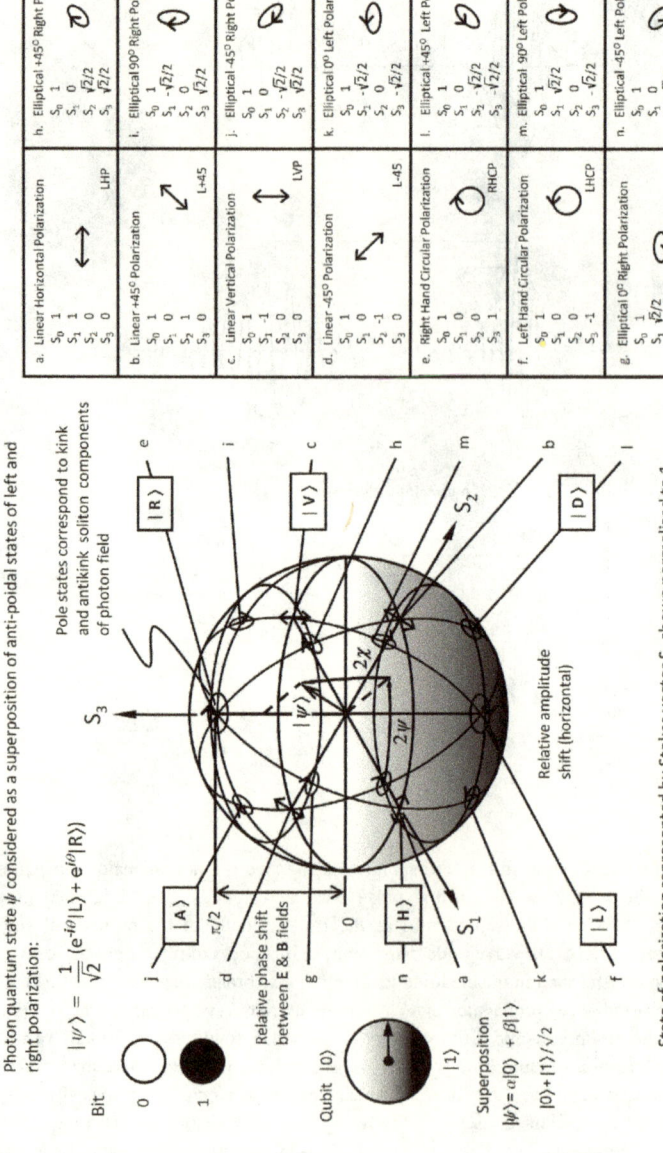

Fig 1-9. Linear, circular and elliptical photon polarization states $|\psi\rangle$ represented by a Stokes vector in a Poincaré sphere.

2. Quantum Vacuum

In relativity, movement is continuous, causality determinate and well defined, while in quantum mechanics it is discontinuous, not causally determinate and not well defined. – David Bohm

You have to say now that space is something. Space can vibrate, space can fluctuate, space can be quantum mechanical, but what the devil is it? – Leonard Susskind

It seems likely that the future theory will be unitary in the sense that the duality of light and matter will disappear. By this I do not claim that we shall necessarily explain one in terms of the other, but perhaps both in terms of some more fundamental concept. – Wolfgang Pauli

If we do get a quantum theory of spacetime, it should answer some of the deepest philosophical questions that we have, like what happened before the big bang? – Michio Kaku

A wave is a propagating disturbance in a medium. Blithe assertions that an electromagnetic wave does not require a medium do not address near-field variation of phase of electric and magnetic field components, radiation impedance, origin of electric charge, retardation, near field effects or finite velocity of propagation. The physical quantum vacuum is not a featureless void, but has discernible, real physical properties, e.g., finite speed of light, c ($= 2.99792458 \times 10^8$ m/s), radiation or characteristic impedance of free space, Z_0 ($= |\mathbf{E}|/|\mathbf{H}| = \sqrt{(\mu_0/\epsilon_0)} = \mu_0 c = 376.73$ Ω) and experimentally observed effects such as Lamb shift, Casimir effect, Scharnhorst effect, Delbrück effect, Davies-Unruh effect, charge screening, birefringence, zero-point energy, etc which may be considered as a manifestation of polarization or vacuum fluctuations. The vacuum exhibits polarization and in the presence of sufficiently intense electromagnetic fields, real matter can be 'materialized' from the vacuum, e.g., electron/positron pair production. Polarizability arises as a result of separable charges – a dipolar irrotational spin torsion field. The Planck aether is somewhat similar to the Dirac model of the electron-positron vacuum as an infinite sea of spin ½ fermion particles of negative energy in which excitations of the vacuum result in formation of particles with positive energy and formation of anti-particles corresponding to vacancy holes in the vacuum sea. Elementary particles (bosons and fermions) represent quasi-particle soliton wave excitations of a vacuum composed of massless Planck dipoles on a scale much smaller than a photon or electron.

In the zero point field (ZPF) representation, the vacuum is an exceedingly stiff, elastic medium filled with harmonic oscillators with frequencies up to the Planck frequency ($f_P = 1.855 \times 10^{43}$ s^{-1}) with each oscillator having a zero-point energy equal to ½$\hbar\omega$. The ZPF is represented as a uniform, isotropic background of electromagnetic radiation that is Lorentz invariant and becomes detectable only when matter is accelerated through the medium. The vacuum appears ultimately to have a discrete or granular nature consisting of ephemeral quantum dipole vortices with characteristic dimensions on the order of the Planck length forming a Bose-Einstein condensate (BEC) superfluid. A non-dissipative, zero-viscosity Bose-Einstein condensate (BEC) superfluid has some resistance to compression and some springiness as a fluid and supports coherent waves. The constancy of the speed of light in deep space indicates a uniform energy density. The dipole vortices are assumed formed by interaction of virtual photon pairs (spinors). The speed of light, in analogy to mechanical wave systems, is a measure of tension, the strength of repulsion between dipoles with speed proportional to the square root of the elastic restoring tension force. The Planck force $F_P = c^4/G = 1.210 \times 10^{44}$ N is repulsive between Planck masses of equal sign and is much larger than the maximum electromagnetic force between electrons $F_e = m^2 c^3/\hbar = 0.21201$ N in part as a result of the large discrepancy in size (Planck wavelength $\lambda_P = \sqrt{(\hbar G/c^3}$ = 1.616 x

2. Quantum Vacuum

10^{-35} m vs. Compton wavelength $\lambda_c = \hbar/mc = 3.8616 \times 10^{-13}$ m). Virtual ZPF pressure is equal and opposite to virtual photon ZPF energy density [ZPF pressure $P = -w \, (dU/dV) > 0$ where w = pressure/energy density = -1]. The cosmological constant Λ is equivalent to a fluid with pressure p and energy density $-\rho_\Lambda$. If purported dark energy has negative pressure, then gravity becomes repulsive rather than attractive and may account for the cosmological constant and expansion of the universe. Dark energy resident in the vacuum has negative pressure causing space to expand such that the dark energy density remains constant at -1. The total amount of dark energy relative to dark matter is theorized to increase with time causing a speed up in the expansion of the universe.

In a Planck aether hypothesis advanced by Winterberg[7], the physical vacuum behaves as a Bose Einstein condensate (BEC) superfluid with a discrete structure composed of positive and negative Planck mass particles. The negative mass particles exhibit negative inertia ($\mathbf{F} = -m\mathbf{a}$) and represent holes similar to the sea of negative electron and positive electron holes proposed by Dirac in 1930. The holes correspond to bubbles in a liquid which move opposite to fluid flow under acceleration. Positive and negative Planck masses each occupy a Planck length volume and may move through each other in effect and form co-rotating or counter-rotating vortices. Cancellation of positive and negative mass results in cosmological constant $\Lambda = 0$ and near zero vacuum energy density. Positive and negative energies likewise cancel accounting for the ~120 order magnitude discrepancy in vacuum energy estimates. Creation of Planck mass generates dissonance driving universe expansion. Fundamental particles such as the photon (boson) and the electron (fermion) are viewed as polarized quasi-particle wave excitations of much smaller Planck particles.

In the context of the physical vacuum composed of Planck dipoles of positive and negative mass, the photon is conceptualized as a soliton spin density wave of massless dipoles. The electric field intensity $\mathbf{E}$ corresponds to the spatial gradient of the scalar potential ϕ plus the temporal rate-of-change in the magnetic vector potential ($\mathbf{E} = -\nabla\phi - d\mathbf{A}/dt$). The variation in scalar potential represents a variation in the volumetric density of Planck dipoles (dM_P/dv). Oscillation in the scalar potential ($d\phi/dt$) corresponds to a mass current (dM_P/dt). The variation in magnetic vector potential represents the differential motion of volumetric regions of Planck dipoles. The photon corresponds to a soliton kink wave in which the amplitude is represented as the electromagnetic vector potential $\mathbf{A}^\mu$. The flow of energy is in the direction of the Poynting vector $\mathbf{S}$ along the least action geodesic line of motion. The flow of energy $\mathbf{S} \, (= \mathbf{E} \times \mathbf{H} = \mathbf{p} \cdot c^2)$ is equivalent to the flow of time ($dt > 0$) which occurs in response to an energy density gradient.

According to SR, time does not elapse for a light wave at a constant frequency (i.e., $d\tau/dt = 0$). Light undergoes a frequency shift in response to an energy density gradient as described by the time dilation expression $\Delta t' = \Delta t/\sqrt{(1 - v^2/c^2)}$ and measured by the index of refraction $K_{PV}(r,\omega,M)$. The volumetric energy density $u \, (= \Sigma n e_i)$ has dimensions of [J/m^3 = m^{-2}·kg·s^{-2}] where n is particle density and e_i is the particle energy. The mass density u/c^2 has dimensions of [kg/m^3]. The momentum density $g = u/c^2 v$ has dimensions [= m^{-2}·kg·s^{-2}] where v = velocity. The energy flux $S \, (= vu = c^2 g)$ has dimensions of [J/m^2·s = kg·s^{-3}]. The estimated quantum mechanical energy density of the vacuum is ~10^{113} J/m^3. For comparison, the energy density of all the matter in observable universe is ~10^{-9} J/m^3.

Tewari[8] postulates the vacuum possesses non-material properties of incompressibility, zero-viscosity, continuity and masslessness of an ideal fluid. The formation of a vortex results in an energyless, fieldless spherical void at the vortex core of a definite volume. Tewari observes it is the creation of this void in which the energy in the void region is displaced outward that is the cause of formation of mass (= void volume x velocity of light). For a spherical electron, the mass of the electron is given as $m_e = (4\pi/3) r_e^3 c$ where the void radius $r_e = c/\omega \cong 4 \times 10^{-13}$ m.

2. Quantum Vacuum

Dyatlov[9] argues the physical vacuum can consist of elementary dipoles, electrical (linked pairs of electrical charge) and gravitational (linked pairs of positive and negative mass) and moments magnetic and spin corresponding to four particle states for a net neutral polarized medium described by Maxwell and Heaviside vacuum equations.

Particle states	
+m, +q	+m, −q
+s, +μ	+s, −μ
−m, +q	−m, −q
−s, +μ	−s, −μ

where m = mass, q = electric charge, s = spin angular momentum and μ = magnetic moment.

This is suggestive of a fluid lattice network similar to spin glasses or charge density waves. The paired matter and antiparticles correspond to solitons associated with topological defects of the spin network.

Urban et al[10] show that the vacuum permeability μ_0 and permittivity ϵ_0 may be attributed to magnetization and polarization of continuously and disappearing fermion pairs. Leuchs and Sanchez-Soto[11] find that the impedance of free space is a function of the square of electrical charges of virtual particle anti-particle pairs. Daywitt[12, 13] makes the argument that the Planck vacuum is a polarizable medium with an effective dielectric constant $\epsilon = 1/\sqrt{\alpha}$, where α = the fine structure constant. Daywitt speculates the Planck particles are massless, the Compton radius is larger than the bare charge (e*) radius and mass arises as a result of random excursions of charge ($r_C = e^{*2}/mc^2$). The vacuum is assumed nonchiral meaning that there is no asymmetry in quantum-vacuum fields resulting in physical differences between right- and left-handed polarization modes.

In Quantum Mechanics (QM) theory, the scalar potential ϕ and magnetic vector potential **A** are the primitives whereas electric **E** and magnetic **H** fields are derivatives. An electromagnetic wave is the exterior derivative of the electromagnetic 4-potential $A^\mu = (\phi/c, \mathbf{A})$. Electromagnetism is associated with spin/torsion of EM fields while gravitation is associated with curvature of geodesics.

- Scalar potential (ϕ). In a quantum polar vacuum, the electric scalar potential ϕ (= V = E/q) is a measure of potential energy/unit charge. Time variation in electromagnetic energy density (U = ½($\epsilon_0 E^2 + B^2/\mu_0$)) results in generation of EM waves. Gravitational potential ϕ_g (= E/m) is a measure of potential energy/unit mass. The gradient in gravitational scalar potential ($\nabla\phi = -\mathbf{g}$) is a measure of wave function curvature and results in the force of gravity ($\mathbf{F} = -m\nabla\phi$).

- Vector potential (**A**). A magnetic field **B** (= $\nabla \times \mathbf{A}$) is defined as the curl of **A** where $\mathbf{A} = \hbar\mathbf{k}/q = (v/c^2)\phi$. The vector potential is proportional to the vector current j_μ which is the source of electromagnetism. The magnetic vector potential A_μ is described by a spin 1 vector potential ($A_\mu = -j_\mu/k^2$). A time-varying electric current produces a time-varying electrokinetic field E_k that is in opposition direction to the inducing current $i(t)$ producing a dragging force that moving electrons exert on neighboring charges. This interaction provides an explanation for Lenz's law ($E_k = -dA/dt$), Einstein-de Haas effect and its inverse the Barnett effect.

2. Quantum Vacuum

- Mass. Mass is a result of wave function interference that obstructs energy flow. An electron acuires rest mass due to an imbalance of electro- and magnetostatic fields inducing spin precession with whirl no. equal to inverse of fine structure constant α^{-1}. A freely- propagating photon in a gravity-free, zero curvature vacuum has no rest mass, but acquires an effective mass when encountering a region of increased energy density. When light is confined within a resonator cavity (a fixed reference frame), the standing wave system acquires rest mass. An electron is equivalent to a confined photon with an electromagnetic mass ≥ 0.511 MeV/c^2. Electron mass $m_e = \pm\sqrt{[(pc)^2 + (mc^2)^2]} = (\hbar\omega + e\phi)/c^2$. Rest mass is a consequence of formation of closed-loop standing waves and is associated with electric charge in accordance with Weber's inertial mass relation $m_W = q\phi_E/3c^2$ where q = electric charge, ϕ_E = electric potential and c = velocity of light. Relative mass varies with velocity according to the SR relation $m = m_0/\sqrt{(1 - v^2/c^2)}$. In the weak field approximation, mass varies with energy density according to the relation $m = m_0\sqrt{(K_{PV})}$ where m = proper mass and K_{PV} is the vacuum refractive index. Energy density is proportional to EM wave nodal density. Proper mass in a gravitational field varies with gravitational gamma Γ (= $1/\sqrt{(1 - 2GM/c^2r)}$ as $m = \Gamma m_0$. Mass results from motion through a region of increasing energy density gradient producing a change in motion due to deceleration. Moving standing waves become contracted as a result of nodal displacement with a change in matter wavelength. Hence, inertial mass equals gravitational mass as both arise from the same causal mechanism. Mass varies with volumetric nodal density and node cross-sectional area which is a function of standing wave ratio. In the context of the quantum vacuum, the Planck mass is given by $m_P = \sqrt{(\hbar c/G)}$. In terms of the vacuum refractive index K_{PV}, the proper Planck mass $m_P = m_{P0}/\sqrt{(K_{PV})}$ and may likewise be attributed to acceleration in a spectral energy density gradient.

- Time. The observed time interval between events is a function of energy flow as a result of a gradient in energy density (∇U). In a Planck vacuum, the scalar potential ϕ is a measure of energy density of Planck masses. In SR/GR, time is said to dilate and for light at constant frequency, time slows to a standstill as there is no fixed reference frame and no position operator. Proper time τ varies as a function of refractive index K_{PV} and gravitational gamma $\Gamma = (dt/d\tau)$ according to the relations $t = t_0\sqrt{K_{PV}} = \Gamma t_0$. Rather than ascribing physical properties to nonphysical abstract dimension, a contrasting view is that apparent slowing of time is a physical effect as a result of a change in wavelength due to EM wave nodal compression associated with increased quantum pressure in a region of increased energy density.

- Electric charge. Electric charge e^-, e^+ is a measure of mass flux flow rate and is related to topological charge or winding number of the magnetic flux quantum (Φ_0 = h/e) where the sign is determined by the winding direction. Electric charge represents a torsion defect (loop closure failure) due to an imbalance of electrostatic and magnetostatic energy. Electric charge is proportional to electron mass and Compton angular frequency and as a result is quantized. In the Kaluza-Klein theory, charge was attributed to rotation in an unseen fifth dimension that corresponds to the Planck scale. The charge polarity is determined by the vortical spin direction. Electrodynamic momentum $\mathbf{p} = \hbar\mathbf{k} = q_e^+\mathbf{A}$ where $\mathbf{A}$ is magnetic vector potential. Applying charge dimensional conversion, electric charge $q_e = \omega_C m_e$ where ω_C = Compton angular velocity and m_e = electron mass. Electric charge results from spin precession of the electron closed-loop standing wave resonance structure (precessing Hopf link) with whirl no. equal to the inverse fine structure constant α^{-1}.

2. Quantum Vacuum

- Electric field. The electric field intensity (strength) **E** corresponds to a measure of the phase alignment of Lorentz contracted dipoles forming Faraday field lines. The **E** field ($E = -\nabla V - dA/dt$) is a measure of accumulation of charge and the time rate change in vector potential **A**. Electromotive force EMF ($= d\phi/dt$) is a result of time rate of change in Planck mass density.

- Magnetic field. The magnetic induction field **B** (magnetic flux density) is the curl of the vector potential **A** field ($B = \nabla \times A$). Magnetic flux lines represent quantized vortex filaments. Vortical motion of Planck masses gives rise to vortical filaments similar to that observed in a Bose-Einstein Condensate (BEC) zero viscosity superfluid such as ^{4}He when stirred while supercooled.

- Voltage. Electromotive force or voltage is equivalent to the time rate of change of potential (EMF $= -d\phi/dt$), a change in Planck dipole density vs. time. Voltage (V = W/q) is equivalent to energy per unit charge where W = work, q = charge.

- Wave function. The wave function ψ describes the shape of the potential field ϕ. EM potential A_μ describes the wave function as a function of position and time. Curvature of wavefronts is a result of localized slowing of EM waves in regions of increased dipolar density. Curvature k of arc length (s) of field wavefronts is equivalent to a gravitational acceleration **g** in the direction of the normal vector **N** and proportional to the radius of curvature of the geodesic described by the tangent vector **T** and binormal vector **B** of the propagating EM potential wave A_μ^a.

- Energy. Energy E is a measure of the time rate of change in curvature ($E = h\nu$) and is equivalent to mass ($E = \pm\sqrt{[(mc^2)^2 + (pc)^2]}$) in motion which is a measure of curvature associated with topological charge. Energy $E = mc^2 = |\text{volts·charge}|/c^2$. In Special Relativity (SR), energy varies with velocity according to the relation $E = m_0c^2/\sqrt{(1 - v^2/c^2)}$. Velocity equals flux/energy density. For small values of v/c, energy varies as $E = m_0c^2[1 + \frac{1}{2}(v/c)^2 + 3/8(v/c)^4 + ...]$. Energy varies with refractive index according to the relation $E = E_0/\sqrt{(K_{PV})} = E_0/\Gamma$. Energy is associated with spatial and/or temporal variation in Planck mass density. Electrostatic potential energy represents a form of stress (resistance to confinement) or volumetric compression of charge elements. Electromagnetism is associated with spin/torsion of EM fields while gravitation is associated with curvature of geodesics, volumetric dipole density and frequency discordance of interacting dipole emitters.

Echoing Gottfried Leibniz's views on relativity, space-time is a mathematical coordinate representation of location of position of coexistent phenomena or events in space and an ordering of events in time in a defined reference frame and is not a physical entity. Space and time are relational properties that depend on the presence of matter and energy. The laws of physics are taken to be invariant, or symmetric, under all changes in reference frames as a result of assumed isotropy. Translations in time are associated with conservation of energy E ($= \gamma mc^2 = \hbar \nu$), translations in space are associated with conservation of linear momentum p ($= \gamma mv = \hbar k$) and rotations in space are associated with conservation of angular momentum J ($= r \times \gamma mv = \hbar m$). Leibnitz developed an expression (*vis viva*) for kinetic energy $E = mv^2$ which anticipated the famous Einstein relation $E = mc^2$. This was elaborated further by Emilie du Châtelet demonstrating energy was proportional to the velocity squared. Thomas Young later identified kinetic energy with a ½ proportionality factor: $E_k = \frac{1}{2} mv^2$ where $[v/c] \ll 1$ such that $E_{total} = E_k + E_P$.

2. Quantum Vacuum

As noted by Tuisku et al [14] the passage of time is associated with energy flow. The hour glass provides a familiar example. Energy density gradients determine flow of energy and mark the passage of time. Geodesic world-lines describe the energy density gradient. Travelling waves and moving standing waves follow the least action path. In the absence of energy density gradients ∇U, time ceases to flow. In a quantum vacuum composed of Planck particles, energy density gradients correspond to volumetric density gradients, a scalar potential ϕ. The potential energy gradient between a region of high energy density and low energy density drives an energy flow down the steepest gradients along space-time geodesics in accordance with principle of least action. Only changes in energy density are observable. For a photon at constant frequency, proper time is equal to zero becoming manifest during a change in frequency as a result of a change in potential. EM wave length distance between nodal points correspond to the metric g_{uv}. Spacetime curvature is equivalent to a curvature of EM wavefronts. With sufficient curvature, an EM wave forms dual closed circulation loops with a spin angular momentum in multiples of Planck's least action constant $\hbar$ (= $h/2\pi$) with each loop corresponding to a fermion. The number of loops corresponds to the Berry geometric phase ϕ. Interactions between particles occur along affine manifold connections defining causal relationships.

The universe is thought to have begun from an initial low entropy (highly ordered) state ~13.8 billion years ago in a Big Bang or Big Uh-Oh and exhibits a dichotomous nature of opposites of curvature and energy. This binary aspect may ultimately be expressed at the Planck scale in the form of Planck dipoles of positive and negative Planck mass m_P. Following the Big Bang, photon collisions resulted in quark-anti-quark and electron-positron formation. In the radiation dominated era, photon collisions overwhelmed gravitational attraction preventing atomic nuclei formation. Approx. 24.4 Kyr after the Big Bang, it is theorized the radiation density in the early universe dropped below that of matter at ~10^{-15} kg/m^3. Both the radiation and matter densities decreased as the universe expanded with radiation density decreasing faster than that of matter as matter was formed. The condensation of matter is thought to have occurred in steps as the universe cooled: Subatomic particles ~10^{13} °K (degrees Kelvin); light elements ~10^9 °K; atoms ~3 x 10^3 °K. The formation of atoms occurring ~3.8 x 10^5 yrs after the Big Bang. The average background microwave temperature is now a very cold ~2.7 °K.

In a model proposed by Sternglass[15] based on an earlier proposal by Lemaître, the universe began as a single rotating electron-positron pair of extreme energy successively dividing into lower energy pairs. The number of divisions since the Big Bang, Sternglass calculates, as exactly twice the inverse of the fine structure constant (2^N where N = $2/\alpha$). The number N of electron masses in the current epoch would be 1.736 x 10^{83} giving the mass of the universe of $[(2/\alpha)] \cdot 2^{\wedge}2/\alpha$ times the mass of two electrons. With each division, the Planck energy density of the electron pair over a spherical volume declines as $(8/\alpha)(e^2/2m_0c^2)^3$. A decline in Planck energy density (U_P = $c^7/\hbar G^2$), dimensionally equivalent to Planck pressure, indicates the Newtonian gravitational constant G increases with time increasing the strength of gravity. The increase in G is accompanied by a decrease in Planck mass m_P (= $\sqrt{(\hbar c/G)}$ and increase in Schwarzschild radius R_S (= $2GM_u/c^2$ = R_m^3/R_c^2) where R_m = mass radius and R_c = spacetime curvature radius. The observable universe appears to consist of a rotating black hole of angular velocity ω (= c/R_S) equal to Hubble's constant H – a closed spinning bubble of positive curvature expanding from an undifferentiated primordial vacuum state. The angular velocity of all matter is derived from the angular velocity of the initial mass pair through successive divisions due to conservation of angular momentum. Radiation emitted in the Big Bang is gradually converted to mass resulting in gravitationally bound, large scale structures such as planets, stars, black holes, galaxies and superclusters as the background radiation

2. Quantum Vacuum

temperature of the universe cools over time. The hierarchical organization of cosmological structures exhibits a linear relation in a logarithmic plot mass expressed in electron pairs ($2m_e$) versus radius in units of the electron Compton radius ($2R_C$). The organizational structures appear to occur in discrete scale groupings of 2^{10} (= 1024) of electron pairs.

The mathematical archetype of a binary geometry may be visualized as a dyad formed by division of a monad, a point or unit circle symbolizing oneness. The dyad is formed of two polarized monads of opposite curvature and energy suggestive of the Big Bang genesis. The transformation of a dimensionless point into a circle of many points in a circuit generated by rotary motion of a point in space and time represents a metaphor of the transcendental creation. A conceptual illustration of this process is shown in Fig. 2-1 in which the dyad consists of a bound rotating electron and positron pair. A bound e^-e^+ pair corresponds to a superposition of 0, 1 states, a quantum bit (qbit). The revolving line of tension separating two points in a unit circle in the complex plane of real and imaginary numbers may be represented as a phasor described by Euler's relation $e^{i\theta} = \cos\theta + i\sin\theta$. In the Sternglass model, all matter consists of electrons and positrons. Evolution of the universe (fireworks model) in conversion of radiation to matter proceeds in stepwise fashion from a single supermassive electron-positron pair with formation of composite structures as the result of 270 successive pair divisions up to the present epoch with half the mass-energy of the preceding parent electron-positron pair.

The present day universe exhibits a dyadic structure of opposites. If, for example, we represent a positive wavefunction curvature as ∪ and negative curvature as ∩, the creation of the universe may be represented as a transformation of the wavefunction monad o with positive and negative curvature into a dyad ∪ + ∩. An arbitrary quantum state vector ψ = a|∪⟩ + b|∩⟩ is a ray in Hilbert space. In bra-ket notation, ⟨∪|∪⟩ = ⟨∩|∩⟩ = 1 or ⟨∪|∩⟩ = ⟨∩|∪⟩ = 0. The creation of the universe corresponds to a wavefunction collapse ⟨ϕ|ψ⟩ from a zero curvature state into an end state of opposite curvatures. Nothingness as the absence of something is equivalent to a summation of opposites.

The Big Bang is theorized as initiating from a singularity event at $t_u = 0$ involving a vacuum decoherence resulting in creation of positive and negative mass Planck dipoles as illustrated in Fig. 2-2. Interference of wave functions of opposing curvature creates dissonance – A repulsive set interaction or 'dark' energy drives universe expansion owing to continuous creation of Planck mass pairs. The phase transition of the Planck vacuum proceeds from an initial spinless, unperturbed, coherent, low entropy state in imaginary time to the present chaotic, frustrated, defect-filled, high spin state with spin wave excitations of Planck dipoles in real time. Physical observables, i.e., light and matter, represent travelling wave and standing spin wave resonant structures, respectively, composed of Planck mass dipoles. Non-spin vacuum states of Planck dipoles with net zero mass represent unobservable states corresponding to dark energy. A Planck mass, in common with a black hole, is characterized by size, mass, charge and spin. The edge of the observable universe (compact, finite manifold M^n of n dimensions) in such a model resembles edge/screw dislocations in a topological insulator. Spatio-temporal wave decoherence provides an explanation for increasing entropy and quantum entanglement. Spatial coherence refers to in-phase alignment of adjacent waves across the wave front. Temporal coherence refers to the extent wave maxima and minima remain in-phase along the wavetrain and may be expressed in terms of coherence length l_c (= $\lambda^2/2d\lambda$). Monochromatic wavetrains correspond to infinite coherence length. A spatial and/or temporal phase shift of coherent quantum waves induces excited vortical spin states and volumetric expansion. Temporal constancy in the speed of light and vacuum pressure

2. Quantum Vacuum

suggests continuous creation of pairs of Planck dipoles associated with volumetric expansion of the universe. The simultaneous creation of positive and negative energy appears consistent with conservation of energy in such a process. Curiously, there seems to have been only one Big Bang rather than a fireworks succession of randomly dispersed Big Bangs in time and space reminiscent perhaps of a highly improbable rogue wave occurrence on an otherwise coherent sea of wavefunctions inducing one-time symmetry-breaking into a degenerate state of opposites. Also, to be explained is why the Big Bang did not immediately self-extinguish. The known laws of physics such as Newton's equations of mechanics, Maxwell's equations of electromagnetism, Einstein's gravitational equations and Schrödinger's wave equations are symmetric in time. Statistical thermodynamics is invoked to explain time's arrow and increase in entropy with time.

The vacuum at the Planck scale has been described as a quantum foam of bubble and void-like structures in constant motion resulting in zero-point energy. Did this quantum foam of positive and negative curvature arise as a result of the Big Bang similar to bubble formation during uncorking of a bottle of champagne or was it pre-existing? Bubbles and voids may be described as positive and negative Planck masses with geometry of inscribed and circumscribed N-polygons with frequency characteristics resembling positive and negative mass. See Fig. 2-3. In the creation of the universe from a singularity event, how does Planck mass arise? What is the nature of the singularity? Given the dyadic nature of the known universe, we might guess that the singularity corresponds to a superimposed source and sink. As these are mutually incompatible states, we might suppose that the singularity represents an alternating source and sink of, say, spherical waves resonantly diverging from and converging toward a point. Considering the Planck vacuum as composed of positive and negative Planck masses, examination of transformation of source/sink doublet flows in the complex plane may provide some clues. For example, consider the complex function f(z) as a power series of infinite terms

$$f(z) = A_0 + A_1 z + A_2 z^2 + A_3 z^3 + A_4 z^4 + \ldots \qquad (2\text{-}1)$$

where the complex number $z = x + iy$ and $i = \sqrt{-1}$.

For a circle centered in the complex plane, the series converges for any value of z within a circle of convergence of radius $R = |z|$ and diverges outside the circle of convergence. The radius of convergence equals the distance to the nearest pole. In the case of the exponential function $e^z = 1 + z/1! + z^2/2! + z^3/3! + z^4/4! + \ldots$, the series converge for all of z. Each term in the series corresponds to a vector with a summation generating a spiral with modulus $|z| = e^z$. Perhaps, at the singularity event @ $t = 0$, the singularity consisting of an alternating source and sink of imaginary mass is transformed into positive and negative real mass by a wavefunction phasor rotation. For example, two infinite power series, G(z) and H(z), each of which have the same interval of convergence $-1 < z < 1$ inside a unit circle centered on the complex plane.

$$G(z) = 1/(1 - z^2) = 1/(1 + z)(1 - z) \qquad (2\text{-}2)$$
$$H(z) = 1/(1 + z^2) = 1/(1 + iz)(1 - iz) \qquad (2\text{-}3)$$

The poles of G(z) lie along the real axis at +z and –z while the poles of H(z) lie along the imaginary axis at +iz and –iz. Viewed on a Riemann sphere, a doublet may be represented a stereo projection of poles of a unit sphere on to the complex plane. Pole rotation of $\pi/2$ radians can convert a singularity (monad) into a pair of singularities (dyad) of either imaginary or real poles. The lines of constant and real parts may be interpreted as a representation of an example of transformation of imaginary mass into real mass.

2. Quantum Vacuum

In the expression for Planck mass

$$m_P = \pm\sqrt{(\hbar c/G)} \qquad (2\text{-}4)$$

the quantum of action $\hbar$ corresponds to a rotation of a mass dipole about a point with a tangential velocity of c. The gravitational constant G represents the strength of mass coupling. Phase transformation corresponds to conversion of counter-propagating massless travelling waves into closed-loop standing waves with real rest mass. In analogy with the electron, the Planck mass may be represented as $m_P = (\hbar\mathbf{k} - q_P)/\mathbf{v}$ where $\mathbf{k}$ = wave propagation vector, q_P = Planck charge (= $\sqrt{(4\pi\epsilon_0\hbar c)}$ = $m_P\,\omega_p$) and $\mathbf{v}$ = velocity.

Magnetic field lines in a Planck vacuum are interpreted as quantized superfluid vortical filaments of spinning Planck dipoles forming closed loops or string-like linear defects terminating in monopole-like point defects or boojums at the boundary. Elementary particles represent spin wave excitations of Planck dipoles. Photons represent travelling wave excitations of Planck dipoles. Electrons and positrons represent the simplest possible resonant, standing wave, closed-loop, topologically bound spin wave structures formed of Planck dipoles in a Planck vacuum. Magnetic field density may be altered by action of compressional Alfvén waves whereas magnetic field line bending may be altered by action of Alfvén shear waves. Filamentary structures in the universe in the Hannes Alfvén cosmological model are attributed to effects of entangled plasma currents of electrons and ions. The universe, itself, appears to be an expanding black hole resonant cavity with a total energy density parameter $\Omega_t = \Omega_b + \Omega_{dm} + \Omega_\Lambda$ where Ω_b = baryonic energy density (~0.04), Ω_{dm} = dark matter energy density (~0.27) and Ω_{de} = dark energy density (~0.73).

A Planck mass represents a standing wave resonator. The interior of bubbles in the Planck vacuum in the model described resembles in some respects the stasis fields of science fiction. The bubbles of space/time of positive and negative energy at the Planck scale are isolated from the entropy gradient of the rest of the universe. The bubble interior is invulnerable to anything externally and reflects incident radiation. The high frequency domain walls form, in effect, an impenetrable barrier to matter and radiation and act as point scattering centers for electromagnetic waves. On collision, the Planck mass bubbles elastically scatter off each other and become immobile in the interior of black holes.

But it is evident that this [Planck] length must be the key to some essential structure. – Arthur Eddington

Our world is so-to-say the simplest of all possible worlds! But all that is music for the future and prior to that it is still necessary to do a lot of mathematics. – Wolfgang Pauli

The more the universe seems comprehensible, the more it seems pointless. – Stephen Weinberg

I am simply fascinated by your [wave equation] and the wonderful new viewpoint it brings. – Paul Ehrenfest

It was not until some weeks later that I realized there is no need to restrict oneself to 2 by 2 matrices. One could go on to 4 by 4 matrices, and the problem is then easily solvable. In retrospect, it seems strange that one can be so much held up over such an elementary point. The resulting wave equation for the electron turned out to be very successful. It led to the correct values for the spin and magnetic moment. This was quite unexpected. The work all followed from a study of pretty mathematics, without any thought being given to these physical properties of the electron. – P.A.M. Dirac

What we observe as material bodies and forces, are nothing but shapes and variations in the structure of space. – Erwin Schrödinger

2. Quantum Vacuum

Electron/positron successive pair doubling cascade

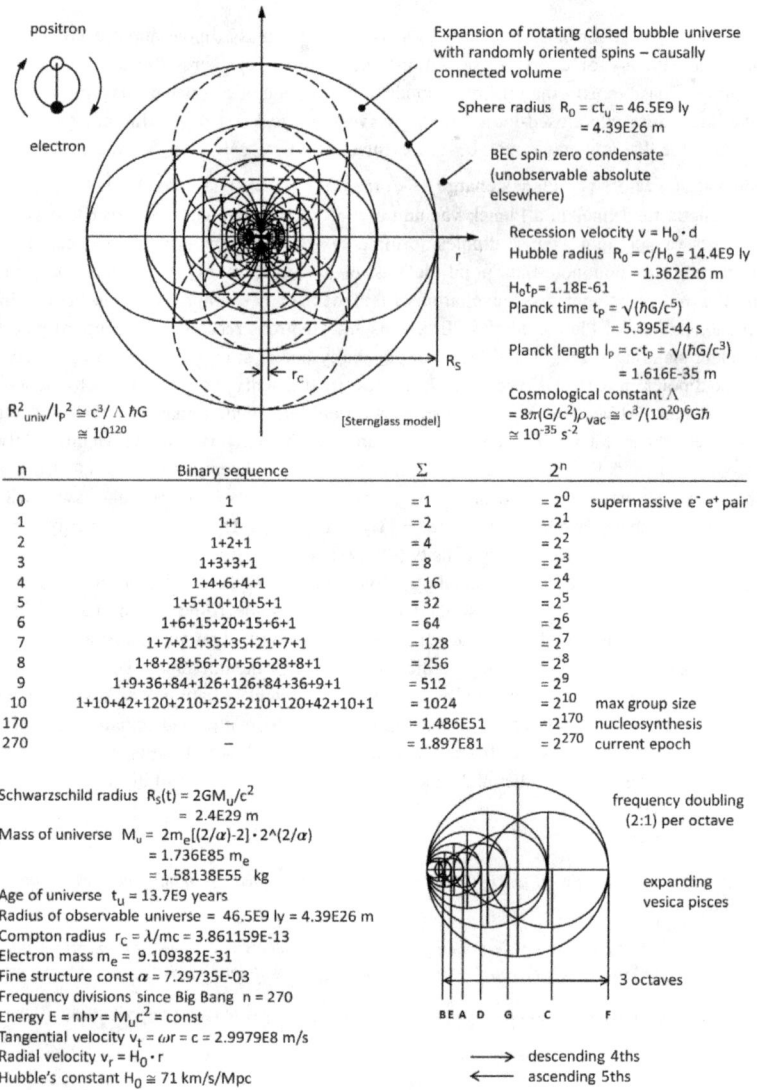

n	Binary sequence	Σ	2^n	
0	1	= 1	= 2^0	supermassive e⁻ e⁺ pair
1	1+1	= 2	= 2^1	
2	1+2+1	= 4	= 2^2	
3	1+3+3+1	= 8	= 2^3	
4	1+4+6+4+1	= 16	= 2^4	
5	1+5+10+10+5+1	= 32	= 2^5	
6	1+6+15+20+15+6+1	= 64	= 2^6	
7	1+7+21+35+35+21+7+1	= 128	= 2^7	
8	1+8+28+56+70+56+28+8+1	= 256	= 2^8	
9	1+9+36+84+126+126+84+36+9+1	= 512	= 2^9	
10	1+10+42+120+210+252+210+120+42+10+1	= 1024	= 2^{10}	max group size
170	–	= 1.486E51	= 2^{170}	nucleosynthesis
270	–	= 1.897E81	= 2^{270}	current epoch

Schwarzschild radius $R_s(t) = 2GM_u/c^2$
 = 2.4E29 m
Mass of universe $M_u = 2m_e[(2/\alpha)-2] \cdot 2^{\wedge}(2/\alpha)$
 = 1.736E85 m_e
 = 1.58138E55 kg
Age of universe t_u = 13.7E9 years
Radius of observable universe = 46.5E9 ly = 4.39E26 m
Compton radius $r_c = \lambda/mc$ = 3.861159E-13
Electron mass m_e = 9.109382E-31
Fine structure const α = 7.29735E-03
Frequency divisions since Big Bang n = 270
Energy $E = nh\nu = M_u c^2$ = const
Tangential velocity $v_t = \omega r = c$ = 2.9979E8 m/s
Radial velocity $v_r = H_0 \cdot r$
Hubble's constant $H_0 \cong$ 71 km/s/Mpc

frequency doubling
(2:1) per octave

expanding
vesica pisces

3 octaves

B E A D   G   C   F

→ descending 4ths
← ascending 5ths

Fig. 2-1. Creation of a polarized dyad from a monad depicting subsequent successive pair doubling. In the Lemaître-Sternglass model, the universe began as a single rotating electron-positron pair of extreme energy successively dividing into lower energy pairs. As the universe cooled, composite structures such as protons and neutrons formed resulting in formation of atoms and molecules. According to this model, anti-matter did not somehow mysteriously disappear during the early formation of the universe, but remains phase-locked to matter in phase quadrature preventing mutual matter/anti-matter annihilation.

2. Quantum Vacuum

Quantum vacuum decoherence

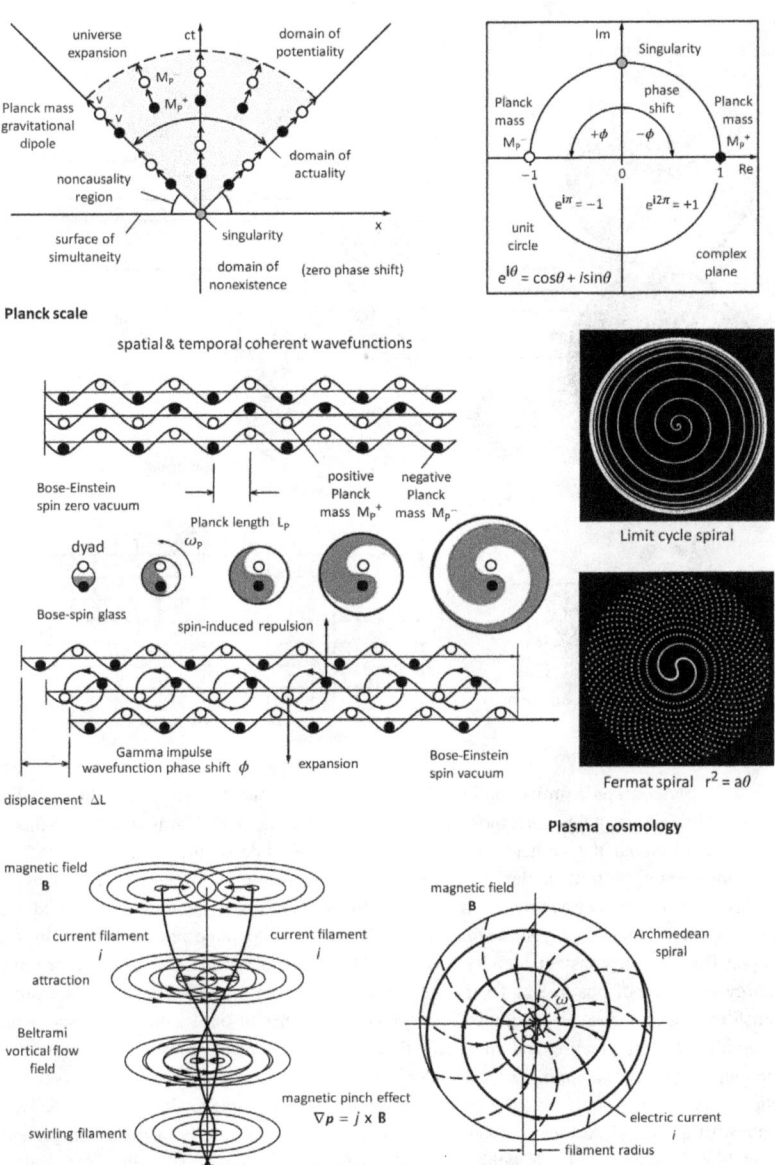

Fig. 2-2. Vacuum expansion (dark energy) associated with Big Bang may be attributed to phase shift of Planck dipoles from coherent, low entropy, spinless state of a Planck vacuum to expanding state with spin wave structures (physical observables). Light and matter represent travelling and standing waves, respectively, composed of Planck mass dipoles.

23

2. Quantum Vacuum

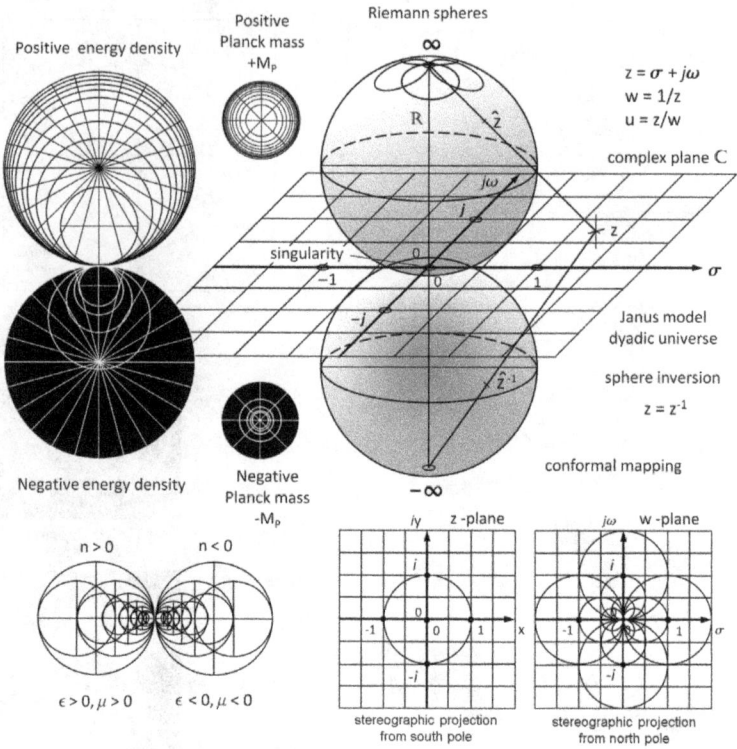

Fig. 2-3. Stereographic projection of poles of a pair of Riemann spheres onto the complex plane. The vertical poles correspond to the range of real numbers from zero to ±infinity. Pole rotation about the vertical axis increases the magnitude of the convergence radius. Vacuum instability from a flat, featureless void may result in formation of regions of positive and negative curvature forming bubble-like regions corresponding to inscribed and circumscribed N-sided polygons which naturally converge to fixed radii described by the Kepler-Bouwkamp constant and its inverse. The nodal patterns correspond to spectral energy densities of positive and negative mass. A vacuum of such Planck mass dipoles constitutes a polarizable medium. The pattern of constant real and imaginary lines resembles that of electric and magnetic fields as well as lines constant curvature and torsion. Conformal mapping of resonator mass likewise resembles that of waveguide impedance. A monopole singularity resembles the electromagnetic dominant mode field pattern of a coaxial AC transmission line viewed end on. Similarly, a dipole singularity resembles the dominant mode field pattern of a parallel wire transmission line. A pulsating singularity may be represented as a wave guide array with sequenced end switching of an AC generator (source) and load (sink) or oscillating transformer. As such, the Big Bang may represent a runaway resonant feedback oscillation. A non-pulsating singularity corresponds to a bimodal degenerate transition from a flat vacuum instability as in a single fold catastrophe manifold expansion from a point with sudden change in curvature into a pairing of opposite curvatures. Expansion is driven by wave function interference.

3. Electromagnetic 4-Potential

In 1856, Maxwell showed that the induced electric field **E** and can be expressed in terms of the magnetic vector potential as $\mathbf{E} = -\delta \mathbf{A}/\delta t$ and deduced that the vector potential **A** represents a potential electromagnetic momentum per unit charge. As defined by Jefimenko[16,17] an electric field $\mathbf{E}_k$ generated by motion of time-variable currents results in an induced electrokinetic force $\mathbf{F}_k$ on nearby electrical charges. The induced electrokinetic force $\mathbf{F}_k$ is proportional to the induced electric field ($= q\mathbf{E}_k = -q\delta \mathbf{A}/\delta t$). The induced current caused by the electrokinetic field is opposite to the inducing current when the current is increasing (positive time derivative) and in the same direction when the current is decreasing (negative time derivative). Jefimenko notes that Faraday induction is not an electromagnetic phenomena – induced currents are not caused by changing magnetic fields, but by an electrokinetic field created by changing electric currents. The induced electrokinetic field $\mathbf{E}_k$ provides an explanation for Lenz's law of electromagnetic induction and Maxwell's displacement current within capacitors as elucidated by Jefimenko[17] and Catt[18].

Magnetic vector potential: $\quad \mathbf{A}^* = -\int \mathbf{E}_k dt + \text{const} \quad$ [V·s·m^{-1}] $\qquad$ (3-1)

where **A*** refers to the retarded magnetic vector potential. Note: For slowly changing fields $\mathbf{A} \approx \mathbf{A}^*$.

The induced electric field $\mathbf{E}_k$ and electrokinetic force are given by

Induced electric field $\qquad \mathbf{E}_k = -\delta \mathbf{A}/\delta t \qquad$ [volts/m] $\qquad$ (3-2)

Induced electrokinetic force $\qquad \mathbf{F}_k = q\mathbf{E}_k = -q\delta \mathbf{A}/\delta t \qquad$ [Newtons] $\qquad$ (3-3)

An electric current *i* has an associated electrodynamic potential momentum **A** which is in opposition to the direction of current. The vector potential **A** represents an interaction of moving charges while the scalar potential ϕ represents interaction between static charges. The vector potential is described as the principal observable in quantum mechanics and is featured prominently in the Aharonov-Bohm effect, super-conductivity, etc. The electromagnetic 4-vector potential A^μ components are illustrated in Fig. 3-1. A notional sketch of a quantized vacuum composed of Planck particles is depicted in Fig. 3-2. In the Planck aether hypothesis advanced by Winterberg[7], the Aharonov-Bohm effect involving a wave function phase shift in the absence of an **E** and **B** field external to a region enclosing a magnetic field may be understood as a result of a net differential in kinetic energy of counter-rotating potential vortices of positive and negative Planck masses which are treated as the source charges that produce the electromagnetic field.

The electromagnetic 4-potential given by $A^\mu = (\Phi/c, \mathbf{A})$ represents energy/charge of the electromagnetic gauge field and is proportional to the 4-current density J^μ. The 4-potential $A^\alpha = (\phi/c, \mathbf{A})$ has a temporal scalar potential component ϕ and three spatial vector potential components A_x, A_y, A_z. The electromagnetic 4-potential is defined as

Electromagnetic potential: $\qquad A^\alpha = (\phi/c, \mathbf{A}) \qquad$ [V·s·m^{-1}] $\qquad$ (3-4)

In the tetrad formalism $A^\alpha = A^{(0)} q_\alpha$ where **A** = 4 x 4 potential matrix of EM potential, q_α = tetrad vectors.

3. Electromagnetic 4-Potential

The scalar potential ϕ and vector potential **A** are related to the potential energy U and potential momentum $\mathbf{p}_{el}$ of a particle of charge q by

Potential energy:	$U = q\phi$	[Joules]	(3-5)
Total energy:	$H = E + q\phi$	[Joules]	(3-6)
Potential electrodynamic momentum:	$\mathbf{p}_{el} = q\mathbf{A}$	[kg·m/sec²]	(3-7)
Generalized canonical momentum:	$\mathbf{P} = \gamma m\mathbf{v} + q\mathbf{A}$	[kg·m/sec²]	(3-8)

The Maxwell electric and magnetic field components represented as a covariant Lorentz tensor of rank 2 may be expressed in matrix form as

$$F_{\mu\nu} = \begin{vmatrix} 0 & c\mathbf{B}_z & -c\mathbf{B}_y & \mathbf{E}_x \\ -c\mathbf{B}_z & 0 & c\mathbf{B}_x & \mathbf{E}_y \\ c\mathbf{B}_y & c\mathbf{B}_x & 0 & \mathbf{E}_z \\ -\mathbf{E}_x & -\mathbf{E}_y & -\mathbf{E}_z & 0 \end{vmatrix}$$

Raising both indices yields the contravariant form of the electromagnetic tensor $F^{\mu\nu}$ which changes the sign of the column and row components.

In terms of the vector potential $A_\mu = (\mathbf{A}, i\phi)$, the Lorentz tensor may be written in compact form as

Induced electrokinetic force: $\quad F_{\mu\nu} = \delta A_\nu/\delta x_\mu - \delta A_\mu/\delta x_\nu \quad$ [Newtons] $\quad$ (3-9)

Electric and magnetic fields are related to the scalar and vector potentials as follows:

Electric field intensity (strength):
$\mathbf{E} = -(\delta\phi/\delta x, \delta\phi/\delta y, \delta\phi/\delta z) - d\mathbf{A}/dt$
$= -\nabla\phi - \delta\mathbf{A}/dt$ [volts/m] (3-10)

Magnetic field (flux density):
$\mathbf{B} = (\delta A_z/\delta y - \delta A_y/\delta z,$
$\delta A_x/\delta z - \delta A_z/\delta x,$
$\delta A_y/\delta x - \delta A_x/\delta y)$
$= -\nabla\Phi_m = \nabla \times \mathbf{A}$ [Webers/m²] (3-11)

In the Evans' tetrad formalism

Electric field intensity (strength):
$\mathbf{E} = -\nabla\phi^a - \delta\mathbf{A}^a/dt$
$\quad - \omega_0{}^a \mathbf{A}^b + \omega^a{}_b \phi^b$ [volts/m] (3-12)

Magnetic field (flux density): $\quad \mathbf{B} = \nabla \times \mathbf{A}^a - \omega^a{}_b \times \mathbf{A}^b \quad$ [Webers/m²] $\quad$ (3-13)

where $\omega_0{}^a$ = scalar spin connection, $\omega^a{}_b$ = vector spin connection.

3. Electromagnetic 4-Potential

Applying a gauge symmetry Lorenz condition on the vector potential gives

$$\delta A_v/\delta x_v = 0 \qquad \text{[Newtons]} \qquad (3\text{-}14)$$

The Lorenz gauge condition ($\delta_a A^a = 0$) represents a partial gauge fixing of the electromagnetic vector potential **A** and is given by

$$\nabla \cdot \mathbf{A} + 1/c^2\, \delta\phi/\delta t = 0 \qquad \text{[V·s·m}^{-2}\text{]} = \qquad (3\text{-}15)$$
$$= \delta A_x/\delta x + \delta A_y/\delta y + \delta A_z/\delta z + (\epsilon_0\mu_0)\cdot dV/dt \qquad \text{[rad}^{-1}\text{]}$$

Gauge fixing relates to an integration constant C which is determined by the limits of integration for a given situation. Only physical quantities remain invariant under gauge transformations. A gauge transformation is unchanged by a transformation of the form **A'** = **A** + $\nabla\phi$ in which a new vector potential **A'** may be defined by addition of an arbitrary scalar function ϕ. A Coulomb gauge is defined by imposing the requirement $\nabla \cdot \mathbf{A} = 0$ while a Lorenz gauge is defined by $\nabla \cdot \mathbf{A} = -\mu_0\epsilon_0(\delta\phi/\delta t)$. Gauge transformations allow changes to the scalar ϕ potential and vector potential **A** without changing the field equations or the **E** and **B** field components. The Lorenz group leaves invariant the quadratic $x_0^2 + x_1^2 + x_2^2 + x_3^2 = 0$ on 4-D Euclidean space, a fundamental postulate of Special Relativity.

The Lorentz force on a positive charge in motion is given by

Lorentz force $\qquad \mathbf{F} = q^+(\mathbf{E} + \mathbf{V} \times \mathbf{B}) \qquad$ [Newtons] $\qquad (3\text{-}16)$
$\qquad\qquad\qquad\quad = q^+[-\nabla(\phi - \mathbf{v}\cdot\mathbf{A}) - d\mathbf{A}/dt]$

An additional dielectric force arises in a nonuniform energy density proportional to gradient of vacuum dielectric:

Dielectric force $\qquad \mathbf{F} = q\mathbf{VB} + \nabla K_{PV}(x,\omega). \qquad$ [Newtons] $\qquad (3\text{-}17)$

As an example of the induced electrokinetic field E_k induced by a changing current flow $i(t)$, consider the movement of electrons within a parallel plate capacitor. The mechanism by which an alternating current is able to flow across a capacitor is described in standard texts in terms of Maxwell's displacement current. The induced electrokinetic force $F_k = qE_k = -q\delta\mathbf{A}/\delta t$ provides an explanation for the mechanism for energy transfer in the absence of electron flow through the intervening space or dielectric separator. During initial charging of the capacitor, current flows radially outward from the connecting lead to the periphery of the conductor plate as voltage is applied. The current is proportional to the rate of change of voltage which varies as $i \cong Cdv/dt$ where C is the capacitance. In a capacitor, current leads voltage and reverses direction with voltage reversal. The phase difference between current and voltage depends on the relative amount of resistance, inductance and reactance in a circuit. The flow of electrons is accompanied by an associated electrodynamic potential momentum **A** which is in opposition to the direction of current. An induced electric field $E_k = -\delta\mathbf{A}/\delta t$ is generated in the adjacent conductor plate(s) resulting in an inward radial current flow towards the connecting terminal.

Spacetime is a manifestation of a physical field. – Carlo Rovelli and Francesca Vidotto

There is no causal relation between the magnetic vector potential and the magnetic field since both are simultaneously produced by the same electric current. – Oleg D. Jefimenko

3. Electromagnetic 4-Potential

Electromagnetic 4-Vector Potential

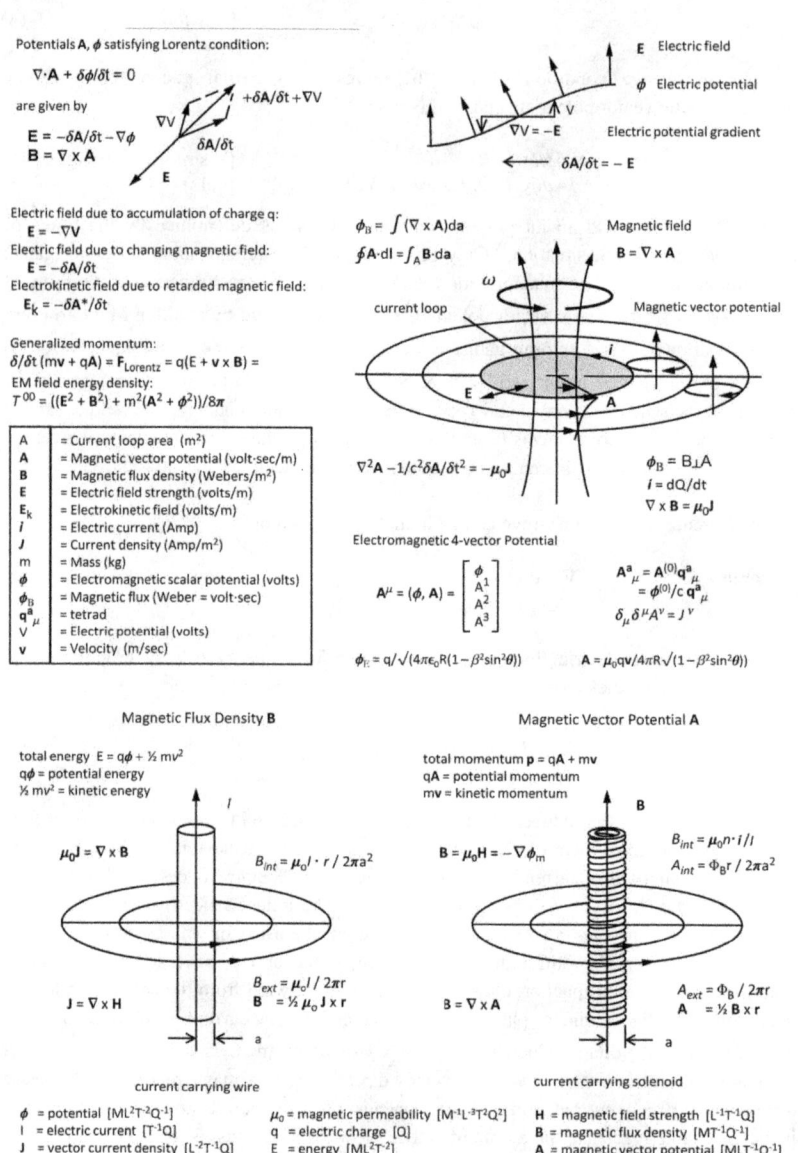

Fig. 3-1. Geometric representation of components of the electromagnetic 4-vector potential $A^\mu = (\phi, \mathbf{A})$ where ϕ = scalar potential and $\mathbf{A}$ = magnetic vector potential.

Force, then, is... Only a Vector. Thy barbèd arrows now have lost their sting. – James C. Maxwell

3. Electromagnetic 4-Potential

Planck energy density ($U_P = c^7\hbar G^2 = m_P c^2/l_P^3$) and corresponding mean free path length $\langle l_m \rangle$ varies in different regions of space. Photon velocity is proportional to the mean free path. The local energy density departures from the local zero-point energy level represent electrical stresses plus or minus electrical potential. Wave speed is controlled by tension, the strength of attraction between neighboring weights (such as a system of weights connected by springs), speed is proportional to square root of tension, e.g., four times tension doubles wave speed. The ratio of effective mass to photon energy-momentum depends on how much it is slowed down in regions of increased electromagnetic energy density. The apparent wave speed limit v_{wp} for Planck dipoles corresponds to the tangential speed of a helical photon or $\sqrt{2}$ times the speed of light c which equals the Weber's constant $c_w (= \sqrt{2}c)$.

The photon, in the model considered, constitutes a propagating torsional wave disturbance of a collection of Planck dipoles composed of positive and negative mass with quantized angular momentum. The rotation of the Planck masses provides the observed photon angular momentum with mass cancellation providing net zero rest mass. A negative mass m_P^- corresponding to a vacancy hole in paired (yin/yang) rotation with a positive mass m_P^+ allows a stable dynamical excitation state without recombination. A photon represented as a kink or topological soliton Planck dipole spin density wave is illustrated in Figs. 3-3 and 3-4. As shown, light waves represent a soliton spin density wave disturbance in a polarized vacuum of variable index of refraction $K_{PV}(r,\omega,M)$. Electric field intensity **E** corresponds to alignment of Planck dipoles while the magnetic field intensity **H** corresponds to a differential rotation of dipoles. The wave function is described as $\Psi = A\sin(2\pi S/\lambda - \omega t)$ where A = amplitude and S = arc length $S = c/v(\sin S_v = c/v_r S_0$. The photon magnetic field **B** represents an alignment of the spin angular momentum vectors of spinning Planck dipoles. The electric field **E** of the photon represents an instantaneous, periodic in-phase alignment of Planck dipoles. The Poynting vector **S** (= **E** x **H**) represents the direction of energy flow. The shape of KdV solitary waves are extraordinarily stable against small distortions and, like solitons, photons may remain unchanged after collision.

Position and momentum fluctuations in spacetime consist of elastic deformations due to curvature and plastic deformations due to torsion. Commutation relations between curvature R and torsion Q may be represented as conjugate variables

$$[Q, R] = (\hbar G/c^3)^{-3/2} \qquad (3\text{-}18)$$

analogous to position and momentum uncertainty relation $[x, p] \geq i\hbar$. At small scales $\Delta Q \Delta R \geq l_P^{-3}$ where l_P is the Planck length. In the de Sabbata/Sivaram spin torsion model, a quantized metric may be constructed of a network of line defects induced by torsion each with an intrinsic spin of $n\hbar/2$ representing a loop closure failure. The torsion defect τ (= $\theta 2\pi/l_p$) is related to $n\hbar = \oint pdq = n \cdot m_P l_P^2/t_P = n \cdot m_P (\theta 2\pi/\tau)^2/t_P$.

Waves from moving sources: Adagio, Andante, Allegro moderato. – Oliver Heaviside

Induction is equal to the number of lines of force passing through the circuit. – Michael Faraday

The electromotive force depends on the change in the number of lines of magnetic induction which pass through a circuit. – James Clerk Maxwell

Mass and spin are two elementary and independent original concepts: as a mass distribution in a space-time is described by energy-momentum tensor, so a spin distribution is described in a field theory by a spin density tensor... We can say that as the mass is responsible for curvature, spin is responsible for torsion. – Venzo de Sabbata and C. Sivaram

3. Electromagnetic 4-Potential

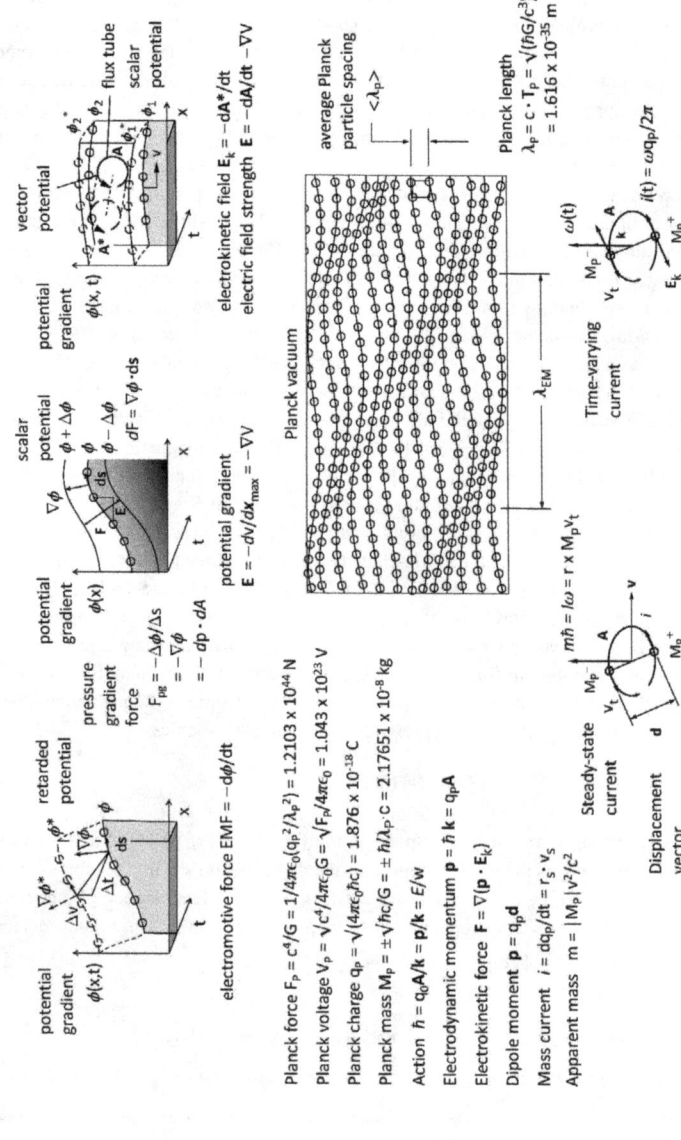

Fig. 3-2. Wave dynamics of quantum vacuum Planck scale (BEC) superfluid composed of Planck particles (M_P^+, M_P^-).

3. Electromagnetic 4-Potential

Soliton kink spin wave

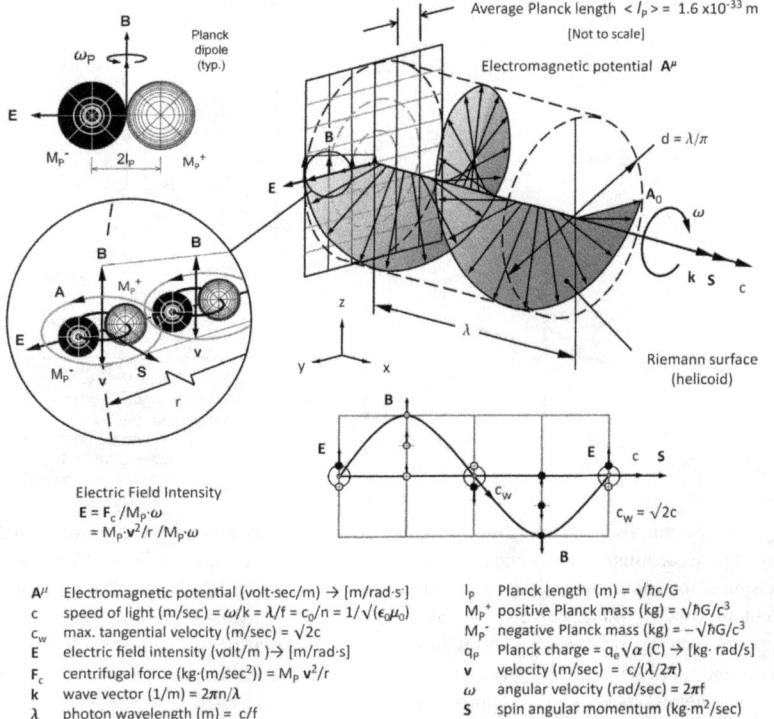

Fig. 3-3. Soliton kink spin density wave in a Planck vacuum formed by synchronous alignment of Planck mass dipoles. In the presence of curvature and torsion, a loop closure failure defect in the Planck mass trajectory arises corresponding to a tetrad rotation and translation in tangent space which yields an uncertainty in position and momentum on the order of the Planck scale. An effective mass results from a change in curvature or torsion.

Waves we cannot have lest they be waves in something. – Sir Oliver Lodge

There must be some fact of which we are entirely ignorant and whose discovery may revolutionize our views of the relations between waves and ether and matter. – Sir William Bragg

However vast the darkness, we must supply our own light. – Stanley Kubrick

"LET THERE BE LIGHT!" And there was light--… – Isaac Asimov (The Last Question)

The dividing line between the wave or particle nature of matter and radiation is the moment "Now". – Lawrence Bragg

This maze of symbols, electric and magnetic potential, vector potential, electric force, current, displacement, magnetic force, and induction, have been practically reduced to two, electric and magnetic force. – George Francis Fitzgerald

The local energy state of the vacuum may be considered to be equivalent to the mass-energy of the object it encapsulates. – Geoffrey S. Diemer

31

3. Electromagnetic 4-Potential

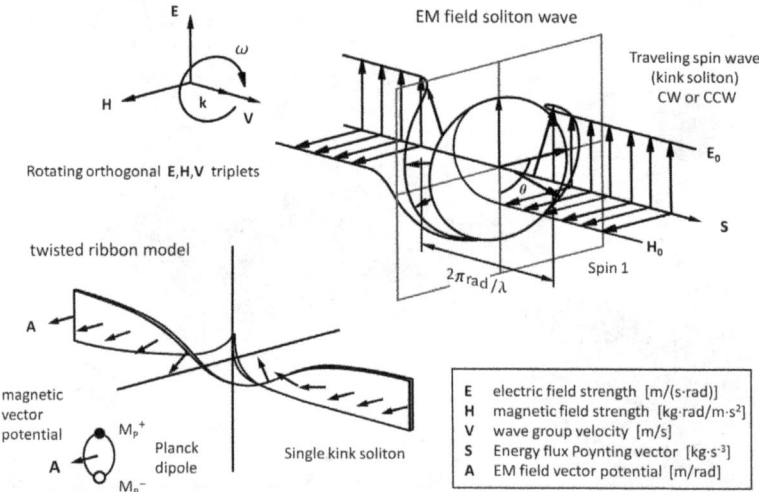

Fig. 3-4. Soliton twisted ribbon model of spin aligned Planck dipoles. A spinning Planck mass dipole constitutes a mass current generating a magnetic vector potential **A**. Rotation of the spin plane about the wave propagation axis creates topological charge. Electric charge [Coulombs/sec] is dimensionally equivalent to mass current rotational flow rate (mass x topological charge rotation velocity) with units of [kg·rad/sec] where rotation refers to spin precession rather than spin. Magnetic vector potential **A** constitutes electromagnetic field momentum/unit charge which is dimensionally equivalent to [m/rad]. A photon wavetrain represents a spin density wave disturbance in a quantized Planck vacuum composed of Planck dipole masses (M_P^+, M_P^-). A pair of coupled rotational solitons exhibit confinement properties similar two quarks interconnected by a string.

The finite time of propagation was thought by field theorists to be sufficient evidence that a field exists in places where there was no matter. The existence of [electromagnetic] waves showed that the propagation of electric and magnetic effects take time, as claimed in the field theory. – William Berkson

The imaginative ideas which scientific work originates depend on a sensitive receptiveness to the oddity of nature, essentially similar to that of the artist. When they are first proposed they often have the same quality of unexpectedness, and perhaps wrongheadedness, as say, cubism, abstract art, or atonal music. – C. H. Waddington

A style of reasoning makes it possible to reason towards a certain kind of propositions, but does not itself determine their true value. – I. Hacking

Ignorance is like a delicate exotic fruit; touch it and the bloom is gone. – Oscar Wilde

Works are of value if they give rise to better ones. – Alexander von Humbolt

The value of an idea lies is using it. – Thomas Edison

One must be prepared to follow up the consequences of theory, and feel one just has to accept the consequences, no matter where they lead. – P.A.M. Dirac

3. Electromagnetic 4-Potential

The velocity of light (c) is given by the Maxwell relation and associated equivalences

$$c = c_o/n = 1/\sqrt{(\epsilon_0 \mu_0)} = Z_0/\Gamma\mu_0 = \sqrt{(v_p v_g)} = c_o/\sqrt{K_{PV}} = c_o/\Gamma = \nu\lambda = \beta/\nu \qquad (3\text{-}19)$$

where:

c_o	Speed of light in vacuo (gravity-free, flat space) $c = \lambda f = \omega/k = \sqrt{(v_g v_p)}$ $= 1/\sqrt{(\mu_0 \epsilon_0)} = c \cdot n = Z_0/\mu_0 = |E|/|H|/\mu_0 = \Gamma c = c\sqrt{K_{PV}} = v/\Delta\phi/\pi = \mathbf{E}/\mathbf{B} = l_P/t_P$		
	Value: 2.997924 x 10^8 (Experimental)	Units: m/s	Dimensions: LT^{-1}
ϵ_o	Electric permittivity $\epsilon_0 = \mu_0/Z_0^2 = \epsilon/K = \epsilon/\sqrt{K_{PV}} = 1/\mu_0 c^2 = \epsilon\sqrt{g_{00}} = \Gamma\epsilon_{00}$		
	Value: 8.854187 x 10^{-12} (Experimental)	Units: Coul^2/nt·m^2 (= farads /meter)	Dimensions: MLQ^{-2}
μ_o	Magnetic permeability $\mu_0 = Z_0/c = \mu/K_m = \mu/\sqrt{K_{PV}} = 1/\epsilon_0 c^2 = \mu\sqrt{g_{00}} = \Gamma\mu_{00}$		
	Value: 4π x 10^{-7} (Defined)	Units: Weber/amp-m (= henrys /meter)	Dimensions: M^{-1}L^{-3}T^{2}Q^2
n	Index of refraction ($c_0/c = \sqrt{\epsilon\mu} = \beta^{-1} = c_0 k/\omega = c_0/(c^2/v_g) = \Gamma = \sqrt{K_{PV}} = 1/\sqrt{g_{00}}$		
	Value: varies	Units: Dimensionless	Dimensions: -
v_p	Phase velocity $v_p = c^2/v_g = c/\beta = E/p = \omega/k = \gamma mc^2/\gamma mv = c/(\Delta\phi/\pi)$ $= c/\sqrt{(1 - (n^2\omega_c^2)/\omega^2)}$		
	Value: varies	Units: m/s	Dimensions: LT^{-1}
v_g	Group velocity $v_g = c^2/v_p = d\omega/dk = c \cdot \beta = c\sqrt{1 - n^2\omega^2/\omega^2} = c \cdot (\Delta\phi/\pi)$; $\omega = \gamma\omega_0$, $k = \gamma\beta\omega_0/c$, $\mathbf{p} = \hbar\mathbf{k} = \gamma m\mathbf{v} = \mathbf{S}/c^2 = \mathbf{v} \cdot \rho_e/c^2$		
	Value: varies	Units: m/s	Dimensions: LT^{-1}
K_{PV}	Polarizable vacuum index of refraction $K_{PV}(r,\omega,M) = \epsilon/\epsilon_0 = \mu/\mu_0 = \Gamma^2 = c_0^2/\phi$ $= 1/g_{00} = g_{11} = g_{22} = g_{33} = n^2$. Index of refraction $n = \sqrt{(\epsilon/\epsilon_0)(\mu/\mu_0)} = \sqrt{(KK_m)}$		
	Value: varies	Units: Dimensionless	Dimensions: -
Γ	Gravitational gamma $\Gamma = dt/d\tau = 1/\sqrt{(1 + (2\phi/c^2))} = 1/\sqrt{(1 - v_e^2/c^2)} = \sqrt{K_{PV}}$		
	Value: varies 1 to ∞	Units: Dimensionless	Dimensions: -
ν	Frequency $\nu = f = c/\lambda = \nu_0/\Gamma = \omega\lambda/k = \omega/2\pi$		
	Value: varies	Units: Hz	Dimensions: T^{-1}
λ	Wavelength $\lambda = c/f = 2\pi/k = \nu \cdot \omega/k = \lambda_0/\Gamma = \lambda_0/\sqrt{K_{PV}}$		
	Value: varies	Units: m	Dimensions: L
Z_0	Vacuum impedance $Z_0 = n|\mathbf{E}|/|\mathbf{H}| = \mu_0 c_0 = \sqrt{\mu_0/\epsilon_0} = 1/\epsilon_0 c_0 = Z_P/4\pi$		
	Value: 376.73	Units: Ohm	Dimensions: ML^{2}T^{-1}Q^{-2}
Z_P	Planck impedance $Z_P = 1/4\pi\epsilon_0 c = \hbar/q_P^2 = 4\pi Z_0$		
	Value: 29.98	Units: Ohm	Dimensions: ML^{2}T^{-1}Q^{-2}
α	Fine structure constant $\alpha = k_e e^2/\hbar c = (1/4\pi\epsilon_0)e^2/\hbar c = e^2 cm_0/2h = R_e/R_C$ $= R_C/R_{em} = \frac{1}{4}(Z_0 \cdot G_0) = e_p^2 = (e/q_P)^2 = g_e^2/4\pi \simeq 1/20\Phi^4$		
	Value: 0.007297352568E-3	Units: Dimensionless	Dimensions: -

3. Electromagnetic 4-Potential

The relatively high, finite velocity of light in vacuo indicates a medium with a high bulk modulus K_B (= $\Delta P/(\Delta V/V$ = $-V(\Delta P/\Delta V)$ with large elastic, but finite stiffness or restoring force. The electric permeability ϵ_0 and the magnetic permeability μ_0 corresponds to a superfluid density and compressibility, respectively. The energy density of the vacuum by some estimates is very high (u_P = $c^7/\hbar G^2$ = 4.632 x 10^{113} J/m^3) whereas the observable energy density of the universe (i.e., energy possessing angular momentum) in the current epoch is of 122-130 orders of magnitude less. With a zero-point energy cutoff at the Planck length l_P, the mass density of the vacuum may be expressed as ρ_{vac} = m_p/r_p^3= $c^5/G^2\hbar$ ≈ 10^{98} kg/m^3. For comparison, this calculated mass of 10^{98} kg is larger than the estimated mass of the visible universe (~1.2-1.7 x 10^{53} kg). One possible resolution for this incredible mismatch has been advanced by Oldershaw[19], by revising the Planck scale based on a scaled gravitational coupling constant G^ψ = $(\Lambda^{1-d})^\psi G_0$ adjusted to an atomic scale where G_{-1} is $\Lambda^{2.174}$ and G_0 is the Newtonian gravitational constant. Using G_{-1} as the appropriate coupling factor, the discrepancy is reduced by 115 orders of magnitude. Macken[20] proposes the discrepancy between the quantum mechanical energy density of the vacuum (~10^{113} J/m^3) and the derived cosmologically observed energy density (~10^{-9} J/m^3) is due the difference in net quantized angular momentum possessed by particles (bosons and fermions) and that of the vacuum modeled as a Bose Einstein Condensate (BEC) superfluid. As explanation for the vacuum energy density mismatch between calculated and observed values, Winterberg[7] hypothesizes that the vacuum must contain a large negative mass cancelling the large positive mass of the zero-point energy. In the Winterberg model, electric charge is associated with a net imbalance of spin direction of vortical flow.

In the Winterberg Planck aether hypothesis, the physical vacuum is ultimately composed of spinning Planck scale mass dipoles of positive and negative energy. Mass may be viewed a result of increased energy density due to increased dipole volumetric density. Gravitationally bound collections of Planck dipoles of positive and negative mass, such as the neutrino or postulated spinor rotons, with net zero particle mass but with positive mass energy of the gravitational field, may account for a portion of dark matter consisting of non-baryonic matter and cold, diffuse baryonic matter. In accordance with the Jefimenko gravitational model, an accumulation of mass results in a polarization of the vacuum with matter consisting of a concentration of positive energy and an outwardly displaced gravitational field of negative energy. One half of the gravitational binding energy lost during mass accretion is converted to heat while the remainder is radiated away. Hence, the total energy of the consolidated mass is less that the mass total of the individual masses prior to contraction.

Electrical permittivity ϵ in units of farads/m with dimensions [$M^{-1}L^{-1}T^2Q^2$], in terms of the derived units kg·rad^2/m^2 or [$M\Theta^2L^{-2}$], is dimensionally equivalent to mass times torsion squared (kg·rad/m·rad/m). Classical action of torsion involves contact interaction between spins. Effective mass m_i is associated with a change in torsion τ due to increased energy density. As change in torsion (rad/m) is related to a change in pitch, mass is analogous to resistance to change in frequency (1/s). The derived dimensions of magnetic permeability μ (hernrys/m) [MLQ^{-1}] becomes (rad·m·sec^2/kg) or [$\Theta M^{-1}LT^2$], which corresponds to angular acceleration/unit mass. The units and dimensions of magnetic vector potential **A** (V·s/m) or [$MLT^{-1}Q^{-1}$] becomes m/rad [$L\Theta^{-1}$]. The derived mechanical dimensions of electric charge in (Coulombs) [Q] are equivalent to (kg·rad/s) [ΘMT^{-1}], a measure of rotational mass motion which appears to represent rate of precession. If neither mass or charge are fundamental, what is the quintessence, the "stuff" of the vacuum from which the Planck dipoles are formed? The ancient philosophical chestnut "Why is there 'something' rather than 'nothing'?" was previously imponderable, while the "What is the something?" is slowly being discovered with ultimate answers to both guided by mathematical physics.

4. Soliton Confinement

Unlike light rays composed of multiple photons, individual photons do not undergo dispersion in a vacuum, but remain confined within a constant radius during propagation in the manner of a soliton wave. The mechanism for soliton confinement is hypothesized to be the result of an abrupt change in permittivity at the amplitude envelope periphery at the speed of light radius creating an optical waveguide. A radial variation in Planck dipole density gives rise to a cylindrical well in the index of refraction K_{PV} of the vacuum which acts as a waveguide and provides topological confinement. The Korteweg de Vries (KdV) equation in soliton theory has been shown to arise in Fermi-Pasta-Ulam nonlinear lattice vibrations as the spacing approaches zero. In a quasi-lattice of a polarizable vacuum composed of Planck dipoles, each Planck mass acts as a harmonic oscillator with coupled harmonic oscillators forming a helical wave and are reflected from the index of refraction boundary similar to an a dielectric resonator or optical ring resonator. At a peak resonant frequency of the Planck dipoles ($\omega_P = \sqrt{(c^5/\hbar G)} = 1.855 \times 10^{43}$ rad/s), a standing wave oscillates within the equivalent of a dielectric resonator in a hybrid (HEM) mode. Assuming a Kerr electro-optic effect due to the alignment of Planck dipoles in an electric field, the refractive index of a polarizable vacuum in analogy to material dielectrics increases with increasing frequency and square of the electric field intensity.

$$n(\omega,|E|^2) = n_0(\omega) + k|E|^2 = n_0 + n_2 I \qquad (4\text{-}1)$$

where n is the index of refraction parallel or orthogonal to the direction of the electric field, ω is the photon frequency, k is a nonlinear susceptibility constant, n_2 is a second order nonlinear refractive index of the vacuum medium, E is the electric field intensity and I is the radiation intensity. For a vacuum, n_0 is ordinarily taken as 1 and $n_2 = 0$. For nonlinear effects with $n_2 > 0$, the nonlinearity is assumed to arise as the result of Planck resonances at the boundary of the optical waveguide with a radial variation in phase velocity v_p. In particular, the phase of the spin wave undergoes a $\pi/2$ change in direction as the Planck dipoles at the phase boundary can no longer remain synchronously in line thus breaking symmetry.

Once polarized, the dipoles are subjected to a torque $\tau = \mu \times \mathbf{E}$. For a circularly polarized photon, the complex wave field may be expected to give rise to torque orthogonal to the direction of propagation. A notional diagram of cylindrical waveguide resulting in topological confinement is illustrated in Fig. 4-1. The waveguide resembles a cladded optical fiber with a refractive index discontinuity at the periphery resulting in total internal reflection confining light within the core when the refractive index of the core n_1 is greater than the substrate/cladding n_2 index. At the cylindrical boundary, the $\mathbf{E}$-field has both radial and tangential components. Outside the boundary, the Planck dipoles become depolarized with random $\mathbf{E}$-field direction. For a hollow waveguide, the wave number is reduced from that in a vacuum resulting in a phase velocity $v_p > c$ and a group velocity $v_g < c$. For a focused beam of photons, even in a vacuum, the axial velocity is reduced below c due to the inclination of the wave vector $\mathbf{k}$ with respect to the axial direction. The group velocity of light $v_g = c/n$ in the case of plane waves. For an individual photon in a vacuum, the group velocity $v_g = c$ with a helical tangential velocity $v_t = \sqrt{2}c = c_W$ where c_W is Weber's constant. In Maxwell's model, the velocity of transverse waves v is the same as the ratio of electrostatic and electromagnetic units ($= c = \sqrt{v_g v_p}$). The velocity of light squared appears as a conversion constant in numerous relations, e.g., $c^2 = E/m = 1/\epsilon_0\mu_0 = \mathbf{E}^2/\mathbf{B}^2 = \phi_E^2/\phi_B^2 = v_p/v_g = \mathbf{S}/\mathbf{p} = V\cdot q/m = Gm_P/l_P = 2GM/r_S = (\mathbf{v} \times \mathbf{g})/\mathbf{K} = \phi_g\cdot\mathbf{v}/\mathbf{A} = (v/\beta)^2 = \omega^2/\mathbf{k}^2 = 4\pi G/\mu_g$.

4. Soliton Confinement

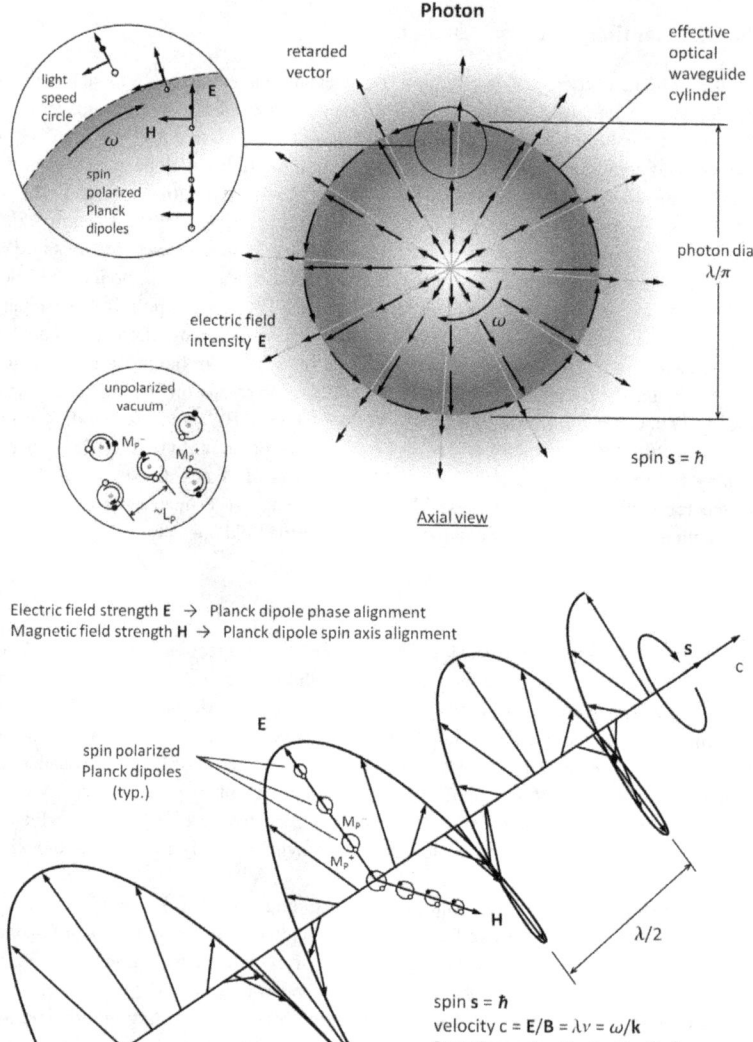

Fig. 4-1. Soliton spin wave confinement in an optical waveguide composed of oriented polarized Planck mass dipoles. At the boundary, Planck dipoles can no longer synchronize.

That he [Einstein] may sometimes have missed the target ... as, for example, in his hypothesis of light quantity, cannot really be held against him. – Max Planck

The task is not to see what has never been seen before, but to think what has never been thought before about what you see everyday. – Erwin Schrödinger

5. Electromagnetic Field Dimensions

The photon has no electric charge, yet is characterized by an electric field component. As conceptualized by Faraday, the electric field represents a tension $(\mathbf{E}^2 + \mathbf{B}^2)/8\pi$ along the field lines and is a measure of the polarization of the vacuum corresponding to a separation of charge. Electric charge q is conventionally taken as a fundamental property and represented as a separate dimension [Q] in the S.I. MKS system of units. However, although related to topological charge, its exact nature has remained a mystery. The conversion of the dimensions of electric charge into equivalent mechanical units is implicit in the Schrödinger definition of electrodynamic momentum **p** of an elemental charge[21]

$$\mathbf{p} = \hbar\mathbf{k} = q\mathbf{A} \tag{5-1}$$

where

p	momentum ($= \mathbf{S}/c^2$); Poynting vector $\mathbf{S} = \mathbf{E} \times \mathbf{H}$		
	Value: Varies	Units: kg·m/s	Dimensions: MLT^{-2}
$\hbar$	Dirac $\hbar$ $(= h/2\pi = p/k = E/\omega = R_C mc = m_P \cdot r_P \cdot c)$		
	Value: 1.0545E-34	Units: J·sec = rad·m^2·kg·s^{-1}	Dimensions: $\Theta L^2 MT^{-1}$
k	wave vector ($\mathbf{k} = k_x\mathbf{i} + k_y\mathbf{j} + k_z\mathbf{k} = m\mathbf{v}/\hbar$; wave no. $k = |\mathbf{k}| = 2\pi/\lambda = 2\pi n/\lambda_0$ $= n\omega/c = k_0/\sqrt{(1-\beta^2)}\cdot\hat{\mathbf{n}}$; $k_0 = \omega\sqrt{(\mu_0\epsilon_0)} = \omega/c$, λ = matter wave wavelength		
	Value: Varies	Units: m^{-1}	Dimensions: L^{-1}
q	electric charge (q = F/E = E/V = V·C/m = $m_e(\omega_C + \omega_p) = \hbar k/A$)		
	Value: $-1.60219E-19$	Units: Coulomb = Amp·sec [kg·rad/sec]	Dimensions: $Q\Theta MT^{-1}$
A	vector potential ($\mathbf{A} = \hbar\mathbf{k}/q = (v/c^2)\phi$; $\mathbf{A}^* = -\int \mathbf{E}_k dt$ + const., $\mathbf{E}_k = -d\mathbf{A}^*/dt$)		
	Value: Varies	Units: volt·sec/m = Wb/m. T·m = N/Amp	Dimensions: $MLT^{-1}Q^{-1}L\Theta^{-1}$

The mechanical dimensions of electric charge [Q] is equivalent to [kg·rad/sec] which corresponds to a rotational mass current. Total momentum of a moving electric charge is the sum of the inertial momentum and electrodynamic momentum

$$\mathbf{p} = \hbar\mathbf{k} = \gamma m\mathbf{v} + q\mathbf{A} \tag{5-2}$$

The velocity of a moving charge or system of charges ($\mathbf{v} = (\hbar\mathbf{k} - q_0\mathbf{A})/\gamma m$) is a function of the imbalance between the total momentum $\hbar\mathbf{k}$ and the electrodynamic momentum $q_0\mathbf{A}$. Momentum **p** times matter wavelength λ_{dB} equals Planck's constant h (= mv·λ_{dB}). In the Planck vacuum model considered here, the mass current is taken as motion of positive and negative Planck masses (m_P^+, m_P^-). For a rotating Planck dipole, the reduced Planck mass $\mu_P = (m_P^+ \cdot m_P^-)/(m_P^+ + m_P^-) = 0$. Hence, the moment of inertia ($I = \mu_P R_P^2 = 0$), angular momentum ($L = \mu_P v R_p = 0$) and the kinetic energy of rotation ($E_r = \mu_P v^2/2 = 0$) are zero.

Pure mathematics is, in its way, the poetry of logical ideas. – Albert Einstein

A strict materialist believes that everything depends on the motion of matter. – James C. Maxwell

5. Electromagnetic Field Dimensions

Conservation of momentum in electrodynamics is realized as both moving charges and attendant electromagnetic fields carry momentum, i.e., total momentum is the sum of mechanical momentum of the particles and the electromagnetic momentum of fields. In electrodynamics, Newton's 3rd law does not hold when considering just the mechanical momentum. In a collection of point charges, mechanical momentum of particles may be interchanged with the electrodynamic momentum of fields. As Mead[21] has shown, the inertia of a collection of interacting charges varies with the square of the number of charges. Hence, the momentum of an electric current in a conducting wire exhibits considerably more momentum than isolated electric charges.

The mechanical dimensions of electric charge may be deduced by setting Newton's expression for inertial mass $F = m_i a$ equal to the electric force on a charged particle $F = q^+E$. At velocities small compared to the velocity of light, inertial mass is essentially constant. Solving for charge: $q = m_i a/E$ where q = electric charge [kg·rad/s], E = electric field intensity [m/s·rad], m_i = mass [kg] and a = acceleration [m/s^2].

Klyushin[22,23,24] has proposed the following dimensionality relating mechanical units for electric units

$$1 \text{ coulomb} \times \text{volt} \times \text{meter/kilogram}^2 = 1 \text{ m}^3/\text{kg·sec}^2 \qquad (5\text{-}3)$$

Spears[25] independently derives the same dimensional units for the gravitational constant G = $G_e = F_g r^2/M_e^2 = G_0 \epsilon_{12}^2/\epsilon_0^2$ [QVL/M^2]. By equating the electrostatic force F_e between two separated electrons to Newton's empirical gravity force $F_{gN} = GM_e M_e/r^2$, a value of G_e in MKS electrostatic units is defined. The calculated numerical value of G = G_e = −6.68541 x 10^{-11} (coulomb·volt·m)/kg^2 is within 0.2% of a widely accepted value of G = −6.67259(85) x 10^{-11} m^3/kg·s^2 and within the range of variations in empirical measurements of G described in the literature (eg., 6.6656(6) x 10^{-11} [*Fitzgerald et al.*] to 6.71540 x 10^{-11} m^3/kg·s^2 [*Michaelis et al.*])[26]. The minus sign is a convention signifying the gravitational force is attractive.

As shown in Fig. 1-2, the E and H field intensity within the photon wave packet increases linearly with radial distance perpendicular to the propagation axis up to the maximum intensity envelope radius at $r = \lambda/2\pi$ and decreasing as $1/r^2$ external to the EHV spin field. The E field intensity (strength) is a vector with a spatial gradient in energy density ($-\nabla\phi$) component and the temporal rate of change of magnetic vector potential ($-dA/dt$) component. The electric field intensity (electric potential gradient) has dimensions of [volts/meter]. The energy gradient results in a difference in the velocity of light in a direction aligned with the E field and that in an opposed direction, hence, Δc has dimension of [meters/sec]. Substituting into the Coulomb electrostatic force equation $F = q^+E$, the corresponding units of electric charge q is (kg·rad/sec). The units of the magnetic field strength H are [Amp·turns/meter]. The corresponding H derived units are (rad^2·kg/(m·sec^2). Similarly, the units of the electromagnetic vector potential A are (volt·sec/m) which corresponds to derived units of (meters/rad).

The mechanical dimensions [kg·rad/sec] of the Coulomb [C] describes a rotational mass flow rate corresponding to a tangential flow through a radial rotating reference plane associated with torsion/spin. The electrical potential (voltage) dimensions [m^2/sec·rad] equivalent to [V] seem to describe an areal velocity of swept area/unit time reminiscent of Kepler's 2nd law which may be possibly be interpreted as Planck dipole spin density waves swept area/unit time. A Coulomb x Volt [kg·m^2/sec^2 = N·m = Joule] seems suggestive of energy corresponding to, say, that of orbital whirl. Charge arises from this precession; mass arises as a result of wave interference. Electrical charge, in this model, represents

5. Electromagnetic Field Dimensions

mass flow in a spiral spin density wave of Planck particles with spin direction corresponding to ± charge. Charge is quantized since spin angular momentum $\hbar$ is quantized. Electrostatic potential energy (= $q\phi$) corresponds to a volumetric compression of Planck dipole charge elements. Planck charge q_P arises from Planck mass precession.

A summary table of mechanical equivalents of electromagnetic quantities is illustrated in Table 5-1. For reference, also shown are Planck scale units. For simplicity, the conversion formulas show only the rotation angle (topological charge) in radians associated with electric charge (Q). The corresponding conversions based on Eqn (5-2) are tabulated in a subsequent section for various electromagnetic parameters illustrating the derived dimensions and MKS SI units conversion into mechanical equivalents.

Units conversion may also be expressed in terms of the electron Compton wavelength λ_C. For example, a length may be represented as the Bohr radius of the hydrogen atom expressed as $a_0 = 4\pi\epsilon_0\hbar/m_e e^2 = \hbar/m_e c\alpha$. A time interval corresponds to the time for light at velocity c to traverse this distance $\Delta t = \hbar/mc^2$. Similarly, the radius of an electron is a length denoted as the reduced Compton wavelength $\bar{\lambda}_C = \lambda_C/2\pi = \hbar/mc$. An angle θ in radians corresponds to an angle subtended by an arc length s equal to the radius r defined by the relation $s = r\theta$. For Noether's theorem for conserved physical quantities undergoing continuous global symmetry transformation in space or time, the matter-energy relations in terms of Planck units are given by

Space – Time Symmetry	Dimension	Conserved Physical Quantity	Matter-Energy Relation
Spatial translation	L	Linear momentum **P**	$= m\mathbf{v} = \hbar\mathbf{k} = \hbar/mc = \lambda_C/2\pi$
Temporal translation	T	Energy E	$= \hbar\omega = h\nu = mc^2 = \lambda_C/c$
Spatial rotation	Θ	Angular momentum **J**	$= \hbar\mathbf{k} = \hbar\lambda_C/2\pi$

where the wave number $k = |\mathbf{k}| = 2\pi/\lambda = 2\pi f/c = n\omega/c$ and the reduced Planck's constant $\hbar = h/2\pi = \mathbf{p}/\mathbf{k} = q_0\mathbf{A}/\mathbf{k} = (S/c^2)\mathbf{k} = E/\omega$.

Space-time reference frames in uniform relative motion known as inertial frames exhibit physical equivalence. Inertial frames are in constant relative motion with respect to each other and are related by Lorentz transformations. A Lorentz transformation $\mathcal{L}$ is a linear transformation of space-time that maps event coordinates (t, x, y, z) in one observer's reference frame to another set of coordinates (t', x', y', z') of the same event in another observer's reference frame. A change in position is a translation; a change in orientation is a rotation; and a change in velocity is a boost. The set of rotations and boosts are known as Lorentz transformations. The set of translations, rotations and boosts are known as Poincaré transformations. Möbius transformations are mappings in the complex plane corresponding to Lorentz transformations. The set of all Lorentz transformations in Minkowski spacetime is named the Lorentz group denoted as Lie group O(1,3). The Lorentz group is a 6-dimensional Lie Group of linear isometries of Minkowski spacetime whereas the Poincaré group is a 10-dimensional Lie group of affine isometries of Minkowski spacetime. Lorentz transformations and rotations may each be represented as 4 x 4 matrices. The subgroup of proper Lorentz transformations, Lorentz group SO(3,1), which is a subset of the Poincaré group, contains the set of all inertial frames and expresses a space-time symmetry or

5. Electromagnetic Field Dimensions

isotropy. Under Lorentz transformations, the form of physical laws remain unchanged in accordance with the general principle of covariance. Lorentz invariants of the electromagnetic field include $\mathbf{E}^2 - \mathbf{B}^2$ and $\mathbf{E} \cdot \mathbf{B}$. The Poincaré group is a group of Minkowski spacetime isometries with ten degrees of freedom in which the interval between events is invariant and includes translations (displacements in space and time), rotations in space and boosts (transformations relating two uniformly moving bodies). The Poincaré group includes two Casimir operators representing rest mass and intrinsic spin associated with elementary particle representations. The gauge group in particle physics represented as U(1) x SU(2) x SU(3) is Lie group of dimension 12 consisting of 1 photonic field + 3 bosonic fields and 8 gluonic fields where $1 + 3 + 8 = 12$ degrees of freedom.

By its very nature the Aether is a vibrating medium. – Isaac Newton

A new fact is battling strenuously for access to your ears. A new aspect of the universe is striving to reveal itself. But no fact is so simple that it is not harder to believe than to doubt at the first presentation. – Lucretius

I regret that it has been necessary for me in this lecture to administer such a large dose of four-dimensional geometry. I do not apologize, for I am really not responsible for the fact that nature in its most fundamental aspect is four-dimensional. – Alfred Norse Whitehead

Space-time does not exist without mass-energy. – D. Lynden Bell

The life contest is primarily a competition for available energy. – Ludwig Boltzmann

Nature has no reverence towards life. Nature treats life as though it were the most valueless thing in the world. – Erwin Schrödinger

There are two striking facts about the universe we live in: (1) The universe is not empty of matter; and (2) the universe is almost empty of matter. It is the task of fundamental physics to understand these facts. – Anthony Zee

In science, one tries to tell people in such a way as to be understood by everyone, something that no one even knew before – But in poetry, it's the exact opposite. – P.A.M. Dirac

Man can know the nature of things. – Henry Kuttner

Imagination will often carry us to worlds that never were. But without it we go nowhere. – Carl Sagan

Self-confidence is an important ingredient that makes for a successful physicist. – Victor F. Weisskopf

...if one chases a light wave with the speed of light, one would have in front of him a time independent wave field. – A. Einstein

In the field of observation, chance favours only the prepared mind. – L. Pasteur

Henceforth, space by itself, and time by itself, are doomed to fade away into mere shadows, and only a kind of union of the two will preserve an independent reality. – Hermann Minkowski

The essential fact is that all pictures which science now draws of nature, and which alone seem capable of according with observational facts, are mathematical pictures. – Sir James Jeans

All attempts to explain the workings of the universe without recognizing the existence of the ether and the indispensable function it plays in the phenomena are futile and destined to oblivion. – Nikola Tesla

The restricted [relativity] theory changed our ideas of space and time in a way that may be summarized by stating that the group of transformations to which the space-time continuum is subject must be changed from the Galilean group to the Lorentz group. – Paul A. M. Dirac

5. Electromagnetic Field Dimensions

Table 5-1. Units Conversion of Physical Quantities.

Quantity	Symbol	Dimension	Planck Formula	Remarks	MKS Units
Length	l	L	$l_P = t_P c = \sqrt{(\hbar G/c^3)}$	$\lambda_C = 2\pi\hbar/mc$, $R_c = \hbar/mc$	meter
Angle	Θ	θ	$\theta = l_P\theta/(\hbar/m_P c)$	$\theta = s/R = s/(\hbar/mc)$	radian
Time	t	T	$t_P = l_P/c = \sqrt{(\hbar G/c^5)} = \hbar/c^2 m_P$	$t_C = \lambda_C/c = \hbar/mc^2$	sec
Mass	m	M QT IT²	$m_P = \sqrt{(\hbar c/G)} = \hbar/cr_P = l_P c^2/G$ $= V_0/c^2 \int \eta(v)\rho_{SED}(v)dv$ $= |V_P \cdot q_P|/c^2$	(Positive energy – negative energy)/$c^2 > 0$. $m_i = \hbar \mathbf{k}/\mathbf{v} =$ |Volt x Charge|/$c^2 = nh v_{dB}/c^2$ $m = 2L/(\mathbf{E}^2 - c^2\mathbf{B}^2) = (e/\mathbf{A})\mathbf{v}$ $m_e = \hbar/c \cdot R_m = (\hbar/c) \cdot \lambda_C/(\alpha^{-1}))\theta_{137}$	kg [Amp·sec²]
Charge	e	Q ΘMT^{-1}	$q_P = \sqrt{(4\pi\varepsilon_0 \hbar c)}$ $= (1/(l_P/Q_P))cT_P$ $= m_P a_P/\mathbf{E}_P \approx (q_e\sqrt{\alpha})/\alpha$ $= m_P \omega_P$	Electric charge equivalent to mass rotational flow rate (Mass x topological charge rotational velocity). Spin precession.	Coulomb Amp·sec [kg·rad/sec]
Volt	V	$ML^2T^{-2}Q^{-1}$ L^2T^{-1}	$V_P = E_P/q_P = (l_P/Q_P)cZ_S$ $= \sqrt{(c^4/4\pi\varepsilon_0 G)}$	Potential difference. A measure of rotational energy. Density waves swept area/unit time.	volt [m²/sec·rad]

cont

5. Electromagnetic Field Dimensions

Table 5-1. Units Conversion of Physical Quantities (cont)

Quantity	Symbol	Dimension	Planck Formula	Remarks	MKS Units
Action	$\hbar$	ML^2T^{-1}	$\hbar = h/2\pi = c^2 T_P^2 Z_s = m_P r_P c$ $= m_P \cdot l_P^2 / t_P = e/\mathbf{A}\mathbf{k}$ $= \mathbf{p}/\mathbf{k}$ $= E/\omega$	Mass x voltage (kg·m^2/sec) Energy x time (J·s) Momentum x radius (N·m·s) = angular momentum	J·s = kg·m^2/sec
Energy	E	ML^2T^{-2}	$E_P = c^2 T_P Z_s = \sqrt{(\hbar c^5/G)}$ $= l_P c^4/G$	Measure of frequency of mass motion or wave curvature. Voltage x charge. $E = \sqrt{((pc)^2 + (mc^2)^2)}$ $E = \frac{1}{2}m(\mathbf{E}^2 + c^2\mathbf{B}^2)$	J = N·m = kg·m^2/sec^2
Force	F	MLT^{-2}	$F_P = E_P/l_P = c^4/G = cZ_s$ $= m_P l_P / t_P^2 = Gm_P^2/l_P^2$ $= (1/4\pi\epsilon_0)q_P^2/l_P^2$	$\mathbf{F}_m = \hbar\omega c^2/c$, $\mathbf{F}_e = \alpha \mathbf{F}_m$ $\mathbf{F}_g = G\hbar^2/c^2\mathbf{r}^4 = 2mc \cdot \Delta v \cdot \mathbf{r}_u$ $F_g/F_e = F_e/F_P$ $F = -(\delta V/\delta x)m = \delta L/\delta x$	N = kg·m/sec^2
Velocity	v	L/T	$v_P = l_P/t_P = c$	$\mathbf{v} = \hbar\mathbf{k}/m$; $\mathbf{v} = (\hbar\mathbf{k} - q_0\mathbf{A})/\gamma m$	m/sec
Frequency	f	T^{-1}	$f_P = \omega/2\pi = c/l_P$	$f = \omega/2\pi = E/h$	1/sec
Angular Frequency	ω	ΘT^{-1}	$\omega_P = 1/t_P = \sqrt{(c^5/\hbar G)}$	$\omega = 2\pi f = E/\hbar = (c/n)\cdot k$	rad/sec

cont

5. Electromagnetic Field Dimensions

Table 5-1. Units Conversion of Physical Quantities (cont)

Quantity	Symbol	Dimension	Planck Formula	Remarks	MKS Units
Scalar potential	ϕ	$\Theta^{-1}L^2T^{-1}$	$V_p = \sqrt{c^4/4\pi\epsilon_0 G} = \sqrt{F_p/4\pi\epsilon_0}$	Energy/Charge $\phi = (1/4\pi\epsilon_0)\Sigma(q_i/r_i) = E/q$	$m^2/(sec \cdot rad)$
Vector potential	$\mathbf{A}$	$L\Theta^{-1}$	$\mathbf{A} = \frac{1}{2}\mathbf{B} \times \mathbf{r} = \Phi_B/2\pi\theta = \hbar\mathbf{k}/q_0$ $= -j_\mu/k^2 = (v/c^2)\phi$	Electromagnetic field momentum/unit charge. $\mathbf{A} = (1/4\pi\epsilon_0)\Sigma(q_i v_i/r_i) = (v/c^2)\phi$ $= \hbar\mathbf{k}/e; \mathbf{A}^* = -\int \mathbf{E}_k\, dt + \mathbf{A}_0$	m/rad m/s
Curvature	k	L^{-1}	$2\pi/l_p$	A measure of imbalance of magnetostatic energy and electrostatic energy	1/m
Torsion	τ	ΘL^{-1}	$\theta 2\pi/l_p$	A change in torsion is a measure of effective mass	rad/m
Temperature	T^0	T	$T_p = E_p/k_B = m_p c^2/k_B = \sqrt{(\hbar c^5/Gk_B^2)}$	Temperature (°K)	°K
Gravitational constant	G	$L^3 M^{-1} T^{-2}$	$G = c^3/Z_S = c^3/(m_p/t_p) = c^4/F_p$ $= \hbar c/m_p^2 = l_p^3/m_p t_p^2 = c^2 l_p/m_p$ $= l_p c^2/m_p = l_p c^4/T_p k_B = l_p c^4/E_p$	$G = F_g r^2/(m_1 m_2) = F_g c^2 r^4/\hbar^2$ $= 3\pi V^2/M_p^2 = G_0 \epsilon_{12}^2/\epsilon_0^2$ $= R_{univ} c^2/M_{univ} = G_0/T^3$	$m^3/kg \cdot sec^2$ $[C \cdot v \cdot m/kg^2]$
Energy/unit mass	c^2	$L^2 T^{-2}$	$c^2 = Gm^p l_p = E_p/m_p$	$c^2 = \phi_E/\phi_B = \mathbf{E}^2/\mathbf{B}^2 = 1/\epsilon\mu = v_p/v_g$	$J/kg = m^2/s^2$

cont

5. Electromagnetic Field Dimensions

Table 5-1. Units Conversion of Physical Quantities (cont)

Quantity	Symbol	Dimension	Planck Formula	Remarks	MKS Units
Coulomb constant	k_C	$\Theta^{-2}M^{-1}L^3$	—	$k_C = 1/4\pi\epsilon_0 = \mu_0 c_0^2/4\pi$ = 8.9875178736 1764E9	$m^3/kg\cdot rad^2$ = $N\cdot m^2\cdot C^{-2}$
Fine structure constant	α	—	$\alpha = e^2/(4\pi\epsilon_0)\hbar c = e_P^2 = k_C e^2/\hbar c$ $= \mu_0 c e^2/2h = e^2/2hc\epsilon_0$ $= \sqrt{(e/q_P)} = Z_0/Z_e$ $= 0.0072973525693(11)$	Charge ratio $(q/q_P)^2$ Impedance ratio (Z_0/Z_e) Force ratio (F_e/F_m) Energy ratio $\frac{1}{2}(E_{so} - E_{mo})/(E_{so} + E_{mo})$	—
Planck's constant	h	ΘML^2T^{-1}	$h = 2\pi\hbar = mv\cdot\lambda = 2\pi E/\omega = 2\pi p/k$ = 6.62607004 E-34	Energy/Frequency $h = E/\nu$ Momentum·wavelength $h = p\cdot\lambda$	$J\cdot s$ = $rad\cdot kg\cdot m^2/s$
Entropy	S_{BH}	$L^{-1}T$	$S_{BH} = c^2 A/4G\hbar$	Bekenstein-Hawking area law $S_{BH} = \text{Area}/(4\hbar G)$	s/m
Flux	ϕ	$L^2\Theta^{-1}T^{-1}$	$\phi = \hbar c/e$	$\Delta\phi\cdot\Delta e \geq \hbar c$ Flux = velocity · energy density	$m^3/s \cdot rad$
Vorticity (mass)	ϕ_m	MLT^{-1}	$\phi_m = \hbar/m$	$\Delta\phi_m\cdot\Delta m \geq \hbar$	kg·m /s
Planck impedance	Z_P	$ML^2T^{-1}Q^{-2}$	$Z_P = 1/4\pi\epsilon_0 c = \hbar/q_P^2 = 29.98$	$Z_P \simeq 29.98$ $Z_P \simeq Z_0/4\pi = 120\pi/4\pi$	ohms

6. Electromagnetic Spectrum

My first thought is always of light. – Galen Rowell

The observed EM frequency spectrum spans more than 140 octaves or ~24 orders of magnitude with no definite upper or lower limit. Visible wavelengths range from ~4.3×10^{-7} m (4,300 Å) to 6.9×10^{-7} m (6,900 Å) with peak eye sensitivity of ~5.55×10^{-7} m (5,550 Å). The visible light spectrum frequencies detectible by the human eye range from ~430 to 770 $\times 10^{12}$ Hz (430-770 THz). The Sun's blackbody spectrum corresponding to a temperature of 5777 °K peaks at about 500×10^{-9} m (500 nm). A tabulation of the higher energy portion of the EM spectrum is shown in Table 6-1. The electromagnetic wave spectrum is shown in Fig. 6-1.

Microwaves (1 mm – 1 meter wavelength) span 300×10^6 Hz (MHz) to 300×10^9 Hz (GHz). X-rays (0.01 – 10 nm) span 30×10^{15} Hz (PHz) to 30×10^{18} Hz (EHz). Gamma rays (< 0.01 nm) have frequencies > 10×10^{18} Hz (EHz). The highest measured gamma cosmic ray frequency is approx. 10^{23} Hz with a corresponding wavelength of 10^{-14} meters and energy of ~100×10^{12} eV. The electron Compton frequency ν_c is ~ 1.2356×10^{20} Hz or a corresponding angular frequency ω_c ~7.7634×10^{20} rad/sec. The calculated Planck frequency f_P is 2.952×10^{42} Hz or a corresponding angular frequency ω_P is 1.855×10^{43} rad/sec appears to represent an upper frequency cutoff limit of the vacuum.

Table 6-1. Electromagnetic Spectrum			
Description	Wavelength	Frequency	Energy
Gamma ray	< 10 pm (1.0×10^{-11} m)	> 10 EHz (> 1.0×10^{19} Hz)	> 1.2 MeV (> 1.2×10^6 eV)
X-ray	0.01 nm – 10 nm (1.0×10^{-11} – 1.0×10^{-8} m)	30 EHz – 30 PHz (3.0×10^{19} – 3.0×10^{16} Hz)	120 eV – 120 KeV (120 eV – 1.20×10^5 eV)
Ultraviolet	10 nm – 400 nm (1.0×10^{-8} – 4.0×10^{-7} m)	30 PHz – 790 THz (3.0×10^{16} – 7.9×10^{14} Hz)	3.1 – 120 eV (1.2×10^3 – 12×10^3 eV)
Visible	390 nm – 750 nm (3.9×10^{-7} – 7.5×10^{-7} m)	405 THz – 790 THz (4.0×10^{14} – 7.9×10^{14} Hz)	1.65 – 3.1 eV
Infrared	750 nm – 1 mm (7.5×10^{-7} – 0.001 m)	300 GHz – 405 THz (3.0×10^{11} – 4.0×10^{14} Hz)	1.20 meV – 1.7 eV (0.0012 – 1.65 eV)
Microwave	1 mm – 1 meter (0.001 – 1 m)	300 GHz – 300 MHz (3.0×10^6 – 3.0×10^9 Hz)	1.2 μeV – 1.2 meV 1.2×10^{-6} – 1.2×10^{-3} eV
Radio	> 1 mm (0.001 m)	> 3 Hz	< 1.2×10^{-5} eV

6. Electromagnetic Spectrum

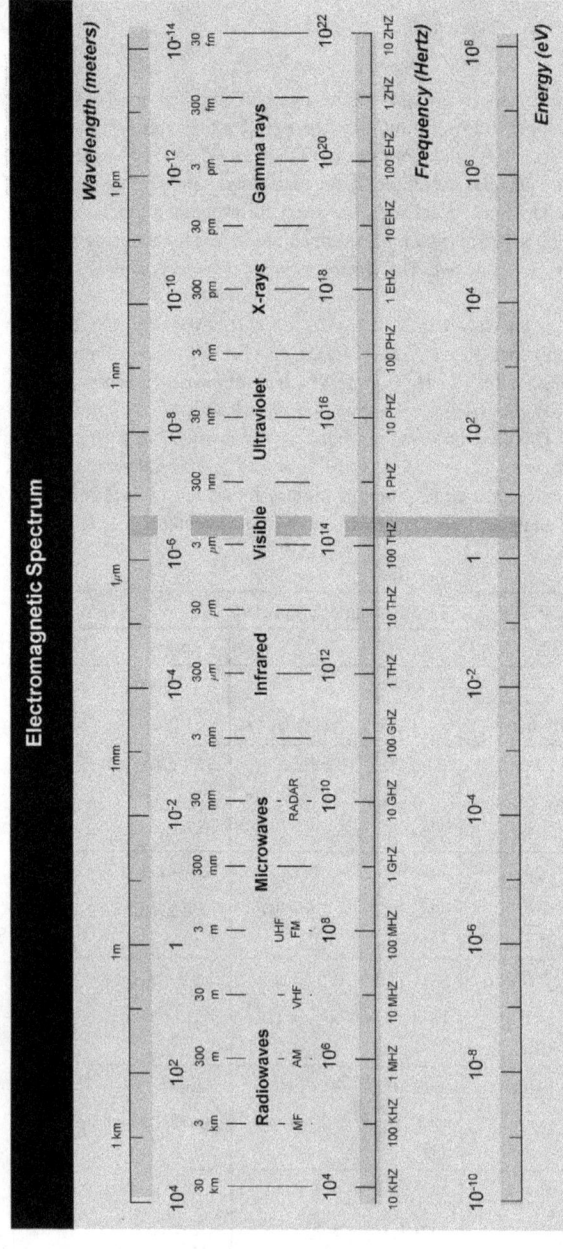

Fig. 6-1. Wavelength, frequency and energy of the electromagnetic spectrum. Wavelength λ is inversely proportional to frequency ($=c/f$) where f = frequency and c = celerity of light. Energy E is a measure of wave curvature and is proportional to frequency ($=h\nu$) where h = Planck's quantum of action. Maximum theoretical vacuum cut-off frequency corresponds to the Planck frequency ($f_P = 2.95 \times 10^{42}$ Hz). The low limit frequency corresponds to the inverse Planck frequency ($1/f_P = 3.38 \times 10^{-43}$ Hz). In principle, as a mathematical limit has two sides, the frequency spectrum conceivably may be infinite and continuous at a Planck subscale inaccessible to observation.

6. Electromagnetic Spectrum

The Planck cut-off frequency ω_c corresponds to the theoretical Planck frequency ω_P (= $\sqrt{(c^5/\hbar G)}$ and defines the maximum possible wave frequency supported by the physical vacuum (Bose-Einstein condensate BEC super-fluid) composed of hypothetical Planck particles. The $\hbar$ term refers to the Planck constant defining the action or minimum quantized value of angular momentum (= $m_P \cdot r_P \cdot c = \Delta p \cdot \Delta q$) where m_P is the Planck mass, r_P is the Planck radius and c is the speed of light. In Planck units, the gravitational constant G = $\hbar c/m_P^2$. The angular cut-off frequency ω_c as a function of wave number k is shown in a dispersion plot in Fig. 6-2.

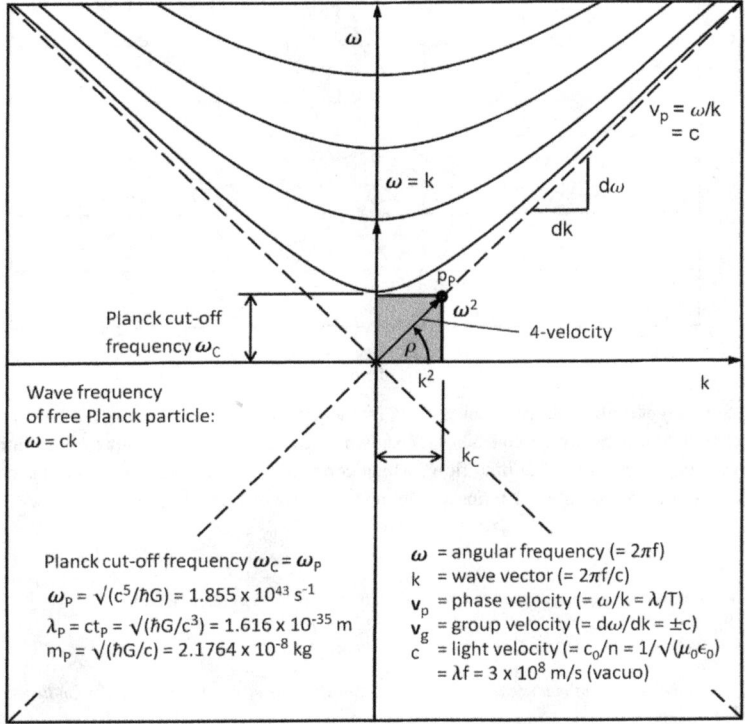

Fig. 6-2. The cutoff frequency ω_{co} of the vacuum is taken as the Planck frequency ω_P (= $\sqrt{c^5/\hbar G}$ = 1.855 x 10^{43} s^{-1}) where the $\hbar$ is the Dirac $\hbar$ (= $h/2\pi$ = 1.055 x 10^{-34} J·s), h is Planck's constant (= 6.626 x 10^{-34} J·s) and G is the universal gravitational constant (= 6.67384 x 10^{-11} m^{3}kg^{-1}s^{-2}).

An inverse (n·1/n = 1) relation is apparent in a number of physical laws such as Newton's 2nd Law: **a** = F/m and Ohm's Law: R = V/I. Similarly, frequency times wavelength = 1·constant = velocity (f·λ = v) or radius times angular velocity = 1·constant = tangential velocity (r·ω = v$_t$). Observed energies, masses, spins, wavelength and frequencies follow a logarithmic spiral progression. A normalized Schauberger plot of 1/r is superimposed on a logarithmic spiral as shown in Figs. 6-3 and 6-4 illustrating the relationship of frequency to wavelength where c = 1.

47

6. Electromagnetic Spectrum

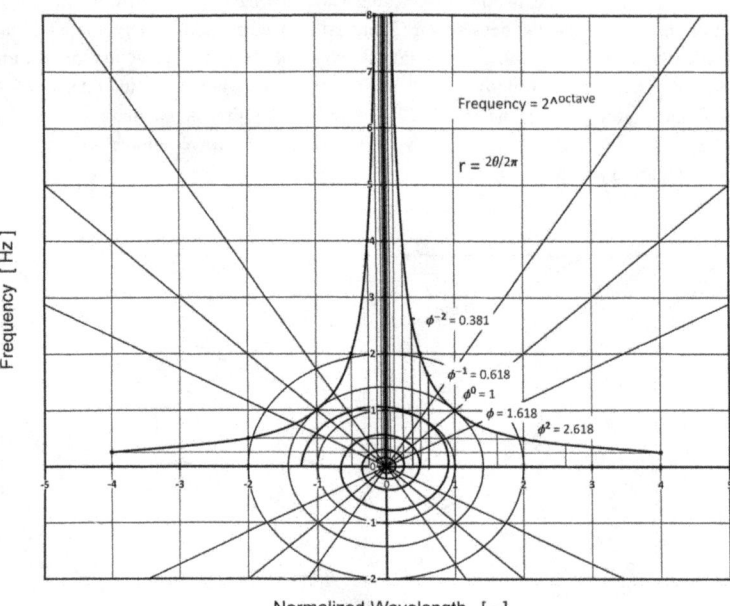

Fig. 6-3. Hyperbolic energy spiral plot of frequency space (spectral energy density) vs. wavelength. Only positive frequencies are shown. Negative frequencies represent negative energy (equivalent to reverse time flow) which correspond to antiparticles. The diagonal lines illustrate Fibonacci Phi Φ ratios. Optimum vortex wavelength $\lambda_{vortex} = 1.618\lambda_{wave}$.

The impulse of all movement and all form is given by Φ. – R. A. Schwaller de Lubicz

Everything is arranged according to number. – Pythagoras

Is it not the essence of mathematics to be conversant with the ideas of number and quantity? – George Boole

Number theorists are like lotus-eaters... they can never give it up. – Leonard Kronecker

Although infinity is needed in mathematics, it occurs nowhere in the physical universe. – David Hilbert

I do not know what I may appear to the world; but to myself I seem to have been only like a boy playing on the seashore, and diverting myself now and then finding a smoother pebble or a prettier shell than ordinary, whilst the great ocean of truth lay all around undiscovered before me. – Isaac Newton

I was born not knowing and have had only a little time to change that here and there. – Richard Feynman

If you want to be a physicist, you must do three things - first, study mathematics, second, study more mathematics, and third, do the same. – Arthur Sommerfeld

6. Electromagnetic Spectrum

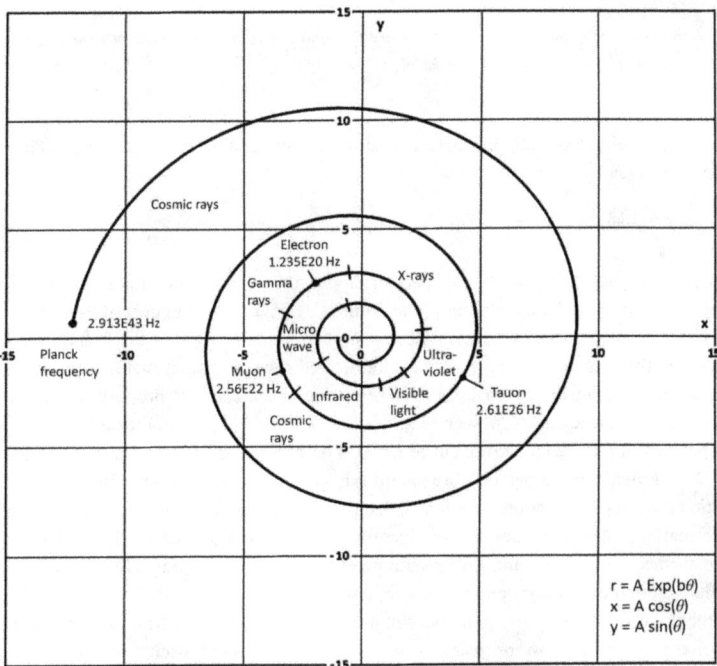

Fig. 6-4. The scaling of frequency and wavelength of matter and energy follows a logarithmic spiral progression. Characteristic frequencies of the lowest mass fundamental particles (electron, muon and tauon) are illustrated in relation to the EM frequency spectrum. The electron Compton frequency f_C (= 1.2356E20 Hz) represents a resonant interaction frequency with the Planck vacuum enabling stability without dissipation. More massive fundamental particles (fermions) represent resonant composite structures of electrons each with a characteristic frequency. The frequency range illustrated extends 143 octaves from 1 Hz to 2.913 x10^{43} Hz corresponding to the Planck cut-off frequency f_P.

There can be no center in infinity. – Lucretius (On the Nature of the Universe)

The nine Indian figures are 9,8,7,6,5,4,3,2,1. With these nine figures, and with the sign 0... any number may be written.... – Leonardo Fibonacci

It is strange that the immense variety of nature can be resolved into a series of numbers. – Sir William Bragg

Information is information. Information is neither matter nor energy. – Norbert Weiner

Napier, Lord of Merchiston, hath set my head and hands at work with his new and admirable Logarithms. – Henry Briggs

Natura non fecit saltum (nature does not make junps). – Gottfried Wilheim Leibnitz

One's ideas must be as broad as Nature if they are to interpret Nature. – Arthur Conan Doyle

7. Curvature and Torsion

The horizon seems quite close to you because the curvature is so much more pronounced than here on earth. It's an interesting place to be. I recommend it. – Neil Armstrong

A circular helix has both intrinsic curvature k and torsion τ components. The total curvature is given by

$$\text{Total Curvature } \kappa_T = \sqrt{(\text{Curvature}^2 + \text{Torsion}^2)} = \sqrt{(\kappa^2 + \tau^2)} \qquad (7\text{-}1)$$

Curvature describes how the tangent spaces roll along the curve (rate of change of direction of tangent vector T in the osculating plane defined by N, T. Torsion describes how tangent spaces twist about a curve under parallel transport (rotation or twist of the Frenet frame $\{T, N, B\}$ about the tangent vector T). For a parametized speed curve, curvature corresponds to acceleration. The intrinsic curvature of an arbitrary space curve is proportional to the arc length raised to an arbitrary power (e.g., $\kappa = s^n$, $\kappa = (s + 1)^n$). For example: A circle corresponds to an arc length exponent of n = 0, a helix n = 2, a log spiral n = −1, a catenary n = −2. For a non-zero curvature manifold, affine transformations allow for contour distortion (scaling, translation, rotation, shearing). A change in torsion represents an affine transformation. An affine connection specifies how tangent spaces at different points on curved surfaces are related and give rise to geodesics. An affine connection ∇ on manifold M is flat where the curvature and torsion are null.

In the conversion of an energetic photon into an electron and positron in pair production, the helix is postulated to be transformed into two tori with each torus of revolution generated by rotation of a Hopf knot (the simplest knot possible), in effect closing the helix end-to-end. The helix is represented in parametric form x = cos v, y = sin v, z = sin u where u and v are parameters defining principle curves. A horn torus, for example, is represented as x = (1 + cos v)cos u, y = (1 + cos v)sin u, z = sin v where u = angle between major radius R and the x-axis; v = angle between major radius R and minor radius r). Conversion from a Frenet-Serret frame to a moving reference Darboux frame defined by unit tangent vector **t**, unit normal vector **u** and unit tangent normal vector **v** is given by $t' = $ **D** x **t**·**n** = **D** x **n**·**b** = **D** x **b** where **D** = τ**t** + κ**b**. The rate of change of the vectors [**t, n, b**] in terms of arc length s is given by $dt/ds = k_n\mathbf{n} - k_g\mathbf{b}$, $d\mathbf{n}/ds = -k_n\mathbf{t} + \tau_g\mathbf{b}$, $d\mathbf{b}/ds = k_g\mathbf{t} - \tau_g\mathbf{n}$. where k_n = normal curvature, k_g = geodesic curvature and τ_g = geodesic torsion. The confined electron topology reflects an imbalance of electrostatic and magnetic energy due to the combination of the toroidal and magnetic fields.

As illustrated in Fig. 7-1 for a circular helix lines of constant curvature κ are shown as horizontally-aligned circles while lines of constant torsion τ are represented as vertically-aligned circles. A freely-propagating photon corresponds to balanced curvature and torsion reflective of a balance of electrostatic and magnetic energy resulting in straight line motion. When a photon is slowed during passage in an optically dense medium, the photon acquires an effective mass which corresponds to decreased torsion τ. The decreased turn height H corresponds to a flatter helix angle. With smaller H, the longitudinal magnetic field decreases. The curvature κ increases as the radius is reduced as a result of conservation of energy and angular momentum. If H is much smaller than radius R, the curvature k increases and approaches 1/R. Increased curvature is equivalent to increased mass.

There is no royal road to geometry. – Euclid

7. Curvature and Torsion

The total photon electromagnetic energy density U may be expressed as a function of the electric field intensity **E** and magnetic flux density **B**

$$U = U_e + U_m = \tfrac{1}{2}\epsilon_0 \mathbf{E}^2 + \tfrac{1}{2}\mu_0 \mathbf{B}^2 \qquad (7\text{-}2)$$

In a non-zero curvature vacuum where the photon acquires an effective mass m with total energy

$$E_t = E_k + E_p = \tfrac{1}{2} mc^2 + GmM/r \qquad (7\text{-}3)$$

Decreased helix inclination angle α corresponds to decreased torsion and increased effective mass. Increased twist angle ϕ and/or increased deflection angle θ corresponds to increased curvature and increased mass.

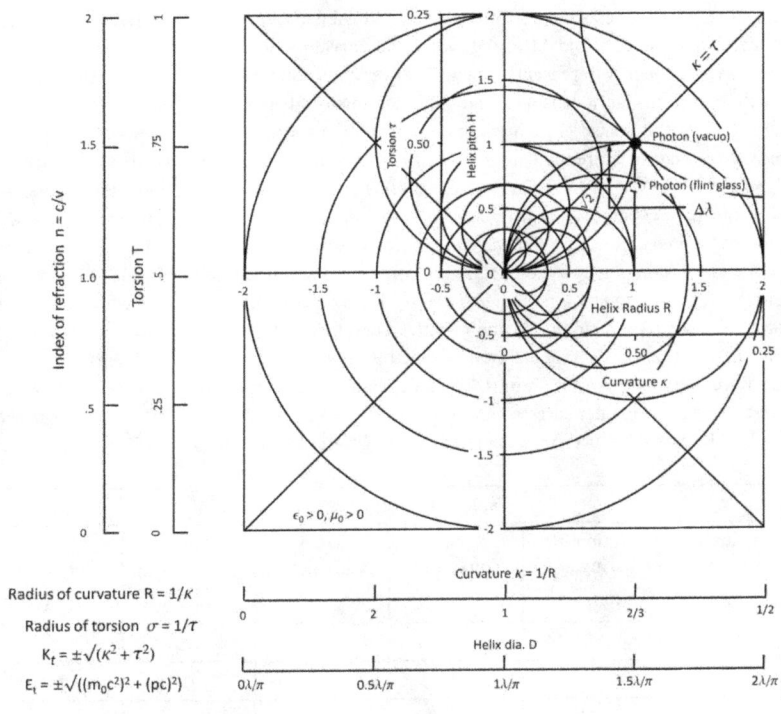

Fig. 7-1. For a freely-propagating photon, the helical curvature and torsion are equal. In a refractive medium, the change in index of refraction n (= ck/ω) produces a corresponding change in torsion τ and wavelength λ. The effective mass of a photon increases in a medium of increased index of refraction while the torsion and velocity of light decreases.

The Frenet-Serret frame refers to stand alone curves without regard to the surface on which the curve resides. With a metric defined, the connection coefficients $\Gamma_{\alpha\beta}{}^\gamma e_\gamma$ (Christoffel symbols) between points on a manifold describe the rate of change of the basis vectors allowing differentiation of tensors. In an affine connection, the torsion and

7. Curvature and Torsion

curvature are invariant. An affine connection is an operator that maps how directions are changed in neighboring tangent spacetimes. Affine torsion relative to a Cartan connection is the source of topological charge and topological spin (3-forms).

Frenet curvature and torsion are scalar 0-forms, affine torsion is a 2-form and topological torsion (3-form).

1-form A – Connection (gauge potential)
2-form dA – Curvature – (0,2) tensor (covariant derivative of connection 1-form)
3-form A^dA – Affine torsion – (1,2) tensor is antisymmetric part of connection.

Torsion is a measure of global rotation of tangent space during parallel transport. Parallel transport along a geodesic changes with rotation. The affine connection specifies how tangent spaces at different points on curved surfaces are related. Only affine connections give rise to geodesics. An affine connection gives a means for transporting tangent vectors to a manifold from one point to another and typically as a covariant derivative which gives a means for taking directional derivatives of vector fields. An affine connection ∇ on manifold M is flat where the curvature and torsion are null. Covariant derivative operator ∇ represents a map from (k, l) tensor fields to $(k, l+1)$ tensor fields which transforms as a tensor on an arbitrary manifold and which reduces to a partial derivative in flat space. A covariant derivative ∇_μ of a vector V^ν consists of the partial derivative and a correction term specified by the matrix $\Gamma_\mu{}^\rho{}_\sigma$ to make covariant: $\nabla_\mu = \delta_\mu V^\nu + \Gamma^\nu_{\mu\lambda} V^\lambda$. A Levi-Civita connection is a Riemannian connection without torsion. An affine connection Γ on a differentiable manifold M with a metric g can be decomposed into Levi-Civita connection Γ plus nonmetricity tensor **S** + torsion tensor **T**.

The electron geometry is conjectured to correspond to a Hopf link embedded in a spindle torus with curvature and torsion generated by 1/3-integer exponents described by a circular torus with positive torsion (2/3) and negative curvature (–1/3) where normal curvature and torsion are constant. For an electron modeled as a horn torus, the Gaussian curvature K (= cos v/a(c + a cos v) varies from 0.5 to 2 depending on slant angle v. The positive outer curvature K_1 of the outer surface cancels the negative inner curvature K_2 of the inner surface ($K_1 A_1 + K_2 A_2 = 0$). Curvature characteristics of a torus are summarized in Table 7-1.

		Table 7-1. Torus Curvature			
Gaussian Curvature K	Azimuth Angle θ	Outer Curvature K_1	Total Curvature $K_1 A_1$	Inner Curvature K_2	Total Curvature $K_2 A_2$
$K > 0$	$\pi/2$	-	-	1/3	1
$K = 0$	0	0	0	0	0
$K < 0$	$-\pi/2$	–1	–1	-	-
Total curvature KA = $2\pi\chi$ where K = curvature, A = surface area, χ = Euler's no. Euler's no. χ = v – e + f where v = no. of vertices, e = no. of edges, f = no. of faces.					

I think the universe is pure geometry – basically, a beautiful shape twisting around and dancing over space-time. – Antony Garrett Lisi

...it is easily deduced from Huygen's principle that light rays propagating at right angles to the gravity field must experience curvature. – Albert Einstein

8. Polarization Effects in a Dielectric Medium

dielectric n. *a medium or substance that transmit electric force without conduction; an insulator*

polarization n. *the action of restricting the vibrations of a transverse wave, especially light, wholly or partially to one direction*

Polarization results from the process of bringing about a partial separation of electric charges of opposite sign in a material by the superposition of an electric field. When a conductor or semiconductor is placed in a static electric field, the free charge carriers move in response to the applied field so as to cancel the field inside the material. Movement of the charge carriers constitutes a time-variable electric current. The induced current is caused by the electrokinetic field E_k. The electrokinetic field is equal to the negative time derivative of the retarded magnetic vector potential A^* where $E_k = -\delta A^*/\delta t$, $A = v\phi_E/c^2 = \hbar k/e$ and $A^* = v'\phi'/c^2$. Once the current ceases, the electric field inside a conductive material or a cavity within a conductive material falls to zero cancelling the external field. The field-free interior cavity of a conductor forms in effect a Faraday cage. Polarization enables storage of electric charge and is dissipated by leakage current. An electret is an insulator or dielectric that retains a permanent electric field analogous to a permanent magnet with a magnetostatic field. The electric induction in a dielectric is given by $D = \epsilon E = \epsilon_r \epsilon_0 E$ where ϵ_r is the relative permittivity of the dielectric of the medium.

An external electric field can induce electric polarization (separation of charges) in a perfect nonconductor with no free charges proportional to the field intensity. At low field intensities, the charge distortion is linear, however, at sufficiently intense field that is comparable to internal electric fields of atoms and molecules, various nonlinear effects can arise. The atomic electric field intensities associated with Coulombic binding are on the order of 10^{11} V/m. Selection of an insulating material for a given application depends in part on the magnitude of the dielectric constant. For high energy density, a relatively large dielectric constant ϵ is needed as the energy density is proportional to the dielectric constant ϵ and the square of the electric field breakdown strength E_B. Conversely, if high transfer rate of energy is desired, an insulator of lower dielectric constant is indicated as the EM wave speed is inversely proportional to $\sqrt{\epsilon}$.

The polarization P is a vector quantity in the direction of flux density D or electric field intensity E equal to $(D - \epsilon_0 E)$. The polarization P (electric dipole moment per unit volume) is associated with bound charges while the electric flux density D is associated with free charges. Dielectric polarization P may be expressed as a function in terms of the magnitude the electric field E and the material susceptibility $\chi = ((\epsilon - \epsilon_0)/\epsilon_0)$. For low field intensity E in a linear medium, the induced polarization $P = \epsilon_0 \chi E$. This is the linear response of the medium. Under a strong electric field, the induced polarization may be expressed by the Taylor series: $P = \epsilon_0[\chi \cdot E_0 + \chi_2 \cdot E^2 + \chi_3 \cdot E^3 + ...]$. This is the nonlinear response. The higher order terms reflect higher wave harmonics. For a sinusoidal external field $E = E_0 \cdot \sin(\omega t)$, the corresponding polarization is $P = \epsilon_0[\chi \cdot E_0 \cdot \sin(\omega t) + \chi_2 \cdot E^2 \cdot (1 - \cos(2\omega t)) + ...]$. Induced polarization effects in a given medium vary with susceptibility and frequency of representative incident electromagnetic radiation is summarized in Table 8-1.

It is contrary to reasoning to say there is a vacuum or space in which there is absolutely nothing.
– René Descartes

8. Polarization Effects in a Dielectric Medium

Table 8-1. Induced Polarization Effects

Susceptibility	Order	Frequency	Polarization Effect	Remarks
$\chi^{(1)}E_0$	1	0	D.C. polarizability (linear effect)	E_0 = external electric field amplitude.
$\chi^{(1)}E_\omega \cos\omega t$	1	ω	refractive index optical polarizability (linear effect)	E_ω = electric field of light wave of frequency ω. Linear absorption/emission/refractive index.
$\chi^{(2)}(E_0)^2$	2	0	D.C. hyperpolarizability	materials lack inversion symmetry.
$\chi^{(2)}E_0 E_\omega \cos\omega t$	2	ω	linear electro-optic Pockels effect	polarization rotation; refractive index change (birefringence) proportional to electric field: $\Delta n = \lambda_0|E_0|$ input frequencies: $0, \omega$.
$\chi^{(2)}(E_\omega)^2$	2	0	optical rectification	induced D.C. voltage across sample. input frequencies: $\omega, -\omega$. Re $\chi^{(2)}(0, \omega, -\omega)$
$\chi^{(2)}(E_\omega)^2\cos 2\omega t$	2	2ω	2nd harmonic generation (SHG)	input frequencies: ω, ω.
$\chi^{(3)}(E_0)^2 E_\omega \cos\omega t$	3	ω	D.C. Kerr effect (quadratic electro-optic effect, QEO)	birefringence proportional to square of electric field; $\Delta n = \lambda_0 K|E_0|^2$ input frequencies: Re $\chi^{(3)}(0, 0, \omega)$

Cont

8. Polarization Effects in a Dielectric Medium

Table 8-1. Induced Polarization Effects (cont)

Susceptibility	Order	Frequency	Polarization Effect	Remarks
$\chi^{(3)} E_0 (E_\omega)^2 \cos 2\omega t$	3	2ω	D.C. induced 2nd harmonic generation	anisotropic materials only. input frequencies: $0, \omega, \omega$
$\chi^{(3)} (E_\omega)^3 \cos \omega t$	3	ω	A.C. electro-optic Kerr effect (degenerate four-wave mixing)	self-phase modulation, self-induced phase and frequency shift, self-focusing. input frequencies: Re $\chi^{(3)} (\omega, \omega, -\omega)$
$\chi^{(3)} (E_\omega)^3 \cos 3\omega t$	3	3ω	3rd harmonic generation	includes isotropic materials. $n = n_0 + 3\chi^{(3)}/8n_0 |E_\omega|^2 = n_0 + n_2 I$. input frequencies: ω, ω, ω.

1. Optical polarization P in terms of applied electric field $E = E_\omega \cos(\omega t)$:
   $P = \epsilon_0 \chi^{(1)} E + \epsilon_0 \chi^{(2)} E^2 + \epsilon_0 \chi^{(3)} E^3 + \ldots$
2. Fourier frequency series in terms of applied frequency:
   $P = P_0 + P_1 \cos \omega t + P_2 \cos 2\omega t + P_3 \cos 3\omega t + \ldots$
3. Index of refraction vector $\mathbf{n} = (c/\omega)\mathbf{k}$; magnitude $n = ck/\omega = \mathbf{H} \cdot Z/\mathbf{E}$. Direction of $\mathbf{n}$ coincides with wave normal $\mathbf{k}$.
   Phase of wave $\phi = \omega((1/c)\mathbf{n} \cdot \mathbf{r} - t)$.

A theory should be favored by far, in which the gravitational field and the electromagnetic field together would appear as a whole. – Albert Einstein

We have two possibilities. Either we use waves in spaces of more than three dimensions...or we remain in three-dimensional space, but give up the simple picture of the wave amplitude as an ordinary physical magnitude, and replace it by a purely abstract mathematical concept... into which we cannot enter. – Max Born

8. Polarization Effects in a Dielectric Medium

The applied electric field and the resultant induced polarization is shown in Figs. 8-1 and 8-2 for a linear and nonlinear medium. As shown, for a linear medium the induced polarization is proportional to the electric field intensity and matches the applied frequency. For a nonlinear medium, the frequency response above the cutoff frequency lags the incident radiation frequency with a mixture of harmonics contributing to distortion of the induced polarization.

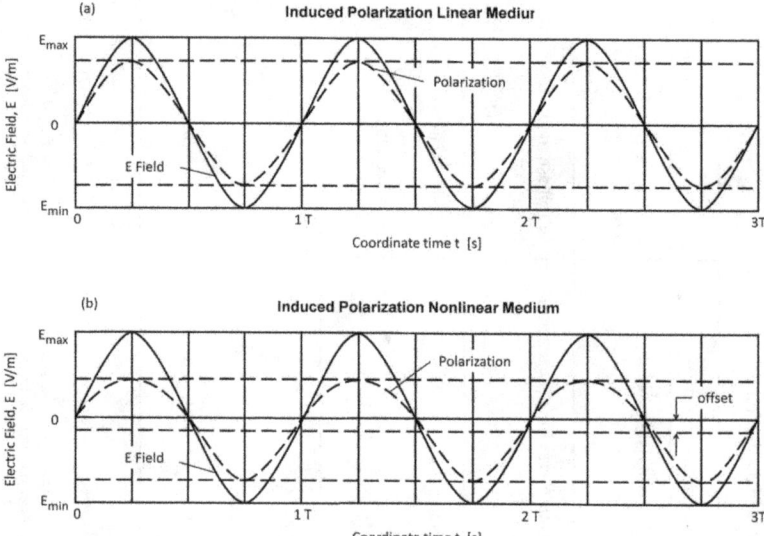

Fig. 8-1. Comparison of the applied electric field and the induced polarization is illustrated for a generalized optical linear and nonlinear medium. As shown, for a nonlinear medium, a d.c. offset bias results in induced polarization waveform asymmetry.

The electric permittivity ϵ of a medium is likewise a function of frequency. For a dielectric medium, the permittivity $\epsilon = \epsilon_0 \epsilon_r$ where ϵ_0 is the electric permittivity of free space (vacuum) and ϵ_r is the relative dielectric constant. At ultraviolet frequencies (~10^{16} Hz) and above, the bound electrons in atoms cannot follow the reversals in the electric field of the incident radiation lessening the polarizability of the medium. In the electronic, atomic, molecular dipole and interfacial absorption bands, the relative dielectric constant ϵ_r increases with decreasing frequency. Energy losses due to light absorption tend to increase with frequency. The refractive index n of an optical medium measured at optical frequencies is related to electronic polarizability associated with molecular polarizabilities (displacement of electrons) of the medium. In Maxwell's theory, charge is associated with a discontinuity in the displacement due to an electric field. This displacement creates a strain in the dielectric that depends on the conductivity of the medium leading to energy dissipation in the form of heat. In wave scattering theory, the forward scattering amplitude is related to the total cross section of the scatterer $\sigma_{tot} = (4\pi/k) \operatorname{Im} f(0)$ where $f(0)$ is the scattering amplitude and k is the wave vector. For a transparent mediums, such as glass, there is a large change in refractive index (anomalous dispersion) and strong absorption at resonance frequencies of bound electrons at infrared and ultraviolet frequencies.

8. Polarization Effects in a Dielectric Medium

The index of refraction n is related to the forward scattering amplitude as follows

$$n = 1 + 2\pi(N \cdot f(0)/k^2) \qquad (8\text{-}1)$$

where N is the volumetric number density of scatterers. For a vacuum, N is the Planck density N_P.

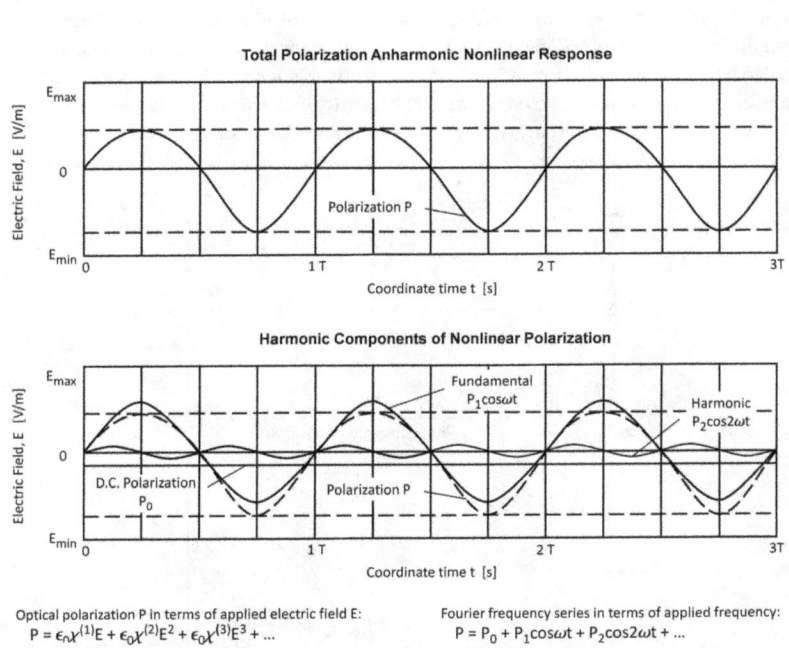

Fig. 8-2. Electric field induced polarization of a nonlinear optical medium results in anharmonic frequency response distortion.

A material medium is susceptible to electronic polarization due to separation of electric charges. Nonlinear effects typically begin to occur at power frequencies with marked absorption peaks at audio, radio, infrared and ultraviolet frequencies. At ultraviolet frequencies (~10^{16} Hz), electrons can no longer follow electric field oscillations of light waves and the atoms become insensitive to polarization, hence, the relative dielectric constant ϵ_r becomes unity. Analogous to matter, a vacuum is susceptible to polarization due to separation of Planck charges. At frequencies approaching the Planck frequency f_P (~10^{42} Hz), the Planck dipoles can no longer follow the electric field oscillations and nonlinear response effects predominate. Above the Planck cutoff frequency, the vacuum becomes saturated and insensitive to polarization. Hence, propagation of electromagnetic waves cannot be supported (group velocity $v_g \rightarrow 0$, phase velocity $v_p \rightarrow \infty$).

Optical resonance manifests as a marked change in the index of refraction and strong absorption of light at or near resonant frequencies. The complex index of refraction may be expressed as $\mathcal{N} = n + ik$ where the real part n is index of refraction (= c/v_p) and k is the extinction coefficient (= $\alpha c/\omega = \omega/v_p = \omega n/c$). The generalized variation in the index of refraction and absorption for a typical resonant peak in the infrared, visible or ultraviolet

8. Polarization Effects in a Dielectric Medium

frequency range is illustrated in Fig. 8-3. Strong resonant peaks occur as the natural frequency in-phase oscillation of electrons couples with the driving excitation frequency of incident optical radiation. Absorption of energy is described by the absorption coefficient α which is related to the extinction coefficient k ($\alpha = \omega/ck$). The sharpness of the resonant peak is measured by the Q ratio which relates resonant bandwidth $\Delta\omega$, resonant frequency ω_0, and absorption coefficient α. The index of refraction of the vacuum is typically taken as $n_0 = 1$ and $n_2 = 0$. For comparison, refractive index of glass n = 1.5. However, in the vicinity of a strong electric field sufficient for $e^- e^+$ production, for example, the vacuum undergoes evident polarization. Confinement of photon solitons suggests vacuum nonlinearity is present for photons as well, particularly, if the wave train is on the order of a few wavelengths where the energy/wavelength is increased over that of longer wave trains.

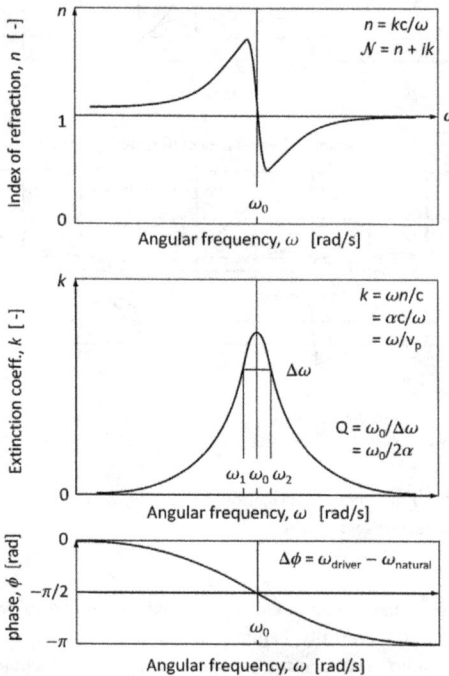

Fig. 8-3. Complex index of refraction, absorption, and phase variation at a resonant peak of a dielectric. At or near resonant peaks, the refractive index n (= c/v_p) undergoes anomalous dispersion in which refractive index decreases with increasing frequency.

Einstein made the qualitative suggestion that photons or light quanta were some kind of singularity of concentration of energy and momentum inside a radiation field. – Leon Rosenfeld

Time cannot be absolutely defined, and there is an inseparable relation between time and signal velocity. – Albert Einstein

In fact, whenever energy is transmitted from one body to another in time, there must be a medium or substance in which energy exists after it leaves one body and before it reaches the other. – James Clerk Maxwell.

8. Polarization Effects in a Dielectric Medium

A dielectric is an insulator that resists the flow of electrons as a result of a full valence band with large band gap and exhibits relatively small polarization due to an external field. The polarization of a dielectric in response to an external electric field results in a variation in the dielectric constant and absorption effects at or near resonance conditions. Irradiation of a dielectric with electromagnetic waves affects bound electrons in atoms altering the equilibrium positions (i.e., polarization) creating an induced electric moment. The degree of polarization is a measure of the refractive index. A wide range of dielectrics, such as glass, exhibit weak damping at optical frequencies allowing transparency. A generalized dielectric frequency response characteristic of a dielectric is shown in Fig. 8-4.

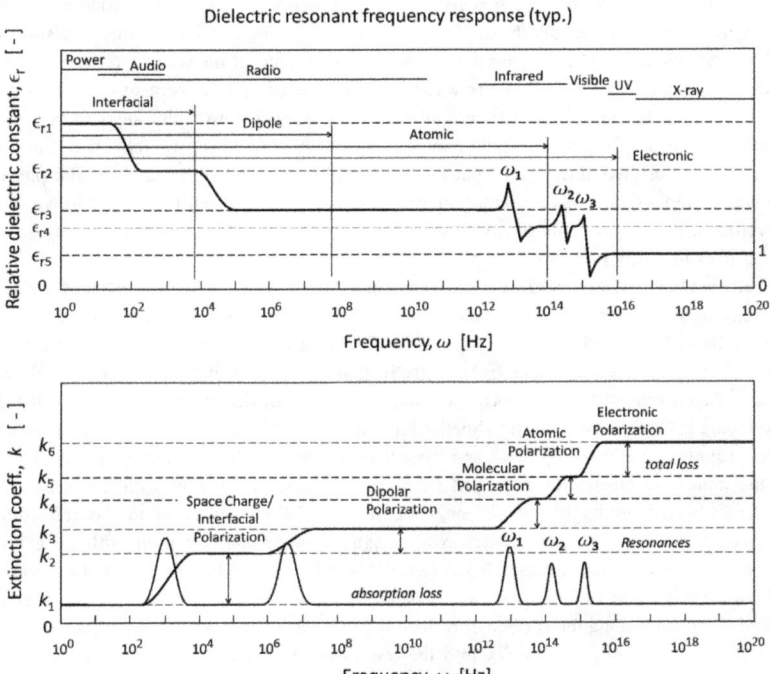

Fig. 8-4. Variation in relative dielectric constant (K) or electric permittivity (ϵ_r) where K = $\epsilon_r = \epsilon/\epsilon_0$ for a typical dielectric is a function of frequency. The total power loss increases with excitation frequency. The index of refraction $n = \sqrt{(\epsilon/\epsilon_0)(\mu/\mu_0)} = \sqrt{(\epsilon_r\mu_r)} = \sqrt{(K \cdot K_m)}$ where $\epsilon_0 = 1/\mu_0 c^2$. For most transparent optical materials, the magnetic permeability is unity (i.e., nonmagnetic), hence, n = $\sqrt{K}$. Dielectric loss tends to increase in materials with higher dielectric constants and is accompanied by temperature rise. A high temperature, high K dielectric ceramic such as TiO_2 has a dielectric constant K of 200 -1000.

The energy of a covalent bond is largely the energy of resonance of two electrons between two atoms. The examination of the form of the resonance integral shows that the resonance energy increases in magnitude with increase in the overlapping of the two atomic orbitals involved in the formation of the bond... – Linus Pauling

9. Reflectors and Dielectric Lenses

To hold, as 'twere, the mirror up to nature. – William Shakespeare

Our inventions mirror our secret wishes. – Lawrence Durrell

Electromagnetic wave reflection need not be limited to conventional mirrored surfaces. Reflectors may take several forms, such as dichroic mirrors, magnetic mirrors, retroreflectors, EM Bragg wave interference nodes, phase conjugate mirrors, gravitational mirrors, etc. In a conventional mirror, a reflected EM spherical divergent wave from a point source remains divergent. A polished surface produces a specular reflection in which the angle of incidence equals the angle of reflection. A rough surface results in a spread reflection while a matte surface produces a diffuse reflection. If the wave is reflected from a phase conjugate mirror (PCM), the wave front is inverted in a convergent beam back to the source in a time-reversed replica. Unlike a conventional mirror, the incident beam is reflected from a PCM in time rather than in space. A comparison of reflection from a conventional polished mirror and a phase conjugate mirror is shown in Fig. 9-1. For phase conjugate mirrors, all wave vector components are inverted resulting in time-reversed wavefronts. Phase conjugate mirrors may be categorized as self-pumped, externally self-pumped, or mutually pumped.

Reflection and refraction of light in an optically dense dielectric medium such as a glass or water may occur in a transverse electric or transverse magnetic mode depending on the polarization of the incident beam. For light within a medium of higher index of refraction than the exterior medium, total internal reflection may occur if the incidence angle is greater than a critical angle θ_c corresponding to the ratio of the index of refraction of the two media given by $\sin\theta_c = n_2/n_1$. Snell's law ($n_1\sin(\theta_1) = n^2\sin(\theta_2)$) holds when there is no relative motion. Refer to Figs. 9-2 and 9-3. With moving media, aberration effects occur. In the transverse electric mode (TEM), the electric vector is parallel with the reflecting surface whereas in the transverse magnetic mode (TMM) is parallel to the reflecting surface. A magnetic mirror interacts with magnetic component of light without a 180 degree phase shift of the reflected E-field as with a ordinary metallic mirror. As the photon helicoid has a finite radius, reflection of single photon wave train is accompanied by a lateral beam shift along the surface of reflection known as the Goos-Hänchen effect. Depth of penetration of the photon helix into the less dense medium represents the evanescent wave.

Retro-reflection of a light wave may be effected by a retroreflector such as a corner cube, cat's eye spherical reflector or phase conjugate mirror which reflects light back to the source with a minimum of diffuse scattering. As shown in Fig. 9-4, an ordinary retroreflector reverses the beam direction and wave front curvature while a phase conjugate mirror reverses the beam direction, but maintains the wave front curvature matching that of the incident wavefront. The phase conjugate mirror acts somewhat similar to a hologram in reconstructing the wavefront spatial and temporal characteristics of the illuminating beam in the signal beam. The reflected signal beam is a phase conjugate of the incident beam. Unlike a reconstruction hologram which produces a converging real image in one first-order wave and a diverging virtual image in a second first-order beam at equal angles to the zero-order beam, a PCM generates a single zero-order return beam. A PCM composed of a photorefractive crystal such as barium titanate requires only one input beam forming a self-pumped phase conjugator enabling modulation of the phase conjugate beam for secure communications.

9. Reflectors and Dielectric Lenses

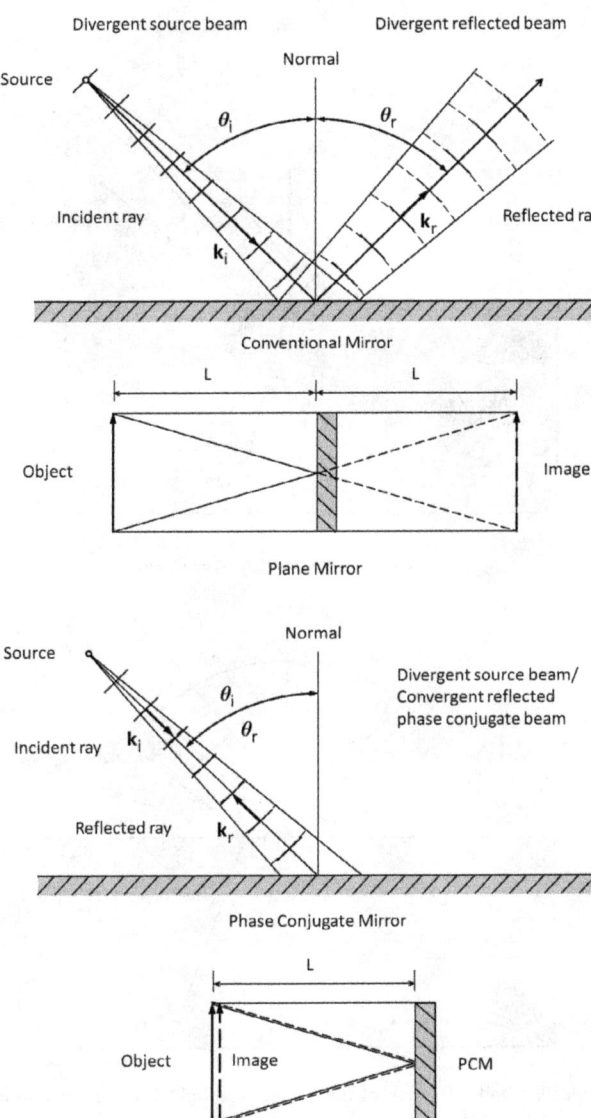

Fig. 9-1. Comparison of reflection from a conventional polished mirror and a phase conjugate mirror. A virtual image corresponds to a negative location in space. A phase conjugate reflection corresponds to a time-reversed image.

"I don't understand...," said Alice. "It's dreadfully confusing!" ... "Living backwards!"– Lewis Carroll (Alice Through the Looking Glass)

61

9. Reflectors and Dielectric Lenses

EM wave reflection and refraction

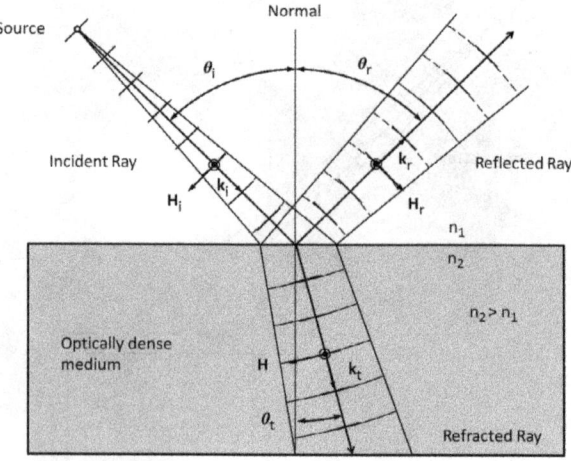

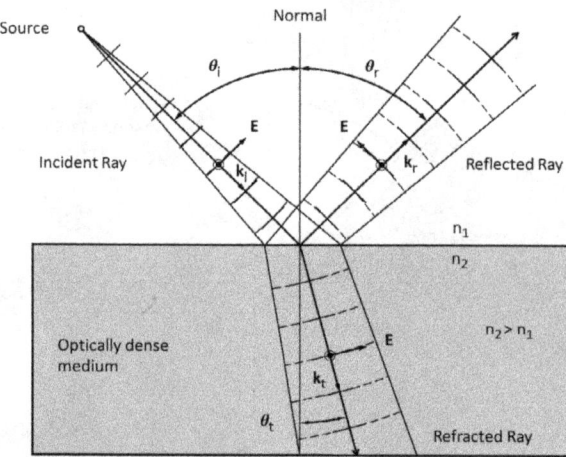

Fig. 9-2. Reflection and refraction of plane polarized light from and within an optically dense medium. High reflectance occurs when incident wavelength is >> plasmonic wavelength of resonant surface electrons.

The mirror reflects all objects without being sullied.. – Confucius

To doubt everything, or, to believe everything, are two equally convenient solutions; both dispense with the necessity of reflection. – Henri Poincaré

Beauty is eternity gazing at itself in a mirror. – Kahlil Gibran

9. Reflectors and Dielectric Lenses

EM wave polarization and internal reflection

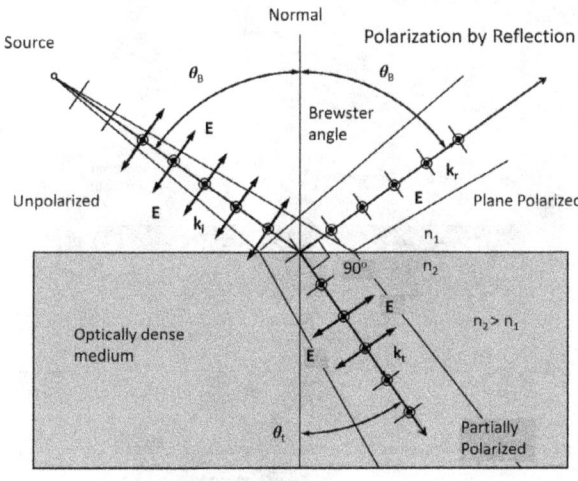

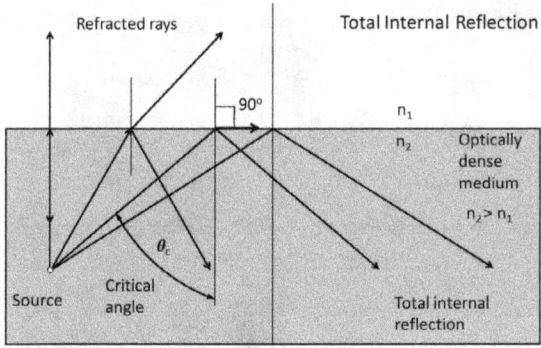

Fig. 9-3. Polarization by reflection and total internal reflection of light. Radiation pressure equals the energy density of the wave incident normally on a perfectly absorbing surface. For a perfectly reflecting surface, radiation pressure ($P_{rad} = \langle S \rangle /c$) doubles.

We drive into the future using only our rearview mirror. – Marshall McLuhan

Our observation of nature must be diligent, our reflection profound, and our experiments exact. ...Observation collects facts; reflection combines them; experimentation verifies the result of that combination. – Denis Diderot

Behavior is the mirror in which everyone shows their image. – Johann von Goethe

Natural beauty takes at least two hours in front of a mirror. – Pamela Anderson

Don't waste your time looking back, you're not going that way. – Ragnar Lothbrok

Sometimes when you stare at the abyss, the abyss stares back at you. – Friedrich Nietzsche

9. Reflectors and Dielectric Lenses

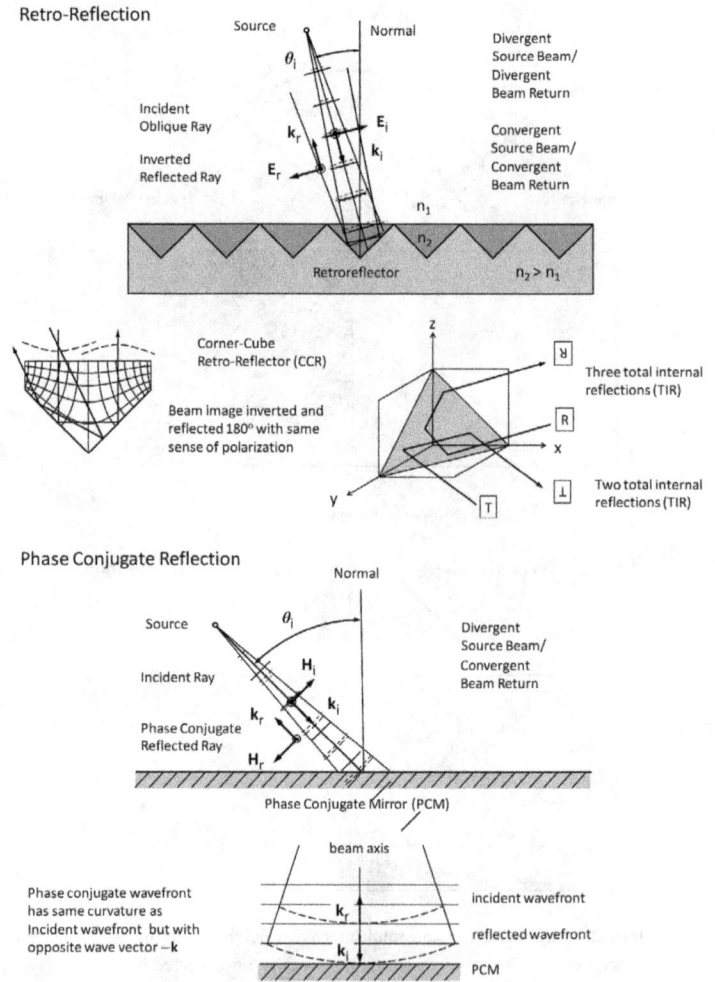

Fig. 9-4. Comparison of retro-reflection from a cube corner reflector and a phase conjugate mirror. All wave vectors are inverted in reflection from a phase conjugate mirror (PCM).

As in time reversal, the wave re-emitted by a phase conjugation mirror will auto-compensate the phase distortion and auto-focus itself on to its initial source, which can be a moving object. – A.P. Brysev et al.

'2001' used a lot of what's called 'front projection.' – Douglas Trumbull

The three Apollo retro-reflector sites form an almost equilateral triangle with sides 1250, 1100, and 970 km, and are almost centered on the nearside of the moon. – NASA

Never fire a laser at a mirror. – Larry Niven

9. Reflectors and Dielectric Lenses

A Fabry-Pérot (F-P) interferometer or étalon is a frequency filter utilizing multiple-beam interference between two flat, planar reflecting surfaces that are parallel or nearly parallel forming an optical standing wave resonator to resolve closely-spaced spectral lines. A parallel-plate étalon is generally comprised of a partially reflecting flat transparent plate with two reflecting surfaces separated by spacers or, in the case of an F-P interferometer, two parallel highly reflecting mirrors, one of which is moveable. Multiple internal reflections give rise to constructive or destructive interference depending on whether the transmitted beams are in phase or out-of-phase. In effect, an F-P étalon transmits only certain narrow wavelength bands while blocking others. Transmission depends on the wavelength of light, incidence angle, refractive index and thickness of the étalon. Sharpness of the rings (finesse) depends on reflectivity of the flats. Resonance frequencies may be tuned by changing the cavity length (distance between mirrors). See Fig. 9-5.

A phase conjugate mirror (PCM) exhibits retro-reflection of light reproducing a time-reversed replica of the incident wave in a backward direction towards the source. A negative index material (NIM) refracts the incident wave such that the group and phase velocities are anti-parallel allowing focusing in the forward direction. Negative refraction results in negative group velocity such that the Poynting vector is in a direction opposite to the wave vector. A Veselago / Pendry flat lens constructed from a NIM meta-material enables near perfect focusing. A comparison of the spatial and temporal characteristics is indicated in Fig. 9-6.

Diffraction of an incident Gaussian beam imaged by an axicon lens gives rise to a Bessel beam within the interference zone. The Bessel beam produces an Airy disc interference pattern. Such beams have been used, for example, as optical tweezers to manipulate particles suspended in the beam. Refer to Fig. 9-7.

Time reversal has a remarkable property of reflected wave fields. Phase conjugation of wave field means the inversion of linear momentum and angular momentum of light. – A. Yu Okulov

When I speak of time, I'm referring to the fourth dimension. Go on, George. The fourth dimension cannot be seen or felt. If you don't mind, will you refresh me on the first three dimensions? Really, Filby. Surely they taught you something in school. Suppose you explain it, doctor. Certainly. For example, when I move in a line, forward or backward…that's one dimension. When I move left or right, two dimensions. When I move up or down…three dimensions. For instance, that box. That box has three dimensions: length, breadth and height. What is the fourth dimension? That's mere theory. No one knows if the fourth dimension exists. On the contrary, the fourth dimension is as real and true as the other three. – H. George Wells (The Time Machine)

Many objects of our three-dimensional perceptual world are not only chiral but appear in nature in two versions, related at least ideally, as a chiral object and its mirror image. An object is chiral if it cannot be brought into congruence with its mirror image by translation and rotation. – Vladimir Prelog

…it was unthinkable that anyone would question the validity of symmetries under 'space inversion', 'charge inversion' and 'time reversal' – Chien Shiung Wu

We have long suspected that phase conjugation to produce specific controlled wave interference is the key to tapping the forces of nature. – Sci com.ru

Whatever you believe has consequences. – Alister McGrath

Nature is a blabbermouth, ask the right question and you'll always get the answer. – James Blish

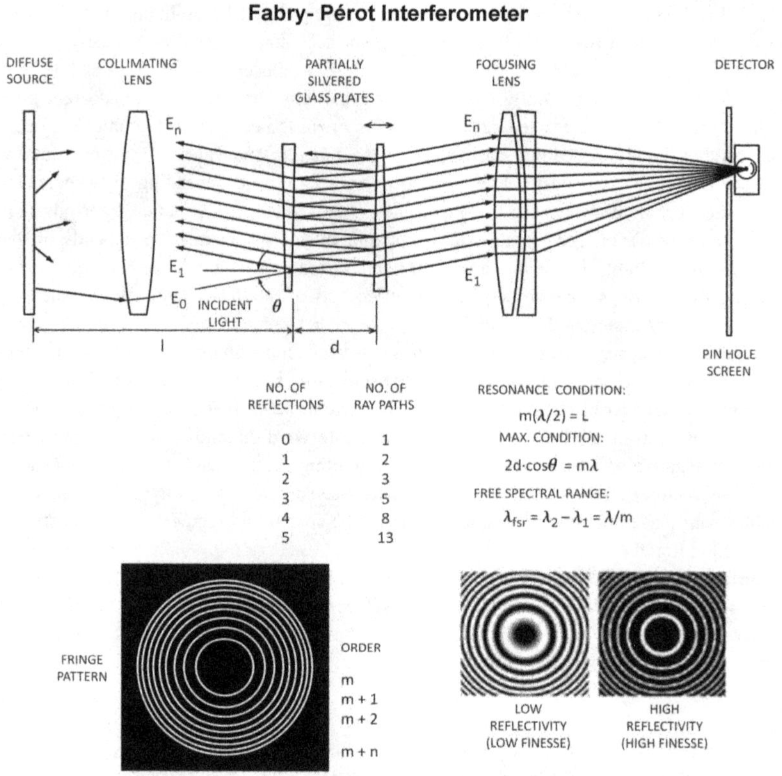

9. Reflectors and Dielectric Lenses

Fabry-Pérot Interferometer

DIFFUSE SOURCE — COLLIMATING LENS — PARTIALLY SILVERED GLASS PLATES — FOCUSING LENS — DETECTOR

E_n, E_1, E_0 INCIDENT LIGHT, θ

PIN HOLE SCREEN

NO. OF REFLECTIONS	NO. OF RAY PATHS
0	1
1	2
2	3
3	5
4	8
5	13

RESONANCE CONDITION:
$$m(\lambda/2) = L$$

MAX. CONDITION:
$$2d \cdot \cos\theta = m\lambda$$

FREE SPECTRAL RANGE:
$$\lambda_{fsr} = \lambda_2 - \lambda_1 = \lambda/m$$

FRINGE PATTERN

ORDER
m
m + 1
m + 2
m + n

LOW REFLECTIVITY (LOW FINESSE) — HIGH REFLECTIVITY (HIGH FINESSE)

Fig. 9-5. Schematic diagram of a scanning Fabry-Pérot (F-P) interferometer. If the mirrors bounding the cavity are moveable with respect to each other, the device is referred to as an "interferometer". If the mirrors are fixed in relation to each other and adjusted for parallelism by spacers then it is said to be an "Etalon". The measure of goodness of an F-P filter is known as "Finesse" which is a ratio of the amount of energy stored in the filter to the amount of energy passed through which is largely determined by the reflectivity of the mirrors. The finesse is a measure of the interferometer's ability to resolve closely spaced spectral lines. Internal absorption of energy reduces the sharpness of the filter transmittance peaks. The distance between intensity peaks vs. frequency in a spectrum plot is termed the "Free Spectral Range (FSR)" of the filter. The FSR is defined as the change in wavelength necessary to shift the fringe pattern by one fringe (FSR = $\lambda^2/(2n \cdot d)$).

> ... at several times we have thought that optics was a finished science, where the last word had been said, or almost. Each time the discovery of new facts, the overthrow or extension of accepted theories, reminded us that science is never finished. – Charles Fabry

> X-rays will prove to be a hoax. – Lord Kelvin

> If I can make this light ray radioactive, it will be the most powerful weapon the world has ever seen. – Dr. Hans Zarkov (Flash Gordon)

> By installing a booster, we can increase the capability on the order of one to the fourth power. – Captain James T. Kirk (Star Trek)

9. Reflectors and Dielectric Lenses

Negative refractive index

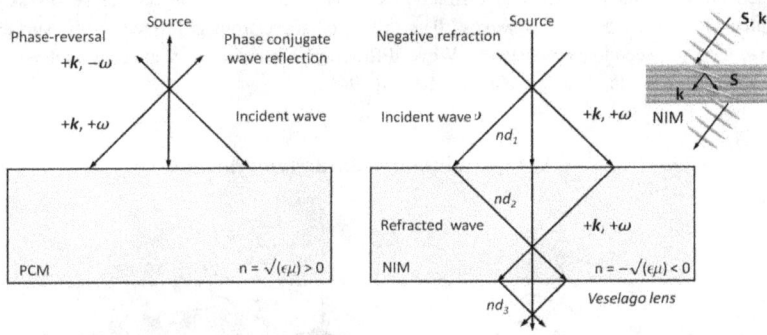

Fig. 9-6. Comparison of a phase conjugate mirror (PCM) and a negative refraction index material (NIM). Negative focusing results in two focal points.

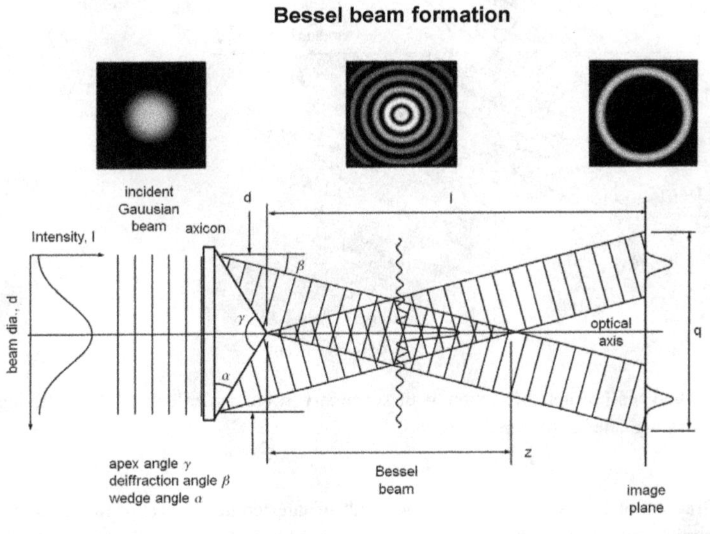

Bessel beam formation

Fig. 9-7. Bessel beam produced by diffraction of incident Gaussian beam by axicon lens.

Yes, Bessel's functions are very beautiful functions, in spite of their having practical applications.
– E. W. Hobson

One cannot, damn it, reduce the whole of philosophy to a screen with two holes. – Jørgen Jørgensen

Time, matter, space – all, it maybe, are no more than a point. – Denis Diderot

It's easy to have a complicated idea. It is very hard to have a simple idea. – Carver Mead

Space happens! – Elle Vianne Sonnet

67

9. Reflectors and Dielectric Lenses

The Huygens-Fresnel principle describes the observed wave pattern of wave propagation based on wave interference and diffraction effects. Each scattering center serves as a source of spherical secondary waves. The diffracted wave front is formed by the surface tangent to the secondary wavelets. Wave diffraction occurs when scattering centers are comparable to the incident wavelength. See Fig. 9-8.

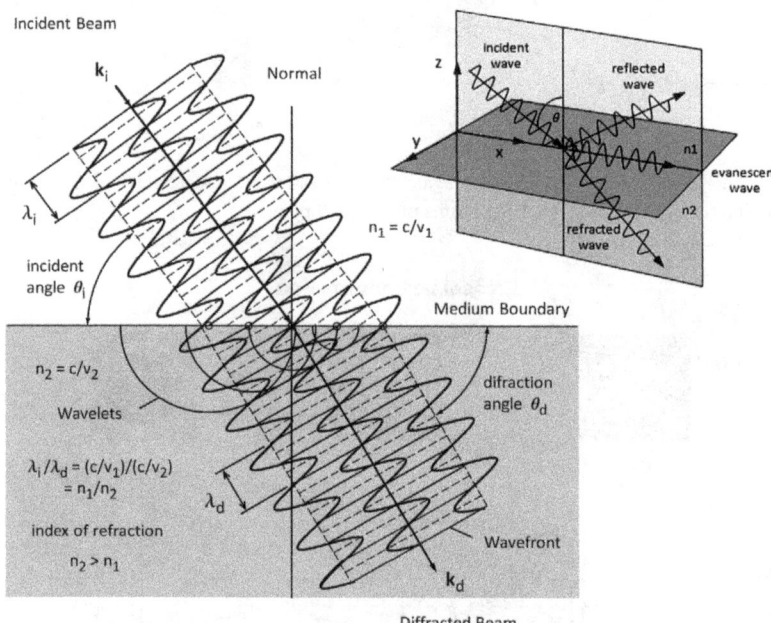

Fig. 9-8. Huygens-Fresnel diffraction with secondary wavelets reflecting off scattering centers generating a deflected wavefront.

Bragg diffraction of EM waves occurs as the result of interference effects of reflection off discrete parallel crystalline planes. The interference is constructive when the phase shift is a multiple of 2π in accordance with Bragg's law $n\lambda = 2d\sin\theta$. See Figs. 9-9 and 9-10. The effect accounts for x-ray diffraction patterns (atomic spacing), iridescent color bands on oil slicks (film thickness), color shifts in Chameleon skin nano crystals (crystal spacing), etc. Refraction occurs when there is a velocity change as measured by the refractive index n.

One may conceive light to spread successively, by spherical waves. – Christiaan Huygens

Simplicity lies concealed in this chaos, and it is only for us to discover it. – Augustin-Jean Fresnel

The concept of a matter wave existing in space is just as valid as a light wave. – Louis de Broglie

Mathematics in general is fundamentally the science of self-evident things. – Felix Klein

9. Reflectors and Dielectric Lenses

Bragg diffraction of EM waves

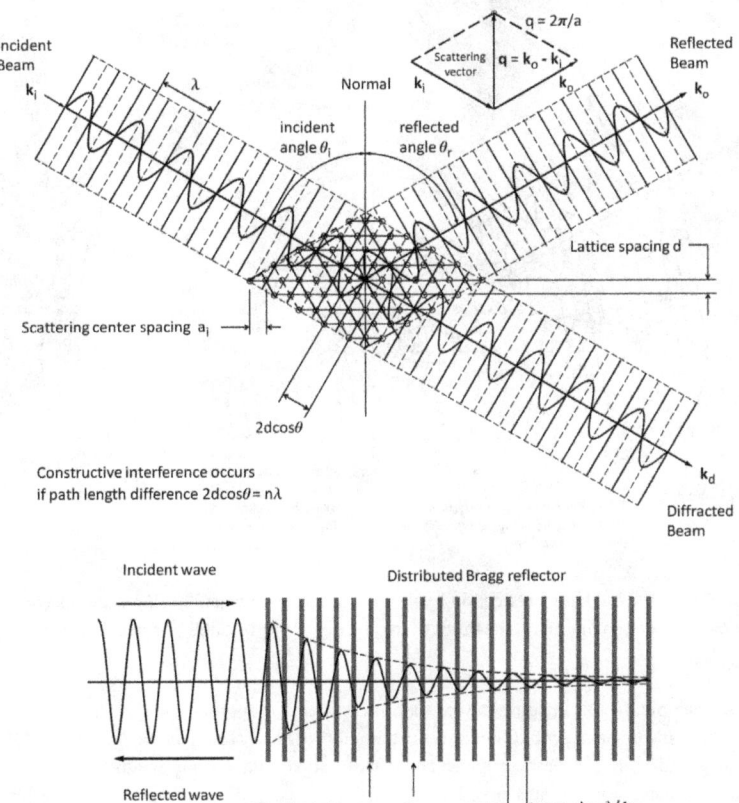

Fig. 9-9. Bragg diffraction of electromagnetic waves by crystalline planes of atomic scattering centers. Bragg diffraction and reflection can also occur by means of Bravais lattice formed by interference nodes of two or more EM wave pump beams. Reflection occurs if the lattice spacing is comparable to the incident beam wavelength. Not only travelling EM waves, but moving standing waves (i.e. particle beams/matter waves), may be reflected and diffracted. This effect was demonstrated in the Davisson-Germer electron diffraction experiment confirming de Broglie's proposal that not only are waves particles, but particles are waves.

The whole subject of the X-rays is opening out wonderfully, Bragg has of course got in ahead of us, and so the credit all belongs to him, but that does not make it less interesting. We find that an X-ray bulb with a platinum target gives out a sharp line spectrum of five wavelengths which the crystal separates out as if it were a diffraction grating. In this way we can get monochromatic X-rays. Tomorrow we search for the spectra of other elements. There is here a whole new branch of spectroscopy, which is sure to tell one much about the nature of the atom. – Henry Moseley

Thus spontaneous radiation is induced radiation of light quanta produced by zero point oscillation of empty space. – Victor Weisskopf

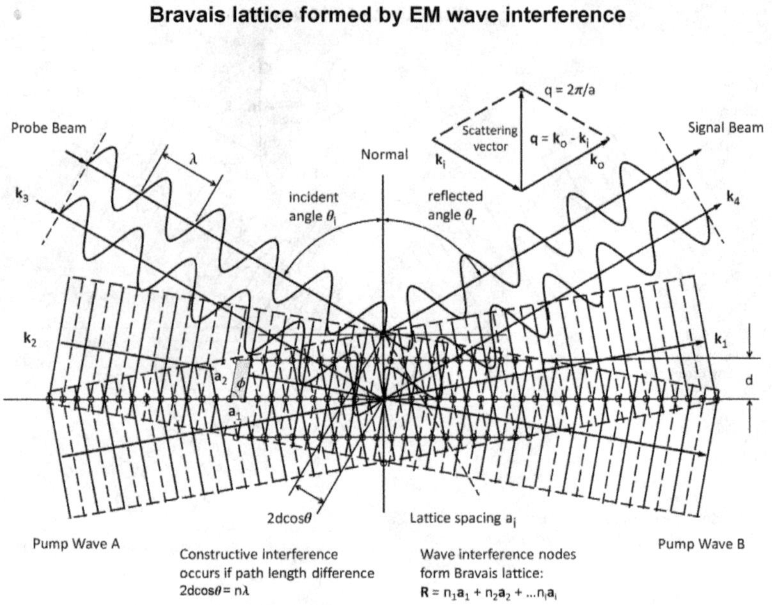

9. Reflectors and Dielectric Lenses

Bravais lattice formed by EM wave interference

Fig. 9-10. Bravais lattice reflection by scattering centers formed by EM wave interference nodes. Lattice spacing in on the order of incident beam wavelength for interference.

A zone plate is an imaging device that has lens-like properties using diffraction, unlike lenses or mirrors using refraction or reflection, to focus EM waves and matter waves. Zone plates consists of concentric circular zones with radially decreasing width that blocks odd Fresnel zones. Zone plates may be utilized across the EM spectrum and, unlike lenses, provide multiple focal points depending on the wavelengths of incident radiation f, $f/3$, $f/5$, $f/7$, etc. Conventional zone plates consist of symmetric, concentric rings with alternating binary transmission of zero and one providing interference to block waves that destructively interfere at the focal point. A Fresnel zone plate has a transparent central zone while a Soret zone plate has an opaque center zone. A Rayleigh-Wood phase-correcting zone plate has an optical thickness of odd zones different than the even zones with a thickness that varies by $\lambda/[2n-1)]$. A zone plate produces convergent/divergent focusing similar to holographic combiner mirrors. A Gabor lens produced holographically exhibits sinusoidal variation. Two displaced overlapped zone plates generate Moiré interference patterns. Refer to Fig. 9-11.

EM waves may be refracted by a graded dielectric lens such as Luneberg or Eaton lens. These are typically used in optical and microwave applications and provide fixed focal length for a given operating frequency. Often the lens are spherical or hemispherical in design with layered construction otherwise known as gradient refractive index (GRIN) lenses. Similar to biologic lenses, GRIN lenses focus light by spatial variation of refractive index as well as surface curvature to efficiently change the direction of light while minimizing aberrations and image blur. Luneberg lenses can focus EM waves to an off-axis point or conversely form a beam of plane waves from a point source. Examples of such lenses are shown in Figs. 9-12 and 9-13. Numerous derivatives have been developed. Dielectric lens are typically used as compact lens antennas and provide advantages over

9. Reflectors and Dielectric Lenses

reflector antennas such as beam shaping, beam steering and sidelobe suppression. Flat surface variants of such spherical lenses may be fabricated by metamaterial structures, quantum wells, or nano-scale antenna resonators layered on a dielectric substrate imparting a phase shift gradient. In addition, such surface layers can be used to produce retroreflection, negative index of refraction and control of amplitude, frequency and polarization. Metamaterials allow synthesis of the gradient index by variation of the dimensions of densely packed periodic arrays of resonant elements enabling phase front control of evanescent EM waves.

Incident EM plane waves in the vicinity of a star are deflected in an Einstein gravitational lens effect producing an Einstein ring or arc focused along a line extending from a minimum focus distance to infinity. In an optical theory of gravity, light deflection is the result of variable refractive index K_{PV} of the vacuum associated with the gravitational field EM energy density. Slow wave dielectric GRIN lenses provide an analogous effect with variable index of refraction of matter and can serve to model various gravitational effects such as gravitational lenses, photonic black holes, etc. An example of simulation of gravitational lensing is shown in Fig. 9-14. According to GR, energy-momentum induces space-time curvature. The Einstein field tensor $G_{\mu\nu}$ is a conversion factor between curvature and energy density. A vacuum exhibits a positive index of refraction such that light bends toward the region of increased energy density. A negative index of refractive material exhibits an analogous 'anti-gravity' effect bending light away from a spherical lens. A gravitational field has negative vacuum energy density and may be represented in anti-de Sitter hyperbolic space with negative curvature.

It was in the investigation of the varying colors of the soap-bubble that Newton... established the fundamental principle of the present generalization of the undulatory nature of light. – Joseph Henry

What de Broglie had found was that momentum should be inversely related to the length of a wave associated with a particle. De Broglie was thus able to show that for electrons moving very close to the speed of light, matter waves would have the same wavelength as those of light or x-rays, and that... matter particles would show the same kind of diffraction and interference effects seen with light waves and X-rays. – Ernest J. Sternglass

Notwithstanding these major arguments the wave theory initially did not meet with complete acceptance. – Max von Laue

There is nothing more deceptive than an obvious fact. – Arthur Conan Doyle

For X-rays, the phenomenon of diffraction by crystals was a natural consequence of the idea that x-rays analogous to light and differ from it only by having a smaller wavelength. – Louis de Broglie

If diffraction or interference phenomena were to be sought it was therefore necessary, in accordance with the basic principles of wave theory, to select for the test arrangement far smaller decisive dimensions than those employed in the corresponding tests with visible light. – Max von Laue

With crystals we are in a situation similar to an attempt to investigate an optical grating merely from the spectra it produces... But a knowledge of the positions and intensity of the spectra does not suffice for determination of the structure. The phases with which the diffracted waves vibrate relative to one another enter in an essential way. To determine a crystal structure on the atomic scale, one must know the phase differences between the different interference spots on the photographic plate, and this task may certainly prove to be rather difficult. – Max von Laue

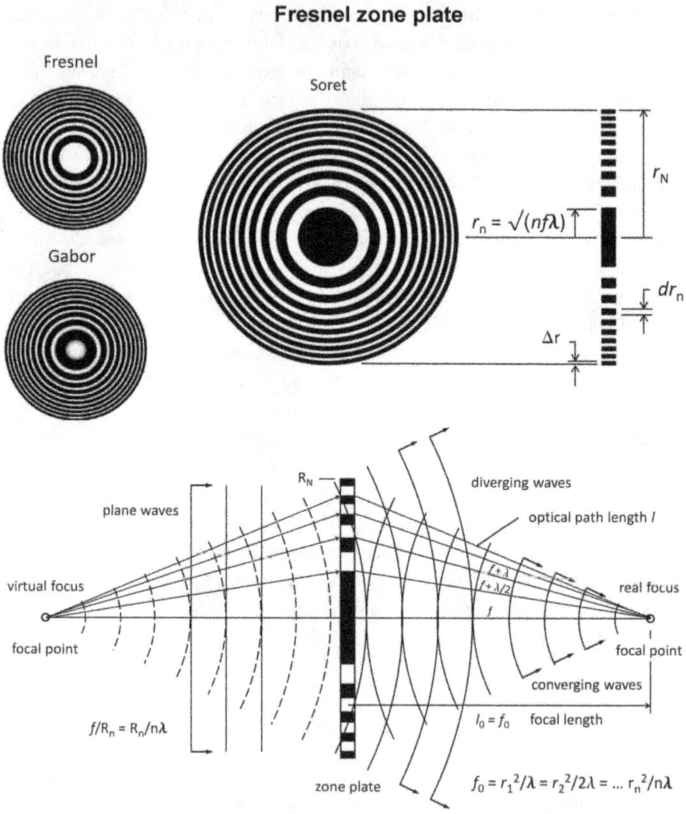

9. Reflectors and Dielectric Lenses

Fresnel zone plate

Fig. 9-11. A Fresnel zone plate has both real and virtual focal points and acts as a converging lens and diverging lens of the same focal length. A zone plate is based on diffraction, rather than refraction or reflection, and selectively blocks incident radiation for focusing. Conventional Fresnel lenses usually consist of all refractive lens ring-shaped segments and sometimes combined with an outer array of reflecting elements. Flat metalenses use arrays of nanostructures such as titanium dioxide nanofins to focus light and eliminate chromatic aberration.

The science of optics...has two different directions of progress...the inductive and deductive method. – Sir William Rowan Hamilton

Mathematics may be defined as the subject in which we never know what we are talking about, nor whether what we are saying is true. – Bertrand Russell

Later generations will regard Mengenlehre (set theory) as a disease from which we recovered. – Henri Poincaré

Mathematics is a more powerful instrument of knowledge than any other that has been bequeathed to us by human agency. – René Descartes

Nature is not embarrassed by difficulties of analysis. She avoids complication only in means. Nature seems to be predisposed to do much with little: it is a principle that the development of physics constantly supports by new evidence. – Augustin-Jean Fresnel

9. Reflectors and Dielectric Lenses

Spherical Luneberg lens

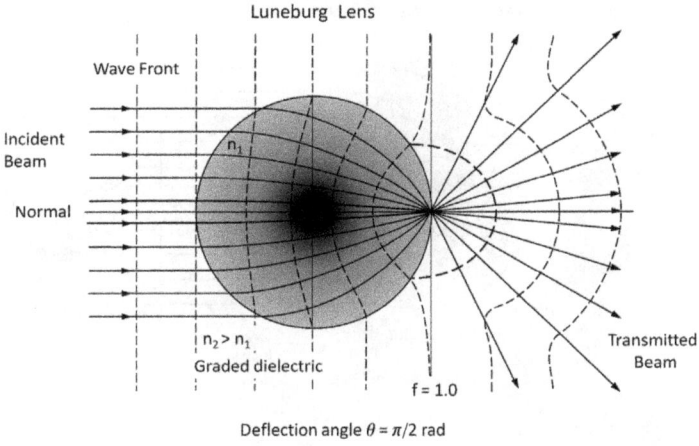

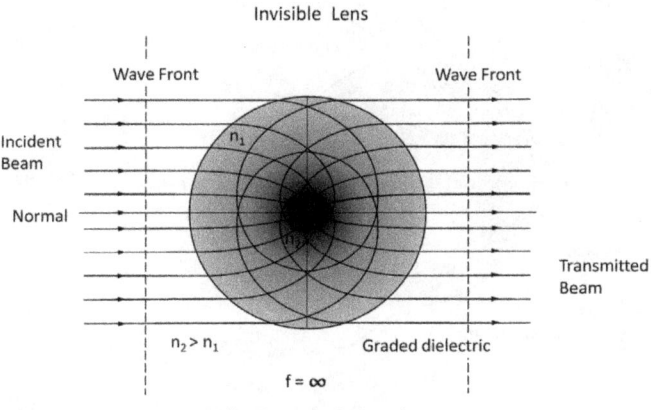

Fig. 9-12. A Luneberg lens is a type of GRIN lens traditionally formed of concentric graded dielectric shells, metallic / ceramic particle matrix or printed circuit metamaterials and able to focus planar wavefronts to a point or convert spherical wave fronts to planar wavefronts. The refractive index is given by $n(r) = \sqrt{(2 - (r/a)^2)}$ where a is the lens radius and r is the radial distance inside the lens. A radar reflector can be made from a Luneberg lens by metallizing parts of its surface and may be used, for example, to augment an aircraft or target drone radar signature.

Never forget the time factor. It always enters the picture in the end. – Robert Morley

We know the past, but cannot control it. We control the future, but cannot know it. – Claude Shannon

The future cannot be predicted, but futures can be invented. – Dennis Gabor

73

9. Reflectors and Dielectric Lenses

Spherical Eaton lens

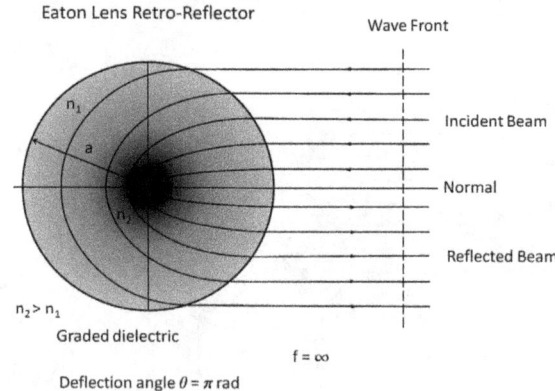

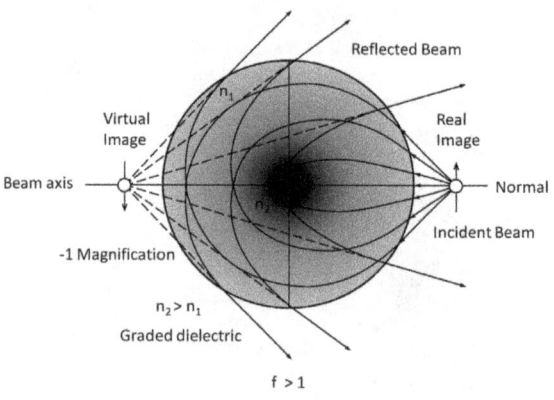

Fig. 9-13. An Eaton lens is another type of GRIN lens that is an omnidirectional retroreflector. This is a spherical lens with radially symmetric refractive index of one at the outer boundary and high refractive index which can approach infinity at the center singularity. The refractive index is given by $n(r) = \sqrt{(2a/r-1)}$ for $r \leq a$ where a is the lens radius and r is the radial distance inside the lens.

I always knew I belonged on the other side of the lens. – Kathy Ireland

Reality is that which, when you stop believing in it, doesn't go away. – Phillip K. Dick

Mathematics is not there till we put it there. – Arthur Eddington

There is no branch of mathematics, however abstract, which may not someday be applied to the phenomena of the real world. – Lobachevsky

One man's 'magic' is another man's engineering. – Robert Heinlein

9. Reflectors and Dielectric Lenses

Gravitational lens

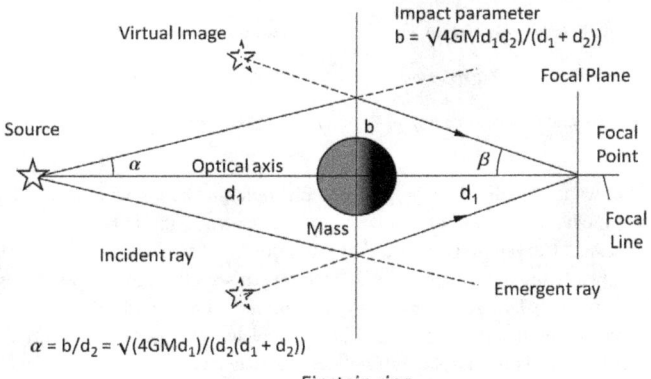

$\alpha = b/d_2 = \sqrt{(4GMd_1)/(d_2(d_1+d_2))}$

Einstein ring

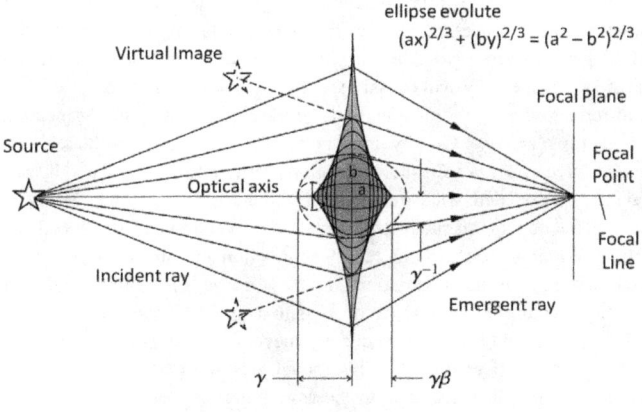

Convergent lens

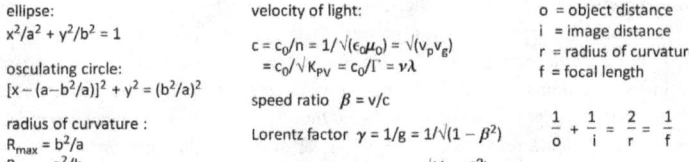

ellipse:
$x^2/a^2 + y^2/b^2 = 1$

osculating circle:
$[x - (a-b^2/a)]^2 + y^2 = (b^2/a)^2$

radius of curvature:
$R_{max} = b^2/a$
$R_{min} = a^2/b$

velocity of light:
$c = c_0/n = 1/\sqrt{(\epsilon_0 \mu_0)} = \sqrt{(v_p v_g)}$
$= c_0/\sqrt{K_{pv}} = c_0/\Gamma = \nu\lambda$

speed ratio $\beta = v/c$

Lorentz factor $\gamma = 1/g = 1/\sqrt{(1-\beta^2)}$

Lorentz contraction $g = \sqrt{(1-\beta^2)}$

o = object distance
i = image distance
r = radius of curvature
f = focal length

$\frac{1}{o} + \frac{1}{i} = \frac{2}{r} = \frac{1}{f}$

Fig. 9-14. Gravitational lensing by an obstructing mass in the line of sight of a radiation source can produce an Einstein ring distortion or arc around the obstruction as seen by an observer. The effect may be simulated by substitution with a concave lens of positive index of refraction. The concave shape corresponds to an evolute of an ellipse.

Provided that our present estimates of the masses of cluster nebulae are correct, the probability that nebulae which act as gravitational lenses will be found practicality a certainty. – Fritz Zwicky

10. Deflection of Light

One may conceive light to spread successively by spherical waves. – Christiaan Huygens

Radiant light consists in Undulations of the Luminferous Ether. – Thomas Young

Light takes the path that minimizes its action. – Piere Louis Maupertuis

Light slows down in a medium due to photon's EM field interaction with the EM energy density of the medium. Wavefront curvature is associated with light deflection and gravitational lensing. Curvature is induced by variation in the vacuum refractive index $K_{PV}(r,\omega,M)$ locally altering the velocity of light resulting in wavefront retardation in regions of increased energy intensity. A general illustration of curved manifolds of a gravitational potential well and vacuum refractive index hill are depicted in Fig. 10-1. In gravity-free region, Minkowski space is isomorphic to Euclidean tangent space. In a gravitational field, the energy density varies resulting in curved manifold parameter spaces while the tangent space of spacetime remains flat. Unlike embedding diagrams associated with illustration of spacetime curvature in GR, spacetime remains Euclidean while variables such as EM nodal distances and frequencies represented in tangent space exhibit dilation, contraction and curvature with an established physical causal means and demonstrable effect.

The index of refraction n ($= c_0/c$) is a measure of the slowing of light in a medium with increased EM energy density. Gravity in a polarizable vacuum may be viewed as a dielectric gradient force effect in a Fresnel zone nonuniform field and mediated by radiation (photons, gravitons) associated with frequency discordance between interacting masses considered as a collection of quantum oscillators. The variation in the speed of light, deflection of light in a gravitational field, length contraction and time dilation effects have been described in terms of local variations in the electric permittivity and magnetic permeability of the vacuum in the presence of mass in a variable polarizable vacuum (PV) model in 1921 by Wilson[27] and subsequently developed by Dicke[28], Puthoff[29,30], Depp[31], Krogh[32] and Storti et al.[33,34] This model was shown to be compatible with General Relativity (GR) without recourse to the assumption of spacetime curvature. The permittivity and permeability appear as scalar fields which, in turn, influence electric field intensity **E** and magnetic flux density **B** in accordance with governing Maxwell equations of electromagnetism. According to the Coulomb equation $F = Q_1 Q_2/4\pi\epsilon r^2$, the Coulomb force between two charges is lower in any dielectric medium than in free space in proportion to the value of the dielectric constant K for the medium. The cause of this decrease in force action is found in the polarization or stretching of the molecules of the dielectric material. The physical vacuum in quantum field theory is a dielectric medium. In the PV model, a variable refractive index K_{PV} is attributed to the quantum vacuum and is proportional to the EM energy density associated with the gravitational field. The variable $K_{PV}(r,\omega,M)$ describes the gradient in the refractive index acting to deflect light. The vacuum permittivity and permeability are not uniformly distributed in space, but form scalar fields. The polarizability of the vacuum in the vicinity of mass differs from the asymptotic far-field value as a result of increased density of EM wave interference antinodes which act as scattering centers slowing and deflecting propagating EM wavefronts.

A gravitational field can be regarded as a medium whose index of refraction is larger than that of a pure vacuum. – Oleg D. Jefimenko

10. Deflection of Light

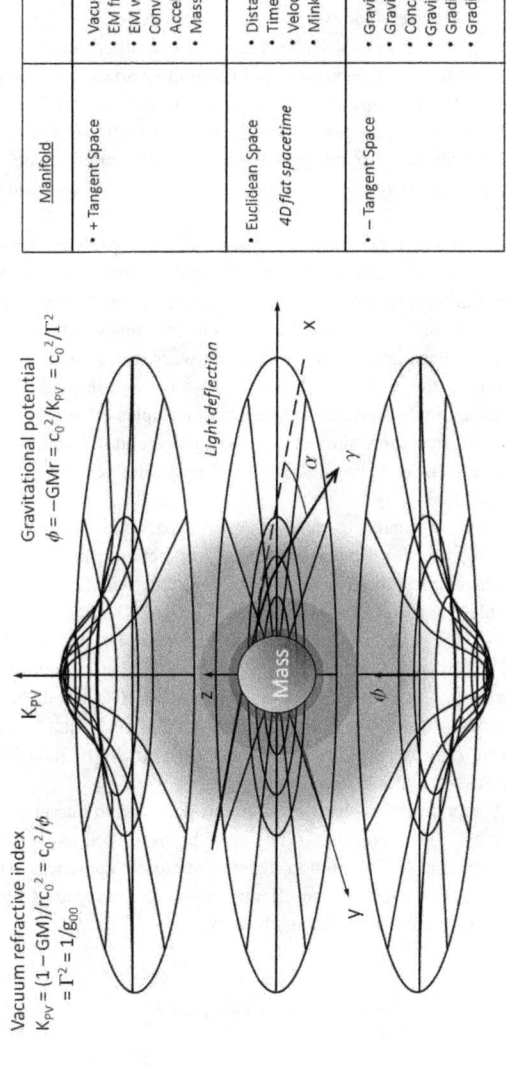

Fig. 10-1. Tangent space representation of EM energy density hill and gravitational potential well in a flat 4D spacetime. A gravitational field consists of a very dense, broad spectrum electromagnetic field with vacuum index of refraction greater than the surrounding far-field medium. This increased field energy density results in a lessening of electromagnetic forces and slowing of electromagnetic interactions.

77

10. Deflection of Light

Unlike the geometric spacetime curvature assumed in the Einstein Theory of General Relativity (GR), gravitation in the Puthoff *et al* PV model[29,30] is described instead by variation in EM wave energy and density due to local variation in the vacuum refractive index K_{PV}.

$$K_{PV} = c^2/c_0^2 = n^2 = (dt/d\tau)^2 = \rho/\rho_0 = \Gamma^2 = (1 - GM/rc_0^2) \quad (10\text{-}1)$$
$$= 1/(1 + 2\phi_g/c^2) = c_0^2/\phi = n^2 = (\epsilon/\epsilon_0)(\mu/\mu_0) \quad (10\text{-}2)$$
$$= 1 + 2GM/rc^2 + (GM/rc^2)^2 + \ldots \quad (10\text{-}3)$$
$$= e^{\wedge}2GM/rc^2 \quad (10\text{-}4)$$
$$= e^{\wedge}r_S/r = 1/g_{00} \quad (10\text{-}5)$$

The refractive index n is a ratio of the phase velocity v in a medium to velocity of light c in a gravity-free, zero curvature vacuum ($n = c/v$). Variable vacuum electric permittivity and magnetic permeability result in a variation in the speed of light providing an explanation of bending of light and gravitational attraction in terms of local scalar ϵ and μ fields (c = $1/\sqrt{\epsilon_0\mu_0}$). Einstein, in 1911, initially proposed a gravitation theory with a variable speed of light as a function of potential, but later, in 1916, adopted an abstract geometrical model in GR[35] without a defined causal physical mechanism for means by which energy was said to curve spacetime.

In lieu of an unexplained mechanism for assumed "bending" of spacetime (e.g., alteration in space (contraction) and time (dilation) in accelerated inertial frames), the Lorentz contraction in the Ivanov-LaFreniere wave model[36,37,38] refers instead to a physical EM wavelength contraction (compression of the nodal distance) and frequency reduction of a standing matter waves in motion. Spacetime remains Euclidean whereas EM waves may undergo distortion. A variation in the vacuum dielectric constant K_{PV} provides the mechanism for alteration in wavelength which occurs in quantized multiples of wavelength. The EM wavelength contraction and frequency shift in a polarizable vacuum accounts for mass in motion and gravitational effects including the energy change, deflection of light, gravitational frequency shift and clock slowing.

Unlike GR (with unexplained mechanism for assumed spacetime distortion), gravitational effects in a polarizable vacuum (including length contraction, time dilation, frequency shift, alteration in the speed of light, etc) are EM wave interaction effects due to local variation in the polarizable vacuum refractive index $K_{PV}(r,\omega,M)$. The vacuum refractive index K_{PV} varies as $e^{\wedge}2GM/rc^2$ or equivalently $1/(1 - 2GM/rc^2)$. The wave vector $k = \omega[1 + GM/rc_0^2]/c_0[1 - GM/rc_0^2] \approx \omega/c_0(1 + 2GM/rc_0^2)$. The deflection of light $\alpha \cong 4GM/rc^2$ in which half of the deflection is due to the scalar potential ϕ ($-GM/r = c_0^2/K_{PV}$) and half due to the vector potential **A** (= $(\mathbf{v}/c^2)\phi$). The time delay dt $\approx [-2GM/c^3]\ln(1 - \cos \alpha)$. The change in vacuum permittivity $\Delta\epsilon = 4GM\epsilon_0/rc^2$. The local velocity of light in a gravitation field c = $c_0/(1 + 2GM/rc_0^2) = c_0/\Gamma = c_0/\sqrt{K_{PV}}$.

Deflection of the wavefront generates an acceleration force proportional to curvature R and is equivalent to that of gravity. The force of gravity may be expressed as $F_g = \frac{1}{2} mc^2(\alpha)/d$ where α = deflection angle and d = separation distance at closest approach. The photon with a zero rest mass in free space acquires an effective mass as the speed of light slows in a refractive medium. The refractive index (n) is given by

$$n = c_0/c = c_0/v_{phase} = c_0/(c^2/v_{group}) = \sqrt{(\epsilon/\epsilon_0)(\mu/\mu_0)} = \Gamma = \sqrt{K_{PV}} = 1/\sqrt{g_{00}} \quad (10\text{-}6)$$
$$= 1 - 2GM/rc^2 + G^2M^2/r^2c^4 = \sqrt{c_0^2/\phi^g} = \sqrt{(\epsilon\mu)}/\sqrt{(\epsilon_0\mu_0)}\sqrt{(\epsilon_r\mu_r)} = \beta^{-1} \quad (10\text{-}7)$$
$$= e^{\wedge}2(GM/c_0^2)/r = e^{\wedge}(2\mu/r) \quad (10\text{-}8)$$

Vacuum voco locum omnem in quo corpora sine resistentia movetur. – Sir Isaac Newton

10. Deflection of Light

The refractive index is a scalar quantity that corresponds to the sum of the absolute values of the local EM field intensities of EM waves existing a given point and time. Inertial (relativistic) mass of a photon is generally found by setting the rest mass to zero in the Einstein energy relation, applying the Planck energy relation and solving for mass m.

$$E = mc^2 = \sqrt{(p^2c^2 + m_0^2c^4)} = h\nu \qquad (10\text{-}9)$$
$$m = h\nu/c^2 = E/c^2 \qquad (10\text{-}10)$$

The total photon energy E_t (= H) may be expressed as a function of the electric field intensity **E** and magnetic flux density **B**. Energy is proportional to wave function curvature.

$$E_t = \tfrac{1}{2} m(\mathbf{E}^2 + c^2\mathbf{B}^2) \qquad (10\text{-}11)$$

In a non-zero curvature vacuum where the photon acquires an effective mass m, the photon energy is

$$E_t = E_k + E_p = \tfrac{1}{2} mc^2 + GmM/r \qquad (10\text{-}12)$$

The quantity $Gm = r \cdot v(r)^2$ where m is gravitational mass, r the radial distance and v(r) the orbital velocity.

Deflection angle of a particle with mass m may be expresses as a ratio of the potential energy to the kinetic energy

$$2\alpha \approx 2\, E_p/E_k = 2(GmM/r)/(1/2mv^2) = 4GM/rv^2 \qquad (10\text{-}13)$$

As described by Krogh[32], the speed of light c (= $c_0(1 - GM/rc_0^2)$) may be represented as a series expansion

$$c = c_0(1 + 2\Phi_g/c_0^2 + 2\Phi_g^2/c_0^4 + \ldots) \qquad (10\text{-}14)$$
$$\text{where } \Phi_g/c_0^2 = -GM/c_0^2 r$$

which gives a first order deflection α_1 in radians in a weak gravitational field of

$$\alpha_1 = 4GM/c_0^2 r \qquad (10\text{-}15)$$

A particle mass passing a central mass such as a star increases velocity as it infalls toward the star. A photon cannot speed up in a region of increasing energy density and, hence, its deflection is twice that of a moving mass. The deflection of light, particulate masses and EM wavefronts in a polarized vacuum of varying refractive index is illustrated in Figs. 10-2 and 10-3 for the example of a non-rotating black hole. As shown maximum deflection occurs at the point of closest approach. Half of the deflection is due to scalar ϕ interaction and half due to vector potential **A** interaction. Wavefront curvature and retardation result in light deflection and gravitational lensing. All force fields are expressed as accelerative curvatures opposing compression to compensate for changes in scale. Spacetime curvature is replaced by variable refractive index $K_{PV}(r,\omega,M)$ which is a measure of EM energy density. Gravity is the result of a gradient in the EM Poynting vector **P** due to variation in K_{PV} ($\nabla \mathbf{P} = k\Delta K_{PV}$). Effective mass is associated with a change in vacuum permittivity ϵ (= $e_0[1 + 4GM/rc^2]$). The metric tensor $d\tau^2 = g_{\mu\nu}dx^\mu dx^\nu$ corresponds to the gravitational potential while the affine connection $\Gamma^\lambda_{\mu\nu}$ represents the gravitational field. A parameter summary is tabulated in Table 10-1.

10. Deflection of Light

The photosphere radius of a black hole corresponds to 1.5 R_S with the critical capture radius for photons lying at $3R_S$ (= $3GM/c^2$) where R_S is the Schwarzschild radius. For a rotating black hole, the outer event horizon expands outward in a radial direction. The ergosphere, an ellipsoidal region in which frame-dragging occurs, extends outward from the outer event horizon out to a static limit and is aligned with the spin axis. For a non-rotating black hole, the minimum stable prograde orbit radius is $2R_S$ while for a rotating black hole the minimum stable orbit radius is R_S.

Curvature of EM wavefronts produced by a gradient in energy density results in a compression of energy. The accelerative force of gravity is a measure of the EM spectral energy density gradient. Opposite curvature of opposing EM wavefronts is associated with gravitational attraction. The Riemann (intrinsic) curvature tensor $R^\rho{}_{\sigma\mu\nu}$ describes acceleration of geodesics with respect to one another. Gravitation corresponds to curvature of the base manifold ($R_{\mu\nu} - \frac{1}{2}g_{\mu\nu}R = -8\pi G/c^2 T_{\mu\nu}$). According to Ivanov[36], the frequency state of the black hole is represented by $\nu_{BH} = m_{BH}c^2/h$ and $R_\nu = R_S/(2Gh/c^4) = Mh/c^2$.

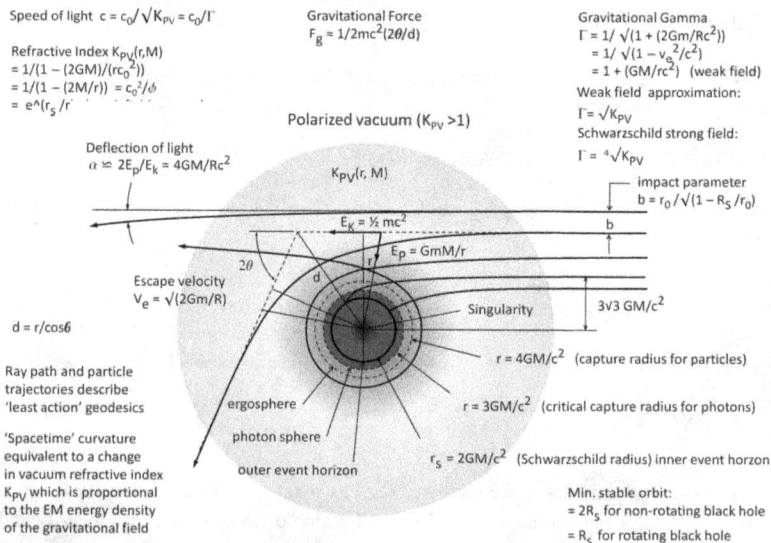

Fig. 10-2. Polarizable vacuum representation of a non-rotating black hole. A static gravitational field only alters the direction and magnitude of the velocity of light whereas the gravitational field of a rotating mass can also alter the rotation of the plane of polarization as a result of interaction of spin and torsion. A rotating black hole, in addition, generates a gravitomagnetic moment as a result of the mass current torsion field interacting with the accretion disc to produce jets along the spin axis. Curvature is due to mass, torsion is due to spin. Acceleration of gravity is a function of the EM energy density gradient.

To the pure geometer the radius of curvature is an incidental characteristic – like the grin of the Cheshire cat. To the physicist it is an indispensible characteristic. – Arthur Eddington

10. Deflection of Light

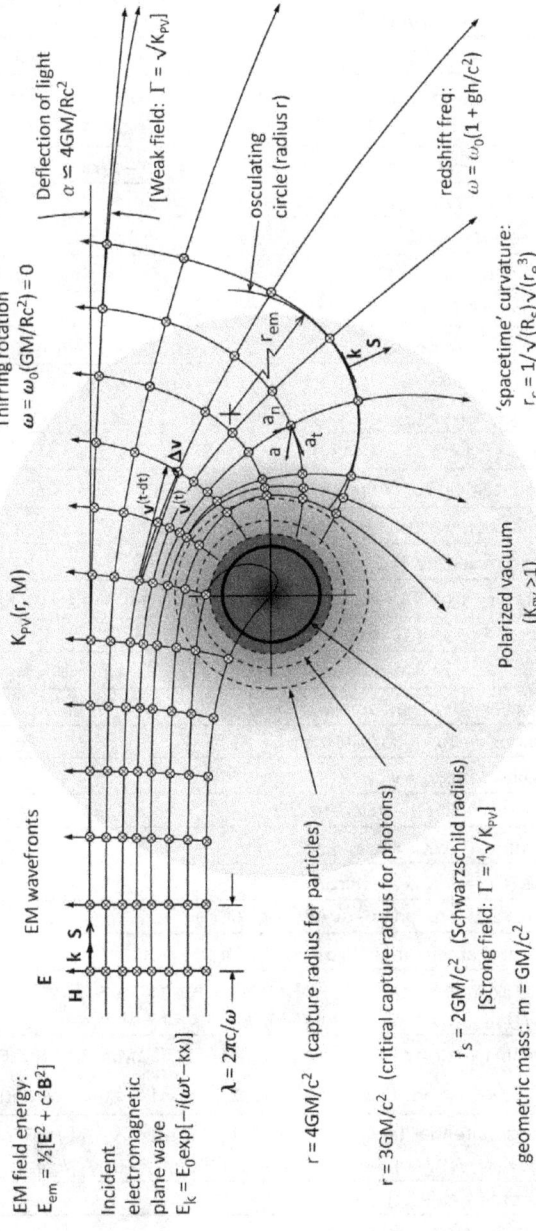

Fig. 10-3. Electromagnetic plane wave wavefront deflection in a gravitational field of a non-rotating black hole. Wavefront curvature κ ($= 1/R$) is produced by a gradient in EM energy density. The vacuum refractive index K_{PV} ($= 1/(1 - (2GM/rc_0^2))$ is a measure of local EM energy density. Deflection of wavefront produces an acceleration proportional to curvature. Close to a black hole there is a radial frame-dragging effect. For a rotating black hole, there is, in addition, a rotational frame-dragging effect. A Kerr black hole possesses mass and momentum but no net electric charge. An extremal black hole is a state of minimum theoretical size for given electric charge and spin angular momentum.

81

10. Deflection of Light

Table 10-1. Polarizable vacuum representative parameter summary

E	Electric field strength $(= -\delta\mathbf{A}/\delta t - \nabla\phi = \mathbf{E}_0 e^{j(\omega t+\theta)} = (k/e_0)\mathbf{D} = Z_0\mathbf{H} = \mathbf{F}/q_0)$
H	Magnetic field strength $(= \phi/\mu A = 1/\mu_0 \nabla \times \mathbf{A} = E/Z_0 = B/K\mu_0 = -(1/\mu_0)\nabla\Phi_m)$
c	Velocity of light $(= \lambda f = \omega/k = c_0/n = K/(\sqrt{\mu_0\epsilon_0}) = \sqrt{(v_g v_p)} = Z_0/\mu_0 = c_0/\Gamma = E/B$ $= \phi_E/\phi_B = v_g/\beta = c_0/\sqrt{K_{PV}} \simeq 2.99792458E8\text{ m/s})$
β	Velocity ratio $(= v/c = \cos\theta = \Delta\phi/\pi = \rho/\gamma = \sqrt{(1-1/\Gamma^2)} = \sqrt{(1-1/K_{PV})} = pc/E)$
v_p	Phase velocity $(= c/n = p/2m = \hbar k/2m = c/\beta = c^2/v_g;$ Wave phase $= kz - \omega t)$
Z_0	Impedance of free space $(= |\mathbf{E}|/|\mathbf{H}| = \sqrt{\epsilon_0\mu_0} = \mu_0 c_0 = Z_P/4\pi = 376.7\ \Omega)$
A	Vector potential $(= \mathbf{p}/q_0 = (\mathbf{v}/c^2)\phi_E = -(1/k^2)\mathbf{j}_\mu = k\hbar/q = (\mu_0/4\pi)(q\mathbf{v}/R)) = \hbar\mathbf{k}/e)$
P_{rad}	Radiation pressure $(\langle T^{33}\rangle) = E_0^2/8\pi = \langle S\rangle/c) = \phi_E/c)$
n	Index of refraction $(= c_0/c = c_0/v_p = c_0\mathbf{k}/\omega = \Gamma = \sqrt{K_{PV}} = 1/\sqrt{g_{00}} = \sqrt{[-g_{11}/g_{00}]})$
K_{PV}	Vacuum refractive index $(= \epsilon/\epsilon_0 = \mu/\mu_0 = \Gamma^2 = c^2/\sqrt{(2GM/R)} = c_0^2/\phi$ $= (\phi_E/\phi_B)^2/\phi_g = 1/g_{00} = g_{11} = g_{22} = g_{33} = 1/(1+2\phi/c^2) = (\epsilon/\epsilon_0)(\mu/\mu_0))$
ρ_e	EM energy density $(= U = W_e = \frac{1}{2}(\epsilon_0 E^2 + B^2/\mu_0) = E_0/\sqrt{K} = S/v = c^2 p)$
T^{00}	EM energy momentum tensor $(= nm/\sqrt{(1-v^2/c^2)} = [(\mathbf{E}^2 + \mathbf{B}^2) + m^2(\mathbf{A}^2 + \phi^2)]/8\pi)$
ρ_g	Gravitational field energy density $(= -g^2/8\pi Gc^2 = (h/c^3)(\omega_{PV}) = -(\nabla\phi^2)/4\pi Gc^2)$
p	EM field momentum density $(= S/c^2 = \rho_e \cdot \mathbf{v}/c^2 = \epsilon_0 \mathbf{E} \times \mathbf{B})$
F_g	Gravitational force $(= m\mathbf{g} = \nabla\mathbf{S} = k\Delta K_{pv}^{-} = -m\nabla\phi = m(2c\cdot\Delta v)\cdot\mathbf{r}_u = m(\mathbf{g} + \mathbf{v} \times \mathbf{K}))$
F_k	Electrokinetic force $(= F_{EP} + F_{DEP} = q\mathbf{E}_k + (\mathbf{P}\cdot\nabla)\mathbf{E})$
M	Mass $_{back\ hole}$ $(= r_S c^2/2G);\ m = m_0/\Gamma = -\phi_g \cdot r = E/\phi = -c_0^2/G\cdot K^{PV} = \mathbf{E}\cdot\mathbf{B}^2/E^2 = \hbar k/v)$
k	Wave vector $(= \lambda/2\pi = 2\pi/\lambda = n\omega/c);\ (= p/\hbar = \sqrt{2ME/\hbar}) = (e/A)/\hbar)$
S	Energy flux Poynting vector $(= (1/\mu_0)\mathbf{E} \times \mathbf{B} = \mathbf{E} \times \mathbf{H})$
$|S|$	Energy flux magnitude $(c\cdot\rho_e = \mathbf{v}\cdot\rho_e = c^2\cdot\mathbf{p})$
r_{em}	Wavefront curvature radius $(=$ arclength $s/\theta)$
κ	Wavefront curvature $(= 1/r_{em} = (v_x a_y - v_y a_x)/\sqrt{(v_x^2 + v_y^2)^3})$
R	Scalar curvature $(= \kappa^2 = 1/r_{em}^2 = -(8\pi G/c^2)T)$
a_n	Wavefront normal acceleration $(= -dV/dt = -g^2/6Gc^2)$
a_t	Wavefront tangential acceleration $(= \alpha r \simeq 4GM/R^2 c^2)$
g	Acceleration of gravity $(= \mathbf{F}/m = -\nabla\phi - \Delta\mathbf{A}_g/dt = \nabla \times \mathbf{A} = a_n + a_t = 1/2mc^2(2\theta/d)$ $= (2c\cdot\Delta v)\cdot\mathbf{r}_u = (\phi/r)\cdot\mathbf{r}_u;\ g_{max} = F_{max}/M = c^4/GM;\ \mathbf{g}_k = \mathbf{v} \times \mathbf{K} = -dAdt)$
Γ	Gravitational gamma $(= dt/d\tau = 1/\sqrt{(1-v_e^2/c^2)} = 1/\sqrt{(1/(2GM/c^2 R))} = H\cdot Z/E$ $= 1/\sqrt{(1-\beta_g^2)} = \sqrt[4]{K_{PV}}$ (strong field) $= \sqrt{K_{PV}}$ (weak field) $= 1/\sqrt{g_{00}} \simeq GM/c^2 R)$
ϕ_g	Gravitational scalar potential $(= -GM/r = c^2(1 - 1/\Gamma)/2 = \mathbf{g}\cdot r = c_0^2/K_{PV} = GE/rc^2$ $= GMc^2/r(\phi_E/\phi_B)^2 = (\phi_E/\phi_B)^2/K_{PV}) = -(\Delta v/v)c^2 = -(\Delta v/v)(E/m) = -(\Delta v/v)/\epsilon_0\mu_0)$
F_i	Inertial force $(= \gamma m a = m d\mathbf{v}/dt)$
F_L	Lorentz force $(= q^+ \mathbf{E} + q^+ (m\mathbf{v} \times \mathbf{B}) = F_e + F_m)$
r_S	Schwarzschild radius $(= 2GM/c^2 = 1/\kappa_S)$

11. Origin of Inertia

Gravitation is not only similar to inertia in its generality, it is also measured by the same number... the mass. – Willem de Sitter

The tendency of modern physics is to resolve the whole material universe into waves, and nothing but waves. These waves are of two kinds: bottled-up waves, which we call matter, and unbottled waves, which we call radiation or light. If annihilation of matter occurs, the process is merely that of unbottling imprisoned wave-energy and setting it free to travel through space. – Sir James Jeans

Named by Kepler and given mathematical form by Newton, the force of inertia remains aloof because it has no obvious local cause. – Peter Graneau

11.1 Trapped waves in a standing wave resonator

Confinement of light results in the creation of mass and inertia. Standing electromagnetic waves within an isolated phase-locked resonator may be understood to explain the origin of mass and inertia while standing wave interactions between coupled resonators provide an explanation for the origin of gravity. The collective inertia of resonators increase when masses move relative to each other via vector potential coupling. Jennison and Drinkwater[39,40,41] have shown that a standing EM wave trapped in a phase-locked cavity exhibits rest mass and intrinsic inertia and classically derived Newton's Second Law ($F = ma$) and the Einstein relation ($E = mc^2$). For a free-floating wave system consisting of two counter-propagating travelling waves in a phase-locked resonant cavity, application of an external force results in an imbalance of radiation pressure of Doppler-shifted waves causing the wave system to move as a whole in a stepwise series of velocity increments. Upon application of an external force to the motive boundary, the blue-shifted incident wave exerts an excess radiation pressure on the reflecting wall and the red-shifted reflected wave exerts a decreased radiation pressure on the motive wall provided the force was applied for an interval equal to or greater than the return of the reflected wave.

Macken[20] observes that coherent light confined in a reflecting box exhibits many of the same properties of fermions including inertia (rest mass), kinetic energy, de Broglie waves, phase velocity, Lorentz contraction, time dilation, etc with the same energy ($E = h\nu = mc^2$). The relativistic contraction due to combination of two Doppler shifts in two opposing propagating waves produces a net decrease in Compton wavelength by factor of $\sqrt{(1 - v^2/c^2)}$. An electron at rest corresponds to a confined photon of energy $E = 0.511$ MeV in a fixed reference frame. As shown by Mead[21], inertial mass $m_i = m_e(1 + qA/m_e v) = \hbar k/v$. LaFreniere[38] maintains that relativistic mass increase with velocity is purely a Lorentz Doppler effect owing to the difference in wave energy of forward and backward propagating waves which corresponds to the Lagrangian interaction energy.

11.2. Confined light

A freely-propagating photon in a zero curvature vacuum has zero rest mass, but when trapped between two Bragg mirrors in a phase-locked optical cavity resonator, light acquires rest mass. A light beam upon reflection from a mirror undergoes momentum reversal cancelling the momentum of the incident wave. Jennison and Drinkwater[39] in 1977 derived Newton's second law for a phase-locked cavity model of a wave system representing a fundamental particle. A trapped standing wave exhibits not only rest mass, but also intrinsic inertia. This effect is illustrated in Fig. 11-1. As shown, a phase-locked resonator in motion represented here as a closed-loop soliton wave exhibits intrinsic inertial

11. Origin of Inertia

properties and illustrates the mechanism of Newton's laws of motion. The effect of induced motion and the cause due to an applied force is accompanied by a delay interval sufficient for a round trip of internal Doppler shifted wave propagation. The periodic rhythmic pulsation of the resonator in motion represents a harmonic oscillator. Wave system motion is in direction of internal standing wave node phase displacement.

Under acceleration, forward and backward propagating waves interact undergoing Doppler shifts resulting in an imbalance of radiation pressure. The total energy E_T of the system consists of the potential energy E_P required to hold the system together plus the forward and backward wave energy $E_{WF} + E_{WB}$. At rest, the wave energy equals the binding energy. Application of an external force $\mathbf{F}$ results in an acceleration $\mathbf{F} = (2/c^2)(E_T - E_P)\mathbf{a} = E_I\mathbf{a}/c^2 = m_i\mathbf{a}$ provided that the force is applied for a duration δt greater than or equal to the time to complete a feedback loop, otherwise the excess incident radiation is re-radiated back into space. The inertial force is a function of only half the total energy of the system as the potential energy is blocked from conversion and makes no contribution. Inertia is the result of internal self-referral dynamics of EM standing waves in an isolated phase-locked cavity oscillator when subjected to an external force. Rest mass is the result of EM momentum transfer at the wall boundaries of the oscillator during wave reflection resulting from wavefront deceleration. A photon propagating in an optically dense medium acquires an effective mass as the velocity c $(= c_0/n = c_0/\sqrt{K_{PV}})$ is slowed. Jennison postulated a tentative model of the electron consisting of two orthogonal spinning EM waves in phase quadrature phase-locked at the Compton wavelength λ_c into a closed-loop system. An equivalent model is that of a toroid generated by two orthogonal spinors representing a photon helix forming a closed-loop soliton wave.

The mechanism of origin of mass from energy of confined light is described by Macken[20] noting that a massless photon acquires rest mass when confined in closed-loop reflecting resonator cavity in a moving reference frame. An illustration of effects of motion on a standing wave system set up by counter-propagating waves in a resonator cavity formed from a laser mirror system is shown in Fig. 11-2. Motion of a confined closed-loop EM wave exhibits inertia characteristics due to self-referral dynamics. Upon application of an external force applied to the motive boundary, the blue-shifted incident wave exerts an excess radiation pressure on the reflecting wall and the red-shifted reflected wave exerts a decreased radiation pressure on the motive wall. Once set in motion, the wave system remains in motion until acted upon by an external force. The de Broglie wavelength represents a matter wave generated by matter in motion, i.e., a modulation of the internal standing wave. Mass is proportional to volumetric EM energy density (m = $4V \cdot \rho_e/3c^2$).

In the example shown, the wave medium corresponds to ordinary positive index of refraction ($n > 0$) resulting in a phase velocity v_{ph} propagating in the same direction as the group velocity v_g. For a negative index of refraction metamaterial, the phase velocity propagates in a direction opposite the group velocity (anomalous dispersion). For a nondispersive medium, the phase velocity v_p $(= \omega/k = c/\beta)$ equals the group velocity v_g $(= d\omega/dk = c \cdot \beta)$. The Bragg mirrors approximate perfectly reflecting conductor surfaces (short circuit) and do not reflect all frequencies.

I was very eager to produce an oscillator for short waves. I was doing science with microwaves, and I would get down to a few millimeters in wavelength, but I wanted to get shorter wavelengths. I wanted to get into the infra-red because I saw there was a lot more to be done there. – Charles H. Townes

The electric masses are nothing more than the places of non-vanishing divergence of the electric field. Light waves appear as undulatory electromagnetic field processes in space. – W. Heisenburg

11. Origin of Inertia

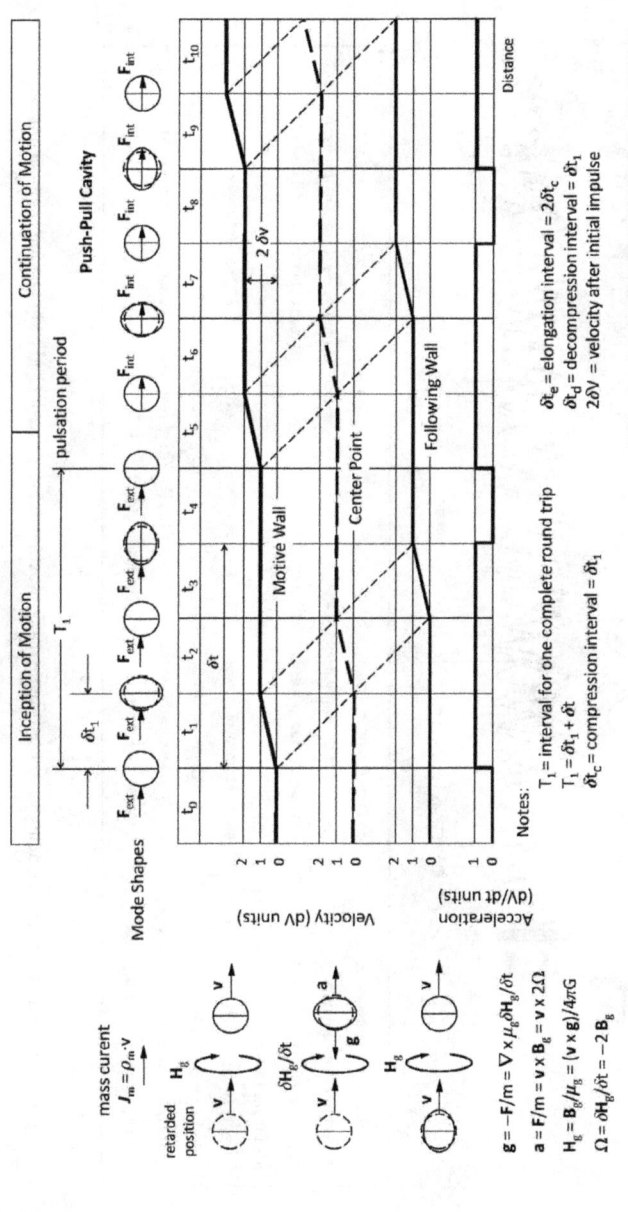

Fig. 11-1. Self-referral dynamics of standing wave radiation trapped in a phase-locked resonator cavity in motion. This push-pull oscillatory action provides an explanation for Newton's three laws of motion in classical physics. A moving mass constitutes a mass current J_m with an accompanying gravitomagnetic field $\mathbf{H}_g$ ($= \mathbf{B}_g/\mu_g$) analogous to an electric current I with an associated magnetic field $\mathbf{B}$. Inertial acceleration $\mathbf{g} = \nabla \times \delta \mathbf{B}_g/\delta t$.

11. Origin of Inertia

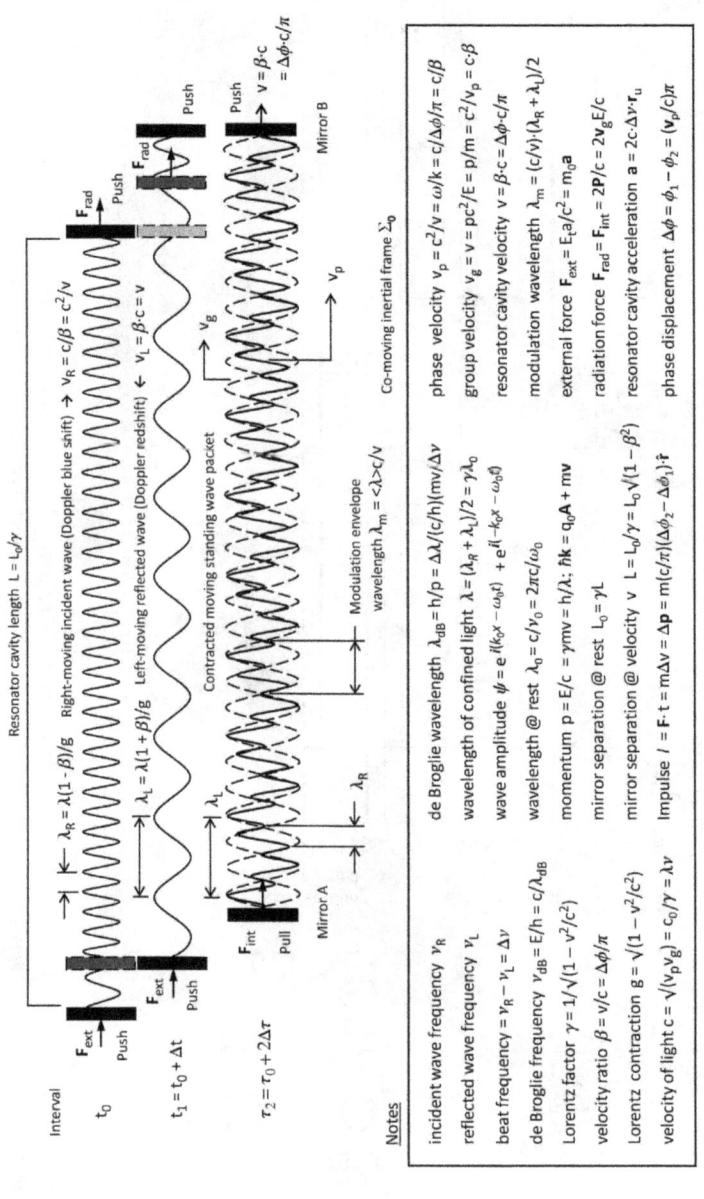

Fig. 11-2. Contracted moving standing wave (de Broglie matter wave) within a phase-locked EM radiation cavity in motion.

11. Origin of Inertia

A summary of variable relations for a phase-locked EM cavity in motion is provided in Table 11-1.

Table 11-1. Phase-locked EM cavity parameters
incident wave frequency $\nu_R = c/\lambda_R = \lambda(1-\beta)/g$
reflected wave frequency $\nu_L = c/\lambda_R = \lambda(1+\beta)/g$
Lorentz-Doppler beat frequency $\Delta\nu = |\nu_R - \nu_L| = (c/h)(mv/\Delta\lambda)\lambda_{dB} = (c/\Delta\lambda)$
de Broglie frequency $\nu_{dB} = E/h = E/(\Delta E/\Delta\nu) = c/\lambda_{dB} = pc/h = (h/mv)(\Delta\nu/\Delta\lambda)$
Lorentz factor $\gamma = 1/\sqrt{(1-v^2/c^2)} = dt/d\tau = 1/g$; Gravitational Gamma $\Gamma = 1/\sqrt{(1-\beta^2)}$
velocity ratio $\beta = v/c = \Delta\phi/\pi = \rho/\gamma = \tan\phi = \tanh\rho = \sqrt{(1-1/\Gamma^2)} = \sqrt{(1-1/\sqrt{K_{PV}})}$
Lorentz contraction $g = \sqrt{(1-v^2/c^2)} = 1/\gamma = \sqrt{(1-v_e^2/c^2)} = 1/\Gamma = 1/\sqrt{K_{PV}} = n^{-1}$
velocity of light $c = \sqrt{(v_p v_g)} = c_0/\sqrt{K_{PV}} = \lambda\nu = c_0/n = c_0/\Gamma = Z_0/\mu_0 = \mathbf{E/B} = \omega/k$
de Broglie wavelength $\lambda_{dB} = h/p = h/\gamma m_0 v = \Delta\lambda/(c/h)(mv/\Delta\nu) = c/\nu_{dB}$
wavelength of confined light $\lambda = (\lambda_R + \lambda_L)/2 = \gamma\lambda_0$
wave amplitude $\psi = e^{i(k_0 x - \omega_0 t)} + e^{i(-k_0 x - \omega_0 t)}$
wavelength @ rest $\lambda_0 = c/\nu_0 = 2\pi c/\omega_0$
momentum $p = \beta E/c = \gamma mv = h/\lambda = eV/c^2 = S/c^2$; $\hbar\mathbf{k} = q_0\mathbf{A} + m\mathbf{v}$
mirror separation @ rest: $L_0 = \gamma L$
mirror separation @ velocity c: $L = L_0/\gamma = L_0\sqrt{(1-\beta^2)}$
Impulse $\mathbf{I} = \mathbf{F}\Delta t = m\Delta v = \Delta\mathbf{p} = m(\mathbf{v}_f - \mathbf{v}_i) = m(c/\pi)(\Delta\phi_2 - \Delta\phi_1)\cdot\hat{\mathbf{r}}$
phase velocity $v_p = c^2/v_g = \omega/\mathbf{k} = c/(\Delta\phi/\pi) = c/\beta$
group velocity $v_g = d\omega/d\mathbf{k} = pc^2/E = p/m = \mathbf{k}\hbar/\nu = c^2/v_p = c\cdot(\Delta\phi/\pi) = c\cdot\beta$
resonator cavity velocity $v = v_g = \beta\cdot c = \Delta\phi\cdot c/\pi = (e\cdot|\mathbf{A}|)/m = v_0/\Gamma$
modulation wavelength $\lambda_m = (c/v)\cdot(\lambda_R + \lambda_L)/2$
external force $\mathbf{F}_{ext} = E_t a/c^2 = m_0 a = m(\Delta v/\Delta t) = \mathbf{I}/t$
radiation force $\mathbf{F}_{rad} = \mathbf{F}_{int} = 2P/c = A\cdot S/c = 2v_g E/c$
resonator cavity acceleration $\mathbf{a} = dv/dt = 2c\cdot\Delta\nu\cdot\mathbf{r}_u = (2c^2/h)(mv/\Delta\lambda)\cdot\lambda_{dB}\cdot\mathbf{r}_u$
phase displacement $\Delta\phi = v_g\cdot\pi/c = (v_p/c)\pi = \pi\cdot\beta$

Mass represents an obstruction to change in energy flow and is analogous to inductance of an AC electrical current. Light exhibits effective mass on slowdown when confined within an optically dense refractive medium or wave-guide transferring some of its momentum to the medium. Light trapped in a mirrored box, likewise, exhibits inertial mass as a result of momentum transfer to the reflecting walls as photons are absorbed and reradiated. A massless photon (boson) acquires rest mass when confined in closed-loop reflecting resonator cavity and demonstrates characteristics of particles of mass such as electrons (fermions). This effect is illustrated schematically in Fig. 11-3 showing the relationship of

11. Origin of Inertia

energy, momentum and inertial characteristics of a phase-locked wave system. The photon acquires effective mass during interaction within the mirrored box via the electromagnetic field. Standing waves store energy in a confined space, and as such, store mass. Partial standing waves correspond to an intermediate state between travelling waves with no rest mass and standing waves with rest mass. A wave system in motion with increased inertial mass corresponds to a contracted partial standing matter wave.

A freely-propagating photon represents a travelling wave. An electron represents a confined high energy photon trapped within a fixed reference frame. An electron at rest with respect to an observer is equivalent to a standing wave confined within a cavity resonator. An electron in motion is equivalent to a standing wave within a Lorentz contracted moving resonator. For a freely-propagating photon in vacuo, the photon travelling wave has no rest mass as there is no fixed reference frame and no defined position operator. In a standing wave resonator, the incident and reflected travelling waves combine to produce a standing wave with cancellation of momentum for a resonator at rest. Motion is the result of difference between the forward and backward matter wavelengths λ_f and λ_b according to the Lorentz Doppler effect and a resulting wave energy imbalance. Once in motion, a phase-locked resonator acquires a relativistic increase in mass $m = \gamma m_0 = m_0/\sqrt{(1 - v^2/c^2)} = m_0/\sqrt{(1 - \beta^2)}$ and corresponding increase in energy $E = E_0/\gamma$.

As illustrated in Fig. 11-1, a phased-locked resonator in motion exhibits an oscillatory rhythmic pulsation and represents a kind of clock or harmonic oscillator. The term resonator refers to standing wave device used to store energy. The oscillatory motion of the boundary walls of a phase-locked resonator form, in effect, a Hertzian dipole antenna generating a dipole radiation field transverse to the direction of motion. The signal radiation is maximized when the standing wave ratio (SWR = E_{max}/E_{min}) is high. The far field radiation pattern corresponds to that of a half-wave dipole and is similar to that of a magnetic loop antenna. If there is no reflection from the walls (SWR = 1), there is no standing wave and the resonator becomes effectively a perfectly matched transmission line (waveguide). For induced motion of a phase-locked resonator, the minimum required force application pulse duration τ or pulse width equals the sum of the compression and elongation intervals (= $\Delta t_c + \Delta t_e$). The interpulse period T for round trip wave propagation within a cavity resonator equals twice the pulse duration τ. The corresponding duty cycle D (= $\tau/T = \tau/2\tau$) equals 0.5. A wave energy phasor diagram for a travelling wave, standing wave and Lorentz contracted moving standing wave of constant total energy is depicted in Fig. 11-4. Motion of a standing wave system corresponds to a phase angle rotation of a phasor $Z\angle\theta$ in the complex plane. Quantum transitions are not discontinuous jumps but are modal wave transitions. Matter waves at high temperature have shorter wavelengths becoming more localized and particle-like. At ultracold temperatures (< 1 ^{0}K), matter waves become coherent forming a Bose-Einstein particle wavefunction condensate.

Inherent force of matter is the power of resisting by which every body, so far as it is able, perseveres in its state either of resting or of moving uniformly straight forward. – Isaac Newton

Although we have ascribed quasiclassical behavior to the heaviness of objects, it would be more accurate to ascribe it to motions associated with sufficiently high inertia. – Murray Gell-Mann

Matter is standing light. – Hans Sallhofer

For X-rays, the phenomenon of diffraction by crystals was a natural consequence of the idea that X-rays are waves analogous to light and differ from it only by having a smaller wavelength. – Louis de Broglie

11. Origin of Inertia

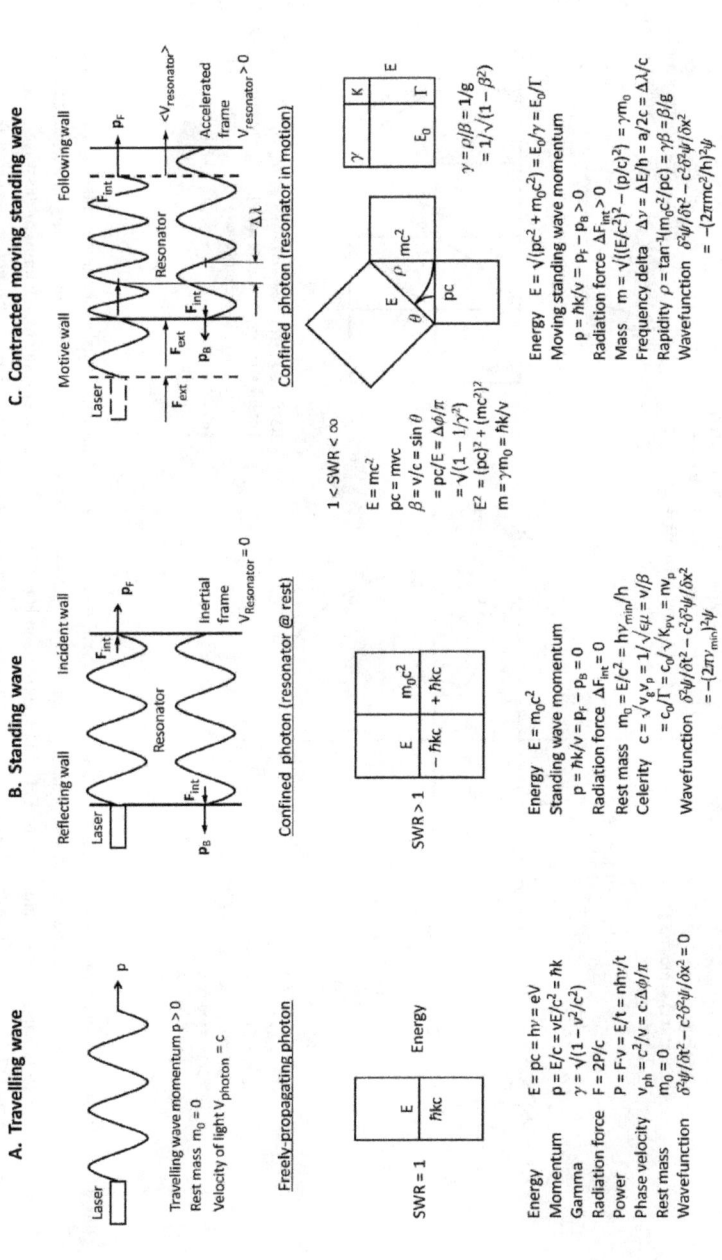

Fig. 11-3. Light confined within reflecting walls of an optical resonator cavity acquires rest mass (m_0). Photon momentum is a fixed ratio of energy.

11. Origin of Inertia

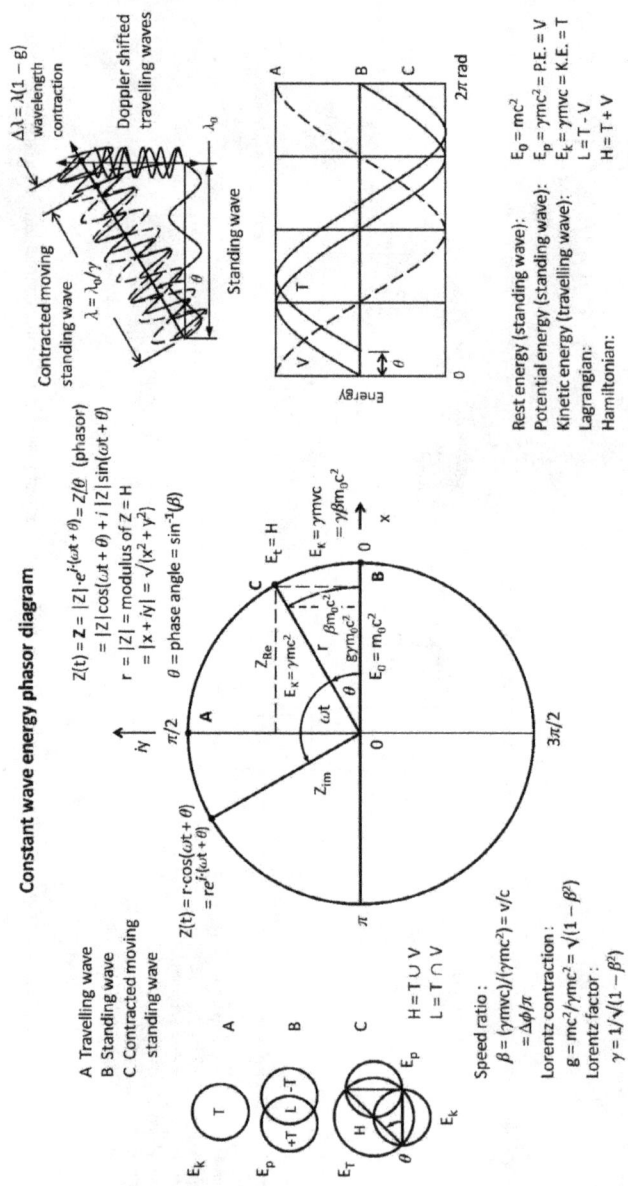

Fig. 11-4. Constant wave energy phasor diagram for a travelling wave, standing wave and Lorentz contracted moving standing wave. Matter wave formation corresponds to a phasor rotation. The Lagrangian $L (= T - V)$ represents the interaction energy or energy difference between kinetic energy $T (= E_k)$ and potential energy $V (= E_p)$. The Hamiltonian $H (= T + V = E_t)$ represents the total energy of the wave system. Mass is a measure of dissonance or destructive interference of out-phase wave functions and represents an obstruction to energy flow.

11. Origin of Inertia

The breathing motion of the resonator resembles a single degree of freedom, undamped, frictionless spring-mass system undergoing simple harmonic motion or a tuned, oscillating LC circuit. A closer analogy is a lossless acoustic, optical or electromagnetic cavity oscillator. A resonant system must contain at least one element in which kinetic energy is stored and another element in which potential energy is stored. For simple harmonic motion, there is a continuous interchange of potential and kinetic energy. For a lumped parameter system, the potential and kinetic energy are stored in separate spatial locations whereas, for a distributed-parameter system, the potential and kinetic energy are not stored separately. An acoustic Helmholtz resonator, common to musical instruments, has one degree of freedom with dimensions of the resonator cavity and aperture that are small compared to the operating wavelength. Resonant cavities, in analogy to waveguides, have various operating modes depending on the resonator geometry, dimensions and cut-off frequency. For a farfield EM wavefront in free space, the transmission mode is a *TEM* mode as the **E**-field and **H**-field are both perpendicular to the propagation direction. However, this mode does not occur in waveguides due to boundary conditions. Hence, field configurations of dominant modes are classified as either transverse electric (*TE*) or transverse magnetic (*TM*). In RF cavities and RLC circuits, energy is stored in and exchanged in electric and magnetic fields. See Fig. 11-5 and Table 11-2. Waveguides are typically rectangular or circular cross-section. Cavity resonators are commonly of cylindrical or pillbox configuration as well as rectangular, spherical, elliptical and toroidal geometries. Oscillation modes for a cylindrical cavity resonator (i.e., E_{mnp} or H_{mnp}) are characterized by subscripts m, n, and p for the number of half-wave patterns of the **E**- or **H**-field along the diameter, circumference and length, respectively. Common modes for a cylindrical cavity resonator include E_{010}, H_{111}, and H_{011}. The TE_{00} mode is often used in connection with specification of laser power output while TE_{10} is common for waveguides.

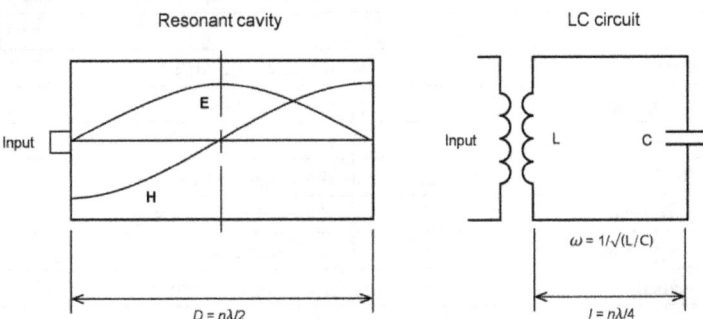

EM cavity resonator equivalent LC circuit

Fig. 11-5. A lossless electromagnetic cavity resonator and equivalent LC circuit. The resonant cavity may be excited by EM radiation introduced via a waveguide or coaxial cable configured as a dipole antenna or loop antenna. The electric and magnetic energy are in phase quadrature. Two such resonators may, in turn, be resonantly coupled to operate out-of-phase. This provides an analogy of coupling of fermions and anti-fermions modelled as resonant standing wave structures of opposing spins. Energy/unit mass $c^2 = \mu\epsilon/LC$.

For it necessarily turns out that inertia originates in a kind of interaction between bodies, quite in the sense of your considerations on Newton's pail experiment. – Albert Einstein

11. Origin of Inertia

Table 11-2. Comparison of mechanical and electromagnetic oscillators

Mechanical Oscillator		Electromagnetic Oscillator	
Spring	Potential Energy $U_P = \frac{1}{2}kx^2$	Capacitor	Electric Energy $U_E = \frac{1}{2}(q^2/C)$
Mass	Kinetic Energy $U_K = \frac{1}{2}mv^2$	Inductor	Magnetic Energy $U_B = \frac{1}{2}Li^2$
System	Total Energy $U_T = U_P + U_K$	System	Total Energy $U_T = U_E + U_B$
Mass current	$i_m = dmv/dt$	Current	$i = dq/dt$
Natural resonant frequency	$\omega = 2\pi v = \sqrt{(k/m)}$, $f_R = 1/(2\pi\sqrt{(mC_M)})$	Natural resonant frequency	$\omega = 2\pi v = \sqrt{(1/LC)}$, $f_R = 1/(2\pi\sqrt{(LC)})$
Impedance	$z_M = r_R + j\omega m\, j/\omega C_M$	Impedance	$z = r + j\omega L - j/\omega C$
Acoustic Cavity Oscillator		*Electromagnetic Cavity Oscillator*	
Cavity	Potential Energy $U_P = \frac{1}{2}kx^2$	Cavity	Electric Energy $U_E = \frac{1}{2}(1/\mu_0)B^2$
	Kinetic Energy $U_K = \frac{1}{2}mv^2$		Magnetic Energy $U_B = \frac{1}{2}\epsilon_0 E^2$
System	Total Energy $U_T = U_P + U_K$	System	Total Energy $U_T = U_E + U_B$
Velocity	$v = \sqrt{(B/\rho_0)} = \delta\varphi/\delta x$	Velocity	$v = c/\sqrt{(\epsilon\mu/\epsilon_0\mu_0)}$
Fundamental resonant frequency	$\omega_1 = \pi v/l$ (cylinder) $f_r = 1/(2\pi\sqrt{(mC_A)})$	Fundamental resonant frequency	$\omega_1 = 2.405c/r$ (cylinder) E_{010} mode
Impedance	$z_A = r_A + j\omega M - j/\omega C_A$	Impedance	$z_E = r_E + j\omega L - j/\omega C_E$
where:			
x = distance		q = electric charge	
k = spring constant		C = capacitance	
m = mass		L = inductance	
B = bulk modulus of elasticity of gas		**B** = magnetic flux density	
ρ_0 = gas mean density		**E** = electric field strength	
ω = angular frequency		ϵ_0 = electric permittivity	
φ = velocity potential		μ_0 = magnetic permeability	
l = resonator cavity length		r = cavity cylinder radius	
C_M = compliance		C_A = acoustical capacitance	
M = inertance / intractance		U = energy	

Using Abraham's theory, Hasenöhrl showed that a cavity with perfectly reflecting walls containing electromagnetic radiation behaves, if set in motion, as it has a mass m given by $m = 4V\rho_e/3c^2$, where V is the volume of the cavity, ρ_e the energy density at rest, and c the velocity of light, — Max Jammer

11. Origin of Inertia

11.3 Contracted standing waves in motion

An object's motion in spacetime is described by a geodesic path (shortest path in units of time between two points) in a curved spacetime manifold. In Lorentz wave transformations, a curved spacetime is replaced by a geodesic path of least resistance describing the wavelength nodal distance (the distance between nodes) in flat spacetime. For example, as shown in Fig. 11-6, consider a moving body consisting of standing EM wave between sources at Points P and E co-moving at a constant velocity v = 0.5c relative to an observer. Let the distance between Points P and E measured simultaneously represent the wavelength of an EM standing wave packet. In tensor notation, the separation between Points P and E is the line element $ds^2 = g_{\mu\nu}dx^\mu dx^\nu = -dt^2 + dx^2 + dy^2 + dz^2$.

Contravariant position 4-vector: $x^\mu = (x^0, x^1, x^2, x^3)$ (11-1)
$= (ct, -x, -y, -z)$
Covariant position 4-vector: $x_\mu = \eta_{\nu\nu}x^\nu = (x^0, -x^1, -x^2, -x^3)$ (11-2)

Contravariant 4-velocity: $v^\mu = dx^\mu/d\tau = (c\gamma, \gamma v)$ (11-3)

In Special Relativity (SR), the Lorentz factor (gamma) is given by $\gamma = 1/\sqrt{(1 - v^2/c^2)} = 1/\sqrt{(1 - \beta^2)}$ which relates transformation of relative motion in flat space of inertial (non-accelerated) reference frames including Lorentz contraction, time dilation and relativistic mass increase effects. Moving bodies (standing matter wave nodal distances) are contracted by a factor of $\gamma = 1/\sqrt{(1 - \beta^2)} = \cosh \beta$ in the direction of motion in accordance with the Lorentz transformation. In normalized coordinates with speed of light c = 1, for the example illustrated, the measured unit length in the x-axis in the S rest frame undergoes a Lorentz contraction of 13.4% in the contravariant S' moving frame of reference at $\beta = v/c = \tanh w = 0.5$. The associated frequency reduction is 14.9% as measured in wavelength units in the covariant S frame, in accordance with the Lorentz Doppler shift.

As shown in Fig. 11-6, lengths L (projected onto the x-axis) and L' (projected onto the x'-axis) are given in wavelength units. The Lorentz contraction in the direction of motion as measured by a fixed observer is $L = L'\sqrt{(1 - v^2/c^2)}$. Proper time is given by $d\tau = 1/c \cdot \sqrt{(dx^\nu dx_\nu)} = dt/\gamma$. Rate of motion or rapidity corresponds to a hyperbolic rotation angle $\rho = \tanh^{-1}(v/c) = \beta\gamma$ where $\gamma = 1/\sqrt{(1 - \beta^2)} = 1/\sqrt{(1 - c \cdot ((\nu_1 - \nu_2)/(\nu_1 + \nu_2))^2)}$. Horizontal lines represent surfaces of simultaneity. The hyperbolic isobars correspond to constant acceleration (Rindler frames) while the 4-velocity straight lines correspond to constant velocity (inertial frames).

According to the Lorentz transformation, motion is equivalent to hyperbolic rotation in Minkowski spacetime. The Lorentz transformation in matrix form is

$$\begin{bmatrix} ct' \\ x' \\ y' \end{bmatrix} = \begin{bmatrix} \cosh \beta & \sinh \beta & 0 \\ \sinh \beta & \cosh \beta & 0 \\ 0 & 0 & 1 \end{bmatrix} \begin{bmatrix} ct \\ x \\ y \end{bmatrix}$$ (11-4)

Relativistic mass increase of matter in motion according to the LaFreniere model (m = γm_0) is attributed to the sum of mass energy of forward and backward matter waves which undergo a Lorentz Doppler effect[38]. Wavelength changes as $\lambda' = \lambda(1 - \beta\cos\theta)$ and frequency varies as $f' = f / (1 - \beta\cos\theta)$ where θ is the angle as measured from the direction of motion. In this model, the total mass is the sum of the active mass ($m_a = g \cdot m_0/(2(1 - \beta))$) corresponding to the forward contracted waves and the reactive mass ($m_r = g \cdot m_0/(2 \cdot (1 + \beta))$) corresponding to the backward expended waves. The active and reactive mass are both positive with a ratio $m_a/m_r = (1 + \beta)/(1 - \beta) = \lambda_b/\lambda_f$. The gain in mass equals $m_a + m_r - m_0$

11. Origin of Inertia

$= \gamma m - m_0$ which corresponds to the increase in kinetic energy. The kinetic energy of a mass in motion may thus be expressed as $E_k = (m_a + m_r - m_0)c^2 = (\gamma m - m_0)c^2$. The total energy is given by $E_t = (m_a + m_r)c^2 = \gamma mc^2$.

The physical basis for wavelength contraction may be understood in the context of the polarized vacuum (PV) model where EM waves undergo a contraction in regions of increased energy density and corresponding change in the dielectric constant $K_{PV}(r,\omega,M)$. The EM wave energy density in uniform motion remains an invariant covariant physical quantity in the Lorentz transformation. Spacetime curvature is replaced by variable refractive index $K_{PV}(r,\omega,M)$ which is a measure of EM energy density. Gravity is associated with a gradient in the EM Poynting vector **P** due to variation in K_{PV} ($\nabla P = k\Delta K_{PV}^2$). A standing wave phase synchronization interaction between oscillators in a nondissipative medium results in a force of attraction equivalent to gravity. The force imbalance is proportional to the difference in wave energy density and inversely to the wave velocity. The instantaneous wave energy density, in turn, is proportional to the square of the wave amplitude. A wave system composed of mass oscillators propagates in the direction of the energy density gradient. This is akin to the "pilot wave" theory in which an object follows a phase wave field path from higher to lower energy density and appears piloted in effect. Motion of matter is accompanied by a localized wave packet (*Wellenpaket*) of de Broglie 'matter' waves described by a wavefunction ψ and associated probability density $|\psi(x)|^2$ as a function of position.

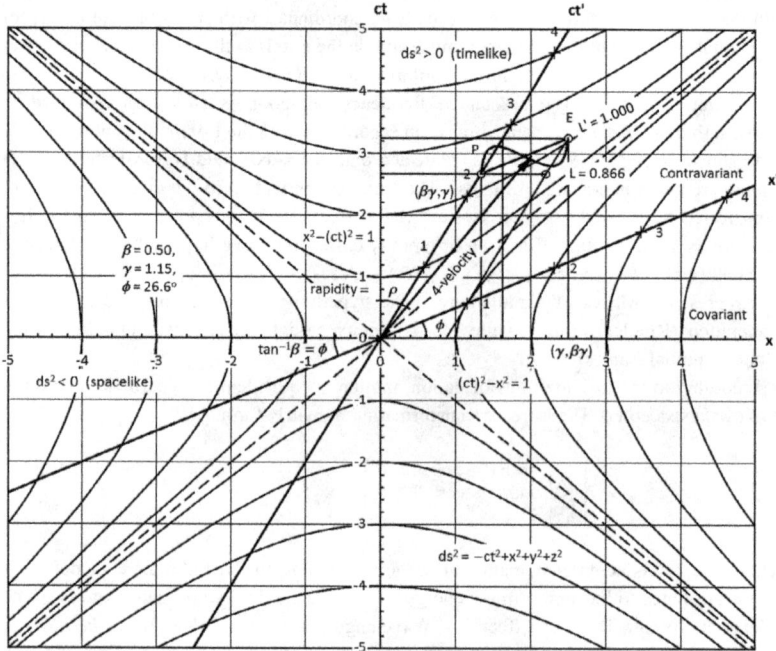

Fig. 11-6. Lorentz wavelength contraction and frequency dilation of a moving contracted standing wave @ velocity v = 0.5c. Positive mass is restricted to $ds^2 > 0$ (timelike intervals). Negative mass is confined to $ds^2 < 0$ (spacelike intervals).

12. Electromagnetic Wave Propagation

We can scarcely avoid the inference that light consists in the transverse undulations of the same medium which is the cause of electric and magnetic phenomena. – James Clerk Maxwell

Light is always propagated in empty space with a definite velocity, "c", which is independent of the state of motion of the emitting body. – Albert Einstein

The theory [of the electromagnetic field] I propose…that in the space there is matter in motion, by which the observed electromagnetic phenomena are produced. – James Clerk Maxwell

It is impossible to travel faster than light, and, certainly not desirable, as one's hat keeps blowing off. – Woody Allen

Light in the form of electromagnetic waves come from accelerated electric quanta. – Johannes Stark

Relative motion of the two reflecting walls of a phase-locked resonator corresponds to an oscillating dipole which emits a transverse radiation field orthogonal to the direction of motion of dipole oscillator. As represented in standard texts, the **E** and **H** wavefront vectors are in-phase far from the source dipole oscillator. Within the evanescent near field region, the **E** and **H** vectors are out-of-phase with a longitudinal polarization component. In the midrange or Fresnel region, the **E** and **H** vectors are partially in-phase. In the near field region, dipole field effects are prominent and drop off as $1/r^2$ whereas EM field drops off as $1/r$. Induction term decays as $1/r^2$ and electrostatic field decays as $1/r^3$ and rapidly decay beyond the evanescent region. The wave phase velocity is superluminal ($v_{ph} > c$) whereas the group velocity is subluminal ($v_g < c$). Refer to Table 12-1 summary below.

The radiation field emitted from a stationary dipole antenna and a rotating dipole antenna are illustrated in Figs. 12-1 and 12-2, respectively. Phase relationship of the electric field intensity **E** and magnetic field intensity **H** as a function of distance from the dipole antenna is shown in Fig. 1-4. An electron may be regarded as a rotating dipole antenna continuously emitting and absorbing pairs of entangled virtual photons. The emission and absorption of photons by separated electrons may be represented as points on a Bloch sphere. The degree of entanglement corresponds to the distance between points. Entangled quantum states represents points lying on the same wavefront, hence, their correlation is causally connected. Non-entangled states are isolated in time or space and are non-causally connected. As a result, instantaneous, non-local entanglement in which action on one particle instantly influences another particle by superluminal 'spooky action at a distance' or pilot wave interaction is a nonphysical illusion.

In the Liénard-Wiechart scalar and magnetic vector potential functions describing the electric and magnetic fields generated by motion of a point charge, electromagnetic radiation arises as a result of acceleration whereas static electric and magnetic fields that result from the particle's uniform motion are associated with the non-radiative EM near-field. The static fields point towards the instantaneous (non-retarded) charge position. The electromagnetic radiation appears to originate at the charge's retarded position (where the charge was when accelerated) reflecting delay due to the finite speed of light.

At the head of these discoveries and insights comes the establishment of the facts that electricity is composed of discrete particles of equal size, or quanta, and that light is an electromagnetic wave motion. – Johannes Stark

12. Electromagnetic Wave Propagation

Table 12-1. Electromagnetic wave phase propagation

Region	E and H Phase Relationship	Remarks
Near field (non-radiative, reactive) region – Dipole wave emission	Induction E-field lags 90 deg out-of-phase with magnetic H field. Near field ($\leq \frac{1}{4}\lambda$). Reactive field $0 < r < \lambda/2\pi$. Radiative to 1 λ. E, H decays as $1/r^2$ and $1/r^3$, respectively. The electric field E is due to charge dipoles and the magnetic field H is due to source currents. $H_0 = E_0/Z$, $Z = \sqrt{(\mu_r \mu_0/\epsilon_r \epsilon_0)} = \Gamma \cdot E/H$. After a period of a ¼ cycle (t = T/4), some of the electric flux lines cross and detach from charges and propagate outward as closed loops carrying away radiant electromagnetic energy. For evanescent waves, phase is independent of distance, so that phase velocity is superluminal. E & H are out-of-phase, hence, Z_0 is not related by 377 ohm characteristic vacuum impedance. Evanescent electromagnetic near-field waves initially propagate faster than the speed of light slowing down to the speed of light within about one wavelength.	Longitudinal E-field component in the near field is partially in the direction of propagation (parallel to k vector). Close to the antenna, the Poynting vector $S = f(r,\theta,\phi)$ is imaginary (reactive), hence energy is not propagating (non-radiating). Energy stored in the field volume is detectable capacitively. Field energy not radiated is alternately returned to the transmitter. Evanescent waves exhibit momentum and spin components orthogonal to the direction of propagation. The transverse momentum varies in proportion to helicity while the transverse spin component does not depend on helicity or polarization. Evanescent waves exhibit decaying amplitude with no oscillations or energy flow.

Cont

Table 12-1. Electromagnetic wave phase propagation (cont)

Region	E and H Phase Relationship	Remarks
Transition (radiative) region (Fresnel zone)	Dynamic E-field consists of induction and radiation field components, the sum total out of phase with the magnetic field. Radiating near-field (Fresnel Zone) extends from $\lambda/2\pi <$ r $< 2D^2/\lambda$ where D = largest antenna dimension. For $D^2/4\lambda <$ r $< 2D^2/\lambda$, E, H fields decay as $1/r$.	Radiating field begins to dominate. Phase angle (90 $< \phi <$ 0 deg).
Far Field (Fraunhofer Zone) – Planar EM Hertzian wave	Radiation E-field in phase quadrature with magnetic H field. E, H radiation fields decay as $1/r$ and dominate over static near-fields; field pattern independent of r. Radiating far-field r $> 2D^2/\lambda$. The antenna impedance $Z_0 = |\mathbf{E}|/|\mathbf{H}| \approx 376.73$ Ω (vacuum free space impedance).	Far field generally regarded to start 2 to 5 wavelengths from source (= $2D^2/\lambda$, where D = dipole element length) or r > 10λ for small antennas. Poynting vector $\mathbf{S} = f(\theta,\phi)$ is real-valued. Radiation resistance of a Hertzian dipole due to electron energy loss via EM radiation: $R_r = (2\pi/3)Z_0(l/\lambda)^2$.

Note: In solid conductors, the magnetic field H lags the electric field E by nearly 45° with $|\mathbf{H}| >> |\mathbf{E}|$. The electromagnetic field energy E (= ½m($\mathbf{E}^2 + c^2\mathbf{B}^2$) is almost entirely magnetic. The E field corresponds to curvature k whereas the B field corresponds to torsion τ. The EM waves propagate with exponential damping $\exp(i\mathbf{k}\cdot\mathbf{x} - i\omega t)$ with penetration depth and exhibit a high frequency current skin effect.

Move like a beam of light; fly like lightning; strike like thunder; whirl in circles around a stable center. – Morihei Ueshiba

Our mind is only a receiver. We just need to tune it to the universe. – Nikola Tesla

Only 4 percent of what's in the universe gives off electromagnetic radiation, so we don't have any handle on the rest. – Barry Barish

12. Electromagnetic Wave Propagation

Contrary to a commonly accepted view that **E** and **B** field create each other to create transverse EM waves, Jefimenko[15] notes that **E** and **B** fields are created by charge density ρ and current density **J** fluctuations at the source and far-field radiation is due to retarded potentials $\phi(r,t)$ and $\mathbf{A}(r,t)$. The retarded scalar and magnetic vector potentials were derived by Jefimenko for electric and magnetic fields in terms of charge and current distributions at retarded times. Special Relativity effects of objects moving at constant velocity are shown to be the result electromagnetic field retardation without recourse to assumption of the same value of velocity of light c, no matter how the source is moving with respect to two independent observers in inertial frames.

An electron, in the ansatz model considered here, may be represented as a EM wave trapped in a phase-locked resonator or, equivalently, a closed-loop, topologically-bound soliton wave. The electron acts as an antenna with emitted or absorbed photon wave vector **k** parallel to electron spin axis **s**. The antenna diameter of the electron corresponds to the electron Compton wavelength. Virtual photons at the Compton frequency are continuously emitted/absorbed from an electron in pairs in opposite directions and helicities. The photon **EHV** dreibein rotates at a electron Compton angular frequency ($\omega_c = 7.763 \times 10^{20}$ rad/s) while the observed photon frequency emitted during electron acceleration is a measure of the overall oscillatory motion of the electron during emission. Although small in size, an electron, by resonant phasing, can couple to much larger wavelengths forming, in effect, a much larger antenna.

Martins and Pinheiro[42] note the induced electrokinetic force $\mathbf{F}_k = q\mathbf{E}_k$ as a function of vector potential **A** is the source of the inertial mass and the radiation force. The radiation force adds to the inertia by energy transfer between the field and the source at a retarded time. Radiation serves as a carrier of momentum and can exert a force on electrons causing motion. Conversely, when a moving electron radiates, the momentum lost is equal to the radiation momentum. When an electron accelerates, the magnetic field increases. The increasing magnetic field gives rise to an electric field producing an electrokinetic force opposing motion. This self-induction effect of the moving electron's interaction with its own electromagnetic field results in an increase in a relativistic electromagnetic mass. Compression of electric field lines reduces electron radius reducing volume occupied by electrostatic energy increasing mass. For an accelerated charge, the induced electric field generated by the time variation of the vector potential $E_k = -d\mathbf{A}/dt$ results in an acceleration of the electric fields in a direction opposite to acceleration vector. The vector potential **A** may be understood as arising from the interaction of Planck masses moving relative to each other. Martins illustrates electric field deformation of a charged particle subjected to a gravitational force, external force, electric force and inertial force. The flux patterns illustrate the same asymmetry as described by Ivanov[36,37] for contracted moving standing wave field interference patterns as discussed in the following sections. These ideas are consistent with the view that all mass is electromagnetic in origin as originally described in essays by Lorentz, Poincaré, Wein and Abraham, and that the Lorentz contraction produces a field asymmetry resulting in a difference in longitudinal and transverse mass.

As in nature, all is ebb and tide, all is wave motion, so it seems that in all branches of industry, alternating currents – electric wave motion – will have sway. – Nikola Tesla

Genius requires solitude. Be alone, that is the secret of invention. – Nikola Tesla

For under certain conditions the chemical atoms emit light waves of a specific length of oscillation frequency – their familiar characteristic spectra – and these can come in the form of electromagnetic waves only from accelerated electric quanta. – Johannes Stark

12. Electromagnetic Wave Propagation

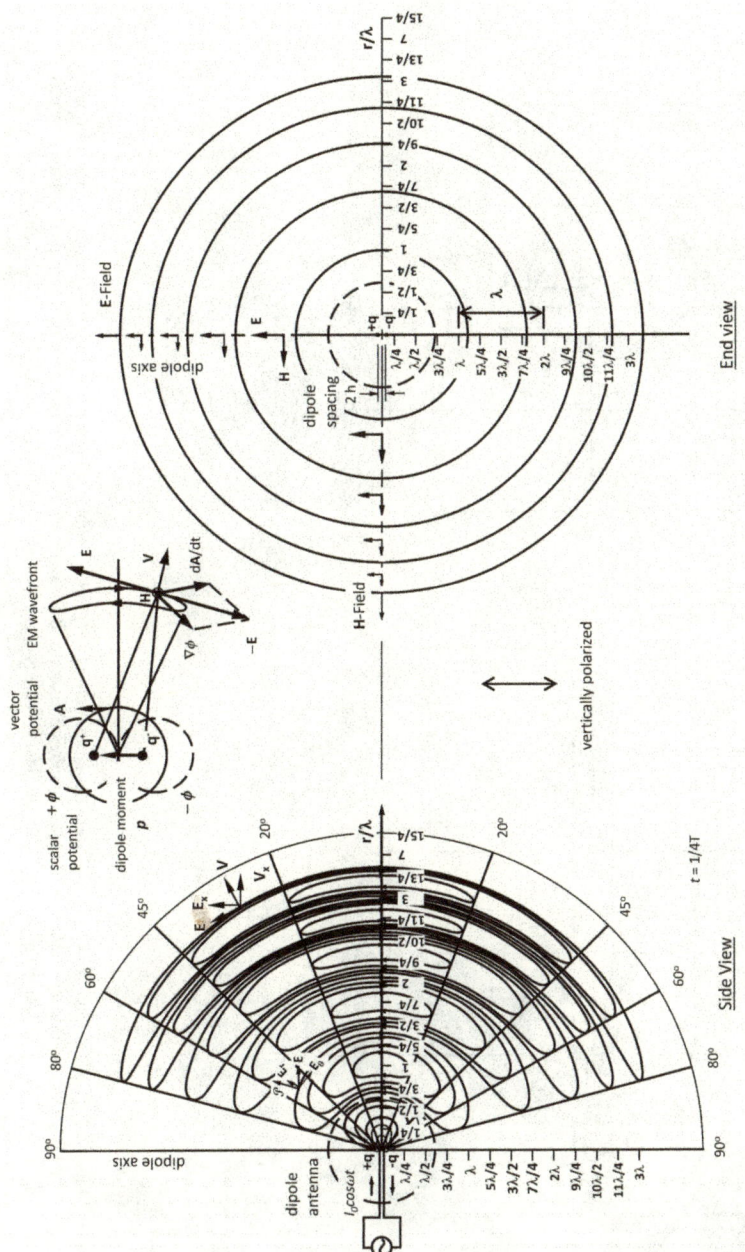

Fig. 12-1. Electric flux lines in the radiation field of a non-rotating dipole antenna (oscillating charge doublet).

12. Electromagnetic Wave Propagation

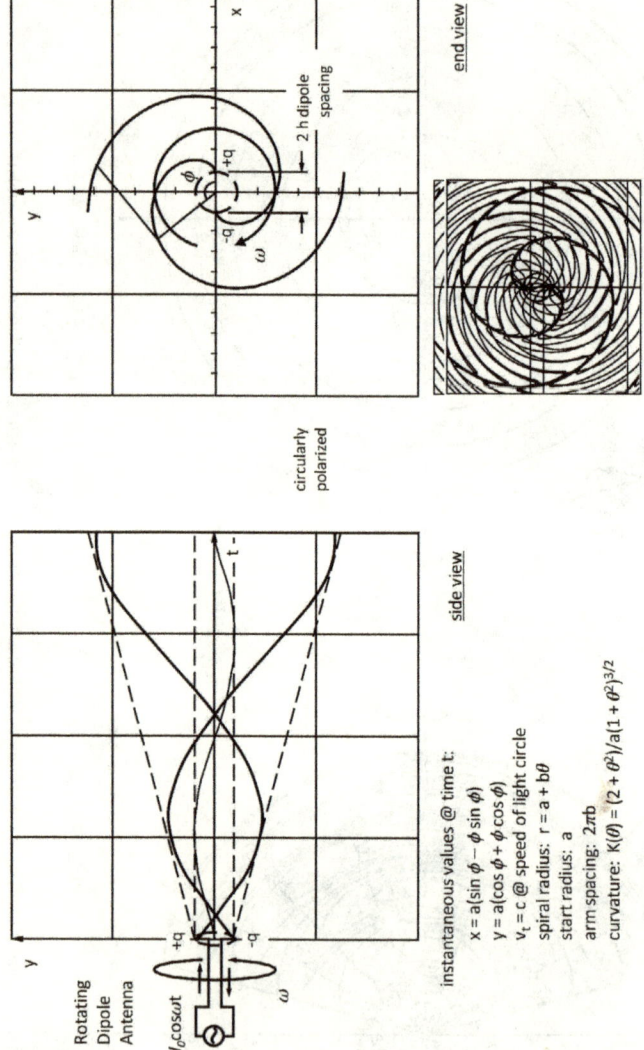

Fig. 12-2. Electric flux lines in the radiation field of a rotating oscillating doublet at relativistic tangential velocity. The pair of wavefront spiral arms represent an entangled, monogamous state. The degree of entanglement corresponds to the distance between points (locality). Entangled quantum states represents points lying on the same wavefront, hence, their correlation is causally connected.

13. Standing Wave Transformations

Ivanov and LaFreniere[36,37,38] have shown that standing waves undergo wavelength (nodal) contraction in the direction of motion. An object in motion relative to a fixed observer undergoes a Lorentz contraction (wavelength compression) in the direction of motion and a Lorentz Doppler shift in frequency (reduction). See Fig. 13-1. The wavelength compression is a physical result of an increase in the vacuum energy density. Moving clocks which are made of standing matter waves undergo time dilation as a result. The EM wavelength contraction and frequency shift in a polarizable vacuum accounts for mass in motion and gravitational effects including the energy change, deflection of light, gravitational frequency shift and clock slowing. The speed of light c appears invariant in all inertial frames due to Lorentz contraction of the measurement apparatus. Spacetime coordinates remain Euclidean. The observed Lorentz space contraction and time dilation are the result of contraction of the nodal distance of the standing wave(s) which constitute the length of measurement. Time dilation is equivalent to a change in the size of the units of measurement which are undetectable to an observer as both the object and the co-moving measurement apparatus undergo Lorentz transformation.

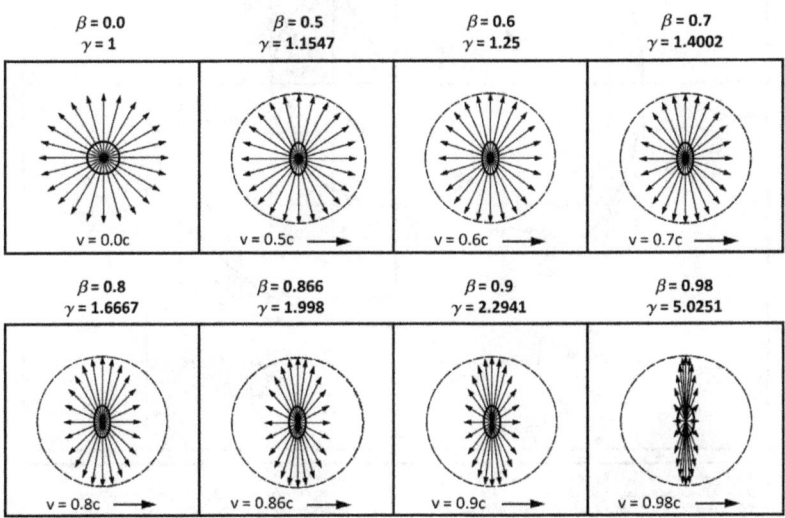

Fig. 13-1. Lorentz contraction of an object in motion in the direction of motion as viewed by a stationary observer. An object and associated gravitational flux field appears contracted referenced to retarded lengths and volumes as measured by a stationary observer. Physical contraction of nodal distance of matter waves occurs in a polarized vacuum as the vacuum dielectric constant K_{PV} (= Γ^2) is equivalent to gamma γ (= $1/\sqrt{(1-(v^2/c^2))}$) which increases with velocity and gravitational gamma Γ (= $1/\sqrt{(1-(v_e^2/c^2))}$) which increases with escape velocity v_e. A co-moving observer will not detect distortion as sensing instruments undergo like contraction and Lorentz Doppler frequency shift.

13. Standing Wave Transformations

LaFreniere[38] derived 'alpha' transformations relating system speed α, arithmetic mean wavelength λ_{am}, geometric mean wavelength λ_{gm}, Lorentz contraction factor g and wavelength compression λ'. The alpha transformations, in terms of standing wave ratios referenced to the mean geometric wavelength, yield results equivalent to the Lorentz 'beta' transformation in terms of β velocity ratios. The relation of Lorentz contraction, Doppler shift and relativistic Doppler shift for a contracted standing wave in motion at relativistic velocity v = 0.5c is illustrated in Fig. 13-2.

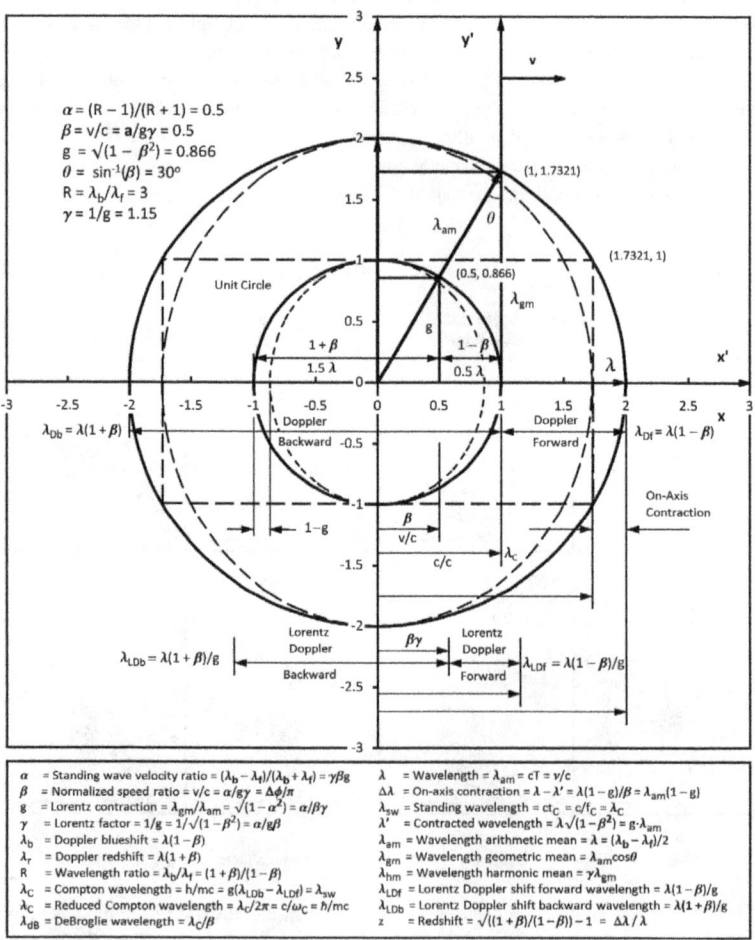

Fig. 13-2. Illustration of Doppler and Lorentz Doppler effect for a contracted moving standing wave moving to the right in the x-axis direction at a velocity of 0.5c. The aberration ϕ angle is a measure of transverse standing wavefront tilt. The electron Compton and de Broglie wavelengths are equivalent to standing and arithmetic mean wavelengths, respectively. The matter wavelength of a moving electron is much larger than its size.

13. Standing Wave Transformations

A standing wave resonator in motion undergoes aberration altering the wavefront propagation direction of transverse standing waves toward the direction of motion. The intersection of crossed wavefronts of counter-propagating transverse waves produces a scissors effect resulting in node progression at the phase velocity ($v_p > c$). This effect is illustrated in Fig. 13-3.

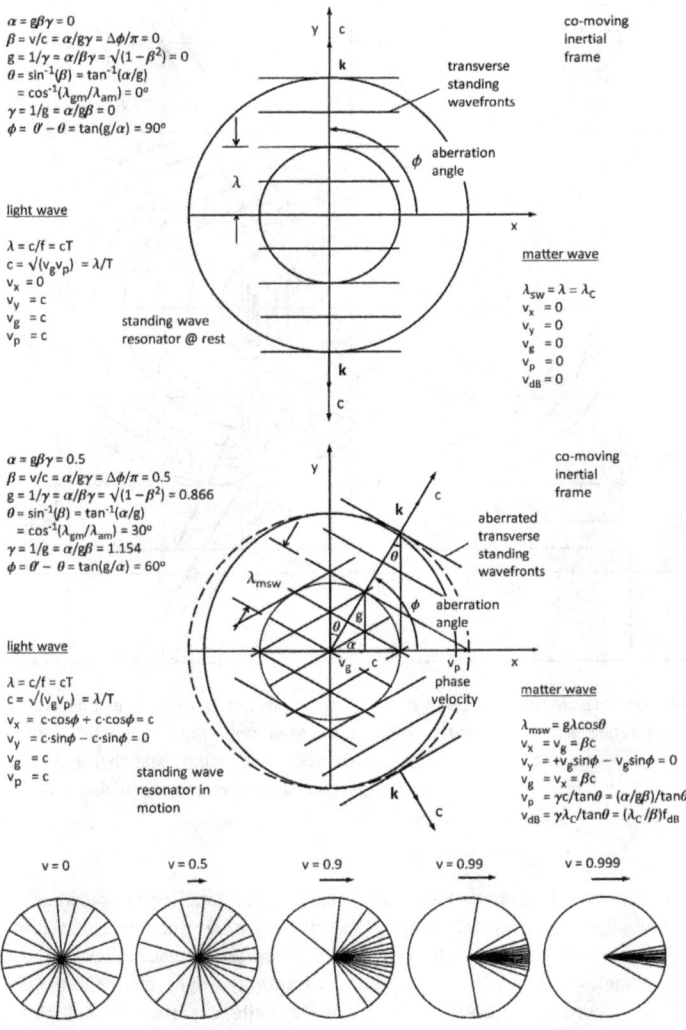

Fig. 13-3. Phase velocity of standing wave system at rest and in motion in a vacuum at ½ velocity of light. A standing wave in motion travels at a velocity < c due to the longer internal zigzag light path travelled than a freely propagating wave at velocity c. Hence, a decrease in resonator size is required for increased velocity, e.g., fermionic matter.

13. Standing Wave Transformations

A comparison of the Lorentz contraction of a moving EM source as observed in a reference frame at rest is shown together with the relativistic aberration as observed in a co-moving inertial frame is depicted in Fig. 13-4.

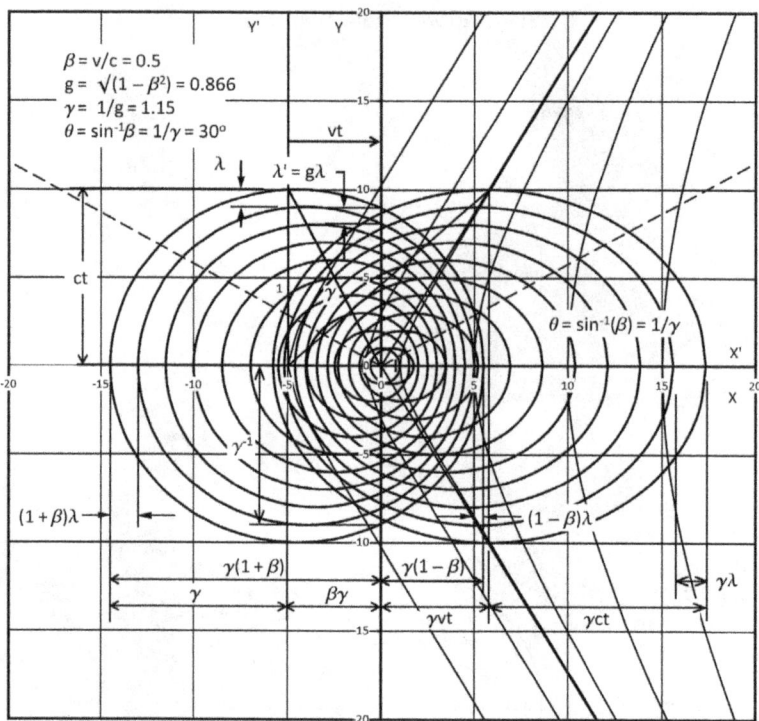

Fig. 13-4. Lorentz Doppler shift of a moving object emitter at v = 0.5c as observed in a co-moving reference frame (primed axes) is compared with the relativistic aberration as observed in a co-moving reference frame. Distances under relativistic aberration scale by a factor $\sqrt{((1 - \beta)/(1 + \beta))}$. The ordinary Doppler shift corresponds to that observed in a reference frame at rest (unprimed axes).

A comparison of the Lorentz-Fitzgerald transformation (applied to spacetime) and the Ivanov-LaFreniere transformations (applied to wavelength) are summarized in Table 13-1. The unprimed coordinates refers to a stationary observer reference frame. The primed system coordinates refer to a reference frame moving to the right relative to the unprimed frame. The x-coordinate denotes number of wavelengths expressed in multiples of λ_{gm}. The t-coordinate denotes number of waves expressed in multiple of period T. A summary of standing wavelength transformation variables for a contracted moving standing wave is shown in Table 13-2.

Physics is simple, but subtle. – Paul Ehrenfest

13. Standing Wave Transformations

Table 13-1. Comparison of Lorentz-Fitzgerald and Ivanov-LaFreniere transformations.[37,38,39]

Description	Lorentz-Fitzgerald	Ivanov-LaFreniere
Velocity of source	$v = v_g = c \cdot \beta = \Delta\omega/\Delta k$	$v = v_g =$ envelope node speed $= c \cdot \beta = \Delta\omega/\Delta k$; matter: $v_{dB} = v_g = c \cdot \alpha$
Velocity of wave	$c = \sqrt{(v_p v_g)} = c_0/n = \omega/k = c_0/\Gamma = \mathbf{E/B}$	$c = \sqrt{(v_p v_g)} = c_0/n = \omega/k = c_0/\Gamma = \mathbf{E/B}$
Normalized speed ratio, Aberration angle	$\beta = v/c = n^{-1} = \tanh\omega = \sin\theta = \rho/\gamma = \Delta\phi/\pi$ $= \sqrt{(1 - 1/\Gamma^2)} = \sqrt{(1 - 1/K_{PV})}$ $= \sqrt{(1 - m_0/m^2)} = pc/E$	$\beta = v/c = n^{-1} = \tanh\omega = \tanh\rho = \alpha/g\gamma = \Delta\phi/\pi$ $= \rho/\gamma = \sqrt{(1 - 1/\Gamma^2)} = \sqrt{(1 - 1/K_{PV})}$ $= \sqrt{(1 - m_0/m^2)} = pc/E$
Standing wave speed ratio	—	$\alpha = (R-1)/(R+1) = (\lambda_b - \lambda_f)/(\lambda_b + \lambda_f)$ $= g\beta\gamma = (1-g) \cdot (1+g)/\beta$
Phase wave speed ratio	$v_p/v_g = c^2/v_g^2$	$v_p/v_g = c^2/v_g^2 = 1/\alpha$
Proper time	$t' = ((t - vx)/(c^2))/\sqrt{(1 - \beta^2)}$ $= \gamma(t - vx/c^2) = (t - \beta x/c)/g$	$t' = gt - \alpha x$ (light) $= gt - \beta x$ (matter)
Lorentz contraction factor	$g = \sqrt{(1 - \beta^2)} = 1/\gamma$	$g = \lambda_{gm}/\lambda_{am} = \sqrt{(1 - \beta^2)} = 1/\gamma = \alpha/\beta\gamma$ $= \alpha/\sinh\omega$

cont

13. Standing Wave Transformations

Table 13-1. Comparison of Lorentz-Fitzgerald and Ivanov-LaFreniere transformations. (cont)

Description	Lorentz-Fitzgerald	Ivanov-LaFreniere
Lorentz factor	$\gamma = 1/\sqrt{(1-\beta^2)} = 1/g = \cosh\omega$	$\gamma = 1/(\lambda_{gm}/\lambda_{am}) = 1/\sqrt{(1-\beta^2)} = 1/g$ λ_{gm} = geometric mean wavelength λ_{am} = arithmetic mean wavelength
Freniere-Lorentz Doppler contraction factor	—	relativistic contraction (matter): $g = \lambda_{gm}/\lambda_{am} = \sqrt{(1-\alpha^2)}$
Ivanov-Freniere acoustic Doppler contraction factor	—	acoustic contraction: $g^2 = (1-\alpha^2)$
On-Axis Wavelength	$\lambda' = \lambda/\sqrt{(1-\beta^2)} = \gamma\lambda$	$\lambda' = \lambda_{am}\sqrt{(1-\beta^2)} = g^2\lambda_{am} = \lambda_{gm}$ $= \lambda\sqrt{((1+\beta)/(1-\beta))}$
Stationary observer coord.	$x = x' + vt'/\sqrt{(1-\beta^2)} = (x' - \beta t)/g$ $y = y'\,;\,z = z'$ $t = (t' + vx'/c^2)/\sqrt{(1-\beta^2)}$	$x = (x' - \alpha t)/\gamma = gx' - \beta t'$ $y = y'\,]\,z = z'$ $t = (t' + \alpha x)/g = g(t' + \alpha x) = gt' - \beta x'$
Proper length|| Direction of motion	$x' = (x - vt)/\sqrt{(1-\beta^2)} = \gamma(x-vt)$ $= \gamma(x - \beta ct) = (x-vt)/g$	$x' = gx + \alpha t$ (light) $= gx + \beta t$ (matter)
Wavelength ratio	$R = \lambda_a/\lambda_r$ Redshift $R_R = \lambda'/\lambda$, Blueshift $R_B = \lambda/\lambda'$	$R = (1+\beta)/g = g/(1-\beta) = \lambda_b/\lambda_f$ $= (1+\beta)/(1-\beta)$; $R_R = \lambda'/\lambda$, $R_B = \lambda/\lambda'$

13. Standing Wave Transformations

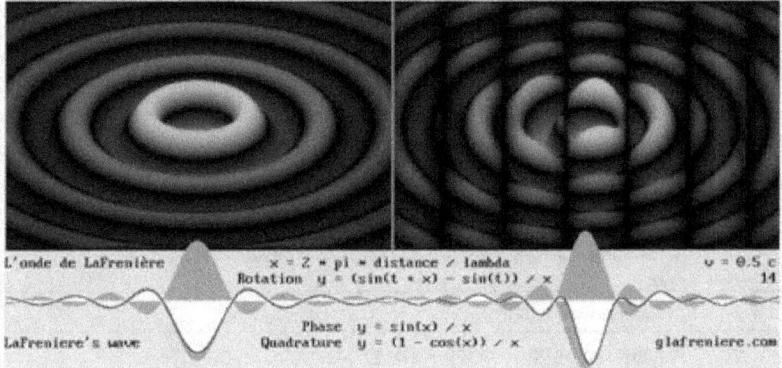

Fig.13-5. Matter is made of confined EM waves. LaFreniere wave model of the electron in motion. [Credit: glafreniere.com; keelynet.wordpress.com/category/gravity control/].

Fig. 13-6. Self-induced wave motion of matter. Propulsion using ultrasonic acoustic moving standing waves has been experimentally demonstrated. Credit: [A New Kind of Propulsion in Space – Proof of Principle – Neila-Tech] https://www.youtube.com/watch?v=jq-kaYVO B4&list= PLjJB6KTdUYEl1-vsh3rWfs1XTvg3eJlDf. Such initial proof-of-concept trials serve as prelude to eventual electromagnetic microwave propulsion demonstration in vacuo.

> *The most beautiful thing we can experience is the mysterious. It is the source of all true art and all science. – Albert Einstein*
>
> *The wave then appears as both a pilot wave and a probability wave... – Louie de Broglie*
>
> *The theory had many virtues and no known vices. – Murray Gell-Mann*
>
> *You have to learn the rules of the game. And then you have to play better than anyone else. – Albert Einstein*

13. Standing Wave Transformations

Matter in motion undergoes wavelength contraction (x' = g·x) and a phase shift (x' = β·t). Moving clocks undergo frequency dilation t' = g·t as well as a time shift (t' = g·t). The phase velocity v_p corresponds to the scissor velocity of the wavefront intersection of two opposed transverse standing waves which varies with wavefront angle θ. In general, $v_g < c < v_p$ although there are exceptions as in anomalous dispersion. For a nondispersive medium, group and phase velocity are equal $v_g = v_p$. Summary equations for Ivanov – LaFreniere contracted standing wave transformations (applied to wavelength) are shown in Table 13-2.

	Table 13-2. Standing Wave Lorentz Transformations[38]
λ, λ_{sw}	Standing wavelength = λ_{am} = cT = c/f = c/ν
λ'	Contracted wavelength (on-axis) $\lambda' = \lambda_{am}\sqrt{(1-\beta^2)}$ = $\lambda\sqrt{((1+\beta)/(1-\beta))}$ = g·λ_{am}
λ'	Doppler wavelength co-moving source and observer $\lambda' = \lambda(\cos(\sin^{-1}(\beta\sin\theta)) - \beta\cos\theta) = \lambda(\sqrt{(1-\beta^2(\sin\theta)^2)} - \beta\cos\theta)$
λ'	Lorentz Doppler wavelength = $\lambda' = \lambda\sqrt{((2/(1+\beta\cos\theta)) - 1)}$
λ_{am}	Arithmetic mean wavelength $\lambda_{am} = (\lambda_b + \lambda_f)/2$ = ct = kωt
λ_{gm}	Geometric mean wavelength $\lambda_{gm} = \sqrt{(\lambda_b \cdot \lambda_f)} = \lambda\cos\theta$
λ_{hm}	Harmonic mean wavelength $\lambda_{hm} = \gamma\lambda_{gm}$
λ_{LDf}	Relativistic Doppler shift forward wavelength $\lambda_{LDf} = \lambda(1-\beta)/g$ = $\lambda\sqrt{((1-\beta)/(1+\beta))}$ (contracted on-axis)
λ_{Df}	Doppler forward wavelength $\lambda_{Df} = \lambda(1-\beta)$ = c/f(1 − β) = $\lambda_{ad}/(1-\beta)$
λ_{LDb}	Relativistic Doppler shift backward wavelength $\lambda_{LDb} = \lambda(1+\beta)/g$ = $\lambda\sqrt{((1+\beta)/(1-\beta))}$ (dilated on-axis)
λ_{Db}	Doppler backward wavelength $\lambda_{Db} = \lambda(1+\beta)$ = c/f(1 + β) = $\lambda_{am}/(1+\beta)$
α	Standing wave velocity ratio $\alpha = (R-1)/(R+1) = (\lambda_b - \lambda_f)/(\lambda_b + \lambda_f)$ = (1 − g)(1 + g)/β = g$\beta\gamma$
β	Normalized speed ratio β = v/c = (R − 1)/(R + 1) = $\sin\theta$ = ρ/γ = n^{-1} = $\alpha/g\gamma$ = tanh ω = tanh ρ = $\Delta\phi/\pi$ = $\sqrt{(1-1/\Gamma^2)}$ = $\sqrt{(1-K_{PV})}$ = pc/E
g	Lorentz contraction g = $\sqrt{(1-\beta^2)} = 1/\gamma = \lambda_{gm}/\lambda_{am} = \sqrt{(1-\alpha^2)} = \cos\theta$ = $\alpha/\beta\gamma = \sqrt{(1-(1-K_{PV}))}$
γ	Lorentz factor $\gamma = 1/g = 1/\sqrt{(1-\beta^2)} = \alpha/g\beta$ = cosh ω
λ_f	Doppler redshift $\lambda_r = \lambda(1+\beta)$
λ_b	Doppler blueshift $\lambda_b = \lambda(1-\beta)$
R	Wavelength ratio R = $\lambda_b/\lambda_f = (1+\beta)/(1-\beta)$; Redshift $R_R = \lambda'/\lambda$, Blueshift $R_B = \lambda/\lambda'$
λ_{aD}	Doppler approaching (source of light) wavelength $\lambda_{aD} = \lambda(1-\beta\cos\theta)$
λ_{rD}	Doppler receding (source of light) wavelength $\lambda_{rD} = \lambda(1+\beta\cos\theta)$
$\langle\lambda\rangle$	Average wavelength $\langle\lambda\rangle = \frac{1}{2}(\lambda_{Red} + \lambda_{Blue}) = \frac{1}{2}(\lambda_{aD} + \lambda_{rD})$

cont

13. Standing Wave Transformations

Table 13-2. Standing Wave Lorentz Transformations (cont)

λ_{aLD}	Lorentz Doppler approaching (source of light) wavelength $\lambda_{aLD} = \gamma\lambda(1 - \beta\cos\theta) = \lambda(1 - \beta\cos\theta)/g$
λ_{rLD}	Lorentz Doppler receding (source of light) wavelength $\lambda_{rLD} = \gamma\lambda(1 + \beta\cos\theta) = \lambda(1 + \beta\cos\theta)/g$
$\Delta\lambda_D$	Doppler wavelength shift $\Delta\lambda_D = (v/c)\lambda = c/\Delta f_D = c/f2\beta$; Redshift $\lambda_R + \Delta\lambda$, Blueshift $\lambda_B - \Delta\lambda$
$\Delta\lambda_{LD}$	Lorentz Doppler wavelength shift $\Delta\lambda_{LD} = \langle\lambda_r\rangle - \lambda = c/\Delta f_{LD} = \lambda 2\beta/g$
c	Velocity of light $c = c_o/n = 1/\sqrt{(\epsilon_0\mu_0)} = \sqrt{(v_p v_g)} = c_o/\sqrt{K_{PV}} = c_o/\Gamma = \nu\lambda$
v_g	Group velocity $v_g = c^2/v_p = d\omega/dk = v_p - \lambda(dv_p/d\lambda) = c\sqrt{(1 - (\omega_{co}/\omega)^2)}$; $v_g = pc^2/E = c\cdot\Delta\nu\cdot\pi = \alpha\cdot c = \beta\cdot c$ (matter wave envelope node speed); $\omega = \gamma\omega_0$, $k = \gamma\beta\omega_0/c$, $\omega_{co} = \omega_p$
v_p	Phase velocity $v_p = c^2/v_g = E/p = \lambda\cdot f = \lambda/T = c/(\Delta\nu\cdot\pi) = c/\beta = \omega/k = c/\sqrt{(1 - \omega_{co}/\omega)^2)}$; $v_p = c/\alpha = \gamma mc^2/\gamma mv$ (matter wave velocity); $p = \hbar k = \gamma mv$, $\omega_p = \omega_{co}$
z	Wavelength shift/Wavelength $z = \Delta\lambda/\lambda = (1 + \beta)/\sqrt{(1 - \beta^2)} = \sqrt{[(1 + \beta)/(1 - \beta)]} - 1 = \sqrt{(R)} - 1$
x	Coordinate length (on-axis) $x = gx' - \beta t'$ (x and x' distance in light-second units, t and t' period in seconds). x' = proper length
t	Coordinate time (on-axis) $t = gt' + \beta x'$ (x and x' distance in light-second units, t and t' period in seconds). t' = coordinate time
x'	Proper length (on-axis) $x' = (x - vt)/g = gx + \alpha t$; Doppler: $x' = g^2x + \beta t$; Lorentz Doppler: $x' = gx + \beta t$ (x and x' distance in light-second units, t and t' period in seconds)
y'	Proper length (transverse axis) Doppler: $y' = gy$; Lorentz Doppler: $y' = y$ (No contraction)
z'	Proper length (transverse axis) Doppler: $z' = gz$; Lorentz Doppler: $z' = z$ (No contraction)
t'	Proper time $t' = (t - \beta x/c)/g = gt - \alpha x$; Doppler: $t' = t - \beta x$; Lorentz Doppler: $t' = gt - \beta x$ (x and x' distance in light-second units, t and t' period in seconds)
θ	Wavefront angle $\theta = \sin^{-1}(\beta) = \tan^{-1}(\alpha/g) = \cos^{-1}(\lambda_{gm}/\lambda_{am}) = \pi/2 - \phi$; Aberration angle $= \phi = \tan(g/\alpha)$
ϕ	Aberration angle $= \phi = \tan(g/\alpha) = \tan(gc/\beta v)$
$\Delta\phi$	Phase shift $\delta\phi = (\phi_2 - \phi_1) = \pi\beta = \pi v/c = \pi v/\lambda_{dB}\cdot\nu_{dB}$
$\Delta\lambda$	Wavelength compression (on-axis contraction: $\Delta\lambda = \lambda - \lambda' = \lambda_{am}(1 - g)$
f_{aD}'	Doppler frequency observer moving towards source $f_{aD}' = f(1 + \beta\cos\theta)$
f_{rD}'	Doppler frequency observer moving away from source $f_{rD}' = f(1 - \beta\cos\theta)$
f_{Df}	Doppler forward frequency $f_{Df} = f(1 + \beta)$

cont

13. Standing Wave Transformations

Table 13-2. Standing Wave Lorentz Transformations (cont)

f_{LDf}	Lorentz Doppler forward frequency $f_{LDf} = \gamma(1+\beta)f$
f_{Db}	Doppler backward frequency $f_{Db} = f(1-\beta)$
f_{LDb}	Lorentz Doppler backward frequency $f_{LDb} = \gamma(1-\beta)f$
Δf_D	Doppler frequency difference $\Delta f_D = f_{Df} - f_{Db} = 2f\beta$
Δf_{LD}	Lorentz Doppler frequency difference $\Delta f_{LD} = f_{LDf} - f_{LDb} = cg/\lambda 2\beta$
f_C	Compton frequency $= c/\lambda_C = mc^2/h = 1.2355E20$ Hz
λ_C	Compton wavelength $= \lambda_{sw} = h/mc = \beta \cdot \lambda_{dB} = g(\lambda_{LDb} - \lambda_{LDf}) = 2.4263E\text{-}12$ m
f_{dB}	de Broglie frequency $f_{dB} = f_C \sqrt{(1-\beta^2)} = g \cdot f_C = c/\lambda_{dB} = (h/mv)\Delta v/\Delta\lambda = pc/h$
λ_{dB}	de Broglie wavelength $\lambda_{dB} = h/mv = h/\gamma m_0 v = h/m\beta c = h/\gamma m_0(c/\pi)\Delta\phi$ $= \lambda_C/\beta = \lambda_{am} = c/v_{dB} = \Delta\lambda/(c/h)(mv/\Delta v)$
λ_{dBt}	de Broglie thermal wavelength $\lambda_{dBT} = \Lambda_T = h/\sqrt{(3mk_BT)}$
f_p	Phase frequency $f_p = v_p/\lambda_p = \beta \cdot c/\lambda_p = c^2 v_g/\lambda_p = (c/\beta)/\lambda_p = (\omega/k)/\lambda_p$
λ_p	Phase wavelength $\lambda_p = \lambda'/\alpha = \lambda \cdot g/\alpha = v_p/f_p = \beta \cdot c/f_p = c/\beta\lambda_p f_p$

A mass object is equivalent to a standing wave resonator. As any physical object has spatial dimensions, a force applied to any point can not simultaneously affect the entire body. As a result, there is a delay time for the wave system to respond. If energy is applied greater than the rate at which the system can absorb the energy input, the excess is radiated away. Power equals the rate of flow of energy. Newton's third law of motion: For every action, there is an equal and opposite reaction assumes instantaneous reaction in one inertial frame and represents a condition of 100% negative feedback. Thus, every action generates an out-of-phase signal of equal and opposite magnitude resulting in self-cancellation. If the phase rotation is 180° out of phase, feedback is positive creating an oscillation alternating between potential and kinetic energy. A resonator, once set into motion, represents a phase rotation such that the energy flow is out of phase. Hence, the resonator under uniform motion becomes a harmonic oscillator and will continue to oscillate unless acted upon by an external force in accordance with Newton's first law. Application of an external force results in a radiation back pressure in proportion to the applied force and, if the external force is applied for twice the critical reaction time, the resonator will accelerate in accordance with Newton's second law. Resonator motion proceeds in hopscotch fashion alternatively switching between dissonance (acceleration jumps) and consonance (constant velocity). Discernible macroscopic quantized effects begin to occur when the temperature is reduced near absolute zero such that the deBroglie thermal wavelengths begin to overlap producing wave function coherence, e.g., atomic Bose-Einstein condensation.

Aether's existence is a very touchy subject in the science world. Speaking about the wave nature of matter without aether is a perfect nonsense. – Gabriel LaFreniere

All my life is a suite of communication failures. – Gabriel LaFreniere

[Re: Ivanov's Standing wave compression vs.Lorentz transformation]. In all my life, I've never read such intelligent and sparkling stuff... I could cry. – Gabriel LaFreniere

14. Phase-locked Phase Conjugate Resonators

A resonant cavity may be formed by closing the ends of a waveguide to make a chamber or cavity. If this cavity is excited in a suitable manner at the proper frequency, it may be made to resonate, as in a laser crystal for example. The reflecting boundaries of a resonator cavity need not be limited to conventional mirrored surfaces. In a conventional mirror, a reflected EM spherical divergent wave from a point source remains divergent. If the wave is reflected from a phase conjugate mirror (PCM), the wavefront is inverted in a convergent beam back to the source in a time-reversed replica. Phase-conjugate wave (PCW) generation may be accomplished via three or four-wave mixing (FWM) in photorefractive crystals, nonlinear optical Kerr media or metamaterials. Four-wave mixing is a nonlinear effect arising from a third-order optical nonlinearity described by a susceptibility $\chi^{(3)}$ coefficient in a Taylor series expansion resulting in induced polarization (P(E) = $\chi_e^{(1)}E(t) + \chi_e^{(2)}E(t)^2 + \chi_e^{(3)}E(t)^3 + ...$) of electric field strength. Nonlinear phenomena that produce phase conjugation include Brillouin scattering, Raman scattering, Kerr FWM, resonant FWM, photon echoes, etc. A Bragg reflector is formed from an interference pattern in the overlap zone of two counter-propagating beams. The effect is similar to conventional Bragg x-ray diffraction from crystals where atomic lattices form periodic scattering centers. A signal beam incident on the interference pattern results in a reflected phase conjugate wave propagating back along the path of the signal beam. A diagram illustrating phase conjugation FWM process is shown in Fig. 14-1 for a nondegenerate case where pump beam frequencies are not equal. Wave interference nodes act as scattering centers of a holographic amplitude grating for incident EM waves to form a reflected phase conjugate beam. Reflection occurs when the incident wavelength is comparable to the Bragg plane spacing.

Four-way mixing can be considered as two simultaneous three-way mixing and scattering processes. In optical mixing, a pump wave mixes with a signal wave generating an interference grating and a second pump wave scatters off the grating generating a phase conjugate wave. An illustration of motion of a wave and phase conjugate wave in the complex plane is shown in Fig. 14-2. In the complex plane, rotation of a phasor $\mathbf{A} = Ae^{i\omega t + \theta}$ and its conjugate $\mathbf{A^*}$ at an angular velocity ω corresponds to a multiplication. Rotation (CCW) by an angle of 90 degrees, for example, results from multiplication of the complex number by $i = \sqrt{-1}$. Modulation of the standing wave (or carrier wave) results in the creation of upper and lower side band frequencies corresponding to harmonics of the modulation frequency centered about the de Broglie frequency ($v_{dB} \pm v_{mod}$). Amplitude modulation produces two sideband signals (S_l, S_u) whenever the amplitude of a signal (f_c) is modulated at a lower frequency (f_m). Sidebands are also produced when the phase or frequency of a carrier signal is modulated. Sideband generation by carrier signal modulation is illustrated in Fig. 14-3. A phase-locked resonator in motion in a phasor diagram may be represented as a cross-section of ellipsoid with oscillating state vector amplitude alternating between an oblate ellipsoid and a prolate ellipsoid rotating with average angular velocity ω. A moving contracted electromagnetic standing wave within the resonator oscillates between negative energy density and positive energy density referenced to a vacuum with a positive time-averaged energy density.

> *The general principal of critical reaction time (CAT) is that the energy of a system can not be changed in zero time... that there is a time in which a system as a whole cannot accept energy input. If the rate of energy input is too great for the system to absorb, the excess energy must be excluded or leave the system, either by transferring the energy to another system and/or changing the form of the input energy so that it may be radiated away. – William O. Davis*

14. Phase-locked Phase Conjugate Resonators

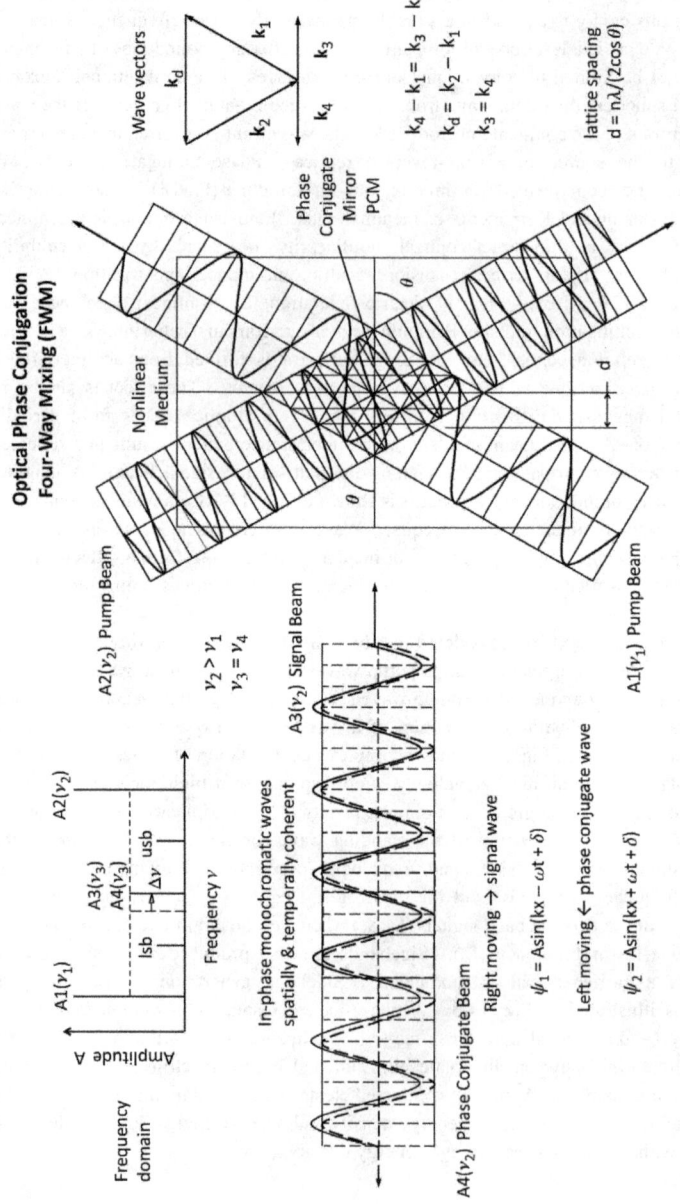

Fig. 14-1. EM wave reflection/diffraction from Bragg planes formed by interference pattern of EM waves.

14. Phase-locked Phase Conjugate Resonators

Under FWM, interference of a pump beam A1 and an opposing pump (or signal) beam A2 create a refractive index grating of alternating grid of variation of refractive index in a nonlinear medium as a result of Kerr/Pockels effects. A signal (or probe beam) A3 reflecting off the interference grating is reflected as a counter-propagating phase conjugate wave A4. The PCW is generated as a third-order nonlinear response of the medium. For non-degenerate FWM (NDFWM), a refractive index modulation at the difference frequency occurs in which two input frequencies v_1 and v_2 (with $v_2 > v_1$) creates two additional frequency components: $v_3 = v_1 - (v_2 - v_1) = 2v_1 - v_2$ and $v_4 = v_2 + (v_2 - v_1) = 2v_2 - v_1$. The frequency v_3 or v_4 can be amplified as a result of parametric amplification. The summation of the A3 signal (or probe beam) and A4 phase conjugate beams forms a standing wave that oscillates in-place if the field amplitudes are equal. Amplitude varies with incidence angle. If the amplitudes are unequal there is a net propagation toward the higher amplitude beam.

With pump beams of sufficiently high amplitude, a portion of the energy in the nonlinear standing waves can transfer to the conjugate wave resulting in amplification. In parametric pumping (3-way mixing), a pump wave at double frequency ($v_p = 2v$) and an incident wave of frequency (v) results in a PCW at frequency v_{pc} (= $2v - v$). Degenerate four-wave mixing (DFWM) phase conjugation involves beams all of the same frequency v. In a 1-channel DFWM process when the two pump frequencies coincide, the idler frequency $v = 2v_1 - v_2$ (where v_1 is the degenerated pump frequency and v_2) is the probe frequency. For nearly degenerate FWM with pump waves of frequency v and incident probe beam ($v + \delta$), the resultant PCW beam frequency v_0 is a difference frequency (= $v - \delta$). In backward NDFWM, the probe beam A1(v_1) and the signal beam A3(v_1) have the same frequency v_1 while beams A2(v_2) and A4(v_2) have a different frequency v_2.

Optical phase conjugation has been a subject of intense study and implemented in a wide variety of applications. Optical resonators with PCMs have been utilized, for example, in laser oscillators with phase conjugation feedback, laser amplifiers with multi-pass gain medium, laser target aiming and auto-focusing implemented with Brillouin enhanced FWM. Phase conjugators provide an alternative to adaptive optics for aberration correction, target aiming, pointing and targeting, interferometry, lensless imaging and optical computing. An unexplored potential is modulation of de Broglie waves or generation of synthesized matter waves utilizing phase conjugation and Doppler shifted pump waves to effect a change in motion of matter.

This all seemed rather childish... to deal properly with waves one should have a wave equation to describe how the wave moves from place to place. – Peter Debye

It is seen that both matter and radiation possess a remarkable duality of character, as they sometimes exhibit the properties of waves, at other times those of particles. – W. Heisenberg

A moving particle is nothing else but the foam on the wave radiation forming the matter of the world. – Erwin Schrödinger

Compounding the difficulties of a rational understanding of matter was the discovery that particles had wave properties and that these matter waves were not of a material nature. In some mysterious way, these waves merely determined the probability of finding a point-like particle in a certain place. – Ernest J. Sternglass

Inertial frames are the reference frames upon which the laws of motion and conservation laws are defined; yet it is still unknown what causes inertial frames to exist and if they have any deeper properties that might prove useful. – Mark Millis

14. Phase-locked Phase Conjugate Resonators

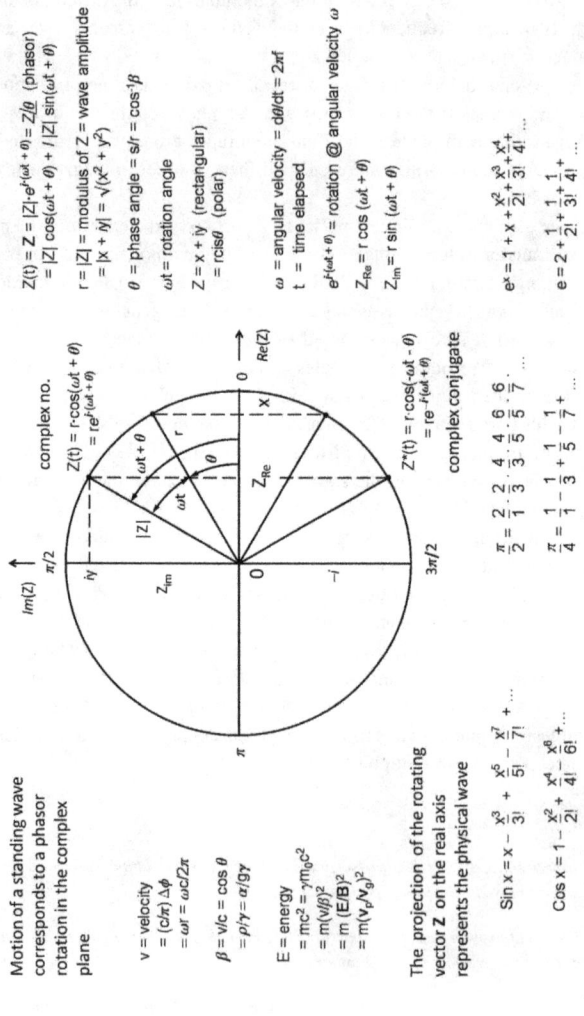

Fig. 14-2. Propagation of a matter wave and its phase conjugate represented in the complex plane. Motion corresponds to rotation (multiplication) of the phasor. Wave system velocity is proportional to a difference in phase. Uniform motion (constant rate-of-change of position) corresponds to constant phasor angular velocity. Time reversal corresponds to a quaternion multiplication in a Wick rotation (t → − ir) which may be simulated with negative refractive index metameterials and phase conjugation.

14. Phase-locked Phase Conjugate Resonators

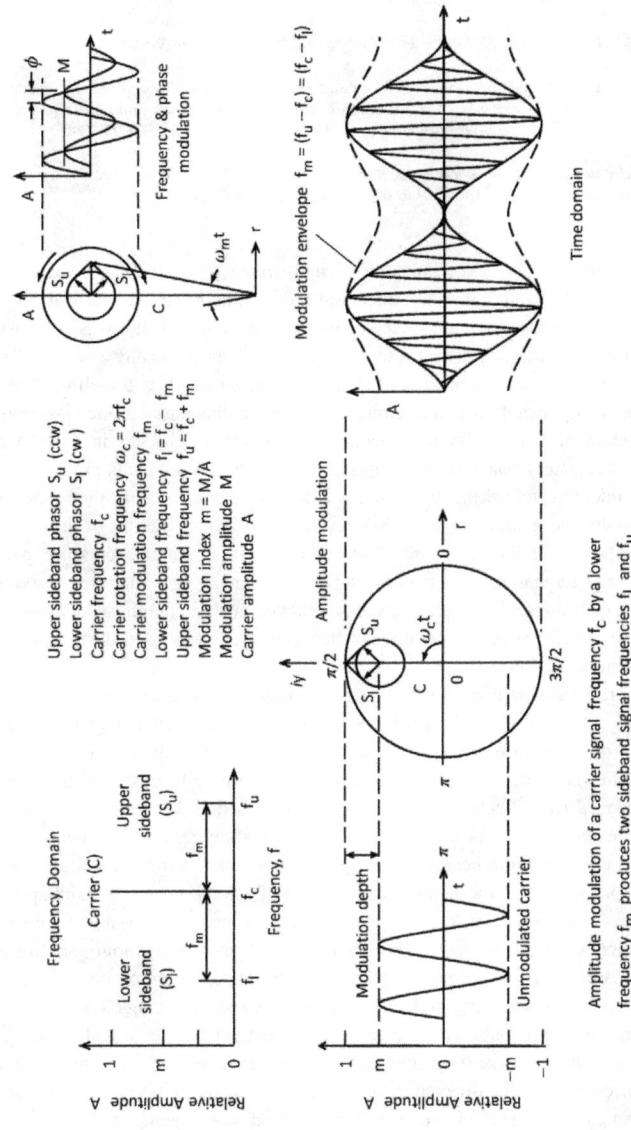

Fig. 14-3. Sideband signals produced by amplitude, frequency or phase modulation of a sinusoidal carrier signal.

15. Phase Conjugate Resonator Experimental Potential

In the environment of an electrically charged body there is a magnetic field which furnishes an apparent contribution to its inertia. Should it not be possible to explain the total inertia of the particles electromagnetically. – Albert Einstein

The inertia of the electron originates in the electromagnetic field. – Max Abraham

For if inertia depends on the zero-point field of the quantum vacuum and if the later can be manipulated, as certain phenomena seem to indicate, then it might not be impossible to control inertia or perhaps even to eliminate it altogether. – B. Haisch, A. Rueda and H. E. Puthoff

If we wish to fully understand and eventually modify gravitational and inertial forces, there is only one thing that we can use (that we know of) to do so: energy. – Geoffrey S. Diemer

In the 1980s, Jennison[39,40] experimentally demonstrated phase-locking effects of free-floating resonators with both light and microwaves. Using a servoed optical etalon on movable trolleys, a fixed wavelength distance between source and reflector was demonstrated. The source emitter and reflector wave system moved as if mechanically connected with a spacing accuracy of > 0.001 wavelength. In another experiment, a travelling EM wave propagating along an oppositely rotating, circular slow wave transmission line was brought to rest and reversed in direction without reflection or refraction resulting in a static dipole electric field in a laboratory rest frame[41]. On-going research developments in metamaterials, nano structures, microelectromechanical systems (MEMS), etc point to further experimental potential of phase-locked, phase conjugate SW resonators. Demonstrations of negative index of refraction, phase conjugation, squeezed /slow light, negative radiation pressure, patterned wave fronts, cross-field/phased array/plasmonic/fractal antennas, etc, suggest phased-locked cavity resonators with unusual properties may be realized such as nonlinear response of the following wall or synchronization interactions between oscillators of different frequencies including self-induced motion.

Generation of rest mass is a result of trapping of radiation in the equivalent of a phase-locked resonator. A phase-locked, frequency modulated, phase conjugate resonator may provide a means for generation of a synthesized matter wave via an inverse Lorentz-Doppler effect. In the Jennison model, inertia arises as a result of a blue or upshifted incident wave and a red or downshifted reflected wave internal to a phased-locked cavity resonator upon application of an external force for an interval sufficient to allow reflection to occur. A phase conjugated resonator with pump beams of controllable frequency acting on a phase conjugate mirror (PCM) is proposed as a means to induce motion of a wave system without application of external force. As shown in Figs. 15-1 and 15-2, it may be possible to generate a synthesized matter wave with controllable self-induced motion. Input conditions are established by pumping of a PCM nonlinear medium by opposing pump beams A1(v_1) and A2(v_2) at frequencies corresponding to the desired Doppler red-shift frequency ($v_1 = v_0 - \Delta v$) and Doppler blue-shift frequency ($v_2 = v_0 + \Delta v$), respectively. The signal beam A3 (v_3) corresponds to the standing wave frequency of the resonator at rest (v_0). The phase conjugate beam A4 (v_4) corresponds to the difference in frequency of the pump beams ($v_1 - v_3$). Mixing of simulated Doppler blue-shifted wave and red-shifted wave pump beams with signal (standing wave) and PC beams is predicted to reproduce a modulated wave phase shift generating an unbalanced radiation pressure resulting in net motion. Radiation pressure p_{rad} equals cyclic time average of energy density (= $\langle u \rangle$). The impulse imparted includes an alternating push/pull force in the direction of motion. The induced phase and frequency shifts

15. Phase Conjugate Resonator Experimental Potential

replicate that produced with application of an external force producing a wave system velocity > 0. The energy associated with motion is $E - E_0 = [2F(v/c)^2]/[1 - (v/c)^2]$ where F = applied force. The amplified pump beam energy input provides the kinetic energy for motion as an inverse Lorentz-Doppler effect.

Addition of energy such as with oscillating wave guide walls or pump waves may allow non-linear response or adjustable gain characteristics. For example, it may be possible to introduce a nonlinearity in response to an applied external force with asymmetric ring resonators with graded index of refraction incorporated between motive wall and the following wall. Substitution of a negative index left hand metamaterial for the wave medium alters the direction of the phase velocity and acts to red shift the incident wave and blue shifts the reflected wave. An inverse Doppler effect may be produced with an array of negative index of refraction resonators such that the incident radiation from the source is red-shifted in the direction of motion while the absorbed radiation is blue-shifted back towards the motive wall emitter. Negative index of refraction (NIM) metamaterials with negative permeability and permittivity exhibit an altered dispersion relation with negative phase velocity and positive group velocity. Negative refraction is ordinarily achieved by a combination of electric and magnetic resonance structures such that both the electric permittivity and magnetic permeability are negative. A chiral material allows a single resonance condition exhibiting a different refractive index for each polarization one of which is negative. EM beam interaction with layers of polaritons within an optical standing wave resonator cavity may result in negative kinetic energy. Force amplification corresponding to a Fresnel coefficient $|F| > 1$ may possibly be realized by parametric pumping of a nonlinear medium at double frequency of the incident wave generating an amplified phase-conjugate wave in a four-wave mixing process. The inertial reaction force may potentially be modulated by application of an exaggerated Doppler frequency shift response augmenting or countering the applied external force. Time reversal of EM waves may be digitally constructed utilizing a transmitter and receiver in a reverberation chamber with sampled signals reversed in time by computer processing. Phase relation between the signal and pump waves determines energy flow, i.e., amplification or deamplification of the signal or phase conjugate (PC) wave. Left-handed metamaterials enable phase wave propagation (wave vector $|\mathbf{k}| = 2\pi/\lambda$) in a direction counter to energy flow (Poynting vector $\mathbf{S} = \frac{1}{2}\text{Re}(\mathbf{E} \times \mathbf{H}) = \mathbf{v} \cdot \rho_e = c^2 \cdot \mathbf{p}$). A phase conjugate cavity resonator with external radiation loss and an increased standing wave ratio corresponds to a state of increased apparent inertia and decreased acceleration response. Parametric pumping of the phase conjugate reflector can result in a decreased SWR corresponding to a state of reduced inertia and increased acceleration response.

The term relativity refers to time and space. ...Briefly it discards absolute time and space and makes them in every instance relative to moving systems. – Hendrik Antoon Lorentz

Nature seems to be delighted with transmutation. – Matvei Petrovich Bronstein

Bring together things that have not yet been brought together and did not seem predisposed to be so. – Robert Bresson

Invention requires both disciplines, strict common sense and wild imagination. – Vanna Bonta

The elementary.particles obey Newton's law of motion for material points. This is the basis on which H. A. Lorentz obtained his synthesis of Newton's mechanics and Maxwell's field theory. The weakness of this theory lies in the fact that it tried to determine the phenomena by a combination of partial differential equations (Maxwell's field equations for empty space) and total differential equations (equations of motion of points), which procedure was obviously unnatural. The inadequacy of this point of view manifested itself in the necessity of assuming finite dimemensions for the particles in order to prevent the electromagnetic field existing at the surfaces becoming infinitely large. – Albert Einstein

15. Phase Conjugate Resonator Experimental Potential

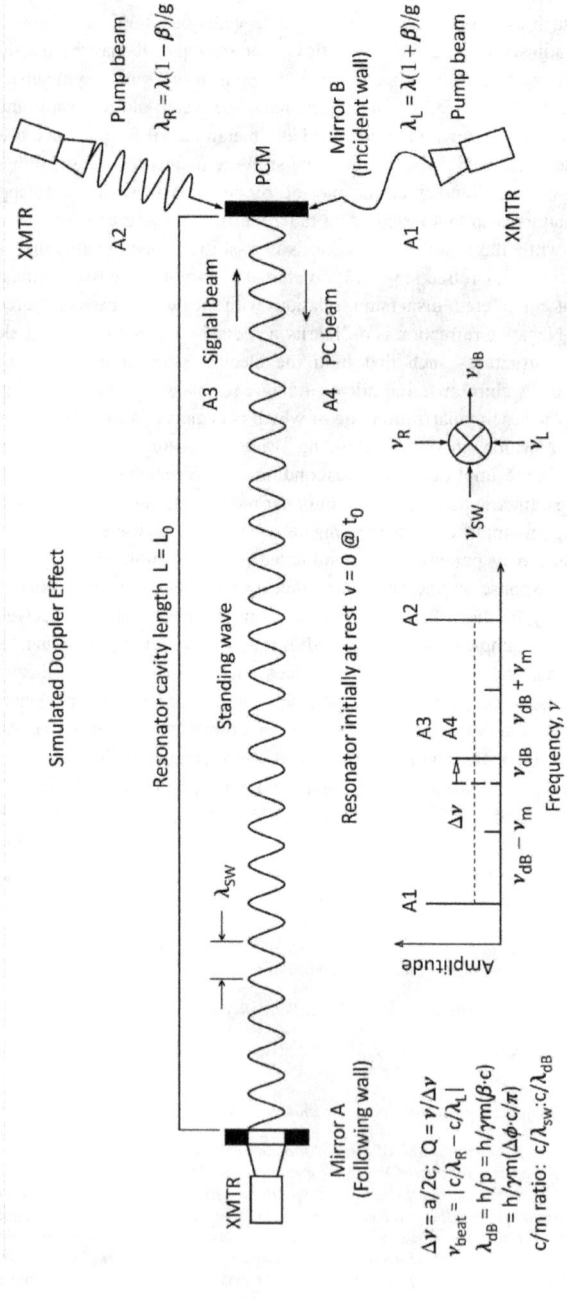

Fig. 15-1. Conceptual diagram for induced motion of a phase-locked resonator irradiated by EM pump waves of imulated relativistic Doppler shifted wavelengths in four-way mixing with an internal standing wave. At rest, the incident signal beam frequency equals the reflected phase conjugate beam frequency. Superposition of a blue-shifted signal beam and and a red-shifted phase conjugate beam results in a Lorentz-Doppler beat frequency $\Delta \nu \, (= c^2/\Delta\lambda \cdot \nu_{dB})$ proportional to de Broglie matter wave frequency $\nu_{dB} \, (= E/h = pc/h = c/\lambda_{dB} = (h/m\nu)(\Delta\nu/\Delta\lambda))$.

15. Phase Conjugate Resonator Experimental Potential

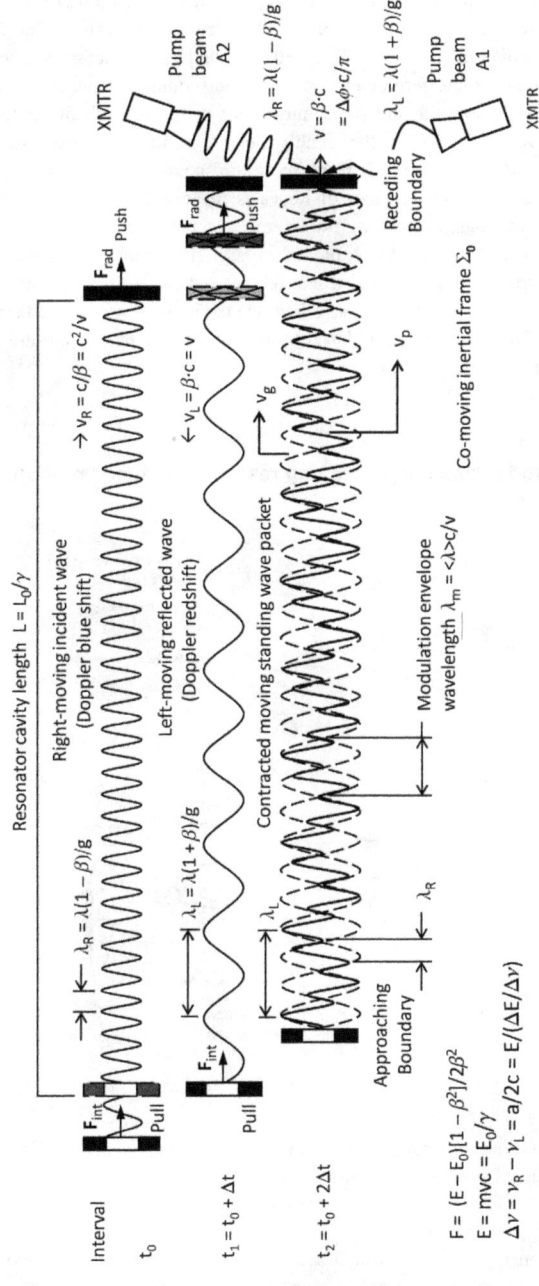

Fig. 15-2. Simulated Lorentz - Doppler effect results in contracted moving standing wave. The internal radiation pressure imbalance results in a net ponderomotive force. No external radiation pressure or expulsion of reaction mass is required for induced motion. A moving standing wave exhibits increased node width where $SWR = |E_{max}|/|E_{min}| = (1 + \Gamma)/(1 - \Gamma)$. Reflection coefficient Γ increases with velocity ratio β ($= v/c$).

15. Phase Conjugate Resonator Experimental Potential

Similar to the previous example for induced linear motion of a phase conjugate resonator with a phase conjugate mirror, angular motion may potentially be realized in a closed loop standing wave resonator in an inverse Sagnac effect. An example concept sketch is indicated in Fig. 15-3. As shown, a synthesized matter wave is generated by pump waves of simulated relativistic Doppler frequencies mixed to produce a phase conjugate mirror. The frequency imbalance creates a radiation pressure differential inducing resonator rotation. For comparison, in a Sagnac effect interferometer, two circulating counter-propagating waves (cw, ccw) produce a standing wave with the same number of nodes in each direction with an integer number of wavelengths within the circular length of the ring resonator. Under rotation, the optical path length is different resulting in a resonant frequency shift of the two standing waves such that the same number of nodes is preserved. The beat frequency is linearly proportional to the angular velocity with respect to the inertial frame in which the center of the loop is at rest. This effect is exploited in laser ring gyroscopes for rotation rate detection by measurement of the spin-induced interference pattern shift of two beams mixed by optical heterodyning. In the reverse Sagnac effect, the cavity platform is fixed (motionless ring laser) and the standing waves rotate induced by motion of an internal aperture along the beam axis shifting node position.

Phase-locked phase conjugate ring resonator induced motion

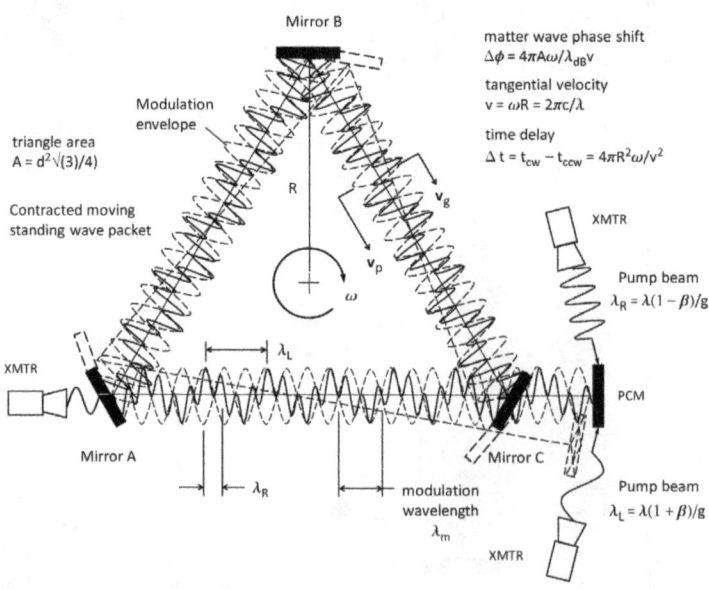

Fig. 15-3. Induced rotation of a phase conjugate ring resonator via inverse Sagnac effect utilizing Doppler modulated synthesized matter waves.

The participation of matter in electromagnetic phenomena has its origin only in the fact that the elementary particles carry unalterable electric charges and on this account are subject ...to the ponderomotive forces and ... possess the property of generating a field. – Albert Einstein

15. Phase Conjugate Resonator Experimental Potential

Acceleration of an oscillator through a vacuum in the Fulling, Davies and Unruh effect is predicted to result in a change of spectrum observed radiation that includes both zero-point radiation of the vacuum and a thermal radiation component proportional to acceleration. For a harmonic oscillator at rest or in a constant velocity frame at absolute zero temperature, the Planck thermal radiation component is suppressed leaving only the ZPF radiation. The ZPF spectrum is Lorentz invariant and, as a result, is not detectable by an observer in motion through space at constant velocity. However, a charged particle accelerated through the ZPF is predicted to experience an electromagnetic force as resistance to acceleration. Supposing a synthesized matter wave may be created in the manner described, how might inertia damping be realized? Recall that the vacuum of Minkowski space is characterized by a spectral energy density $U(\omega)d\omega = k(\hbar\omega^3/c^3)d\omega$ which, in an inertial frame, is uniquely Lorentz invariant. A physical vacuum composed of Planck dipoles is represented as having a zero-point energy $E = \frac{1}{2}\hbar\omega_P$. Inertia has been conjectured by Haisch, Rueda and Puthoff[43] to be a consequence of a magnetic component of the Lorentz force arising in accelerated motion of a charged particle though such a zero-point field (ZPF). In other words, inertia is attributed to ZPF electromagnetic resistance to acceleration. The frequency spectrum remains unchanged by a transformation from one constant velocity inertial frame to another at rest, but becomes distorted during relative acceleration with respect to the ZPF background.

The intensity of the observed electromagnetic spectrum at any frequency is greater in an accelerated rest frame than one at rest and is proportional to the cube of the frequency ω^3. The ZPF radiation spectrum is illustrated in Fig. 15-4 together with the thermal radiation effect of acceleration. Observed acceleration induced thermal radiation spectrum of a moving observer is shown compared with a simulated zero-point radiation spectrum. Classical radiation in a resonant cavity in an inertial frame is equivalent to empty-space zero point radiation. Pump beam frequencies with simulated zero-point radiation spectrum in a resonator inertial frame compensates for observer motion by changes in frequency vs. intensity such that no detectable Doppler shift by a co-moving observer is produced at constant velocity.

The zero-point spectrum is independent of observer velocity due to compensating changes in frequency and intensity. As a result, the laws of physics remain conformally invariant. Inertial mass varies with change in EM mode density. This seems to suggest that a synthesized matter wave with a modulated intensity that is proportional to ω^3 frequency characteristic may allow for mass motion without Doppler induced inertial effects in an accelerated (non-inertial) Rindler frame. The observer at rest and an observer in motion in a co-moving reference frame of the resonator see the same zero-point radiation spectrum with no detectable Doppler shift. The intensity of the simulated Doppler red and blue shifted pump waves may be modulated by a parametric amplifier that is controlled to vary with the cube of the wave frequency oscillation. The simulated ZPF spectrum need not necessarily replicate a ω^3 manifold. For example, addition of one or more control factors may be introduced to generate a local minima/maxima in the simulated spectrum or even a catastrophic fold or cusp to alter system behavior such as bimodality or divergence. An abrupt transition from a constant velocity inertial frame to a constant acceleration Rindler frame may be described as a discontinuous transition state resembling a catastrophic jump from one system state pathway to another. To effect such a transition, control factors may possibly include, for instance, switch from Doppler to inverse Doppler mode, normal to anomalous dispersion, asymmetric Fano resonance detuning, upshift/down-shift Doppler gain adjustment or addition of higher order, non-linear frequency components. Negative energy may be created by a moving mirror with increasing acceleration generating a very small negative energy flux in front of the mirror. Quantum squeezing of light in optical resonators has been experimentally demonstrated with ultra-high intensity femtosecond chirped-pulse amplification (CPA) lasers generating pulsed positive and negative energy. Strongly focused ultrashort light pulses

15. Phase Conjugate Resonator Experimental Potential

generate quantum fluctuations in the E-field intensity above and below the vacuum noise level. A squeezed vacuum state temporarily increases the local speed of light whereas a stretched vacuum lowers said velocity. A material object disposed between such regions will experience a net force in the direction of the field gradient. A phased-locked resonator irradiated with alternating positive/negative energy pulses may allow modulation of acceleration. Modulation of phase and/or intensity may be realized by addition of an array of coaxial Fano cavity resonators deployed along the optical axis of a coupled phase-locked resonator. Parametric pumping of the Fanno resonators may be used to amplify optical effects enhancing the interaction of light and matter allowing generation of a nonlinear Fanno profile. Altering the phase relation by interference of the central beam resonance frequency and Fano resonances may yield a nonuniform acceleration gradient along the beam axis.

Nonlinear polarizable materials, such as piezoelectric crystals used in accelerometers, typically exhibit reciprocity over a portion of the operating range. Application of voltage results in mechanical strain and vice versa. However, application of voltage to an accelerometer does not result in acceleration. Symmetry is broken. An additional conversion step is required to induce motion, eg., Modulation of EM waves using nonlinear materials such as piezoelectric materials, Gunn diodes, varactors, Kagome lattices, metamaterials, etc with asymmetric response may be utilized to generate an difference in frequency and/or phase of oppositely directed traveling waves within a resonator to induce motion via inverse Lorentz-Doppler effect.

Since nature is a principle of motion and change, and since our inquiry is about nature, we must not overlook the question of what motion is. For without understanding motion, we could not understand nature. – Aristotle

Fundamental research is obliged to provide mankind with such knowledge about the world that it allows it to survive in space under any circumstances, even if the planet is lost. – Yuri N. Ivanov

My interest is in the future because I am going to spend the rest of my life there. – Charles Kettering

I like the dreams of the future better than the history of the past. – Thomas Jefferson

The present is saturated with the past and pregnant with the future. – Gottfried Leibnitz

We are called upon to be the architects of the future, not its victims. – R. Buckminster Fuller

I imagine having 25th century science this century. – Harold E. Puthoff

Science is magic that works. – Kurt Vonnegut

The world is rational. There are systematic methods for solutions of all problems. – Kurt Gödel

So many fail because they don't get started – They don't go. They don't overcome inertia. They don't begin. – Ben Stein

A new scientific truth does not triumph by convincing its opponents and making them see the light, but rather because its opponents eventually die and a new generation grows up that is familiar with it. – Max Planck

Our fight must be not only to the stars but into the nature of our own beings. Because it is not merely where we go, to Alpha Centauri or Betelgeuse, but what we are as we make our pilgrimage there. Our natures go with us. – Phillip K. Dick

From an incandescent mass we have originated, and into a frozen mass we shall turn.– Nikola Tesla

It is not in the stars to hold our destiny but in ourselves. – William Shakespeare

The concern for man and his destiny must always be the chief interest of all technical effort. Never forget it among your diagrams and equations. – Albert Einstein

15. Phase Conjugate Resonator Experimental Potential

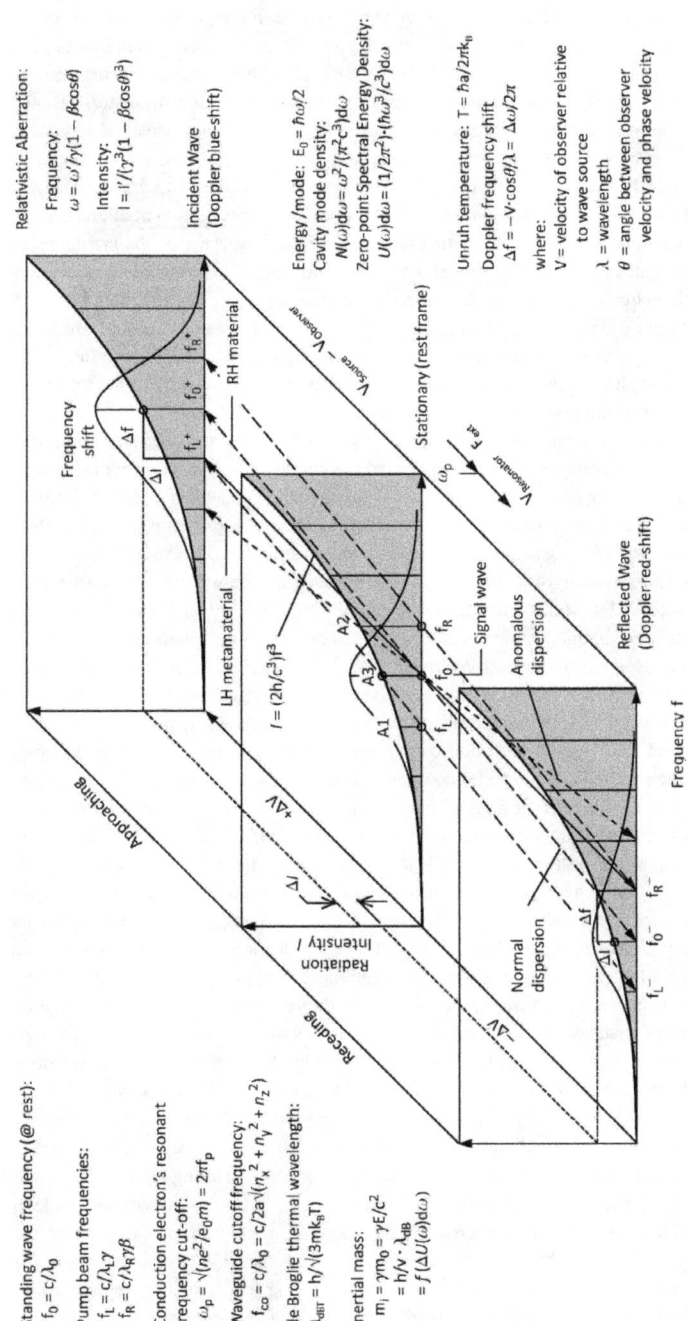

Fig. 15-4. Acceleration induced thermal radiation spectrum of a moving observer compared to a zero-point radiation spectrum.

123

15. Phase Conjugate Resonator Experimental Potential

A resonator with an EM sources, physical walls, sensors, etc made of matter necessarily consist of inherent atomic oscillators with mass and inertia formed from collections of electrons. Matter consists of a collection of synchronized mass oscillators of electrons effectively bound to nuclei linked by EM wave fields. Standing wave resonant coupling occurs only at discrete frequencies. Wave energy trapped within a manmade resonator is miniscule compared to the energy content of constituent atoms of the resonator. Hence, complete elimination of inertia and associated matter waves would not appear theoretically possible without total matter annihilation. Acceleration of electrons results in an opposing electrokinetic force $F_k = qE_k = -q(dA/dt)$ with an induced electric field E_k which varies with the time rate of change of vector potential (dA/dt). Hence, this opposition to acceleration may be identified with Newton's 2nd law of motion ($F = m_i a = \hbar ka/v$) where m_i is the inertial mass. Generation of a cancelling vector potential $-dA/dt$ would require an equivalent number of electrons simultaneously sloshing in the opposite direction to the acceleration vector **a**. Superposition of the wave phasor and its phase conjugate results in cancellation of imaginary (vertical) phase component. Superposition of a wave phasor Z(t) and its negative phase conjugate $-Z^*(t)$ results in cancellation of the real (horizontal) component and the imaginary (vertical) wave energy components.

Referring to the wave phasor diagram of Fig. 15-5, from a symmetry argument, cancellation may conceivably be realized with a phase conjugate resonator combined with a mirror image phase conjugate resonator that simultaneously generates a negative phase conjugate of the wave phasor. During acceleration or deceleration, a negative complex phase conjugate phasor $-Z^*\angle\theta$ is generated countering the positive $Z\angle\theta$ wave phasor. Wave cancellation by superposition of a time varying synthesized matter wave Z(t) and a negative complex conjugate $-Z^*(t)$ seems suggestive of a possibility to effect inertia modification of a man-made resonator (without destruction of matter and conversion into photons).

Motion of a phase-locked cavity resonator consists of an oscillatory sequenced series of accelerative jumps in which the phase conjugate wave phasor is generated and a subsequent coasting (constant velocity) interval allowing system recovery and dimensional stabilization. See Figs. 15-6 and 15-7. For a push-pull cavity, upon application of an impulsive external force, the resonator undergoes an initial compression as the Doppler blue-shifted travelling wave propagates in the direction of motion towards the incident wall. Absorption of the wave energy and back reflection imparts a radiation pressure force of twice the magnitude of the incident force due to momentum reversal. Wave reflection of the red-shifted wave towards the motive wall results in an energy imbalance and consequent pulling force. Such a push-pull resonator cavity is equivalent to a mass spring system. Adding additional energy to the synthesized matter wave via pump beams may allow modification of the apparent effective mass of the overall wave system altering the acceleration response. For example, superposition of an injected out-of phase signal matched to the red-shifted reflected wave and an out-of-phase signal matched to the blue-shifted incident wave would result in a net zero amplitude simulated de Broglie matter wave. Increasing the energy density u within a resonator cavity by irradiation of pump beams during compression by application of an external force increases the internal radiation pressure p ($= u/3$). For a laser optical cavity resonator with collimated EM radiation or a phase-locked, phase conjugate resonator cavity, the radiation pressure equals the energy density ($p = \langle u \rangle$). During compression, the increased energy density and the reduced volume results in increased energy content, temperature, stiffness and increased inertial mass. Likewise, absorption of energy from the resonator cavity during expansion, results in decreased energy density, temperature, stiffness and decreased inertial mass. The effect is similar to response to that of a dynamically-controlled, variable shock absorber or of two coupled absorbers operating out-of-phase.

15. Phase Conjugate Resonator Experimental Potential

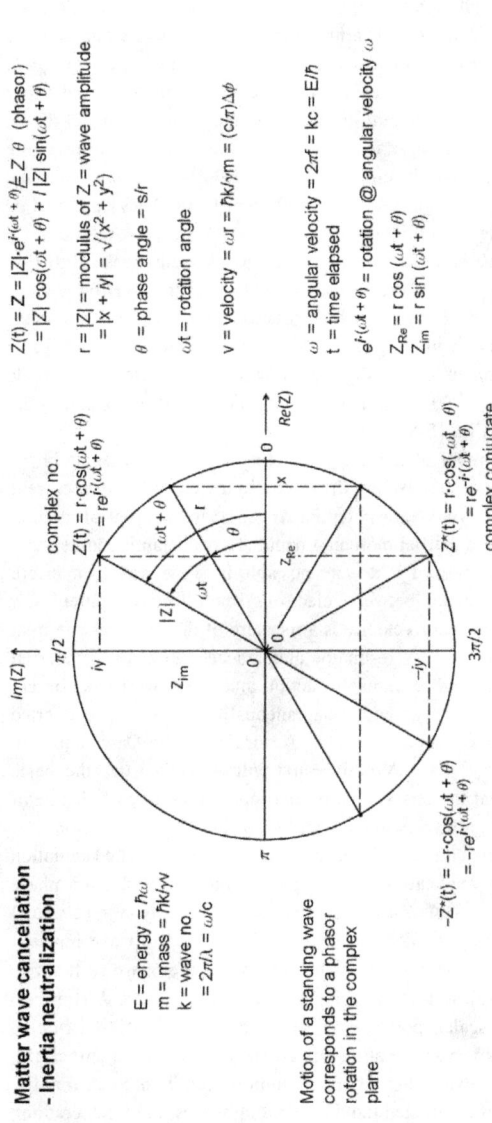

Matter wave cancellation - Inertia neutralization

E = energy = $\hbar\omega$
m = mass = $\hbar k/\gamma v$
k = wave no.
 = $2\pi/\lambda = \omega/c$

Motion of a standing wave corresponds to a phasor rotation in the complex plane

Superposition of phasors $Z(t)$ and $-Z^*(t)$ cancels real and imaginary wave components

$Z(t) = Z = |Z| \cdot e^{i(\omega t + \theta)} = \angle Z \; \theta$ (phasor)
 $= |Z|\cos(\omega t + \theta) + i|Z|\sin(\omega t + \theta)$

$r = |Z|$ = modulus of Z = wave amplitude
 $= |x + iy| = \sqrt{(x^2 + y^2)}$

θ = phase angle = s/r
ωt = rotation angle
v = velocity = $\omega r = \hbar k/\gamma m = (c/\pi)\Delta\phi$
ω = angular velocity = $2\pi f = kc = E/\hbar$
t = time elapsed
$e^{i(\omega t + \theta)}$ = rotation @ angular velocity ω
$Z_{Re} = r\cos(\omega t + \theta)$
$Z_{Im} = r\sin(\omega t + \theta)$

The projection of the rotating vector Z on the real axis represents the physical wave

Fig. 15-5. Illustration of wave cancellation of synthesized matter wave phasor and negative phase conjugate. A standing wave is created by superposition of a traveling wave and a counter-propagating out-of-phase traveling wave. Motion of a standing wave results in blue- and red-shifted traveling wave components creating a de Broglie matter wave. The matter wave exhibits a Lorentz-Doppler frequency shift and Lorentz contraction. The resonator, while in motion, undergoes an oscillatory contraction and elongation. The phasor angle undergoes incremental rotation during accelerative jumps and remains constant during uniform motion.

125

15. Phase Conjugate Resonator Experimental Potential

During decompression while the resonator is elongated, the energy flow is reversed. The energy transfer to and from the phase conjugate resonator is accomplished by parametric amplification/deamplification of the phase conjugate reflector timed to the natural resonator frequency. The energy transfer to or from the resonator acts to oppose or augment the application of external force with internal radiation pressure altering the induced acceleration $a = (F_{ext} - F_{int})/m_i$. A somewhat similar waveguide amplification technique is a tunnel diode reflection amplifier. A basic waveguide circuit consists of an ac signal source, a waveguide circulator, waveguide, tuner, tunnel diode and load. Input power from the ac source flows along through the waveguide, through the circulator, through the tuner into the tunnel diode, the bias of which is adjusted to provide negative resistance. The reflected voltage travelling wave travels through the waveguide, through the circulator which isolates the source from the load to the load. The magnitude of the voltage reflection coefficient Γ_L ($= (V_2/V_1)e^{2\gamma l} = (Z_L - Z_0)/(Z_L + Z_0)$) is a measure of the voltage amplifier gain.

One way to illustrate apparent wave cancellation is to apply a Lissajous phase shift to a phasor diagram. The phase shift corresponds to the difference in phase of the de Broglie wave and a counterpropagating synthesized de Broglie wave. Rotation of $\pi/2$ radians about the imaginary axis of the phasor reduces the magnitude of the $Z(t)$ R_e wave component to zero. In the example shown, the real component corresponding to rest mass energy appears to disappear at a phase shift angle $\phi = 90°$ and $270°$ as the horizontal component lies perpendicular to the image plane. See Fig. 15-8.

The inertia of a collection of N number of moving interacting electrons varies as N^2 as shown by Mead[21] which explains the relatively low drift velocity of electrons in a current carrying wire. Moving charges interact via vector potential **A**. Total momentum of moving electric charges equals the sum of the the inertial momentum and electrodynamic momentum. Inertia modification may, perhaps, be realized if it were possible in some way to interfere with the electromagnetic waves exchanged between electrons or collectively alter their momentum **p** ($= \hbar k - q_0 A = S/c^2$). Resonator velocity is a measure of the difference in total momentum and electrodynamic momentum. A resonator under acceleration by an inertial force represents a state of inertial and Rindler frames as action and subsequent reaction are not simultaneous as the energy cannot be absorbed instantaneously but must be converted within the critical reaction time of the system as a whole. According to the Davis equation for applied force in one direction $F = kx + Vdx/dt + md^2x/dt^2 + Dmd^3x/dt^3$, the basic Newtonian equation is revised such that there is a force proportional to the rate-of-change of acceleration (jerk or surge) in addition to the Newton's force proportional to acceleration.

For a resonator composed of matter, conceivably, irradiation of a phase-locked radiation loop by external phased radiation and application of voltage may induce an electron phase shift á la the Aharonov-Bohm effect $\Delta\phi = q\Phi_B/\hbar$ to effect a change in velocity ($v = (c/\pi)\cdot\Delta\phi$) of the resonator such that $\mathbf{v} = (c/\pi)\cdot q\Phi_B/\hbar = (\hbar k - q_0 A)/\gamma m_i = \hbar k/m_i$. As the wave function ψ of the electron varies with scalar potential ϕ and mass m ($= |V \cdot q|/c^2$), the phase shift varies as $\Delta\phi = -qVt/\hbar$. Hence, a similar result to that of vector potential **A** (magnetic A-B effect) may be realizable with change in scalar potential ϕ (electric A-B effect). Disruptive stress/strain effects of rapid acceleration/deceleration on composite molecular structures, biologic organisms, etc is otherwise known as the 'Humpty Dumpty' problem. Such fragility may be reduced by pulse modulation of contracted moving standing waves, in effect, cradling of atoms and molecules at nodal zones analogous to eggs held in an egg crate with a uniformly applied force. The nodal wavelengths are adjusted to correspond to average molecular spacing intervals in a solid by pulse width compression/modulation thereby lessening excessive bond stress and strains associated with relative molecular movement induced by progressive acceleration-induced compressional shock waves.

15. Phase Conjugate Resonator Experimental Potential

Resonator velocity staircase

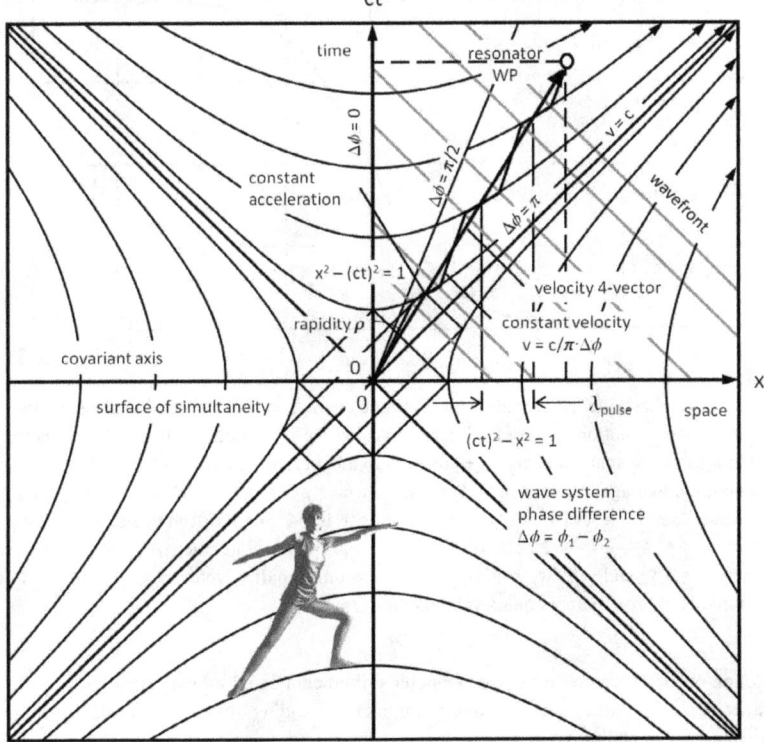

Fig. 15-6. A Minkowski spacetime diagram illustrating a phase-locked standing wave resonator in motion. The velocity 4-vector designates a world point in R^4 with real coordinates $x = (x_0, x_1, x_2, x_3)$ which follows a velocity staircase with periodic acceleration jumps. 4-momentum equals mass x 4-velocity. Hyperbolic rotation represents motion in spacetime with rapidity $\rho = \tanh^{-1} \beta = \beta\gamma$. Proper acceleration is the rate of change of rapidity with respect to proper time. A Lorentz boost (translation without rotation) describes the transformation from the moving time-line of one co-moving body to another in uniform motion (for motion in the x–direction: $x' = \gamma(x - vt)$, $t' = \gamma(t - vx/c^2)$). Uncertainty in position remains invariant under Lorentz boosts (area = $2(ct^2 - x^2)$). The difference between the initial and final endpoints in a gravitational field are related to elastic deformations due to curvature and plastic deformations due to torsion. The relative acceleration between initial and final endpoints due to space-time distortion reflects the rate of change of momentum due to curvature k and change in position due to torsion τ corresponding to **E** and **B** fields.

It is this idea of 'phase-displacement', which when coupled with a distance or 'space-displacement' gives rise to a travelling wave. – Eric Laithwaite

A resonator can not gain energy or lose energy in a continuous manner. It can not gain a fraction of a quantum, it must acquire a whole quantum or none at all. – Henri Poincaré

15. Phase Conjugate Resonator Experimental Potential

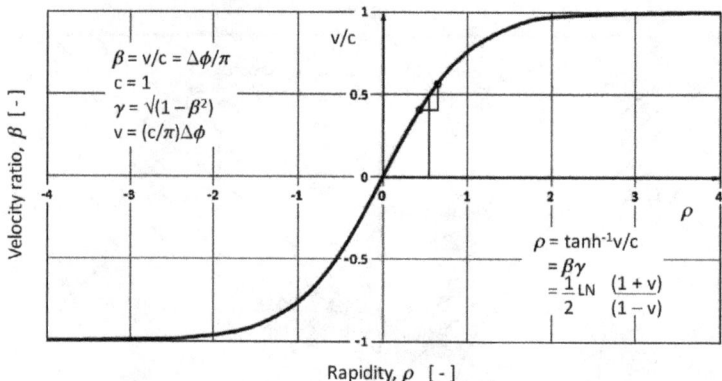

Fig. 15-7. A phase-locked resonator at constant average velocity exhibits an oscillation in rapidity ρ (rate of motion). Rapidity corresponds to the magnitude of the velocity vector or speed in hyperbolic spacetime and is equivalent to a hyperbolic rotation angle. In the example illustrated, the average velocity ratio $\beta = v/c = \Delta\phi/\pi = \rho/\gamma = \sqrt{(1-1/\Gamma^2)} = \cos\theta = \tanh\rho = 0.5$, Lorentz factor $\gamma = \sqrt{(1-v^2/c^2)} = 1/g = \cosh\rho = 1.154$, phase difference between coupled oscillators $\Delta\phi = \phi_2 - \phi_1 = \pi/2$ radians, velocity of light $c = 1$ in geometrized units, velocity of resonator $v = 0.5c$ and rapidity $\rho = \beta\gamma = 0.577$. At unit rapidity, velocity $v = 0.7616c$. For $v \ll c$, rapidity approximately equals velocity ratio $\rho \cong \beta$.

Motion of matter creates a Lorentz Doppler shifted contracted moving standing wave, i.e., a matter wave. Synthesis of a moving contracted standing wave from a standing wave consisting of a signal wave and counter-propagating phase conjugate wave at rest may be generated by nondegenerate four-wave mixing (NDFWM) with simulated Lorentz Doppler shifted red- and blue-shifted waves to produce the inverse effect of induced motion of matter. Parametric amplification of the pump beams provides the energy of motion. This effect is illustrated in Fig. 15-9.

Unlike naturally occurring materials, metamaterials composed of periodic resonant microstructures may be designed to react resonantly out-of-phase with external excitation. Effective negative mass of a coupled resonant system may be realized by careful tuning over a given frequency range as in anti-vibration systems such that a passive damper mass m moves out-of-phase with a spring-coupled active support mass M. A similar effect may be potentially realized with a negative index material (NIM) in a phase-locked standing wave resonator as shown in Fig. 15-10. Application of an external force to the resonator shell results in a Doppler blue-shifted impulse wave. Interaction of the impulse wave with a NIM element exhibiting an inverse Doppler effect results in secondary emission of a Doppler red-shifted wave in the direction of propagation and a reflected Doppler blue-shifted wave in the counter-propagating direction with a 180° phase shift. The resulting displacement of the NIM mass element m is out-of-phase with the PIM resonator mass M. A zero effective mass effect may similarly be produced with zero-index metamaterial (ZIM) with zero permitivitty ϵ and zero permeability μ yielding zero phase variation (rigid body motion) and zero Doppler shift. Effective zero mass of electrons in a Dirac-cone state have been experimentally demonstrated in a superconducting graphene lattice resulting in high speed electron motion without collision. Zero inertial mass corresponds to a condition where $m = (\hbar\mathbf{k} - q_0\mathbf{A})/\mathbf{v}_g = 0$

15. Phase Conjugate Resonator Experimental Potential

Phasor with phase shift rotation

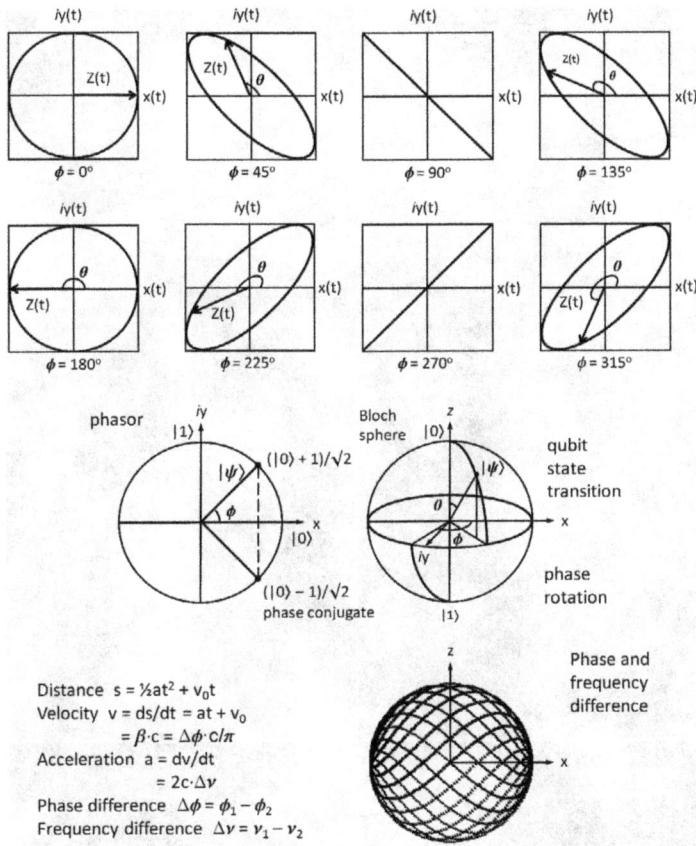

Fig. 15-8. Lissajous-phase shift of a complex wave energy phasor $Z(t)$ with components potential energy $V(t)$ and kinetic energy $T(t)$ where the in-plane phase angle θ represents motion of the standing wave system and the out-of-plane phase angle ϕ is the difference in phase between the de Broglie matter wave and a synthesized counterpropagating de Broglie wave. The interaction between the real and imaginary parts of the wavefunction each with half of the total energy of the wave system describes excitations of the matter wave and energy propagation. The Lagrangian L(t) represents the interaction energy (= T(t) − V(t)).

If the flying arrow is at every instant of its flight at rest in space equal to its length, when does it move? – Zeno of Elea

Mathematics is not real, but it feels real. Where is this place? – Richard Feynman

All fundamental processes are reversible. – Richard Feynman

Electrons are timid little things but notional; you have to let them know who's boss. – R. Heinlein

15. Phase Conjugate Resonator Experimental Potential

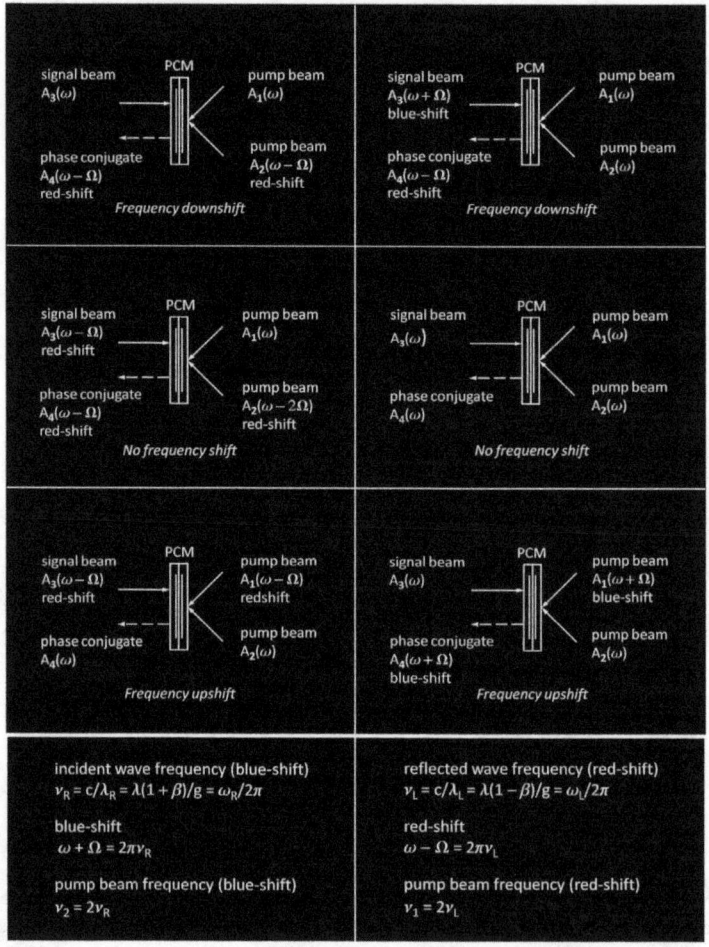

Fig. 15-9. Phase conjugation by degenerate four-wave mixing requires a signal, phase conjugate and two pump beams. Pump waves at simulated Doppler red- and blue-shift frequencies acting on a phase conjugate reflector in a nonlinear medium may be used to modulate a standing wave resulting in a contracted moving standing wave as a means to induce motion of a phase conjugate standing wave resonator. This effect is the inverse of motion of matter generating de Broglie 'matter' waves. Synthesis of matter waves is a straight forward application of phase conjugate 4-way mixing using amplified Lorentz Doppler pump beams to enable self-induced motion without expulsion or reaction mass.

Life is travelling to the edge of knowledge, then a leap taken. – D. H. Lawrence

It is necessary; therefore it is possible. – Giuseppe Antonio Borgese

To fly and find pure etheral substances that are not matter. – Dejan Stoznovic

15. Phase Conjugate Resonator Experimental Potential

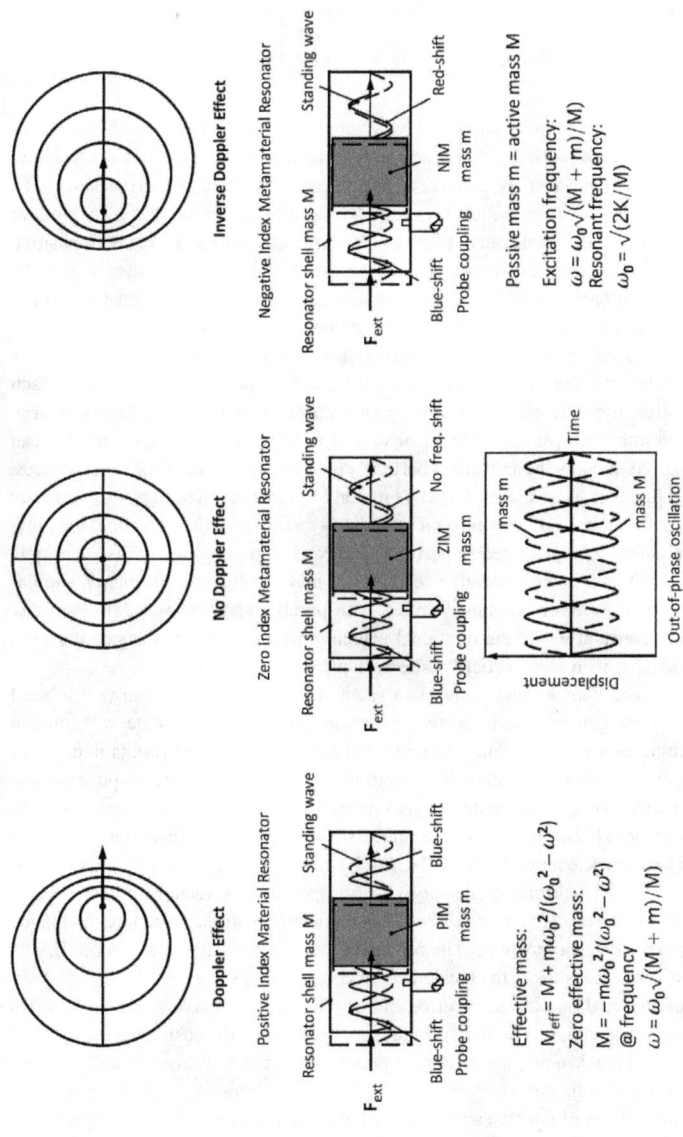

Fig. 15-10. Coupled phase-locked standing wave resonators with positive, negative and zero refractive index with corresponding Doppler effects. Motion is a result of an imbalance of momentum and electrodynamic momentum.

16. Planck Aether

If you would understand the invisible, look carefully at the visible. – Talmud

To deny the ether is ultimately to assume empty space has no physical qualities whatever. The fundamental facts of mechanics do not harmonize with this view. – Albert Einstein

Quite undeservedly, the ether has acquired a bad name. – Frank Wilczek

In the Planck aether hypothesis advanced by Winterberg[7], the ultimate building blocks of matter are Planck mass particles. The vacuum is posited as a two-component superfluid composed of an equal number of positive and negative Planck mass particles with an average volumetric density of one Planck mass (M_P) per Planck volume (l_P^3). A positive mass attracts both positive and negative masses, while a negative mass repels both positive and negative masses. Like positive mass Planck mass M_P^+ repel each other as do like negative mass Planck particles M_P^-. A negative gravitational mass is expected to exhibit negative inertial mass and a negative moment of inertia. For negative mass particles, the direction of acceleration is opposite to the direction of force resulting in repulsion. As described by Winterberg, opposite Planck mass particles, M_P^+ and M_P^- initially attract each other, but on collision decelerate and pass through each other such that the kinetic energy and linear momentum are restored. The positive and negative mass Planck particles can combine into gravitationally bound quasi-particle co- and counter-rotating ring vortices. The vortex core filament when closed forms a toroidal loop (roton). See Figs. 16-1 through 16-6. As shown, the polarized vacuum represented as a quantized BEC superfluid spinning Planck scale dipoles. A quantized superfluid vortex is composed of a Planck dipole consisting of a positive M_P^+ and negative M_P^- Planck mass rotating at the Planck angular frequency ω_P. For a Rankine combined vortex with purely tangential flow, the vorticity within the core is constant where the rotational velocity increases with r. Outside the core, the flow is irrotational with zero vorticity. The vorticity has a discontinuity at r_0.

In the hypothesized Planck mass vortex in a vacuum BEC superfluid, energy displaced from the vortex core generates a high energy density (false) vacuum state with broken symmetry resulting in mass formation. A similar negative energy vortex results in negative mass formation. The Planck vacuum is composed of an equal number of positive and negative mass particles (m_P^+, m_P^-) with net zero mass. The reduced exterior energy density results in gravitational attraction as a result of energy density (quantum pressure) differential. The Planck energy E_P is 1.9561E9 J with an average energy density u_P of 4.63E113 J/m^3. The Planck energy density U_P corresponds to a vacuum pressure P_v (= $U_P/3$) of 1.47 E113 N/m^2. This energy density is associated with aggregations of Planck dipoles with spin angular momentum. The refractive index of a polarizable vacuum K_{PV} (= $e^{\wedge}2GM/rc^2$) is a scalar measure of the local vacuum energy density.

A vacuum of an equal number of positive and negative Planck masses retains the zero-point energy of the vacuum, hence, the average vanishes making the cosmological constant Λ exactly zero. In Planck units, the estimated present age of the visible universe is ~10^{60} Planck time (t_P) and of a size of ~10^{60} Planck length (l_P). The present mass of the universe is ~10^{60} (m_P). Summation of positive and negative Planck masses of the vacuum equals zero ($\sum(m_P^+ + m_P^-) = 0$). The expansion of the universe attributed to dark energy may be understood as the result of continuous creation of Planck dipoles from the vacuum each of which form a gravitationally bound coupled pair of positive and negative mass with a net translation of motion owing to the attraction and repulsion effects of mass polarity.

16. Planck Aether

The energy density (pressure) of the far-field zero curvature polarizable Planck vacuum modeled as a Bose-Einstein superfluid is

$$\rho_P = m_P c^2 / l_P^3 = c^7 / \hbar G^2 \qquad (16\text{-}1)$$

where

ρ_P	= Planck energy density (= 4.63298E113 J/m^3 = kg/m·s = N/m^2) = Planck pressure (= N/m^2)
m_P	= Planck mass (= $\sqrt{(\hbar c/G)}$ = 2.17455E-08 kg)
c	= velocity of light (= $l_P/t_P = c_0/\Gamma = \Gamma \cdot v_p = c_0/\sqrt{(\epsilon/\epsilon_0)(\mu/\mu_0)}$ = 2.99792458E8 m/s)
l_P	= Planck length (= $\sqrt{(\hbar G/c^3)} = mG/c^2 = ct_P$ = 1.616199E-35 m)
$\hbar$	= Dirac constant (= $\Delta p/\Delta mv = h/2\pi$ = 1.0545E-34 J·s = m^2·kg·s^{-1})
G	= Newton's gravitational constant (= $c^3/Z_S = \hbar c/m_P^2$ = –6.6726E-11 N·m^2·kg^{-2})

The mass density (ρ_m) of matter equals the energy density E within a volume V expressed as $\rho_m = (mc^2)/V$ with positive mass energy with respect to the Planck energy and equal to the negative gravitational field energy. In a black hole of radius equal to the Schwarzschild radius R_S (= $2Gm/c^2$), the volume is maximally compressed at Planck pressure. A black hole corresponds to a maximally rigid arrangement of impenetrable Planck masses.

The probability amplitude of finding a mass particle at a given location **r** is represented as the square of the wave function $|\psi|^2$ where $\psi(r) = \langle r|\psi \rangle = Ae^{i\phi}$. For mass particles taken as density fluctuations of Planck dipoles in a BEC superfluid then $A^2(\mathbf{r})$ corresponds to the local mean density of dipoles $\langle \rho_P \rangle$. In the Sakharov-Puthoff model[43], it is the perturbation of the zero-point field (ZPF), generated by Planck oscillators in the presence of matter, that results in gravitational force. Owing to the extreme vacuum energy density (pressure), the perturbation distance is on the order of the dynamic Planck length L_P (= 1.616 x 10^{-35} m). The ZPF itself does not gravitate and does not contribute to the overall curvature of the universe. The energy density of the vacuum is given as $\rho_E = (\hbar/8\pi^2 c^3)\omega_{co}^4$ where ω_{co} is an interaction cutoff frequency taken as on the order of the Planck frequency ω_P (= $\sqrt{(c^5/\hbar G)}$ ≈ 1.855 x 10^{43} s^{-1}).

A Planck dipole pair corresponds to a superposition of quantum mechanical states that can be treated as binary digital bit 0 or 1 or both at the same time. This property represents a quantum bit or qbit. Information represents the number of distinguishable wave patterns of dipoles in a given volume of space. One expression of information is I = k_B log 2 = (E/T) log 2 where k_B = Boltzman's constant, E = Energy and T = Temperature. Planck dipoles seemingly are immutable and inseparable with individual masses of opposite polarity that cannot merge into one. An ensemble of multiple dipoles can collapse into 2^n states. Interactions represent an information encoding and decoding process. The electric field **E** is conjectured to represent a momentary phase alignment of Planck dipoles while the magnetic field **B** represents a coaxial spin axis alignment. The presence of an electric or magnetic field corresponds to transformation into aligned states. Planck particles of positive and negative mass are irreducible and having opposite curvatures and energy. The complementary yin/yang geometry of a coupled rotating dipole corresponds to a gamma pulse in the complex plane. Rotation in phase space follows a path of constant Hamiltonian with evolution described by a Fermat limit cycle spiral. The dyadic yin/yang geometry reflects fundamental similarities to long held philosophical theories of Zou Yan, Pythagoras, Leibnitz, Tesla and others. A rotating source/sink pair is suggestive of a ouroboros flux tube link oriented along the Planck mass spin axis closing each circuit as two interconnected loops with a full twist in the form of a Hopf link fiber bundle.

16. Planck Aether

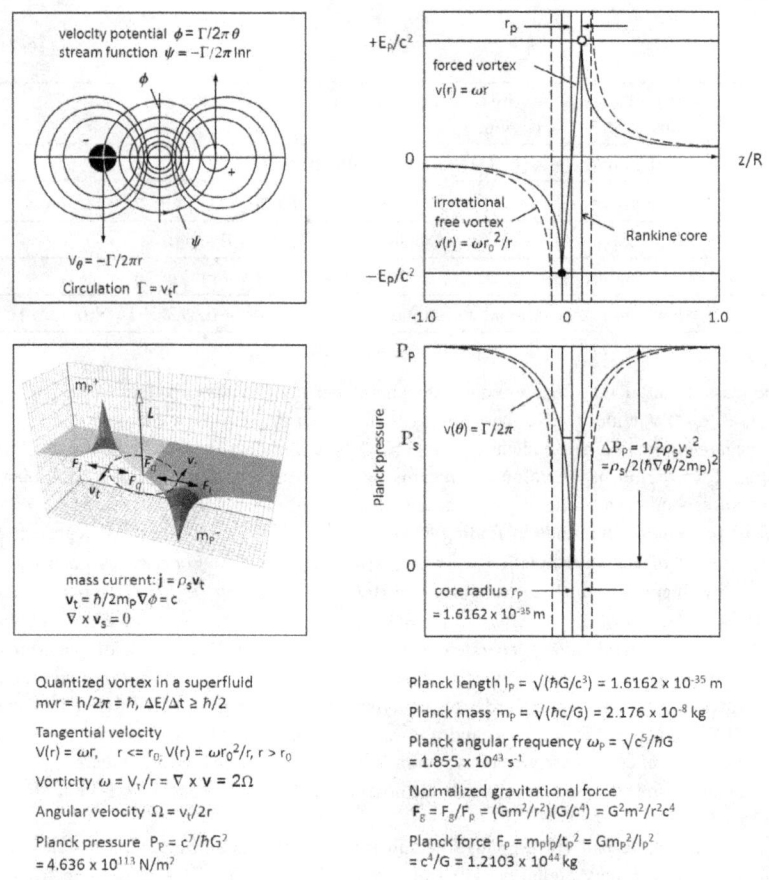

Fig. 16-1. Polarized vacuum represented as a quantized BEC superfluid spinning Planck scale dipoles. Chirality reversal is equivalent to mass inversion (m → −m).

How did the universe start? Based on statistical thermodynamics theory, Ludwig Boltzman proposed that the universe began as a chance fluctuation in a state of minimal entropy. How did this happen? We might consider the expansion of the universe as a transient response describable in terms of a transfer function T(s) characterizing the spatial or temporal frequency of the system's natural response Y(s) to an input forcing function R(s). A true vacuum or absolute nothingness may represent an unstable condition as a resonant oscillation between two untenable states of +∞ and −∞ in an n-dimensional Hilbert space and in a runaway positive feedback loop ultimately splitting into two opposite extremes (+∞ to 0, 0 to −∞) of wavefunction energy and curvature (Cornu spiral), i.e., the Big Bang creation of the existing universe of dualities. This sublimation sequence of a unity of opposites somewhat resembles the Hegel dialectic of thesis, antithesis → synthesis. Positive and negative Planck masses may be considered as energy storage elements of the quantum vacuum. A Planck dipole corresponds to a critically damped two-element system

16. Planck Aether

with a natural frequency response described by a second order differential equation with real roots. The primordial universe before the Big Bang in this model corresponds to an RLC circuit with zero resistance. In the complex frequency plane with R = 0 and damping coefficient $\alpha = 0$, the response is oscillatory with no damping. The characteristic equation roots are located on the imaginary axis with roots $\pm\omega_n$. The Big Bang may, perhaps, represent a transition from an open loop system with open loop poles to a closed loop system with closed loop poles. The root locus is a curve of the location of the poles of a transfer function in the complex frequency s-plane as some parameter, typically gain K, is varied. Stability is normally described when the real part of the pole is to the left of the imaginary axis ($-\sigma_x$). At the singularity event, in this model description, the universe may have undergone a $\pi/2$ phase shift with poles rotated to the real axis with circular arc radius ω_n. The natural frequency ω_n is taken as the Planck cut-off frequency ($\omega_n = \omega_P = \sqrt{(c^5/\hbar G)}$). Continued expansion of the universe may involve continuous creation of positive and negative Planck mass dipoles without further energy input (1 = −1 or 2 = 0).

As shown, for a positive m^+ and negative mass m^- dipole, the center of mass of the two-body system lies at the center of rotation of the orbiting pair system. The orbital path traces out a circular helix corresponding to Dirac spinors. Vortex rings constitute a current of Planck masses. Planck charge is a spin precession effect. The spin precession frequency ω_p (= q_P/m_P = 8.6193E25 rad/s) is much less than the Planck angular frequency ω_P (= $\sqrt{(c^5/\hbar G)}$ = 1.855E43 rad/sec. Planck charge to electric charge ratio $q_P/q_e \approx 11.709$. Overall there is an equal number of positive and negative Planck charges with the consequence that the critical energy density parameter $\Omega_\Lambda = 1$ and the cosmological constant $\Lambda = 0$ (= $\Omega_\Lambda - 1$).

An electronic example of a yin/yang geometry is found in analysis of L-type circuits employing two reactive elements for transforming a complex load impedance Z_l into a pure resistance Z_0 as represented in Smith charts of network impedance. The impedance path is restricted to one comma-shaped portion in a complex plot of resistance Z_0 vs. complex frequency $j\omega$. The complementary comma-shaped portion indicates a forbidden area where the load impedance cannot be transformed into a pure resistance. Representative LC networks for transforming complex impedances are shown in Fig. 16-7. A positive Smith chart which is used for evaluation of networks and waveguides with positive resistance has a real axis of +Re/Z_0 extending from 0 to ∞. A negative Smith chart used for negative resistances is a mirror image with resistance −Re/Z_0 ranging from ∞ to 0.

The most puzzling aspect of space is its very existence. – Lee Smolin

There is still a difference between something and nothing, but it is purely geometrical and there is nothing behind the geometry. – Martin Garner

There is nothing in the world except empty curved space. – John Archibald Wheeler

Because there is a law such as gravity, the Universe can and will create itself from nothing. – Stephen Hawking

Some months ago we discovered that certain light elements emit positrons under the action of alpha particles. – Frederic Joliot-Curie

[From uranium] there are present at least two distinct types of radiation one that is very readily absorbed, which will be termed the α radiation, and the other of a more penetrative character, which will be termed the β radiation. – Sir Ernest Rutherford

[Radium emits electrons with a velocity so great that] one gram is enough to lift the whole of the British fleet to the top of Ben Nevis; and I am not quite certain that we could not throw in the French fleet as well. – Sir William Crookes

16. Planck Aether

Planck mass vacuum vortex

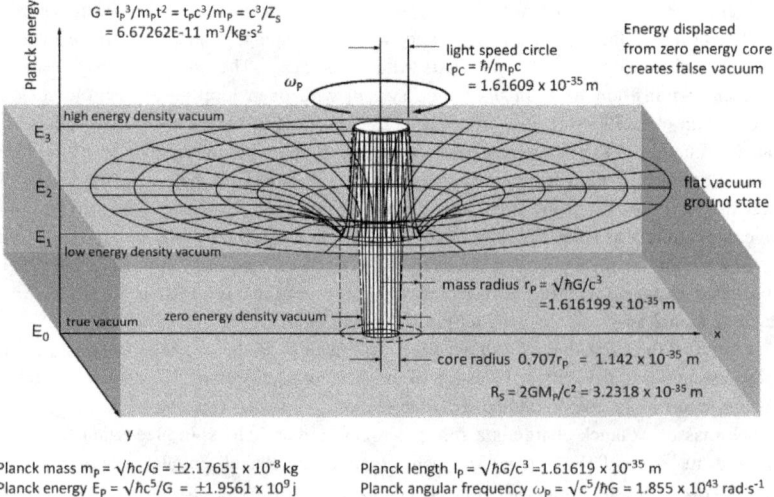

Planck mass $m_p = \sqrt{\hbar c/G} = \pm 2.17651 \times 10^{-8}$ kg
Planck energy $E_p = \sqrt{\hbar c^5/G} = \pm 1.9561 \times 10^9$ j
Planck length $l_p = \sqrt{\hbar G/c^3} = 1.61619 \times 10^{-35}$ m
Planck angular frequency $\omega_p = \sqrt{c^5/\hbar G} = 1.855 \times 10^{43}$ rad·s^{-1}

Fig. 16-2. Planck mass potential vortex in a Planck vacuum BEC superfluid.

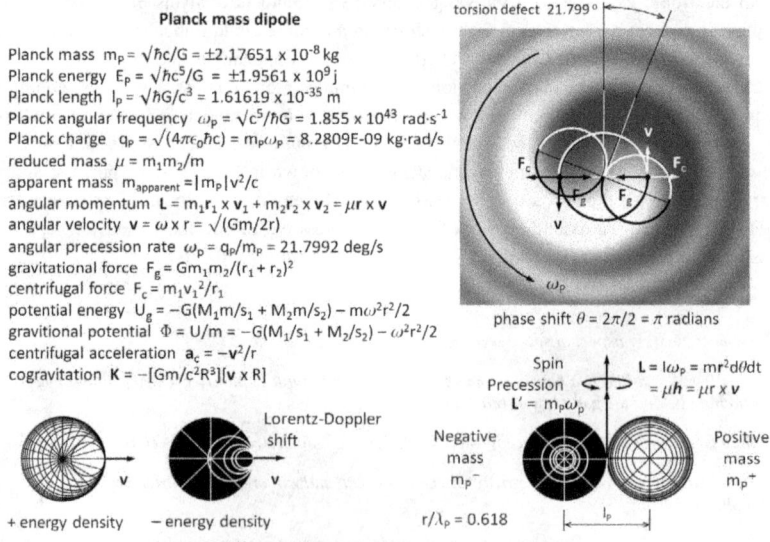

Planck mass dipole

Planck mass $m_p = \sqrt{\hbar c/G} = \pm 2.17651 \times 10^{-8}$ kg
Planck energy $E_p = \sqrt{\hbar c^5/G} = \pm 1.9561 \times 10^9$ j
Planck length $l_p = \sqrt{\hbar G/c^3} = 1.61619 \times 10^{-35}$ m
Planck angular frequency $\omega_p = \sqrt{c^5/\hbar G} = 1.855 \times 10^{43}$ rad·s^{-1}
Planck charge $q_p = \sqrt{(4\pi\epsilon_0 \hbar c)} = m_p \omega_p = 8.2809\text{E-}09$ kg·rad/s
reduced mass $\mu = m_1 m_2/m$
apparent mass $m_{apparent} = |m_p|v^2/c$
angular momentum $\mathbf{L} = m_1 \mathbf{r}_1 \times \mathbf{v}_1 + m_2 \mathbf{r}_2 \times \mathbf{v}_2 = \mu \mathbf{r} \times \mathbf{v}$
angular velocity $\mathbf{v} = \omega \times r = \sqrt{(Gm/2r)}$
angular precession rate $\omega_p = q_p/m_p = 21.7992$ deg/s
gravitational force $F_g = Gm_1m_2/(r_1 + r_2)^2$
centrifugal force $F_c = m_1 v_1^2/r_1$
potential energy $U_g = -G(M_1m/s_1 + M_2m/s_2) - m\omega^2 r^2/2$
gravitional potential $\Phi = U/m = -G(M_1/s_1 + M_2/s_2) - \omega^2 r^2/2$
centrifugal acceleration $\mathbf{a}_c = -v^2/r$
cogravitation $\mathbf{K} = -[Gm/c^2 R^3][\mathbf{v} \times R]$

Fig. 16-3. Rotating coupled Planck mass dipole of interacting positive and negative mass. Faraday electric and magnetic field lines are composed of strings of Planck mass dipoles.

The magnetic energy is kinetic energy of a medium occupying the whole of space, and that electric energy is the energy of strain of the same medium. – William H. Thomson

16. Planck Aether

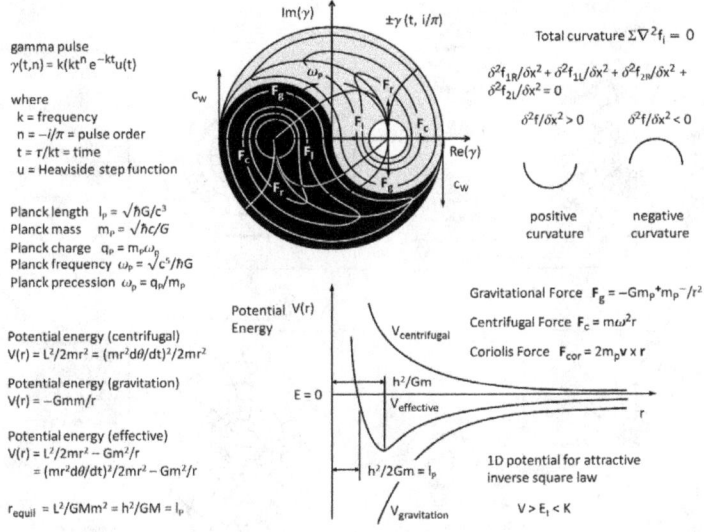

Fig. 16-4. Spin dynamics of a rotating Planck dipole of positive and negative mass. Planck charge q_P arises as a result of spin precession ω_p where $\omega_p \ll$ Planck angular frequency ω_P.

Everything [∞] is nothing [0], with a twist. – Kurt Vonnegut

0 is not the immediate successor of any number. – Giuseppe Peano

I hold in fact
(1) That small portions of space are in fact of a nature analogous to little hills on a surface which is on the average flat, namely, that the ordinary laws of geometry are not valid in them.
(2) That this property of being curved or distorted is continually being passed on from one position to another after the manner of a wave.
(3) That this variation of curvature of space is what really happens in that phenomena which we call the motion of matter, whether ponderable or etheral.
(4) That in the physical world nothing else takes place but this variation, subject possibly to the law of continuity. – William Kingdon Clifford

Under the influence of elastic forces, the electrons can vibrate about their positions of equilibrium. In doing so, and perhaps also on account of other more irregular motions, they become the centres of waves that travel outwards in the surrounding ether and can be observed as light if the frequency is high enough. In this manner we can account for the emission of light and heat. – Hendrik Lorentz

I cannot but regard the ether, which can be the seat of an electromagnetic field with its energy and vibrations, as endowed with a certain degree of substantiality, however different it may be from all ordinary matter. – Hendrik Lorentz

By being set in movement in this fluid, the ether, becomes gross matter. Its movement arrested, the primary state reverts to its normal state. It appears, then, possible for man through harnessed energy of the medium and suitable agencies for starting and stopping ether whirls to cause matter to form and disappear. To cause the birth and death of matter would be man's greatest deed, which would give him mastery of physical creation, make him fulfill his ultimate destiny. – Nikola Tesla

Ere many generations pass, our machinery will be driven by power obtainable at any point in the universe. It is a mere question of time when men will succeed in attaching their machinery to the very wheelwork of nature. – Nikola Tesla

It seems the stability of matter itself depends on the zero point energy sea. – H. E. Puthoff

16. Planck Aether

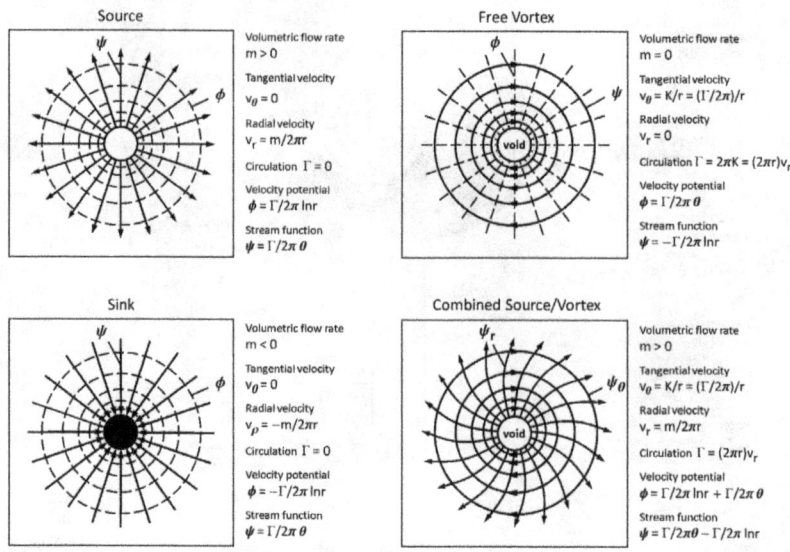

Fig. 16-5. Illustration of source, sink, free vortex and source/vortex characteristics.

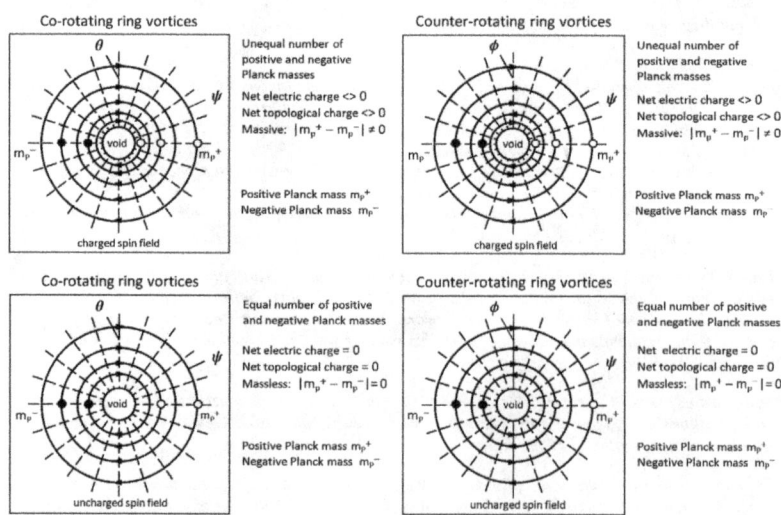

Fig. 16-6. Comparison of source sink flow and a pair of co-rotating and contra-rotating Planck mass vortices in a model of the Planck aether advanced by Winterberg[7].

... the physical origin of "space curvature" is associated with matter in Einstein's General Theory of Relativity is explained by the internal circulation of a fluid in a vortex ring. – Ernest J. Sternglass

16. Planck Aether

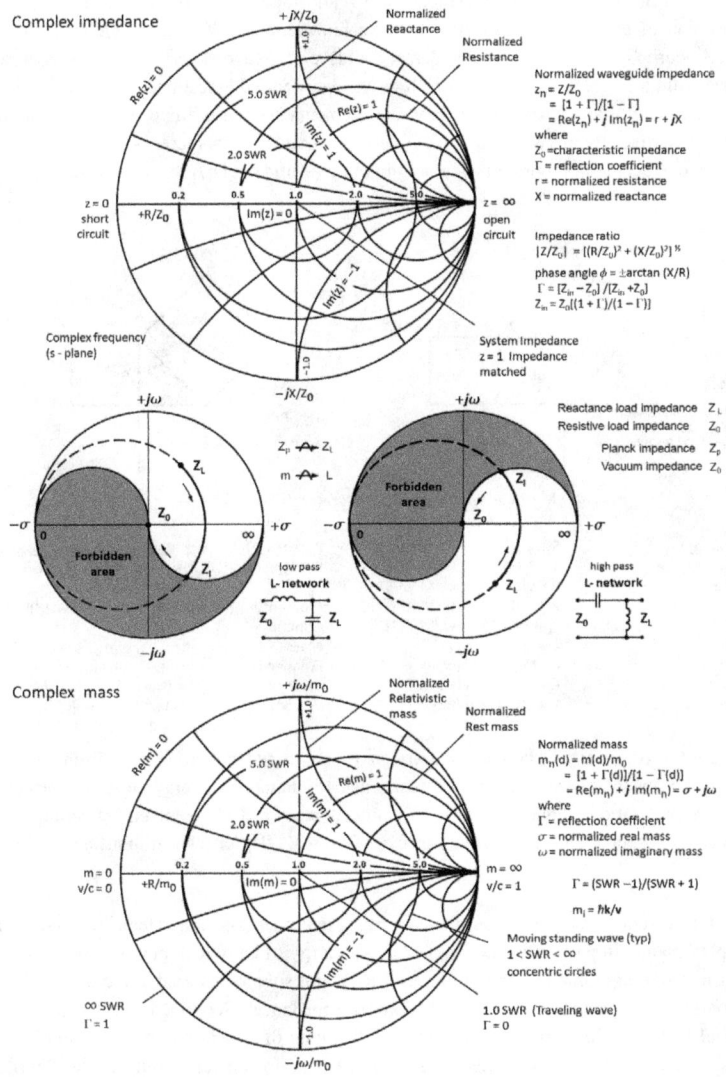

Fig. 16-7. Smith chart representation in the complex frequency plane of impedance matching characteristics of an L-type circuit. Transformation of complex load impedance Z_L into a resistive load impedance Z_0 is indicated by the load impedance vector. The forbidden area of transformation is denoted by the shaded area. An LC circuit is a lumped parameter equivalent to an electromagnetic cavity resonator. Inductive reactance X_L (= $2\pi fL$) is a measure of opposition to a change in energy flow as is inertial mass m_i (= $\hbar k/v$ = E/c^2 = |volt x charge|/c^2). Ergo, resonator mass is analogous to waveguide impedance.

Nothing takes place in the world whose meaning is not that of some maximum or minimum. – Leonard Euler

139

16. Planck Aether

An energy triangle bears resemblance to an impedance triangle. Mass represents obstruction of change in energy flow as does impedance. Altering the environment of a wave system by increasing energy density adds effective mass. Reconfiguring a wave system from a travelling wave to a standing wave system adds rest mass. Reconfiguring an RLC circuit by adding resistive or reactive elements increases impedance and collective electron inertia. Relativistic mass increase corresponds to an increase in impedance and resistance to acceleration proportional to gamma γ (= $dt/d\tau = 1/\sqrt{(1 - v^2/c^2)}$). See Fig 16-8.

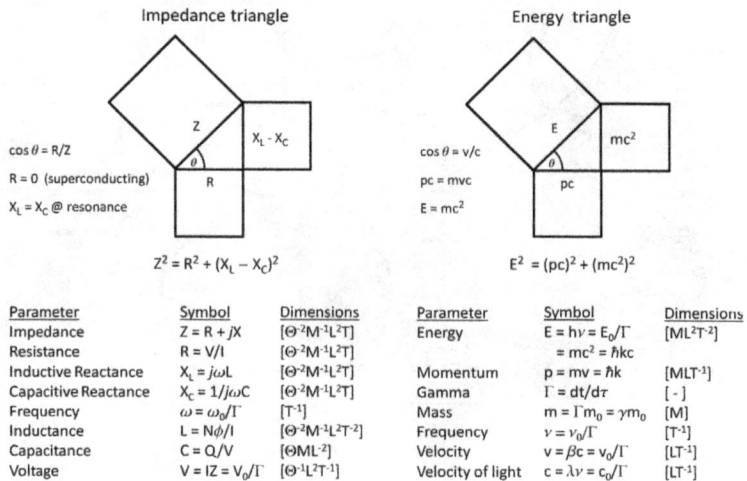

Parameter	Symbol	Dimensions	Parameter	Symbol	Dimensions
Impedance	$Z = R + jX$	[$\Theta^{-2}M^{-1}L^2T$]	Energy	$E = h\nu = E_0/\Gamma$	[ML^2T^{-2}]
Resistance	$R = V/I$	[$\Theta^{-2}M^{-1}L^2T$]		$= mc^2 = \hbar kc$	
Inductive Reactance	$X_L = j\omega L$	[$\Theta^{-2}M^{-1}L^2T$]	Momentum	$p = mv = \hbar k$	[MLT^{-1}]
Capacitive Reactance	$X_C = 1/j\omega C$	[$\Theta^{-2}M^{-1}L^2T$]	Gamma	$\Gamma = dt/d\tau$	[-]
Frequency	$\omega = \omega_0/\Gamma$	[T^{-1}]	Mass	$m = \Gamma m_0 = \gamma m_0$	[M]
Inductance	$L = N\phi/I$	[$\Theta^{-2}M^{-1}L^2T^{-2}$]	Frequency	$\nu = \nu_0/\Gamma$	[T^{-1}]
Capacitance	$C = Q/V$	[ΘML^{-2}]	Velocity	$v = \beta c = v_0/\Gamma$	[LT^{-1}]
Voltage	$V = IZ = V_0/\Gamma$	[$\Theta^{-1}L^2T^{-1}$]	Velocity of light	$c = \lambda \nu = c_0/\Gamma$	[LT^{-1}]

Fig. 16-8. An energy triangle comparison to an impedance triangle. Both mass and electrical impedance are measures of resistance to change in energy flow. Fibonacci and Pythagorean ratios represent non-resonant (damped) and resonant standing waves conditions, respectively. The ubiquitous c^2 (= E/m = $\mathbf{E}^2/\mathbf{B}^2$) term has dimensions [J/kg].

A further comparison may be made by representing mass rather than impedance on a complex plane. In particular, the mass or mass-energy of an object may be represented as a point in a stereographic projection of a unit Riemann sphere centered at the origin. This is analogous to projection of a 3-D Smith chart of impedance. See Fig. 16-9. Latitude circles parallel to the complex plane represent loci of points of constant rest mass. For positive mass spheres, rest mass is limited to a range of 0 to $+\infty$ corresponding to the South and North poles, respectively. Applying a rotation $z \mapsto e^{i\alpha}z$, corresponds to an elliptic Möbius transformation. Expansion or contraction away from or towards the poles along longitude lines equivalent to mass expulsion or accretion corresponds to a hyperbolic Möbius transformation. A family of invariant circles centered at N = ∞ is equivalent to constant velocity motion and corresponds to a parabolic Möbius transformation. A rotation and expansion generates spiral curves which converge to the N pole (acceleration) or to the S pole (deceleration) is equivalent to relativistic acceleration corresponding to a loxodromic Möbius transformation. The stereographic image of a line on the complex plane is a circle on the sphere passing through the N pole = ∞. The circle marks the intersection of a cutting plane at an angle to the complex plane. The line corresponds to the wave vector $\mathbf{k}$ of a freely-propagating travelling wave in a zero curvature vacuum. Similarly, the projection of

16. Planck Aether

the intersection of a cutting plane parallel to the complex plane corresponds to the wave vector **k** of a closed-loop standing wave which has positive rest mass. Circles on the Riemann sphere denote a standing wave resonator. Latitude circles on a Riemann sphere correspond to resonators at rest while inclined circles on a rotating Riemann sphere correspond to moving standing wave resonators. Negative real mass in analogy to a negative Smith chart may be represented as $-Re/M_0$. Similarly, imaginary mass is identified as $+Im/M_0$ while negative imaginary mass is identified as $-Im/M_0$. The characteristic wavelength corresponds to the center point of a given mass circle.

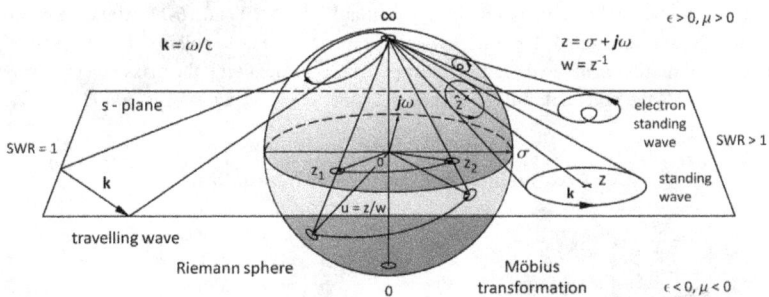

Fig. 16-9. Complex mass represented on a Riemann sphere projected onto a complex plane. Mass oscillators have associated characteristic wavelengths that may be represented in the complex frequency plane. Matter in motion generates de Broglie matter waves with defined propagation characteristics that may be similarly plotted. Complex mappings in the complex plane in the form of Möbius transformations correspond to Lorentz transformations. A spherical radiation pattern of photons of light rays emanating from an isotropic point source correspond to a Riemann sphere with rays of light that may be represented as a complex number. Motion corresponds to rotation in the complex plane.

Moving particles generate matter waves characterized by the de Broglie wavelength $\lambda_{dB} = 2\pi\hbar c/mv = h/p$ with a frequency $\nu_{dB} = E/2\pi\hbar$. Matter waves propagate with a wave phase velocity $v_{dB} = c^2/v$ and an angular frequency of $\omega_{dB} = mc^2/\hbar$. De Broglie waves are generated as a result of contraction of a standing wave in motion. In terms of the Planck aether hypothesis, the de Broglie wave is a propagating phase wave formed by rotation of Planck dipoles of positive m_P^+ and negative mass m_P^- in the wake of the moving particle.

The expansion of the universe has been theorized to be the result of a repulsive gravity effect of purported dark energy. Neither Newtonian or relativistic gravitation precludes the existence of negative mass, however, discrete physical non-void negative mass objects are not observed. Negative mass in nature appears, for example, in the form of mass deficit of nuclei due to mass binding energy, in fluid bubbles due to mass displacement, low density 'dark' solitons in ultracold superfluids, and polaritons (combinations of ½ light and ½ matter) created by interaction of standing light waves in optical microcavities with quasi-particle excitons (electrons and electron holes). Assuming potential flow, the added inertial mass of a moving bubble is ½ the mass of displaced liquid. Objects corresponding to potential energy holes such as quasi-particle bubbles in a liquid, phonons in a crystal lattice, laser pulses in a nonlinear optical mesh lattice of two coupled fiber optic loops, exhibit effective negative mass $(1/(d^2E/dp^2) = \hbar/(d^2\omega/dk^2))$. In a semiconductor crystalline lattice, for example, such as silicon, the effective mass of the electron is ~14% less than in a

16. Planck Aether

vacuum and in gallium arsenide ~7% less. Conceptually, in Newtonian mechanics, three types of masses may be defined: inertial mass m_i, active gravitational mass m_a and passive gravitational mass m_p. A positive active gravitational mass attracts both positive and negative masses whereas a negative active gravitational mass repels both positive and negative masses. The Gravitation constant is positive as mass density is positive and gravitation is attractive. A negative inertial mass reacts with opposite momentum in accordance with Newton's second law. A proximate positive effective mass and a negative effective mass, if equally mobile, move in tandem as if mechanically coupled through action-reaction symmetry breaking. The initial separation upon phase-lock remains constant but decreases at relativistic velocities. Conceptual diagrams of equivalence of positive and negative gravitational and inertial mass is shown in Figs. 16-10 and 16-11. Relative motion and forces are illustrated for various combinations of mass polarity (++,+−,−−,−+). Direction of motion is in the direction of the potential gradient with the mass object or wave source towards the potential minima or maxima (standing wave anti-nodes or nodes).

Positive and negative mass interaction

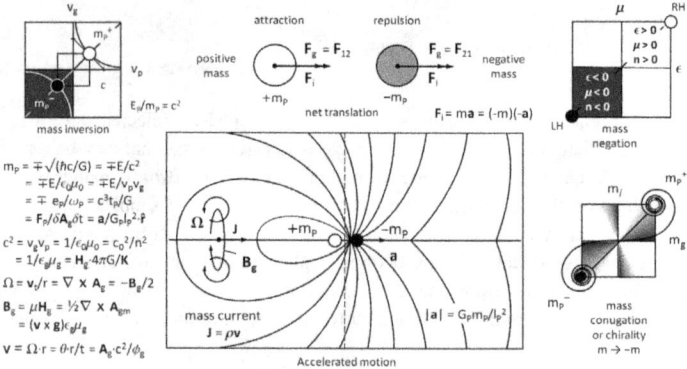

Element	Gravitational Force	Inertial force	Momentum
Positive mass m_P^+	$F_g = Gm_P^2\,\hat{r}/r^2$	$F_i = +m_P^+ a$	$P = +m_P^+ v$
Negative mass m_P^-	$F_g = -Gm_P^2\,\hat{r}/r^2$	$F_i = -m_P^- a$	$P = -m_P^- v$

Fig. 16-10. Hypothetical positive and negative Planck mass interaction forming a gravitational dipole with zero net gravitational mass and inertia. Since energy and mass are equivalent, negative energy implies negative mass ($= -m_0 c^2$). For negative mass, acceleration is opposite to applied force. The inertial mass of the negative mass remains positive, i.e., $m_i = |m_g|$. For a freely falling body near the surface of the Earth, inertial mass is equal to the gravitational mass ($m_i \mathbf{a} = m_g \mathbf{g}$). Negative gravitational mass is tantamount to negative inertial mass. Energy and momentum remain conserved in a push/pull interaction.

Only the future can reveal the ultimate use to which humans will put the remaining fire of the gods, the quantum fluctuations of empty space. – Harold E. Puthoff

There are very few things which we know, which are not capable of of being reduc'd to a Mathematical Reasoning; and when they cannot it's a sign our knowledge of them is very small or confus'd... – John Arbuthnot

16. Planck Aether

According to the JANUS bimetric cosmological model by Jean-Pierre Petit, the possibility of a diffuse negative mass energy surrounding positive mass energy objects appears as a plausible explanation for the observed expansion of the universe and appearance of globular voids in clusters of stellar clusters as well as observed galactic rotation velocity curves. Remarkably, this model obviates need for dark energy and dark matter ad hoc hypothesis.

Positive and negative gravitational and inertial mass comparison

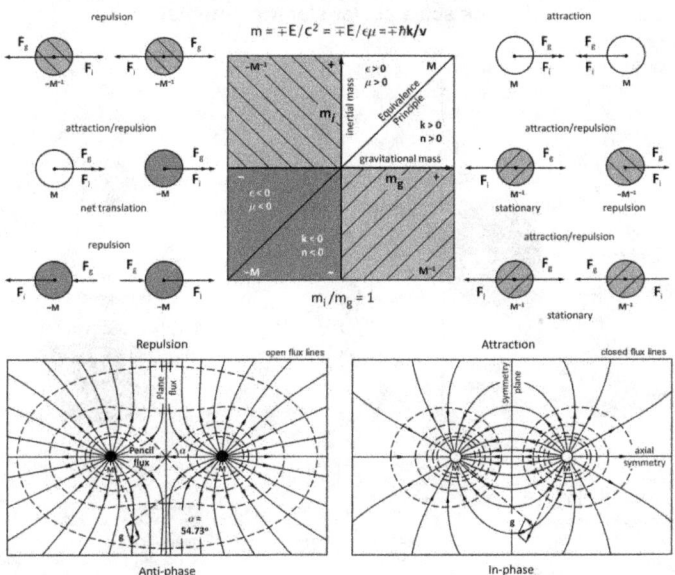

Fig. 16-11. Conceptual diagram of equivalence of positive and negative gravitational and inertial mass. Different combinations of mass dipoles yield net attraction (positive gravitational mass and positive inertial mass + positive gravitational mass and positive inertial mass), net repulsion (negative gravitational mass and negative inertial mass + negative gravitational mass and negative inertial mass) or net translation (positive gravitational mass and positive inertial mass + positive gravitational mass and negative inertial mass; positive gravitational mass and positive inertial mass + negative gravitational mass and negative inertial mass). A negative mass is equivalent to a negative energy wave packet in uniform motion with oppositely directed velocity and momentum. Negative mass corresponds to a volumetric mass deficit in the surrounding medium (lattice defect). Negative mass energy = negative Lagrangian ($L < 0$) where the electrostatic energy $E_{s0} <$ magnetostatic energy $E_{m0} \rightarrow \mathbf{E}^2 < c^2\mathbf{B}^2$. An inverse whispering gallery mode (decreasing radial frequency gradient) exhibits negative pressure analogous to repulsive gravity.

All mass is interaction. – Richard P. Feynmann

The notion of mass although fundamental to physics, is still shrouded in mystery. – Max Jammer

Negative pressure produces repulsive gravity, and that's the secret of what makes inflation possible. – Alan Guth

16. Planck Aether

Electric flux lines ϕ_E are interpreted as a phase alignment of Planck dipoles formed from positive and negative Planck masses. Alignment results in tension along the flux lines due to attraction of positive and negative masses and compression perpendicular to the flux lines. The strength of the electric field intensity **E** is a function of the product of Planck electric dipole moment μ_{qP} and Planck mass m_P. Electric flux lines are an emergent result from superposition of potentials. Vector potentials **A** arise from motion of scalar potentials ϕ. Spinning dipole motion results in a tight spiral. This effect is illustrated in Fig. 16-12.

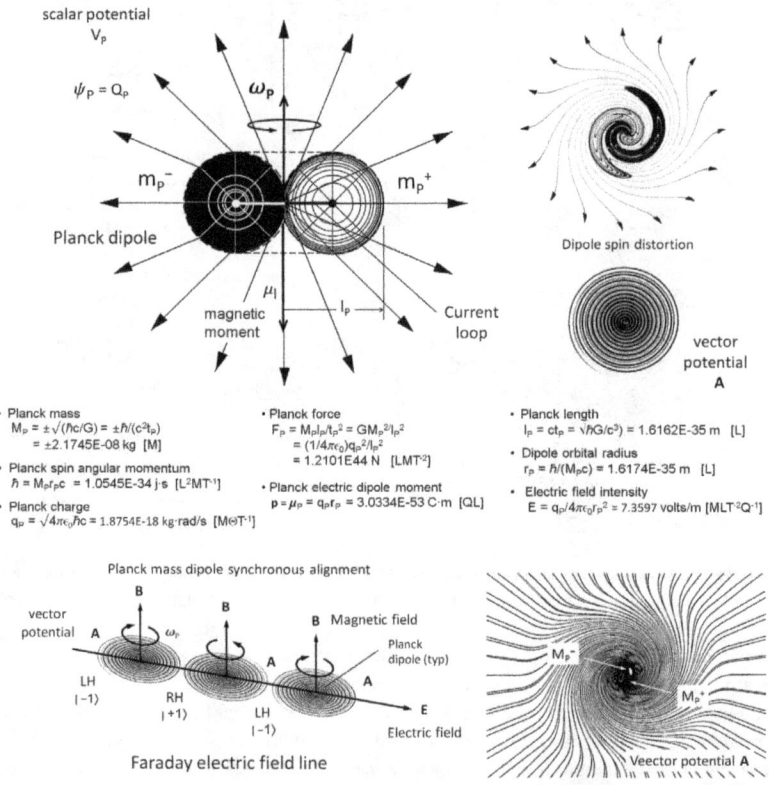

Fig. 16-12. Faraday electric field lines **E** are postulated to result of synchronous alignment of multiple Planck dipoles with charge dipole moment (= $\mu_P m_{qP}$).. Electric field strength is a measure of the difference in number of charges between two points. Planck charge q_P arises a result of Planck dipole spin precession, a torsion defect. Only scalar potentials exist below the Planck limit whereas electric **E** and magnetic **B** field lines are only manifest above the Planck limit. A positive Planck mass m_P^+ has positive energy density while a negative Planck mass m_P^- has negative energy density and are mutually incompatable. Opposite Planck electric charges attract while opposite Planck masses repel. Superposition of opposite Planck masses and charges results in net zero charge and mass in the far field.

16. Planck Aether

Magnetic flux lines ϕ_B are interpreted as vortex filaments of rotating Planck mass dipoles in the Planck vacuum. The magnetic field intensity **H** is a measure of the vortex strength of entrained dipoles. See Fig. 16-13. The **E** field and **H** field vectors are mutually perpendicular and orthonormal to the direction of energy propagation represented by the Poynting vector **S** = **E** x **H** according to the right hand rule.

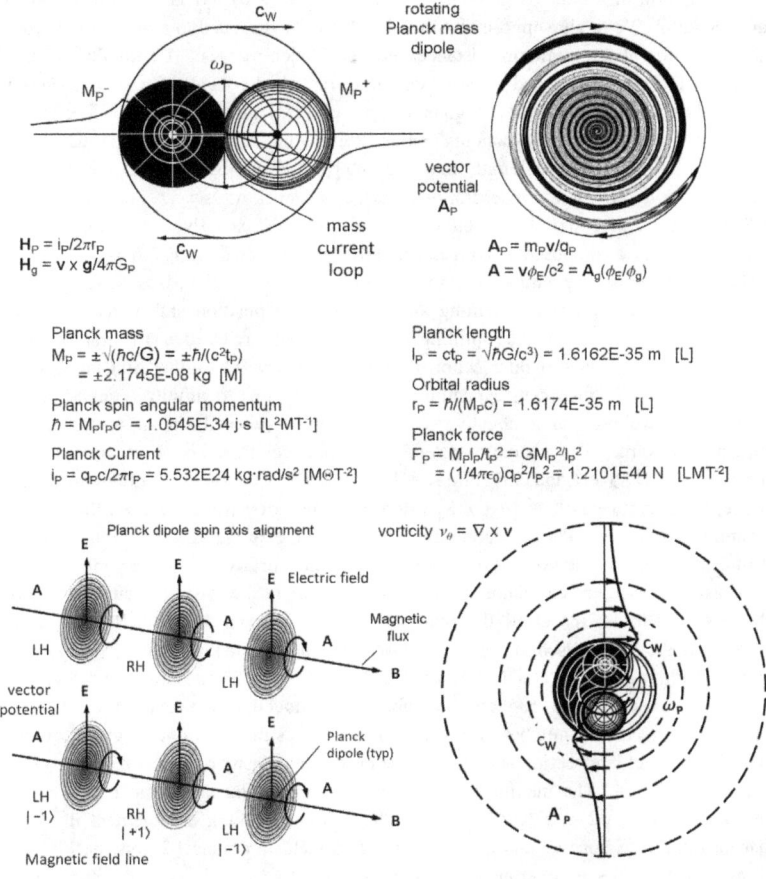

Fig. 16-13. Magnetic field strength **H** is postulated to be an effect of a vortex rotation of Planck mass dipoles. Magnetic field lines are the result of spin axis alignment of multiple Planck dipoles. Magnetic field strength is a measure of movement of charges through a cutting plane. Magnetic flux lines represent spin-aligned closed loop vortical filaments. Planck dipoles may spin in either direction. An arrangement of alternating spins (LH/RH chirality) is required for electric field polarization. Magnetic field lines are generated by vector potential. Planck charge Q_P responsible for scalar potential ϕ arises as a result of spin precession. Electric field lines are generated by superposition of scalar potentials.which couple synchronously. Curl of the magnetic field **B** results from rotation.

16. Planck Aether

The quantum vacuum is conjectured to consist of a densely packed quantum foam of Planck masses with positive and negative curvature and energy. The characteristic size of a Planck mass dipole is taken to be on the order of the Planck length L_P (= 1.616199 x 10^{-35} m). The vacuum is hypothesized to have spontaneously degenerated from a flat, featureless of void into a bimodal equipartition of positive and negative curvature and energy. As the universe expands, the energy density decreases. If the vacuum consists of a quantum foam of Planck masses, how do these evolve with time? Are new Planck dipoles continuously formed to fill the void? As positive energy equals negative energy, no additional energy input appears required to form new bubbles ($E^+ = m_P^+ c^2 = E^- = m_P^- c^2 = 0$). What sets the size of the Planck length? We might suppose that the Planck length is related to a radius of curvature and may represent the minimum distance between Planck masses. A naturally occurring frequency limit of positive and negative curvature radii is the Kepler-Bouwkamp constant associated the limit of inscribed polygons, i.e., $\rho = \cos(\pi/3) \times \cos(\pi/4) \times \cos(\pi/5) \times \cos(\pi/n)$ = 0.1149420448... The corresponding radius of curvature $K = 1/\rho$ = 8.700366252... In such a model, the Planck length may be interpreted as the sum of the two radii corresponding to the minimum separation distance of Planck masses. A notional sketch of the Planck vacuum composed of bubbles of positive and negative curvature including magnetic field lines composed of vortical flux tubes is illustrated in Fig. 16-14.

The evolution of a positive curvature bubble may be described as starting with a degenerate singularity point exhibiting an uncertainty in position and momentum $\Delta x \Delta p$ expanding into a circle (monad) splitting into two linked circles (dyad) forming a vesica piscis. The nodal points of intersection form triangles (triad). Further division creates a square (tetrad) and subsequent expanding series of polygons to infinity (Aleph-null) and beyond. The evolution of a negative curvature bubble proceeds in an inverse fashion contracting inward. The geometry resembles a spherical fractal such as an Apollonian sphere. For a 3D sphere, the curvature $K = 1/R^2$ where R is the radius. For a non-spherical surface, the curvature is $K = 1/(R \times S)$ where R is the rafter (distance from the center of curvature to point P on the osculating circle) and S is the swing (distance from point P to the directrix or axis). The rafter is a measure of intrinsic (surface) curvature while the swing is a measure of extrinsic curvature. Gravity is a measure of the in-surface curvature. For a spherical mass in 4D spacetime the swing $S = \sqrt{(2R_S^3/M)}$ and the rafter $R = \sqrt{(R_S^3/2M)}$ where R_S is the Schwarzschild radius (= $2GM/c^2$). For a Planck mass m_P (= $\sqrt{\hbar c/G}$), the Schwarzschild radius $R_S = 2L_P$ where L_P is the Planck length. For a spin-polarized vacuum, the minimum diameter of a magnetic flux tube in the model depicted equals $4L_P$.

In 2D, the maximum number of equal diameter circles that can touch a center circle of equal size without intersection is 6. In 3D, the maximum density of sphere packing (kissing number) equals 12. The maximum density of sphere packing in 3D for a face-centered cubic lattice is $\pi/3\sqrt{2}$ = 0.7404. For the Planck mass packing arrangement illustrated, adjacent regions of positive and negative curvature yields a compact 2-zone packing such that no positive masses overlap (repulsion) but negative mass envelopes may overlap (attraction). The high frequency limit sets up in effect an impenetrable force field barrier preventing overlap of positive masses. The mutual exclusivity of positive and negative curvature suggests a possible explanation for the high stiffness of the vacuum and the resultant high velocity of light. The vacuum acts as a perfect fluid without dissipation losses or resistance to motion. The net mass of a Planck dipole is zero ($\Sigma(m_P^+ + m_P^-) = 0$), hence, it is concluded that Planck dipoles may freely move about or rotate with respect to each other allowing propagation of waves or spin waves. The Planck frequency f_P (= $\sqrt{(c^5/\hbar G)}/2\pi$ = 2.949E42 Hz) is taken as the cutoff frequency of the quantum vacuum. The corresponding cutoff wavelength λ_P is the Planck length (= $l_P = \sqrt{(\hbar G/c^3)} = ct_P$). The cutoff wavelength is the maximum wavelength that will propagate without attenuation. This wavelength is taken to represent the longest wavelength possible within the spherical volume of the Planck mass which corresponds to a standing wave resonator, in particular, a whispering gallery mode resonator. The quantum vacuum represented as a Bose-Einstein superfluid composed of Planck masses appears globally to be isotropic indicating an overall uniform density. The positive and negative Planck masses in this model are somewhat separable and able to slide past each other in a superfluid-like motion. The estimated quantum mechanical energy of the vacuum ($\sim 10^{113}$ J/m^3) resides in the spin angular momentum of the Planck mass dipoles. The nature of the sub-Planckian vacuum remains a mystery – Just what is the composition of the Planck mass? The inertial mass of the Planck mass ($m_P = \hbar/cR_P$) varies with change in EM mode density, and may be represented as the summation of spectral energy energy modes within an enclosed volume over the range of allowable frequencies up to the Planck frequency ν_P, i.e. $m_i = V_0/c^2 \int \eta(\nu) \rho_{SED}(\nu) d\nu$ where $\eta(\nu)$ = no. of modes, V_P = Planck mass volume (= $4/3\pi r_P^3$) and ρ_{SED} = spectral mode energy density. What is the physical nature of the wavefunction existing at the subPlanck scale where curvature K ≤ l_P^{-2} ? Are we to imagine yet some still finer granularity or discretization of a surreal subquantal vacuum as an infinite regress of an ensemble of states of contingencies á la nesting Matryoska dolls in a succession of mother Nature's hat tricks?

Planck quantum vacuum model

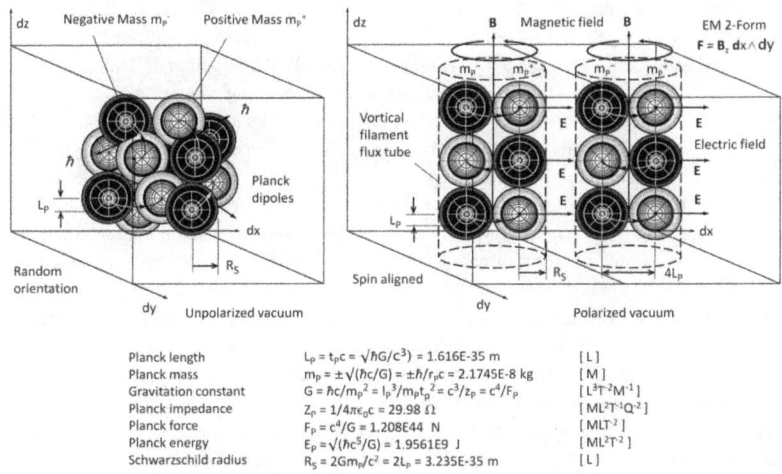

Fig. 16-14. Illustration of Planck vacuum composed of positive and negative curvature of spherical zones in a foam-like quasi-lattice. The minimum separation distance between Planck masses equals the Planck mass Schwarzschild radius. Velocity of light varies with Planck mass volumetric density. Creation of positive and negative mass from the vacuum obeys 0, 0 → 1, −1. Flattening of the vacuum is the converse process 1, −1 → 0, 0 corresponding to absence of curvature. Positive Planck masses may be expected to exhibit substantial local gravitational squeezing of the vacuum (negative energy density) with ZPF mode wave lengths $\lambda \geq 8\pi R_S$ (≥ 8.50 E-34 m). The squeezed vacuum state results in asymmetric amplitude of travelling EM waves oscillating between regions of positive and negative energy density, but with positive time averaged spectral energy density. The Planck masses act as impenetrable point scattering centers for incident EM waves.

Section 2 - Electricity

17. Electron Model

The electron can no longer be conceived as a single, small granule of electricity; it must be associated with a wave, and this wave is no myth; its wavelength can be measured and its interferences predicted. – Louis de Broglie

There was a time when we wanted to be told what an electron is. The question was never answered. No familiar conceptions can be woven around the electron; it belongs to the waiting list. – Arthur Eddington

We can therefore say that we have now reached a theoretical understanding of the existence of the electron, but in no way that of its constitution. – Pascual Jordan

The magnetic cleavage of the spectral lines is dependent on the size of the charge of the electron, or, more accurately, on the ratio between the mass and charge of the electron. – Pieter Zeeman

The bridge between the electron and other elementary particles is provided by the fine structure constant, $\alpha \sim 1/137$, as manifested in the factor-of-137 spacings between the classical electron radius, electron Compton radius, and the Bohr radius... – Malcolm H. MacGregor

A number of torus ring electron models have been previously proposed including those by Compton[44], Allen[45], Thomas[46], Jennison[47], Bergman/Wesley[48,49], Hestenes[50], Williamson / van der Mark[51], Kanarev[52,53], Winterberg[7], Ginzburg[54], Carroll[55], Heaston[56], Lucas[57], Rivas[58], Gauthier[59], Klyushin[60] and others in addition to a variety of disc and spherical wave models by Crane, Macken, Wolf, Haselhurst, Tomes, Cabala, LaFreniere, MacGregor, Tewari, Ghosh, etc. In superstring theory, the electron is represented as a closed loop with size on the order of the Planck scale with no explanation of electric charge. What discriminators may be applied to determine which of the proposed models best represents the observed reality? The ability to account for pair production and annihilation presents a critical test. It is experimentally observed that energetic photon(s) and electrons/positrons may be interconverted in pair production/annihilation processes. The geometrical transformation of an energetic photon into an electron/positron pair must be mathematically demonstrable. Likewise, the emission of a pair of photons with opposite momenta resulting from the annihilation of an electron and positron must be shown to be geometrically possible for a plausible model. The creation of electric charge and rest mass during electron/positron formation must be explained and calculable. The observed physical properties of the photon, electron and positron must be accounted for and quantified.

A photon wave-train is described as a helicoid in a twisted ribbon travelling wave geometrical model. A closed-loop double loop Hopf strip may be formed from a ribbon with a full twist. An eccentric hula-hoop motion of a Hopf strip generates a swept volume toroidal envelope corresponding to a closed-loop sanding wave. See Fig. 17-1. An example of a torus ring model is depicted in Fig. 17-2 illustrating an electron consisting of two orthogonal spinors generated by a rotating Hopf link corresponding to a poloidal and toroidal current loop. The charge trajectory is described by a precessing Hopf link, the simplest form of knot, embedded in a torus manifold created by eccentric motion.

It has to do with electrons. You know, the little things that whirl in space. Never heard of them. – Dr. Stanton (Riders to the Stars)

You know, it would be sufficient to really understand the electron. – Albert Einstein

17. Electron Model

Electron-positron pair production requires electric fields greater than the Schwinger field critical value $E_{cr} = m^2c^3/e\hbar \cong 1.3 \times 10^{18}$ V/m sufficient to provide the required rest mass energy. Pair production occurs by decay of a sufficiently energetic photon near an atomic nucleus $\gamma \rightarrow e^- + e^+$ or through photon-photon interaction via Breit-Wheeler decay $\gamma\gamma \rightarrow e^+e^-$ such as a probe photon propagating through a polarized short-pulsed electromagnetic field. Photons (gamma rays) may be generated in a reverse annihilation process by collision of an electron and positron $e^+e^- \rightarrow \gamma\gamma$. An inelastic collision of a photon with a free electron results in Compton scattering with a result momentum transfer altering the photon wavelength. Acceleration or deceleration of an electron results in photon emission such as EM radio wave emission at low energy or Bremsstrahlung emission, synchrotron radiation, or Cherenkov radiation at high energy with photon energy proportional to frequency.

The classical electron radius r_e also known as the Thompson scattering length is derived by assuming the electron is a sphere with uniform charge density and equating the potential energy of charged sphere $E_p = c^2/\pi c_0 r^2$ with the electron's rest mass energy $E = m_0 c^2$ and solving for radius r. The resultant value of this estimate is 2.8179×10^{-15}m. For comparison, the calculated Compton wavelength λ_C of 2.4263×10^{-12}m as determined from scattering experiments yields a reduced Compton radius R_C ($= \lambda_C/2\pi$) of 3.8616×10^{-13}m.

Remarkably, an electron, a spin $\frac{1}{2}\hbar$ fermion with quantized electric charge (e^-) and positive rest mass ($m_e = 0.511$ MeV/c^2), may be created from an energetic photon, a spin ± 1 $\hbar$ boson with no electric charge or rest mass. A toroid configuration of the electron has long been posited to account for the observed physical properties such as quantized electric charge, magnetic moment, g-factor, spin angular momentum, etc. The electron's wavefunction ψ may be represented as consisting of two spinor components one with right-hand and one with left-hand helicity. An example of a model electron formed from a single wavelength helical photon topologically confined into a torus configuration is shown in Figs. 17-3 through 17-9. In this model, the helical path of a photon is in the form of Hopf link which under rotation traces out a trajectory path of toroidal geometry. The electron has a magnetic dipole moment μ_s as a result of its intrinsic spin angular momentum S ($=\hbar[\frac{1}{2}(\frac{1}{2}+1)]^{\frac{1}{2}}$). In the ansatz model considered here, the electron has both a toroidal spin and poloidal spin component. If there were no spin-spin interaction of the toroidal spin angular momenta S_o and the poloidal spin angular momentum S_r, the orientation of one vector would be independent of the other.

The electron in the toroid model illustrated has both toroidal rotation and poloidal rotation such that the **EHV** triplet charge path develops two internal rotations for each toroidal rotation. The toroidal radius corresponds to the reduced Compton radius ($R_C = \lambda_C/2\pi$). The tangential velocity of rotation v_t equals the velocity of light c ($v_t = R_C\omega_C$). Due to the increased magnetic field, the orbital charge velocity internally varies from superluminal at the orbital periphery to sublight velocity at the spin center. The $\frac{1}{2}\hbar$ spin characteristic of the electron arises from the ratio of the Compton and Zitterbewegung rotational frequencies ($\omega_C/\omega_{zbw} = \frac{1}{2}$) resulting in the observed net spin in a reference frame at rest with the observer.

The fundamental fact of electron theory, the existence of discrete electrical particles, thus manifests itself as a characteristic quantum phenomena, namely as equivalent to the fact matter waves only appear in discrete quantized states. – Pasqual Jordan

Nobody has ever seen an electron. Nor a thought. – Robert A. Heinlein

Zitterbewegung: what happens when you take too much coffee. – Prof. Rabindra Mohapatra

17. Electron Model

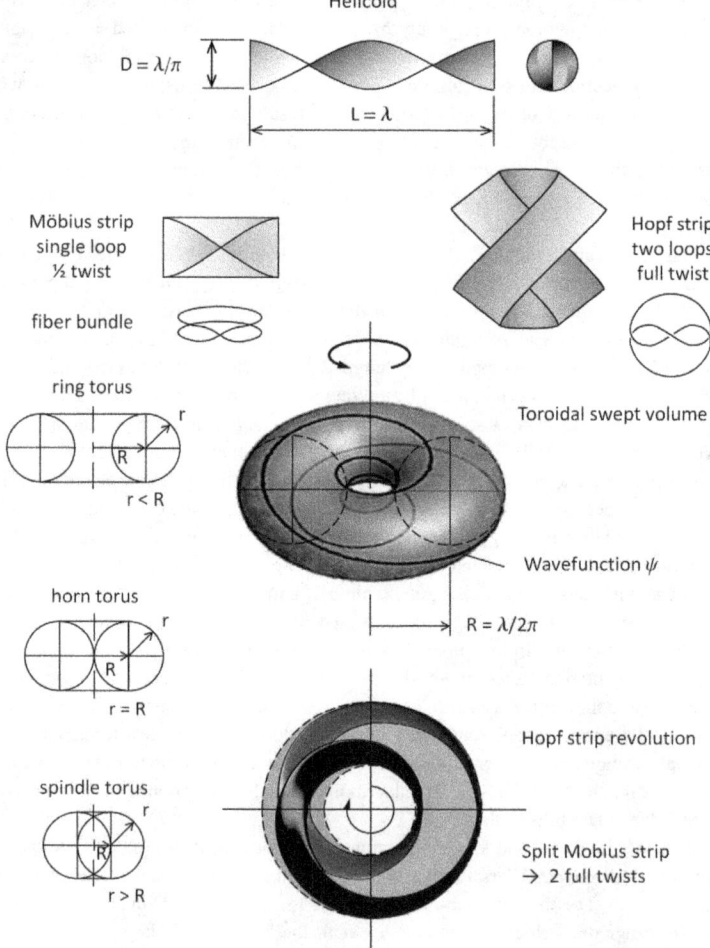

Fig. 17-1. Illustration of a rotating closed-loop Hopf link formed from a twisted ribbon (helicoid) embedded in a torus of revolution. Such a geometry exhibits a ½ spin characteristic, i.e., a 720° rotation is required to return to the starting position. Similarly, a spin 1 photon (a travelling wave of helical geometry) may be transformed into a spin ½ electron (a closed-loop standing wave of toroidal geometry). Electric charge (spin precession) corresponds to a torsion field dislocation defect (loop closure failure).

The only thing you can say about the reality of the electron is to cite its mathematical properties. – Martin Gardner

As far as the laws of mathematics refer to reality, they are not certain; and as far as they are certain they do not refer to reality. – Albert Einstein

17. Electron Model

Toroidal electron model

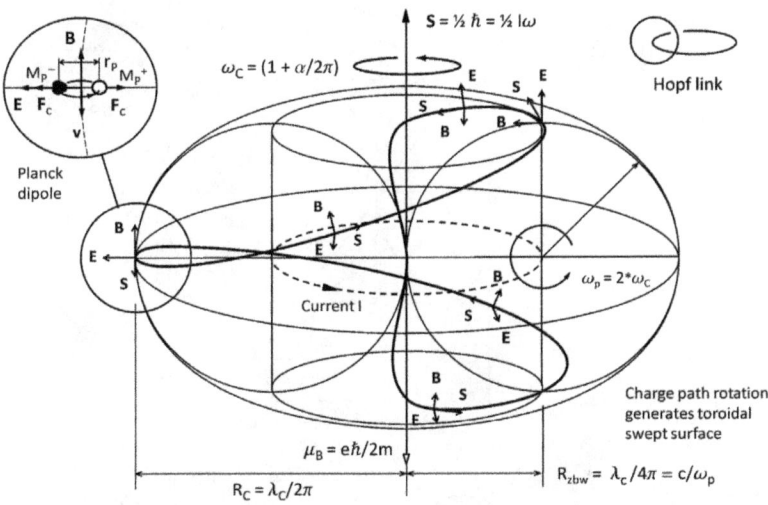

Parameter	Symbol	Relation	Value	Units
Electric charge	e	$= F/E = F_k/E_k = \hbar k/A =$ $= E/V = p/k \cdot A = \sqrt{(\alpha q_P^2)}$	1.60217E-19	Coul
		$= m(\omega_C + \omega_p) \simeq m\omega_C$	7.019E-10	kg·rad/s
Mass	m_e	$= E/c^2 = \hbar/c \cdot R_C = |V \cdot q|/c^2$ $= \hbar k/v = e/(\omega_C + \omega_p) \simeq e/\omega_C$	9.10938E-31	kg
Compton wavelength	λ_C	$= h/mc = h/p$	2.4263E-12	m
Compton frequency	f_C	$= mc^2/h = c/\lambda_C = \omega_C/2\pi$	1.2355E20	Hz
Compton radius:	R_C	$= \lambda_C/2\pi = \alpha a_0 = \sqrt{(E/m\omega^2)}$	3.8616E-13	m
Compton angular frequency	ω_C	$= c/R_C = mc^2/\hbar \simeq e/m$	7.7634E20	rad/s
Zitterbewegung angular frequency	ω_{zbw}	$= 2\omega_C$	1.5527E21	rad/s
Precession frequency	ω_p	$= \omega_{e/m} = q/m = \alpha\omega_C$	1.7588E11	rad/s
Spin angular momentum	s	$= \tfrac{1}{2}\hbar = i\omega = \tfrac{1}{2}m_e R_C^2 \omega$	5.2725E-5	J·s
Bohr magneton	μ_B	$= e\hbar/2m$	9.274E-24	J/T
Rest Energy	E	$= \hbar c/R_C = m_0 c^2 = m\phi$	8.187E-14	J

Fig. 17-2. Toroidal electron is formed by a topologically confined photon inside the Compton radius as a result of imbalance of electrostatic and magnetostatic energy. The charge path represents a spin wave phase alignment of entrained Planck dipoles, the rotation of which generates a toroidal form. The propagation of the rotating spin wave describes a current loop of radius equal to one half the Compton radius.

17. Electron Model

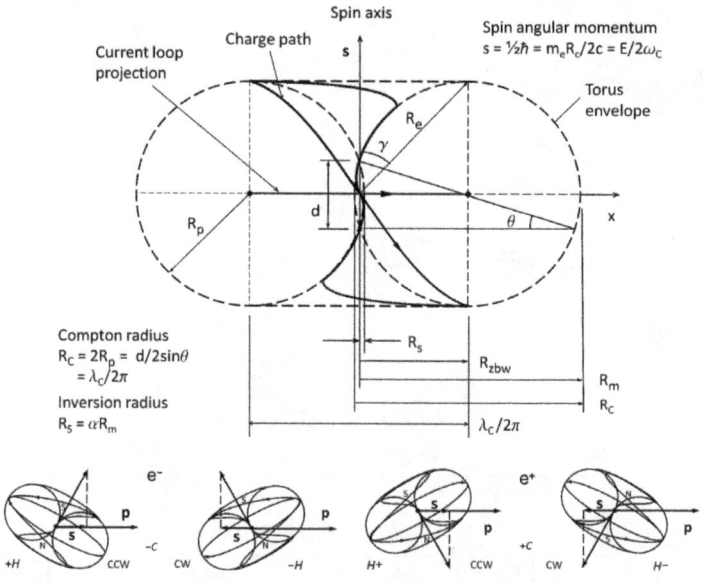

Parameter	Symbol	Relation	Value	Units
Charge radius	R_e	$= \sqrt{(2\lambda_C/4\pi)} = R_{zbw}/\sin\gamma$	2.7305E-13	m
Classical electron radius	R_0	$= (1/4\pi\epsilon_0)\, e^2/m_e c^2$ $= k_e(e^2/m_e c^2)$	2.8179E-15	m
Compton radius	R_C	$= \lambda_C/2\pi = \alpha a_0 = h/2\pi mc$	3.8616E-13	m
EM radius	R_{em}	$= \hbar/mc^2 = R_m$	1.2880E-21	m
Helix radius	R_{photon}	$= R_{helix} = \lambda_C/2\pi$ $= \sqrt{(E/m_e\cdot\omega^2)}$	3.8616E-13	m
Mass radius (max)	R_m	$= \hbar c/E = \hbar/mc$	3.8613E-13	m
Poloidal radius	R_p	$= R_m/2$	1.9308E-13	m
Spindle (inversion) radius	R_s	$= R_C - R_m = \alpha R_m$	2.6983E-17	m
Toroidal radius	R_t	$= R_C(1 + \alpha/2\pi)$	3.8616E-13	m
Zitterbewegung radius	R_{zbw}	$= \langle R_m \rangle = \lambda_C/4\pi = \hbar/2mc$ $= c/2\omega_C = R_e \sin\gamma$	1.9308E-13	m

Fig. 17-3. Characteristic dimensions of electron torus model. The toroidal circumference corresponds to the Compton wavelength λ_C of the confined photon. The electron Compton radius is $R_C = \lambda_C/2\pi = (\lambda_C/\alpha^{-1})/\theta_{137}$. Electron charge (a torsion defect due to loop closure failure) is a result of a slight spin precession with whirl no. $= \alpha^{-1}$ where α = fine structure constant. Electron mass results from wave function interference due to spin precession.

17. Electron Model

Electron ring configuration

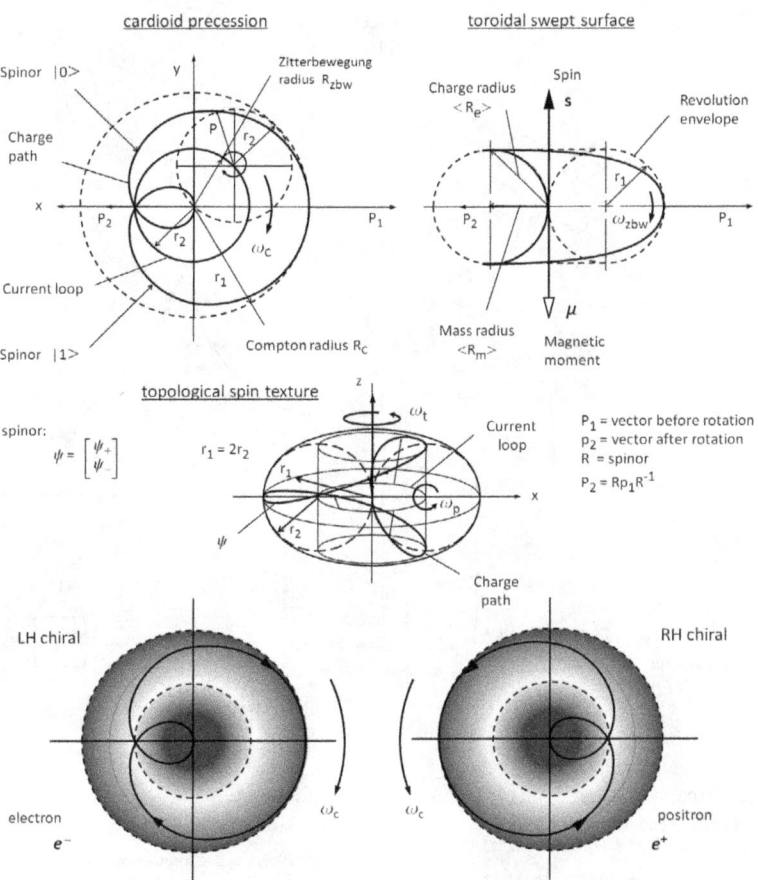

Fig. 17-4. Electron depicted as precessing epitrochoid charge path composed of two orthogonal spinors describing a rotating Hopf link with toroidal and poloidal current loop components of 2:1 rotary octave. Rotation of the charge path generates the torus geometry. The spin ratio of the Compton angular frequency ω_C and the Zitterbewegung angular frequency ω_{zbw} (= $2\omega_C$) corresponds to observed spin of ½. Precession induces an internal magnetic field that is stronger by a factor of α^{-1} than the spin induced magnetic field external to the Compton radius aligned along the spin direction.

How electrons are made has not yet been determined. – Oliver Heaviside

To the electron...May it never be of any use to anybody. – J. J. Thomson (Cavendish Lab toast)

I now believe in the idea of the self-rotating electron. – Wolfgang Pauli

How can one look happy when he is contemplating the anomalous Zeeman effect? – Wolfgang Pauli

17. Electron Model

Electron spin precession

Homoclinic zoom-whirl orbit with eccentric epitrochoid rossette and quasi circular whirls

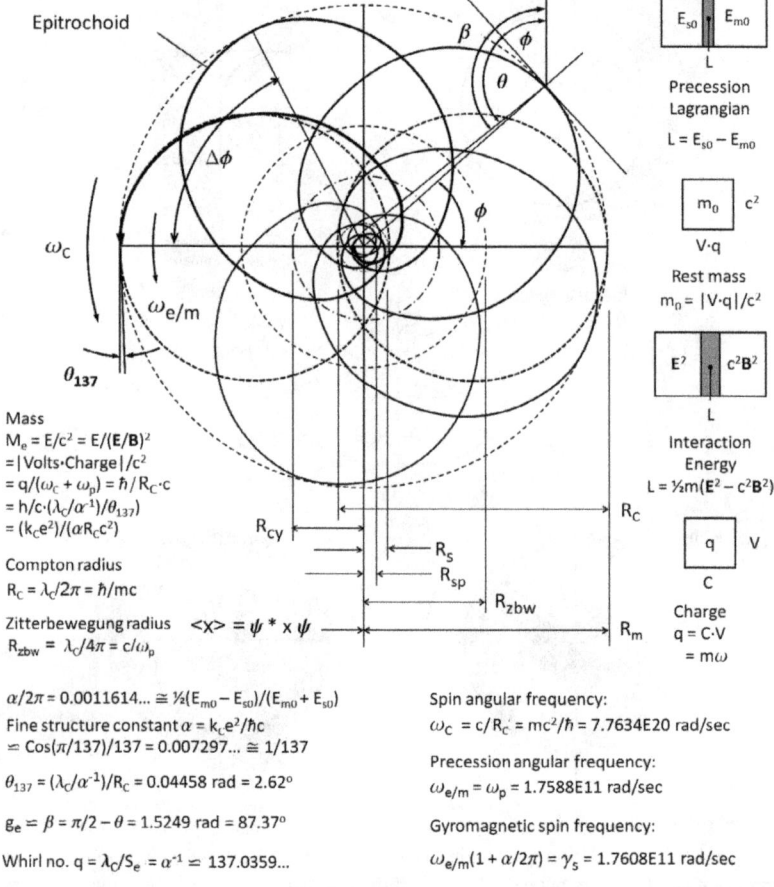

Mass
$M_e = E/c^2 = E/(E/B)^2$
$= |\text{Volts·Charge}|/c^2$
$= q/(\omega_C + \omega_p) = \hbar/R_C \cdot c$
$= h/c \cdot (\lambda_C/\alpha^{-1})/\theta_{137}$
$= (k_c e^2)/(\alpha R_C c^2)$

Compton radius
$R_C = \lambda_C/2\pi = \hbar/mc$

Zitterbewegung radius $\langle x \rangle = \psi^* \times \psi$
$R_{zbw} = \lambda_C/4\pi = c/\omega_p$

Precession Lagrangian
$L = E_{s0} - E_{m0}$

Rest mass
$m_0 = |V \cdot q|/c^2$

Interaction Energy
$L = \frac{1}{2}m(E^2 - c^2B^2)$

Charge
$q = C \cdot V$
$= m\omega$

$\alpha/2\pi = 0.0011614... \cong \frac{1}{2}(E_{m0} - E_{s0})/(E_{m0} + E_{s0})$
Fine structure constant $\alpha = k_c e^2/\hbar c$
$\cong \cos(\pi/137)/137 = 0.007297... \cong 1/137$

$\theta_{137} = (\lambda_C/\alpha^{-1})/R_C = 0.04458$ rad $= 2.62°$

$g_e \cong \beta = \pi/2 - \theta = 1.5249$ rad $= 87.37°$

Whirl no. q $= \lambda_C/S_e = \alpha^{-1} \cong 137.0359...$

Spin angular frequency:
$\omega_C = c/R_C = mc^2/\hbar = 7.7634E20$ rad/sec

Precession angular frequency:
$\omega_{e/m} = \omega_p = 1.7588E11$ rad/sec

Gyromagnetic spin frequency:
$\omega_{e/m}(1 + \alpha/2\pi) = \gamma_s = 1.7608E11$ rad/sec

Fig. 17-5. In the electron model illustrated, electron spin precession $\omega_{e/m}$ is due primarily to imbalance of electrostatic and magnetostatic energy resulting in an eccentric whirl orbit of the charge path about the spin axis. The precession follows a zoom-orbit whirl with a periapsis advance θ_{137} that is a function of the fine structure constant α and the Compton radius R_C. Synchronization occurs every α^{-1} revolutions. The Thomas precession frequency ω_T is approximately equal in magnitude to Compton angular frequency ω_C. Electron rest mass m_0 equals E/c^2 ($= |V \cdot q|/c^2 \simeq \hbar/a_0 c a = e/\omega_C$). Electric charge $q = \hbar k/A = C \cdot V = p/k \cdot A$ where k = wave no., A = vector potential, C = capacitance, V = voltage, p = momentum.

In quantum theory, mass corresponds to the periodicity of waves. – Arthur Eddington

The rigid electron is no working hypothesis, but a working hindrance. – Hermann Minkowski

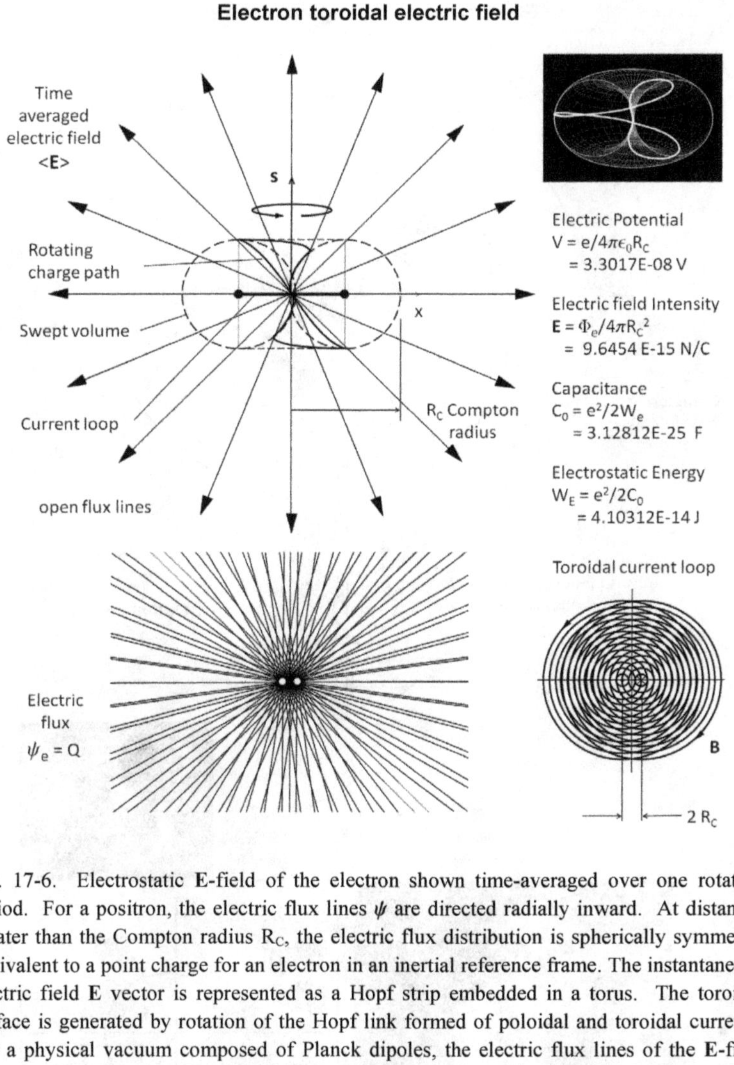

17. Electron Model

Electron toroidal electric field

Electric Potential
$V = e/4\pi\epsilon_0 R_C$
= 3.3017E-08 V

Electric field Intensity
$E = \Phi_e/4\pi R_C^2$
= 9.6454 E-15 N/C

Capacitance
$C_0 = e^2/2W_e$
= 3.12812E-25 F

Electrostatic Energy
$W_E = e^2/2C_0$
= 4.10312E-14 J

Labels: Time averaged electric field $\langle E \rangle$; Rotating charge path; Swept volume; Current loop; open flux lines; R_C Compton radius; Electric flux $\psi_e = Q$; Toroidal current loop; $2 R_C$

Fig. 17-6. Electrostatic E-field of the electron shown time-averaged over one rotation period. For a positron, the electric flux lines ψ are directed radially inward. At distances greater than the Compton radius R_C, the electric flux distribution is spherically symmetric equivalent to a point charge for an electron in an inertial reference frame. The instantaneous electric field **E** vector is represented as a Hopf strip embedded in a torus. The toroidal surface is generated by rotation of the Hopf link formed of poloidal and toroidal currents. For a physical vacuum composed of Planck dipoles, the electric flux lines of the E-field represent momentary synchronous alignment of positive M_P^+ and negative M_P^- Planck masses.

Charge is certainly a well-characterized attribute, but the question: "What is charge?" is presently unanswerable. – Riccardo C. Storti

A full theory of the dynamics of the electron depends upon assumptions concerning its structure, an unsolved problem. – William Berkson

I believe that the development of the theory along the correct lines will then lead to a numerical value of the fine-structure constant $\alpha = e^2/\hbar c = 1/137$, and to an explanation of the fact that arbitrarily high masses do not appear concentrated in a given region of space region in nature. – Wolfgang Pauli

17. Electron Model

Electron as a rotating spin wave

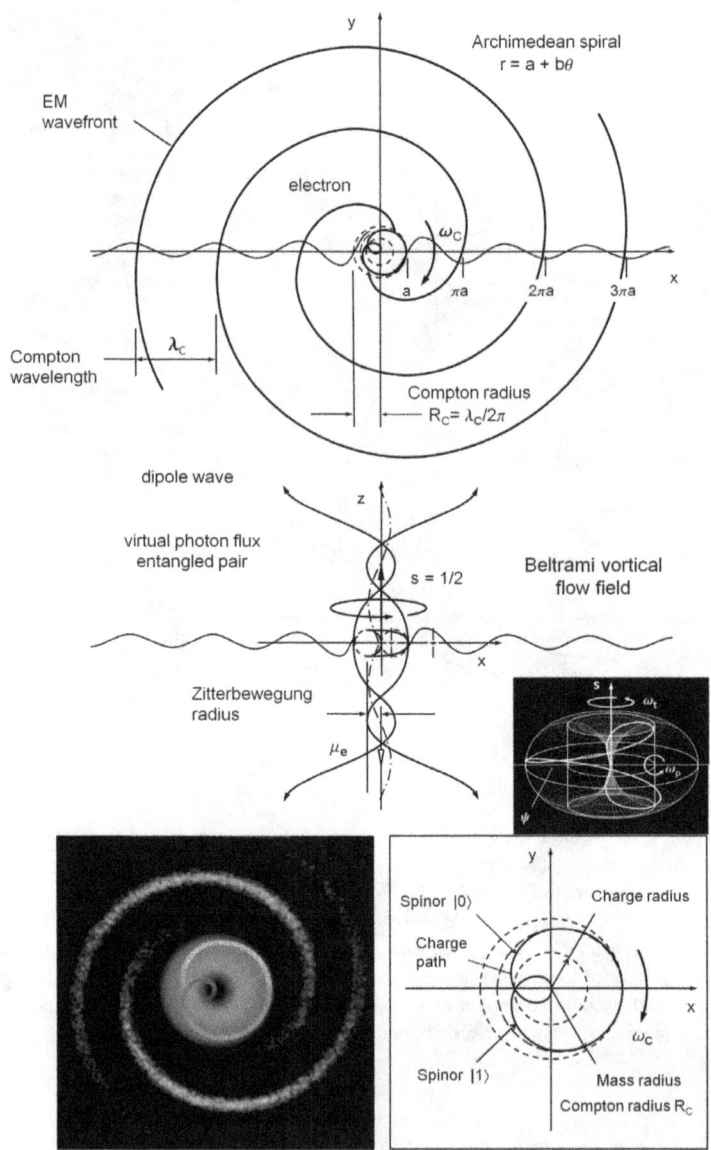

Fig. 17-7. The electron continuously generates a dipolar spin wave at the Compton frequency f_C (= mc^2/h = 1.236E20 Hz). The electron acts as a spinning dipole antenna with virtual radiation emission of a pair of entangled wavefronts along the spin axis. Viewed along the spin axis, the dipole waves describe an Archimedean spiral. Virtual photons are continuously emitted and absorbed and, like unobservable Cheshire cats, appear to wink in and out of existence – but have a measurable zero point energy effect en masse.

17. Electron Model

Electric field of an oscillating electron

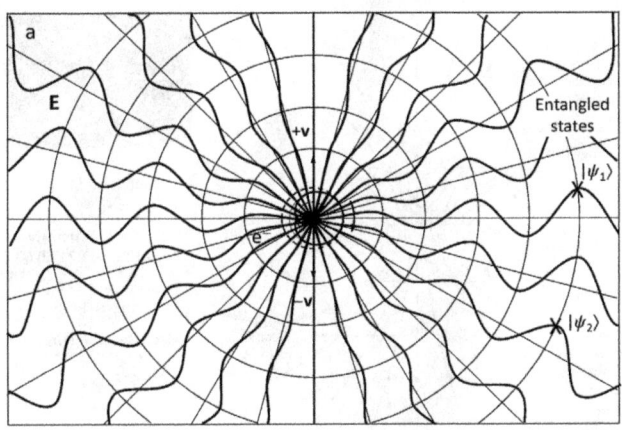

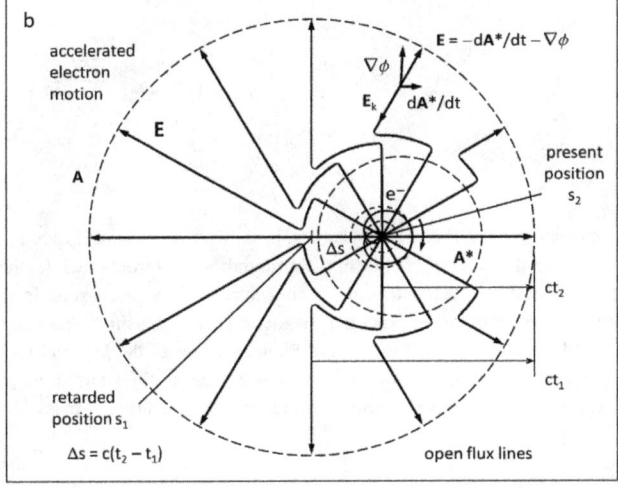

Fig. 17-8. Oscillation of the electron at frequencies less than the Compton frequency f_C in response to an external electromagnetic field results in generation of observed EM waves in synchronicity with the imposed frequency. In a) is depicted vertical sinusoidal oscillation. Horizontal acceleration over an interval Δt is illustrated In b) showing a propagating spherical disturbance with farfield flux pointing in the direction of the retarded initial starting position. Entangled states represent different points on the same wavefront and, hence, share properties of energy, momentum and polarization. Measurements of particle states are thus correlated as they are sampling the same wavefront. Hence, no superluminal influence or "spooky action at a distance" is involved in perceived entanglement. An 'observer' is simply an interacting object being acted upon – No consciousness is required.

All matter is simply undulations in the fabric of space. – William Clifford

All forces are consequences of geometry. – Timothy Ferris

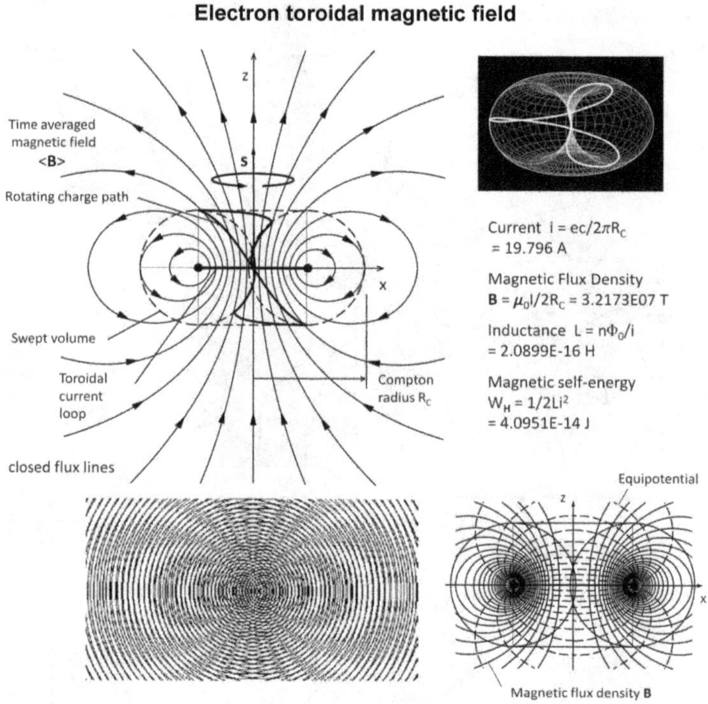

17. Electron Model

Electron toroidal magnetic field

Current $i = ec/2\pi R_C$
  $= 19.796$ A

Magnetic Flux Density
$B = \mu_0 I/2R_C = 3.2173\text{E}07$ T

Inductance $L = n\Phi_0/i$
  $= 2.0899\text{E-}16$ H

Magnetic self-energy
$W_H = 1/2Li^2$
  $= 4.0951\text{E-}14$ J

Fig. 17-9. Magnetostatic **B**-field of the electron is shown time-averaged over one rotation period. The external magnetic field flux configuration is toroidal while the internal magnetic field is poloidal. Magnetic flux is concentrated in a pinch zone in the central region creating a concentration of potential magnetic energy density. The magnetic flux lines represent filaments of vortical rotation of Planck dipoles of the physical vacuum. The external magnetic field corresponds to the interference pattern of the current loop magnetic flux. Spin precession results in wave field interference and generation of mass.

The electron, in a sense, may be considered as a form of a primitive electrical machine consisting of an electric circuit linked with a magnetic circuit similar to a transformer or electric motor. See Fig. 17-10. The electromotive force (e.m.f.) in an electric circuit equals current x resistance. The magnetomotive force (m.m.f.) in a magnetic circuit equals flux x reluctance which is proportional to the product of current and number of turns or ampere-turns. Electrical power produced in the electric current loop is the product of electromotive force ($\mathcal{E}$ = e.m.f.) and current (I). In an electron, the estimated electrical power (P = V·I) is $\cong 5.1096 \times 10^5$ watts. As shown by Macken[20], the circulating electrical power ($P_c = E_i \omega_C = \omega_C^2 \cdot \hbar$) in an electron is $\cong 6.355 \times 10^7$ watts. Maximum current induced by the magnetic flux for a given e.m.f. occurs when electric circuit resistance (R) and magnetic circuit reluctance ($\mathcal{R}$) are both minimized. Maximum magnetic flux (ϕ_B) is realized for a given current (I) with minimal reluctance. The electric current I (= dQ/dt) or equivalently, the magnetic vector potential **A**, represents mass motion of Planck dipoles in a spin density wave. Electric charge e [Q] is dimensional equivalent to mass•radians/unit time [MΦT^{-1}].

17. Electron Model

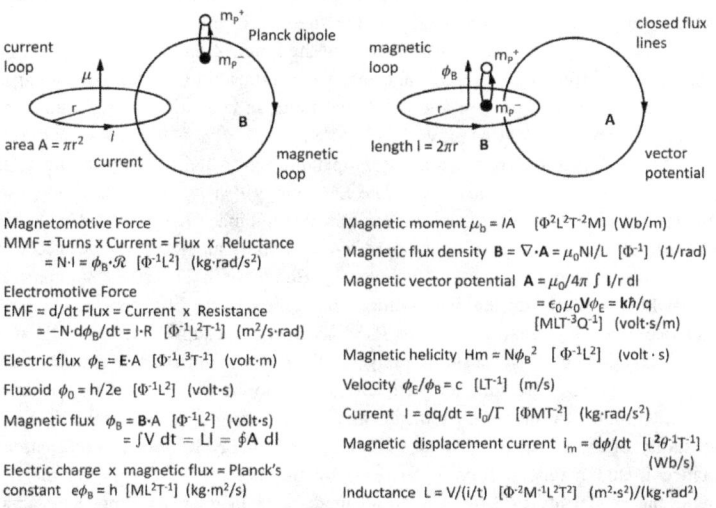

Fig. 17-10. An electron represented schematically as a primitive electrical machine linking a super-conducting electric circuit with a magnetic circuit resembles a transformer. Two interconnected loops (fiber bundles) form a Hopf link. The magnetic flux density $\mathbf{B}$ ($= -\nabla\phi_B = \nabla\cdot\mathbf{A}$), the gradient of the magnetic scalar potential ϕ_B ($= LI = \int V dt = \oint \mathbf{A}\cdot d\mathbf{l}$), represents vortical motion of spin chains of Planck dipoles with dimension of inverse radians [Φ^{-1}]. Broken magnetic flux lines reconnect to form closed loops.

A single electron represents a vast store of internal energy with a rest energy of 0.511 MeV/c^2, an electric current of 19.7 Amperes and an electric power estimated at 5.109E05 Watts. The electron shrinks in size with increasing kinetic energy and, thus, appears as a point object when probed at high energy. The change in poloidal diameter is a measure of the individual temperature of an electron[61]. With spin-oriented electrons trapped in metamaterial quantum wells, might it be possible to synthesize an electron-like artificial spin wave with an excess of magnetostatic energy over electrostatic energy that precesses to generate electric charge? For example, if a surfeit of electrons is confined in a superlattice of oriented, layered graphene sheets and subjected to rotation, an expanding and contracting Moiré pattern results changing the energy density in a periodic fashion. The diffraction patterns form an adjustable fractal antenna allowing coupling to external photonic fields. Connected to an external resonant LC circuit, the electron oscillations may be converted to EM waves or conversely, as in a dielectric, electrons may be excited in resonance with incident EM waves. To reproduce Maxwell's results of electrical disturbances in dielectrics producing transverse waves, Helmholtz deduced that the vacuum must be polarizable electrically and magnetically. If there were any resistance to vacuum polarization, longitudinal waves would result in addition to transverse waves. In a material dielectric, the movement of electrons result in self-induction interaction and magnetic interference thus decreasing the local velocity of light and increasing the index of refraction.

17. Electron Model

The size and energy content of the electron varies with motion as a function of the Lorentz factor γ. As illustrated in Fig. 17-11, the electron represented as a standing wave resonator undergoes a Lorentz contraction in the direction of motion. An electron shrinks in size as energy of motion increases as a function of the Lorentz factor $\gamma \; (= 1/\sqrt{(1 - (v/c)^2)})$. The electric permittivity $\epsilon \; (= \epsilon_0/\sqrt{g_{00}})$ and magnetic permeability $\mu \; (= \mu_0/\sqrt{g_{00}})$ vary with the GR metric coefficient g_{00}. In a gravitational field, an equivalent relation holds as the gravitational gamma $\Gamma = dt/d\tau = 1/\sqrt{(1 - (2GM/c^2 R))} = 1/\sqrt{g_{00}} = 1/\sqrt{(1 - (v_e/c)^2)}$ where t = coordinate time (stationary metric), τ = proper time, R = radial distance and v_e is the escape velocity of the central mass M. The EM energy density in a polarizable vacuum may be represented by a scalar vacuum index of refraction index $K_{PV} \; (= \Gamma^2 = 1/g_{00})$.

The mechanism for storage of energy and re-radiation by an electron under acceleration is elucidated in a theory of forces developed by Bergman[48]. To briefly summarize, the electric field amplitude during acceleration undergoes relativistic contraction and the electron decreases in size (radius $R = R_0/\gamma$) as energy acquired increases ($E = \gamma E_0$). Magnetic induction ($E = -\phi/dt$) stores energy in the surrounding electromagnetic radiation field. See Fig. 17-12. The radiation field corresponds to the induced EM field asymmetry under acceleration. A measure of the local field distortion is represented by the non-orthogonality of the **E** and **H** vectors. The magnetic energy E_m, electrostatic energy E_s, inductance L and magnetic flux ϕ_m increase by the Lorentz γ factor. The increase in magnetic energy density is equivalent to an increase in magnetostatic pressure acting to reduce the size of the electron. The change in electron radius, mass, flux, inductance, capacitance and energy as a function of velocity ratio β and Lorentz factor γ is shown in Figs. 17-13 through 17-18. The Schwinger correction[61] for spinning mass is given by $m_s = m(1 - \alpha/2\pi)$ in which the electromagnetic mass $\Delta m = m \cdot \alpha/2\pi$ where α is the fine structure constant. In the toroid electron model illustrated, the additional mass due to the magnetic field ($\Delta m = m_e \alpha/2\pi$) results in a reduced Compton radius ($R_C' = \alpha R_C = \alpha \hbar/mc$). The magnetostatic energy E_{m0} is reduced $E_{m0}' = E_{m0}(1 - \alpha/2\pi)$ while the electrostatic energy is increased $E_{s0}' = E_{s0}(1 + \alpha/2\pi)$. Langragian energy $L = E_{s0}(1 + \alpha/2\pi) - E_{m0}(1 - \alpha/2\pi)$.

The combined matter wave of a coupled pair of electrons (superconducting Cooper pairs) corresponds to a partially overlapping superposition of wave functions with length and peak amplitude greater that that of a single electron in isolation. This effect enables tunneling of pairs of electrons through an insulating barrier of a few nanometers thick between two superconducting wires that would otherwise impede current flow of single electrons. This effect is utilized in a Josephson junction superconducting tunneling current switch. On-off current flow is electromagnetically controlled via an external magnetic field in an adjacent control wire enabling very fast switching speeds of up to ~10 terahz (THZ).

De Broglie's ideas were for a free electron by itself, and Schrödinger extended them to apply to an electron moving in an electromagnetic field. – P.A.M. Dirac

Einstein suggested that mass might be an interrelation between electromagnetic and gravitational fields. – A.R. Weyl

No, It's quite impossible for the electron to have spin. – Hendrik Lorentz

I would rather know what an electron is. – Albert Einstein

Light travels with the velocity of light. That's a nice sentence. – Prof. Rabindra Mohapatra

Time reversal goes with inversion of energy and mass. Consequently, there are two kinds of antimatter of opposite mass. – Jean-Pierre Petit

17. Electron Model

Lorentz contraction of a standing wave resonator in motion

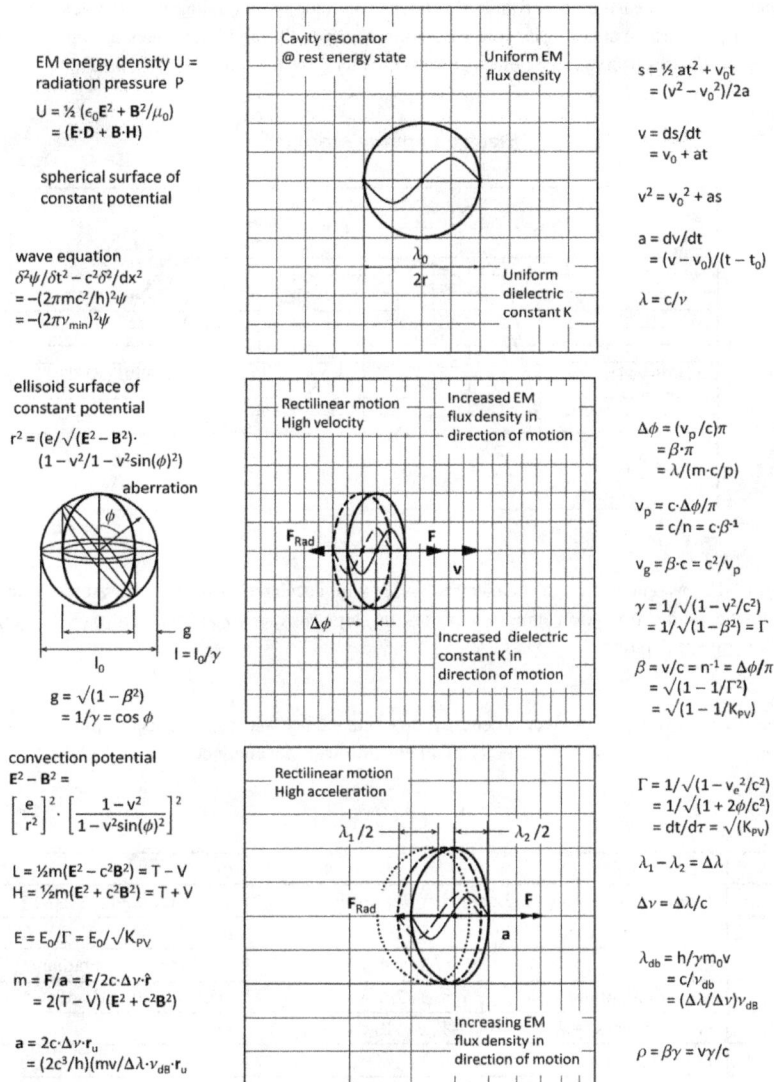

Fig. 17-11. Matter in motion undergoes a velocity reduction and Lorentz contraction in the direction of motion as a result of increased EM flux density where $g = 1/\gamma = 1/\Gamma = 1/\sqrt{K_{PV}}$.

Inertial frames are uniformly moving (unaccelerated) frames that obey Newton's first law. In Einstein's special theory of relativity, which describes motion of objects in flat spacetime in the absence of gravity where acceleration $g = 0$, inertial frames are known as Lorentz frames. Bergman demonstrated that an inertial frame is one in which both acceleration and radiation field are absent. During acceleration of a charged particle, the

17. Electron Model

electromagnetic field becomes distorted (breaking symmetry) where additional kinetic energy is stored in the accompanying magnetic field. During and subsequent to deceleration as measured in a frame of reference at velocity coinciding with the velocity of the charged particle at rest prior to acceleration, the radiation field dissipates in radiation of energy as the electromagnetic field regains symmetry.

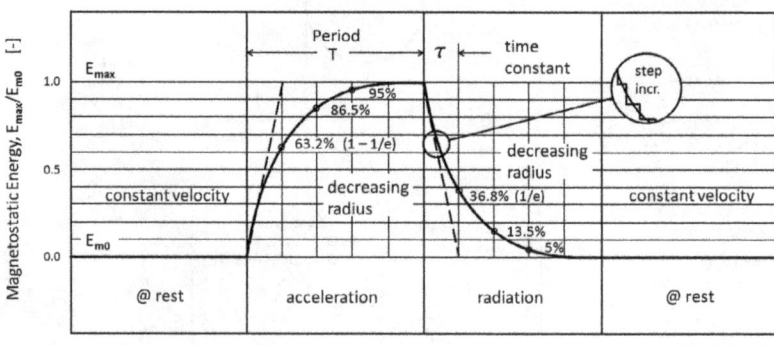

Fig. 17-12. Magnetic flux of the electron varies with acceleration with energy stored in the magnetic field. Radiation emission occurs in stepwise quantized increments ($E = nh\nu$) following acceleration.

The effect of velocity on electron energy, size, inductance and capacitance is illustrated in Figs. 17-13 through -18. The velocity of an electron within a conductor is

$$\mathbf{v} = (\hbar\mathbf{k} - q_o\mathbf{A})/m \qquad (17\text{-}1)$$

where

v	electron velocity	m/s
ℏ	Planck's constant ($= h/2\pi = p/k = S/c^2k = E/\omega$)	J·s (= rad·kg·m²/s)
k	propagation vector ($= E/\hbar c$)	m⁻¹
q_o	electric charge ($= F/E = E/V = \hbar\mathbf{k}/\mathbf{A} = \mathbf{p}/\mathbf{k}\cdot\mathbf{A}$)	Coulomb (= kg·radian/s)
A	vector potential ($= \hbar\mathbf{k}/q = \mathbf{p}/q = \phi_B/2\pi r = (v/c^2)\phi_E$)	Weber/m (= m/rad)
m	electron mass ($= \hbar k/v = E/\phi = k_C e^2/\alpha R_C c^2 = \gamma m_0$)	kg
Δm	radiative mass shift ($= e^2\langle A\rangle/2m_0$)	kg

The electromagnetic mass of a collection of electrons in a current carrying wire varies as the square of the number of charges ($= (nq_0)^2$), hence, the overall drift velocity is much less than the individual electron velocity. The velocity of an electron is reduced to zero when the total momentum $\hbar\mathbf{k}$ equals the electrodynamic momentum $q_o\mathbf{A}$.

Something unknown is doing we don't know what. – Sir Arthur Eddington

The concept of electric charge is as mysterious in its fundamental reality as is the concept of mass. – Malcolm H. Mac Gregor

17. Electron Model

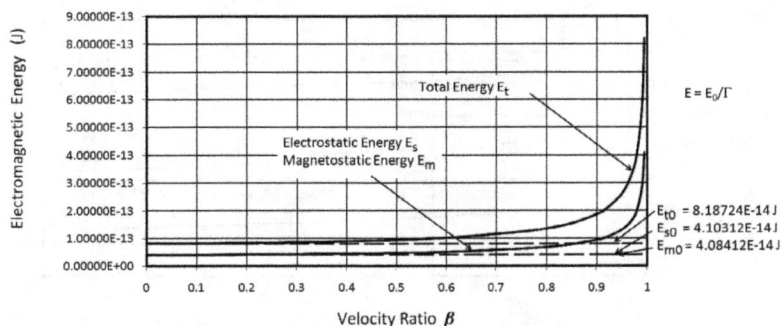

Fig. 17-13. Variation in electron energy as a function of velocity ratio β (= v/c).

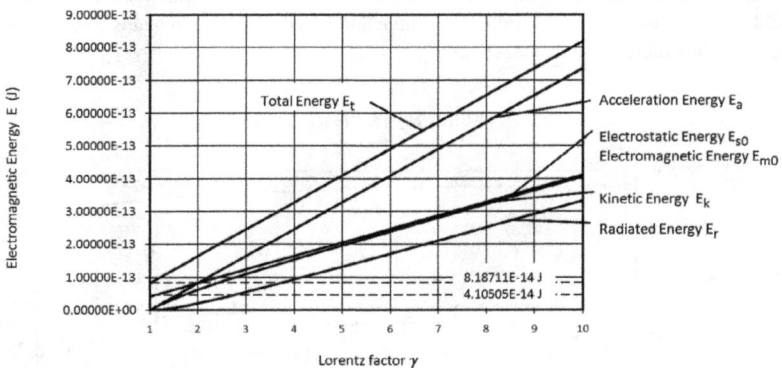

Fig. 17-14. Electromagnetic energy as a function of Lorentz factor γ. [After Bergman]

No familiar conceptions can be woven around the electron. – Sir Arthur Eddington

Einstein nodded, indicating that he had thought about this often himself, but that the whole problem of the origin of mass had remained a puzzle, and that the present theoretical efforts in quantum theory that treated the electron as a point source of an electromagnetic field led to infinite masses, requiring questionable mathematical subtraction procedures to arrive at experimentally observed values... he agreed that one must try to develop detailed descriptions of phenomena on the atomic scale that were anschaulich or visualizable. – Ernest J. Sternglass

After long reflection in solitude and meditation, I suddenly had the idea, during the year 1923, that the discovery by Einstein in 1905 should be generalized by extending it to all material particles and notably to electrons. – Louis de Broglie

... By which strange coincidence could a representation of probabilities propagate in space through time like a physical wave able to be reflected, refracted and diffracted. – Louis de Broglie

17. Electron Model

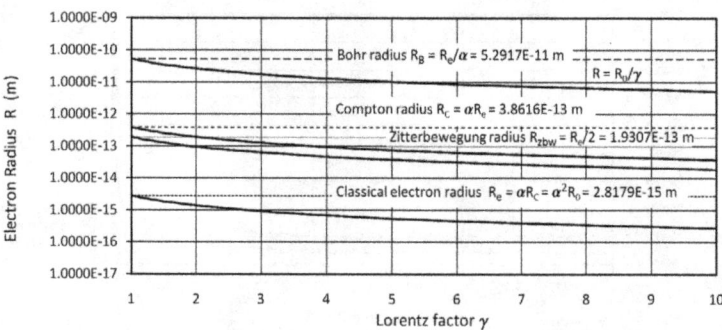

Fig. 17-15. Variation in electron radius as a function of Lorentz factor $\gamma\ (=1/\sqrt{(1-\beta^2)})$. The absorption of energy causes electrons to contract in size as a function of Φ^2 accordingly increasing the wave function curvature, kinetic energy, and volumetric energy density. The Compton radius reflects an equilibrium between torsion and the gravitoelectromagnetic field. Absorption of energy results in a contraction in radius as spin remains constant and torsion is insufficient to prevent collapse.

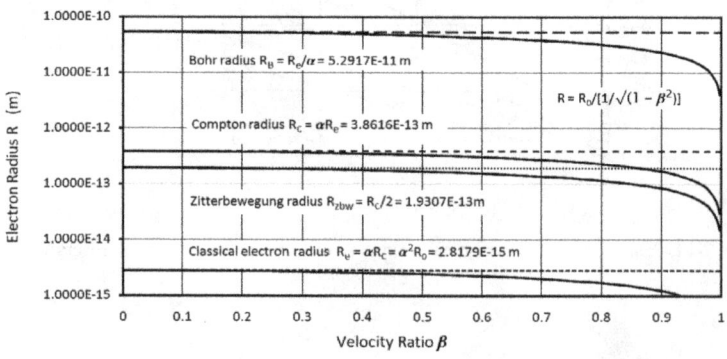

Fig. 17-16. Variation in electron radius as a function of velocity ratio $\beta\ (=v/c=\sqrt{(1-1/\Gamma^2)})$.

In 1923, the French physicist Louis de Broglie theorized that electrons would also have to be regarded as a wave-like pulse that became smaller as the electron was made to move faster. – Ernest J. Sternglass

One should not expect that ordinary quantum mechanics which treats the electron as a point-charge could hold under these conditions [i.e., when the electron has energy greater than 137 mc^2, and its wavelength is smaller than the classical radius of the electron]. – H. Bethe and W. Heitler

The theoretical determination of the fine structure constant is certainly one of the most important of the unsolved problems in modern physics. – Wolfgang Pauli

17. Electron Model

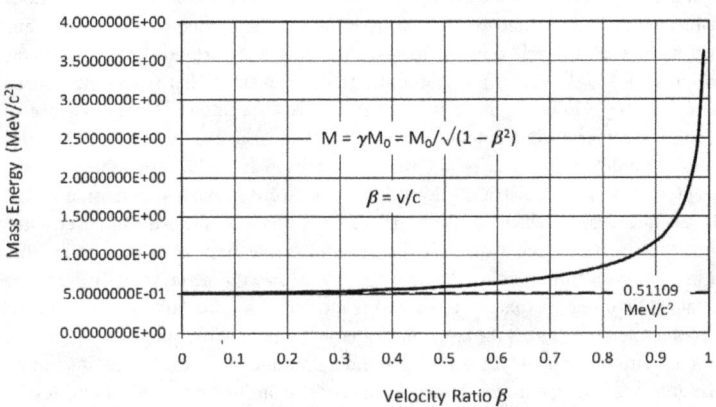

Fig. 17-17. Electron mass energy as a function of velocity ratio β.

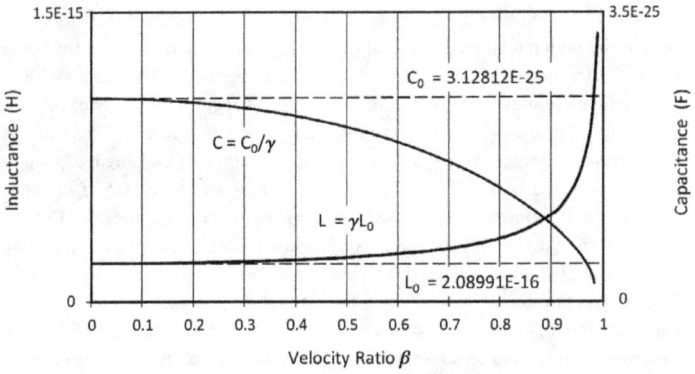

Fig. 17-18. Electron self-inductance and capacitance as a function of velocity ratio β (= $v/c = \Delta\phi/\pi = \rho/\gamma = \alpha/g\gamma = \cos\theta = \tanh\rho = pc/E = \sqrt{(1 - 1/\Gamma^2)} = \sqrt{(1 - 1/K_{PV})}$).

In the early 1890s, the energy that it took to establish an electric field in the space surrounding a spherical charge was shown by the Dutch theoretical physicist Hendrik Lorentz to make it possess inertia or resistance to a change in motion, a property we associate with mass. Thus, mass might be only a manifestation of energy in the region around a source of an electric field or "charge," with nothing but empty space in the center. – Ernest J. Sternglass

The total mass of an electron is merely an electromagnetic phenomena. – W. Kauffman

If alpha [the fine structure constant] were bigger than it actually is, we should not be able to distinguish matter from ether [the vacuum, nothingness]. – Max Born

In order to make progress, one must leave the door to the unknown ajar. – Richard Feynman

17. Electron Model

J. J. Thompson discovered the electron as a fundamental particle in 1897 and determined the charge to mass (e/m) ratio by measuring the deflection of an electron beam in a cathode ray tube subjected to both an electric and magnetic field. The motion of the electron and the electrodynamics of moving objects, in general, finds a description in terms of the Lorentz factor γ (= $1/\sqrt{(1 - (v/c)^2)}$) in Einstein's four dimensional flat spacetime Special Theory of Relativity. However, in regards to the curved spacetime General Theory of Relativity, as Einstein writes in a note in Ehrenfest in 1920, "The electric field still remains unconnected. Overdeterminism does not work. Nor have I produced anything for the electron problem." Kaluza, circa 1919, hypothesized a five dimensional extension of GR and which was developed further by Klein in 1926 in a quantum mechanical description. Klein identified electric charge with periodic standing wave motion in an unseen fifth dimension in a compact 'rolled-up' cylindrical space with size on the order of 10^{-31} m. The origin of Planck's constant h (quanta of action) is attributed to circular periodicity of the fifth dimension. Attempts to treat the electron as a point particle led to singularity problems in Quantum Electrodynamics (QED) theory yielding infinities of calculated quantities. Renormalization was attempted to subtract infinities from infinities to yield finite results consistent with measured results. Dirac commented in 1975 lectures on the renormalization problem that "This is just not sensible mathematics. Sensible mathematics involves neglecting a quantity when it turns out to be small – not neglecting it just because it is infinitely great and you do not want it." The much-hyped string theory posits the electron is a one dimensional closed loop string existing a multi-dimensional universe of 26 dimensions. Extra dimensions have yet to be observed nor have testable predictions been verified experimentally. No explanation is advanced to explain orgin of electric charge.

The electron may be understood as a finite object which varies in size according to its energy of motion. An electron corresponds to a standing wave resonator with a topologically confined photon with a minimum rest mass energy of 0.511 MeV/c^2. A standing wave resonator corresponds to an electromagnetic waveguide with closed ends. In motion, a standing wave resonator undergoes a Lorentz contraction and a Lorentz Doppler shift resulting in a contracted moving standing wave with red- and blue-shifted counter-propagating waves. Such a moving contracted standing wave constitutes a de Broglie matter wave with wavelength λ = h/p where h is Planck's constant and p is momentum. The quantization of wavelength and, hence, energy is the result of bounding or limiting wavelength to an integral number of waves within the resonator. For motion of an electron in a gravitational field, the analogous gamma factor $\Gamma = dt/d\tau = 1/\sqrt{(1 - (v_e/c)^2)}$ where dt = observed time interval by a remote observer, $d\tau$ = proper time interval, and escape velocity $v_e = \sqrt{(2GM/R)}$.

The electron exhibits a Zitterbewegung "trembling" motion at an angular frequency of twice the Compton angular frequency ($\omega_{zbw} = 2\omega_C$). Zitterbewegung, in stochastic electrodynamics (SED), is attributed to EM zero-point fluctuations. The electron Compton frequency represents a stable energy resonance condition with surrounding ZPF vacuum. The Compton frequency of the electron ω_C remains constant as does the orbital whirl α^{-1}. As such, the electron represents a spin density wave of Planck dipoles with phase-locked gyration. Similar to quantized vortices in superconducting BEC superfluid such as ^{4}He, the electron contains quantized angular momentum while the surrounding vacuum fluctuations do not. The total internal energy E_{int} of the electron is 8.146E-14 J of which half is kinetic energy and half self-energy. The internal rotation generates an outward inertial force countering the vacuum quantum pressure force. For a spherical volume of radius R_C, the corresponding energy density is 3.39E23 J/m^3 (= $E_{int}/4/3\pi R_C^3$). The internal electric field provides an outward radial tension force countering the inward compression force of the magnetostatic field. To resist collapse and remain stable, the electron's electrostatic

17. Electron Model

pressure must counterbalance the external magnetostatic pressure. The internal electrostatic energy density of the electron u_e is 4.1278E20 J/m^3 (= $\alpha \hbar c/(8\pi R_C^4)$) and the magnetostatic energy density u_m is 4.118E20 J/m^3 (= $B^2/2\mu$) for a total internal EM energy density of 8.235E20 J/m^3. Refer to Fig. 17-19. A comparison of the calculated characteristics for the illustrated electron model and experimentally observed physical properties is summarized in Table 17-1.

What is electricity was good for? I do not know, but I'm pretty sure that Her Majesty's government will soon tax it. – James Clerk Maxwell

A theory is a supposition which we hope to be true, a hypothesis is a supposition which we expect to be useful; fictions belong to the realm of art; if made to intrude elsewhere, they become either make-believes or mistakes. – George Johnstone Stoney

Nature presents us, in the phenomenon of electrolysis, with a single definite quantity of electricity which is independent of the particular bodies acted on... If we make this our unit quantity of electricity, we shall probably have made a very important step in our study of molecular phenomena. – George Johnstone Stoney

An electron would not know how large it ought to be unless there existed independent lengths in space for it to measure itself against. – Sir Arthur S. Eddington

What is the structure which can appear under probing with electromagnetic fields as a point charge, yet as far as spin and wave properties are concerned exhibits a finite size on the order of the Compton wavelength? – Asim Barut

Bertrand Russell had given a talk to the then new quantum mechanics, of whose wonders he was most appreciative. He spoke hard and earnestly in the New Lecture Hall. And when he was done, Professor Whitehead, who presided, thanked him for his efforts, and not least for "leaving the vast darkness of the subject unobscured". – J. Robert Oppenheimer

It didn't take long for people to latch on to the idea that Pauli's fourth quantum number described the electron's spin which could be thought as pointing either up or down, giving a nice double-valued quantum number. – John Gribbin

Whether we consider electrons, light quanta, benzol molecules, or stones, we shall always come up against these two characteristics, the corpuscular and the undular. – Werner Heisenberg

Essentially all models are wrong, but some are useful. – George Box

In a vacuum, a charged particle acquires mass because of its interaction with its own field, that is, its self-energy. – Abraham Pais

To a request to explain what an electron really is supposed to be we can only answer, "It is the ABC of physics." – Arthur Eddington

No language which lends itself to visualizability can describe quantum jumps. – Max Born

Those wonderful instruments the new accumulators... absorb and condense the energy coming from whatever source... the transformer, a more wonderful contrivance still, which takes... from the accumulator... gives it back to space in whatever form. – Jules Verne (In the Year 2889)

A mass-object of any type may be represented by its PV [polarizable vacuum] spectrum. – Geoffrey S. Diemer

You work on renormalization theory for years and years, and then you decide: I'd better work on something else. – Prof. Rabindra Mohapatra

17. Electron Model

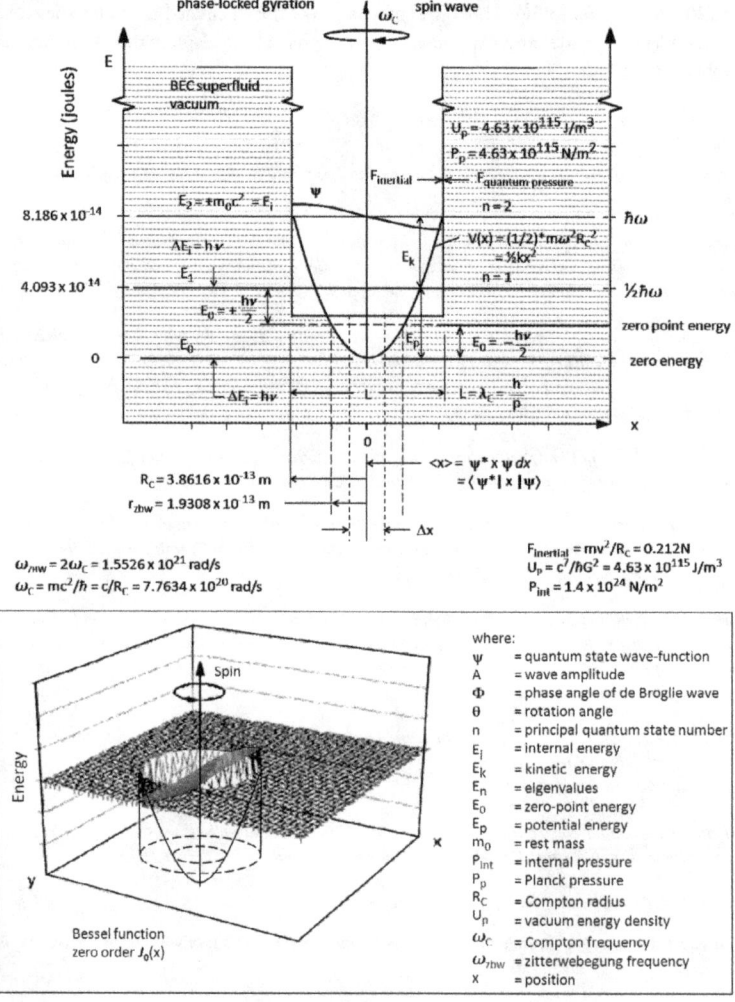

Fig 17-19. The electron represented as a resonant spin density wave confined in an oscillating deep potential well of the quantum vacuum. Zitterbewegung corresponds to the motion of the center of charge aound the center of mass at twice the Compton frequency. Spin precession with whirl no. q = α^{-1} results wave interference → electric charge and mass.

I spin on the circle of a wave upon wave of the sea. – Pablo Neruda

The ordinary operators of algebra suffice to resolve problems in the theory of curves. – Joseph-Louis Lagrange

Just because something appears plausible, it doesn't mean that it is. – David Leinweber

17. Electron Model

Table 17-1. Electron Physical Characteristics

Parameter	Symbol	Relation	Value (calc./exp.)	Units	Dimensions	Remarks
Compton wavelength	λ_C	$= h/m_ec = h/p$	2.42631058E-12	m	L	Typ. range of electron-photon interaction
		$= hc/eV = c/f_C$	0.0243	Å		
		$= (2\pi/\alpha)R_C = e^2/2\pi\varepsilon_0 m_0 c^2$	0.00243	nm		
Reduced Compton wavelength	$\lambdabar_C$	$= \lambda_C/2\pi = a_0\alpha$	3.81123974E-12	m	L	
Compton frequency	f_C, ν_C	$= \omega_C/2\pi$	1.2356E20	Hz	T^{-1}	
		$= c/\lambda_C = mc^2/h$				
Compton angular frequency	ω_C	$= mc^2/\hbar = \sqrt{(k/m)} = 2\pi f_C = c/R_C$	7.7634E20	rad/s	T^{-1}	
	ω_C'	$= \omega(1 + \alpha/2\pi)$	7.7724E20	rad/s	T^{-1}	Non-resonant. Bergman[48]
Resonant frequency	f_{LC}	$= 1/\sqrt{(LC)}$	1.2355E20	1/s	T^{-1}	LC circuit
	ω_{LC}	$= 2\pi/\sqrt{(LC)}$	7.7709E20	rad/s	T^{-1}	LC circuit
		$= 2\pi c/\sqrt{(4\pi^2 R^2)} = c/R$				
Classical electron radius	R_0, r_e	$= (1/4\pi\varepsilon_0)(e^2/m_ec^2)$	2.81792E-15	m	L	Uniformly charged sphere
		$= \alpha\lambda_C/2\pi = \alpha R_C$				
Compton radius, Quantum radius	R_C, R_q	$= \hbar/mc = c/\omega_C = \hbar c/E = R_e/\alpha$	3.8616E-13	m	L	Mackcn[20] Schilling[62]
		$= \sqrt{((2\Phi_0\mu m)/(c^2m_{e_-}))}$				$\lambda_C \sim a_0/137$
Reduced Compton radius	R_C	$= \lambda_C/2\pi = \lambdabar_C$	3.81123974E-12	m	L	
Reduced Compton wavelength	Δr	$= \lambda_C/2\pi - R_C$	3.42507974E-12	m	L	
Classical electromagnetic radius	R_{em}	$= \hbar/mc^2$	1.2880E-21	m	L	
Compton radius Phi ratio	$R_C\phi$	$= 2\pi R_C/\Phi$	3.74886E-13	m	L	$\Phi = 1.618033...$
de Broglie radius	r_{dB}	$= r_C/\beta\gamma = r_C/(v/c)(1/\sqrt{(1-(v/c)^2)}$	varies	m	L	
Lorentz contracted radius	r_L	$= r_C/\gamma = r_C/(1/\sqrt{(1-(v/c)^2)}$	varies	m	L	
Cyclotron radius	r	$= (mv^2)/qBV = v/\omega_C = mv/qB$	varies	m	L	

cont

17. Electron Model

Table 17-1. Electron Physical Characteristics (Cont)

Parameter	Symbol	Relation	Value (calc./exp.)	Units	Dimensions	Remarks
Cyclotron frequency	ω_c	$= qB/m_e = (e/m)B$	varies	Hz	T^{-1}	
Thomson scattering cross-section	σ_e	$= (8\pi/3)(e^2/4\pi\epsilon_0 mc^2)^2$ $= (8\pi/3)(\alpha\hbar c/mc^2)^2$ $= (8\pi/3)(\alpha\lambda_c/2\pi)^2$ $= (8\pi/3)r_e^2$	0.6652459 2E-28 0.6652460 6E-28 0.6515543 9E-28 0.6652459 2E-28 0.6652461 6E-28	m^2	L^2	$r_e < 4.60168$ E-28 m
Bohr radius	a_0, R_0	$= 4\pi\epsilon_0 \hbar^2/(m_e e^2)$ $= \hbar/(m_e c\alpha)$ $= \alpha^{-1}\lambda_c/2\pi = R_0/\alpha \cong \langle r_n\rangle/n^2$ $= \hbar/(m_e c\alpha)$	5.2910525E-11 5.2914124E-11 5.2917721E-11 0.529	m	L	quantum state n = 1 $1\text{Å} = 1.0 \times 10^{-10}$ m
Electric charge	e^-, q_e, q	$= F/E = m\omega = \sqrt{(\alpha q_p^2)} = \sqrt{\alpha}(q_p)$ $= (\hbar r_0/c)\omega = (s/c)q_0$ $= m_e(\omega_c + \omega_s) = -\phi/V$	1.6021764E-19 7.019E-10	C = A·s kg·rad/s	Q	$q_e = -e, q = ne$ 1C = 6.242E18e, $e = 2\pi\hbar/\mu_0 q_D$ $V = Q/4\pi\epsilon_0$
Bare charge	e^*	$= e/\sqrt{\alpha}$	1.8755E-18	C	Q	Daywitt[63,64]
Planck charge	q_P	$= \sqrt{4\pi\epsilon_0\hbar c} = \sqrt{(c^2/\alpha)} \approx 11.6q$	1.876E-18	C	Q	$e \approx 0.0854245 q_P$
Electromagnetic mass, Mass energy	m_{em}	$= m_e c^2$	0.5109990 6	MeV/c^2	M	$1\ eV/c^2 = 1.763 \times 10^{-36}$ kg
	m_{em}	$= \sqrt{(100^2)}10 = \sqrt{\alpha_{gs}}/10$	0.5116672 74	MeV/c^2	M	$\Phi = 1.61803398 7498...$
	m_{em}	$= m_e c^2$	8.1871043 8E-14	J = N·m	ML^2T^{-2}	
	Δm	$= m_e \cdot \alpha/2\pi$	1.0579772 6E-33	MeV/c^2	M	electromagnetic mass
	Δm	$= m_e \cdot (g/2 - 1)$	1.0563 5E-33	MeV/c^2	M	
	$m_0, \Delta m$	$= \alpha m_e$	6.6474376 5E-33	MeV/c^2	M	Bare mass

cont

17. Electron Model

Table 17-1. Electron Physical Characteristics (Cont)

Parameter	Symbol	Relation	Value (calc./exp.)	Units	Dimensions	Remarks
Mass	m_e	$= (1/4\pi\epsilon_0)(e^2/r_e \cdot c^2)$	9.1093-4E-31	kg	M	Mass radius $= R_m$
		$= E/c^2 = eV/c^2 = h/c \cdot R_m$	9.10941E-31			$= qB/\omega = \eta B/v$
		$= 2\pi\Phi_0 i_H/\pi R^2 \cdot c^2$	9.1095E-31			
		$= h\omega/c^2 = e(E/B)^2$	9.10876E-31			
		$= q \cdot E/(2c \cdot \Delta v \cdot r^2)$				
		$= |V \cdot q|/c^2 = q(\omega_c + \omega_p)$	9.109826E-31			
		$= h/R_c \cdot c = h/c \cdot (\lambda_c/a^2)/\theta_{137}$	9.109386E-31			
		$= 2(e^2/2C - LF/2)/(E^2 - c^2B^2)$				
		$= h/c\lambda = (k_e e^2)/(\alpha R_c c^2)$	9.109386E-31			
Planck mass ratio	m_e/m_P	$m_e/m_P = m_e/\sqrt{(hc/G)}$	4.185E-23	-	-	$= H_\beta$
Charge-to-mass ratio	e/m, g, $\omega_c/e/m$	$e/m = \omega \cdot e/m$	1.75882008E11	C/kg rad/sec	Q/M ΘT^{-1}	$\rho = m/e$ $= 5.68563\text{E-}12$ kg/C
Change radius to mass radius ratio	r_e/r_m	$r_e/r_m = 1/\sqrt{2}$	0.7070715	-	-	
Spin quantum no.	s, p_s	$m_s = S_z/\hbar$	±1/2	-	-	Projection of m_s $|S| = (\sqrt{3}/2)\hbar$
Angular Moment of Inertia	I, I_z	$= 2/3 (mR^2)$	9.0559E-56	kg·m²	ML²	Hollow sphere $R = R_C$
		$= 2/5 (mR^2)$	5.4335E-56	kg·m²	ML²	Uniform solid sphere $R = R_C$
		$= 2/5 (m(R^5 - r^5)/(R^3 - r^3))$	6.0157E-56	kg·m²	ML²	Thin spherical shell $r = R_C/2$, $R = R_C$
		$= \frac{1}{2} MR^2$	6.7919E-56	kg·m²	ML²	Uniform solid disc $R = R_C$
		$= \frac{1}{2} m(r^2 + R^2)$	8.4899E-56	kg·m²	ML²	Annular cylinder $r = R_C/2$, $R = R_C$
		$= 1/8(m(4R^2 + 5r^2)$	8.9144E-56	kg·m²	ML²	Solid torus $r = R_C/2$, $R = R_C$
		$= 1/8(m(4R^2 + 5r^2)$	6.791935E-56	kg·m²	ML²	$R = R_C$, $r = \sqrt{(e^2/(16\pi 4\epsilon_0^2 R^2 E^2))}$
		$= mR^2$	1.3583E-55	kg·m²	ML²	Thin ring $R = R_C$
		$= mR^2$	6.78398E-56	kg·m²	ML²	Thin ring $R = R_C\sqrt{2}$
Rotational Inertia	I_z	$= L/\omega = (\hbar/2)/\omega$	6.79148E-56	kg·m²/rad	ML²Θ^{-1}	$\omega = \omega_C$

cont

17. Electron Model

Table 17-1. Electron Physical Characteristics (Cont)

Parameter	Symbol	Relation	Value (calc./exp.)	Units	Dimensions	Remarks
Spin angular momentum of charged spinning ring	P_S	$= m_m cR$	5.2607E-35	J·s kg·m²/s	ML^2T^{-1}	Bergman[48,49]
		$= (e^2/8\pi^2\epsilon_0 c)\ln(8R/r) = \hbar/2$	5.2725E-35	J·s kg·m²/s	ML^2T^{-1}	Thin ring r = 8R exp(−π/α) = 3.2303E-199 m
		$= (e^2/8\pi^2\epsilon_0 c)\ln(8R/r)$	5.2725E-35	J·s kg·m²/s	ML^2T^{-1}	Thin ring $r = \sqrt{(e^2/16r^4\epsilon_0^2R^2E^2)}$ = 7.64714E-20 m
Intrinsic spin angular momentum	S, L_S	$= \hbar\sqrt{(s(s+1))}$ $= ((\sqrt{3}/2)\hbar$	9.1322E-35 9.1324E-35	J·s kg·m²/s	ML^2T^{-1}	s = 1/2
Spin angular momentum Z component	S_z	$= \pm 1/2\hbar = m_s\hbar = I\omega$ $= (|m|\hbar)/(\mu_B g)$ $= \hbar\omega/\omega_S = \hbar/2$	5.2725E-35 5.2664E-35 5.2725E-35	θJ·s θkg·m²/s	ΘML^2T^{-1}	secondary spin quantum no. $m_s = \pm 1/2$
Orbital angular momentum	L L_z	$= r \times p = mvR = m_e v R_C$ $= I\omega = m\hbar = j\hbar$ $= m_e vr = g\hbar/2$ $= (\mu_L/-e)(2m_e)$ $= mR_C v + eR_{C0}^2 B/2c$ $= \hbar\sqrt{(l(l+1))}$	1.0545887E-34 1.05457E-34 1.05457E-34 1.05458E-34 varies varies	θJ·s θkg·m²/s	ΘML^2T^{-1}	$r = R_C, v = c$
Angular momentum (homogeneous sphere)	L	$= 3/5 m_e R^2 \omega_C$	6.32742E-35	θJ·s θkg·m²/s	ΘML^2T^{-1}	
Total angular momentum	J	$= S + L$	varies	θJ·s θkg·m²/s	ΘML^2T^{-1}	

cont

17. Electron Model

Table 17-1. Electron Physical Characteristics (Cont)

Parameter	Symbol	Relation	Value (calc./exp.)	Units	Dimensions	Remarks
Total momentum	p	$= \hbar k = mv + eA$	varies	kg·m²/s	ML^2T^{-1}	Single electron. A = vector potential
		$= L(ne)^2 v$	varies	kg·m²/s	ML^2T^{-1}	Collection of electrons. Ref: Mead[20] L = inductance
Velocity of light	c	$= \lambda f = \omega/k$	2.99792458E08	kg·m²/s	ML^2T^{-1}	
		$= \sqrt{(k/\tau)} = 1/\sqrt{(\epsilon_0 \mu_0)}$				
		$= \sqrt{((k/\mu_0)} = c_0/\sqrt{(\epsilon_r \mu_r)}$				
		$= E/B = c_0/n$				
		$= \sqrt{(v_p v_g)}$				
		$= c_0/T$				
		$= l_P/t_P$				
Tangential velocity	v_{ct}	$= L/mr = S_z/m_e R = (\hbar/2)/(m_e R_C)$	1.498857E08	m/s	LT^{-1}	$r = R_C$
	v_t	$= c/(1 + \alpha/2\pi)$	2.9944680E08	m/s	LT^{-1}	@ R_m
	v_{cm}	$= \alpha r c = e/\sqrt{(4\pi\epsilon_0 m_e r)}$	2.187691263E06	m/s	LT^{-1}	Bohr atom
Velocity ratio, Bohr atom	β_B	$= \sqrt{((\lambda_C(4\lambda + \lambda_C))/(2\lambda + \lambda_C))}$	1.16138653E-03	-	-	Michaud[65] λ = wavelength of orbiting electron = 4.5633525E-8 m, λ_C = Compton wavelength = 2.42631021E-12 m
		$= v/c = \alpha$				
	β_B	$= \sqrt{(4EK + K^2)/(2E + K)} = v/c = \alpha$	1.16138653E-03	-	-	Michaud[65] E = electron rest mass energy = 8.18710414E-14 J K = Bohr gyroradius energy = 4.35974380E-18 J

cont

17. Electron Model

Table 17-1. Electron Physical Characteristics (Cont)

Parameter	Symbol	Relation	Value (calc./exp.)	Units	Dimensions	Remarks
Electric field strength (electric potential gradient)	E	$= -\nabla V = -\delta A/\delta t$ $= D/\epsilon_0$ $= F_e/e$	varies	volts/m = Nt/Coul = J/Coul·m	$MLT^{-2}Q^{-1}$	
Electric flux	ϕ_E, ψ	$= Q/\epsilon_0 = E \cdot A$ $= EA\cos\theta = |E \cdot L| \cdot A_{loop}$ $= \Sigma_i E_i \cdot \Delta A i$	1.8095127E-08	v·m = N·m^2·C^{-1} = kg·m^3·s^{-3}·A^{-1}	$L^3MT^{-2}Q^{-1}$	Q no. of lines of electric field E through area A $A_{toms} = 1.87604E\text{-}24 \, m^2, R = R_C$
Electric displacement (electric flux density, induction)	D	$= Q/A = e/\pi R^2$	3.4199E05	Coul/m^2	$L^{-2}Q$	$R = R_C$
Surface charge density (sphere)	σ	$= e/(4\pi R^2)$	8.5500E04	Coul/m^2	$L^{-2}Q$	$R = R_C$
Current	I	$= q/t = ev'_c$ $= ev/2\pi r = e\omega/2\pi$ $= \phi/L = n\epsilon_0 \psi L$ $= (\gamma\phi_0)/(\gamma L_0)$ $= I_0$ $= \mu/\pi R^2$ $= e/(\hbar/mc^2)$	19.7963 19.796 19.773 19.795 19.796	A = C/s	$T^{-1}Q$	magnetic moment $\mu = 9.274E\text{-}24 \, J/T$ Flux $\phi = h/2e$ Turns/unit length n
	I_e	$= e\omega/2\pi R_C$ $= (c^2/4\pi)((em/(h/2))$ $= (2\pi cB)/\mu_0$	19.796 19.798 19.773	A = C/s	$T^{-1}Q$	$r = R_C$
	I'	$= e\omega^3/2\pi = e\omega(1 + a/2\pi)/2\pi$ $= ec/2\pi R'$	20.023	A = C/s	$T^{-1}Q$	

cont

17. Electron Model

Table 17-1. Electron Physical Characteristics (Cont)

Parameter	Symbol	Relation	Value (calc./exp.)	Units	Dimensions	Remarks
Current density	j	$= 1/\mu_0 \nabla \times B = I/\pi R^2$	9.396E-25	A/m²	$L^{-2}T^{-1}Q$	
Current area	A_c	$= \lambda^2/4\pi = \mu_l/i$	4.6847E-25	m²	L^2	
Bohr magneton	μ_B	$= (2e/2m_e)\hbar/2 = e\hbar/2m_e$	9.27400968E-24	$J/T = m^2 \cdot A$	$L^2T^{-1}Q$	
		$= e\hbar/2m_e = S_z e/m$	5.7883826E-5	eV/T	$L^2T^{-1}Q$	
Magnetic dipole moment (sphere)	μ	$= (3/10)eR^2\omega$	-5.5644E-24	J/T $m^2 \cdot A$	$L^2T^{-1}Q$	
Magnetic dipole moment, Magnetic dipole moment, Spin magnetic moment	$\mu, \mu_e,$ m, M_z	$= i\cdot A = (2e\nu)\pi r^2 = e|s|/m_e$	-9.28483 2E-24	J/T $m^2 \cdot A$	$L^2T^{-1}Q$	$g_s = -2$
		$= \pi R^2 \omega e/2\pi = e e R/2 = (e/2m_e)L$	-9.2847637E-24			
		$= -\gamma e \hbar/2 = \gamma_e \hbar/2$	-9.2841E-24			
		$= a_\mu \mu_B$	-9.284764E-24			
		$= -(1/2)g_s\mu_B/\hbar = -(g/2)\mu_B(S/(\hbar/2))$	-9.2847701E-24			
		$= e\hbar/2m_e(1 + \alpha/2\pi)$	-9.2847636E-24			
		$= ecR_\alpha/2 = (evr)/2$	-9.273 79E-24			
		$= (ecR_c/2) + (ecR_0/4\pi)$	-9.273 79E-24			Free
		$= \gamma_s S$	-9.2734E-24			Bound
		$= e\hbar/(2m_e)$	-9.2734E-24			
		$qv r/2 = (q/2m)L$	-9.274E-24			
		$= -g_s(e/2m_e)S = g_s(e/2m)(\hbar/2)$	-9.2734E-24			
		$= -(g/2)\mu_B$	-9.274E-24			
		$= -g_s\mu_B S$	-9.274E-24			
		$= (g_s\mu_B/\hbar)S$	-9.274E-24			
		$= -g_s\mu_B S + L$	varies			

cont

17. Electron Model

Table 17-1. Electron Physical Characteristics (Cont.)

Parameter	Symbol	Relation	Value (calc./exp.)	Units	Dimensions	Remarks
Spin magnetic Moment Z-component	μ_{sz}	$= -g_s(e\hbar/2m_e)m_s$ $= -g_s\mu_B m_s$ $= -2m_s\mu_B$ $= \pm\frac{1}{2}g_s\mu_B$ $= \pm\mu_B$ $= ceR_\infty/2$	$-9.273379E-24$ $-9.274010E-24$ $-9.274010E-24$ $\pm 9.274010E-24$ $\pm 9.274009E-24$ $-9.274027E-24$	J/T $m^2 \cdot A$	$L^2T^{-1}Q$	$g_s = -2$, $m_s = \pm 1/2$
		$= \pi \bar{a}^2(I/c)k$	$5.8870E-31$	$A \cdot m \cdot s$	LQ	
		$= \mu_0/\gamma = \mu_{orl}/\gamma$	varies	J/T	$L^2T^{-1}Q$	$\gamma = 1/\sqrt{1-\beta^2}$
Orbital magnetic Moment	μ_L	$= \gamma_L L$ $= -g_L(e/2m_e)L$ $= -(e/2m)L$ $= -((g_L\mu_B)/\hbar)L$ $= i \cdot A$ $= (-e/2m_e)\sqrt{l(l+1)}\hbar$ $= \sqrt{l(l+1)}\mu_B$	$\pm 9.263951E-24$	J/T $m^2 \cdot A$	$L^2T^{-1}Q$	$g_L = -1$
Orbital magnetic moment Z-component	μ_{Lz}	$= -g_L m_l \mu_B = -\mu_B(L/\hbar)$	$\pm 9.274009E-24$, $\pm 1.854801E-24$ $\pm 2.782202E-23...$	J/T $m^2 \cdot A$	$L^2T^{-1}Q$	$m_l = 0, \pm 1, \pm 2$
Total magnetic moment	μ_J	$= \mu_L + \mu_S = g_J\mu_B(J/\hbar)$ $= -(e/2m_e)(L + gS)$	varies	J/T $m^2 \cdot A$	$L^2T^{-1}Q$	
Anapole moment Charge moment	T	$= (1/4\pi c) \cdot i \cdot (2\pi^2 \cdot t_\theta^2 \tau_\phi)$ $= (\pi \cdot e \cdot \tau_\theta \cdot \tau_\phi)/2\pi$	$2.9475E-44$ $1.8784E-44$	$C \cdot m^2$	QL^2	i = current torus vol. $= 2\pi^2 \cdot t_\theta^2 \tau_\phi$

cont

17. Electron Model

Table 17-1. Electron Physical Characteristics (Cont)

Parameter	Symbol	Relation	Value (calc./exp.)	Units	Dimensions	Remarks
g-factor/2	$g/2$	$= (\mu/\mu_B)/(S/\hbar)$	1.00115965218	-	-	Free electron
		$= 2\mu/\mu_B = \hbar\omega_S/B\mu_B$				
		$= 2\omega_S/\omega_C$				
		$= \nu_S/\nu_C = \nu_S(eB/2\pi m)$	1.0011595	-	-	
		$= 1 + (g-2)/2$	1.0011614	-	-	
		$= 1 + \omega_a/\omega_C$	1.0011614	-	-	
		$\approx 1 + \alpha/2\pi$	1.0011595	-	-	
		$= 1 + C_2(\alpha/\pi) + C_4(\alpha/\pi)^2 + C_6(\alpha/\pi)^3 + \ldots$	1.0011287	-	-	
		$= 1 + (\alpha/2\pi) + ((2\alpha^2)/3)(\alpha/2\pi) - (4/3)(\alpha/2\pi)^2$	1.00111674	-	-	$\Phi = 1.6180339\ldots$ Max. damping ratio $(5/3) - (7/12^2)$
		$= 1/\sin(\Phi) = 1/\sin(1.6180339)$				
		$= <a_c^2>/<a_m^2>$	1	-	-	Non-relativistic point electron
Spin g-factor	g_e, g_S	$g_e = 2\mu_e/\mu_B$	-2.002319043	-	-	Free electron
		$= (\|\mu\|\hbar)/(\mu_B\|s\|)$				
		$= 2m_e/(e\hbar/2m_e)$				a_e = anomalous moment
		$= -2(1 + a_e)$				
		$= (\mu_e-1)/((e/2m\hbar))$	-2.00245539170	-	-	
		$g_S = \|g_e\| = -g_e$	2.0023193 0437	-	-	
		$g = 2(1 + \alpha/2\pi + \ldots)$	$\cong 2.002319304\ldots$	-	-	QED, $C_1 = 0.5$, $C_2 = -0.326478966$, $C_3 = 1.1765$
		$= 2 + C_1\alpha + C_2\alpha + C_3\alpha + \ldots$				

cont

17. Electron Model

Table 17-1. Electron Physical Characteristics (Cont)

Parameter	Symbol	Relation	Value (calc./exp.)	Units	Dimensions	Remarks
Spin g-factor (cont)	g_e, g_S	$g = 2 \cdot (\omega_L/\omega_C) \cdot ((q/M)/(e/m_e))$	varies	-	-	Bound electron. Larmor freq. $\omega_L = g_S(\mu_B/\hbar)B$ Cyclotron freq. $\omega_C = 2(\mu_B/\hbar)B$
		$g_e = -2/\sin(\Phi)$	-2.00233473	-	-	$\Phi = 1.618033...$
	g_{Dirac}	$g_{Dirac} = \mu_S(2m)/eS$	2	-	-	Free electron.
Orbital g-factor	g_L	$g_L = -(\mu_L \hbar)/\mu_B L$	-1	-	-	
Total (Landé) g-factor (atomic)	g_J	$g = 1 + [J(J+1) + S(S+1) - L(L+1)]/2J(J+1)]/2L(L+1)]$ For free electron $g = 2$, $\mathbf{J} = \mathbf{S}$	varies	-	-	S = spin angular momentum L = orbital angular momentum, J = total angular momentum
		$g_J = -(\mu_J \hbar)/\mu_B L$	varies	-	-	
	$g_{J\,Dirac}$	$= 2 - 2/3(Z\alpha^2) + ...$	varies	-	-	
Anomalous moment	a_μ, a_e	$a_\mu = |(g-2)/2|$ $= \alpha/2\pi$ $= 1/(200^4)/2\pi$ $= \mu_e/\mu_B$	$\cong 0.0011614$ $= 0.0011613...$ $= 0.00116102...$ $= 0.00115965858...$ (experimental)	-	-	
Gyroradius magnetic orbit radius	r_0	$= \gamma m_0 v/q B_0$	varies	m	L	

cont

17. Electron Model

Table 17-1. Electron Physical Characterisitics (Cont)

Parameter	Symbol	Relation	Value (calc./exp.)	Units	Dimensions	Remarks
Gyromagnetic ratio, Spin gyromagnetic ratio	$\gamma_s, \gamma_e, \gamma$	$= -e/m_e = 2e/(2m_e)$	$-1.7582201E11$	C/kg	QM^{-1}	spin
		$= -2\mu_e/\hbar$	$-1.7608597E11$	$T^{-1} \cdot s^{-1}$		
		$= -g_e(e/2m_e)$	$-1.760859E11$	C/kg		
		$= g_e \mu_B/\hbar$	$-1.760979E11$	$T^{-1} \cdot s^{-1}$		
		$= 2\pi g_e \mu_B/h$	$-1.7608592E11$	$T^{-1} \cdot s^{-1}$		
	γ_{el}	$= \mu_s/I_s$	$-8.793486E10$	A·s/kg	QM^{-1}	orbital
		$= e/2m_e$	$-8.7941007E10$	A·s/kg		
	γ'	$= \gamma_0/\gamma$	varies	rad/sec	ΘT^{-1}	$\gamma = 1/\sqrt{(1-\beta^2)}$
Gyromagnetic factor ratio	g_{bound}/g_{free}	$= 1 - (Z\alpha)^{2/3} + \alpha(Z\alpha)^2/4\pi$	varies	–	–	Direction of electron intrinsic spin angular momentum
Magnetic quantum no. (electron)	m_s	$= \mathbf{l} \cdot \omega/\hbar = \mu_s/g_l\mu_B$	$\pm \frac{1}{2}$	–	–	
	m_l	$= \mathbf{L}_z/\hbar = \mu_{lz}/g_l\mu_B$	Orbital	l	m	No. of orbitals and orbital orientation
		Range $-l$ to $+l$	s	0	0	
			p	1	$0, \pm 1$	
			d	2	$0, \pm 1, \pm 2$	
			f	3	$0, \pm 1, \pm 2, \pm 3$	
	m_j	$= m_l + m_s$	varies $0, \pm 1/2, \pm 3/2, \ldots$	–	–	
Magnetic Flux Density (Induction)	B_0	$= E/2\mu_B = (\mu_0\pi e c)/(\alpha^3\lambda^2)$	235051.735	T = Webers/m²	$MT^{-1}Q^{-1}$	Bohr orbit Michaud[65]
		$= \gamma\mu_0 v \hbar\omega_q$				

cont

17. Electron Model

Table 17-1. Electron Physical Characteristics (Cont)

Parameter	Symbol	Relation	Value (calc./exp.)	Units	Dimensions	Remarks
Magnetic Flux Density (Induction)	B	$= \mu H = \mu_0 \mu_r H = \nabla \times A = -\nabla \phi_m$ $= \mu_0(H + M)$ $= \mu_0 \cdot I/(2R_C)$	3.217E07	T = Webers/m^2	MT^{-1}Q^{-1}	external toroidal field
	B	$= L/ct^2 = jh/ct^2$	2.2064E09	T	MT^{-1}Q^{-1}	$r = R_C$
	B_p	$= \mu_0 \cdot N \cdot i/(2\pi r_0)$ $= \mu_0 \cdot N \cdot i \cdot \pi \cdot \text{tr}/(\text{m}_e c)$	3.263606	T	MT^{-1}Q^{-1}	internal poloidal field
	B_r	$= \pi a^2(i/c) 2\cos\theta/r^3$	1.074E07	T	MT^{-1}Q^{-1}	$\theta = 0, r = R_C, a = r$
	B_θ	$= \pi a^2(i/c) \sin\theta/r^3$	0	T	MT^{-1}Q^{-1}	$\theta = 0, r = R_C, a = r$
	B_r	$= \pi a^2(i/c) 2\cos\theta/r^3$	6.581E-11	T	MT^{-1}Q^{-1}	$\theta = \pi/2, r = R_C, a = r$
	B_θ	$= \pi a^2(i/c) \sin\theta/r^3$	5.372E05	T	MT^{-1}Q^{-1}	$\theta = \pi/2, r = R_C, a = r$
	B_z	$= (\mu_0 \cdot I \cdot a^2)/(2(a^2 + z^2)^{3/2})$	1.230E07	T	MT^{-1}Q^{-1}	$\theta = 0, a = R_C, z = 0$
	B_z	$= (\mu_0 \cdot I \cdot a^2)/(2(a^2 + z^2)^{3/2})$	1.965E07	T	MT^{-1}Q^{-1}	$\theta = 0, a = R_C, z = R_C/2$
Magnetic Flux Density (Point dipole)	B_r	$= (2\mu_0\mu_\text{s}\cos\theta)/(4\pi r^3)$	1.31412E18	T	MT^{-1}Q^{-1}	$\theta = 0$
	B_θ	$= (\mu_0\mu_\text{s}\sin\theta)/(4\pi r^3)$	0	T	MT^{-1}Q^{-1}	$\theta = 0$
	B_r	$= (2\mu_0\mu_\text{s}\cos\theta)/(4\pi r^3)$	2.1276E14	T	MT^{-1}Q^{-1}	$\theta = \pi/2$
	B_θ	$= (\mu_0\mu_\text{s}\sin\theta)/(4\pi r^3)$	1.7366E30	T	MT^{-1}Q^{-1}	$\theta = \pi/2$
Magnetic Flux Density (Thin Ring)	B	$= e^2 4\pi^2 \epsilon_0 c\pi R$ $= \sqrt{(\mu_0 \epsilon_0 E^2)}$	3.21735E07	T	MT^{-1}Q^{-1}	$r = R_C/10$

cont

17. Electron Model

Table 17-1. Electron Physical Characteristics (Cont)

Parameter	Symbol	Relation	Value (calc./exp.)	Units	Dimensions	Remarks
Magnetic Flux Density (rotating spherical shell)	B	$= ((2\mu_0 R_c\omega r)/3)(\cos\theta\hat{r} - \sin\theta\hat{\theta})$	2.14735E07	T	$MT^{-1}Q^{-1}$	$r < R_c$, $\theta = \pi/2$ rad
		$= ((2\mu_0 R_c\omega r)/3)(\cos\theta\hat{r} - \sin\theta\hat{\theta})$	2.14735E07	T	$MT^{-1}Q^{-1}$	$r < R_c$, $\theta = \pi/2$ rad
		$= ((\mu_0 R^4\omega r)/3)(1/r^3)(2\cos\theta\hat{r} + \sin\theta\hat{\theta})$	3.20212E-18	T	$MT^{-1}Q^{-1}$	$r \geq R_c$, $\theta = 0$ rad
		$= ((\mu_0 R^4\omega r)/3)(1/r^3)(2\cos\theta\hat{r} + \sin\theta\hat{\theta})$	1.60106E-18	T	$MT^{-1}Q^{-1}$	$r \geq R_c$, $\theta = \pi/2$ rad
Internal Average Electric Field Strength	$\langle E \rangle$	$= \sqrt{((6\hbar c)/(\pi\varepsilon_0\lambda^4))}$	3.51619E16	volts/m	$MLT^{-2}Q^{-1}$	
Electric Field Strength	E	$= \Omega H$ $=\Phi_c/4\pi R_c^2 = e/4\pi R_c^2$ $= -\sqrt{(B^2/\mu_0\varepsilon_0)}$	9.6454E15	volts/m	$MLT^{-2}Q^{-1}$	
Induced Electric Dipole Moment	μ	$= qr$	varies	Coul·m	LQ	Separation of charges in an external field
de Broglie wavelength	λ_{dB}	$= h/p = h/\gamma m_0 v = 2\pi c^2/(\gamma v \omega_c)$ $= h/\sqrt{(2mE_k)} = h/\sqrt{(2meV)}$ $= hc/\sqrt{((m - mc^2)(E + mc^2))}$ $= 2d\sin\theta/n$ $= \sqrt{(2\pi\hbar^2/mk_B T)}$	varies	m	L	Bragg relation $\lambda = 2d\sin\theta/n$
de Broglie frequency	f_{dB}	$= pc/h = 2\pi c/\lambda_{dB}$ $= (c/h)\sqrt{(m r c^2 + p_x^2 + p_y^2 + p_z^2)}$	varies	Hz	T^{-1}	
Magnetic Field Strength	H	$= (\mu_0\mu)\cdot 1/2R_c$ $= B/\mu_0$	2.56E13 2.56E13	Amp·turn/m	$L^{-1}T^{-1}Q$	

cont

17. Electron Model

Table 17-1. Electron Physical Characteristics (Cont)

Parameter	Symbol	Relation	Value (calc./exp.)	Units	Dimensions	Remarks
Magnetic Field Strength	H_r	$= -\delta\psi_m/\delta r = (1/4\pi)(2\mu\cos\theta/r^3)$	1.6279E06	Amp·turn/m	$L^{-1}T^{-1}Q$	$\theta = 0$ rad; μ = magnetic dipole moment
			9.9671E-11	Amp·turn/m	$L^{-1}T^{-1}Q$	$\theta = \pi/2$ rad
	H_θ	$= -(1/r)\delta\psi_m/\delta\theta = \mu\sin\theta/(4\pi r^3)$	0	Amp·turn/m	$L^{-1}T^{-1}Q$	$\theta = 0$ rad
			-1.2815E13	Amp·turn/m	$L^{-1}T^{-1}Q$	$\theta = \pi/2$ rad
Magnetic Flux Linkage	Λ	$= N\Phi$	2.0899E-16	Wb = volt·sec	$ML^3T^{-2}Q^{-1}$	
		$= L \cdot I$	2.0921E-16			
Electric dipole moment (EDM)	p, d_e	$= q\mathbf{d} = q\lambda_c = q\hbar/(2m_e)$	1.2373E-31	Coul·m	LQ	$\mathbf{d}$ = distance between equal and opposite charges
		$= -U/E$	8.4777E-30			
		$= erv/c$	Exp. < 0.07E-28			
Inertial mass of magnetic field of spinning ring	m_m	$= (3\mu_0 e^2)/(32\pi^2 R)$	7.93478E-34	kg	M	$R = R_c$
Magnetostatic mass	m_m	$= E_m/c^2 \approx E_f/2c^2 \approx m_e/2$	4.54420E-31	kg	M	Bergman[48]
Mass of confined photon	m	$= U/c^2 = (hc/\lambda)/c^2$	9.109413E-31	kg	M	$\lambda = \lambda_c$
Energy of confined photon	U	$= hc/\lambda$	8.187132E-14	J	$MT^{-2}L^{-1}$	$\lambda = \lambda_c$
Internal Energy	E_i	$= mc^2 = \hbar\omega_c$	8.1871E-14	J	$MT^{-2}L^{-1}$	
		$= \gamma E_0$	varies	J	$MT^{-2}L^{-1}$	
Kinetic Energy Self Energy	E_0, U_0	$= mc^2/2$	4.0935524E-14	J	ML^2T^{-2}	Free electron
	E_k	$= E_0(\gamma v^2/2c^2)$	varies	J	ML^2T^{-2}	$\gamma = 1/\sqrt{1-\beta^2}$

cont

17. Electron Model

Table 17-1. Electron Physical Characteristics (Cont)

Parameter	Symbol	Relation	Value (calc./exp.)	Units	Dimensions	Remarks
Acceleration Energy	E_a	$F \cdot d = E_0(\gamma - 1)$	varies ($\hbar$ increments)	J	ML^2T^{-2}	Bergman[48] Acceleration distance $d = E_0(\gamma - 1)/F$
Radiated Energy	E_r	$= E_a - E_k = (E_0(\gamma-1)^2)/2\gamma$	varies ($\theta/2$ incr.)	J	ML^2T^{-2}	Bergman[48]
Potential energy	E_p	$\cong -e^2/4\pi\varepsilon_0 r_n$	varies	J	ML^2T^{-2}	n^{th} quantum state, r_n = electron and proton separation.
Total energy	E_t	$= \hbar\omega + e\phi$ $\pm \sqrt{((pc)^2 + (mc^2)^2)}$	varies	J	ML^2T^{-2}	
Energy Density	U_q	$= E_t/R_q^3 = (\hbar\omega c^4)/c^3$	1.4217E24	J/m³	$MT^{-2}L^{-1}$	Macken[20]
Internal Vacuum Pressure	P_q	$= U_q/3$	0.4739E24	N/m²	$MT^{-2}L^{-1}$	
Electrostatic Energy	W_e, U, E_s	$= e^2/4\pi\varepsilon_0 r_0$ $= m_e c^2$	8.1871E-14	J	ML^2T^{-2}	$r_0 = R_C$
	E_{s0}	$= e^2/2C_0$	4.0935E-14	J	ML^2T^{-2}	Bergman[48] Ring capacitance = C_0 = 3.12812E-28 F
	E_{s0}'	$= E_{s0}(1 + \alpha/2\pi)$	4.0983E-14	J	ML^2T^{-2}	
Electrostatic Energy external to radius r	E_{ext}	$= \alpha\hbar c/2r = 1/2\varepsilon_0 E_i$ $= e^2/(8\pi\varepsilon_0 R)$	2.9870E-16	J	ML^2T^{-2}	$R = R_C$
Electric field energy density	u_e, η_E	$= \varepsilon_0 E^2/2$	4.1281E20	J/m³	$MT^{-2}L^{-1}$	
Electrostatic Energy (Charged Sphere)	E	$= q^2/8\pi\varepsilon_0 r$	2.98720E-16	J	ML^2T^{-2}	$r = R_q$
		$= (3/5)(1/4\pi\varepsilon_0)(e^2/r_e)$	3.5846E-16	J	ML^2T^{-2}	$r = R_q$

cont

17. Electron Model

Table 17-1. Electron Physical Characteristics (Cont)

Parameter	Symbol	Relation	Value (calc./exp.)	Units	Dimensions	Remarks
Electrostatic Energy (Charged Spinning Ring)	E_{cr}, E_e	$= e^2/C$	4.0935E-14	J	ML^2T^{-2}	Capacitance $C = e^2/2E_e$
Electrostatic Energy (Thin Charged Spinning Ring)	E_{cr}	$= (e^2/8\pi^2\epsilon_0 R)\ln(8R/r)$ $= \alpha mc^2 \ln(R_c/R_g)$	3.727119E-14 3.79711E-14	J J	ML^2T^{-2} ML^2T^{-2}	$C = 4\pi^2\epsilon_0 R/(\ln(8R/r)$ $R_g =$ charge radius
Hydrogen Atom Binding Energy	E	$= -(GMm_e)^2 m_e/(2\hbar^2)$	-13.6	eV	ML^2T^{-2}	
Centrifugal Force, Curvature Force	F_c	$= m c^2/R_C = mv^2/r$ $= E/R_C$	2.210106E-01	$N = kg \cdot m/s^2$	MLT^{-2}	$r = R_C$ Energy/Radius
Electric Coulomb Force	F_e	$= qE = (1/4\pi\epsilon_0)(e^2/r^2)$	varies	$N = kg \cdot m/s^2$	MLT^{-2}	Macken[20]
Max Repulsive Force	F_{MR}	$\approx \alpha F_m = \alpha \hbar c/R_c^2$ $\alpha(l_P^2/r^2)F_P = \alpha(l_P^2/r^2)(c^4/G)$	1.5470E-03 varies	$N = kg \cdot m/s^2$	MLT^{-2}	
Relativistic Force	F_r	$= \hbar c/R_c^2 = m^2 c^3/\hbar$ $= P_c/c = \hbar c/R_c^2 = E_I^2/\hbar c$ $= PR_c^2 = F_m$	2.211998E-01 2.211998E-01	$N = kg \cdot m/s^2$ $N = kg \cdot m/s^2$	MLT^{-2} MLT^{-2}	Force between electrons @ distance R_c. Macken[20] Force required to deflect circulating power. Macken[20]
Magnetic Force	F_m	$= evB$	varies	$N = kg \cdot m/s^2$	MLT^{-2}	
Lorentz Force	F_L	$= q(E + (v \times B))$	varies	$N = kg \cdot m/s^2$	MLT^{-2}	
Magnetic Compression Force	F_H	$= (1/2)\mu_0\Phi_I^2$	2.4623E-04	$N = kg \cdot m/s^2$	MLT^{-2}	$\Phi_I = I/2, I = e\omega/2\pi$
Electric Power	P	$= V \cdot I$	5.1096E05	$W = J/s$	ML^2T^{-3}	
Circulating Power	P_c	$= E_I\mu_C = E_I/\hbar = \omega_c^2\hbar = \hbar c^2/R_c^2$	6.6356E07	$W = J/s$	ML^2T^{-3}	Macken[20]

cont

17. Electron Model

Table 17-1. Electron Physical Characteristics (Cont)

Parameter	Symbol	Relation	Value (calc./exp.)	Units	Dimensions	Remarks
Magnetostatic Energy, Magnetostatic Mass Energy	W_m, E_m, U_B	$= \frac{1}{2} LI^2 = \frac{1}{2} I\phi = \frac{1}{2} L(ec/2\pi R_C)^2$	4.08406E-14	J	ML^2T^{-2}	R_m = magnetic field radius ~ R_C
		$= 2\mu r^2/3Rm^3 = 2(eh/2mc)^2/3Rm^3$	4.09512E-14			
		$E_{mo} = \phi_0^2/2L_0 = I_0\phi_0^2/2$	4.08412E-14	J	ML^2T^{-2}	Bergman[48]
		$= ((\mu_0 N^2 \Gamma^2 (R-r))/4\pi)\ln(R/r)$	1.174E-14	J	ML^2T^{-2}	$R = R_C$, $r = \sqrt{(e^2/16\pi^4 \epsilon_0^2 R^2 E^2)}$
		$m_m c^2 = 2\phi^2/L = 2(L \cdot \dot{r})^2/L$	4.09198143E-14	Wb²/H, J	ML^2T^{-2}	O.F. Shilling[62]
		$= -\mu \cdot B = -I \cdot A \cdot B$	varies	J	ML^2T^{-2}	B = Ext. B Field
		$= \gamma E_{m0}$	varies	J	ML^2T^{-2}	$\gamma = 1/\sqrt{(1-\beta^2)}$
	E_{m0}'	$= E_{m0}(1 - \alpha/\pi)$	4.0746333E-14	J	ML^2T^{-2}	Bergman[48]
Magnetic self-energy	W_H	$\approx (\alpha/2\pi)mc^2 = 0.0011641 mc^2$	9.5085318E-17	J	ML^2T^{-2}	Schwinger correction[61] Irrotational.
		$\Delta mc^2 = 593$ eV				
Magnetic field energy density	u_m, u_H, η_B	$= B^2/2\mu_0 = (\mu_0 N^2 I^2)/(8 s^2 r^2)$	4.11867E20	J/m³	$MT^{-2}L^{-1}$	
Electric/Magnetic field energy density ratio	U_m/U_e	$= U_m/U_e$	1.002289574	-	-	
Potential Energy	U	$= -\mu B \cos\theta = -\mu \cdot B$	varies	J	ML^2T^{-2}	
Magnetic Mass	m_m	$= 2\phi^2/Lc^2 = 2(Li)^2/Lc^2$	4.552943E-31	kg	M	O.F. Schilling[62]
		$= \pi\phi_0 i_B/\pi R^2 c^2$	4.55995E-31			
		$= p_\theta/cR$	4.54419E-31			

cont

17. Electron Model

Table 17-1. Electron Physical Characteristics (Cont)

Parameter	Symbol	Relation	Value (calc./exp.)	Units	Dimensions	Remarks
Magnetic mass fraction	m_e/m_m	$= m_e/m_m$	0.49885	-	-	
Ratio of electrostatic to magnetostatic energy	W_e/W_m	$= W_e/W_m$	1.0047	-	-	
Total Internal Energy	W_I, E_I, U	$= m_e c^2$ $= W_e + W_m$	8.18711E-14 8.18711E-14	J	ML^2T^{-2}	
Magnetostatic pressure	P_m	$= -U_m/3$ $= (B^2/2\mu_0)/3$	1.3910E19 – 1.3728E20	$N \cdot m^{-2}$	$ML^{-1}T^{-2}$	Outer surface pressure equivalent to exterior energy density
		$= E_{m0}/Vol_{sphere}$	1.6964E23	$N \cdot m^{-2}$	$ML^{-1}T^{-2}$	$R = R_C$
		$= E_{m0}/Vol_{torus}$	1.4398E23	$N \cdot m^{-2}$	$ML^{-1}T^{-2}$	Horn torus $R = R_C$, $r = R_C/2$
			9.1786E35	$N \cdot m^{-2}$	$ML^{-1}T^{-2}$	Thin ring, R_C $r = \sqrt{(e^2/(16\pi^4\epsilon_0^2 R^2 E^2))}$
		$= F_H/A_{sphere}$	8.3653E19	$N \cdot m^{-2}$	$ML^{-1}T^{-2}$	$R = R_C$
		$= F_H/A_{torus}$	8.3563E19	$N \cdot m^{-2}$	$ML^{-1}T^{-2}$	Thick ring A = Ring surface area $= 4\pi^2 rR = 1.8760$E-24 m^2
		$= -\sigma vB$	2.5602E13	$N \cdot m^{-2}$	$ML^{-1}T^{-2}$	Charged ring σ = Surface charge density $= 2.6544$E-03 C/m^2, $v = c$, $B = 3.21735$ T

cont

17. Electron Model

Table 17-1. Electron Physical Characteristics (Cont)

Parameter	Symbol	Relation	Value (calc./exp.)	Units	Dimensions	Remarks
Magnetostatic pressure (cont)	P_m	$= -B/\mu_0$	2.5602E13	$N \cdot m^{-2}$	$ML^{-1}T^{-2}$	$B = 3.21735$ T
Electrostatic pressure	P_e	$= U_e/3$	1.37604E20	$N \cdot m^{-2} = J \cdot m^{-3}$	$ML^{-1}T^{-2}$	Inner surface pressure equivalent to interior energy density
		$= E_{s\phi}/Vol_{sphere}$	1.6971E23	$J \cdot m^{-3}$	$ML^{-1}T^{-2}$	$Vol_{sphere} = (4/3)\pi r^3 = 2.4121E\text{-}37$ m^3, $r = R_C$
		$= E_{s\phi}/Vol_{torus}$	8.2563E26	$J \cdot m^{-3}$	$ML^{-1}T^{-2}$	Thick ring, $R = R_C$, $r = R_C/2$
		$= E_{s\phi}/Vol_{torus}$	5.2633E33	$J \cdot m^{-3}$	$ML^{-1}T^{-2}$	Thin ring, R_C, $r = \sqrt{(e^2/(16\pi^4\epsilon_0^2 R^2 E^2))}$
		$= F_E/A_{sphere}$	1.37604E20	N	MLT^{-2}	$R = R_C$
		$= F_E/A_{torus}$	6.48444E20	N	MLT^{-2}	Thick ring, $R = R_C$, $r = R_C/2$
		$= \sigma E$	8.2373E20	$N \cdot m^{-2}$	$ML^{-1}T^{-2}$	Charged ring $E = e/A\epsilon_0 = e/(4\pi r^2 \epsilon_0 R)$ volts/m
		$= e^2/(16\pi^4\epsilon_0^2 R^2)$	5.2633E33	$N \cdot m^{-2} = kg/ms^2$	$ML^{-1}T^{-2}$	$R = R_C$, $r = \sqrt{(e^2/(16\pi^4\epsilon_0^2 R^2 E^2))} = 7.6471E\text{-}20$ m
		$= e^2/(16\pi^4\epsilon_0^2 R^2)$	8.2563E26	$N \cdot m^{-2} = kg/ms^2$	$ML^{-1}T^{-2}$	Thick ring $r = R_C/2$, $R = R_C$
Wavelength/rev	r_ϕ	$= \sqrt{(E/m_e \cdot \omega^2)} = c/f_C$	2.426E-12	m	L	
Radial Strain	ϵ_e	$= \Delta L/L = \sqrt{\alpha L_p/R_C}$	3.5753E-24	-	-	Macken[20]

cont

17. Electron Model

Table 17-1. Electron Physical Characteristics (Cont)

Parameter	Symbol	Relation	Value (calc./exp.)	Units	Dimensions	Remarks
Toroidal radius	R_t	$= \hbar/mc = c/\omega_C = \lambda_C/2\pi$ $= R_C/\sqrt{(1 + R_C^2\omega^2/c^2)}$	3.8616E-13	m	L	
Poloidal radius	R_p	$= R_C/2 = \lambda_C/4\pi$ $= e/4\pi^2\epsilon_0 c B R_C$ $= e/4\pi^2\omega E R_C$	1.9308E-13 1.2305E-13 1.2291E-13	m	L	Thick ring
		$= \sqrt{(e^2/(16\pi^4\epsilon_0^2 R^2 E^2))}$	7.64714E-20			Thin ring
Mass radius	R_m	$= \hbar c/E$	3.8613E-13	m	L	
Charge radius, Effective charge radius	R_c, R_q	$= \lambda_C/2\pi - (\sqrt{2})^*\lambda_C/4\pi$ $= \sqrt{(A_c/\pi)}$	3.86159E-13	m	L	A = current loop area
		$= \sqrt{((\sqrt{A_c/\pi^2})^2 + (\sqrt{A_c/\pi^2})^2)}$	4.7295E-13			
	R_E	$= \alpha^2 R_0$ $= \alpha^3 R_C$	1.5005914E-19 1.5005942E-19	m	L	
Magnetic radius, Magnetic field radius	R_m, R_H	$= \sqrt{((2\pi\theta_\omega)/(\pi mc^2))}$ $= R_C(1 + \alpha^2/2\pi)^{2/3}$	3.8616E-13 3.86458E-13	m	L	O.F. Shilling[62]
Berry phase	γ	$= (e / \hbar)\phi = (\pi / \hbar)L = j\pi$	varies	rad	Θ	
Mass deficit angle	α	$= 2\pi(1 - \sin\gamma)$	π	rad	Θ	= 180 deg
Loop radius	$R_C/2$	$= \lambda_C/4\pi$	1.90561987E-12	m	L	
Classical electron charge radius	r_e	$= \alpha\lambda_C/2\pi$ $= \alpha^2 a_0 = \alpha^2\hbar/m_0 c^2$	2.7812E-13	m	L	

cont

17. Electron Model

Table 17-1. Electron Physical Characteristics (Cont)

Parameter	Symbol	Relation	Value (calc./exp.)	Units	Dimensions	Remarks
Electromagnetic radius, Classical electron radius	R_0	$= \alpha \hbar / m_e c$ $= (1/4\pi\epsilon_0)(e^2/m_e c^2)$ $= k_0 e^2/mc^2$ $= \alpha R_C$ $= \alpha \cdot e/(\omega_C + \omega_e/m)$ $= e^2/8\pi\epsilon_0 U_e$	2.8177488E-15 2.8179403 2E-15 2.8179404E-15 2.8179 4566E-15 2.8179 4566E-15 2.8224 448E-15	m	L	typ. range of electron-electron interaction $R_C \sim \lambda_C/137$
Electromagnetic radius	R_{em}	$= R_e/\alpha$	5.29178214E-11	m	L	
Classical electron radius (charged spherical shell)	R_{0s}	$= (1/2)e^2/mc^2$	1.56 76908E-25	m	L	Surface charge distribution
Classical electron radius (charged solid sphere)	R_{0ss}	$= (3/5)e^2/mc^2$	1.88 12289E-25	m	L	Uniform charge distribution
Thin ring radius	r	$= 8R \exp(-\pi/\alpha)$	3.3203E-199	m	L	Bergman[48] Thin ring $r \ll R$, $R = R_C$
		$= \sqrt{(e^2/(16\sigma^4 \epsilon_0 R^2 E^2))}$	7.64714E-20	m	L	Thin ring $r \ll R$
Torix inversion radius (spindle torus)	R_i	$= R_C - R_m$	2.698327E-17	m	L	
Thermodynamic charge	h	$= \lambda_0 \omega m_e = \lambda_0 T_0 S_0 m_e$	1.045E-34	kg·m²/s	ML^2T^{-1}	Klyushin[60] Poloidal rotation. $m_e \lambda_0 = 7.072$E-10 kg/s

cont

17. Electron Model

Table 17-1. Electron Physical Characteristics (Cont)

Parameter	Symbol	Relation	Value (calc./exp.)	Units	Dimensions	Remarks
Temperature	T_0	$= \omega_0/S_0$ $= \hbar/\lambda_0 T_0 m_e$	5.9299E9 1.147E-4	°K m²/s	T^0 L^2T^{-1}	Klyuschin[60] $\omega_0 = 7.7634$E20 rad/s $S_0 = 6.7061$E24 rad/m² 1 °K = 1.9523E-11 m²/s
Entropy (wave no.)	S_0	$= 1/\lambda_0 = 1/r_0^2$ $= 1/(c^2/i\omega_0^2) = \omega_0/T_0$	6.7061E24	rad/m²	ΘL^2	Klyuschin[60] $r_0 = 3.8616$E-13 $T_0 = 5.9299$E9 °K
X-ray wavelength	λ_{min}	hc/Ve	varies	m	L	Duane-Hunt relation
Self-Capacitance	C	$= e^2/2E_c$	3.12812E-25	Farads	$M^{-1}L^{-2}T^2Q^2$	
		$= e/V$	3.13560E-25	Farads	$M^{-1}L^{-2}T^2Q^2$	
		$= (4\pi^2\varepsilon_0 R)/\ln(8R/r)$	3.12817E-25	Farads	$M^{-1}L^{-2}T^2Q^2$	Bergman[48] Thin ring $r \ll R$, $R = R_c$ $r = \sqrt{(e^2/(16\pi^4\varepsilon_0^2R^2E^2))}$ $= 7.64714$E-20 m
		$= C_0/\gamma$	varies	Farads	$M^{-1}L^{-2}T^2Q^2$	$\gamma = 1/\sqrt{(1-\beta^2)}$
Self-Inductance	L	$= n\Phi_0/i = n(h/2e)/i = \Lambda/i$ $= 2\Phi^2/mc^2$ $= n\mu_0\pi R^2/\mu = L_\bullet$	1.0446E-16 (n=1) 1.0446E-16 2.0899E-16 (n=2)	Henry = Wb/A	ML^2Q^{-2}	n = turns/unit length. O.F. Shilling[62]
		$= \mu_0 R[\ln(8R/r)-2]$ $\approx \mu_0 R[\ln(8R/r)]$	7.5285E-18	Henry = Wb/A	ML^2Q^{-2}	Bergman[48] Thin ring $r \ll R$, $R = R_c$, $r = 7.64714$E-20 m

cont

17. Electron Model

Table 17-1. Electron Physical Characteristics (Cont)

Parameter	Symbol	Relation	Value (calc./exp.)	Units	Dimensions	Remarks
Self-Inductance (cont)	L	$= 2U_B/I^2$	2.0899E-16	Henry	ML^2Q^{-2}	$R_2 = R_{C,} R_1 = 7.64714E\text{-}20$ m
		$= (\mu_0 N^2 R_2 - R_1)/2\pi)\ln(R_2/R_1)$	2.9660E-17			
		$= (V/f_C)/I$	2.0897E-16	Henry	ML^2Q^{-2}	
		$= \gamma L_0$	varies			$\gamma = 1/\sqrt{(1-\beta^2)}$
Inductance Thin Ring	L	$\mu_0 R(\ln(8R/r) - 2)$	2.08991E-16	Henry	ML^2Q^{-2}	$R = R_C, r = 7.64714E\text{-}20$ m
Inductance Thick Ring	L	$B_p \pi r_p^2/i$	1.9308E-20	Henry	ML^2Q^{-2}	r_p = poloidal radius
						B_p = poloidal magnetic field
Magnetic Flux Quantum	Φ_0	$= (h/2e) = (\pi R_C^2 c^2 m)/(2n)$	2.067833E-15	Weber (Wb)	$ML^2T^{-1}Q^{-1}$	
		$= 1/k_J$				
Magnetic Flux	Φ_B, Φ_m	$= n(h/2e) = n\Phi_0$	2.067833E-15	Weber (Wb)	$ML^2T^{-1}Q^{-1}$	n = turns/unit length = 2
		$= \pi R_C^2 c^2 m/2\mu$	2.0677E-15	= volt-sec		
		$= L \cdot i$	2.089E-16	$= T \cdot m^2$		
		$= \pi r^2 B = (\pi/e)L = j\pi\hbar/e$	varies	Wb	$ML^2T^{-1}Q^{-1}$	
		$= \gamma \Phi_0$	varies	Wb	$ML^2T^{-1}Q^{-1}$	$\gamma = 1/\sqrt{(1-\beta^2)}$
Quantized Vorticity	Φ_m	$= \hbar c/e$	1.9731E-07	m^3/rad-sec	$L^3T^{-1}\theta^{-1}$	
Flux/Vorticity	$\Phi e/\Phi_m$	$= e^2/(\hbar c/e)$	8.12E-13	$C/m^3 \cdot kg \cdot s^{-2}$	$\Theta^2 ML^{-3}$	
No. of Fluxoids	n	$= \Phi_B/\Phi_0$	1	—	-	
		$= (mc^2 \pi R^2)/(2\Phi_0 \mu)$				

cont

17. Electron Model

Table 17-1. Electron Physical Characteristics (Cont.)

Parameter	Symbol	Relation	Value (calc./exp.)	Units	Dimensions	Remarks
Electromotive Force (EMF)	ε	$= -Nd\phi_m/dt = E/q$ $= W/q = V = IR$ $= -(\sigma/e)dL/dt$ $= -(\sigma/e)\tau$ $= -N \cdot A \cdot (dB/dt)$	5.1096E05	volt	$ML^2T^{-2}Q^{-1}$	Open circuit potential. Force induced on charge ϕ_m = enclosed magnetic flux, $E = W$ = Energy, τ = torque
Magnetomagnetic Force (MMF)	$\mathcal{F}, F_m$	$= nI = 1*I$	1.97963	A	$\Theta M T^{-1}$	n = no. of turns i = current
		$= \phi \mathcal{R}$	1.97963	A	$\Theta M T^{-1}$	$\mathcal{R}$ = Reluctance
		$= HL$	1.97736	A	$\Theta M T^{-1}$	L = Toroid Coil Length
Reluctance	$\mathcal{R}$	$\mathcal{F}/\Phi = l_m/\mu A_m$	4.78393E05	Amp turns/m turns/H	$M^{-1}L^{-2}Q^{-2}$	L = Coil Length A = loop area
	R_m	$= V_m/\Phi = n \cdot I/\phi$	4.78393E05	Amp turns/m turns/H	$M^{-1}L^{-2}Q^{-2}$	V_m = magnetic potential difference Φ = magnetic flux
Resistance	R	$= \rho_c l_c / A_c = Z_c l^2$ $= V/I$	2.5811E04 2.5802E04	Ω	$ML^2T^{-1}Q^{-2}$	
Toroidal loop area	A_t	$= \pi R_c^2 = \lambda_c^2/4\pi = \mu/\hat{i}$	4.6847E-25	m^2	L^2	
Poloidal loop area	A_p	$= \pi (R_c/2)^2$	1.1712E-25	m^2	L^2	
Cardioid area	A_c	$= (3/2)\pi a^2$	1.7568E-25	m^2	L^2	$a = R_c/2$

cont

17. Electron Model

Table 17-1. Electron Physical Characteristics (Cont)

Parameter	Symbol	Relation	Value (calc./exp.)	Units	Dimensions	Remarks
Precession area	A_a	$= \frac{1}{2} S_e \cdot R_m$	3.4183E-27	m^2	L^2	
Schwarzschild radius (electron), Gravitational radius	R_S, R_G	$= 2Gm_e/c^2$	1.355E-57	m	L	$\sim 10^{-22} l_p$
	r_q	$= \sqrt{(q^2 G/(4\pi\epsilon_0 c^4))}$	9.152E-37	m	L	Charged black hole
	R_{min}	$= Gm_0/c^2$ $= Gm_p/c^2$ $= l_p^2/R_q$	6.7760E-58 6.7635E-58 6.7643E-58	m	L	
Orbital precession	$\Delta\phi$	$= 6\pi G(m_1 + m_2)/(c^2 a(1-\epsilon^2))$ $= 3\pi Gm^2(1 + m_1/(m_1+m_2))/(ac^2(1-\epsilon^2))P$	varies	rad/s	ΘT^{-1}	a = ellipse semi-major axis = $R_e/2$, b = semi-minor axis = $R_e/4$. ϵ = eccentricity = $\sqrt{(a^2-b^2)}/a$. P = orbital period. Prograde precession in direction of orbital motion.
Orbital frequency	f_L, ω_L	$= \gamma/\beta = q\beta/m_e$	1.76086285E11	1/s = Hz	T^{-1}	Hestenes[50] β = effective magnetic field = 1.000 T, $\gamma = m_e/e$
Spin precession	g, γ_S, $\omega_{e/m}$	$= (e/m_e)(1 + \alpha/2\pi)$ $= \omega_e/m$	1.76086285E11	C/kg rad/s	Q/M ΘT^{-1}	Same as gyro-magnetic spin ratio γ_S.

cont

17. Electron Model

Table 17-1. Electron Physical Characteristics (Cont)

Parameter	Symbol	Relation	Value (calc./exp.)	Units	Dimensions	Remarks
Thomas-Wigner rotation (circular orbit), Thomas precession	$\Delta\theta_T$, Ω_T, TP, TWR	$= 2\pi(v^2/c^2)$ $= 2\pi(\gamma - 1)$ $= (\Delta v/v)(1 - \sqrt{(1-v^2/c^2)})$ $= (a\Delta t/v)(1 - \sqrt{(1-\beta^2)})$ $= \omega_T T$ $= \pi v^2(1 + (3/4)v^2 + (5/8)v^4 + (35/64)v^4 + (63/12)v^8 + ...)$ $= -\Delta\phi(1 + (v^2/1 - v^2)\cos(\phi)^2)$	-0.0003346 -0.0001673	rad	Θ	Lorentz factor $\gamma = 1/\sqrt{(1 - v^2/c^2)} = $ N/(N – 1). No. of nutations N = no. of orbital waves. Total angle of rotation = $2\pi\gamma$. Retrograde precession to orbital revolution.
Thomas precession angular velocity, precession rate	ω_T, $\delta\Omega/\delta t$, ω_T	$-d\theta/dt = -(\gamma - 1/v^2) v \times a$ $= (1/2c^2) v \times a = -(v^2/c^2)(eE/m_e)$ $= -\gamma(\beta \times a)/c\beta^2$ $= -(\gamma^2/\gamma + 1)(v \times a)/2c^2$ $= (2\pi/\gamma T)(\gamma - 1)$ $= -(e/m_ec)(1/\gamma + 1) \beta E_0 $ $= -(e/mc^2) v \times E$ $= (g_e - 1)eB/2m$ $= \gamma^2 \omega t$ $= \omega_c(\gamma - 1) = \omega - \omega_T$ $= -\omega[1/\sqrt{(1-\beta^2)} - 1]$	-2.0671E16	rad/s	ΘT^{-1}	$\omega_T/\omega = \gamma - 1 = (\Delta\theta/T)/(2\pi/T)$. $\beta = v/c$, T = orbital period (interval). v/R = a/v. Relativistic gyroscopic precession opposite to ½ Larmor rotation. $\omega_T = \omega(1 - r^2\omega^2/c^2)^{-0.5} = -\omega_L/2$
Thomas time dilation	$d\tau$	$= (1/\sqrt{(1 - v^2/c^2)}dt = \gamma dt$	varies	s	T	

cont

17. Electron Model

Table 17-1. Electron Physical Characteristics (Cont)

Parameter	Symbol	Relation	Value (calc./exp.)	Units	Dimensions	Remarks
Larmor precession frequency	ω_L, ω_e	$= eB/mc$ $= m\mu_B B/\hbar = (g_s/4\omega/\hbar)H$ $= (\mathbf{v} \times \mathbf{a})/c^2 = -2\omega_r$	varies	rad/s	ΘT^{-1}	m = magnetic quantum number
Total precession frequency	ω_t	$= \omega_L + \omega_r = (\mathbf{v} \times \mathbf{a})/2c^2$ $= -(e^2/2mc^2)\gamma \mathbf{v} \times \mathbf{E}$	varies	rad/s	ΘT^{-1}	excl. geodetic precession
Whirl no.	q	$= \lambda_C/S_e = (2\pi c/\omega_C)/S_e = \alpha^{-1}$	$= 137.0359...$	—	—	Arc length $S_e = R_C \theta = 1.77056$E-14 m
Betatron frequency	ω_β	$= \omega_r/\sqrt{(2\gamma)}$	varies	rad/s	ΘT^{-1}	ω_p = plasma freq.
Cyclotron frequency	ω	$= qB/m$	varies	rad/s	ΘT^{-1}	For $\omega = \omega_C$, $B = 4.4139$E9 T
Geodetic (de Sitter) precession	Ω_G	$= (3/2)^*GM/2c^2 r^2)\sqrt{(GM/r)}$ $= 3\gamma^2 \omega m (1 - a\omega)/r$	2.9969E8	rad/s	ΘT^{-1}	Prograde, $r = R_C$, $\omega = d\phi/dt$
Geodetic precession angle	$\Delta\Omega_G$	$= \Omega_G/f_C$	2.4247E-12	rad	Θ	
gravitomagnetic precession	$\Delta\phi_g$	$= \pi((2\pi GM)/(c^2 r))$	1.1025E-44	rad	Θ	Prograde, $r = R_C$
gravitomagnetic precession rate	ω_g	$= 2\pi(v^3/2\pi c^2) = \pi((2\pi GM)/(Tc^2 r))$ $= \Delta\phi_g/T$	3.7618E-17	rad/s	ΘT^{-1}	T = period = $2\pi r/v$
Stiffness	k	$= m_e \omega c^2$	5.4902E11	kg·rad^2/s^2	$ML^2 T^{-2}$	
Energy density (spherical)	U	$= (3/4\pi)E_E/R_q^3$	3.3942E23	J/m^3	$ML^{-1}T^{-2}$	Macken[20]
Quantized acceleration	a_q	$= m_e c^3/\hbar$ $\mu \cdot c\hbar \cdot \mathbf{L}$	8.1871E-14	m/s^2	LT^{-2}	
Gravitational acceleration @ electron center	A_q	$= L_p \omega c^2 = \sqrt{(m^4 c^5 G/\hbar^3)}$	9.7404E6	m/s^2	LT^{-2}	Macken[20]

cont

17. Electron Model

Table 17-1. Electron Physical Characteristics (Cont)

Parameter	Symbol	Relation	Value (calc./exp.)	Units	Dimensions	Remarks
Gravitational acceleration @ R_q	g_q	$= Gc^2m^3/\hbar^2$	4.0764E-16	m/s²	LT^{-2}	Macken[20]
Magnetic scalar potential, magnetostatic potential, Magnetic potential	ϕ_m, Ω	$\phi_m = -i \cdot \Omega/(4\pi)$	varies	A = J/Wb = C/s	QT^{-1}	Ω = solid angle sub-tended by loop contour C at point p. $\Omega = -2\pi(1-z/(R/\sin\theta))$ $= 2\pi(1-\cos\theta)$. For Pt p @ center of current loop, $(z=0), \Omega = 2\pi$ sr.
	ψ_m	$= \mathbf{m} \cdot \mathbf{r}/4\pi r^3 = m \cdot \cos\theta/4\pi r^2$	-4.954	A	QT^{-1}	$\theta = 0$ rad **m** = dipole moment $\mathbf{H} = -\nabla\psi_m$
		$= \mathbf{m} \cdot \mathbf{r}/4\pi r^3 = m \cdot \cos\theta/4\pi r^2$	-3.035E-16	A	QT^{-1}	$\theta = \pi/2$ rad
	ϕ_t	$= i\Omega = (ev/2\pi r)/2)$	9.89816	A	QT^{-1}	$r = R_c, v = c$
Magnetic scalar potential (magnetic dipole)	U	$= ((\mu_0 m)/(4\pi r^2))(\cos\theta)$	-3.592E-14	A	QT^{-1}	**m** = dipole moment, θ = half angle subtended by loop. $r = R_c, \theta = 0$
Magnetic scalar potential (rotating spherical shell)	U	$= ((\mu_0 R^4\omega\sigma)/(3r^2))\cos\theta$	varies	A	QT^{-1}	$r > R_c$ ω = angular vel. σ = surface charge density
		$= -((2\mu_0 R_c\omega r)/3)r\cos\theta$	varies	A	QT^{-1}	$r < R_c$

cont

17. Electron Model

Table 17-1. Electron Physical Characteristics (Cont.)

Parameter	Symbol	Relation	Value (calc./exp.)	Units	Dimensions	Remarks
EM scalar potential	ϕ, V_{ab}	$=\int_{ab} E\cdot d\ell$	varies	volts	$ML^2T^{-2}Q^{-1}$	Potential drop. $V_{ab} = V_a - V_b$
Magnetic vector potential	A	$A = p/q_0 = \hbar k/q_0$	1.7044E-16	volt-sec/m = Wb/m	$MLT^{-1}Q^{-1}$	Magnetic field strength/unit length. Wave no. $k = \omega_E/c$
		$A = (\mu_0 i/4\pi)\oint d\ell \cdot i_r$				
		$A = v\phi_E/c^2 = \epsilon_0\mu_0 v\phi_E$		T·m = N/A		
		$A_\varphi(r,\theta) \cong (\mu_0/4\pi)(I\pi R^2)\sin\theta/r^2$	1.1835E-04 for $\theta = 90°$, $r = R_C$	volt-sec/m	$MLT^{-1}Q^{-1}$	Current loop ext $r \gg R$
		$A_\varphi(r,\theta) \cong (\mu_0 I/4)(r/R)\sin\theta$	8.6368E-07 for $\theta = 90°$, $r = R_C$	volt-sec/m	$MLT^{-1}Q^{-1}$	Current loop int $r \ll R$
		$A_\varphi(r,\theta) = \pi a^2(I/c)\sin\theta/r^2$	3.9478E-06 for $\theta = 90°$, $a = R_C$	volt-sec/m	$MLT^{-1}Q^{-1}$	Current loop ext
		$A_\varphi(r,\theta) = (\mu \cdot 1 \cdot a^2 r \sin\theta)/(2(a^2 + r^2)^{3/2})$	4.1844E-05 for $\theta = 90°$, $a = R_C$	volt-sec/m	$MLT^{-1}Q^{-1}$	$r = R_C$
		$A_\varphi(r,\theta) = (\mu_0 \cdot 1 \cdot a^2 r \sin\theta)/(2(a^2 + r^2)^{3/2})$	1.1835E-04 for $\theta = 90°$, $a = R_C$	volt-sec/m	$MLT^{-1}Q^{-1}$	$r = 0$
		$A = \Phi_m/2\pi \hat{\varphi} = \nabla\phi_m$ $A' = A + \nabla\phi_m$	varies	volt-sec/m	$MLT^{-1}Q^{-1}$	$\phi_m(r)$ = magnetic scalar potential $(T \cdot m^2) \neq$ magnetic flux $\phi_m (T \cdot m^2)$.
		$A = \frac{1}{2} B \times r$	6.21206E-06	volt-sec/m	$MLT^{-1}Q^{-1}$	$r = R_C$
		$A_{tot} = \Phi_B v/2\pi a^2 = \Phi_B/2\pi R_C = \Phi_B/\lambda_C$	1.70254E03	volt-sec/m	$MLT^{-1}Q^{-1}$	$r = R_e$, $a = R_e$
		$A_{ext} = \Phi_B/2\pi r$	1.70254E03	volt-sec/m	$MLT^{-1}Q^{-1}$	$r = R_C$

cont

17. Electron Model

Table 17-1. Electron Physical Characteristics (Cont.)

Parameter	Symbol	Relation	Value (calc./exp.)	Units	Dimensions	Remarks
Magnetic vector potential (rotating solid sphere)	A	$A = \mu_0 J r_0^2/3$	6.5978E-07	volt-sec/m	$MLT^{-1}Q^{-1}$	Sphere of radius r_0, Current density J
Magnetic vector potential (rotating spherical shell)	A	$A(r,\theta,\phi) = (1/3)\mu_0 R\omega\sigma\sin\theta\hat{\phi}$	varies	volt-sec/m	$MLT^{-1}Q^{-1}$	$r < R_c$
		$A(r,\theta,\phi) = ((\mu_0 R^4 \omega\sigma)/3)(\sin\theta/r^2)\hat{\phi}$	varies	volt-sec/m	$MLT^{-1}Q^{-1}$	$r \geq R_c$
Magnetomotive force (MMF)	$\mathcal{M}, \mathcal{F}, F_m$	$= I \cdot N$	varies	Amp-turns A·N	$T^{-1}Q\Theta$	
Critical electric field	E_{crit}	$= m_e^2 c^3/e\hbar$	1.323375E18	V/m	$MLT^{-2}Q^{-1}$	Max. strength of vacuum
Electric dipole moment	p	$= qd$	varies	Coul-m	LQ	d = displacement vector
Zitterbewegung angular velocity	ω_{zbw}	$= 2\omega_c$ $= 2m_e c^2/\hbar$	1.55268E21 1.55279E21	rad/s	ΘT^{-1}	
Zitterbewegung frequency	f_{zbw}	$= \omega_{zbw}/2\pi$	2.4712E20	$1/s$ ($= Hz$)	T^{-1}	
Zitterbewegung radius	r_{zbw}	$= (\lambda_c/2\pi)/2 = R_c/2$ $= 0.5\hbar/mc$ $= c/2\omega$	1.930800E-13 1.930665E-13 1.930806E-13	m	L	
Average EM Energy Imbalance	$\Delta E/E$	$= (E_{s0} - E_{m0})/(E_{s0} + E_{m0})$ $\approx 2(\alpha/2\pi)$	2.0023206	-	-	
Dynamical mass	m_z	$= (\hbar/2)\omega$	4.09325E-14	$m^2 \cdot kg/s^2$	L^2MT^{-2}	Hestenes[50] Energy stored in zitter motion.
Electrostatic orbital radius	r_{ee}	$= \varepsilon_0 h^3/4\pi^2 m_e^2 m_e q$ $= \sqrt{(q/4\pi\varepsilon_0 B v_0)} = q/4\pi v_0 m_e$	5.2856E-11	m	L	Bohr atom

cont

17. Electron Model

Table 17-1. Electron Physical Characteristics (Cont)

Parameter	Symbol	Relation	Value (calc./exp.)	Units	Dimensions	Remarks
Breaking torque	P	$= \tau \cdot \omega_c = \mathcal{E} \cdot I$	1.0115E7	N·m = kg·m²/s²	ML²T⁻²	
Magnetic orbital radius	r_{mo}	$= r_B$ = Bohr radius	5.2917E-11	m	L	
Euler characteristic (torus)	χ	$= 2 - 2g$	0	-	-	genus g = 1, Total curvature = 0
Planck/Compton ratio	$x_C/x_P, \varepsilon_P$	$= L_P/R_C$ $= m_e/m_P$ $= E/E_P$ $= \omega_C/\omega_P$	4.1851E-23	-	-	
Reduced Compton radius Phi ratio	x/Φ	$= 2\pi R_C/\Phi$	3.74886E-13	m	L	$\Phi = 1.618033...$
Electric permittivity	ε_0	$\varepsilon_0 = \mu_0/Z_0^2 = c/K = 1/Z_0 c_0$	8.84025E-12 - 8.854187E-12	C²/N·m² = F/m	MLQ⁻²	
Magnetic permeability	μ_0	$\mu_0 = Z_0/c$	4πE-7	Wb/A·m = N/A² = H/m	M¹L⁻³T⁻²Q²	
Potential energy	V	$= Q/4\pi\varepsilon_0$	3.3017E-08	volts	ML²T⁻²Q⁻¹	
Electrostatic potential	φ	$= -e \cdot V$	-5.2898E-27	kg·m²/s²	ML²T⁻²	
Cyclotron impedance	Z_c	$= V/I = \pi r^2 B/N \cdot e$ $= (n/N)(h/2e^2)$	varies	Ohm (Ω) = Volt/Amp	ML²T⁻¹Q⁻²	N = no. of electrons, n = quantum state I = cyclotron current = Neω/2π
Electron impedance	Z_e	$= Z_0/\alpha$	51,649	Ohm (Ω) = Volt/Amp	ML²T⁻¹Q⁻²	orbital electron in the 1st orbital of a hydrogen atom

cont

17. Electron Model

Table 17-1. Electron Physical Characteristics (Cont)

Parameter	Symbol	Relation	Value (calc./exp.)	Units	Dimensions	Remarks
Vacuum Impedance	Z_Θ	$=\|E\|/\|H\|$ $=\sqrt{(\varepsilon_o/\mu_o)}$ $=\mu_o c_0$ $=\sqrt{(\mu_0/\varepsilon_0)}$ $=1/\varepsilon_0 c_0$ $=4\pi Z_P$	376.7303135	Ohm (Ω) = Volt/Amp	$ML^2T^{-1}Q^{-2}$	Free space
Planck Impedance	Z_P	$=V_P/I_P$ $=\hbar/q_P^2$ $=1/4\pi\varepsilon_0 c$ $=Z_\Theta/4\pi$ $=1000(\alpha/2\pi)R_K$	29.979245	Ohm (Ω) = Volt/Amp	$ML^2T^{-1}Q^{-2}$	
Frequency horizon radius	R_ν	$=(2Ghv/c^4)\nu$	6.778307E-58	m	L	
Coulomb constant	k_e, k_C	$=1/4\pi\varepsilon_0$	8.98755179E09	$N\cdot m^2/C^2$	$ML^{-3}T^{-2}Q^2$	
Josephson const.	K_J	$=2e/h=1/\Phi_0$	4.835978E14	Hz/V	$M^{-1}L^{-2}TQ$	Φ_0 = magnetic flux quantum
Verdet constant	V	$=\theta/Bd=(e/2mc)\lambda dn/d\lambda$	varies	rad^2/m	ΘL^{-1}	Wavelength and material dependent
von Klitzing const.	R_K	$=h/e^2$ $=Z_P/1000(\alpha/2\pi)$	25.812807449	Ohm (Ω) = Volt/Amp	$ML^2T^{-1}Q^{-2}$	
mass moment constant	k_S^g	$=mR^2/\mu$	1.462E-32	s^2/rad	$T^2\Theta^{-1}$	$\mu=-ge\hbar/2=g_e\mu_B/2$. Schilling[62]
Mass charge spin constant	G	$=(c^2/4\pi)(cm/h/2)$	4.949419674	rad·kg	ΘM	S.Ghosh et al[66]
Hall resistance	R_H	$=h/e^2$	25812.8056(12)	Ohm (Ω)	$ML^2T^{-1}Q^{-2}$	$\sigma=ve^2/h$ where $v=1,2,3$, etc

cont

17. Electron Model

Table 17-1. Electron Physical Characteristics (Cont)

Parameter	Symbol	Relation	Value (calc./exp.)	Units	Dimensions	Remarks
Mass number	n	$= M/(\alpha^{-1} m_e)$	varies	-	-	M = particle mass.
Mass quantum no. bosons	m_b	$= m_e/\alpha = (\hbar c/e^2) m_e = (2/3) m_f$	70.02526	MeV/c^2	M	Q_L. Nambu[67]
Mass quantum no. fermions	m_f	$= (3/2) m_e/\alpha$	105.0379	MeV/c^2	M	$Q_{3/2}$. MacGregor[68]
Schwarzschild Radius/Mass ratio constant	G/c^2	$= R_S/M$	7.4260489 4E-28	m/kg	LM^{-1}	$R_S = 2GM/c^2$ MacGregor[68]
Planck's constant	h	$= R_g/c^2 = E/\nu$	6.626093E-34	J·s (= kg·m²/s)	$ML^2 T^{-1}$	
Reduced Planck's constant, Dirac h	$\hbar$	$= h/2\pi = e^{*2}/\alpha c = q_P^2/c = e^2/\alpha$ $= E/2\pi\nu$ $= R_g/c^2$	1.0545E-34	J·s (= kg·m²/s)	$ML^2 T^{-1}$	QM angular momentum operator
Planck length	l_P	$= \sqrt{(\hbar G/c^3)} = c T_P$	1.616199E-35	m	L	
Planck time	t_P	$= \sqrt{(\hbar G/c^5)}$	5.3955E-44	s	T	
Planck angular frequency	ω_P	$= \sqrt{(c^5/\hbar G)}$	1.855E43	rad/s	ΘT^{-1}	
Planck frequency	f_P	$= \omega_P/2\pi$	2.949E42	Hz	T^{-1}	
Planck mass	m_P	$= \sqrt{(\hbar c/G)}$	2.17651E-8	kg	M	$\lambda_{dB} = R_S$
Planck force	F_P	$= Gm_P m_P/r_{min}^2 = c^4/G$ $= m_P l_P t_P^{-2} = G(m_P^2/l_P^2)$ $= (1/4\pi\epsilon_0)(q_P^2/l_P^2)$	1.2084E44	N	MLT^{-2}	

Cont

17. Electron Model

Table 17-1. Electron Physical Characteristics (Cont)

Parameter	Symbol	Relation	Value (calc./exp.)	Units	Dimensions	Remarks
Planck power	P_P	$= c^5/G$	3.6222E52	W (= J/s)	ML^2T^{-3}	
Planck energy	E_P	$= \sqrt{(\hbar c/c^5)}$	1.9561E9	J	ML^2T^{-2}	
Planck energy density	U_P	$= c^7/\hbar G$	4.6329E113	J/m³	$ML^{-1}T^{-2}$	
Planck pressure	p_P	$= F_P/l_P^2$	4.6329E113	N/m²	$ML^{-1}T^{-2}$	
Planck mass density	ρ_P	$= m_P/l_P^3$	5.1509E96	kg/m³	ML^{-3}	
Planck dynamic viscosity	η_P	$= \sqrt{((\rho_P c^4)/(8\pi G))} = \tau_{ij}/\gamma_{ij}$	4.9762E69	kg/m·s	$ML^{-1}T^{-1}$	stress tensor: τ_{ij}; displacement rate tensor: γ_{ij}
Planck voltage	V_P	$= \sqrt{(c^4/4\pi\epsilon_0 G)} = \sqrt{(F_P/4\pi\epsilon_0)}$	1.043E27	volts	$ML^2T^{-2}Q^{-1}$	
Planck charge	q_P	$= \sqrt{(4\pi\epsilon_0 \hbar c)}$	1.8754E-18	Coul	Q	
Planck current	i_P	$= q_P c/(2\pi R_P)$	5.5322E24	A (= Coul/s)	$T^{-1}Q$	
Planck electric field strength	E_P	$= c^4 G \sqrt{}/(4\pi\epsilon_0 \hbar c)$	6.4423E61	volts/m	$MLT^{-2}Q^{-1}$	
Planck magnetic flux density	B_P	$= c^3 G\sqrt{}/(4\pi\epsilon_0 \hbar c)$	2.1489E53	T = webers/m	$MT^{-1}Q^{-1}$	
Planck magnetic field strength	H_P	$= i_P/2\pi r_P$	5.4432E53	webers/m²	$L^{-1}T^{-1}Q$	
Planck linear acceleration	a_P	$= c^4/(Gm_P)$	-5.5562E51	m/s²	LT^{-2}	
Planck temperature	T_P	$= m_P c^2/K_B = \sqrt{(\hbar c^5/G K_B)}$	1.4168E32	°K	-	

cont

17. Electron Model

Table 17-1. Electron Physical Characteristics (Cont)

Parameter	Symbol	Relation	Value (calc./exp.)	Units	Dimensions	Remarks
Rydberg constant	R_∞	$= (m_e e^4)/(8\varepsilon_0^2 h^3 c)$ $= \alpha^2/2\lambda_c$ $= \alpha/4\pi a_0$	1.0973731 E7	m^{-1}	L	
Einstein constant	k	$= 8\pi G/c^4$	−2.0765E-43	1/N	$T^2M^{-1}L^{-1}$	
Gravitational coupling ratio	a_g	$= Gc^{-2} m_e^3 / \hbar^2$	−4.0845E-16	m/s^2	LT^{-2}	
Gravitational coupling constant	α_g	$= Gm_e^2 / \hbar c$ $= (m_e/m_p)^2$ $= (t_p \omega_c)^2$	1.7518E-45	-	-	
Strong gravitational constant	Γ	$= e^2/(4\pi\varepsilon_0 m_e n_e)$	1.514E29	$m^3 kg^{-1} s^{-2}$	$L^3 M^{-1} T^{-2}$	
Ratio of electrostatic & gravitational coupling constants	α/α_g	$= e^2/(4\pi\varepsilon_0 Gm_e^2)$	4.17E42	-	-	
Electrogravitic length	L_S	$= \sqrt{(ke^2 G)/(c^5)}$	1.38E-36	m	L	
Euler's identity	$e^{i\pi}$	$e^{i\pi} = -1 = \Phi^{-1} - \Phi$	−1	-	-	$\Phi = 1.61803398...$ $\pi = 3.141592655...$ $e = 2.71828182...$ $i = \sqrt{(-1)}$
Graviton exchange coupling constant	a_g	$= (Gm_p^2 2\pi)/\hbar c$	5.82E-39	-	-	

cont

17. Electron Model

Table 17-1. Electron Physical Characteristics (Cont)

Parameter	Symbol	Relation	Value (calc./exp.)	Units	Dimensions	Remarks
Fine structure constant	α	$= v_e/c$ $= Z_0/Z_\infty$ $= e^2/4\pi\epsilon_0\hbar c = (1/4\pi\epsilon_0)e^2/\hbar c$ $= k_e e^2/\hbar c = 2\pi k e^2/(hc)$ $= e^2 c\mu_0/2h$ $= \frac{1}{2}(Z_0 \, G_0)$ $= (e/q_P)^2$ $= c_0{}^2$ $= \lambda_0/a_0 = R_0/\lambda_C$ $= R_0/R_C$ $= R_e/R_{em}$ $= F_e/F_m$ $= \lambda_0/r_{dB}$ $= \cos(\pi/137)/137$ $= \omega_F/\omega$	7.2973525E-3 $\approx 1/137$	-	-	scaling in velocity, force, impedance, lifetimes; elementary charge to Planck charge; Bohr radius to Compton radius; Compton radius to electro-magnetic radius; de Broglie/Compton wave number ratio; orbital precession to spin. G_0 = conductance quantum = $2e^2/h$
		$= 1/2 0 \phi^4$ where $\phi = 1.6180...$	0.007294902...		-	$\Phi = (1+\sqrt{(5)})/2 = 1.6180339...$
Inverse fine structure constant	α^{-1}	$= 1/\alpha = 2\pi q/\phi = q = 1C/Se = (\hbar c)/(k_e e^2)$ $= 2h/(e^2\sqrt{(\epsilon_0/\mu_0)}) = \pi(3-\sqrt{5})$	137.0359917...	-	-	scaling in mass, whirl no.
Reduced fine structure constant	$\alpha/2\pi$	$= \alpha/2\pi$ $= \frac{1}{2}(E_{m0} - E_{\infty})/(E_{m0} + E_{\infty})$	0.0011641	-	-	scaling in radius, angular velocity; magnetic/electrostatic energy imbalance

Quantum electrodynamics gives us a complete description of what an electron does; therefore in a sense it gives us an understanding of what an electron is. – F. J. Dyson

It is only in a theory with electron spin that one can see why the wave function is complex. – David Hestenes

17. Electron Model

The electron external magnetic field is generated by a loop current by toroidal circulation of electric charge. The poloidal magnetic field is confined internally and is not directly observable. The magnetic moment μ_e is aligned with the toroidal axis and opposite in direction to the spin angular momentum vector **s**. The *g*-factor (g_s) of the electron is given as the ratio $g_s = -g_e = 2\mu_e/\mu_B = 2.002319304$. The anomalous magnetic moment of the electron ($\alpha_e = 1.15965$) represents the excess over the Dirac calculated moment of $g_s = 2.0$. The anomalous moment is associated with the fine structure constant α (i.e., $a_\mu = |(g-2)/2| = \alpha/2\pi$). The measured *g*-factor indicates the spinning charge radius and mass radius are not equal. For $g > 1$, the charge radius $\langle R_c^2 \rangle = g \cdot \langle R_m^2 \rangle$. A charge path following a cardioid trajectory can account for this asymmetry as well as the $\hbar/2$ spin characteristic. Rather than the postulated interaction with virtual particles of the QED theory, the increased magnetic moment is attributed to the poloidal (anapole) magnetic field of the toroidal electron which adds an additional internal magnetic field. The external electric field of the rotating electron does not self-interact. The magnetic field inside does not rotate with the spin. For an electron at rest, the intrinsic magnetic field is given by

$$\mathbf{B}_r = -2(-e\hbar/2m_e c)(g/2)(\cos\theta/r^3) \tag{17-2}$$
$$\mathbf{B}_\theta = -(-e\hbar/2m_e c)(g/2)(\sin\theta/r^3) \tag{17-3}$$

where *g* is the electron gyromagnetic factor (≈ 2.0023).

The intrinsic magnetic moment μ_B may be calculated by considering a charge equivalent to $-2e$ (e^- charge circulated at $2\omega_c$) circulating in an eccentric orbital motion opposite to the toroidal spin motion ($-e$ charge circulated at ω_c) as follows:

$$\mu = IA = (2ev)/(2\pi r)\pi r^2 = evr = e|s|/m_e \tag{17-4}$$
$$= e\hbar/2m_e = \mu_B$$

where:

I		current
A		current loop area
e		electron charge
v		velocity of charge (varies from superluminal at extrema and subluminal at center of spin)
s		intrinsic angular momentum of double charge
m_e		mass of electron
$\hbar$		spin angular momentum
μ_B		Bohr magneton

In the model shown, the electron magnetic moment, $\mu_e = \mu_B(1 + \alpha/2\pi)$, has both toroidal spin and orbital spin components. The spin component is associated with toroidal spin about the center-of-mass whereas the orbital spin magnetic moment arises as a result of the cardioid loop around the center of charge. The orbital precession is twice the rotation frequency (Zitterbewegung = $2\omega_c$) resulting in an Epitrochoid (limacon) charge trajectory path. The $\alpha/2\pi$ term represents the orbital precession. The number of turns for a complete rotation (whirl no. q) $= \lambda_c/s_e = \lambda_C/R_C \cdot \theta = 137.036$ ($\cong 1/\alpha$).

The zitterbewegung is a local circulatory motion of the electron. – David Hestenes

17. Electron Model

For a single electron with spin ($S_z = \hbar/2$) directed along the z-axis, the magnetic moment

$$\mu_{spin,\,z} = (-e\hbar/2m_ec)(g/2) \approx -e\hbar/2m_ec \qquad (17\text{-}5)$$

The magnetic moment of the electron μ_s is slightly greater than the Bohr magneton μ_B. The calculated magnetic dipole moment μ is greater than that of sphere of equivalent radius (−9.27E-24 vs. −5.56E-24 J/T) indicating that the electron is not perfectly spherical. The electron gyromagnetic ratio γ_s for an equivalent double charge may be expressed as

$$\gamma_s = e/m_e = 2e/(2m_e) \qquad (17\text{-}6)$$

The intrinsic spin magnetic moment $\mu_s = -\gamma_s \mathbf{s}$ where the negative sign indicates the spin angular momentum $\mathbf{s}$ is oppositely directed. For a positron, both vectors are aligned. For comparison, the orbital magnetic moment μ_L for a bound electron in a hydrogen atom

$$\mu_L = \gamma_e \mathbf{L} = -(\mu_B/\hbar)\mathbf{L} \qquad (17\text{-}7)$$

where:

γ_e	gyromagnetic ratio ($= \sqrt{(1 - \alpha(Z-1))}$; α = fine structure const ($= k_e \cdot e^2/\hbar c$); Z = nuclear charge
μ_B	Bohr magneton
L	orbital angular momentum

Electron spin magnetic moment μ_s for the hydrogen atom is

$$\mu_s = \gamma_e \mathbf{s} = g_s \gamma_e \mathbf{s} = -g(\mu_B/\hbar)\mathbf{s} \qquad (17\text{-}8)$$

where:

γ_s	gyromagnetic ratio
s	spin angular momentum
g_s	gyromagnetic factor

The magnetic moment of an electron modeled as a sphere with homogeneous charge distribution is $\mu_{sphere} = -5.644\text{E-}24$ J/T. Experimental results indicate a larger magnetic moment μ than this with a difference expressed as in terms of Landé g-factor: $g_s = \gamma/(e/2m) \cong 2.0023$. The ratio of the magnetic moment M to the angular momentum J of the shell in Pauli's equation is

$$\mu_L = M/J = \gamma_e/2m_ec \qquad (17\text{-}9)$$

where:

M	magnetic moment
J	total angular momentum

Dirac's equation not only accounted for the spin of the electron and its observed magnetic moment, but also correctly explained the fine structure constant of the hydrogen atom. – Slyvan S. Schweber

17. Electron Model

The intrinsic magnetic moment μ_e (= $\mathbf{M}_z$) is slightly greater than the Bohr magneton μ_B. The difference $\mu_e - \mu_B$ is a consequence of Zitterbewegung (trembling motion). The total magnetic moment $\mu_T = \mu_s + \mu_L$ where μ_T = magnetic moment created by volumetric current (spin) and μ_L = Impulse moment due to rotation (orbital). The impulse moment μ_L is attributed to the Barnett effect.

The anomalous gyromagnetic ratio (Δa_f ~0.0011596521811) arises as the result of interaction of spin with the Lorentz force in curved spacetime. Anomalous magnetic moment arises from radiative corrections owing to interaction of the electron with its own EM field. In the presence of an external magnetic field $\mathbf{B}$, the magnetic dipole precession is given by $\omega_L = d\theta/dt = (e/2mc)\mathbf{B}$ and is termed the Larmor frequency. Thomas precession is retrograde motion due to length contraction. The Larmor precession is twice the Thomas precession and in the opposite direction: $\omega_L = (\mathbf{v} \times \mathbf{a})/c^2 = -\omega_{Th}$. The spin precession frequency is twice as large as the orbital angular momentum precession: $\omega_L = g(e/2mc)\mathbf{B}$ where g = 2. Experimentally, g is found to be 2.00232 resulting in a slightly higher precession frequency. The spin magnetic moment μ_s for the electron is collinear but oppositely directed from the spin angular momentum $\mathbf{S}$ and precesses at the same angular frequency.

Electromagnetic fields are described by Maxwell's equation while matter fields are described by Schrödinger's equation. The existence of particles in a unitary view of nature may be understood as a quantization of these fields. Photons represent quantization of travelling waves (bosons) in accordance with the Planck relation. Electrons represent quantization of standing waves (fermions) in accordance with the Einstein relation. Polaritons represent quantization of a mixed state of bound photons and quasi-particle excitons (electrons plus electron holes).

I really believe that the electron moment is 12% greater than the Bohr magneton. This seems to be in almost exact agreement with Schwinger's calculation and accounts completely for the hyperfine structure in hydrogen and deuterium... – I. I. Rabi

A hole, if there were one, would be a new kind of particle, unknown to experimental physics, having the same mass and opposite charge to the electron. – P.A.M. Dirac

Mass is a very important thing, something whose origin we must explain. But it's also the most difficult problem in physics. Just look at the range of masses. You start with an electron and go to the Planck mass, which is twenty-two orders of magnitude larger. – Abdus Salam

This is Dr. Compton who is with us from the United States to discuss his work on "The Size of the Electron". I hope you will listen attentively, but you don't have to believe him. – Ernest Rutherford

We have found a strange footprint on the shores of the unknown. – Arthur Eddington

A theory has only the alternative of being right or wrong. A model has a third possibility: it might be right but irrelevant. – Manfred Eigen

Everything should be as simple as possible, but no simpler. – Albert Einstein

The electron is not as simple as it seems. – Sir Lawrence Bragg

If you want to do serious physics, sometime you just have to learn it. – Prof. R. Mohapatra

By studying nature, man can overtake his imagination; he can discover and understand what he is even unable to imagine. – Lev Landau

18. Pair Production and Annihilation

A theory with mathematical beauty is more likely to be correct than an ugly one that fits some experimental data. – P. A. M. Dirac

A geometrical model of the photon should account for pair production of an electron and positron from an energetic photon as well as production of gamma rays from collision and annihilation of an electron and positron. An electron/positron model must provide an explanation for the origin of observed electric charge and mass. As illustrated by Feynman diagrams, pair production and annihilation represents the transformation of EM waves into matter and vice versa. See Fig. 18-1. In an annihilation process indicated by $e^- + e^+ \rightarrow \gamma + \gamma$, the collision of an electron and positron, each with a rest mass of 0.511 MeV results in the production of two photons each with frequency of 1.25×10^{20} Hz and opposite polarization and momentum. In pair production, an inverse process $\gamma + \gamma \rightarrow e^- + e^+$ occurs in which the interaction of two photons, each with a frequency of at least 1.25×10^{20} Hz results in the creation of an electron and positron with opposite charge, spin and momenta. Pair production can also occur in an unsymmetrical process $\gamma \rightarrow e^- + e^+$ in which a single photon with an energy of at least 1.02 MeV corresponding to a frequency of 2.50×10^{20} Hz on interaction with an atomic nucleus results in creation of an electron and positron. The minimum threshold photon energy $E = h\nu$ required for pair production is $2m_0c^2$ (1.02 MeV) or twice the rest mass (0.511 MeV) of an electron (positron). In each case, electric charge and momentum is conserved. For annihilation of an electron and positron with non-relativistic velocities producing two gamma rays $e^+e^- \rightarrow \gamma\gamma$, the cross-section as derived by Dirac is $\pi r_0^2 (c/v)$. The cross-section for pair production by head-on collision of gamma rays $\gamma\gamma \rightarrow e^+e^-$ is $\sigma = \pi r_e^2 (1 - (m e^2 c^4)/\sqrt{(E_1 E_2)}) \cong \pi r_e^2$ where r_e is the classical electron radius.

Quantum Electrodynamics (QED) theory, while providing a mathematical formalism to describe electromagnetic interactions, conveys little insight into the nature or structure of the electron. Feynman diagrams, a graphical representation of probability amplitudes, yield scant information on the internal details of the interaction processes at the vertex points or geometrical description of the propagator. As the quantum numbers of the constituent particles before and after interaction are known, even though the interaction is unobservable, it may nonetheless be possible to infer details of the interaction process and deduce the likely geometrical structures of the photon and electron that enable conversion. While various models of photons and electrons have been proposed which can account for some of the observed characteristics, an additional screening is necessary to discriminate viable candidates. A process description must account for not only for the initial and final configuration states but also the transition path between demonstrating convertibility and conservation of quantum mechanical properties. Pair production and annihilation processes provide a severe test of any proposed geometrical models of such particles.

The conservation of charge in pair production and annihilation exhibits a certain temporal asymmetry:

photon (0) → electron (−1) + positron (+1)
electron (−1) + positron (+1) → photon (0) + photon (0)

Mother nature is a tyrannical bookkeeper, yet the even/odd quantum arithmetic progression here seems a bit curious: $0 = +1 -1 = 0 + 0$. Might this little bang of pair production provide insight into the Big Bang in which the universe arises out of seeming nothingness

18. Pair Production and Annihilation

into a pairing of opposites in a degenerate duality state. The arithmetic may be more understandable in terms of set theory. The null or empty set $\phi = \{\ \}$ represents nothingness – a set with no elements. If we define a set consisting only of the null set, we get a set with one element $\{\phi\}$. If we further define a paired set consisting of two elements, we get $\{\phi, \{\phi\}\}$. In order theory, a Hasse diagram may be used to represent such an ordered set: $\{\ \} \rightarrow \{x\},\{y\} \rightarrow \{x,y\}$. In essence, such a set represents something created *ex nihilo* – a mathematical metaphor for creation of the cosmos. The origin of the universe in terms of ordered parings in set theory will be examined in more detail in a later section.

Quantum Electrodynamics (QED) Diagrams

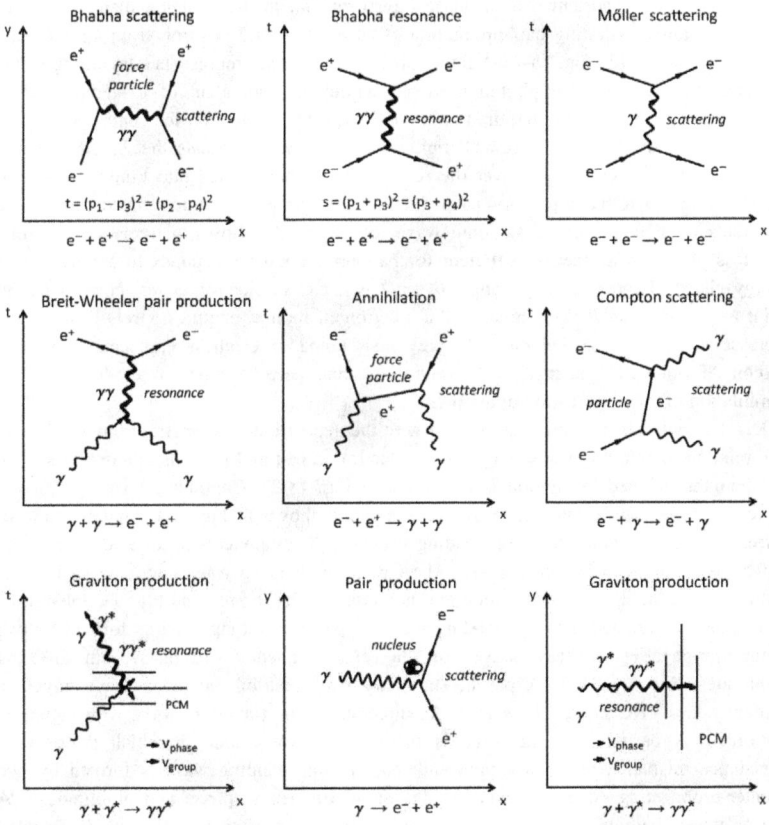

Fig. 18-1. Quantum electrodynamics (QED) diagrams of Bhabha, Compton and Möller scattering, Breit-Wheeler e^-/e^+ pair production, electron/positron e^-/e^+ annihilation, electron/positron e^-/e^+ pair production, e^- Compton scattering, and graviton $\gamma\gamma^*$ production. A spin 2 graviton $\gamma\gamma^*$ represents a momentary resonance between a photon γ and a phase conjugate photon γ^* with oppositely directed momentum.

209

18. Pair Production and Annihilation

A freely-propagating photon may be represented as an electromagnetic travelling wave. An electron at rest corresponds to a photon topologically confined within the equivalent of a standing wave resonator forming a closed loop electromagnetic standing wave. An electron in motion may be modeled as the equivalent of a photon confined within a Lorentz contracted moving standing wave resonator. The kinetic energy of a travelling wave and potential energy of a standing wave may be represented in a wave phasor diagram. Motion of a standing wave system may be represented as a function of phase angle on a complex phasor diagram with real and imaginary energy components corresponding to potential and kinetic energy, respectively. A photon (spin 1 boson) in a zero curvature vacuum is massless whereas an electron (spin ½ fermion) is massive. In Supersymmetry theory, a boson and a fermion may be interchanged in a hypothetical supersymmetry transformation. Repeated supersymmetry transformation of a fermion into a boson and back into a fermion is theorized to result in a spatial translation. The conversion of an electron into a photon may be similarly represented as a phase transformation in the complex plane. A wave phasor diagram illustrating pair production of an electron and positron from an energetic photon is illustrated in Fig. 18-2. A freely-propagating photon represents a travelling wave. An electron consists of a photon confined within an equivalent of a standing wave resonator. The minimum required photon frequency required for pair production corresponds to twice the electron Compton frequency which equals the Zitterbewegung frequency. Any excess energy over the rest energy is transformed into kinetic energy of motion of the electron and positron pair. An electron (or positron) in motion corresponds to a Lorentz contracted moving standing wave resonator. As shown, a freely propagating, massless photon with energy sufficient for pair production corresponds to a point on an energy phasor diagram at phase angle $\theta = \pi/2$ whereas an electron at rest corresponds to point at phase angle $\theta = 0$ radians. An electron in motion acquires a relativistic mass increase as a function of velocity. As previously noted, an electron represents a confined photon of equivalent energy. Conversion of one particle type to another involves annihilation of one for creation of another.

Kinetic energy of motion is associated with the generation of a de Broglie matter wave. The relation of standing wave length of an electron at rest and relativistic Doppler shifted wavelengths induced by motion is illustrated in Fig. 18-3. Contracted moving standing wave components of a standing wave resonator are shown in Fig.18-4. A tabulation of corresponding contracted moving standing wavelength components of an electron moving at 0.5c is reproduced in Table 18-1. The Compton standing wave undergoes a Lorentz contraction in the direction of motion and is modulated by the red and blue Doppler shifts producing a contracted moving standing wave. A partial standing wave is formed by two counter-propagating coherent waves of the same frequency and phase, but differing amplitudes. See Fig 18-5. A partial stationary wave exhibits an open wave envelope without nodes. Net energy flow is in the direction of the travelling wave with increased amplitude. A partial standing wave is formed in a waveguide in which there is an impedance mismatch between source and load. A full standing wave is formed by two counter-propagating coherent waves of the same frequency, phase and amplitudes. A phasor wave diagram for an electron moving at 0.5c depicting Compton, de Broglie, Doppler, Lorentz Doppler, standing wave, and contracted moving standing wave components is shown in Fig 18-6.

> *Many theorists suspect that space has an intrinsic structure – that it is 'grainy'- but this structure is on a much finer scale than any known subatomic particle. The structure could be of an exotic kind: extra dimensions over and above the three that we are used to... – Martin Rees*

18. Pair Production and Annihilation

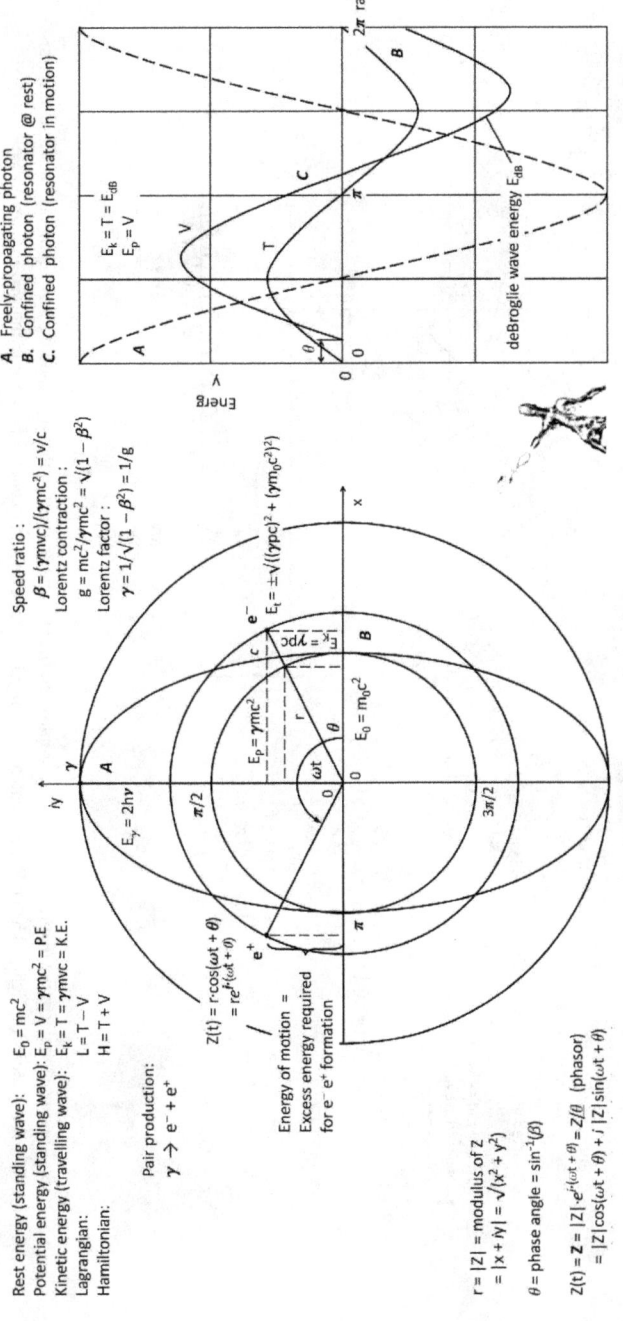

Fig. 18-2. Wave energy phasor diagram of electron and positron pair production from annihilation of an energetic photon.

211

18. Pair Production and Annihilation

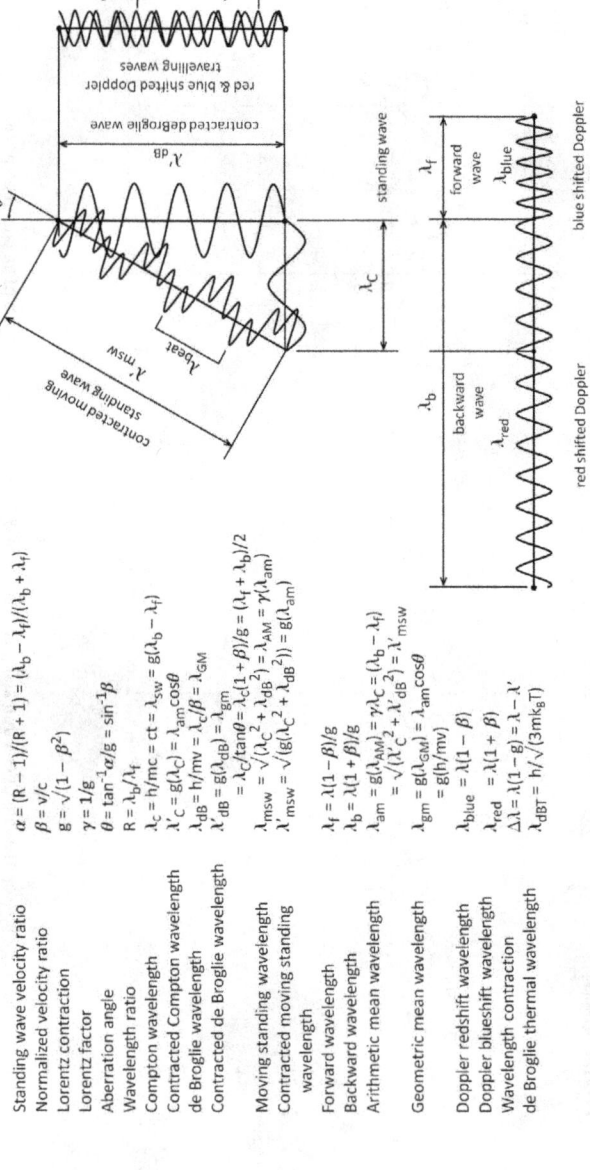

Standing wave velocity ratio	$\alpha = (R-1)/(R+1) = (\lambda_b - \lambda_f)/(\lambda_b + \lambda_f)$
Normalized velocity ratio	$\beta = v/c$
Lorentz contraction	$g = \sqrt{(1-\beta^2)}$
Lorentz factor	$\gamma = 1/g$
Aberration angle	$\theta = \tan^{-1}\alpha/g = \sin^{-1}\beta$
Wavelength ratio	$R = \lambda_b/\lambda_f$
Compton wavelength	$\lambda_C = h/mc = ct = \lambda_{sw} = g(\lambda_b - \lambda_f)$
Contracted Compton wavelength	$\lambda'_C = g(\lambda_C) = \lambda_{am}\cos\theta$
de Broglie wavelength	$\lambda_{dB} = h/mv = \lambda_C/\beta = \lambda_{GM}$
Contracted de Broglie wavelength	$\lambda'_{dB} = g(\lambda_{dB}) = \lambda_{gm}$
	$= \lambda_C/\tan\theta = \lambda_C(1+\beta)/g = (\lambda_f + \lambda_b)/2$
Moving standing wavelength	$\lambda_{msw} = \sqrt{(\lambda_C^2 + \lambda_{dB}^2)} = \lambda_{AM} = \gamma(\lambda_{am})$
Contracted moving standing wavelength	$\lambda'_{msw} = \sqrt{(g(\lambda_C^2 + \lambda_{dB}^2))} = g(\lambda_{am})$
Forward wavelength	$\lambda_f = \lambda(1-\beta)/g$
Backward wavelength	$\lambda_b = \lambda(1+\beta)/g$
Arithmetic mean wavelength	$\lambda_{am} = g(\lambda_{AM}) = \gamma\lambda_C = (\lambda_b - \lambda_f)$
	$= \sqrt{(\lambda'_C{}^2 + \lambda'_{dB}{}^2)} = \lambda'_{msw}$
Geometric mean wavelength	$\lambda_{gm} = g(\lambda_{GM}) = \lambda_{am}\cos\theta$
	$= g(h/mv)$
Doppler redshift wavelength	$\lambda_{blue} = \lambda(1-\beta)$
Doppler blueshift wavelength	$\lambda_{red} = \lambda(1+\beta)$
Wavelength contraction	$\Delta\lambda = \lambda(1-g) = \lambda - \lambda'$
de Broglie thermal wavelength	$\lambda_{dBT} = h/\sqrt{(3mk_BT)}$

Fig. 18-3. Lorentz-Doppler wave diagram illustrating Compton, de Broglie and contracted moving standing wave components for an electron moving at ½ the velocity of light. The de Broglie matter wave results from a beat frequency difference in internal Lorentz-Doppler shifted foreward and backward traveling wave components of elemental mass oscillator standing waves.

212

18. Pair Production and Annihilation

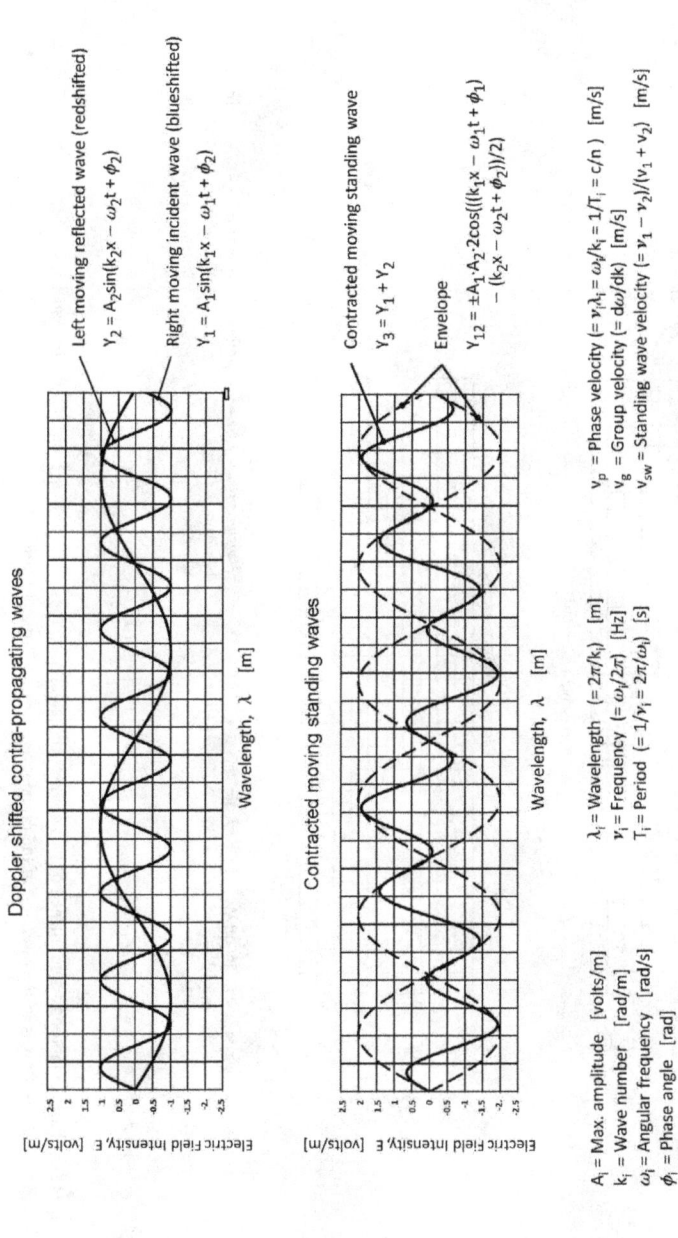

Fig. 18-4. Contracted moving standing wave within a standing wave resonator in motion. A traveling wave has a standing wave ratio SWR = 1, whereas a standing wave has SWR > 1. Hence, for a moving standing wave, $1 < \text{SWR} < \infty$.

18. Pair Production and Annihilation

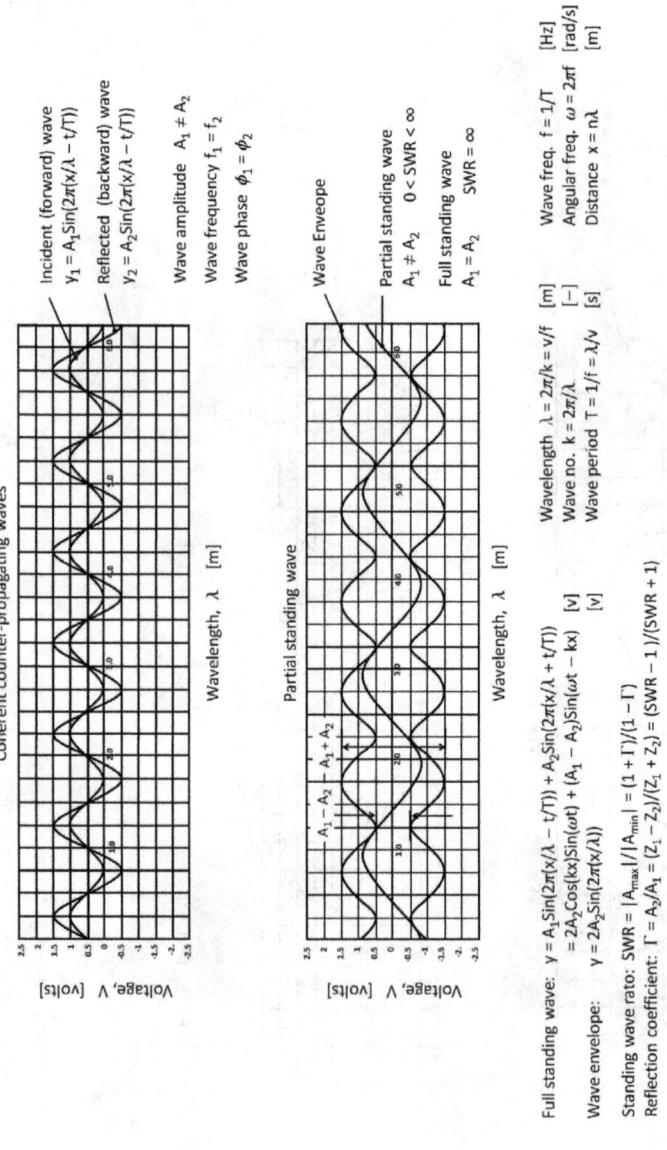

Fig. 18-5. Partial standing wave formed of two counter-propagating waves of equal phase and frequency, but with different amplitudes. Resultant energy flow is in the direction of the propagating wave of greater amplitude.

18. Pair Production and Annihilation

Table 18-1. Example of contracted moving standing wave components for an electron moving at a velocity of 0.5c.

Description	Symbol	Relation	Value	Units
Standing wave velocity ratio	α	$= (\lambda_b - \lambda_f)/(\lambda_b + \lambda_f)$ $= (R - 1)/(R + 1)$ $= g\beta\gamma$	0.5	-
Normalized speed ratio	β	$= v/c = \alpha/g\gamma = \Delta\phi/\pi$ $= \rho/\gamma = \sqrt{(1 - 1/\Gamma^2)}$	0.5	-
Velocity of light	c	$= \lambda \cdot f = \omega/k = 1/\sqrt{(\epsilon_0\mu_0)}$	2.99792E8	m/s
Lorentz contraction	g	$= \lambda_{gm}/\lambda_{am} = \sqrt{(1 - \alpha^2)}$ $= \alpha/\beta\gamma$	0.866	-
Lorentz factor	γ	$= 1/g = 1/\sqrt{(1 - \beta^2)}$ $= \rho/\beta = \alpha/\gamma\beta = \cosh w$	1.15	-
Doppler blueshift	$1 - \beta$	$= 1 - v/c = 1 - \Delta\phi/\pi$	0.5	-
Doppler redshift	$1 + \beta$	$= 1 + v/c = 1 + \Delta\phi/\pi$	1.5	-
Doppler blueshift wavelength	λ_{blue}	$= \lambda(1 - \beta) = \alpha \cdot \lambda$	1.2131E-12	m
Doppler redshift wavelength	λ_{red}	$= \lambda(1 + \beta)$	3.6334E-12	m
Wavelength ratio	R	$= \lambda_b/\lambda_f$ $= (1 + \beta)/(1 - \beta)$	3	-
Doppler factor	λ_o/λ_s	$= \sqrt{(\lambda_b/\lambda_f)}$ $= \sqrt{((1 + \beta)/(1 - \beta))}$	1.7320	-
Relativistic wavelength (observer)	λ	$= \lambda' \sqrt{(\lambda_b/\lambda_f)}$ $= \lambda' \sqrt{((1 + \beta)/(1 - \beta))}$ $= \lambda_o$	4.2024E-12	m
Source wavelength	λ'	$= \lambda_s = \lambda/\sqrt{(\lambda_b/\lambda_f)}$ $= \lambda/\sqrt{((1 + \beta)/(1 - \beta))}$	2.4263E-12	m
Wavelength contraction	$\Delta\lambda_{LD}$	$= \lambda(1 - g) = \lambda - \lambda'$	3.2506E-13	m
Transverse Doppler shift	$\Delta\lambda_{LD}$	$= \lambda - \lambda' = \gamma \cdot \lambda - \lambda$ $= \lambda_o - \lambda_s$	3.7535E-13	m
Compton wavelength	λ_C	$= h/mc = \lambda = ct = \lambda_{sw}$ $= g(\lambda_b - \lambda_f) = \beta\lambda_{db}$	2.4263E-12	m
Contracted Compton wavelength	λ'_C	$= g \cdot \lambda_C$	2.1012E-12	m
Reduced Compton wavelength	R_C	$= \lambda_C/2\pi = \hbar/mc = c/\omega_C$	3.8615E-13	m
de Broglie wavelength	λ_{dB}	$= h/mv = \lambda_C/\beta$	4.8526E-12	m

Cont

18. Pair Production and Annihilation

Table 18-1. Example of contracted moving standing wave components for an electron moving at a velocity of 0.5c. (cont)

Description	Symbol	Relation	Value	Units
Contracted de Broglie wavelength	λ'_{dB}	$= g \cdot \lambda_{dB} = \lambda_{GM}$ $= \lambda_C/\tan\theta = g(h/mv)$	4.2024E-12	m
Doppler approaching wavelength	λ_{aD}	$= \lambda(1 - \beta\cos\theta)$	1.3757E-12	m
Doppler receding wavelength	λ_{rD}	$= \lambda(1 + \beta\cos\theta)$	3.4769E-12	m
Lorentz Doppler forward wavelength	λ_{LDf}	$= \lambda(1 - \beta)/g$ $= \gamma\lambda(1 - \beta)$	1.4008E-12	m
Lorentz Doppler backward wavelength	λ_{LDb}	$= \lambda(1 + \beta)/g$ $= \gamma\lambda(1 + \beta)$	4.2025E-12	m
Standing wavelength	λ_{sw}	$= ct = c/f = \lambda_C$	2.4263E-12	m
Contracted standing wavelength	λ'_{sw}	$= \lambda\sqrt{(1 - \beta^2)} = g \cdot \lambda$ $= g \cdot \lambda_{am} = \lambda'_C$	2.1012E-12	m
Moving standing wavelength	λ_{msw}	$= \sqrt{(\lambda_C^2 + \lambda_{dB}^2)} = \lambda_{AM}$ $= \gamma\lambda'_{msw} = \gamma\lambda_{am}$	5.4254E-12	m
Contracted moving standing wavelength	λ'_{msw}	$= \sqrt{((g \cdot \lambda_C)^2 + (g \cdot \lambda_{dB})^2)}$ $= \lambda_{am}$ $= \sqrt{((\lambda'_C)^2 + (\lambda'_{dB})^2)}$	4.6985E-12	m
Arithmetic mean wavelength	λ_{am}	$= \lambda'_{msw} = g \cdot \lambda_{AM}$	4.6985E-12	m
Geometric mean wavelength	λ_{gm}	$= \lambda_{am}\cos\theta = \lambda'_{sw}$	2.1012E-12	m
Harmonic mean wavelength	λ_{hm}	$= \lambda_{gm}\cos\theta = \lambda_{GM}^2/\lambda_{AM}$	1.8197E-12	m
Redshift	z	$= (\lambda_o - \lambda_s)/\lambda_s = \Delta\lambda/\lambda_s$ $= \sqrt{((1 + \beta)/(1 - \beta))} - 1$	2.7320	–
Source velocity (electron)	v	$= \beta \cdot c = (\hbar k - qA)/m$	1.4989E8	m/s

Based on de Broglie's relation between momentum and wavelength of a free electron, Schrödinger developed the relativistic wave equation for motion of an electron moving in an electromagnetic field. Quantum transitions are not discontinuous quantum jumps but a modal wave transition. The Klein-Gordon equation was used in an attempt to model the electron which proved unsuccessful as the electron has a spin characteristic. The Klein-Gordon equation solution for electron plane waves in the vicinity of a solenoid predicts a phase shift now known as the Aharonov-Bohm effect.

18. Pair Production and Annihilation

Phasor wave diagram for a relativistic electron

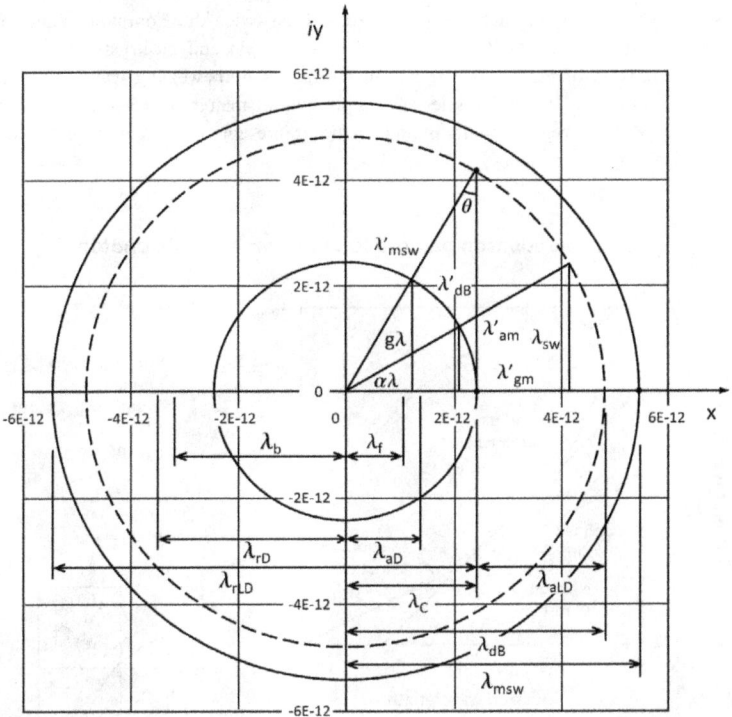

$\lambda_C = h/mc = \lambda_{sw} = 2.4263\text{E-}12$ m
$\lambda_{dB} = h/mv = 4.8526\text{E-}12$ m
$\lambda_{msw} = \sqrt{(\lambda_C^2 + \lambda_{dB}^2)} = 5.4254\text{E-}12$ m
$\lambda_f = \lambda(1-\beta)g = 1.0506\text{E-}12$ m
$\lambda_b = \lambda(1+\beta)g = 3.1518\text{E-}12$ m
$\lambda_{aLD} = \gamma\lambda(1-\cos\theta) = 1.5885\text{E-}12$ m

$\lambda'_C = g \cdot \lambda_C = 2.1012\text{E-}12$ m
$\lambda'_{db} = \lambda(1+\beta)g = 4.2025\text{E-}12$ m
$\lambda'_{msw} = \sqrt{(\lambda'^2_C + \lambda'^2_{dB})} = 4.6985\text{E-}12$ m
$\lambda_{aD} = \gamma(1-\beta\cos\theta) = 1.3756\text{E-}12$ m
$\lambda_{rD} = \gamma(1+\beta\cos\theta) = 3.4769\text{E-}12$ m
$\lambda_{rLD} = \gamma\lambda(1+\cos\theta) = 4.0148\text{E-}12$ m

Fig. 18-6. Phasor wave diagram for an electron moving at a velocity of 0.5c depicting Compton, de Broglie, Doppler, Lorentz Doppler, standing wave and contracted moving standing wave components. A phase wave, often referred to as a "pilot" wave, appears to accompany the particle. The phase wave reflects a relativistic transformation in the internal frequency state of the particle. The phase wave of the electron is responsible for quantization of atomic orbits and spectra.

Physics once again is very wrong. – Wolfgang Pauli

Truth is stranger than fiction. Get your facts first, then you can distort them as you please. – Mark Twain

Each photon then only interferes with itself. Interference between two different photons never occurs. – Paul Dirac

Every great discovery I ever made, I gambled that the truth was there, and then I acted in faith until I could prove its existence. – Arthur H. Compton

18. Pair Production and Annihilation

A comparison of a free photon of Compton wavelength with a confined photon corresponding to an electron is shown in Fig. 18-7. In the electron/positron pair production process ($\gamma \rightarrow e^- + e^+$), an energetic photon of at least twice the Compton frequency is converted into an electron and a positron of opposite spin and electrical charge. The geometrical relation of the de Broglie wavelength and Zitterbewegung wavelength of an electron is illustrated. The de Broglie wave represents a 'matter' wave of an electron in motion. The Zitterbewegung or 'trembling' motion represents a spin component at twice the electron Compton frequency.

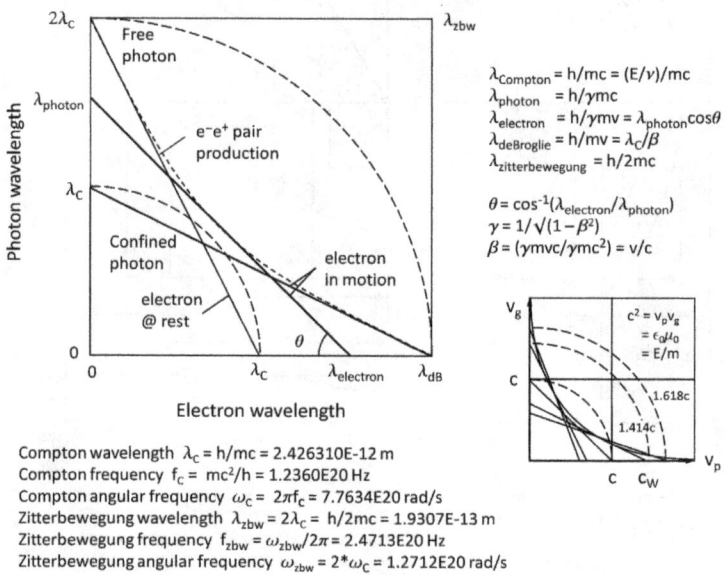

Fig. 18-7. Electron Compton wavelength, zitterbewegung wavelength and de Broglie wavelength is compared to the wavelength of an energetic photon required for pair production of an electron and positron.

An example of the relation of energy (E) and momentum (pc) of a relativistic electron in motion is depicted in Fig. 18-8 for the case of a velocity equal to 0.5 times the velocity of light. For comparison, the same condition is also represented on a tangent space diagram of angular frequency (ω) vs. wave no. (k). The magnitude of wave no. k otherwise known as the propagation vector or wave vector is a measure of spatial frequency (k = $2\pi/\lambda$ = ω/c where λ = wavelength, ω = angular frequency and c = speed of light). The resonance ratio ϵ_P represents the ratio of the electron Compton angular frequency ω_C and the Planck angular frequency ω_P. (= ω_C/ω_P = 7.76E20/1.855E43 = 4.185 E-23).

The electron as it leaves the atom, crystallizes out of Schrödinger's mist like a genie energy from the bottle. – Arthur Stanley Eddington

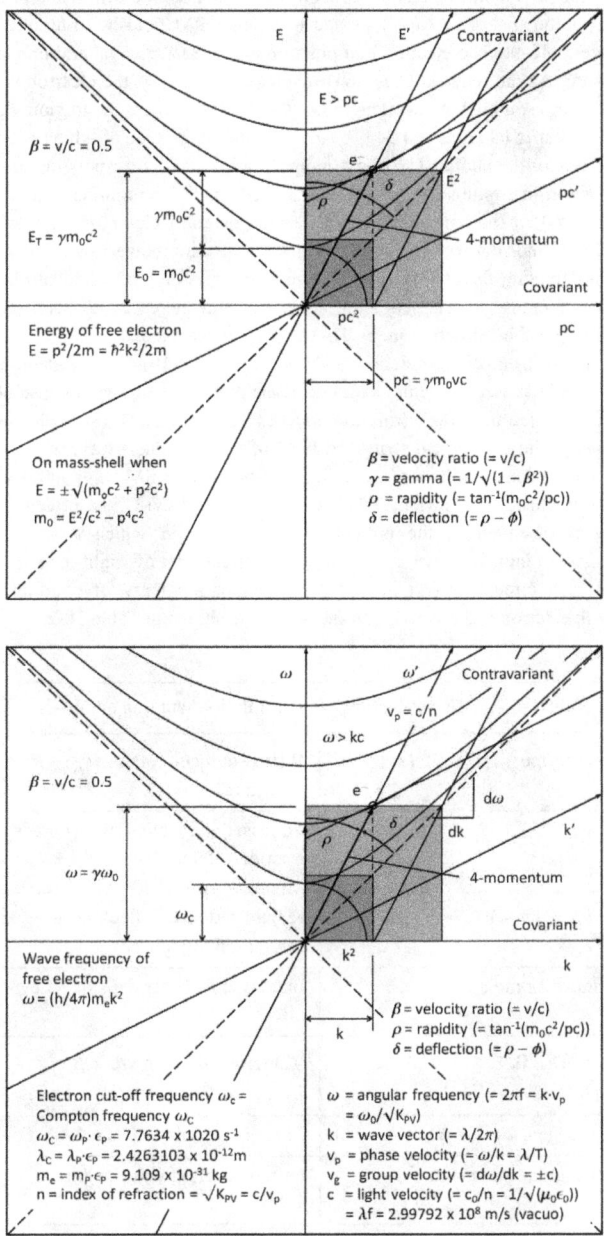

Fig. 18-8. Energy and momentum of a freely-propagating, high speed electron ($\beta = v/c = n^{-1} = 0.5$). In this example, the kinetic energy E_k slightly exceeds the rest mass energy E_0 and the corresponding wave frequency ω_k slightly exceeds the cut-off frequency ω_c.

18. Pair Production and Annihilation

For a freely-propagating photon, the energy E equals the momentum p times c. As a consequence, the photon has no rest mass. Pair production of an electron and positron from an energetic photon (E ≥ 1.02 MeV) slowed in a dense EM field of an atomic nucleus is illustrated in Fig. 18-9. The electron and positron emitted have equal and opposite charge and momentum. As the reduced Compton wavelength ($\lambda_c/2\pi$) of the electron and positron is 3.811×10^{-12} m vs. a photon wavelength (λ_{pp}) of 1.21×10^{-12} m, the torsion κ of electron and positron at rest is taken as $\pm 1.311 \times 10^{11}$ m^{-1}, the magnitude of which is equal to the initial curvature κ of the photon λ_{pp}. The principal curvature κ_1 corresponding to the inverse of the electron toroidal radius (= $2\pi/\lambda_c$) is 2.623×10^{11} m^{-1}. The minimum photon angular frequency required for pair production is twice the electron Compton angular frequency ($2\omega_C = 1.5527E21$ rad/sec) which is equal to the electron Zitterbewegung frequency ω_{zbw}.

A diagram depicting the change in curvature k and torsion τ of a helical photon wave train due to an increase in magnetic flux density **B** and increased electromagnetic field density denoted by the refractive index $K_{PV}(r,\omega,M)$ in the vicinity of an atomic nucleus during electron/positron pair production is shown in Fig. 18-10. During electron/positron pair creation, the increased curvature k and decreased torsion τ during collapse of a helical photon wave train due to Faraday rotation and Levi-Civita effects results in formation of two loops of opposite topological charge. A twist of $\theta = 2\pi$ radians/wavelength is required for charge quantization. The compression and splitting of the EM wave into two loops of opposite charge and spin is attributed to the Faraday and Levi-Civita effects. A similar spin-splitting is observed in the optical Magnus effect in which a wave of mixed polarization splits into two waves in opposite directions of right and left circular polarizations in a circular waveguide. A parameter summary of winding and twist associated with electron and positron pair production is shown in Table 18-2.

Table 18-2. Pair production and annihilation winding and twist	
$\Delta\phi_F = (e^3\lambda^2)/(2\pi/mc^2)^2) \int n_e(z)B_{\parallel}(z)dz$	**B** field-induced rotation $\theta_F \rightarrow$ Increased curvature κ
$\theta_K = \pi/2 \cdot (1/\sqrt{K_1} - 1/\sqrt{K_2})$ $= \pi/2 \cdot (1 - v/c) = \pi/2 \cdot (1 - \beta)$	Increased magnetic field – Faraday rotation effect
$\Delta\theta_K = \pi/2 \cdot (v_1/c - v_2/c)$	K field-induced deflection $\theta_K \rightarrow$ Decreased torsion τ
θ twist per loop: 2π radian/λ	Increased EM flux density = Levi-Civita effect
$K_{PV} = \Gamma^2 = c_0^2/(\mathbf{E_C}/\mathbf{B_C})$ $\beta = \sqrt{(1 - 1/\Gamma^2)}$	Velocity reduction v/c $\simeq \beta =$ $1 - (4\pi G \cdot B^2/\mu_0 c^4)((L/2)^2 - \lambda^2)$
Winding no.: 1 (γ), 2 (e$^-$, e$^+$)	Impedance: $Z_0 = |\mathbf{E}|/|\mathbf{H}|$; $Z_C = |\mathbf{E_C}|/|\mathbf{H_C}|$
Photon spin angular momentum: $\hbar$	Pair production wavelength: $\lambda_C/2 =$ 1.21E-12 m
Electron/Position spin angular momentum: $\hbar/2$	Pair production energy threshold: 1.02 MeV

18. Pair Production and Annihilation

Electron/Positron Pair Production

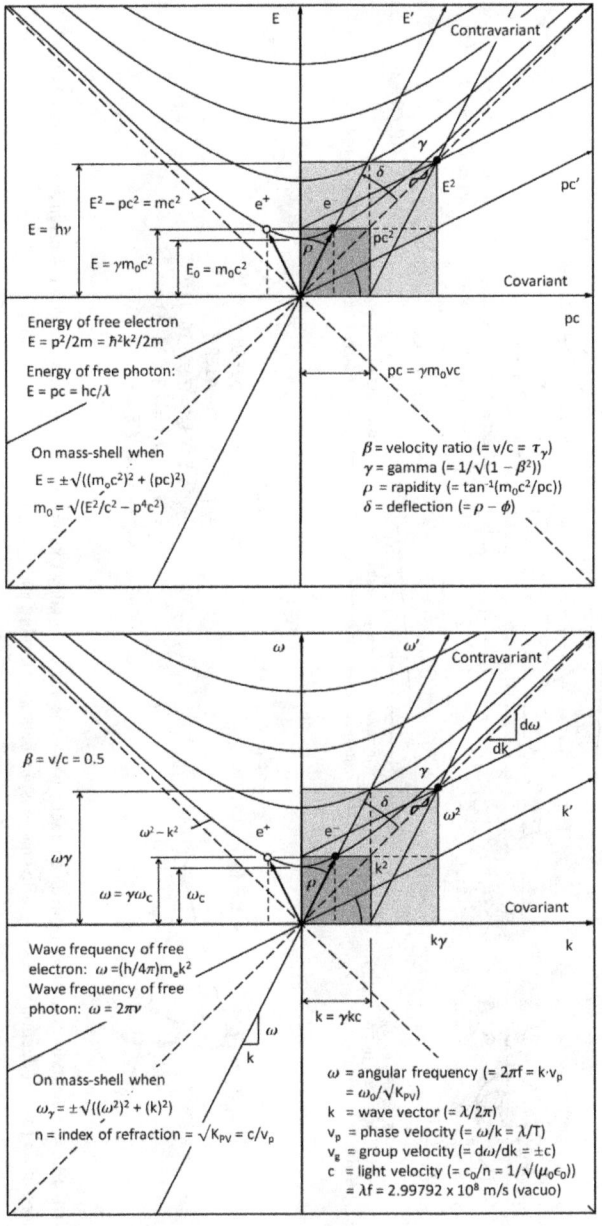

Fig.18-9. Representation of an electron and positron formation from an energetic photon in energy E vs. momentum pc and angular frequency ω vs. wave vector k diagrams. Pair production only occurs at or above the cut-off frequency ω_c (= $2 \cdot \omega_C$ = 2.50E20 Hz).

18. Pair Production and Annihilation

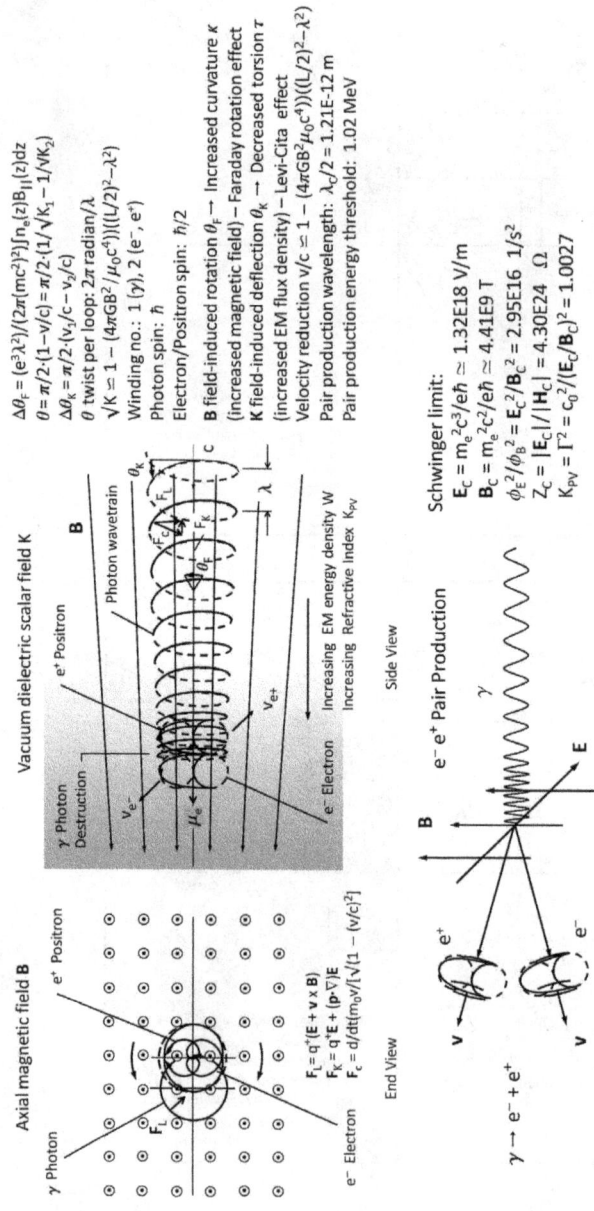

Fig. 18-10. Electron/positron formation from an energetic photon is due to Faraday rotation and Levi-Civita velocity effects in an intense EM field. Electron/positron pair production yields opposite charge, spin, helicity, chirality and magnetic moments.

The wave function ψ of an electron in an atom does not describe a smeared-out electron with a smooth charge density. The electron is either here, or there, or somewhere else, but wherever it is, it is a point charge.– Richard P. Feynman

18. Pair Production and Annihilation

The process of pair production from an energetic photon in the vicinity of an atomic nucleus is illustrated in Fig. 18-11. The photon is compressed due to the increased electromagnetic energy density U (= ½($\epsilon_0 \mathbf{E}^2 + \mathbf{B}^2/\mu_0$)) encountered near the nucleus. The electron is represented as a rotating Hopf link space curve which generates a spindle toroidal form corresponding to a circular helix connected end-to-end. Transformation of a photon γ into an electron e^- and positron e^+ involves conversion of linear momentum into rotary momentum to absorb energy of a collision.

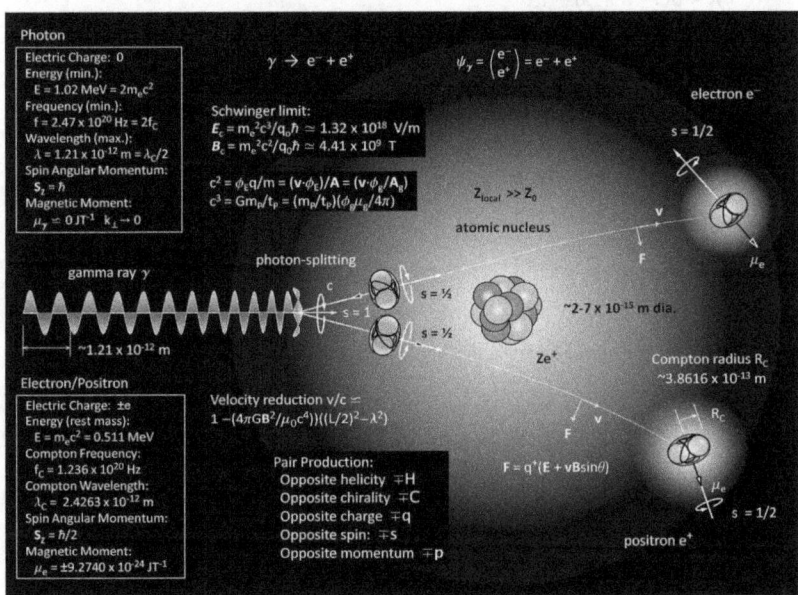

Fig. 18-11. Illustration of electron/positron pair production from an energetic photon in the vicinity of an atomic nucleus (~30 x 10^{-15}m). Photon conversion is triggered by interaction with the intense EM field of the nucleus with field strength exceeding the Schwinger limit for nonlinear vacuum polarization. The decrease in photon torsion corresponds to an increase in effective mass as a result of encountering increased EM energy density where local vacuum impedance $Z \gg Z_0$. Pair production involves transformation of propagating photon traveling wave into an electron and positron topologically confined standing waves. The electron e^- and positron e^+ are each depicted as a torus of revolution (Hopf link embedded on an annular manifold) with opposite spins and magnetic moments.

Electrons and positrons have opposite charge, but have the same rest mass. Positrons are the antimatter counter part of electrons. Photons have zero electric charge and zero rest mass. Quantum Electrodynamics (QED) is silent as to explanation for creation of electric charge or origin of rest mass. However, we do know something of the initial and final states, hence, we ought to be able to deduce the likely geometrical conversion process. Rest mass and electric charge are intimately related, i.e., electric charge does not arise independent of mass. Electric charge of the electron results from loop closure failure due to precession while rest mass results from quantum wave interference due to precession.

223

18. Pair Production and Annihilation

As the electric field is a gradient of potential, the electric field associated with an atomic nucleus provides a local change in light velocity. The nucleus provides intense EM energy density altering the local refractive K index slowing the speed of light and disrupts the photon's apparent self-focusing effect. As demonstrated by the Levi-Civita effect, electric and magnetic fields alter the velocity of light. The following relations for the Levi-Civita effect in terms of refractive index K were derived by Puthoff[29]

$$\nabla^2 \sqrt{K} = -(\sqrt{K}/4\lambda)[1/2(\mathbf{B}^2/K\mu_0 + K\epsilon_0 \mathbf{E}^2) - \lambda/K^2(\nabla K)^2] \quad (18\text{-}1)$$

$$\sqrt{K} \approx \alpha(1 - \beta^2 z^2/2) = \alpha\cos\beta z \quad (18\text{-}2)$$
$$\sqrt{K} \approx 1 - (4\pi G \cdot \mathbf{B}^2/\mu_0 c^4)[(L/2)^2 - z^2] \quad (18\text{-}3)$$

where:
- K = Vacuum refractive index (= $K_{PV}(r,\omega,M) = c^2/v^2$)
- L = Photon path length in z direction
- α, β = Integration constants $\alpha^2\beta^2 = 4\pi G \mathbf{B}^3/\mu_0 c4$

The velocity of light is slowed within an axially aligned magnetic field (e.g., solenoid)

$$v_L(z)/c \approx 1/K \approx 1 - (4\pi G \cdot \mathbf{B}^2/\mu_0 c^4)[(L/2)^2 - z^2] \quad (18\text{-}4)$$

The transit time τ of light for axially aligned **B** and **E** fields is given by [29]

$$\tau \approx L/c[1 + (2\pi G \mathbf{B}^2 L^2)/3\mu_0 c^4] \quad (18\text{-}5)$$
$$\tau \approx L/c[1 + (2\pi \epsilon_0 G \mathbf{E}^2 L^2)/3c^4] \quad (18\text{-}6)$$

In the presence of an axial magnetic field, light undergoes a rotation (Faraday effect) and undergoes a decrease in velocity (Levi-Civita effect). The optical Faraday rotation is characterized by the Verdet constant ($\nu = \theta/Bd$) relating the rotation angle (θ) per propagation distance (d) and the axial magnetic flux density (**B**). The increase in K index acts as a damping term ($-M_s/dt$) decreasing the photon's longitudinal magnetic field (photo-magneton) and decreasing wavelength ($\lambda = -K \cdot \gamma/M_s$) causing the photon magnetic moment to precess, spiralling outward and rotating perpendicular to direction of travel. Substituting a wavelength of the reduced Compton wavelength for the path length L and calculated values for **E** and **B** field strengths in Eqns 17-5 and 17-6 both yield an estimated duration $\tau \approx 1.28809111\text{E-}21$ s which corresponds to a rotation of $\theta/3$ rad for dissolution of the incident photon. The corresponding velocity threshold is 1.1671E-48 c.

Electron wave vortices in the presence of longitudinal magnetic field have been experimentally demonstrated to undergo a rotation analogous to the Faraday rotation of optical vortices[69]. The Faraday rotation of electrons in a vacuum arises as a result of Zeeman interactions in propagation parallel to an external uniform longitudinal magnetic field in the absence of a Lorentz force. The circulating charge in the electron vortex beam generates a magnetic moment μ_B within the beam waist parallel to the symmetry axis which interacts with the external magnetic field.

You don't want any self-respecting vacuum to break gauge invariance. – Prof. R. Mohapatra

I'm no scientist, but couldn't we use that Ben Franklin thing? – Col. Jack O'Neill

So you understand everything... except those minor vacuum fluctuations. – Prof. R. Mohapatra

19. Coulomb's Law

19.1 Electrostatics

The electrostatic force of attraction between two stationary point electric charges in scalar form is defined by Coulomb's law published in 1785 by Charles Augustin de Coulomb

$$F_e = (1/4\pi\epsilon_0) q_1 q_2 / r^2 = k q_1 q_2 / r^2 \qquad (19\text{-}1)$$

where:

F_e	Coulomb force (= Source x field strength)		
	Value: Varies	Units: Newton (N) [= kg(m/sec^2)]	Dimensions: MLT^{-2}
k	Coulomb's constant (= k_C = ¼$\pi\epsilon_0$)		
	Value: 8.98755178736817E9	Units: N·m^2/C^2 [= kg·m^3/s^2· Coul^2 = kg·m^3/s^4·A^2]	Dimensions: ML^{3}T^{-2}Q^{-2}
q_1	charge on object 1		
	Value: Varies	Units: Coulomb (C)	Dimensions: Q
q_2	charge on object 2		
	Value: Varies	Units: Coulomb (C)	Dimensions: Q
ϵ_0	electric permittivity of free space		
	Value: 8.854187817E-12	Units: m^{-3}·kg^{-1}·s^4·A^2 (= Farads/m)	Dimensions: M^{-1}L^{-3}T^{2}Q^2
r	Radius vector (distance between charge centers = $|r_2 - r_1|$)		
	Value: Varies	Units: m	Dimensions: L

In vector notation, the above may be written as $\mathbf{F}_e = (k q_1 q_2 / r^2)\mathbf{r}_u$ where $\mathbf{r}_u$ is the unit vector.

Coulomb's law is analogous to Newton's law of gravitation. The magnitude of the Coulomb proportionality constant k (= 9.0 x 10^9 N·m^2/C^2) is much greater than the Newton's gravitational constant G (= 6.67 x 10^{-11} N·m^2/kg^2), hence, a unit electrical charge will attract a unit electrical charge with much greater force that a unit mass will attract a unit mass. Under a Lorentz transformation, Coulomb's law becomes a function of time.

$$\begin{aligned} F_e &= (1/4\pi\epsilon_0)(q_1 q_2 / r^2)(\sqrt{(1 - v^2/c^2)}) \\ &= k(q_1 q_2 / r^2)(\sqrt{(1 - \beta^2)}) \end{aligned} \qquad (19\text{-}2)$$

A moving charge constitutes an electric current generating a magnetic field as described by Oersted's law: $\nabla \times \mathbf{B} - 1/c^2\ \delta\mathbf{E}/\delta t = \mu_0 \mathbf{J}$. Both sides of the equation change in the same way such that the equation is said to be covariant.

By Ohm's discovery a large part of the domain of electricity become annexed by Coulomb's discovery of the law of inverse squares, and completely annexed by Green's investigations.
– Oliver Heaviside

...two electrons feel an electric force between them to compensate for a shift in the electron waves. – Ron Mallett

19. Coulomb's Law

19.2 Generalization of Coulomb's Law

For the case of moving charges, Wilhelm Weber in 1846 derived the following relation for two charges in relative motion[70] which reduces the electrostatic force due to effects of relative velocities and accelerations

$$\mathbf{F}_w = \mathbf{F}_{21} = q_1 q_2 \mathbf{r}_{21}/4\pi\epsilon_0 r^3 - [1/8\pi\epsilon_0 (dr^2/dt^2)] + q_1 q_2 \mathbf{r}_{21}/4\pi\epsilon_0 r^2 c^2 (d^2r/dt^2) \quad (19\text{-}3)$$
$$= q_1 q_2 \mathbf{r}_{21}/4\pi\epsilon_0 r^3 (1 - \dot{r}^2/2c^2 + r\ddot{r}/c^2)$$
$$= q_1 q_2 \mathbf{r}_{21}/4\pi\epsilon_0 r^3 (1 - \dot{r}^2/c_w^2 + r\ddot{r}/2c_w^2)$$

where charges q_1 and q_2 are separated by a distance r with a relative velocity $v = \dot{r} = dr/dt$ and a relative acceleration $dv/dt = \ddot{r} = d^2r/dt^2$. The above relation, now known as Weber's Law, reduces to Coulomb's Law in the limit where velocities and accelerations are zero. With relative motion, the Weber's law of electrical action accounts for Ampere's Law of electrodynamics. Weber deduced that at a relative limit velocity of $\dot{r} = \pm c_w$, such that $\ddot{r} = 0$, the force of attraction between opposite charges would fall to zero. In other words, the electrostatic force represented by the 1st term on the right-hand side of Eqn (19-3) is cancelled by the electrodynamic force represented by the 2nd term with the charges moving a constant relative velocity c_w. The Weber constant, $c_w = \sqrt{2}c = 4.395 \times 10^8$ m/s, is the ratio of electrodynamic to electrostatic units of charge where c = the velocity of light. The value of c_w is defined as the relative velocity for which two electrical masses do not interact at all. Weber also derived a velocity dependent potential energy $U_w = q_1 q_2/4\pi\epsilon_0 r_{12}[1 - \dot{r}^2_{12}/2c^2] = Kq_1 q_2/r)(1 - \dot{r}^2/2c^2)$ showing the Weber force was consistent with the conservation of energy. Electrostatics is recovered from Weber's force when there is no motion between charge, i.e., $\dot{r} = 0$. Utilizing the experimental results by Weber and Kohlrausch for the measured value of c_w for the ratio of electrostatic to electromagnetic forces and relating the rate of propagation v of transverse waves in an elastic medium as the square root of the ratio of the torsion modulus m to mass density of the medium ρ_m (i.e., $v = \sqrt{m/\rho_m}$), Maxwell, conceptualizing light as a transverse torsion wave in an elastic medium, deduced the value of the speed of light ($c = 1/\sqrt{\epsilon_0 \mu_0}$).

A variety of similar equations have been subsequently derived to describe the dynamic effects. Klyushin[22,23,24] provides a concise summary of various of the proposals for comparison as follows:

Neumann
$$\mathbf{F}_{21} = +q_1 q_2/4\pi\epsilon_0 c^2 r^3 [\mathbf{r}_{21}(\mathbf{v}_1 \cdot \mathbf{v}_2)] \quad (19\text{-}4)$$

Grassman
$$\mathbf{F}_{21} = -q_1 q_2/4\pi\epsilon_0 c^2 r^3 [\mathbf{v}_2(\mathbf{r}_{21} \cdot \mathbf{v}_1) - \mathbf{r}_{21}(\mathbf{v}_1 \cdot \mathbf{v}_2)] \quad (19\text{-}5)$$

Ampere
$$\mathbf{F}_{21} = -q_1 q_2/4\pi\epsilon_0 c^2 r^5 [3(\mathbf{r}_{21} \cdot \mathbf{v}_1)(\mathbf{r}_{21} \cdot \mathbf{v}_2) - 2(\mathbf{v}_1 \cdot \mathbf{v}_2)r^2] \, \mathbf{r}_{21} \quad (19\text{-}6)$$

Whittaker
$$\mathbf{F}_{21} = -q_1 q_2/4\pi\epsilon_0 c^2 r^3 [\mathbf{v}_1(\mathbf{r}_{21} \cdot \mathbf{v}_2) + \mathbf{v}_2(\mathbf{r}_{21} \cdot \mathbf{v}_1) - \mathbf{r}_{21}(\mathbf{v}_1 \cdot \mathbf{v}_2)] \quad (19\text{-}7)$$

For the case of a negative charge q_1 rotating at constant velocity around a positive charge q_2 at rest, Klyuschin[24] provides the following relation r

$$\mathbf{F}_{21} = q_1 q_2 \, \mathbf{r}_{21}/4\pi\epsilon_0 r^3 + q_1 q_2 \, v_1^2 \mathbf{r}_{21}/4\pi\epsilon_0 r^3 c^2 \quad (19\text{-}8)$$

Ampere was the Newton of electricity. – James C. Maxwell

19. Coulomb's Law

This formula indicates a radial force that augments the Coulomb force. Because of centripetal acceleration, the rotating charge q_1 does not radiate. Radiation emission occurs only if q_1 is accelerated tangentially. Klyuschin derives the following generalized relation for the force between moving charges

$$\begin{aligned}\mathbf{F}_{21} = &\; q_1q_2\,\mathbf{r}_{21}/4\pi\epsilon_0 r^3 + q_1q_2/4\pi\epsilon_0 r^3 c^2 \cdot \{[\mathbf{r}_{21}(\mathbf{v}_1\cdot\mathbf{v}_2) - \mathbf{v}_1(\mathbf{r}_{21}\cdot\mathbf{v}_2) - \mathbf{v}_2(\mathbf{r}_{21}\cdot\mathbf{v}_1) + \\ & 3\mathbf{r}_{21}/r^2[(\mathbf{r}_{21}\cdot\mathbf{v}_1)(\mathbf{r}_{21}\cdot\mathbf{v}_2)] + [\mathbf{r}_{21}(\mathbf{v}_1-\mathbf{v}_2)^2 - (\mathbf{v}_1-\mathbf{v}_2)[\mathbf{r}_{21}\cdot(\mathbf{v}_1-\mathbf{v}_2)]] - \\ & 3\mathbf{r}_{21}(\mathbf{v}_1-\mathbf{v}_2)/r^2[\mathbf{r}_{21}[\mathbf{r}_{21}\cdot(\mathbf{v}_1-\mathbf{v}_2)] - (\mathbf{v}_1-\mathbf{v}_2)r^2] + [\mathbf{r}_{21}[\mathbf{r}_{21}\cdot(\mathbf{a}_1-\mathbf{a}_2)] - \\ & (\mathbf{a}_1-\mathbf{a}_2)r^2] + 1/c(\mathbf{v}_2-\mathbf{v}_1)[\mathbf{r}_{21}\cdot(\mathbf{v}_1\times\mathbf{v}_2)] + \mathbf{r}_{21}/c[(\mathbf{r}_{21}\times\mathbf{v}_2)\cdot\mathbf{a}_1 - \\ & (\mathbf{r}_{21}\times\mathbf{v}_1)\cdot\mathbf{a}_2] + 3\,\mathbf{r}_{21}/r^2 c[\mathbf{r}_{21}\cdot(\mathbf{v}_1-\mathbf{v}_2)]\cdot[\mathbf{r}_{21}\cdot(\mathbf{v}_1\times\mathbf{v}_2)]\}\end{aligned} \quad (19\text{-}9)$$

The first term on the right-hand side of the equation is the gradient of scalar products. The dynamic components of the Neumann, Grassmann, Ampere, and Whittaker formulas are reflected in the first square brackets. The 2nd bracket is a product of scalar and dynamic components valid if at least one of the components is charged. The 3rd bracket describes the radiation field of accelerated charges. The last three terms in braces relate to electroweak interactions proportional to $1/c^3$.

A generalized Weber's law for electromagnetism advanced by Assis[71] has the form

$$\mathbf{F}_{21} = q_1q_2\,\mathbf{r}_{21}/4\pi\epsilon_0 r^2 \cdot (1 - \alpha(\dot{r}^2 - 2r\ddot{r}/c^2) - \beta(\dot{r}^4 - 4r\dot{r}^2/c^4) \\ - \gamma(\dot{r}^6 - r\dot{r}^2/c^4) - \dots) \quad (19\text{-}10)$$

When applied to two neutral dipoles this relation for $\beta > 0$ yields a non zero attractive resultant force equivalent to Newton's law of gravitation as a fourth order electromagnetic effect due to the terms of fourth and higher orders of $1/c$. Setting $\gamma/\beta = -7/3$ is said to reproduce precession of perhelion advance.

Following the method used by Assis, Lucas[57] derives an electrodynamic universal force law as

$$\mathbf{F}_{21} = q_1q_2/4\pi\epsilon_0 r^2 \cdot \{(1 - \beta^2)\mathbf{r} + 2\mathbf{r}\mathbf{a}/c^2/[1 - \{\mathbf{r}\times(\mathbf{r}\times\mathbf{v})/r^4\}^2]^{1/2} - (1 - \beta^2)\cdot[(\mathbf{r}\cdot\beta)\mathbf{r}\times(\mathbf{r}\times\beta) + r^2/c^2\,(\mathbf{r}\times(\mathbf{r}\times\mathbf{a})]/[1 - \{\mathbf{r}\times(\mathbf{r}\times\mathbf{v})/r^4\}^2]^{3/2}] \quad (19\text{-}11)$$

where $\mathbf{r} = \mathbf{r}_2 - \mathbf{r}_1 = \mathbf{r}_{21}$, $\beta = (1 + v^2/c^2)$, $\mathbf{a}$ = acceleration, $\mathbf{v}$ = velocity

The 1st term in the non-relativistic limit is described as spherical symmetry while the 2nd term represents chiral symmetry due to the triple cross product vector. A combination of spherical and chiral symmetry is said to produce left and right hand symmetry with a spiral pattern based on the prime number sequence 1, 3, 5, 7, 11, 13... The statistically averaged v/c, v^2/c^2, v^3/c^3 terms average to zero while the v^4/c^4 terms results in an average force of a magnitude on the order of that of gravitation. As vibrating neutral electric dipoles radiate energy over time, the force of gravity decays (GM/r decreases with time). This effect provides an explanation for Hubble redshift, expansion in size of planets and moons, and high velocity of outer arms of spiral galaxies. The radiated intensity of neutral dipole radiation corresponds to cosmic microwave background radiation. The 2nd term is proportional to $\mathbf{r}\times(\mathbf{r}\times\mathbf{v})$ results in a spiral motion and is said to account for the Titius-Bode quantization law ($A_n = R_0 A_0^n$) of planetary orbits. A non-radial term $(\mathbf{r}\cdot\mathbf{v})\mathbf{r}\times(\mathbf{r}\times\mathbf{v})$ results in an equatorial tilt.

Scientists advance their careers by sharing what they know. Engineers largely advance their careers by hiding what they know. – Robert Laughlin

20. Origin of the Electron Fine Structure Constant α

20.1 Background

The fine structure constant α is dimensionless quantity described as a fundamental physical constant characterizing the strength of the electromagnetic interaction. Introduced by Sommerfeld in 1916 to describe the spacing of splitting of spectral lines in multi-electron atoms in terms of electric charge (e), speed of light (c) and Planck's constant (h), the α constant represents a relativistic correction term to the Bohr theory of the energy level of an electron:

$$E_{n,j} = [\mu e^4/(4\pi\epsilon_0)^2 2\hbar^2 n^2][1 + \alpha^2/n(1/(j + \tfrac{1}{2}) - \tfrac{3}{4}n)] \qquad (20\text{-}1)$$

where $\alpha = 2\pi e^2/hc$, reduced electron mass $\mu = mM/(m + M)$, and n and j are quantum numbers.

Besides being a measure of the fine structure of spectral lines, the fine structure constant measures the coupling strength of the interaction of two electrons. The dimensionless fine structure constant may be expressed as

$$\alpha = \tfrac{1}{4}\pi\epsilon_0(e^2/\hbar c) = k_C e^2/\hbar c = e^2/q_P^2 \qquad (20\text{-}2)$$
$$= 0.0072973552\ldots \cong 1/137$$

where:
- e = electric charge (= q = F/E = $\hbar k/A$ = E/V = $V\cdot C$) = −1.6021E-19 Coul
- q_P = Planck charge (= $\sqrt{(4\pi\epsilon_0 \hbar c)}$ = $q_e\sqrt{(\alpha \hbar c/2\pi)}$ ≃ $m_P \omega_P$ = 8.619E25 kg·rad/s)
- ϵ_0 = permittivity of free space (= 8.854187817E-12 F/m = 1.7251E8 kg·rad^2/m^3)
- $\hbar$ = reduced Planck constant (= $h/2\pi$ = p/k = $(q_0 A)/k$ = $(S/c^2)k$ = mc^2/ω)
- c = speed of light (= $1/\sqrt{(\mu_0\epsilon_0)}$ = E/B = ω/k = $l_P t_P$ = $m_P^2 G/\hbar$ = 2.99792458 m/s)
- k_C = Coulomb constant (= $1/4\pi\epsilon_0$ = 8.98755E9 kg·m·s^{-2}·C^{-2})

The fine structure constant α has several interpretations including ratio scaling in tangential velocity v to velocity of light c, electromagnetic force F_e to max. force F_m at Compton radius, impedance of free space Z_0 to electron impedance Z_e, elementary unstable lifetimes τ_i/τ_π, Bohr radius R_0 to Compton radius R_C, classical electron radius R_e to Compton radius R_C; Compton radius R_C to electromagnetic radius R_{em}, electric charge e to Planck charge q_P, resonance frequency ω_p/ω, Phi ratio Φ, etc. For example:

$$\alpha = v_t/c; \ = Z_0/Z_e; \ = R_0/R_C; \ = R_e/R_C; \ = R_C/R_{em}; \qquad (20\text{-}3)$$
$$= (e/q_P)^2; \ = \omega_p/\omega; \ \cong 1/20\Phi^4$$

The inverse fine structure constant α^{-1} (= 137.035999...) is associated with scaling in mass ratios while the reduced fine structure constant $\alpha/2\pi$ (= 0.0011614...) is associated with scaling in angular velocity and energy ratios. For example, orbital angular momentum of hadrons is found to vary in proportion to mass squared $L \propto (\alpha/2\pi)m^2$. The electron fine structure constant α is ubiquitous in numerous characterizations of the size of the electron. Various calculated electron radii as a function of multiples of the fine structure constant α are illustrated in Fig. 20-1. The fine structure const. α appears related to the inverse Kepler-Bouwkamp constant $K \simeq \phi^2/2\sqrt{(\pi\alpha)} \simeq 8.7000366\ldots$ where ϕ is a Fibonacci ratio.

When I die, my first question to the devil will be "What is the meaning of the fine structure constant?"
– Wolfgang Pauli

20. Origin of the Electron Fine Structure Constant α

Eddington, Pauli, Born, Heisenberg, Feynman, etc have famously written about the mysterious magic number 137 in quantum theory. In a letter to Heisenberg in 1934, Pauli wrote "Everything will become beautiful when [1/137] is fixed."[72] Feynman wrote: "It has been a mystery ever since it was discovered more than fifty years ago, and all good theoretical physicists put this number on their wall and worry about it. Immediately you would like to know where this number comes from... Nobody knows. It's one of the greatest mysteries of physics..."[73] Penrose observes "Many of today's physicists might be less optimistic than their predecessors about finding a direct mathematical 'formula' for α, or other 'constants of Nature'. Nowadays, physicists tend to regard these quantities as functions of the *energy* of the particles involved in an interaction, rather than simply as numbers."[74] The fine structure constant α is approx. equal to 0.007 or $1/20\Phi^4$ expressed in terms of Phi resonant damping ratios. In quite another numerological context, the mysterious number 007 was said to have been used by mathematician Dr. John Dee in the sixteenth century for eyes only communication with Queen Elizabeth and later famously adopted by author Ian Fleming.

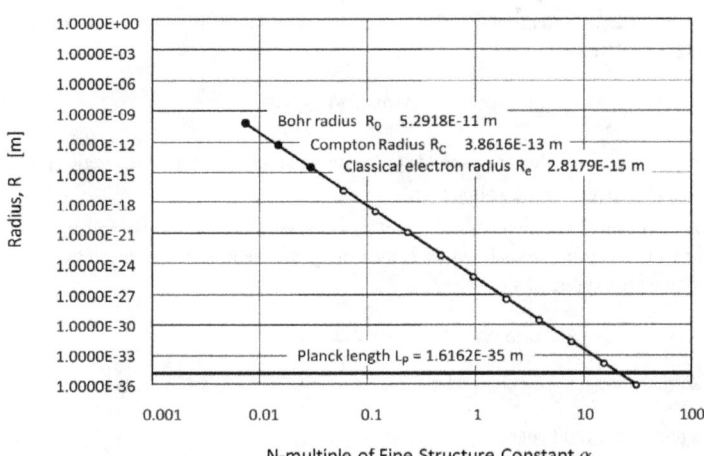

Fig. 20-1. Electron radii as a function of the fine structure constant α.

Besides the electron fine structure constant α, other dimensionless coupling constants are

Coupling constant	Symbol	Relation	Relative strength	n-gon 1/n	Precession angle (radians)
Strong	α_s	$= g_s^2 E/\hbar c$; $g_s^2/4\pi$	1	1	2π
Electromagnetic	α	$= e^2/4\pi\epsilon_0\hbar c$	1/137	137	0.01459
Weak	α_ω	$= G_F m_p c/\hbar^3$; $g_w^2/4\pi$	3×10^{-7}	3.333E6	6×10^{-7}
Gravity	α_G	$= G m_p^2/\hbar c$	5.9×10^{-39}	1.69E38	-

Alpha sets the scale of nature... it controls everything we see. – Frank Close

20. Origin of the Electron Fine Structure Constant α

20.2 Electron charge-to-mass ratio

The electric charge-to-mass e/m_e ratio of the electron is conventionally expressed in dimensions of [Q/M] and in SI units has been experimentally measured: $e/m_e \cong$ 175,882,008,800... C/kg [CODATA 2010]. From measured deflection of electron beams in a magnetic field, the ratio may be calculated from the applied magnetic field B, electron velocity v, electric potential V and path radius r.

$$e/m = v/Br = 2V/(B^2 r^2) \qquad (20\text{-}4)$$

What is the origin of this quotient and why does it have this value? Substituting the charge conversion dimensions ($Q = \theta M T^{-1}$) as derived by Klyushin[25], the charge-to-mass ratio e/m has dimensions of θT^{-1}:

$$\omega_{e/m} = e/m = \gamma_s = 1.758820088\text{E}11 \text{ rad/sec} \qquad (20\text{-}5)$$

This angular velocity is quite small in comparison to the Compton angular frequency ω_C (= 7.76343E20 rad/sec) and is interpreted as a precession frequency. If we add the Schwinger ($\alpha/2\pi$) correction to the g-factor where α = fine structure constant ($\cong$ 7.297352568E-03):

$$\omega_{e/m}(1 + \alpha/2\pi) = 1.76086\text{E}11 \text{ rad/sec} \qquad (20\text{-}6)$$

The above calculated value corresponds to measured value of the gyromagnetic spin ratio γ_s ($\cong$ 1.706859708E11 C/kg = rad/(s·T)).

What is cause of this precession? Let us consider the electrostatic and magnetostatic energy of the electron. As shown by Bergman[48], the electrostatic field energy E_{s0} of the electron may be expressed as

$$E_{s0} = e^2/2C = 4.103212\text{E-}14 \text{ J} \qquad (20\text{-}7)$$

where C = capacitance of a charged spinning ring = C_0/γ, C_0 = 3.128125E-25 F.

The magnetostatic field energy E_{m0} is given as

$$E_{m0} = LI^2/2 = 4.08412\text{E-}14 \text{ J} \qquad (20\text{-}8)$$

where L = self-inductance of a spinning ring = γL_0, L_0 = 2.08991E-16 (H) and I = current = $e\omega/2\pi$ = 19.979 (A). The total energy $E_t = \gamma E_{t0}$ where $E_{t0} = E_{s0} + E_{m0}$ = 8.18724E-14 J and γ = Lorentz factor = $1/\sqrt{(1 - v^2/c^2)}$.

If we compare the ratio of the magnetostatic energy to the electrostatic energy

$$E_{m0}/E_{s0} = 4.10312\text{E-}14 \text{ J}/4.08412\text{E-}14 \text{ J} = 1.004652165 \qquad (20\text{-}9)$$

and calculate the ½ average energy difference (kinetic plus potential):

$$\tfrac{1}{2} (E_{m0} - E_{s0})/(E_{m0} + E_{s0}) = 0.001160342 \qquad (20\text{-}10)$$

we find this value is slightly less than the fine structure constant ratio $\alpha/2\pi$ (= 0.00116140973). The difference $\alpha/2\pi - \tfrac{1}{2}(E_{m0} - E_{s0})/(E_{m0} + E_{s0}) \cong 0.0000010654$. The ½ factor represents the kinetic energy portion of the energy differential – the potential energy

20. Origin of the Electron Fine Structure Constant α

having null effect as the potential energy is blocked from conversion. Based on the Virial theorem of R.E. Clausius, in a central force field of a bound system, the average kinetic energy $\langle T \rangle$ $(= \frac{1}{2}mv^2)$ is half the average potential energy (virial) $\langle V \rangle$ $(= -2\langle T \rangle)$.

The Thomas precession frequency ω_T $(\cong (\mathbf{v} \times \mathbf{a})/2c^2)$ for a relativistic velocity v = c at the Compton radius R_C equals the orbital frequency. The Thomas precession is in a retrograde direction to the electron spin. The prograde geodetic (de Sitter) precession or gravito-magnetic effect is assumed to account for a portion of the slight difference between calculated $\alpha/2\pi$ and the average energy imbalance ratio. Hence, the imbalance of magnetostatic and electrostatic energy appears responsible for nearly all of the observed precession.

20.3 Thomas precession

The Thomas precession is an effect proposed by L. H. Thomas in 1926 which causes a precession with respect to a laboratory reference frame as a result of acceleration perpendicular to the rotation tangential velocity[46]. The Thomas precession, a kinematic relativistic effect, results in a precession angle of $-(v^2/c)$ radians/orbital radian ($\Delta\theta_T = \omega_T T = 2\pi v^2/c^2$). The Thomas angular velocity ω_T $(\gamma-1)((\mathbf{v} \times \mathbf{a})/v^2) = (\gamma-1)/\omega \cong (\mathbf{v} \times \mathbf{a})/2c^2$. Davies and Jennison[75] have shown that Thomas precession is due to the difference between the angular velocity of a rotating disc as observed in the laboratory frame and as observed at a point on the rotating disc. Ashworth[76] derives the following expression:

$$\omega_T = \omega - \omega_r = -\omega[(1 - v^2/c^2)^{-1/2} - 1] \qquad (20\text{-}11)$$
$$= \omega - \omega(1 - \beta^2)^{-1/2}$$
$$= -V^{-1}(dV_1/dt_1)[1 - \beta^2)^{-1/2} - 1]$$

where V = relative velocity between the two reference frames and β = v/c = $\Delta\phi/\pi$.

Interaction of the electron with an external magnetic field results in a Larmor precession ω_L which yields a total precession rate of $+(v^2/2)$. Total precession is the sum of Larmor plus Thomas precession:

$$\omega_{tot} = \omega_L + \omega_T = (\mathbf{v} \times \mathbf{a})/c^2 + (\mathbf{v} \times \mathbf{a})/2c^2 \qquad (20\text{-}12)$$
$$= \frac{1}{2}(\mathbf{v} \times \mathbf{a})/c^2$$

In the model illustrated, the cycloid eccentric motion undergoes a Thomas-Wigner precession rotation $\Delta\theta_T$ with each rotation at the Compton frequency ω_C. The circular path may be approximated by an N-sided polygon with sides of length $l = 2\pi r/N$. In an inertial laboratory rest frame, the rotation angle at each vertex is $\theta = 2\pi/N$ where N = 137. After N successive rotations, the net rotation is $\Delta\theta = 2\pi(\gamma-1)$ in the local object frame while the rotation in the laboratory frame is 2π radians. For an inscribed 137-sided polygon related to a consequence of Thomas precession, the fine structure constant $\alpha = ke^2/\hbar c \approx \cos(\pi/137)/137 = 0.007297$ where k = Coulomb constant. The periphery of a disc with radius r rotating at ω has length $l' = 2\pi r/\sqrt{(1 - r^2\omega^2/c^2)}$. The Zitterbewegung radius R_{zbw} corresponds to the radius required to make the l' circumference equal to R_C at a angular frequency ω_c.

$$R_{zbw} = (\lambda_c/2\pi)/2 = 0.5\hbar/mc = c/2\omega = 1.9301\text{E-}13 \text{ m} \qquad (20\text{-}13)$$

Hestenes[50] derives the same result as above for R_{zbw} and derives an expression for dynamic mass $m_d = \hbar\omega/2$ with units [m^2·kg·rad/s^2]. Similar to that experimentally observed

20. Origin of the Electron Fine Structure Constant α

for a steadily precessing gyroscope, the effect of retrograde precession is to reduce the net rotational velocity and, hence, apparent inertial mass m_i.

The mass radius R_m is equal to the relativistic contraction of the reduced Compton radius $R_C = \lambda_c/2\pi = \hbar/mc$

$$R_m = R_C/\sqrt{(1 + R_C^2\omega^2/c^2)} = 3.8419\text{E-}13 \text{ m} \qquad (20\text{-}14)$$

The relativistic contraction of the reduced Compton radius R_C results in a retrograde Thomas precession with Thomas-Wigner rotation angle $\Delta\theta = \lambda_C/\alpha^{-1} = 4.585\text{E-}02$ rad (= 2.627 deg). The rate of precession $\omega/\omega_p = (\Delta\theta/T)/(2\pi/T) = \gamma - 1$. The phase-locking boundary radius $R = c/\omega$ for a disc rotating with angular velocity ω corresponds to the Compton radius R_C. The mass radius R_m for an electron may be expected to lie within $0 < r < c/\omega$. For a frame in a disc rotating relative to the observer at an angular velocity $\omega + \Delta\omega$, the disc appears to be rotating at different angular velocities at different radii – slower at hub, ω at disc rim.

The time dilation in terms of the Lorentz factor is:

$$t_0 = t_v/\sqrt{(1 - (v/c))^2} \qquad (20\text{-}15)$$

substituting v and c in terms of r

$$c = \omega * r_{rim}$$

$$v = \omega * r$$

gives the time dilation in terms of radius:

$$t_0 = t/\sqrt{(1 - (r/r_{rim})^2)} \qquad (20\text{-}16)$$

Consequently, at the rim, time would not appear to be progressing at all whereas, at the hub, time would be progressing at the normal rate. As indicated in Fig. 20-2, the radius *r* in a rotating (non-inertial) reference frame is equal to the circumference $C/2\pi$. Inertial radius appears contracted in a rotating disc system at relativistic velocities. At relativistic velocities, in the stationary inertial frame, the circumference C_K appears greater than $2\pi r$. A contracted radius may be defined to correspond with the circumference C_{Kp} as perceived by a comoving observer. The contracted radius as measured by a revolving observer in frame K_p is $r_r = r\sqrt{(1 - r^2\omega^2/c^2)}$. Refer to Fig. 20-3.

It is one of the greatest damn mysteries of physics; a magic number that comes to us with no understanding by man. – Richard P. Feynman

There are no arbitrary constants... Nature is so constructed that it is possible logically to lay down such strongly determined laws which only contains logically deduced constants. – Albert Einstein

Can you explain the fine structure constant? No? So see you again when you have done it. – Paul A. Dirac

The fine structure constant is undoubtably the most fundamental pure dimensionless number in all of physics. It relates the basic constants of electromagnetism (the charge of the electron), relativity (the speed of light), and quantum mechanics (Planck's constant). – David J. Griffiths

There are considerable mysteries surrounding the strange values that Nature's actual particles have for their mass and charge. For example, there is the unexplained 'fine structure constant'... governing the strength of electromagnetic interactions. – Roger Penrose

20. Origin of the Electron Fine Structure Constant α

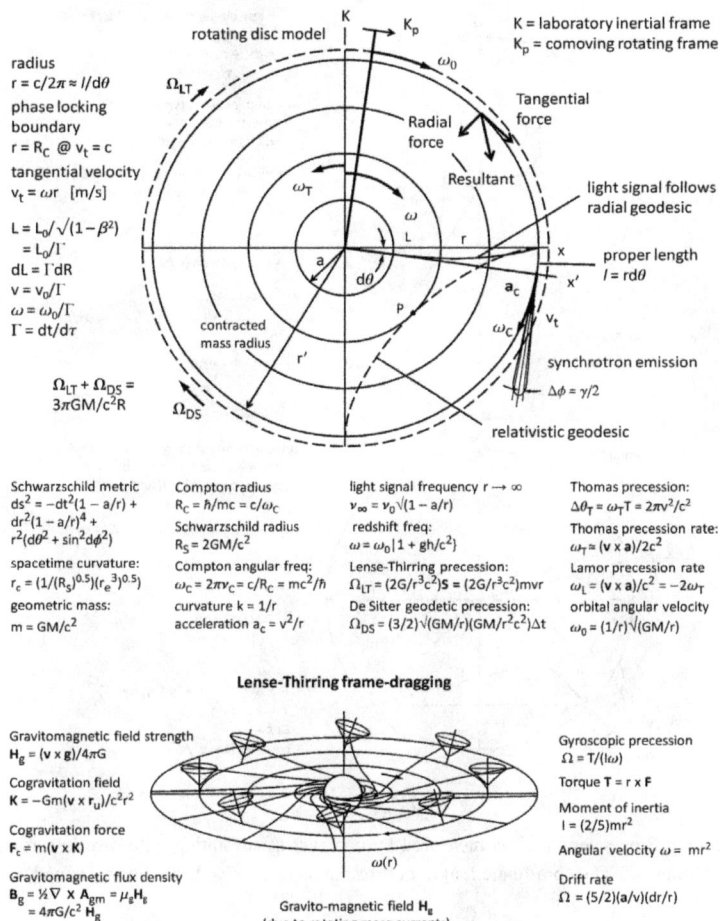

Fig. 20-2. Relativistic effects apparent in a comoving rotating reference frame for a rotating disc are shown in the upper diagram. The gravitomagnetic field effects of a rotating mass is shown in the lower diagram. The gravitomagnetic permeability μ_g (= $4\pi G/c^2$) is very low due to c^2 term in the denominator, hence, reduced gravitomagnetic flux.

The whole Galileo-Newton system thus sank to the level of first approximation, becoming progressively less exact as the velocities concerned approached that of light. – Hendrik Antoon Lorentz

Come forth into the light of things, Let Nature be your teacher. – William Wordsworth

After long reflection in solitude and meditation, I suddenly had the idea, during the year 1923, that the discovery made by Einstein in 1905 should be generalized by extending it to all material particles and notably to electrons. – Louis de Broglie

20. Origin of the Electron Fine Structure Constant α

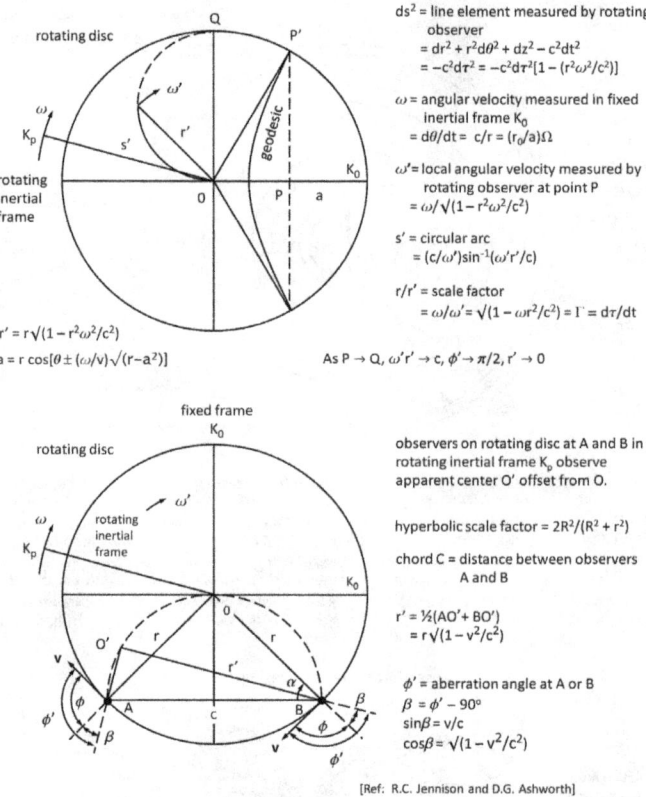

Fig. 20-3. Proper length r' as measured by observer(s) corotating with disc in a rotating inertial frame K_p. The coordinate length r corresponds to the fixed, stationary frame K_0.

J. J. Thomson, in 1881, theorized that a charged particle acquires an effective (electromagnetic) mass as a result of its interaction with its own electric field, i.e., self-energy[77]. Of the total rest mass energy of the electron E ($= mc^2$), half is kinetic energy and half is potential self-energy. Rest mass results from wavefunction interference due to precession and appears associated with a radial contraction. The contraction results in a positive mass deficit angle. Refer to Fig. 20-4. Based on a dimensional argument as indicated, the mass of the electron m_e may be expressed as a function of the Schwarzschild radius R_S ($= 2GM/c^2$) divided by the Einstein constant k ($= 8\pi G/c^4$). A radius of $4R_S$ ($= 4GM/c^2$) corresponds to a black hole particle capture radius. The radial contraction corresponds to $2R_S$ ($= k \cdot m_e/2\pi$). For zero mass (deficit angle $\alpha = 0$), the ½ semi-angle γ would equal $\pi/2$ rad. The radial contraction of $2R_S$ is equal to two times the electron strain amplitude ϵ_P ($= \sqrt{Gm^2/\hbar c} = 4.1854E-23$) times the Planck length l_P. A circumferential contraction of length $2R_S$ is 43 orders of magnitude smaller than orbital precession distance ($= \lambda_C/137$) at the Compton radius and 32 orders of magnitude smaller than the Planck length. Referenced to the Compton radius R_C, the deficit angle is vanishingly small.

20. Origin of the Electron Fine Structure Constant α

Electron mass deficit angle

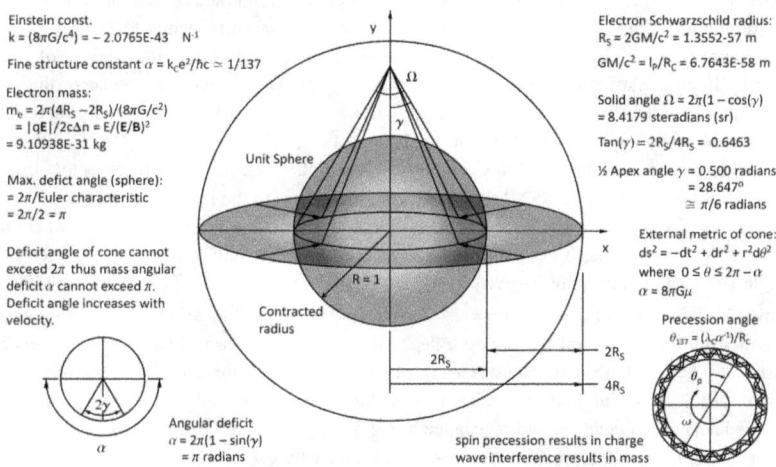

Fig. 20-4. Electron mass (confined energy) appears associated with a radial contraction resulting in an associated deficit angle. For an electron at rest, the calculated mass angular deficit equals π radians. Angular deficit is related to curvature $\Delta\theta = \int R \, dA$ where Gaussian curvature $R = 8\pi G\rho/3c^2$ and ρ = mass density. Spacetime impedance $Z_s = m_P/t_P = c^3/G$.

20.4 Electron stability

What accounts for the electron's indefinitely long lifetime? How is the configuration maintained during energy absorption and release? Why is it that the electron's energy is not radiated away as it spins? The stability of the electron may be attributed to continuous influx of energy compensating for energy losses generating a standing wave with net neutral energy flow. In a Planck vacuum, the space-time continuum exhibits a fine scale granularity at the Planck scale. Planck dipoles formed of positive and negative Planck masses constitute Archimedean spiral antenna emitters and absorbers embedded in a lossless medium. Electromagnetic waves radiated from the electron as it spins are absorbed by the surrounding Planck dipoles and re-emitted generating spherical Huygens wavelets. Half of the energy absorbed by each dipole is reflected back toward the electron where it is reabsorbed. See Fig. 20-5. The radiation pattern of an Archimedean spiral antenna with no end loss in a lossless medium is hemispherical about the bore axis ($\theta = 0$) with an amplitude ($\cos^2\theta/2$) in the forward and reflected direction. The interaction of the electron and the vacuum occurs at a unique resonance ratio of the electron Compton frequency f_C and the Planck frequency f_P (= 1.2360E20 Hz /2.9497E42 Hz = 4.19019E-23).

All the laws of physics in their elementary forms are reversible. – Léon Brillouin

I am busy just now again on electro-magnetism, and I think I have got hold of a good thing, but I can't say. It may be a weed instead of a fish that, after all my labor, I may at last pull up. – Michael Faraday

It is more important that a proposition be interesting than it be true…But of course a true proposition is more apt to be interesting than a false one. – Alfred North Whitehead

20. Origin of the Electron Fine Structure Constant α

The electron itself may be viewed as equivalent to an Archimedean spiral dipole antenna. An Archimedean spiral antenna radiates from an active region where the circumference equals the wavelength which for an electron corresponds to the Compton wavelength λ_C (= 2.4283E-12 m). The characteristic free space impedance of the surrounding medium is $Z_0 = 367.7$ Ω. The internal impedance of the electron according to Babinet's principle for a 2-arm spiral is $Z_0/2 = 188.5$ Ω. The low frequency operating point is determined by the outer radius r_2 and is given by $f_{low} = c/2\pi r_2 = c/(2\pi R_C) = 1.256$E20 Hz which corresponds to the electron Compton frequency f_C. The high frequency cutoff limit $f_{high} = c/2\pi r_1 = c/(2\pi\lambda_P) = 2.9522$E42 Hz which corresponds to the Planck frequency f_P.

The spiral of Archimedes spiral is described by the polar equation $r = a + b\theta$ where a defines the start radius, b defines the distance between turns and θ describes the rotation angle. The spiral curvature is given by $K(\theta) = (2 + \theta^2)/a(1 + \theta^2)^{3/2}$. The arc length $s(\theta) = \frac{1}{2}a[\theta\sqrt{(1 + \theta^2)} + \ln\sqrt{(1 + \theta^2)}] = a(\theta + 1/6\ \theta^3 - 1/40\ \theta^5 + 1/112\ \theta^7 - 5/1152\ \theta^9 + \ldots)$. For a 2-arm Archimedean spiral frequency independent antenna, the arm spacing between each turn is $s = [(r_2 - r_1)]/4N = w$ where N = no. of turns = $\theta/2\pi$, r_1 = inner radius = $s = b/2\pi$, r_2 = outer radius of spiral midline, w = arm width = s. For such an antenna, at high frequencies, the impedance and resistance are relatively constant. For small x, the electric field strength $E = K \sin x/x$ where $x = \pi(L/\lambda)\sin\theta$ where L = antenna diameter. The radiated power (energy per unit time) P varies as $K'(\sin x/x)^2$. The reflected wave has opposite polarization compared to the outward traveling wave.

The maximum electrostatic repulsive force between two electrons if held stationary at a separation distance equal to Compton's radius R_C according to Coulomb's equation is $F_e = (1/4\pi\epsilon_0)e^2/R_C^2 = 1.5471$E-03 N. Macken[20] shows that this force $F_e = \alpha F_m$ where α = fine structure constant and F_m is the maximum theoretical force possible if all of the circulating power is deflected ($F_m = \hbar c/R_C^2 = 0.2119$ N). See Fig. 20-6. For two electrons separated by a distance of 1 mm, the Coulomb repulsion force $F_{elec} = 2.307$E-22 N. For comparison, the gravitational force of attraction of two electrons at the same distance is $F_{grav} = 5.547$E-65 N which is ~10E43 times weaker.

When you want to know how things really work, study them when they're coming apart. – William Gibson

You will get your difficulties with the point electron. – Paul Ehrenfest

Let us begin with the fine-structure constant... The fine-structure constant is really the ratio of two natural units or atoms of action....We obtain action when we multiply energy by time. ...We are challenged to find a unified theory of electric particles and radiation in which the electrostatic type of action and the quantum type of action are traced to their source. – Arthur Stanley Eddington

Jung and Pauli's mutual effort to discover the cosmic number or fine structure constant, which is a fundamental physical constant dealing with electromagnetism, or, from a different perspective, could be considered the philosopher's stone of the mathematical universe. – Todd Hayen

The fact however that alpha has just its value 1/137 is certainly no chance but itself a law of nature. It is clear that the explanation of this number must be the central problem of natural philosophy. – Max Born

QED [Quantum electrodynamics] reduces... "all of chemistry and most of physics,: to one basic interaction, the fundamental coupling of the photon to electric charge. The strength of this coupling remains, however, as a pure number, the so-called fine-structure constant, which is a parameter of QED that QED itself is powerless to predict. – Frank Wilczek

20. Origin of the Electron Fine Structure Constant α

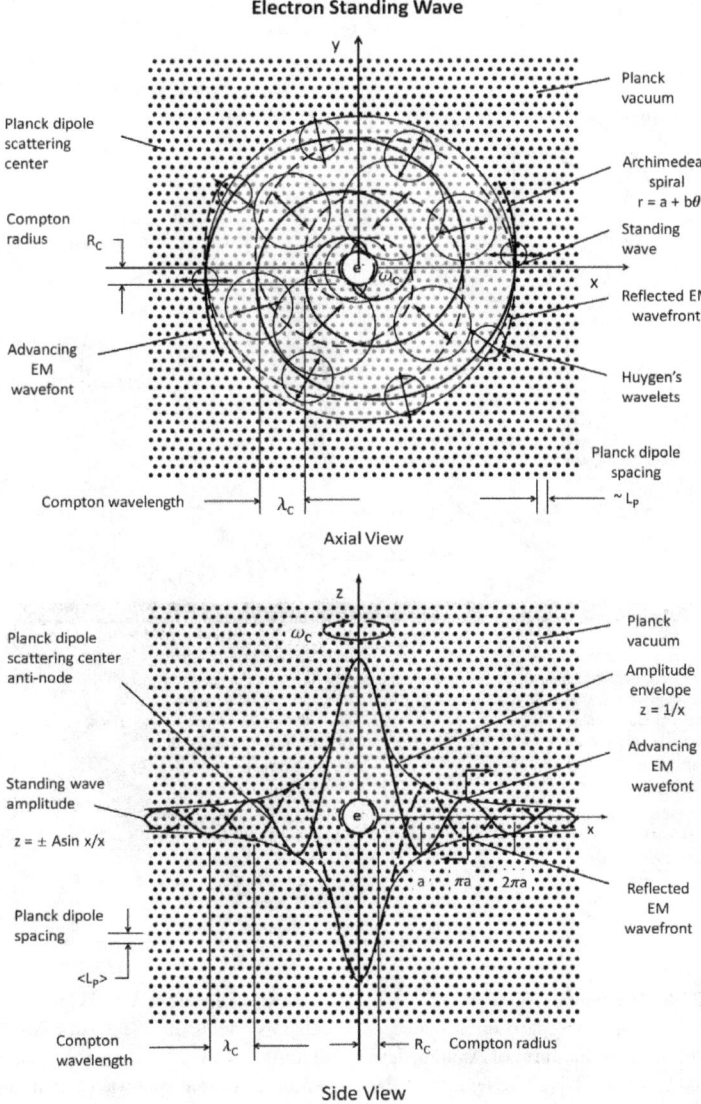

Fig. 20-5. Electron standing wave pattern described as a rotating Archimedean spiral (axial & side view). An external EM standing wave resonance is formed by an outward propagating travelling wave and an inward reflected travelling wave from the surroundings external to the Compton radius. The internal electron standing wave structure is formed by a photon topologically confined within the quantum vacuum light speed circle boundary at the Compton radius.

The actual state of our knowledge is always provisional and... there must be, beyond what is actually known, immense new regions to discover. – Louis de Broglie

20. Origin of the Electron Fine Structure Constant α

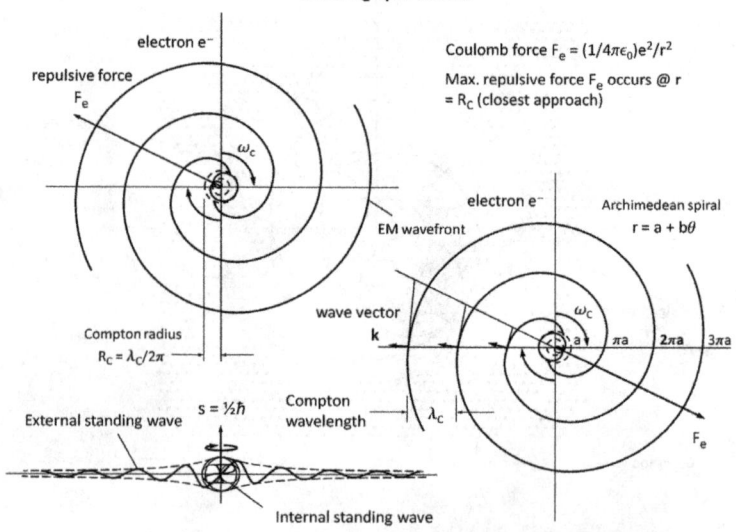

Parameter	Symbol	Relation/Value
Charge	q^-, e^-	$= 1.6021 \times 10^{-19}$ Coulombs $= 7.019 \times 10^{-10}$ kg·rad/s
Compton frequency	f_C	$= 1.24 \times 10^{20}$ Hz $(= c/\lambda_C = mc^2/h = \omega_C/\pi)$
Compton angular frequency	ω_C	$= 7.76 \times 10^{20}$ s^{-1} $(= mc^2/\hbar = c/Rq = E_i/\hbar)$
Compton wavelength	λ_C	$= 2.43 \times 10^{-12}$ m $(= h/mc = h/p = hc/eV)$
Reduced Compton wavelength	$\lambdabar_C$	$= 3.86 \times 10^{-13}$ m $(\lambdabar_C = \lambda_C/2\pi = c/\omega_C = \hbar/mc)$
Classical electron radius	r_e	$= 2.82 \times 10^{-15}$m $(= (1/4\pi\epsilon_0)e^2/m_ec^2))$
Planck length	L_P	$= 1.616 \times 10^{-35}$ m $(= ct_P = \sqrt{(\hbar G/c^3)})$
Planck/Compton ratio	r_{PC}	$= 4.18 \times 10^{-23}$ $(= L_P/\lambdabar_C = L_P/R_C)$
Internal energy	E_i	$= 8.1871 \times 10^{-14}$ J $(= mc^2 = \hbar\omega_C = m(E/B)^2)$
Rest mass	m_e, m_0	$= 9.1094 \times 10^{-31}$ kg $(= E_i/c^2 = \hbar/R_Cc = \omega_C\hbar/c^2)$
Rest mass energy	E	$= 0.511$ MeV/c^2 $(= k_Ce^2/\alpha R_C)$
Planck force	F_P	$= 1.210 \times 10^{44}$ N $(= E_P/L_P = c^4/G = F_m/\epsilon_\beta^2)$
Max force F_m @ R_C	F_m	$= 0.21201$ N $(= m^2c^3/\hbar = \hbar c/R_C^2 = \epsilon_\beta^2 F_P)$
Coulomb force	F_e	$= e^2/4\pi\epsilon_0 R_C^2 = \alpha\hbar c/R_C^2 = q^+/E = \alpha F_m$

Fig. 20-6. Maximum Coulomb repulsive force between electrons occurs at closest proximity equal to separation distance of Compton's radius R_C. The radiated EM wavefronts are in the form of Archimedean spiral arms. Two electrons on approach are repelled by constructive interference of EM wavefronts. Positron spin waves rotate in a direction opposite to that of electrons. Electron and positron interaction give rise to destructive wave interference. Spin precession (torsion field dislocations) of like helicity repel while those of unlike helicity attract.

If we start with two electrons we can change the scale of the wave of one of the electrons by shifting its wave relative to the other electron. The shifting of the wave leads to a compensating conversion factor that we call the electric force between electrons. – Ronald Mallett

Forces of nature are conversion factors needed to compensate for changes in scale. – Ronald Mallett

21. Electric Charge

The day when we shall know exactly what electricity is will chronicle an event probably greater, more important, than any other recorded in history the human race. – Nikola Tesla

Electric charge is a fundamental, physical property of matter similar to, and inseparable from mass, and, like mass, is a conserved quantity. Electric charge, a scalar quantity, remains invariant under a Lorentz transformation. The electrostatic energy of a moving charge decreases with velocity by a Lorentz factor $g = 1/\sqrt{(1 - (v/c)^2)}$ while increasing in magnetic energy such that the total energy remains constant.

A particle with charge undergoes Lorentz contraction and relativistic mass increase with motion with mass becoming infinite at light speed. Like mass and spin angular momentum, charge is quantized. Mass variation with velocity appears as a radiation resistance effect. Electric charge (e) is a two-valued scalar quantity parameterized by a integer quantum number $Q = \pm 1$ associated with leptons and fractionally $Q = \pm 1/3, \pm 2/3$ in the case of quarks. Planck charge q_P is a function of the speed of light c and Planck's constant $\hbar$, a unit of action with dimensions of spin angular momentum. In meter kilogram second (MKS) units, electric charge [Q] and mass [M] are considered as separate fundamental dimensions.

21.1 Dimensions of electric charge

Charge has been likened to French perfume – Particles cloaked with it make opposites attract. Electric charge has variously been described as a source or sink, a broken symmetry caused by spin, the end of a broken string, a vacuum density stress or strain, a measure of vacuum curvature, a form of imaginary energy, a complex part of a field, a measure of localized time distortion, a dimensionless proportionality factor associated with angular momentum around an unseen fifth dimension, a topological property of a spin or torsion field, a measure of chirality and apparent time flow, a coupling constant relating vector current to the vector potential, etc. However, the fundamental nature has remained elusive and in standard texts in adopted units of measurement, electric charge is described following Benjamin Franklin's characterization as a fundamental dimension represented simply, however, opaquely as [Q] with ± polarity. In 1939, Nikola Tesla lamented '…Day after day I have asked myself "what is electricity?" and have found no answer. 80 years have gone by since that time and I still ask the same question, unable to answer it.'

Charge has its origin in rotational motion of energy. In the Kaluza-Klein theory, electric charge is proportional to the angular momentum (= energy/frequency) of motion around an unseen curled up fifth dimension. The charge/mass ratio suggests energy of motion at charge radius > energy at mass radius (e/m = 1.758820088E11 C/kg). Electric charge is related to topological charge associated with spin precession.

$$e = m_e\omega = m_e(\omega_C + \omega_p) = 1.6021\text{E-}19 \text{ Coulomb} \qquad (21\text{-}1)$$
$$= 7.0719\text{E-}10 \text{ kg·rad/sec}$$

where $\omega_p = d\theta/dt = e/m = \omega_{e/m} = 1.758820088\text{E}11$ rad/sec, $\omega_C = 7.7634\text{E}20$ rad/sec.

The topological charge rotation angle $d\theta$ represents the incremental precession

$$d\theta = S_e/R_C = \lambda_C/q = 4.585\text{E-}02 \text{ rad } (= 2.627°) \qquad (21\text{-}2)$$

where

q = whirl number = $\alpha^{-1} \cong 137.03599967$

S_e = Arc length = $R_C\theta = \lambda_C/\alpha^{-1} = \lambda_C/q \cong 1.77056437$ m

239

21. Electric Charge

Anomalous magnetic moment (fractional part of spin) is related to formation of charge.

Toroidal models of the electron have been previously advanced which involve toroidal and poloidal rotations. Klyushin[24] posits that the toroidal (or equatorial) rotation determines electric charge while the poloidal (or meridional) rotation determines spin. The electric charge is given as

$$q = m\omega = 7.0719 \times 10^{-10} \text{ kg·rad/sec} \qquad (21\text{-}3)$$

where m is mass ($m_e = eV/c^2 = k_C e^2/\alpha R_C c^2 = h/c \cdot (\lambda_C/\alpha^{-1})/\theta_{137} = 9.109 \times 10^{-31}$ kg) and ω is the toroidal rotation angular velocity ($\omega_C = mc^2/\hbar = 7.76 \times 10^{20}$ rad/sec). Electric charge results from spin precession ω_p and rest mass m_e arises from wave interference due to precession. Hence, electric charge and mass are intimately related.

For a toroidal geometry, the open electric flux lines (time averaged) extend in a radial direction about the center of rotation with closed magnetic flux lines associated with the toroidal current loop aligned with the loop spin axis. The poloidal magnetic flux forms about the poloidal loop current.

The ratio of electric repulsion to gravitational attraction force between two electrons is

$$F_{elec}/F_{grav} = \tfrac{1}{4\pi\epsilon_0}[q^2/r^2]/[Gm^2/r^2] = 4.17 \times 10^{42} \qquad (21\text{-}4)$$

Substituting for q yields an expression in terms of ω

$$\omega^2/4\pi G\epsilon_0 = 4.17 \times 10^{42} \qquad (21\text{-}5)$$

Substituting the electron Compton angular velocity $\omega_C = mc^2/\hbar = 7.7633 \times 10^{20}$ rad/s yields a value of electric charge

$$e = q = m\omega = 9.109 \times 10^{-31} \text{kg } (7.7633 \times 10^{20} \text{ rad/sec}) = 7.0719 \times 10^{-10} \text{ kg·rad/s} \qquad (21\text{-}6)$$

Correspondingly,

$$1 \text{ Coulomb} = q/e = [7.0719 \times 10^{-10} \text{ kg·rad/s}]/[1.602 \times 10^{-19} \text{ Coul per electron}] \qquad (21\text{-}7)$$
$$= 4.414 \times 10^9 \text{ kg·rad/s}$$
$$\epsilon_0 = 1.725 \times 10^8 \text{ kg·rad}^2/\text{m}^3 \qquad (21\text{-}8)$$
$$\mu_0 = 6.449 \times 10^{-26} \text{ m·s}^2/(\text{kg·rad}^2) \qquad (21\text{-}9)$$
$$e/m_e = 1.7588 \times 10^{11} \text{ rad/s} \qquad (21\text{-}10)$$

The electric permittivity of free space ϵ_0 (= $8.854187817 \times 10^{-12}$ F/m) corresponds to an aether mass density. The magnetic permeability μ_0 (defined as $4\pi \times 10^{-7}$ H/m) corresponds to an aether compressibility.

Electrodynamic momentum qA where q is electric charge [Q] and **A** is the magnetic vector potential [MLT^{-1}Q^{-1}] has the corresponding mechanical dimensions [MLT^{-1}] or MKS units of kg·m/sec which is equivalent to linear momentum (mass x velocity). The canonical total momentum (sum of kinetic momentum and potential momentum) is thus represented as **P** = m**v** + q**A** whereas the observed momentum is given by **P** − q**A** = $\hbar$**k** − q**A**. Electric charge q = $\hbar$k/A = P/A = (dP/dt)/(dA/dt).

The vacuum free space impedance $Z_0 = \sqrt{(\mu_0/\epsilon_0)} = \mu_0 c = E/H = 4\pi Z_P = 376.73 \, \Omega = 1.75 \times 10^{-17} \text{ m}^2 \cdot \text{s}/(\text{kg·rad}^2)$.

Hence,

$$1 \text{ Ohm} = 4.65 \times 10^{20} \text{ m}^2\cdot\text{s/kg·rad}^2 \qquad (21\text{-}11)$$
$$1 \text{ Ampere} = 1 \text{ Coul/sec} = 4.44 \times 10^9 \text{ kg·rad/s}^2 \qquad (21\text{-}12)$$
$$1 \text{ Volt} = 1 \text{ Ohm} \times 1 \text{ Ampere} = 2.07 \times 10^{-10} \text{ m}^2/\text{s·rad} \qquad (21\text{-}13)$$
$$1 \text{ Farad} = 1 \text{ Coul}/1 \text{ Volt} = 2.144 \times 10^{19} \text{ rad·kg/m}^2 \qquad (21\text{-}14)$$

21. Electric Charge

The local permittivities ϵ_1, ϵ_2 and path permittivity ϵ_{12} in an accelerated region are increased over the far field permittivity ϵ_0 as a result of increased density of hypothetical vacuum quantum Planck particles which increase the capacitance similar to the effect of fine metallic particles distributed between poles of an air-dielectric capacitor.

A summary table of dimensional conversions is shown in Table 21-1. A more extensive list of conversions of various physical quantities is provided in a section 54.

The phenomena of magneto-electric induction forced me to postulate the principle of (special) relativity. – Albert Einstein

Maxwell makes the important assumption that the magnetic vortices have a mass depending on the magnetic permeability of the medium. – William Berkson

Faraday induction is not caused by changing magnetic fields, but by electrokinetic fields produced by changing electric currents. – Oleg. D. Jefimenko

Faraday's theory of induction by lines of force could equally well be formulated in terms of the vector potential. – Sir Edmund T. Whittaker

Since Neumann's vector potential is the potential of the force between current elements, the space derivative will give the force acting between current elements... The force between current elements should be distinguished from the electromotive force due to electromagnetic induction, which is equal to the time derivative of the vector potential. – William Berkson

There are two main reasons to believe that the lines of force have an independent physical existence. The first is that they can be curved and the second is that they take time to propagate. – Michael Faraday

The distribution of the magnetic lines of force could be determined by assuming that there is a tension along the lines of force and a pressure between them. – Michael Faraday

Before I began the study of electricity I resolved to read no more mathematics on the subject till I first read through Faraday's 'Experimental Researches in Electricity.' – James Clerk Maxwell

There is nothing quite as frightening as someone who knows they are right. – Michael Faraday

I am somewhat exhausted; I wonder how a battery feels when it pours electricity into a nonconductor? – Sherlock Holmes

I am an expert of electricity. My father occupied the chair of applied electricity at the state prison. – W. C. Fields

Almost all of the material phenomenon which occur under terrestrial conditions are recognized as quantum mechanic consequences of the electrical attraction between electrons and nuclei and of the gravitational attraction between massive objects. – Victor Weiskopf

I take it that a monograph of this sort belongs to the ephemeral literature of science. – Gilbert Newton Lewis

...it should be remembered that the atomicity of electron charge has already found its expression in the specific numerical value of the fine structure constant, a theoretical understanding of which is still missing today. – Wolfgang Pauli

A contradiction – free union of the condition of quantum theory with the corresponding prediction of field theory is only possible in a [theory] that provides a particular value of Sommerfeld's constant e^2/hc. – Wolfgang Pauli

21. Electric Charge

Table 21-1. Units Conversion of Physical Quantities

Quantity	Symbol	Dimension	Planck Formula	Remarks	MKS Units
Length	l	L	$l_P = t_P c = \sqrt{(\hbar G/c^3)}$	$\lambda_C = 2\pi\hbar/mc$, $R_C = \hbar/mc$	meter
Angle	Θ	–	$\theta = l_P \theta/(\hbar/mpc)$	$\theta = s/R = s/(\hbar/mc)$	radian
Time	t	T	$t_P = l_P/c = \sqrt{(\hbar G/c^5)}$ $= \hbar/c^2 m_P$	$t_C = \lambda_C/c = \hbar/mc^2$	sec
Mass	m	M, QT, IT²	$m_P = \sqrt{(\hbar c/G)} = \hbar/c^2 t_P$ $= E_P/c^2$	(Positive energy – negative energy)/c^2 > 0. m = $\hbar k$∼qA/v m = |Volt·Charge|/c^2 = **W/g** = $k_C e^2/\alpha R_C c^2$ = E/($\mathbf{E}/\mathbf{B}$)^2 = $2L/(\mathbf{E}^2 – c^2\mathbf{B}^2)$ = m_0/T	kg
Charge	e	Q, $\Theta M T^{-1}$	$q_P = \sqrt{(4\pi\epsilon_0 \hbar c)}$ $= (1/(I_P/Q_P))cT_P$ $\approx (q_e\sqrt{\alpha})/\alpha$ $= m_P \omega_P$	Electric charge equivalent to mass rotational flow rate (Mass x precession rate). Spin torsion defect - topologicalcharge	Coulomb Amp·sec [kg·rad/sec]
Volt	V	$ML^2T^{-2}Q^{-1}$ L^2T^{-1}	$V_P = E_P/q_P = (I_P/Q_P)cZ_s$ $= \sqrt{(c^4/4\pi \epsilon_0 G)}$	Potential difference. A measure of rotational energy. Density waves swept area/unit time.	volt [m²/sec·rad]

cont

Table 21-1. Units Conversion of Physical Quantities (cont)

Quantity	Symbol	Dimension	Planck Formula	Remarks	MKS Units
Action	$\hbar$	ML^2T^{-1}	$\hbar = c^2 T_P^2 Z_S = m_P l_P c$ $= m_P \cdot l_P^2 / t_P$	Mass × voltage ($kg \cdot m^2/sec$) Energy × time ($J \cdot s$) Momentum × radius ($N \cdot m \cdot s$) = angular momentum.	$J \cdot s =$ $rad \cdot kg \cdot m^2/sec$
Energy	E	ML^2T^{-2}	$E_P = c^2 T_P Z_S = \sqrt{(\hbar c^5/G)}$	Measure of frequency of mass motion or wave curvature. |Voltage × charge| $E = \pm\sqrt{(pc)^2 + (mc^2)^2}$ $E = \frac{1}{2}m(\mathbf{E}^2 + c^2\mathbf{B}^2)$	$J = N \cdot m =$ $W \cdot s =$ $kg \cdot m^2/sec^2$
Force	$\mathbf{F}$	MLT^{-2}	$F_P = E_P/l_P = c^4/G = cZ_S$ $= m_P l_P/t_P^2 = Gm_P^2/l_P^2$ $= (1/4\pi\varepsilon_0)q_P^2/l_P^2$	$\mathbf{F}_m = \hbar\omega c^2/c,\ \mathbf{F}_e = \alpha\mathbf{F}_m.$ $\mathbf{F}_g = Gh^2/c^2\mathbf{r}^4$ $F_g/F_e = F_e/F_P$	$N =$ $kg \cdot m/sec^2$
Pressure	p	MLT^{-2}	$p_P = F_P/l_P^2 = m_P/t_P^2 l_P$	Force/area	J/m^2
Velocity	v	L/T	$v_P = l_P/t_P = c$	Translational velocity v $= (\hbar k - qA/\gamma m$ $\mathbf{v}_g = \hbar \mathbf{k}/m_i = d\omega/d\mathbf{k} = c^2/v_p$ $= c \cdot \Delta\nu/\pi = \beta \cdot c;\ v_p = \omega/\mathbf{k}$ $= c^2/v_g = c/\Delta\nu/\pi = c/\beta$	m/sec

cont

21. Electric Charge

Table 21-1. Units Conversion of Physical Quantities (cont)

Quantity	Symbol	Dimension	Planck Formula	Remarks	MKS Units
Frequency	f	T^{-1}	$f_P = \omega/2\pi = c/l_P$	$f = \omega/2\pi = E/h$	1/sec
Angular Freq.	ω	ΘT^{-1}	$\omega_P = 1/t_P = \sqrt{(c^5/\hbar G)}$	$\omega = 2\pi f = c \cdot k/n$	rad/sec
Scalar potential	ϕ	$\Theta^{-1} L^2 T^{-1}$	$V_P = \sqrt{c^4/4\pi\epsilon_0}G = \sqrt{F_P/4\pi\epsilon_0}$	Energy/Charge	$m^2/(sec \cdot rad)$
Vector potential	$\mathbf{A}$	$L\Theta^{-1}$	$\mathbf{A} = \frac{1}{2}\mathbf{B} \times \mathbf{r} = \Phi_B/2\pi\theta = mv/e$ $= \mathbf{k}\hbar/q_0 = (\mathbf{v}/c^2)\phi_E$	Electromagnetic field momentum/unit charge.	m/rad
Curvature	k	L^{-1}	$2\pi/l_P$	A measure of imbalance of magnetostatic energy and electrostatic energy	1/m
Torsion	τ	ΘL^{-1}	$\theta 2\pi/l_P$	A change in torsion is a measure of effective mass	rad/m
Gravitational constant	G	$L^3 M^{-1} T^{-2}$	$G = c^3/Z_S = c^3/(m_P/t_P)$ $= c^4/F_P = t_P c^3/m_P$ $= \hbar c/m_P^2 = l_P^3/m_P t_P^2$ $= l_P c^2/m_P = l_P c^4/t_P k_B$ $= l_P c^4/E_P = c^2 l_P/m_P$ $\approx 6.67261E\text{-}11$	$G = \mathbf{F}_g r^2/(m_1 m_2)$ $= \mathbf{F}_g c^2 \tau^4/\hbar^2$ $= 3\pi V^2/M_P^2$ $= G_0 \epsilon_{12}^2/\epsilon_0 = G_0/T^3$ $= R_S c^2/2M = c^2 R_{univ}/M_{univ}$	$m^3/kg \cdot sec^2$ [$C \cdot v \cdot m/kg^2$]

cont

244

21. Electric Charge

Table 21-1. Units Conversion of Physical Quantities (cont)

Quantity	Symbol	Dimension	Planck Formula	Remarks	MKS Units
Coulomb constant	k_e	$\Theta^{-2}ML^3$	—	$k_e = 1/4\pi\epsilon_0 = \mu_0 c_0^2/4\pi$ $= 8.987551\text{E}9$ $= 4.6129\text{E-}10$	$kg \cdot m \cdot s^{-2} \cdot C^{-2}$ $m^3/kg \cdot rad^2$
Fine structure constant	α	—	$\alpha = e^2/(4\pi\epsilon_0)\hbar c = e_p^2$ $= k_e e^2/\hbar c = \mu_0 c e^2/2h$ $= \sqrt{(e/q_P)} = Z_0/Z_e$ $= 0.0072973525693(11)$	Velocity ratio (v_{et}/c), Impedance ratio (Z_0/Z_e) Force ratio (F_e/F_m) Energy ratio ($\pi(E_{s0}-E_{m0})/(E_{s0}+E_{m0})$)	—
Planck's constant	h	ΘML^2T^{-1}	$h = 2\pi\hbar = 6.62607004\text{E-}34$	Energy/Frequency $h = E/\nu$	$rad \cdot kg \cdot m^2/s$
Reduced Planck's constant	$\hbar$	ΘML^2T^{-1}	$\hbar = h/2\pi = E/2\pi\nu = e^2/\alpha$ $= 1.0545\text{E-}34$	Dirac h	$rad \cdot kg \cdot m^2/s$
Planck charge	q_P	$M\Theta T^{-1}$	$q_P = \sqrt{(4\pi\epsilon_0 \hbar c)} = \sqrt{(e^2/\alpha)}$ $= m_P \omega_P = 11.70963$ q	Torsion defect precession $\omega_P = q_P/m_P = 8.619\text{E}25$ rad/s	$kg \cdot rad/s$
Temperature	T^0	T	$T_P = E_P/k_B = m_P c^2/k_B$ $= l_P c^4/Gk_B = Gm_P/c^3$	Temperature (°K)	°K
Entropy	S_{BH}	$L^{-1}T$	$S_{BH} = c^2 A/4G\hbar$	Bekenstein-Hawking area law $S_{BH} = $ Area $/ (4\hbar G)$	s/m

245

21. Electric Charge

Electrical charge equivalence to mass angular flow rate may represent mass flow in a spiral spin density wave of Planck dipoles with spin direction corresponding to ± charge. Charge is quantized since spin angular momentum is quantized. As shown, electric charge (e) has dimensions of kg·rad/sec [MΘ/T] equivalent to [Q]. Electric charge is a measure of spin precession which has the same dimensions as spin angular momentum. In this respect, electric charge is a dual to vortex charge. The topological charge of an optical vortex, for example, is an integer number, either positive or negative, that represents the integral number of turns per wavelength of light or strength of the vortex (orbital angular momentum). The sign of the topological charge is dependant on the direction of twist. The spin angular momentum of an electron is a constant $|S_z| = \frac{1}{2}\hbar$. In a free (irrotational) vortex, the radial velocity $v_r = 0$ while the tangential velocity varies inversely with radius $v_\theta = \Gamma/2\pi r$. The circulation/unit area or vorticity $\Gamma = 2\pi C$ is a constant. The angular momentum/unit mass is constant: $rv_\theta = \Gamma/2\pi$. Electrons with opposite spin couple to form a pair of charge 2e (phonon mediated attraction) forming a charge = vortex composite each made up of charge q and vorticity Φ. The action picks up an imaginary contribution $\pm i(q_i + i\Phi)$. The intrinsic statistics angle $\theta_1 = 0 \pmod{2\pi}$ for bosons and $\theta_1 = \pi \pmod{2\pi}$ for fermions. If charged particles are fermions and $q\Phi = \pi$ then the composite has Bose statistics. In general, $q\Phi$ need not be an integer multiple of π. Massless particles can have no charge as there is no torsion field defect (failure of loop closure due to precession).

As previously noted, the electron exhibits a spin precession which is characterized by the fine structure constant α in addition to its quantum spin angular momentum $s = \hbar/2$. See Fig. 21-1. Josephson junction oscillations and quantum Hall resistance steps are characterized by α. Superconductivity – a pairing of bound electrons (Cooper pairs interacting through exchange of phonons) of opposite spins travelling together in a supercooled conductor acting as a spin 0 boson – may be associated with charge density waves (CDW) with amplitude and/or phase fluctuations. High temperature superconductivity is thought to the result of spin fluctuations rather than phonon lattice vibration interactions in low temperature superconductors. With paired spins, a possibility for enhanced coupling arises if, in addition, the spin precession of each electron is in phase alignment. Such a coupled phase-lock relationship producing an attractive interaction may possibly help explain high temperature conductivity effects such as the postulated short-range spin waves. A quantum-mechanical knot soliton may be constructed from a set of nested tori of interlinked loops in a Hopf fibration. Such a stable topological knot in a quantum field if induced to precess by an external magnetic field may provide a means to synthesize a multi-loop electron analog with increased charge.

I used to wonder how it comes about the electron is negative. Negative-positive – these are perfectly symmetric in physics. There is no reason whatever to prefer one to the other. Then why is the electron negative? I thought abought this for a long time at last all I could think was "It won the fight." – Albert Einstein

You know my methods. Apply them. – Sherlock Holmes

Assuming that our explanation of the lines of force by molecular vortices is correct, why does a particular distribution of vortices indicate an electric current? – Michael Faraday

Brunettes are full of electricity. – Villers de L'Isle-Adam

I wonder if I could make an electric bass. – Leo Fender

21. Electric Charge

Spin wave precession

Electric charge $q_e = m_e \omega_p$ = 1.60217E-19 C = 7.0719E-10 kg·rad/s
Fine structure constant $\alpha = ke^2/\hbar c$ = 7.29735E-3
Gyromagnetic spin frequency $\omega_{e/m}(1 + \alpha/2\pi)$ = 1.76086E11 rad/s
Inversion radius $R_s = R_C - R_m$ = 2.6983E-17 m
Precession angle $\theta_{137} = (\lambda_C/\alpha^{-1})/R_C$ = 0.04458 rad

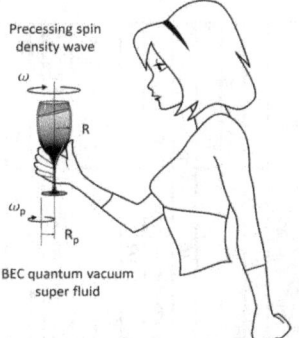

Precessing spin density wave

BEC quantum vacuum super fluid

Parameter	Symbol	Value	Units
Speed of light	c	≅ 2.9979E08	m/s
Electron charge	e	≅ 7.0710E-10	kg·rad/s
Electron mass	m_0	≅ 9.1093E-31	kg
Charge/mass	e/m_0	≅ 1.7588E11	rad/s
Compton frequency	ω_C	≅ 7.7634E26	rad/s
Compton radius	R_C	≅ 3.8616E-13	m
Zitterbewegung radius	R_{zbw}	≅ 1.9306E-13	m
Precession frequency	ω_p	≅ 1.7588E11	rad/s
Whirl no.	α^{-1}	≅ 137.0359	-

Fig. 21-1. Electric charge is intrinsic to matter and exhibits similarities to topological and vortical charge, i.e., loop closure failure. The electric charge of the electron has dimensions of spin angular momentum and appears to be the result of a slight spin precession (torsion defect) corresponding to the inverse fine structure constant α^{-1}.

21.2 Electrical charge characteristics of elementary particles

Elementary particles of matter exhibiting electric charge are illustrated in Fig. 21-2. Each generation consists of one lepton doublet and one quark doublet in the Standard Model. Electric charge and color charge occurs only with particles with rest mass. Leptons including electrons and positrons carry integer electric charge whereas quarks carry 1/3rd fractional charge. Particles with half-integral spin (measured in units of h/2π, where h = Planck's constant ~6.626 x 10^{-34} joule sec) obey Fermi-Dirac statistics and, hence, are called fermions. Those with integral spin obey Bose-Einstein statistics and are called bosons. Fermions behave according to the exclusion principle formulated by Wolfgang Pauli which precludes occurrence of two identical fermions in a given system to occupy the same state. Bosons, in contrast, may be brought together in virtually unlimited numbers in a superposition state with some scattering. Fermions and antifermions exhibit a chiral (left- and right-hand mirror image) symmetry. Each generation corresponds to a different helicity (left-handed or right-handed) state. Charge (C) conjugation symmetry transforms a particle into its anti-particle. Parity (P) symmetry transforms a particle or wave into its mirror image. Charge parity (CP) violation is associated with the weak force and observed in decays of K, B and D-mesons. An illustration of symmetry relations of electric charge, particle type, helicity, and color charge is illustrated in Fig. 21-3. Not shown in this static diagram are the spin, velocity or orbital angular momentum vectors describing dynamic effects.

One could call electrons with negative mass donkey electrons (which behave oppositely). A donkey electron has rest energy $-m_0 c^2$, and its motion will result in additional negative energy. – George Gamow

Mists where the electron behaves and misbehaves as it will, where the forces tie themselves up into knots of atoms and come untied... – D.H. Lawrence

247

Elementary Particles of Matter

	Fermions			Anti-Fermions			
Spin	1/2	1/2	1/2	1/2	1/2	1/2	**Spin**
Mass	<2.2 eV/c^2	<0.17 MeV/c^2	<15.5 MeV/c^2	<15.5 MeV/c^2	<0.17 MeV/c^2	<2.2 eV/c^2	**Mass**
Charge	0	0	0	0	0	0	**Charge**
Leptons	electron neutrino, ν_e	muon neutrino, ν_μ	tau neutrino, ν_τ	tau neutrino, $\bar{\nu}_\tau$	muon neutrino, $\bar{\nu}_\mu$	electron neutrino, $\bar{\nu}_e$	**Anti-Leptons**
Spin	1/2	1/2	1/2	1/2	1/2	1/2	**Spin**
Mass	0.511 MeV/c^2	105.7 MeV/c^2	1.777 GeV/c^2	1.777 GeV/c^2	105.7 MeV/c^2	0.511 MeV/c^2	**Mass**
Charge	−1	−1	−1	+1	+1	+1	**Charge**
Leptons	electron, e^-	muon, μ^-	tau, τ^-	tau, $\bar{\tau}^+$	muon, $\bar{\mu}^+$	positron, $\bar{e}^+$	**Anti-Leptons**
Spin	1/2	1/2	1/2	1/2	1/2	1/2	**Spin**
Mass	2.3 MeV/c^2	1.275 GeV/c^2	173.07 GeV/c^2	173.07 GeV/c^2	1.275 GeV/c^2	2.3 MeV/c^2	**Mass**
Charge	2/3	2/3	2/3	−2/3	−2/3	−2/3	**Charge**
Color	rgb	rgb	rgb	$\underline{\text{rgb}}$	$\underline{\text{rgb}}$	$\underline{\text{rgb}}$	**Color**
Quarks	up, **u**	charm, **c**	top, **t**	top, $\bar{t}$	charm, $\bar{c}$	up, $\bar{u}$	**Anti-Quarks**
Spin	1/2	1/2	1/2	1/2	1/2	1/2	**Spin**
Mass	4.8 MeV/c^2	0.5 MeV/c^2	4.18 GeV/c^2	4.18 GeV/c^2	0.5 MeV/c^2	4.8 MeV/c^2	**Mass**
Charge	−1/3	−1/3	−1/3	+1/3	+1/3	+1/3	**Charge**
Color	rgb	rgb	rgb	$\underline{\text{rgb}}$	$\underline{\text{rgb}}$	$\underline{\text{rgb}}$	**Color**
Quarks	down, **d**	strange, **s**	bottom, **b**	bottom, $\bar{b}$	strange, $\bar{s}$	down, $\bar{d}$	**Anti-Quarks**
Generation	I	II	III	III	II	I	**Generation**
	Bosons						
Spin	1	2	1	1	0	0	**Spin**
Mass	0	0	91.2 GeV/c^2	80.4 GeV/c^2	<0.0002 GeV/c^2	126 GeV/c^2	**Mass**
Charge	0	0	0	±1	0	0	**Charge**
Force Carriers	photon, γ	graviton, $\gamma\gamma^*$	Z boson, Z^0	W boson, W^+, W^-	gluon, **g**	Higgs boson, **H**	**Force Carriers**
Quantum Field	Electromagnetism	Gravitation	Weak Nuclear	Weak Nuclear	Strong Nuclear	Higgs Field	**Quantum Field**

Fig. 21-2. Mass, charge and spin characteristics of fundamental particles and antiparticles.

21. Electric Charge

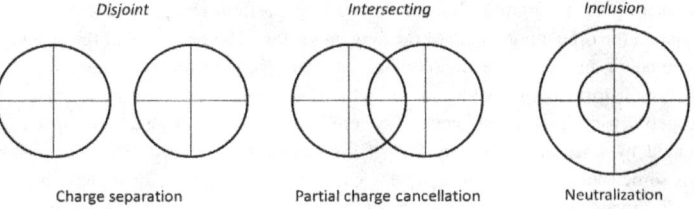

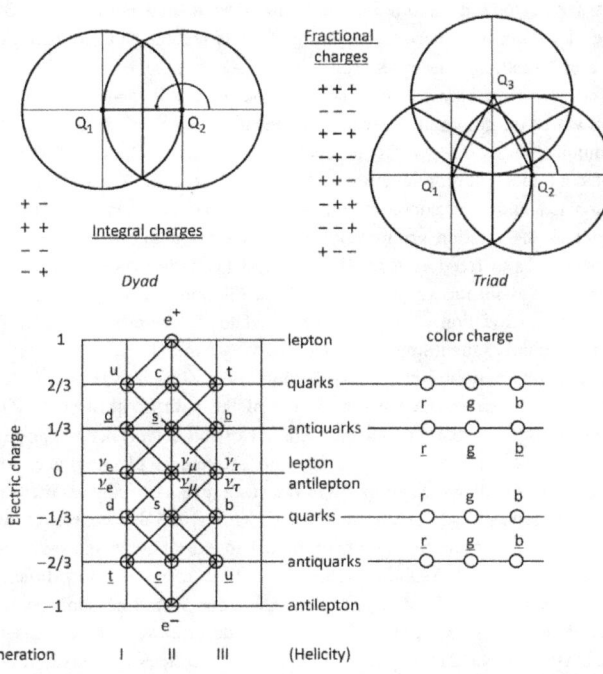

Fig. 21-3. When there is symmetry, there is conserved charge. Electric charge symmetry is apparent, but symmetry alone does not provide an obvious dynamic causal mechanism for origin of electric charge or define underlying fundamental dimensionality.

21.3 Relation of electric charge to topological charge

Electric charge and topological charge or twist have several mathematical properties in common (e.g., ± polarity, quantized, conserved, localized, scale invariant, etc) and it is has been long speculated that the mysterious property of electrical charge has in part a geometrical origin. Consider a physical situation in which upon completion of a loop in a space of parameters, a selected quantity, s, representative of the system state fails to return to its original state. The difference between the initial value of s and the final value is the geometric phase or anholonomy. A simple example is the parallel transport of a vector

249

21. Electric Charge

around a loop on the surface of a sphere. See Fig. 21-4. Transport of a tangent vector along a longitude great circle by a polar angle θ and then along a latitude circle by an angle ϕ and back along a longitude line to the starting position results in accumulated phase difference in the orientation angle of the tangent vector. The curvature of the sphere results in a geometric phase. The anholonomy is a measure of intrinsic curvature. Local symmetry transformations (distortion) reflect introduction of forces between points.

The geometric phase of a quantum system in n-dimensional parameter space may be determined by integrating the two-form (flux) threading a closed circuit. A geometric phase is somewhat different than a topological phase in that geometrical phase includes the effect of curvature (twisting of the fiber bundle connecting the base manifold to the parameter tangent space). An example of an anholonomy geometric phase was discovered by S. Pancharatnam in connection with polarization states of light. The phase of a light wave with a wave vector k transported around a loop designated as states $|A\rangle$, $|B\rangle$, and $|C\rangle$ on the surface of a Poincaré sphere representing the polarization state exhibits a difference in phase. The difference in phase between $|A\rangle$ and $|A'\rangle$ is equal to $-1/2$ x solid angle Ω subtended by a spherical triangle ABC. The order sequence of rotations is of importance with operators which do not commute which relates to physical variables which cannot be determined simultaneously as per the Heisenberg Uncertainty Principle. For operators which do not commute, the commutator $[A,B] = AB - BA \neq 0$. Non-Abelian symmetry groups of elements a and b are non-commutative if ab $\neq$ ba. An example of a non-Abelian symmetry group is the rotation group SO(3). An Abelian symmetry group $AB - BA = 0$ which has less degrees of freedom than SU(2) has a U(1) form. As shown in Fig. 21-5, a rotation of α radians about the x-axis followed by a rotation of β radians about the y-axis shows a difference in location of a point compared to the same rotations performed in opposite order. Similarly, quantum operations involving two distinct paths in Hilbert space of quantum states when applied in different orders yield non-equivalent transformations. The geometric phase is determined by the shape of the path as illustrated in Fig. 21-6. Electric charge corresponds to a loop closure failure (torsion defect) due to spin precession. Electron rest mass arises as a result of wave function interference of spin precession.

A SU(3) x SU(2) x U(1) gauge symmetry in flat spacetime has a 5-D geometry. The GR Minkowski spacetime has a 4-D geometry. The Lorentz group that conserves the interval ds is O(1,3). The Kaluza-Klein 5-D geometry added an extra unseen space-like dimension to describe electromagnetism. A similar notion is to add an imaginary time dimension in an orthogonal complex plane to the 4-D spacetime geometry. Adding additional rotational degrees of freedom allows for spin/torsion fields to describe electromagnetism, angular momentum, etc. An n-dimensional space is a manifold – a collection of a set of points with connections (topology). A 10- or 11-dimension geometry has been proposed in superstring theory by Green and Schwarz. A mathematical representation of allowable rotations (6 degrees of freedom) in a 10-dimensional spacetime manifold is shown in Fig. 21-7 illustrating the Poincaré symmetry group of 4D translations and 6D rotations (Lorentz group O(1,3). Rotary coordinates are referenced to a 6-D tangent manifold (K' frame) coupled to a 4-D spacetime base manifold (K frame). A Minkowski metric $ds^2 = -(dt^2) + dx^2 + dy^2 + dz^2$ is equivalent to a flat Euclidean 4-D metric $ds^2 = d\tau^2 + dy^2 + dx^2$ if substitute an imaginary time variable for time $t = -i\tau$ (Wick rotation). Curvature in Minkowski spacetime = 0 compared to de Sitter space > 0, anti-de Sitter space < 0 metrics.

It is through science we prove, but through intuition we discover. – Henri Poincaré

I used to think information was destroyed in a black hole. – Stephen Hawking

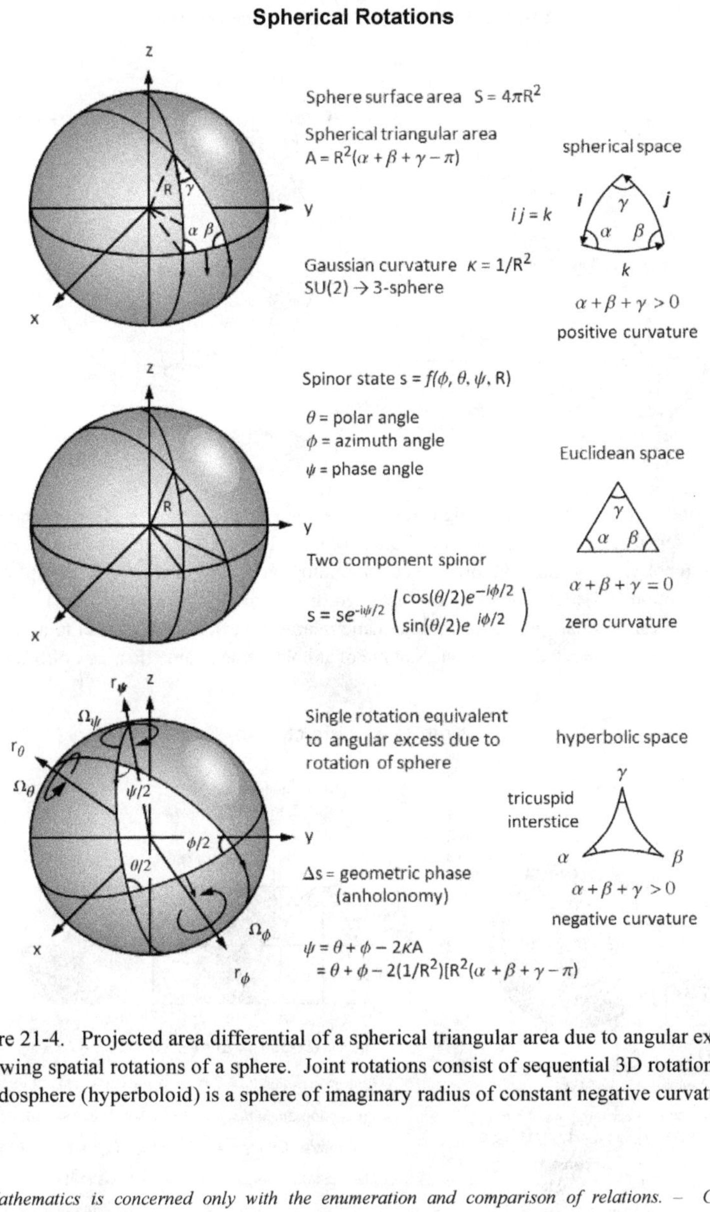

21. Electric Charge

Spherical Rotations

Sphere surface area $S = 4\pi R^2$

Spherical triangular area
$A = R^2(\alpha + \beta + \gamma - \pi)$

spherical space

$ij = k$

Gaussian curvature $\kappa = 1/R^2$

SU(2) → 3-sphere

$\alpha + \beta + \gamma > 0$
positive curvature

Spinor state $s = f(\phi, \theta, \psi, R)$

θ = polar angle
ϕ = azimuth angle
ψ = phase angle

Euclidean space

Two component spinor

$s = se^{-i\psi/2} \begin{pmatrix} \cos(\theta/2)e^{-i\phi/2} \\ \sin(\theta/2)e^{i\phi/2} \end{pmatrix}$

$\alpha + \beta + \gamma = 0$
zero curvature

Single rotation equivalent to angular excess due to rotation of sphere

hyperbolic space

tricuspid interstice

Δs = geometric phase (anholonomy)

$\alpha + \beta + \gamma > 0$
negative curvature

$\psi = \theta + \phi - 2\kappa A$
$= \theta + \phi - 2(1/R^2)[R^2(\alpha + \beta + \gamma - \pi)]$

Figure 21-4. Projected area differential of a spherical triangular area due to angular excess following spatial rotations of a sphere. Joint rotations consist of sequential 3D rotations. A pseudosphere (hyperboloid) is a sphere of imaginary radius of constant negative curvature.

Mathematics is concerned only with the enumeration and comparison of relations. – Carl Friedrich Gauss

Geometry should be ranked not with arithmetic, which is purely aprioristic, but with mechanics. – Carl Friedrich Gauss

Finally, a piece of pure mathematics that will not be sullied by applications. – Godfrey H. Hardy

Mathematics is like childhood diseases. The younger you get it, the better. – Erwin Schrödinger

I counted to ten slowly, using binary notation. – Robert A. Heinlein

21. Electric Charge

Anholonomy geometric phase difference

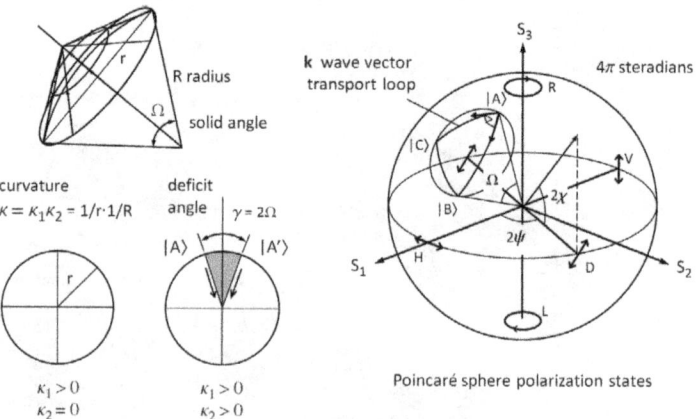

Fig. 21-5. Polarization states of light represented on a Poincaré sphere. Parallel transport of a vector around a closed path on the surface results in a geometric phase difference which represents the anholonomy. The holonomy group represents the group of all transformations of the tangent space obtained from parallel transport around different closed curves. Thomas precession, a kinematic relativistic effect with respect to an inertial frame over an accelerated path, is an example of anholonomic geometric phase difference.

Origin of electric charge

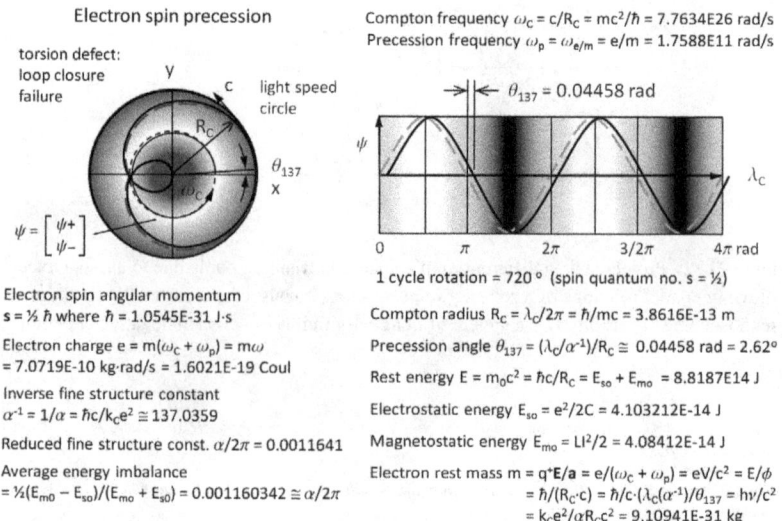

Fig. 21-6. Electron charge arises as a result of slight spin precession of the electron with whirl no. q equal to the inverse fine structure constant α^{-1}. Precession represents a loop closure failure (torsion defect) resulting in wave function interference (decoherence).

21. Electric Charge

Translational coordinates x_1, x_2, x_3 are holonomic whereas rotational coordinates φ_1, φ_2, φ_3 are anholonomic. By definition, translational motion at constant velocity is unaccelerated whereas inertial rotation is accelerated motion by virtue of continuous change in direction. In translational motion of a nonrotating gyroscope, the inertial mass and gravitational mass are equal and is isotropic with respect to translation. A rotating gyroscope, however, exhibits inertial mass which is anisotropic with respect to rotation. Shipov[78, 79] maintains that, as a consequence, Newton's 3rd law of motion no longer applies to the center of mass of a rotating system as part of the translational energy is transformed into internal rotational energy and vice versa. Shipov derives the following expression of inertial mass M_i of a rotating gyroscope

$$M_i = (M + 2m)[1 - 2m/(M + 2m)\sin^2\phi] \qquad (21\text{-}15)$$

where
 M = gimbal mass
 m = rotating mass = $m_i(\phi, v, t)$
 ϕ = rotational angle in spin plane

The equation of motion for a system with variable inertial mass (e.g., by varying the angular frequency of internal rotation) and conservation of total energy (E_t = constant) for a free gyroscope becomes

$$M_i dv_c/dt = - (dM_i/dt)v_c. \qquad (21\text{-}16)$$

where
 v_c = velocity of center of mass $B\omega_0\sin\phi_0 = \omega_0$ = initial angular velocity @ t = 0
 E_t = total energy = $(M + 2m)v^2/2 + mr^2\omega^2 - 2mrv\omega\sin\phi$
 L = kinetic energy + rotational energy = $\frac{1}{2}Mv^2 + \frac{1}{2}2mv_t^2$

As an example of conversion of rotary motion to translation, Laithwaite & Dawson have demonstrated linear momentum may be developed from pairs of oppositely precessing gyroscopes (GRB Patent 2289757, U.S. Patent 5860317). An example of a gyroscopic inertial drive using a single gyroscope is U.S. Patent 3404854 by di Bella. Generally, such devices involve mass motion in a three-dimensional, closed-loop Viviani curve with varying velocity or acceleration. The space curve describes a figure-8 pattern formed by the intersection of a sphere and a cylinder. Unlike linear momentum in which the acceleration vector is opposite in direction to the velocity vector or uniform rotary motion in which the acceleration vector is orthogonal to the velocity vector, in rotary motion under varying acceleration, the acceleration vector may lag the velocity vector in different portions of the curved trajectory path. For gyroscopes following a Viviani curve trajectory at the top and bottom of the figure-8, the angular acceleration is oppositely directed at the top and bottom of the loops (translational dominated portion). Precession occurs in the central portion of the figure-8 (precession dominated portion) resulting in precessional forces with reaction forces developed parallel to the horizontal plane bisecting the figure-8 along the equator of the Viviani window sphere. See Fig. 21-8. The precession rate is inversely proportional to the rotor spin rate, thus for a fast spin rate the precession is slow.

It's fascinating to be able to watch the Einstein warping of spacetime directly in the tilting of these GP-B gyroscopes. – Francis Everitt

I again committed, in regards to gravity, something which puts me in danger of being shut up in an insane asylum. – Albert Einstein

21. Electric Charge

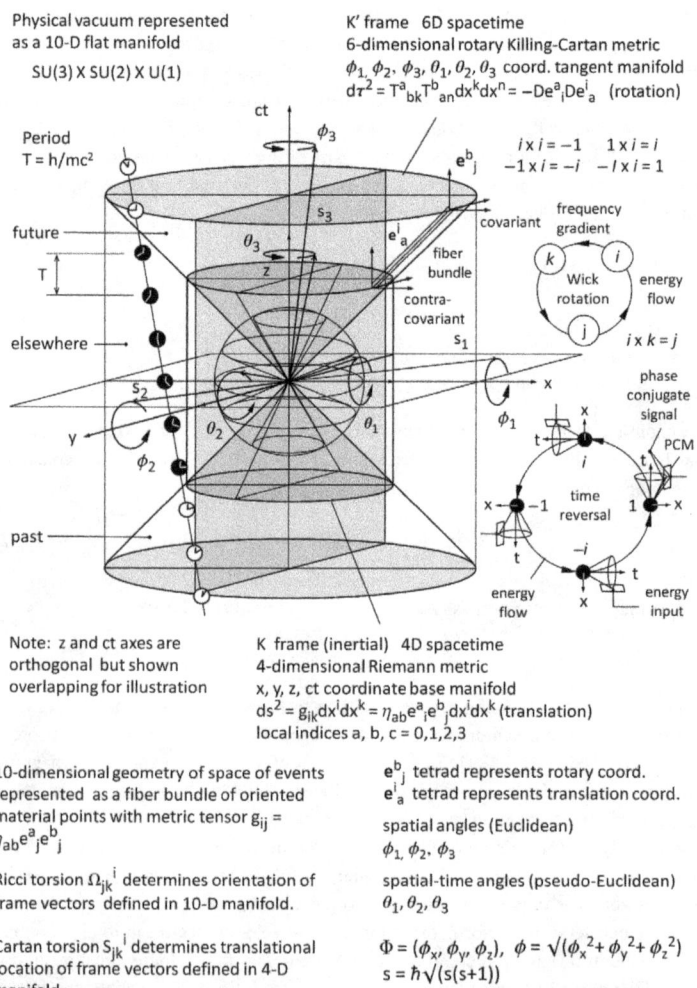

Fig. 21-7. Illustration of 4-D spacetime metric plus a 6-D rotations metric in a 10-D dimensional manifold. Energy flow proceeds in the time direction. Time travel corresponds to a quaternion multiplication in a Wick rotation (t → −iτ) and requires energy input. An analogy is an amplified phase conjugate reflected 'time-reversed' wave signal or action of waveguide circulators on signal propagation. In Clifford Geometric Algebra, time is a bivector, i.e., a rotation. Dual operator i ($= \sqrt{-1}$) transforms cross product into a bivector.

Time machines? Even we know that's nothing but trouble. – Col. Jack O'Neill

The Mannschenn drive moved forward in space and backward in time using temporal precession and matter phasing. – Arthur Bertram Chandler

21. Electric Charge

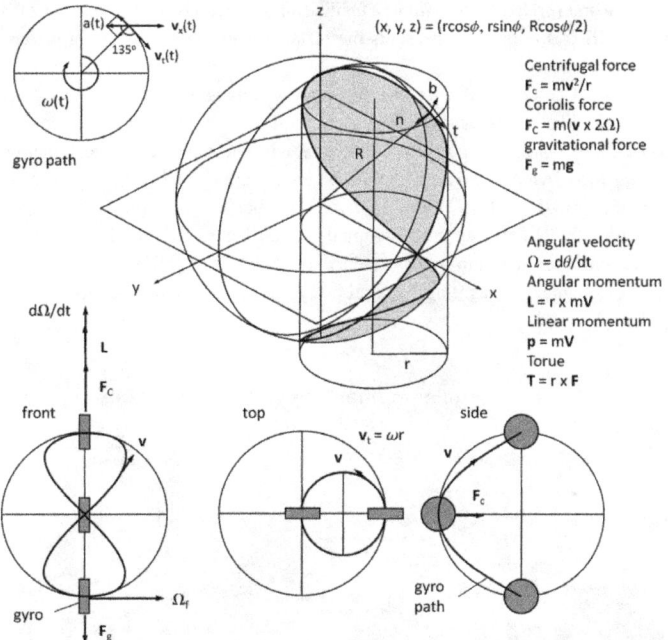

Fig. 21-8. Gyroscope trajectory path following a Viviani's curve described by the intersection of a sphere and cylinder results in a combination of linear and angular momentum. Effects of nonplanar orbital motion and precession produces periodic translation. The trajectory path corresponds to a Mobius manifold spinor fiber bundle.

The rotary coordinates may likewise be applied to represent elementary particle dynamics such as angular momentum interactions. A geometrical representation in the form of a vertex interaction is illustrated in Fig. 21-9. As shown, angular momentum bivectors are interpreted geometrically as oriented planes. The attitude or orientation of each plane is defined by the rotation angles. The magnitude of the bivector is equivalent to the exterior product of two component vectors **a^b**. The intersection of surfaces with edges represent momentum trajectories. Connected vertices represent causality. The vertices may be applied to describe particle interactions in momentum space. Given the particle momentum and spin of two interacting particles, the resultant may be calculated from the interaction simplex. Such interactions form connection elements of spin network diagrams commonly employed in loop quantum gravity theories.

> *You see that if the precession is hurried, it is more than sufficient to balance gravity and the gyrostat rises. – John Perry*

> *Now – think hard! -- suppose we put a gyroscope in a frame, then impress equal forces at all three spatial coordinates at once; what would it do? – Robert Heinlein*

> *Reductio ad absurdum, which Euclid loved so much, is one of a mathematician's finest weapons – Godfrey H. Hardy*

21. Electric Charge

In a twist ribbon model, if electric charge $\pm Q$ is taken as a measure of mass flow rate with sign corresponding to spin rotation (phase difference), then by extension, quark color charge may be interpreted as a 3-finned fusilli ribbon analogous to an electrical 3-phase wave-form. RGB Color charge represents the instantaneous phase current in increments of $1/3Q$. See Fig. 21-10.

Similarities of 3-color quark charges are likewise apparent in experiments of colliding vortical optical soliton beams of opposite topological charge. For example, head-on collisions of counter-propagating Gaussian beams with the same topological charge tend to form standing waves while collisions of vortex beams with opposite topological charge tend to breakup into three rotating beamlets each with phase shifts of $2\pi/3$. Repulsive or attractive force effects between cross-coupled soliton beams due to changes in refractive index of a nonlinear medium may result depending on relative beam direction (co-propagating or counter-propagating), spatial and/or temporal coherence, and relative phase between them.

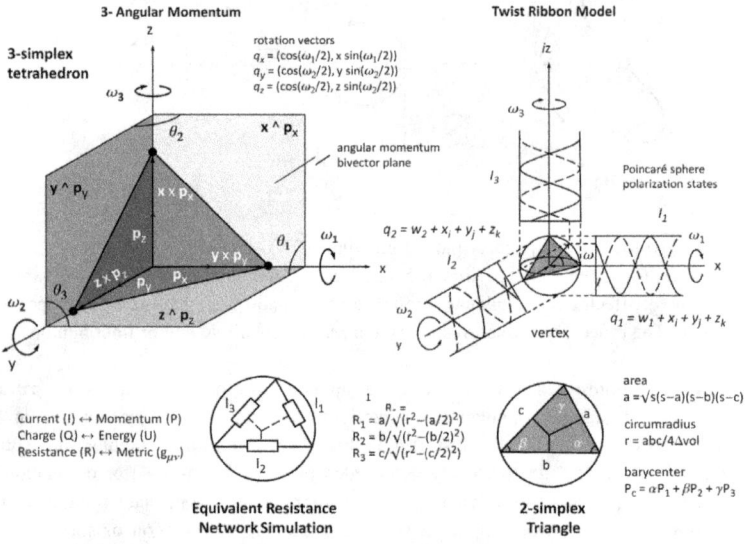

Fig. 21- 9. Spin network vertices representation of quark dynamic interaction.

Mathematicians do not study objects, but the relations between objects. – Henri Poincaré

There is something magical about three you know – a trio is tight and nicely economical. – Ian Williams

And there is the fact that what matters is not how things are but rather how they interact... manifesting themselves in these interactions as spin networks. – Carlo Rovelli

One can take the colour-electric field as...a string stretched between quarks as a consequence of space having properties that make it something like the electric version of a superconductor. – Lee Smolin

21. Electric Charge

Quark Color Charge

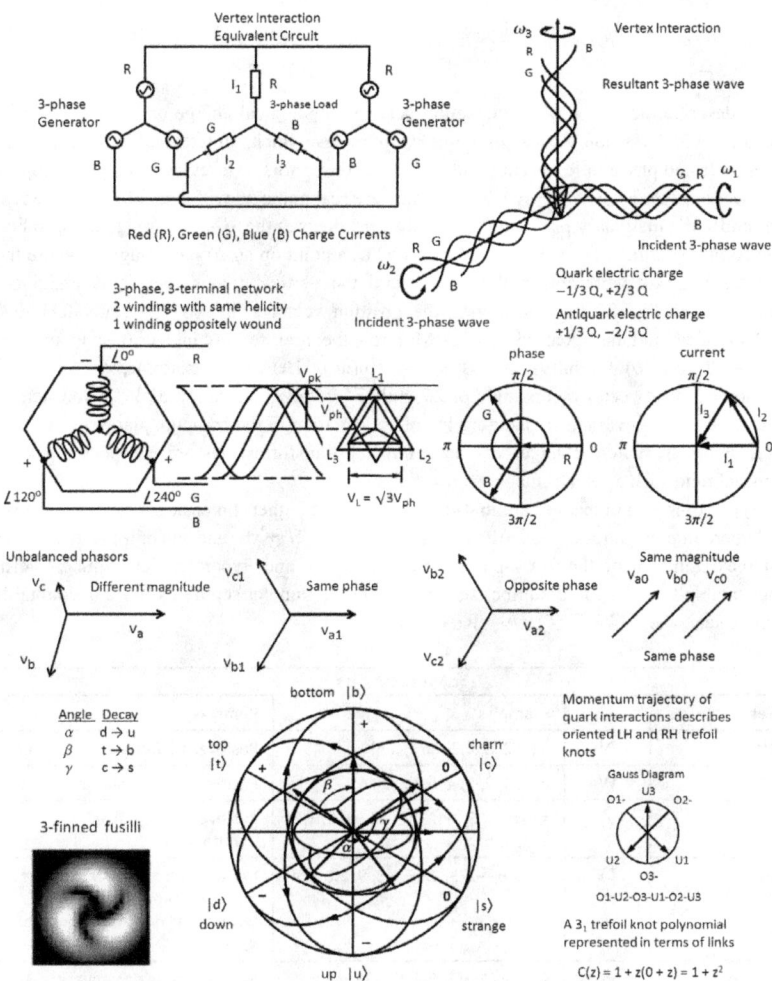

Fig. 21-10. Twist ribbon representation of quark color charge interaction and equivalent 3-phase circuit. Quark momentum trajectories correspond to trefoil knots.

The basic idea of current algebra was to assume the existence of quarks, whose motion gives rise to currents, such as electric current. – Frank Close (The Infinity Puzzle)

"Three quarks for Muster Mark!" might be "Three quarts for Mister Mark," in which case the pronunciation "kwork" would not be totally unjustified. In any case, the number three fitted perfectly the way quarks occur in nature. – Murray Gell-Mann

I get all tangled in your ribbons. – Consuelo de Saint-Exupéry

Arithmetic! Algebra! Geometry! Grandiose trinity! Luminous triangle! – Comte de Lautreamont

One chord is fine. Two chords are pushing it. Three chords and you're into jazz. – Lou Reed

257

22. Complex Numbers

There is nothing to that mathematics – it's all imaginary. – W. Yergen

To describe the notion of a connection between topological charge and electric charge, recourse will be made to use of complex numbers which, unlike real numbers, have magnitude and phase to represent points and position vectors. To review briefly, a complex number is of the form $z = x + iy$ where x and y are real numbers with x identified as the real part and y the imaginary part and the symbol 'i' denotes the $\sqrt{-1}$, an imaginary number. The complex number $z = x + iy$ may be plotted as a point on an Argand diagram where the horizontal x-axis represents real numbers and the vertical y-axis represents imaginary numbers. See Fig. 22-1. Multiplying the position vector Z by $\sqrt{-1}$ is equivalent to a rotation of $\pi/2$ radians. According to DeMoivre's theorem, for any integer (n) and any real number θ: $(\cos(\theta) + i\sin(\theta))^n = \cos(n\theta) + i\sin(n\theta)$. Hence, for example, Z^2 involves a rotation of the Z vector by an angle of 2θ while Z^3 involves a rotation of 3θ. A system is said to be gauge-invariant if changing the phase (rotation in the complex plane) results in a local symmetry where changes due to symmetry transformations are compensated by a physical field such as electromagnetism.

Hypercomplex numbers relate to 4-dimensional and higher dimensional spaces. The set of hypercomplex numbers are of the form $a + bi + cj + dk$ and encompass the set of complex numbers of the 2D z-plane. Both complex and hypercomplex numbers form uncountable sets. The real number set and the surreal number set are likewise uncountable infinite sets where $\{N \subset Z \subset Q \subset A_R \subset R \subset C \subset H\}$.

		Number sets	
Set	Symbol	Examples	Remarks
Natural	N	= {1, 2, 3, 4, 5, 6, 7, …n}	Positive. Countable set.
Whole	W	= {0, 1, 2, 3, 4, 5, 6, 7, …n}	Positive. Countable set.
Integer	Z	= {−n,…−5, −4, −3, −2, −1, 0, 1, 2, 3, 4, 5, …n}	Positive and negative. Countable set.
Rational	Q	½, 5/6, −2/3, 2.25, −5, 9, 1.0, etc	n/d where n and d are integers and d ≠ 0. Repeating and terminating decimals. Countable set $\leq \aleph$.
Irrational	$\bar{Q}$	$\sqrt{2}, -\sqrt{3}, \sqrt{5}$, 0.6180339…, e, π, e^n, ln2, sin(π/3), Φ, etc	Non-repeating decimals R/Q.
	A_R	$\sqrt{2}, -\sqrt{3}, (1+\sqrt{5})/2$, etc	Roots of a non-zero polynomial equation w/ rational coefficients.
		e, π, e^π, γ, Ω, ln2, sin(π/3), etc	Transcendentals R/A.
Real	R	−5, −$\sqrt{3}$, 0, $\sqrt{5}$, 8/3, e, π, −2π, ϕ, Φ, etc	All numbers represented as points on the real number line.
Surreal	S	{ | }, {0|}, {0| }, { |0}, {−2|0}, {0|1}, [(1,0), (2,0)], etc	ordered set pair x = {L|R} where $L \leq R; \epsilon < \mathbf{R} < \omega$
Complex	C	$0 + \pi i$, $1.5 + \pi i$, $5 - 2i$, $-1 - 3\pi/2 i$, $e + \pi i$, $\pi + \sqrt{2} i$, etc	$a + bi$ where a and b are real numbers and $i = \sqrt{-1}$. 2D. SO(2)
Hypercomplex	H	−5 + ei +2j + 0k, 0 + πl + 3j−6k, etc	$a + bi + cj + dk$. 3D quaternions, SU(2)

22. Complex Numbers

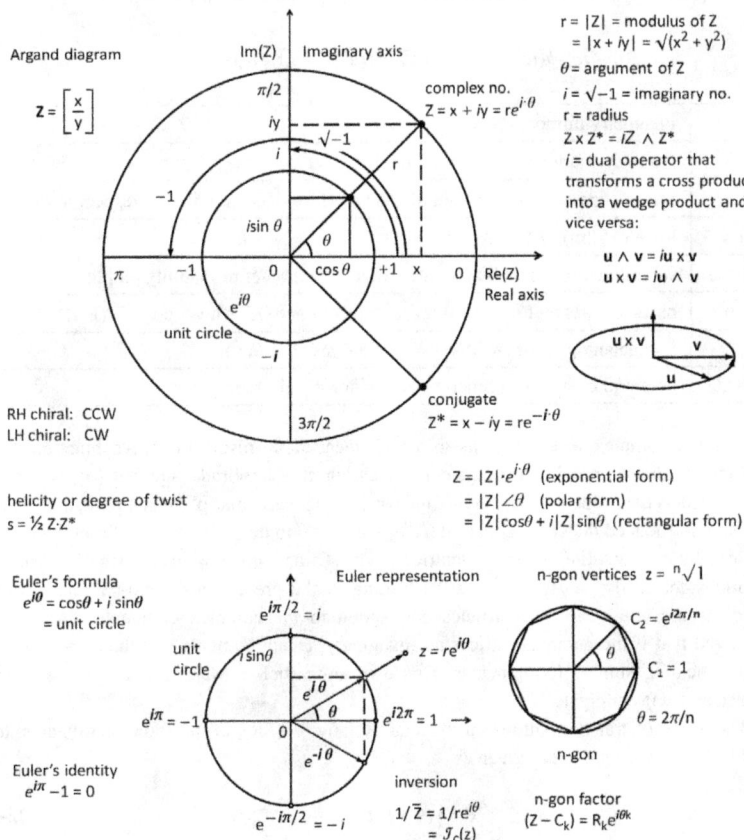

Fig. 22-1. A complex number Z plotted as a point and a position vector in a 2D complex plane formed by a real number axis and an orthogonal imaginary number axis.

Rotations in the complex plane describe changes in phase. Forces are the result of symmetry breaking and local gauge transformation. In Maxwell's theory, the electrodynamic potentials and electromotive forces are interrelated – EMF is a gradient of a scalar function. The EMF on any element of a conductor is measured by the instantaneous rate-of-change of the vector potential **A**. Electromagnetism is, in essence, a $U(1)$ fiber-bundle theory. Weyl introduced the concept of gauge transformation in Quantum Electrodynamics (QED) theory and discovered that Maxwell's equations are Lorenz gauge invariant reflecting a gauge symmetry ($\nabla \cdot \mathbf{A} = -(1/c^2)\delta\phi/\delta t$). The change in phase of the electrodynamic field is associated with a change in the four-dimensional vector potential $A_\mu = (\phi, \mathbf{A})$. These transformations involve at every point in spacetime, a phase transformation, i.e., a Lie Group $U(1)$ rotation in the complex plane.

"$i^2 + j^2 + k^2 = ijk = 1$" – Sir William Rowan Hamilton

259

22. Complex Numbers

A famous example of application of complex numbers is the non-relativistic Schrödinger wave equation which is a time-dependant complex, partial differential equation

$$i\hbar(\delta/\delta t)\psi(r,t) = -(\hbar^2/2m)\nabla^2\psi(r,t) + V(r,t)\psi(r,t) \tag{22-1}$$

where

i	imaginary number ($= \sqrt{-1}$)
$\hbar$	Dirac constant ($= h/2\pi = \Delta p/\Delta mv$; $h = E/\nu$)
$i\hbar\delta/\delta t$	Energy operator ($= E$); Rate of change of wave equation with respect to time
$E\psi$	$E\psi = (\hbar^2/2m)k^2\psi$; $\delta^2\psi/\delta x^2 = (2m/\hbar^2)[V(x) - E]\psi = k^2\psi = 0$
$\psi(r,t)$	wave function describing matter wave evolution & probability amplitude $|\psi|^2$
m	particle mass ($= \mathbf{F}/\mathbf{a} = \hbar\mathbf{k}/\mathbf{v} = E/c^2 = (e\mathbf{A})/\mathbf{v} = m_0/\Gamma = h\cdot\nu_{dB}/c\cdot v = E/(\mathbf{E}/\mathbf{B})^2$)
∇^2	Laplacian operator ($= \nabla\cdot\nabla = \delta^2/\delta x^2 + \delta^2/\delta y^2 + \delta^2/\delta z^2$)
$V(r,t)$	potential energy as a function of position r and time t

To convert complex wavefunctions to a real, measurable result, the wave function ψ is multiplied by its complex conjugate ψ^* yielding the amplitude squared $|\psi|^2$ which is interpreted as the probability of finding the particle in a particular place at a particular time.

Dirac introduced the Dirac equation ($\iota\gamma^\mu\delta_\mu - m)\psi = 0$ to describe charged fermions where γ_μ are anticommuting Gamma matrices. The Dirac equation gives negative energy solutions for $E(p) = \pm\sqrt{(pc)^2 + (mc^2)^2)}$ leading to the prediction of anti-particles. Dirac fermions are complex. Anti-particles correspond to the complex conjugate. Majorana theorized that if the gamma matrices are imaginary, ψ can be made real thus describing a charge neutral, spin ½ fermion that is its own antiparticle. Majorana fermions are real valued and self-conjugate.

The Lorentz transformations of Special Relativity (SR) of position coordinates for motion in the x-direction are given as

$$x' = \gamma(x - \beta ct)$$
$$y' = y$$
$$z' = z$$
$$ct' = \gamma(ct - \beta x) \tag{22-2}$$

where $\gamma = 1/\sqrt{(1-\beta^2)}$, $\beta = v/c$, $\beta^2 = 1 - 1/\gamma^2$, $\gamma^2 - \beta^2\gamma^2 = 1$.

The Lorentz transformations may be viewed as rotations in a four-dimensional spacetime where the fourth dimension is ict. In this complexified spacetime, the Lorentz transformations of event coordinates is described in terms of an imaginary Wick rotation angle φ

$$x' = \gamma x + i\beta\gamma(ict)$$
$$y' = y$$
$$z' = z$$
$$ict' = -i\beta\gamma x + \gamma(\iota ct) \tag{22-3}$$

where $\gamma = \cos\varphi$, $i\beta\gamma = \sin\varphi$, $I = \sqrt{-1}$, $\gamma^2 + (i\beta\gamma)^2 = 1$, $\cos^2\varphi + \sin^2\varphi = 1$.

The true metaphysics of the square root of minus one is elusive. – Carl Friedrich Gauss

22. Complex Numbers

In four-dimensional spacetime, real numbers are traditionally used. Using properties of complex numbers, Penrose[74] has developed an eight-dimensional spacetime model known as twistor theory in which the building blocks are twistors. The eight quantities of twistor theory describe three dimensions of twistor position in space, two angles that describe the direction of twistor motion, energy, spin and polarization of motion. Elementary particles may be constructed from combinations of twistors. Forces correspond to different deformations of twistor space. Spin networks may be modeled in 8-D twistor space.

Yet another illustrative example of complex numbers is found in conformal mapping in hydrodynamics (2-D potential fluid flows) and electrostatics (2-D electrostatic fields). Analytic complex functions that can be reduced to the Laplace equation in two dimensions enable conformal transformations of difficult boundary conditions into simpler ones. The solution of Laplace's equation, subject to the specified boundary conditions, yields the pattern for electrostatic, magnetostatic and gravitational fields. Any complex analytic function of two real functions $u(x,y)$ and $v(x,y)$

$$\omega(z) = u(x,y) + iv(x,y) \qquad (22\text{-}4)$$

satisfies the Cauchy-Riemann equations

$$\delta u/\delta x = \delta v/\delta y \quad \text{and} \quad \delta u/\delta y = -\delta v/\delta x \qquad (22\text{-}5)$$

For given values of (x,y) in the z-plane, corresponding values of (u,v) in a given region s may be determined and plotted in the w-plane. The field may be plotted in the z-plane taking the values of the u and v parameters as $\phi(x,y) = u$ and $\psi(x,y) = v$. The field may likewise be plotted in the w-plane by taking x and y as parameters. The function $u(x,y) = c_1$ and $v(x,y) = c_2$ solution curves are orthogonal and therefore may be used to represent mutually orthogonal stress lines and equipotentials of a force field. For example,

$w(z)$ complex field	$u(x,y)$ real component	$v(x,y)$ imaginary component
Electrostatic	electric potential	flux lines of force
Magnetostatic	magnetic potential	flux lines of force
Inviscid fluid flow	velocity potential	stream function
Heat flow	isothermals	heat flow lines
Gravitational	gravitational potential	flux lines of force
Elastic	strain	stress lines

Various practical problems may be modeled as combinations of sources and sinks, dipole flows, parallel flows and circulation flows. As a simple example of the analogous force field effects is the attraction and repulsion of bodies immersed in a potential flow and the attraction and repulsion of electrical charges in an electrostatic field. This effect is illustrated in Fig. 22-2 in a 2-D cross-section of a potential fluid flow.

If, for example, +1, −1 and the square root of −1 had been called direct, inverse and lateral units, instead of positive, negative and imaginary (or even impossible), such an obscurity would have been out of the question. − Carl Friedrich Gauss

Poles arise whenever there is a sudden change in reluctance about a magnetic circuit. − Eric R. Laithwaite

22. Complex Numbers

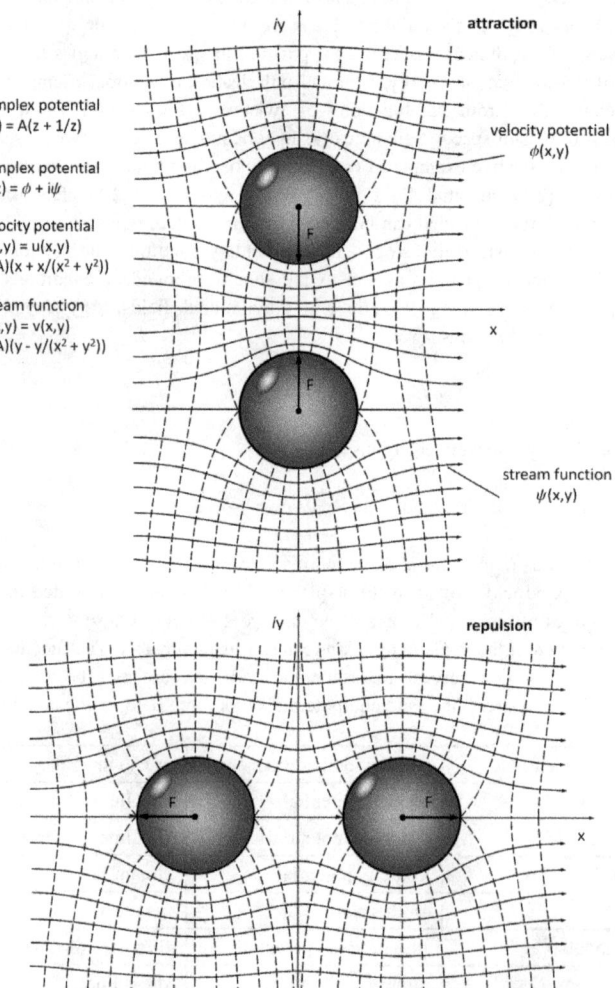

Fig. 22-2. Potential flow around a pair of spheres in a fluid without viscosity. The pressure differential results from the velocity gradient described by the equation of Bernoulli for a homogeneous, incompressible, inviscid, steady-state fluid flow. For the case where the line connecting the center of the spheres is perpendicular to the flow direction, the spheres attract whereas if the spheres are aligned such the connecting line is parallel to the flow direction, the spheres repel. Each sphere may be simulated by a doublet flow $\Omega(z) = 1/z$.

Discontinuous two-dimensional fluid motion in hydrodynamics can be deduced by interpreting functional relations between complex variables. – Andrew Forsyth

I recoil in fear and loathing from that deplorable evil: continuous functions with no derivatives. – Charles Hermite

23. Phasors

Do not imagine that mathematics is hard and crabbed, and is repulsive to common sense. It is merely the etherealization of common sense. — William Thomson (Lord Kelvin)

A complex number may also be used to represent a vector as a function of time $Z(t)$ and is termed a phasor $Z\angle\theta$. If the position vector rotates at a angular velocity ω, as shown in Fig. 23-1, the vector may be used to represent a sinusoidally varying quantity where the amplitude (A), angular frequency (ω) and phase angle (θ) are time invariant. Pairs of phasors may be combined to produce Lissajous figures describing complex harmonic motion.

Phasors are commonly employed in electrical engineering applications to represent voltage and current sinusoids. For example, if an instantaneous voltage as a function of time is described as

$$v(t) = V\cos(\omega t + \theta) \tag{23-1}$$

then by Euler's theorem, the voltage can be represented as the real part of a complex function

$$v(t) = \text{Re}\{V e^{j(\omega t + \theta)}\} = \text{Re}\{(V e^{j\theta})(e^{j\omega t})\} \tag{23-2}$$

The first term is a transform of the voltage $v(t)$ defined as a phasor where

$$\mathbf{V} = V e^{j\theta} = V\angle\theta \tag{23-3}$$

The second term denotes rotation $e^{j\theta} = \cos\theta + j\sin\theta$. Like vectors, phasors have magnitude and direction and may be resolved into components. A phasor diagram of electromotance $\mathcal{E}$ (= $-d\phi_m/dt = \mathcal{E}_0\cos\omega t$) is the sum of three components: 1) electromotance $j\omega L I_0$ in the inductor which leads current by 90°, 2) $I_0 R$ component which is in phase with current, and 3) the electromotance $-jI_0/(\omega C)$ component which lags current by 90°. The voltage of a series RLC circuit is

$$\mathbf{V} = R\mathbf{I} = j\omega L\mathbf{I} + \mathbf{I}/j\omega C = (R + j\omega l + 1/j\omega C)\mathbf{I} \tag{23-4}$$

Impedance is the ratio of the voltage phasor to the current phasor $\mathbf{Z} = \mathbf{V}/\mathbf{I} = Z\angle\phi_Z$ where $\mathbf{Z}_R = R\angle 0°$, $\mathbf{Z}_L = \omega L\angle 90°$ and $\mathbf{Z}_C = 1/\omega C\angle -90°$. For a circuit consisting of several impedances in series, $\mathbf{Z} = \mathbf{Z}_1 + \mathbf{Z}_2 + \mathbf{Z}_3 + \ldots \mathbf{Z}_n$, the corresponding phasor is taken as a vector sum rather than an algebraic sum. For impedances in parallel, the circuit admittance is the reciprocal of impedance and is given by $\mathbf{Y} = 1/\mathbf{Z} = 1/\mathbf{Z}_1 + 1/\mathbf{Z}_2 + 1/\mathbf{Z}_3 + \ldots 1/\mathbf{Z}_n$. The reciprocal of complex number $re^{j\theta}$ is $e^{-j\theta}/r$.

Ratios of pairs of complex numbers may be geometrically represented on a Riemann sphere of unit radius centered in the complex plane. Two-state systems described by a complex projective Hilbert space such as spin states of an electron may be represented as a point on a Riemann sphere where the general spin state of magnitude $\tfrac{1}{2}\hbar$ is a combination of up and down spin states $|\psi\rangle = \{w, z\} = w|\uparrow\rangle + z|\downarrow\rangle$. Similarly, the photon polarization state, described by a Stokes vector of magnitude $q = \sqrt{(z/w)} = w|+\rangle + z|-\rangle$, is a combination of + and − helicity states.

Local symmetry transformations of the unitary group U(1) of a complex variable are related to the conservation of charge. Transformations of phase in 2π radian multiples leaves phase unchanged. Symmetry is preserved if the phase of the electron wavefunction

263

23. Phasors

is matched by changes in the electromagnetic field. Gauge invariance of the phase of the electron wavefunction generates force field changes counteracting changes in phase thereby conserving electric charge. Symmetry-breaking of the U(1) gauge field of electromagnetism is also associated with superconductivity in which photons in the gauge field acquire an effective mass as a result of the Meissner effect such that the field cannot penetrate the superconductor interior.

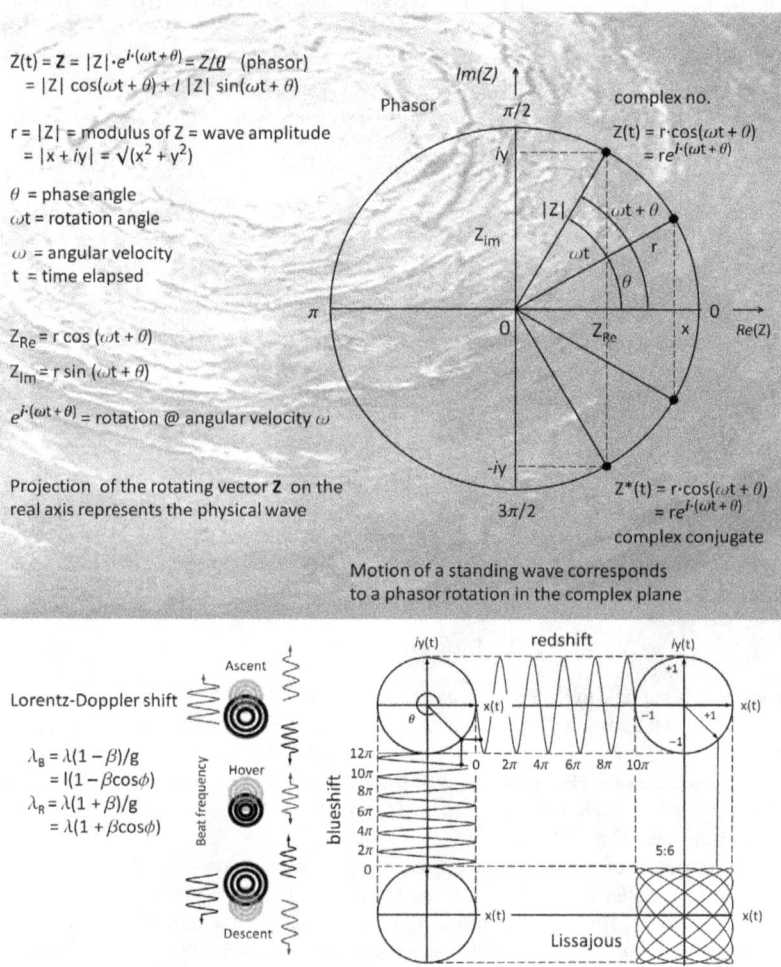

Fig. 23-1. A complex number $Z(t)$ as a function of time plotted as a rotating vector or phasor $\mathbf{Z} = Z\underline{/\theta}$. Quantities expressed as ratios, percentages, motion, energy, musical notes, etc varying in time may be represented as angles on a phasor diagram.

264

24. Quaternions

I have yet to see any problem, however complicated, which, when you looked at it the right way, did not become still more complicated. – Poul Anderson

It seemed (and still seems) to me natural to connect this extra-spatial unit with the conception of time. – William Rowan Hamilton

Quaternions represent a generalization of complex numbers that provides a description of rotations in three dimensions just as complex numbers do in two dimensions. Representations of rotation in 3D are not a 3D vector space using vector addition but may be described in higher dimensional 4D quaternion space. A quaternion is set of four numbers of the form $q = [\mathbf{v}, w] = w + x_i + y_j + z_k$ which represents a sum of a scalar w and a vector $x_i + y_j + z_k$ with three imaginary components where w, x, y, and z are real numbers. The magnitude of $q = |q| = \sqrt{(w^2 + x^2 + y^2 + z^2)}$. Just as $i = \sqrt{-1}$ in complex numbers, with quaternions $i = j = k = \sqrt{-1}$. The vector component $\mathbf{v}$ represents a rotation about an axis and the scalar component w represents the amount of rotation about this axis. See Fig. 24-1. A rotation of a point with coordinates x, y, z into another point in R^3 can be written as the operation

$$(x_i + y_j + z_k) \rightarrow q(x_i + y_j + z_k)q^{-1} \quad (24\text{-}1)$$

where q is the quaternion. Like rotations, quaternion multiplication is not commutative. Unlike quaternions, octonion multiplication is not associative.

Example operations include

Addition $q + q' = [\mathbf{v}, w] + [\mathbf{v}', w]$ $\quad = [\mathbf{v} + \mathbf{v}', w + w']$	Norm $N(q) = \sqrt{qq^*} = \sqrt{q^*q}$ $\quad = w^2 + \mathbf{v}\cdot\mathbf{v} = w^2 + x^2 + y^2 + z^2$
Multiplication $qq' = [\mathbf{v}, w][\mathbf{v}',w']$ $\quad = [\mathbf{v} \times \mathbf{v}' + w\mathbf{v}' + x'\mathbf{v}, ww' - \mathbf{v}\cdot\mathbf{v}']$	Inverse $q^{-1} = q^*/N(q)$ $q^{-1}q = qq^{-1} = 1$
Conjugation $q^* = [\mathbf{v},w]^* = [-\mathbf{v}, w]$	Rotation $q = q_2 q_1$

In contrast to complex numbers in which ' i ' represents rotation by 90 deg, ' i ' represents rotation by 180° in quaternion numbers. Rotation by 1 or −1 corresponds to 360° rotation which represent the same rotation. 4D quaternions can represent 3D reflections, rotations and scaling. Quaternions provide an alternative to represent rotations as 3 Euler angles (3D rotations as 3 numbers) which, although more intuitive, are limited by gimbal lock when the rotation angle approaches ± 90 deg. Conversion of quaternions to Euler angles is dependent on the order of rotations. Quaternions are associated with rotations in SO(3) and SO(4) in 3- and 4-dimensions, respectively. Like tensors, quaternions can be added together or multiplied by a scalar. Unlike tensors, quaternions are restricted to four dimensions and may provide explanation for the four fundamental forces.

It is unfair to call a vector a quaternion as to call a man a quadruped. – Oliver Heaviside

24. Quaternions

Maxwell adopted quaternions for describing electric and magnetic field interactions which included the effect of scalar potentials on electric charges. Maxwell's equations represented in standard texts are a stripped-down vectorial force field description by Heaviside. Vector notation was invented by Heaviside and Gibbs following Maxwell's early death. The four Maxwell equations can be expressed as a single bi-quaternion equation in which the real numbers x_0, x_1, x_2 and x_3 are replaced by complex numbers representing an 8D spacetime. Quaternions find application in physics, computer graphics, inertial navigation, and robotics for example, as coordinates for rotations and orientations. Scalar and 3-vector potential gauge fields of electromagnetism and gravitomagnetism of the 4-vector form $\mathbf{A}_\mu = (\phi/c, -\mathbf{A}_x, -\mathbf{A}_y, -\mathbf{A}_z) = \mathbf{A}_0, -\mathbf{A}_1, -\mathbf{A}_2, -\mathbf{A}_3$ may be represented in terms of a 4-dimensional quaternion gauge field. A set of events forms a coordinate pattern described by time-like scalar potential $\mathbf{A}_0$ and space-like vector potential $\mathbf{A}_1$, $\mathbf{A}_2$, $\mathbf{A}_3$. Changes in potential creates a field gradient generating forces equal to the source times the field strength. Forces represent conversion factors for changes in scale (gauge). The SU(3), SU(2), U(1) symmetry groups correspond to strong, weak and electromagnetic forces. Particle generations may be described in hyperdimensional spinor space by Dixon division algebra $T := C \otimes H \otimes O$ product where C = complex numbers, H = quaternions and O = octonions.

Quaternion representation

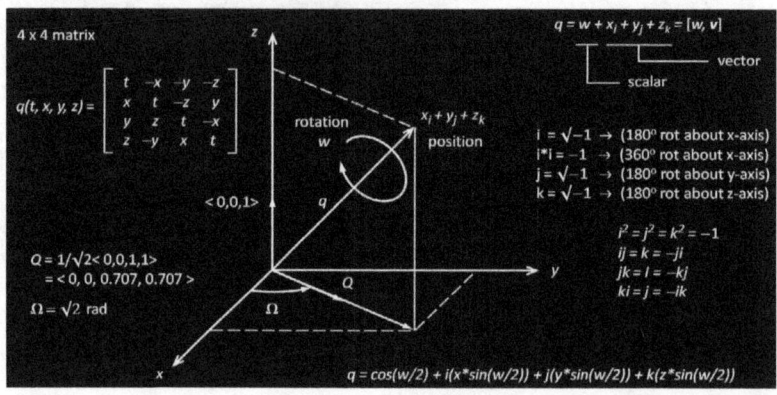

Fig. 24-1. Quaternions provide a description of orientation and rotation of objects in 3D space. The Lorentz group, a Poincaré subgroup, in 4D spacetime is associated with the quaternions and complex number spaces ($H \otimes C$). Octonions are formed of pairs of quaternions defining coordinates in 8-D space and may be represented as a Fano cube.

This topsy turvy system [vectors] is earnestly recommended to physicists as being precisely what they want. Not a bit of it. They have said so by their silence. Common sense of fitness of things revolts against the algebra of vectors. Nothing could be more unnatural. – Glasgow Herald, 17 Dec review: Heaviside's Electrical Papers (1894).

Octonions are to physics what the Sirens were to Ulysses. – Pierre Ramond

I fear that, to a mind of sufficient intellectual power, the whole of mathematics would appear trivial. – Bertrand Russell

25. Spinors

No one fully understands spinors. Their algebra is formally understood but their general significance is mysterious. – Michael Atiyah

In Quantum Mechanics (QM), quaternions can represent particle spin described in Lie Group theory as Spin(n) of dimension n which has elements known as spinors. Spinor rotation operators were introduced by Élie Cartan in 1913. A rank 1 spinor may be represented as a generalization of a vector by a two-component complex vector with a spatial part consisting of a vector (flagpole) with two additional features: a rigid 'flag' indicating the plane in space containing the vector and a sign. Unlike vectors or tensors, spinors change sign when rotated through 360°. Spinors were found indispensible to describe the spin angular characteristic of the electron and other spin ½ fermions.

A spinor inverts negative upon a 360° rotation and returns to the original position upon a 720° rotation, such as for example, rotation of a tangent or orthonormal vector on a Möbius strip. Refer to Figs. 25-1 and 25-2. Three-dimensional spinors of Group Spin(3) = SU(2). Quaternions are equivalent to spinors in 3D. For an SO(3) rotation, the spinor has 2 complex components. Spinors can represent rotations in any number of dimensions. In quantum physics applications, spinors provide mathematical representations of spin angular momentum properties of fermions (spin ½ objects), energy transport in EM fields, loop quantum gravity phase space, etc. In 3D, j = ½ spinor representation of the Spin(3) rotation group is derived from the Pauli spin matrices, a set of three 2 x 2 complex matrices (σ_1, σ_2, σ_3) which are generators for SU(2) where the spin operator is given by $J = \hbar/2\ \sigma$. The spin angular momentum observable corresponds to the spin operator which in ket notation has the form $S = \hbar/2\ (|\uparrow\rangle\langle\uparrow| - |\downarrow\rangle\langle\downarrow|)$ where the ↑ and ↓ arrows represent the direction of the z-component of angular momentum eigenvectors and $|*\rangle$ represents the S_z eigenstates. The 4-component Dirac spinors ω_p are solutions to the relativistic Dirac equation and provide a connection of the quantum state of the electron and the Lorentz group. The Dirac spinor is an element of a 4D complex vector Hilbert space of combinations of Dirac 4 x 4 gamma matrices γ^μ, μ = 0, 1, 2, 3 that are the spin and charge operators. The four components of the wave function ψ correspond to the spin-up and spin down electron and spin-up, spin-down positron. The Dirac spinor can be decomposed into positive and negative energies E = $\pm\sqrt{(p^2c^2 + m^2c^4)}$. An example of a physical interpretation of negative energy is propagation of a wave in a medium with negative index of refraction (metamaterial).

Zitterbewegung has been described as an interference effect between wave function positive and negative energy states with frequency ω_{zbw} increasing linearly with mass, decreasing with positive and negative spinor energy state overlap separation and amplitude A_{zbw} decreasing with mass. The Majorana spinor ψ is an element of a 4D real vector Hilbert space of linear combinations of Majorana matrices $i\gamma^\mu$, μ = 0, 1, 2, 3. Unlike Dirac fermions, neutrally charged Majorana fermions are their own antiparticle ($\psi = \psi_c$) and may provide a model for neutrinos. The seemingly mysterious property of 720° (= 4π radians) rotation of fermion spin to return to the initial state may be understood to be a consequence of simultaneous cyclic motion in two orthogonal directions, e.g., toroidal and poloidal rotation, otherwise known as a bispinor with ½ spin.

Let us think of a wave group... which in some way gets into a small closed 'path', whose dimensions are of the order of a wave length. – Erwin Schrödinger

25. Spinors

Spinor representation

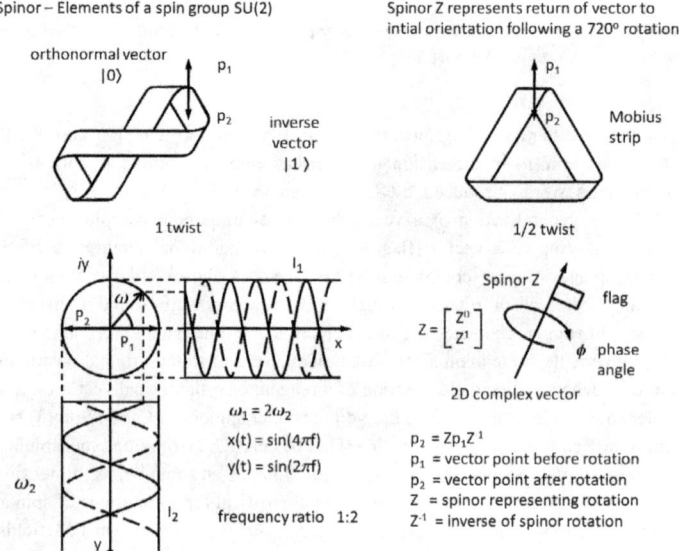

Fig. 25-1. Spinor Z rotation operation is representative of a fermion ½ spin characteristic in which 2 x 360° rotations are required to return to initial position and restoring chirality. A 4π rotation of a phasor traces out a Möbius band. SU(2) spin group is noncommutative.

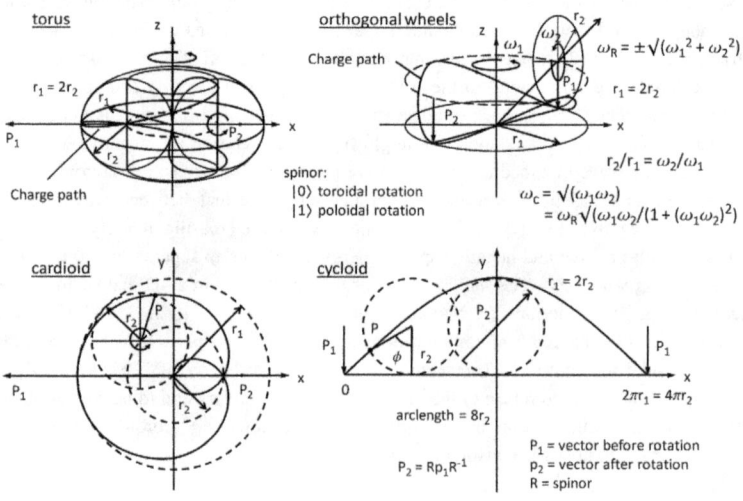

Fig. 25-2. Spinor examples of 720 degree rotation to return to initial orientation involving two cyclic operations with rotational frequencies differing by a factor of 2 (rotor equiv.).

26. Topological Charge

Topology is precisely the mathematical discipline that allows the passage from local to global...
— René Thom

I get the same charge from juxtaposition of colors as I do from juxtaposition of chords. — Joni Mitchell

Topological charge is expressed as a dimensionless quantum number that takes on only one of a discrete set of values. Topological charge results from a quantization procedure rather than simply a symmetry of a given system. Knots in a twisted rubber band, for example, correspond to topological charge as a result of differential rotation. The number of twists in a kink soliton is its topological charge. When two kinks collide, the number of twists remains unchanged. A kink and anti-kink annihilate on collision. Such properties are reminiscent of electrical charge. Geometric charge is a related concept that involves curvature. The Aharonov-Bohm (AB) effect is an example of a physical effect of a geometric phase change which produces a diffraction angle shift in electron beams. Helicity of an EM wavefront is an example of a topological phase change.

The electric charge states (+1, −1) may be represented on a unit circle in the complex plane Z corresponding to the nth roots of unity $Z^n = 1$ where n = 2. The nth roots of unity correspond to n equal eigenvalues for an element of SU(n). Square root of unity as an analogous representation of electric charge (q^+, q^-) is shown in Fig. 26-1.

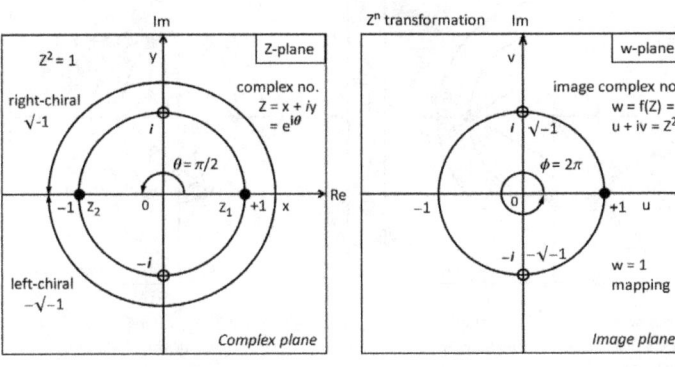

Fig. 26-1. Electric charge represented as square roots of a complex number Z.

But don't you see what this implies? It means that there is a fourth degree of freedom for the electron. It means that the electron has a spin, that it rotates. — Samuel Goudsmit

26. Topological Charge

Fractional electric charge states of quarks (+1/3, −2/3) and antiquarks (−1/3, + 2/3) may be similarly represented as cubic roots of a complex number mapped onto a contracted w image plane. In Fig. 26-2, the cubic root of unity mapped onto a complex image plane illustrates an analogous representation of quark charge. One set of roots correspond to quarks and mirror image set corresponds to antiquarks.

The topological charge represents a twist angle θ corresponding to the number of turns N about a point ($\Phi = 2\pi N$). The winding number n describes the number of loops around the origin and is defined as

$$n = \tfrac{1}{2\pi} \oint_\Gamma \nabla\phi \cdot \mathbf{dr} \qquad (26\text{-}1)$$

where ϕ = phase of the complex field $\psi = |\psi|e^{i\phi}$.

A vortex considered as a topological object may be represented as a wave function

$$\psi(\mathbf{r}) = e^{i\theta}(\mathbf{r} - \mathbf{r}_0) = \langle \mathbf{r}|\psi\rangle \qquad (26\text{-}2)$$

Topological charges of vortices and skyrmions (mass currents) are similar to electric charges and are conserved quantities. The creation of a vortex requires the simultaneous creation of an anti-vortex. Pair annihilation results if the vortices subsequently collide cancelling the charges.

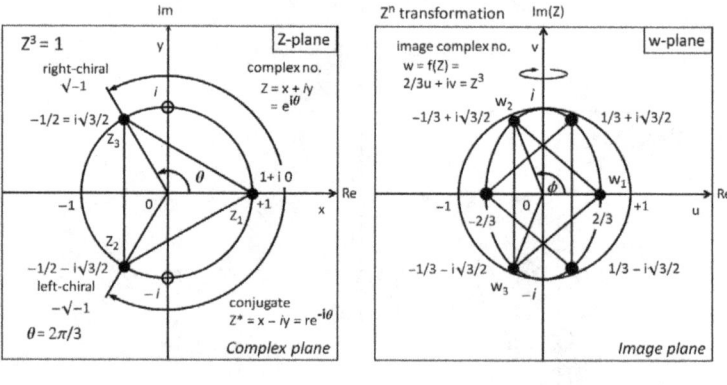

Fig. 26-2. Cubic roots of a complex number Z corresponds to a representation of quark charge.

It is by logic we prove, it is by intuition we discover. – Henri Poincaré

I am a practitioner of the science of deduction, of using known facts in a case to unveil the unknown. – Sherlock Holmes

I am not interested in proofs, but only in what nature does. – P.A.M. Dirac

Push on blindly, but we still have our highly developed sense of smell. – Col. Jack O'Neill

26. Topological Charge

A geometric depiction of relation of electric charge (e⁻, e⁺) to topological charge Φ is illustrated in the form of a tusi couple (2-cusp hypocycloid). A 2-cusp hypocycloid corresponds to SU(2). As shown in Fig. 26-3, a Tusi couple is a 2-cusped hypocycloid obtained by rolling a circle of radius b inside a circle of radius a = 2b resulting in a line segment. A tusci couple illustrates quantization of lepton electric charge modeled in relation to topological twist due to rolling motion. Rotation of the inner circle results in oscillation of a trace of fixed point on the circumference of the smaller circle in linear motion along a diameter of the outer circle R = 2r. In the complex plane the square roots of unity ($z_n = 1$ where order n = 2) lie on the unit circle at $0 \pm i$. A similar construction (not shown) with a circle of radius a rolling on the outside of another larger circle of radius 2a results in a 2-cusped epicycloid known as a nephroid.

Electric charge vs. topological charge

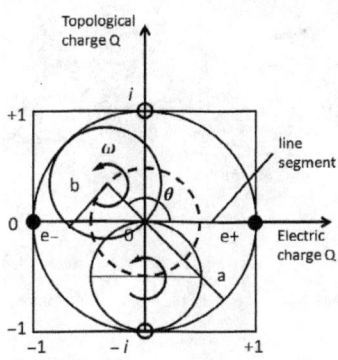

Tusi couple (2-cusped hypocycloid)

Rolling rotation of a small circle of radius b inside a larger circle of radius 2b results in a linear oscillation of a trace generated by fixed point on the smaller circle along a diameter of the larger circle.

n-cusp hypocycloid:

$x = (a/n)[(n - 1)\cos\phi - \cos[(n - 1)\phi]$

$y = (a/n)[(n - 1)\sin\phi + \sin[(n - 1)\phi]$

For a = 2b   n = a/b = 2

$x = a \sin\phi$
$y = 0$

Fig. 26-3. Electrical charge vs. topological charge illustrated as a Tusi couple – a mathematical device that generates a linear motion of a point along the diameter of a circle that is twice the diameter of the smaller inner circle rolling inside the larger circle. Linear oscillation of the point along the major diameter of the circle requires two revolutions of the inner circle similar to the ½ spin characteristic of the electron and positron.

Fractional quark electric charge (−1/3, +2/3) may be produced in a similar fashion by a 3-cusp hypocycloid. A 3-cusp hypocycloid (deltoid) is formed by rolling of inner circle of 1/3rd radius of unity circle inside outer circle. See Fig. 26-4. Anti-quarks with electric charge (+1/3, −2/3) are related to chirality – left- and right-hand mirror-image symmetry which is associated with a radius of convergence k = −Q/6. A 3-cusp hypocycloid corresponds to Lie Group SU(3). For vertices of a regular *n*-gon inscribed in a unit circle in the complex plane with Z_n at the origin of the complex plane, the vertices are given by $C_{k+1} = e^{ik(2\pi/n)}$. The vertices of the n-gon are the nth roots of unity $z = (1)^{1/n}$. Cusps $C_n = e^{2\pi ki/n}$ lie on the unity circle in incremental rotations of $2\pi/n$.

Facts cannot be observed without the guidance of some theory. – Auguste Comte

If I knew something about it, I wouldn't lecture on it! – Arnold Sommerfeld

26. Topological Charge

The complex function $h(z) = 1/(1 + z^2)$ has two singularities in the complex plane at $z = i, -i$. The center of group SU(n) has n eigenvalues that lie on a circle that circumscribes the hypocycloid which intersect at the cusps. Cusps lie on quark charge circle and correspond to cube roots of unity circle. Topological charge = no. of times phase changes from 0 to 2π per wavelength. Turning no. = total curvature/2π. Angular displacement of the rolling circle corresponds to topological charge twist angle $e^{iq} = 2\pi/n$ radians. Integer charge $Q = 2\pi$ radians. Electric charge Q is a spin precession effect (torsion defect).

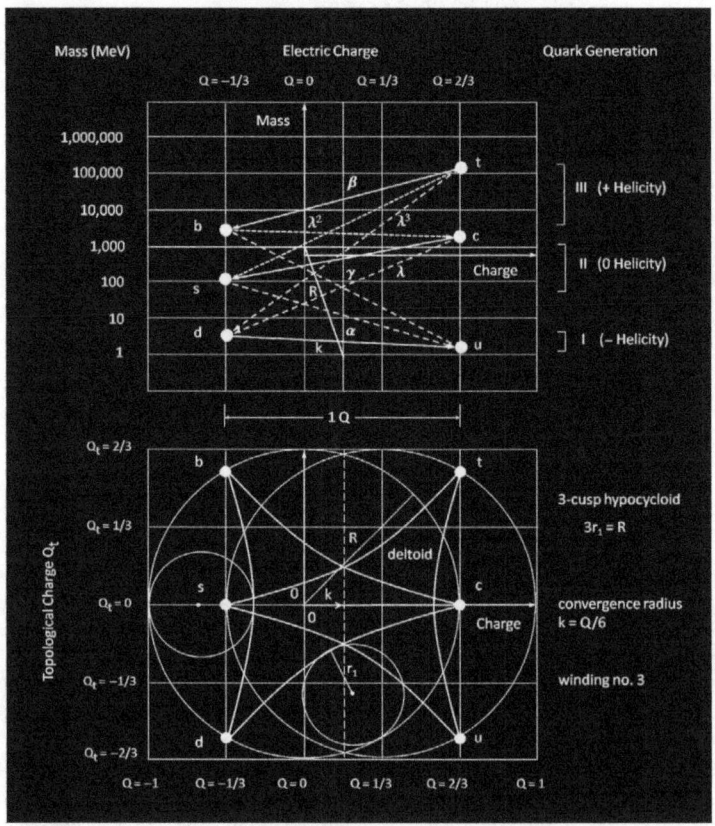

Fig. 26-4. A 3-cusped hypocycloid illustrating fractional quantization of quark electric charge in terms of topological twist due to rolling motion. Rotation of the inner circle results in 3-cusp deltoid. In the complex plane the cubic roots of unity correspond to singularities analogous to quarks and antiquark charges. Antiquarks (not shown) exhibit a mirror image symmetry with $k = -Q/6$. Electric charge $Q \simeq 7.0719$ kg·rad/sec (precession).

Why are charges of elementary particles limited to the values +1, −1 and 0? These and many other such puzzles seem to lie entirely beyond the power of our present theories. – Murray Gell-Mann

26. Topological Charge

Observed quark decay mode trajectories from most to least massive are illustrated in Fig 26-5. Quark flavor type and mass are associated with the direction and degree of twist or helicity. Quark topological and electrical charge relationship to helicity may be described by a 3-cusp hypocycloid (deltoid) as shown in Fig 26-6.

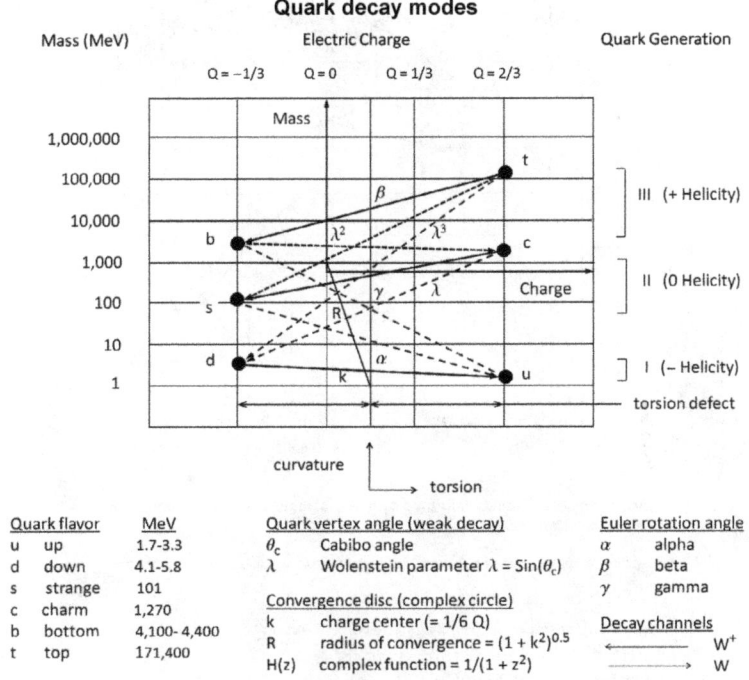

Fig. 26-5. Quark decay modes illustrating relation of mass and helicity. Up and down quarks (Generation I) correspond to negative helicity. Strange and charmed quarks (Generation II) exhibit zero helicity. Bottom and top quarks (Generation III) correspond to positive helicity. Increased mass is associated with increased curvature. Increased charge is associated with increased torsion. Mixing angles are a measure of the ratio of strength of interaction forces. The weak mixing angle θ_W connects the photon (carrier of the electromagnetic force) and the Z boson (carrier of the weak force). In a weak neutral current involving particle scattering by the weak nuclear force, energy is exchanged but charge is not. Strange particles only decay via the weak force.

> *Quarks came in a number of varieties – in fact, at first, only three were needed to explain all the hundreds of particles and the different kinds of quarks – they are called u-type, d-type, s-type. – Richard P. Feynmann*

> *We called the new [fourth] quark the "charmed quark" because we were pleased, and fascinated by the symmetry it brought to the subnuclear world. "Charm" also means a "a magical device to avert evil," and in 1970 it was realized that the old three quark theory ran into very serious problems. ... As if by magic the existence of the charmed quark would [solve those problems]. – Sheldon Lee Glashow*

273

26. Topological Charge

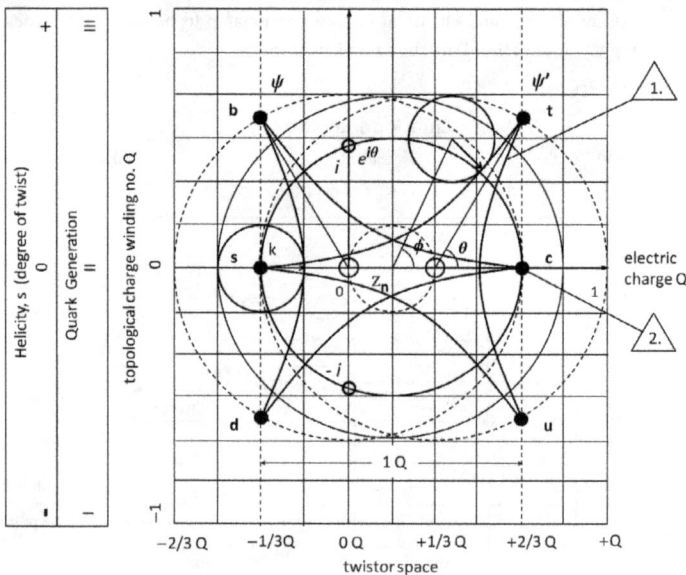

1. 3-cusp hypocycloid (deltoid) formed by rolling of inner circle of $1/3^{rd}$ radius of unity circle inside outer circle.
2. Cusps lie on quark charge circle and correspond to cube roots of unity circle.
3. Charge $Q = 2\pi$ radians; Helicity $s = \frac{1}{2} Z \cdot Z^*$
4. Angular displacement of the rolling circle corresponds to topological charge twist angle $e^{i\theta} = 2\pi/n$ radians.
5. Turning no. = total curvature/2π
6. Charge asymmetry convergence radius: Quarks: $k = +Q/6$; Antiquarks: $k = -Q/6$
7. State vector $|\psi\rangle$ is invariant under rotation (unitary transformation). $\psi = A\exp(\phi) = |A|e^{i\phi}$
8. Topological charge = no. of times phase changes from 0 to 2π per wavelength.

Fig. 26-6. Quark charge and type relationship to helicity depicted in a diagram of topological charge vs. electrical charge. Electric charge is expressable in terms of $SU(2)_L$ and $U(1)_Y$ couplings (weak isospin g and weak hypercharge g') such that $e = g \sin\theta_W = g'\cos\theta_W$ where θ_W is the Weinberg angle (= $\arctan g_1/g_2 = \cos^{-1}(m_W/m_Z) = \arcsin\sqrt{(1/\phi^3)}$)) which is a ratio of the mass of the W and Z bosons.

Mathematics is the only true metaphysics. – Baron William Thomson Kelvin

Strangeness is a quantum number introduced to describe certain hadrons first observed in the 1950's and called strange particles because of their anomalously long lifetimes.... Like isotopic spin, strangeness depends on the properties of the multiplet, but it measures the distribution of charge amomg the particles rather than their number. – Sheldon Lee Glashow

26. Topological Charge

In quantum chromodynamics (QCD) theory of strong nuclear interactions, quarks have a color charge of red (R), green (G) or blue (B) and antiquarks have a color charge of antired ($\bar{A}$), antigreen ($\bar{G}$) or antiblue ($\bar{B}$). Baryons are theorized to consists of quark triplets and mesons of quark pairs. Baryonic states may be represented as $\psi(r,g,b) = \psi(r) \times \psi(g) \times \psi(b)$ and antibaryonic states as $\bar{\psi} = \psi(\bar{r}) \times \psi(\bar{g}) \times \psi(\bar{b})$. Mesonic states may be described as $\psi(r,\bar{r}) = -\psi(\bar{r},r)$, $\psi(g,\bar{g}) = -\psi(\bar{g},g)$ and $\psi(b,\bar{b}) = -\psi(\bar{b},b)$. Other combinations such as tetraquark (two quarks and two antiquarks) and pentaquark (4 quarks and one antiquark) combinations have been tentatively identified. Quarks are bound together by the strong force such that color charges cancel. In a twisted ribbon model, quark RGB color charge interactions may be visualized as 3-fined fusilli corkscrew with net colorless combinations (ternary neutral states). Binary interactions between quarks and antiquarks correspond to a 2-fined corkscrew (helicoid). Anticolors (cyan, magenta, yellow) correspond to a counter-propagating phase conjugate wave. See Fig. 26-7. Baryons of red, green and blue quarks are conjectured to spin in one direction (CCW) generating effective mass while antibaryons composed of antired, antigreen and antiblue spin in the opposite direction (CW) generating effective antimass.

Quark color charge interactions

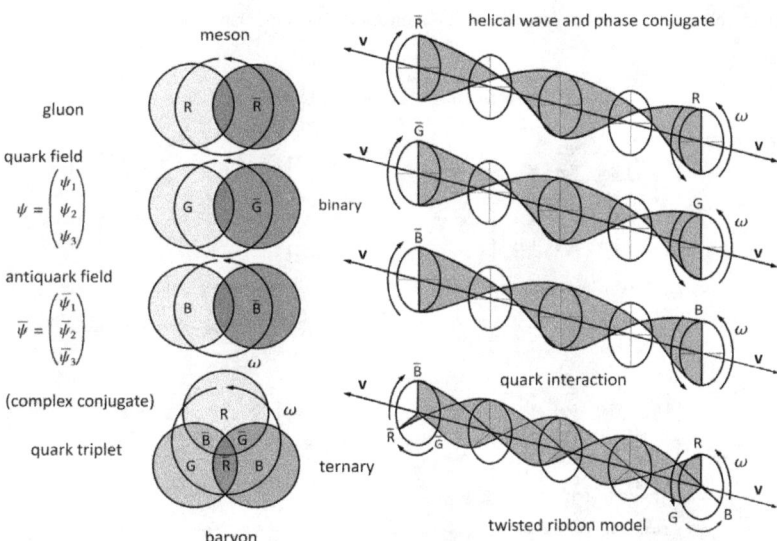

Fig. 26-7. Quark color charge interactions represented as helical wave and counter-propagating phase conjugate waves (rotating, multi-fin, twisted ribbon configurations).

With the new views advocated by Riemann...the texture, structure or geometry of space is defined by the metrical field, itself produced by the distribution of matter. – A. D'Abro

Science walks forward on two feet, namely theory and experiment. Sometimes it is one foot which is put forward first, but sometimes the other, but continuous progress is only made by the use of both. – Robert Milikan

26. Topological Charge

The relation of electric charge and color charge in matrix form resembles spatial and temporal components of 4D Riemannian spacetime metric representations of translation, twist (topological charge) and rotation rate. Riemannian space is a subset of Euclidean space in which distances, angle and curvature are represented as tensors. A Riemannian manifold is a space with a Riemannian metric and a Levi-Civita connection. The geometry of spacetime in Riemann curved manifolds is described by a 4 x 4 symmetric matrix. The Riemann curvature tensor R_{abcd} can be decomposed into a Ricci curvature R_{ab} and a Weyl curvature tensor C_{abcd}. The trace of the Ricci tensor $R = R_{00} + R_{11} + R_{22} + R_{33}$ is the sum of the diagonal components. Electric charge Q as previously noted has mechanical dimensions of [MΘT^{-1}] corresponding to a mass rotational flow rate in time. By analogy as shown in Fig. 26-8, color charge appears to correspond to spatial variation of mass flow with dimensions [MΘL^{-1}]. The RGB color charge Riemann manifold surface domain for the complex cube root function $w = f(z) = z^{1/3}$ resembles a spiral ramp (helix). The Ricci curvature (volume changing) may perhaps be associated with a massive Higgs-like scalar excitation (wave function amplitude $|\psi|$) while the Weyl curvature tensor (volume preserving) may represent tidal distortion effects (wave phase ψ/θ) of massless longitudinal and transverse radiation fluctuations. A Ricci torsion may possibly be associated with pseudo-scalar phase excitations. Lattice-like displacement of Planck dipoles of the vacuum similar to polarons in condensed matter physics where dragging the distortion field effectively increases mass resembling the conjected Higgs mechanism for mass generation.

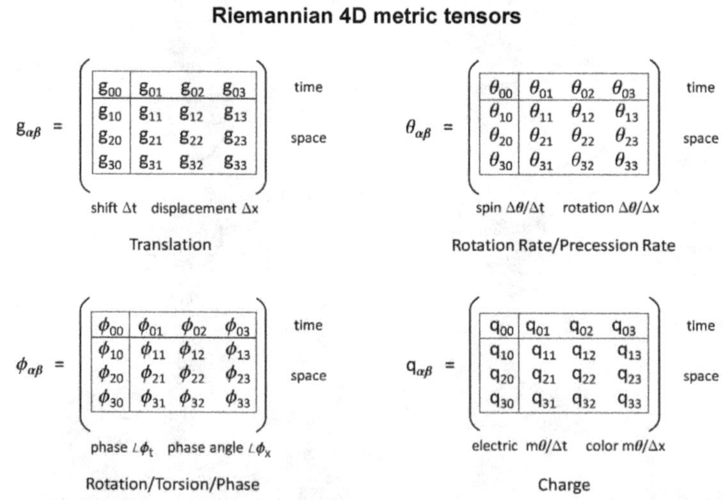

Riemannian 4D metric tensors

Fig. 26-8. Riemann metric comparison of spatial and temporal components for translation, rotation, rotation rate, and electric charge.

Therefore either the reality on which our space is based must form a discrete manifold or else the reason for the metric relationships must be sought for, externally, in the binding forces acting on it. – Bernhard Riemann

Phyiscists came to understand that each of the quantities c, G and h plays the role of conversion factors. – Frank Wilczek

26. Topological Charge

Electric charge Q is related to Isospin I_3 and Hypercharge Y: $Q = I_3 + Y/2$. Hypercharge is defined as $Y = s + c + b + t + B$ where s = strangeness, c = charm, b = bottomness, t = topness and B = baryon number. The 3rd component of Isospin is given as $I_3 = Q - Y/2$. Isospin according to the Gell-Mann Nishijimá rule is $q/e = I_3 + (S + B)/2$ where q is particle charge, e is electron charge, I_3 is projection of Isospin, S = strangeness and B is baryon number. Baryon number is a function of the difference between the number of quarks and antiquarks: $B = (n_q - n_{\bar{q}})/3$ where q denotes quark and $\bar{q}$ denotes anti-quark. Hypercharge is related to electric charge and weak isotopic charge: $Y = 2(Q - I_3)$. Weak hypercharge is defined as $Y_W = 2(Q - T_3)$ where T_3 is the third component of weak isospin. Isospin is related to the differences between the numbers of up and down quarks and up and down antiquarks: Isospin $I_3 = \frac{1}{2}[(n_u - n_{\bar{u}}) - (n_d - n_{\bar{d}})]$. A comparison of quark charge quantum numbers is summarized in Table 26-1.

Table 26-1. Particle type quantum number comparison.

Particle Type	Symbol	Spin	Electric Charge	Color Charge	Isospin	Isospin 3rd projection	Hyper-charge	Baryon No.	Lepton No.
		S	Q	RGB	I	I_3	Y	B	L
Quarks	u	1/2	2/3	RGB	1/2	1/2	1/3	1/3	0
	d	1/2	-1/3	RGB	1/2	-1/2	1/3	1/3	0
	c	1/2	2/3	RGB	0	0	4/3	1/3	0
	s	1/2	-1/3	RGB	0	0	-2/3	1/3	0
	t	1/2	2/3	RGB	0	0	4/3	1/3	0
	b	1/2	-1/3	RGB	0	0	-2/3	1/3	0
Leptons	ν_e ν_μ ν_τ	1/2	0	0	0	1/2	-1	0	1
	e μ τ	1/2	-1	0	0	-1/2	-1	0	1
Bosons	W^+	1	1	0	1	1	0	0	0
	W^0	1	0	0	0	0	0	0	0
	W^-	1	-1	0	1	-1	0	0	0
	Z^0	1	0	0	0	0	0	0	0
	g	1	0	*	0	0	0	0	0
	H	0	0	0	0	0	0	0	0
	γ	1	0	0	0	0	0	0	0
	$\gamma\gamma$*	2	0	0	0	0	0	0	0

* color + anti-color: $R\bar{R}$, $G\bar{G}$, $B\bar{B}$.

Col. O'Neill: *I need a seven-letter word. "Up, Down, Charm... Blank."*
Dr. Jackson: *"Strange".* Col. O'Neill: *Yeah, well, thanks anyway.*

On quantum theory I use up more brain grease than on relativity. – Albert Einstein

27. Bound Particle States

*Nature uses a certain number of themes over and over, if you just have the sense to look for them.
– Aristides Yayanos*

There are two kinds of worlds we line in a material world, bound by laws of physics, and the world inside our mind, which is just as important. – Alan Moore

Man cannot make principles; he can only discover them. – Thomas Paine

Obey the principles without being bound by them. – Bruce Lee

The ground we walk on, the plants and creatures, the clouds above constantly dissolving into new formations – each gift of nature possessing its own radiant energy, bound together by cosmic harmony. – Ruth Bernhard

A vacuum is a hell of a lot better than some of the stuff that nature replaces it with. – Tennessee Williams

27.1 Multiple electron and positron states

Electrons and positrons can temporarily form bound composite particle states which are unstable and decay into gamma rays. Metastable electron-positron bound states can exist in different ground state configurations depending on relative spin states. Examples of such metastable states and example decay modes are summarized in Table 27-1 below.

The maximum repulsive force F_m between electrons at the closest separation distance equal to the Compton radius R_C is

$$F_m = \hbar c / R_C^2 = m^2 c^3 / \hbar \qquad (27\text{-}1)$$

where $R_C = \hbar/\gamma mc = R_{C0}/\gamma = R_{C0}\sqrt{(1 - v^2/c^2)}$.

The electric current I associated with an orbiting electron with orbital period T and velocity v

$$I = e/T = ev/2\pi r \qquad (27\text{-}2)$$

For an electron with negative charge, the electric current is opposite to the direction of motion (mass current).

The magnetic moment μ associated with the current loop of area A is

$$\mu = iA = evr/2 = (e/2m_e)\, L \qquad (27\text{-}3)$$

where L = orbital angular momentum.

The electron model considered here represents an extension of geometrical orbital spin models of electron-positron composite particles previously proposed by Sternglass[15]. Example composites, in terms of the ansatz, including the pion, muon and tauon are illustrated in Figs 27-1 through and 27-4. The minimum approach distance corresponds to the Compton radius R_c where the maximum repulsive force is developed. This force is responsible for electron degeneracy pressure within stars counteracting gravity. At relativistic orbital velocities, the composite particles undergo a Lorentz contraction in which R_C varies with Lorentz factor γ.

27. **Bound Particle States**

Table 27-1. Properties of Electron and Positron Composite Particles

Particle Type	Symbol	Configuration	Charge Radius	Spin s, Moment μ	Decay Modes	Mass Energy	Lifetimes*
Positronium							
Para-positronium	p-Ps	electron $e^- \uparrow$ + positron $e^+ \downarrow$ (anti-parallel spins) ($S=0$, $M_S=0$) Singlet $^1S_0$, ^SP_s	$R_{psq} = 1.9308\text{E-}13$ m $R_{ps} = R_C/(m_{ps}/m_e)$ $= 1.4193\text{E-}15$ m $R_{ps} = v/\omega = \alpha c/2\omega$ $= 1.4089\text{E-}15$ m Positronium Bohr radius: $R_B = 2\hbar^2/m e^2 = 2 a_0 =$ $R_C/2 = 1.058\text{E-}10$ m $\approx 1\text{Å}$	$S_{pP} = 0\hbar$ $\mu_{Ps} = \mu_e/2$ $= -4.636\text{ E-}24$ J/T	$e^- + e^+ \to 2\gamma$ $e^- + e^+ \to 4\gamma$ $e^- + e^+ \to 6\gamma$	1.021 MeV/c^2 $= 2m - m\alpha^2/4$ $\approx 2m_e$	$1.244\text{E-}10$ s
Ortho-positronium	o-Ps	electron $e^- \uparrow$ + positron $e^+ \uparrow$ (parallel spins) ($S=1$, $M_S = -1, 0, 1$) Triplet $^3S_1$, ^TP_s	$R_{psq} = 1.9308\text{E-}13$ m $R_{ps} = R_C/(m_{ps}/m_e) =$ $1.4193\text{E-}15$ m $R_{ps} = v/\omega = \alpha c/2\omega$ $= 1.4089\text{E-}15$ m $R_B = 2\hbar^2/m e^2 = 2 a_0 =$ $R_C/2 = 1.058\text{E-}10$ m $\approx 1\text{Å}$	$S_{oP} = 1\hbar$ $\mu_{Ps} = \mu_e/2$ $= -4.636\text{E-}24$ J/T	$e^- + e^+ \to 3\gamma$ $e^- + e^+ \to 5\gamma$	1.021 MeV/c^2 $= 2m - m\alpha^2/4$ $\approx 2m_e$	$1.386\text{E-}7$ s

Cont

27. Bound Particle States

Table 27-1. Properties of Electron and Positron Composite Particles (cont)

Particle Type	Symbol	Configuration	Charge Radius	Spin s, Moment μ	Decay Modes	Mass Energy	Lifetimes*
Pion (Pi-Meson)							
Neutral Pion	π_0	electron $e^-\downarrow$ + positron $e^+\uparrow$ ($u\bar{u}$ - $d\bar{d}$)/$\sqrt{2}$ (Magnetic moments antiparallel)	$R_{\pi q}$ = 1.9308E-13 m = $R_C/(m_\pi/m_e)$ = $m_e(1+(3/2)\alpha(9))$	$S_\pi = 0\,\hbar$ $\mu_p = 0$	$\pi_0 \to \gamma + \gamma$ $\pi_0 \to e^+ + e^- + \gamma$ $\pi_0 \to e^+ + e^-$ $\pi_0 \to 3\gamma$ $\pi_0 \to 4\gamma$ $\pi_0 \to e^+ + e^- +$ $e^+ + e^-$	134.97 MeV/c^2 $\cong$ 264.2 m_e $\cong 2m_e/\alpha$	0.83E-16 s
	π_1^0	electron $e^-\uparrow$ + positron $e^+\uparrow$ (Magnetic moments antiparallel)	$R_{\pi q}$ = 1.9308E-13 m = $R_C/(m_\pi/m_e)$ = $m_e(1+(3/2)\alpha(9))$	$S_\pi = 1\,\hbar$		134.97 MeV/c^2 $\cong$ 264.2 m_e $\cong 2m_e/\alpha$	

Cont

27. Bound Particle States

Table 27-1. Properties of Electron and Positron Composite Particles (cont)

Particle Type	Symbol	Configuration	Charge Radius	Spin s, Moment μ	Decay Modes	Mass Energy	Lifetimes*
Pion (Pi-Meson) (cont)							
Neutral Pion	$\overline{\pi}^0_1$	electron $e^-\uparrow$ + positron $e^+\uparrow$ (Magnetic moments parallel)	$R_{\pi q} = 1.399\text{E-}15$ m $= R_C/(m_\pi/m_e)$ $\cong m_e(1+(3/2)\alpha(9))$	$S_\pi = 1\,\hbar$		141.03 MeV/c² $\cong 276\, m_e$ $((2/\alpha)-2))m_e$ $= 272.072\, m_e$	
Positive Pion	π^+	$e^-\downarrow + e^+\uparrow + e^+\uparrow$ $\pi_0 + e^+\uparrow$ $u + \overline{d}$	$R_{\pi q} = 1.4138\text{E-}15$ m $= R_C/(m_\pi/m_e)$ $\cong m_e(1+(3/2)\alpha(9))$	$S_{\pi\pm} = 0$	$\pi^+ \to \mu^+ + \nu_\mu$ $e^+ + \nu_e + \overline{\nu_\mu} + \nu_\mu$ $\pi^+ \to \pi_0 + e^+ + \nu_e$ $\pi^+ \to e^+ + e^+ + e^- + \nu_e$	139.57 MeV/c² $\cong 273.13\, m_e$ $\cong m_{\pi 0} + 9 m_e$ $\cong 2m_e/\alpha$	2.60E-8 s
Negative Pion	π^-	$e^-\downarrow + e^+\uparrow + e^-\uparrow$ $\pi_0 + e^-\uparrow$ $d + \overline{u}$	$R_{\pi q} = 1.4138\text{E-}15$ m $= R_C/(m_\pi/m_e)$ $\cong m_e(1+(3/2)\alpha(9))$	$S_{\pi\pm} = 0$	$\pi^- \to \mu^- + \overline{\nu_\mu}$ $e^- + \overline{\nu_e} + \nu_\mu + \overline{\nu_\mu}$ $\pi^- \to \pi_0 + e^- + \overline{\nu_e}$ $\pi^- \to e^- + e^- + e^+ + \overline{\nu_e}$	139.57 MeV/c² $\cong 273.13\, m_e$ $\cong m_{\pi 0} + 9 m_e$ $\cong 2m_e/\alpha$	2.60E-8 s

Cont

27. Bound Particle States

Table 27-1. Properties of Electron and Positron Composite Particles (cont)

Particle Type	Symbol	Configuration	Charge Radius	Spin s, Moment μ	Decay Modes	Mass Energy	Lifetimes*
Muon (Mu-meson)							
Positive Muon	μ^+	electron e^- ↑ + positron e^+ ↑ + positron e^+ ↓	$R_{\mu q}$ = 1.8676E-15 m = $R_C/(m_\mu/m_e)$ = $(2/3)(e^2/m_e c^2)$ $R_\mu = (3/2)\hbar/m_\mu c)$ = 2.8011E-15 m $R_B = a_B^* = \hbar^2/m_\mu e^2$ = 255.927E-13 m	S_μ = ½ $\hbar$ μ_μ = -4.49045E-26 J/T = $g_\mu(e/2m_\mu)S$ $\cong e\hbar/2m_\mu c$ $\mu_\mu \cong 1.0011 \mu_e$ $\cong 1.001165916$ $e\hbar/2m_\mu$ $\mu_\mu \cong 8.9\mu_N$ $\cong 3.18 \mu$	$\mu^+ \to e^+ + \overline{\nu}_e + \nu_\mu$ $\mu^+ \to e^+ + \gamma$ $\mu^+ \to e^+ + 2\gamma$ $\mu^+ \to e^+ + e^- + e^+$	105.658 MeV/c^2 (0.106 GeV/c^2) $\cong$ 206.67 m_e $\cong 3m_e/2\alpha$ = $(\mu_e/\mu_\mu)m_e$	2.197E-6 s
Negative Muon	μ^-	electron e^- ↓ + positron e^+ ↑ + electron e^- ↑	$R_{\mu q}$ = 1.8676E-15 m = $R_C/(m_\mu/m_e)$ = $(2/3)(e^2/m_e c^2)$ = $R_e = (2/3)R_0$ = $R_C/(1+3/2\alpha)$ $R_\mu = (3/2)\hbar/m_\mu c)$ = 2.8011E-15 m $R_B = a_B^* = \hbar^2/m_\mu e^2$ = 255.927E-13 m	S_μ = ½ $\hbar$ μ_μ = -4.49045E-26 J/T = $g_\mu(e/2m_\mu)S$ $\cong e\hbar/2m_\mu c$ $\mu_\mu \cong 1.0011 \mu_e$ $\cong 1.001165916$ $e\hbar/2m_\mu$	$\mu^- \to e^- + \overline{\nu}_e + \nu_\mu$ $\mu^- \to e^- + \gamma$ $\mu^- \to e^- + 2\gamma$ $\mu^- \to e^- + e^+ + e^-$	105.658 MeV/c^2 (0.106 GeV/c^2) $\cong$ 206.67 m_e $\cong 3m_e/2\alpha$ = $(\mu_e/\mu_\mu)m_e$	2.197E-6 s

Cont

Table 27-1. Properties of Electron and Positron Composite Particles (cont)

Particle Type	Symbol	Configuration	Charge Radius	Spin s, Moment μ	Decay Modes	Mass Energy	Lifetimes*
Muonium							
Positive Muonium	$M_{\mu+}$	$\mu^+\uparrow + e^-\uparrow$ $\mu^+\uparrow + e^-\downarrow$ paramagnetic	$R_{\mu q} = 1.8676\text{E-}15$ m $R_B = a_{Mu,0}$ $= 4\pi\epsilon_0(\hbar^2/e^2)(1/M_e*)$ $= 0.5317\text{E-}10$ m $= 1.0044 a_0$	$S_\mu = 0\,\hbar\;(^S\text{Mu})$ $S_\mu = 1\,\hbar\;(^T\text{Mu})$	$\mu^+ e^- \to \mu^- e^+$	$\cong 106$ MeV/c^2 $\cong 207.8\,m_e$ $\cong 206.768\,M_e$	2.197E-6 s
Negative Muonium (Anti-muonium)	$M_{\mu-}$	$\mu^-\downarrow + e^+\downarrow$ $\mu^-\downarrow + e^+\uparrow$	$R_{\mu q} = 1.8676\text{E-}15$ m $R_B = a_{Mu,0}$ $= 4\pi\epsilon_0(\hbar^2/e^2)(1/M_e*)$ $= 0.5317\text{E-}10$ m $= 1.0044 a_0$	$S_\mu = 0\,\hbar\;(^S\text{Mu})$ $S_\mu = 1\,\hbar\;(^T\text{Mu})$	$\mu^- e^+ \to \mu^+ e^-$	$\cong 106$ MeV/c^2 $\cong 207.8\,m_e$ $\cong 206.768\,M_e$	2.197E-6 s

* Relativistic lifetime $\tau = \gamma\tau_0$ where τ_0 is at rest lifetime.
** Alpha ratios: Ref: MacGregor[69]

We thus see that a nonlinear interaction causing attraction can generate a self-organizing system. – Norbert Weiner

The history of the cosmos is the history of the struggle of becoming. – D.H. Lawrence

The principle of the constancy of the speed of light... can be valid only for regions of constant gravitational potential. – Albert Einstein

27. Bound Particle States

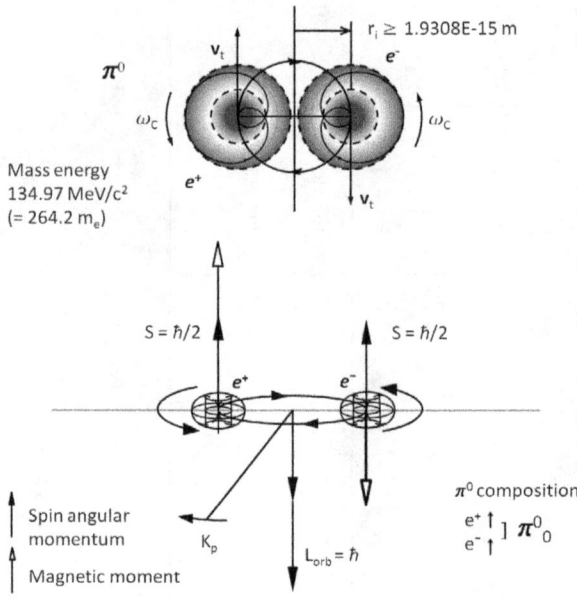

Fig. 27-1. Neutral Pion π^0 consisting of a spin 0 neutral pi-meson composed of a positron-electron pair (Sternglass model). The positron and electron pair each have spin angular momentum oriented either parallel or anti-parallel to each other. The orbital angular momentum is equal to twice the spin angular momentum (L = $\hbar$). A spin 0 neutral pion has a lifetime of ~0.83E-16 sec and can decay into 2 gamma rays. Spin 1 neutral pions are more energetic with a mass of ~264 m_e electron masses. Binding energy is decreased when magnetic moments are parallel due to mutual repulsion increasing the net mass energy. The strong force mediated by pions which decay results in an intensity which varies as Aexp($-\lambda/r$)/r^2 where $\lambda = \alpha/v = m_Uc/\hbar$ and v = meson velocity. The exponential factor accounts for the effect of particle decay on effective range of the strong force.

The pi meson, or pion, the true mediator of the nuclear force, was discovered in 1947. – Anne Rooney

The pion decays to the less heavy muon, emitting a quantum of radiation that carries away the difference in energy and angular momentum in the form of a neutrino. – Ernest J. Sternglass

Both the pion and muon have almost the same mass creating confusion among physicists for a decade in the 'ten-year' joke. – J. Robert Oppenheimer

Yukawa's conjecture was fully confirmed... The particle he predicted has been found and it is called the pi meson, or pion. It weighs about 270 electron masses, and comes in three forms: positive, negative and neutral. – Murray Gell-Mann

"How hot?" she asked, toying with her skirt. "Ionizing radiation hot," I said. "Neutral pion decay hot." Elena snorted, "You're such a romantic," she said. – David Walton

When one jumps over the edge, one is bound to land somewhere. – D. H. Lawrence

27. Bound Particle States

Positively charged Mu-meson (Muon)

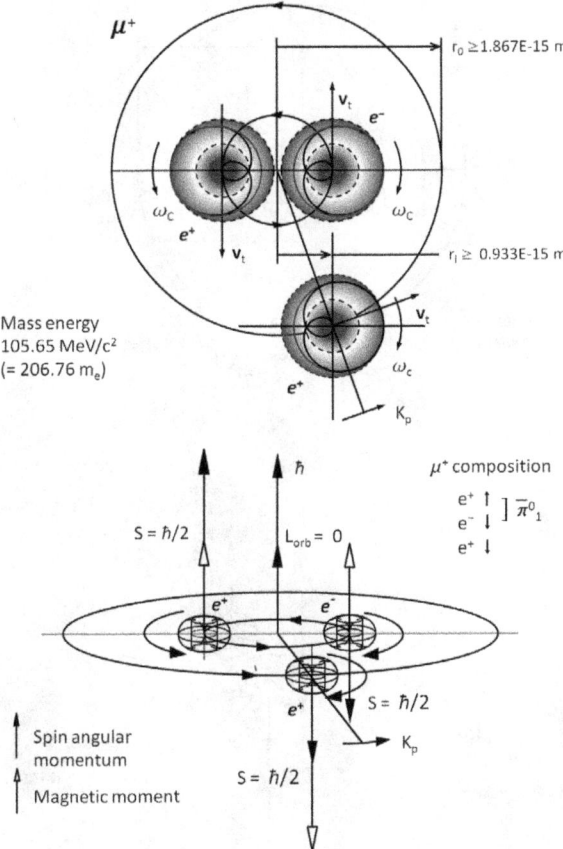

Fig. 27-2. Positive Muon μ^+ consisting of a positron-electron and positron. Positron e^+ is at rest relative to the precessing frame K_P of central pair system consisting of a spin 1 neutral pi-meson π^0_1 (Sternglass model).

Muon: Who ordered that? – Isidor Isaac Rabi

Why is the muon, some dumb particle, 200 times heavier than the electron? Why is the proton about 2,000 times heavier than the electron? Why is the electric charge of the electron what it is? – Sheldon Lee Glashow

Don't believe any calculation in meson theory which uses Feynman diagrams. – R. P. Feynman

There was much breathless talk of new elements, bizarre optical properties, and other things which puzzled men of science are wont to say when faced by the unknown. – H. P. Lovecraft

Science has done it, they've finally found the elusive g-factor. Women everywhere are rejoicing. – Rico Pablo

285

27. Bound Particle States

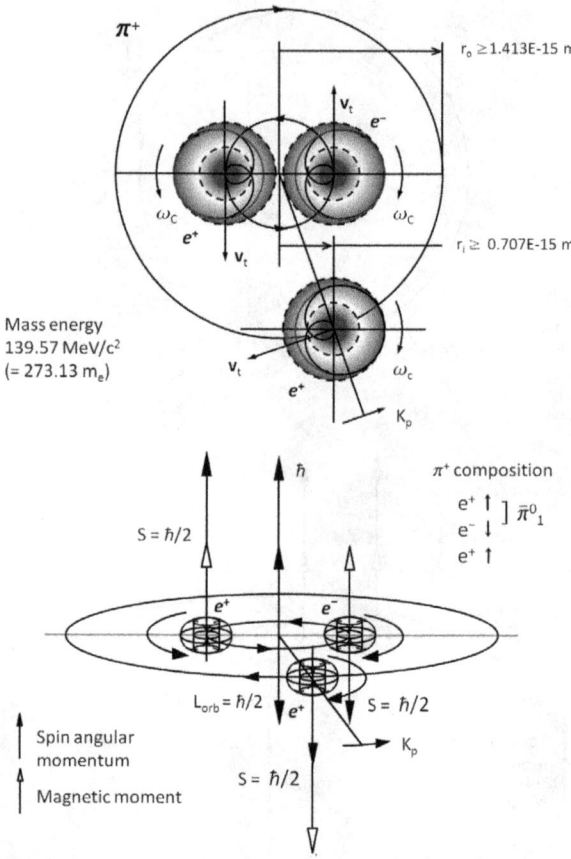

Fig. 27-3. Positive Pion π^+ consisting of a positron-electron and positron. A positron e^+ is in orbital motion relative to the precessing frame K_P of a central pair system consisting of a spin 1 neutral pi-meson π^0_1. A π^+ is equivalent to an $u\bar{d}$ quark pair with opposite spins [↑↓] and +1/3 + 2/3 = +1 electric charge. (Sternglass model).

Are there charged leptons with masses greater than that of the muon? – Martin Perl

With Mie's view of matter there is contrasted another, according to which matter is a limiting singularity in the field, but charges and masses are force-fluxes in the field. – Herman Weyl

Interestingly, the value of muon anomalous magnetic moment $a_\mu = (g_\mu - 2)/2$ appears to be larger than the Standard Model value by greater than three standard deviations. – B. Lee Roberts

When truth is discovered by someone else, it loses something of its attractiveness. – Aleksander Solzhenistsyn

27. Bound Particle States

Negatively charged Tau-meson (Tauon)

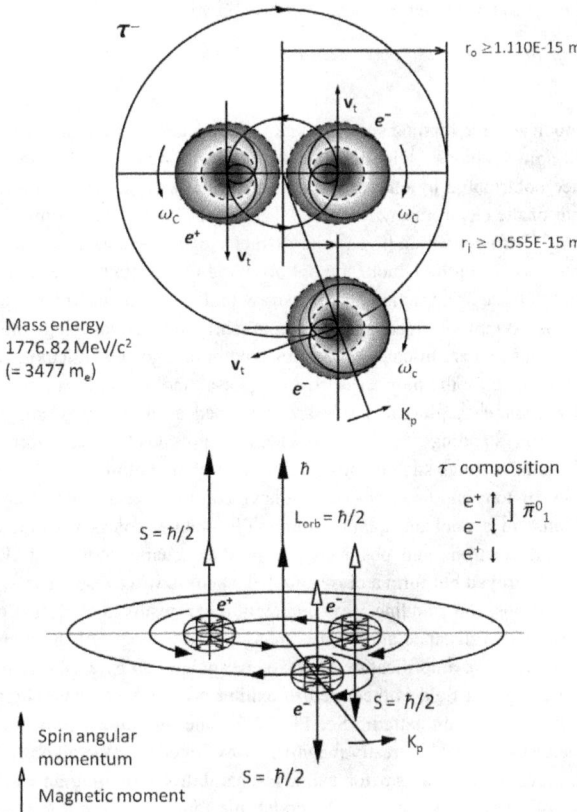

Fig. 27-4. Negative Tauon τ^- consisting of a positron-electron and electron. Electron e^- is in orbital motion relative to the precessing frame K_P of central pair system consisting of a spin 1 neutral pi-meson $\bar{\pi}^0{}_1$ (Sternglass model).

The letter τ is from the Greek τριτον for third – the third charged lepton... Is the tau simply a standard model lepton... or outside the standard model? – Martin Perl

The U (Unknown) particle was estimated to be 1800 mass-energy units, twice as heavy as a hydrogen atom! – Nigel Calder

Thus by 1978 all confirmed measurements agreed with the hypothesis that τ was a lepton which was produced by electromagnetic interaction and, at least in its main modes, decayed through the conventional weak interaction. – Martin Perl

When the waves are taken into account..., as in the case of the electron, waves have to be accommodated within a certain distance. – J. J. Thomson

Inertia is the property of matter by which it resists force. – Klaatu

287

27. Bound Particle States

27.2. Gluon fields

gluon, *n. a massless gauge boson that binds quarks together to form baryons, mesons and other hadrons and is associated with the strong nuclear force.*

glueball, *n. hypothetical composite particle composed of gluons*

The strong nuclear force binding spin ½ quarks is attributed in the standard QM model to an exchange of spin 1 gluons. Gluons account for approximately half of the proton spin with the balance contributed by quarks and antiquarks. The strong force is ~137 times (= α^{-1}) the strength of the electromagnetic force. The strong interaction is limited in range to about a fermi (10^{-15} m) corresponding to the diameter of a nucleon or approx. 1% of the electron Compton wavelength. Gluons are not observed in isolation and exhibit an unusual property in that the force of attraction is increases with distance described as a confinement property known as asymptotic freedom. In quantum chromodynamics gauge theory (QCD), there are eight gluons that are bi-colored carrying a color charge and anticolor. The number of gluons is determined by the number of SU(3) color-anticolor combinations. In a model proposed by LaFreniere[38], gluonic fields are described as a superposition of a pair of electron or positron standing waves. An electron and positron can form temporary resonances as orbital pairings such as positronium and pions before annihilating. Leptons (electrons, muons, tauons and neutrinos) do not have color charge and are insensitive to the strong interaction. In a nucleon, quarks continually change colors via gluon exchange. Collisions between electrons and positrons can produce quarks indicating electrons and positrons are not destroyed but form a close coupled standing wave resonance.

An electron and positron standing wave interaction can result in attraction or repulsion depending on separation distance and relative phase. The strength of the gluonic field is attributed to standing wave amplification of wave energy emitted by each electron/positron. The intensity of the gluon field is theorized to exhibit a cross-section variation similar to that of an Airy disc diffraction pattern. See Fig 27-5. The flux tube area is proportional to the square of field strength. The greatly magnified wave energy of the gluon field flux tube is viewed as a plausible mechanism for quark inseparability providing an explanation for the 'bag' model of quark confinement. The nodal interference pattern, for example, in the case of a quarter period offset results in a Fresnel interaction zone of alternating wave curvature with a unidirectional wave velocity in opposition to the Coulomb force of attraction. The cross-section of the co-axial Fresnel zones form in effect a diffractive lens or zone plate concentrating the wave energy towards each electron and/or positron. Refer to Fig. 27-6. The nested ellipsoidal Fresnel zones act like mirrors focusing in-phase waves at the focii. The energy contained in each Fresnel zone is proportional to its enclosed volume, increasing the separation distance of the foci requires additional energy input. When stretched, the flux lines become confined within a narrow flux tube approximating a one-dimensional string (field). Moving standing waves are compressed in wavelength with an increase in frequency and amplitude. In the LaFreniere model, an e^-e^+ pair close together within a few Compton radii ($R_C \sim 10^{-13}$ m) interact via this diffractive lens effect creating a gluonic field of plane standing waves within the first innermost Fresnel zone (gluon tube) to bind together to form a quark. Compressed moving EM waves transform into gluonic fields with greater increased wave energy. Charge cancellation of like charges in close proximity is theorized to be due to an effective charge screening by an interference zone of depolarized Planck dipoles. In-phase constructive wave interference results in repulsion while out-of-phase destructive interference results in attraction.

27. Bound Particle States

The gluon is characterized as having spin 1 with zero electric charge. For a quark composed of an electron and positron each with spin ½ and −1 and +1 electric charge indicates the spins are aligned with opposite rotation. Thus two geometric possibilities arise: 1) The e^-e^+ orbit as a pair in the same spin plane or 2) the e^-e^+ pair are coaxially aligned. A orbital pairing of an electron and positron model previously described with Archimedean dipole spiral arm wavefronts with a $\pi/2$ phase quadrature is illustrated in Fig. 27-7. The wavefront interference pattern as shown also exhibits an Airy disc intensity profile similar to that of spherical waves. To avoid self annihilation, the electron and positron need to retain soliton-like properties to allow interpenetration. The constant $\pi/2$ phase locking appears required to prevent electron and positron self annihilation as in the case of the π meson. Electric field lines represent a synchronous alignment of Planck dipoles. At close approach of the electron and positron ($< \frac{1}{2} R_C$), the Planck dipoles would need to continuously change orientation by π radians as each wavefront sweeps by and there is a limit to how fast the spin flips can occur. Beyond this limit, the Planck dipoles spins depolarize assuming random orientations and thus form a temporary separation boundary between counter-rotating spiral arms allowing each arm to retain integrity. However, it is not apparent how precise phase quadrature can be readily accomplished as the e^- and e^+ spins remain constant without phase-locking synchronization.

Shown in Fig. 27-8 is the second case of an e^-e^+ in a close proximity coaxial arrangement. Here again, the spins are aligned and net zero electric charge with opposed spin precession. However, there is no necessity for centrifugal force to offset the Coulomb force of attraction. At close range ($< \frac{1}{2} R_C$), again there is a limit to how fast the intervening Planck dipoles can spin flip in synchronicity with the counter-rotating e^-e^+ spiral wavefronts due to the elastic limits of the vacuum as measured by the speed of light. Beyond this limit, the Planck dipoles depolarize forming, in effect, a charge screen thus isolating the e^- and e^+ fields from each other. Hence, the electrostatic Coulomb attraction is neutralized and the electron and positron are no longer attracted at close range. A rotation $\pi/2$ phase difference allows closer approach. The e^- and e^+ pair would tend to oscillate along the spin axis alternating between a near state where charge screening occurs and at further distances where charge screening is ineffective. The electric dipole moment at such small separations would be very small. Of these two possibilities, coaxially coupling of a an e^-e^+ with aligned spins resulting in a massless spin 1 gluon seems simpler and perhaps is the more likely. Phase locking involves conversion of kinetic energy into potential energy stored in gluon fields. Electric charge with dimensions [$M\Theta T^{-1}$] as previously noted corresponds to mass spin precession.

Color charge corresponds to mass winding number with dimensions [$M\Theta L^{-1}$]. Color charge interaction is held as the mechanism for strong nuclear force binding quarks together in the nucleus. In the twist ribbon model, topological charge increases with twist. Increasing the separation distance, increases the number of turns/sec increasing charge and, hence, Coulombic force of attraction. Color charge appears to arise when the spin axes of paired electrons (or positrons) are aligned and may represent a coaxial spin alignment configuration rather than a orbital spin axis alignment.

What physics tells us is that everything comes down to geometry and the interaction of elementary particles. – Antony Garrett Lisi

'Excellent,' I cried. 'Elementary,' said he. – Arthur Conan Doyle

That's too sophisticated. We'll do simple field theory, where everything I say is correct.
– Prof. Rabindra Mohapatra

27. Bound Particle States

In QCD theory, two up quarks (u) and a down quark (d) constitute a proton. Two down quarks (d) and an up quark (u) form a neutron. See Fig. 27-9. The up quark has ½ spin, 2/3 charge and mass of ~2-5E-3 GeV/c^2 while the down quark has ½ spin, −1/3 charge and mass of ~4−9E-3 GeV/c^2. The three quarks account for ~1% of the mass of the proton with the balance due to interactions of quarks and gluons. In the Sternglass orbital model, a neutral pion π_0 consists of either an uu̲ quark or dd̲ quark pair. An (u) quark in this description corresponds to an $e^-\uparrow e^+\uparrow$ orbital pair with aligned spins while an anti-up quark u̲ corresponds to an $e^-\uparrow e^+\uparrow$ orbital pair with opposite orbital spin ↓↓. Likewise, a (d) quark corresponds to an $e^-\downarrow e^+\downarrow$ orbital pair with aligned spins while an anti-up quark u̲ corresponds to an $e^-\downarrow e^+\downarrow$ orbital pair with opposite orbital spin ↑↑. In the LaFreniere model, a neutron consists of three pairs of electrons in a tetrahedral arrangement interlinked with 18 gluon fields while a proton consists of a neutron with a central positron interlinked with 15 gluon fields. The mass of the thee component quarks of a nucleon (proton or neutron) accounts for ~1% of the mass of the nucleon. The balance of the mass of the nucleon resides in the potential energy in the gluon fields of the strong force binding the quarks together. Potential energy represents stored energy of position or arrangement whereas kinetic energy represents energy of motion. Mass appears as a result of interaction energy of wave function interference – a spin precession effect and, notably, is not due to some magical Higgs field interaction. An electron/positron model of a proton and neutron illustrating QCD equivalence is shown Figs. 27-10 and 27-11. An electron situated between two positrons in a string configuration are matched in phase and form an up quark. Similarly, a positron disposed between two electrons phase matched form a down quark. The electric dipole moments (EDM) of each pair cancel out resulting in a net zero EDM and conserving parity. The residual strong force between quarks is responsible for "gluing" the nucleon together overcoming the repulsive Coulomb force.

As Sternglass has shown, the pi-meson model may be extended to include excited nucleon states such as hyperons or strange baryons with strangely longer lifetimes. Additionally, composite particles such as the J/Psi with very high energy states of relativistically rotating e^-e^+ coupled pairs are also successfully described accounting for the longer lifetime. In the case of the J/Psi meson, a spin 1 neutral pion is bound in the center of four sets of spin 1 Rho mesons. K-mesons or kaons exhibit a hypercharge characteristic. The neutral K_0 has a +1 hypercharge while the $\overline{K}_0$ antiparticle has a −1 hypercharge. The amplitude combination of ($K_0 + \overline{K}_0$) results in a K_L or "K-long" with a longer decay time than K_S or "K − short" formed by the combination ($K_0 - \overline{K}_0$). The K_L decay modes appear to exhibit violations of time-reversal and charge invariance. An apparent violation of charge-exchange, parity and time (CPT) symmetry in which a neutral kaon K_0 decay into two pions proceeds at a different reaction rate than combining of two kaons into a neutral pion as well as direct decay into two pi mesons ($\pi_0 + \pi_0$, $\pi^+ + \pi^-$) may also be explained in terms of spin-orbit coupling by the Sternglass model.

The total energy of the universe is constant; the total entropy is continually increasing. – Rudolf Clausius

Unfortunately, QCD has nothing whatsoever to say about the quark mass spectrum, nor, for that matter does any existing theory. – Kurt Gottfried

Squalid-state physics. – Murray Gell-Mann

It all seems so arbitrary in the ridiculous collection of fundamental particles, the lack of patterns to their masses. – Leonard Susskind

27. Bound Particle States

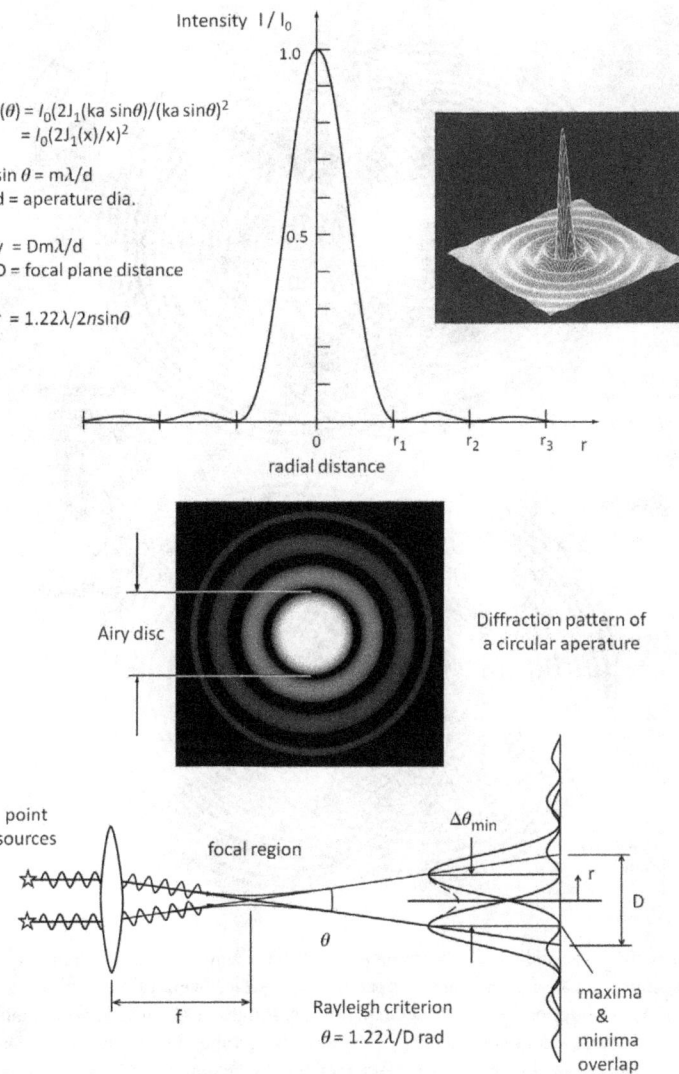

Fig. 27-5. Airy diffraction pattern from a circular aperture and Rayleigh criterion for minimum separation.

In the hands of Science and indomitable energy, results the most gigantic and absorbing may be wrought out by skillful combinations of acknowledged data and the simplest means. – Sir George Biddell Airy

Perceptions determines what goes on. Your answer is encoded in your question. – Greg Bishop

27. Bound Particle States

Gluon field Fresnel zone patterns

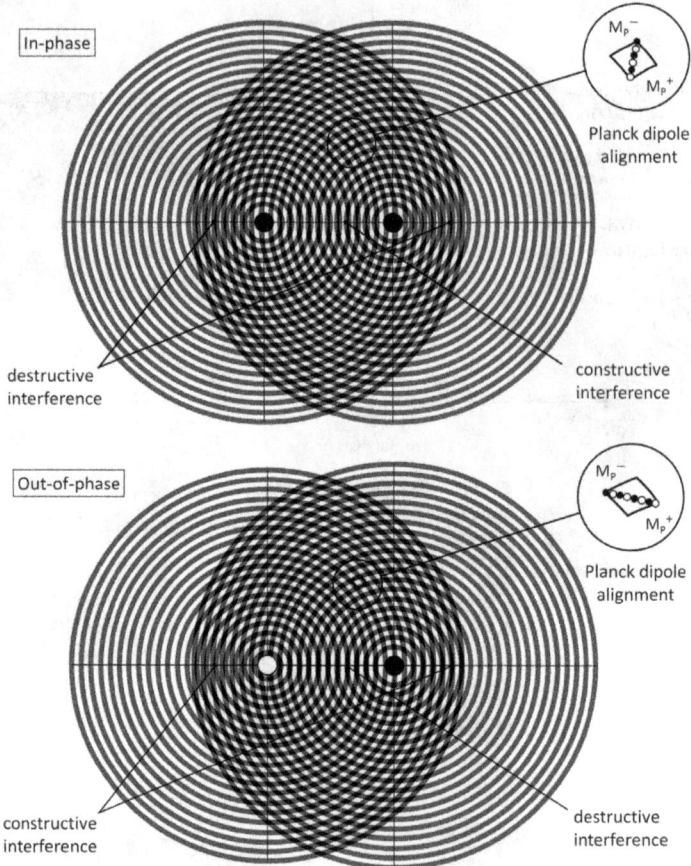

Fig. 27-6. In-phase and out-of-phase gluon field Fresnel zone patterns with Airy diffraction cross-section. As an example, quark separation in a proton with a Compton radius R_{CP} of 2.1031E-16 m, from proton scattering experiments, suggest a gluon field with a minimum distance of ~0.33E-15 m or ~1.56 R_{CP}. Quark mass is a measure of interaction energy of destructive wave function interference corresponding to the number of interference nodes.

A mathematician may say anything he pleases, but a physicist must be at least partially sane. – J. Willard Gibbs

There is a plan, it seems to me, that reaches out of the electron to the rim of the universe... and beyond my feeble intellect. – Clifford D. Simak

The last level of metaphor in the Alice books is this: that life viewed rationally and without illusion, appears to be a nonsense tale told by an idiot mathematician. – Martin Gardner

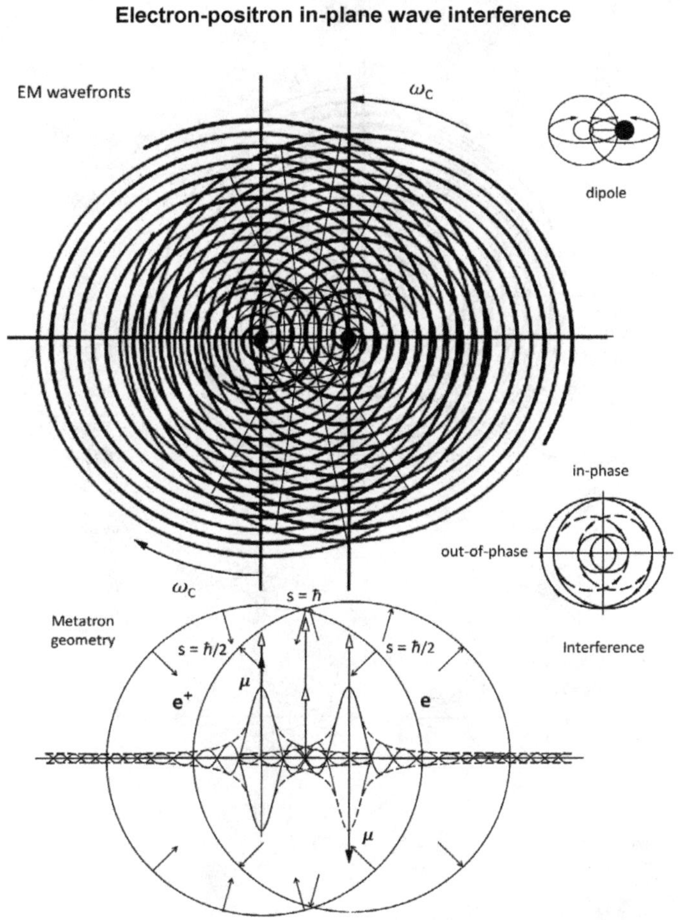

27. Bound Particle States

Electron-positron in-plane wave interference

Fig. 27-7. Hypothetical model of a quark with a massless, spin 1 gluon field illustrated as a close coupled out-of-phase electron/positron pair in orbital relationship with spins aligned. Interactions depend on an overlay and phase combination of their electronic wavefunctions. An out-of-phase quadrature phase shift of $\pi/2$ is assumed to prevent self-destruction. Spin alignment results in increased mass and spin angular momentum than if spins opposed

The general laws of Nature are not, for the most part, immediate objects of perception. – George Boole

... for all these speculations regard the atom as a foreign substance – a sort of 'grit' in the aether... - Oliver Lodge

Outstanding among the areas of physics which have been left out of recent theories of elementary particles are gravitation and cosmology. – Freeman Dyson

Physicists try to decrypt physical reality to understand time & space and beginning & end of things. – Julian Assange

In order to have local symmetry, we must introduce the gluon fields. – Frank Wilczek

27. Bound Particle States

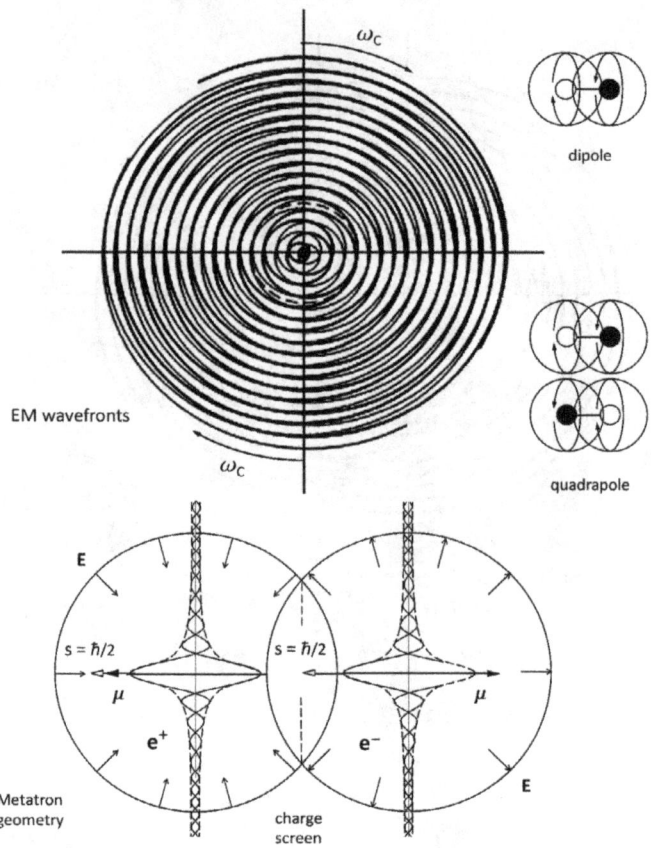

Fig. 27-8. Coaxial coupling of an e^-e^+ pair with aligned spins as a hypothetical quark model with a massless, spin 1 gluon field. A charge screen formed by depolarized Planck dipoles prevents Coulombic attraction. The electron and positron oscillate along the spin axis alternating between a state where charge screening is effective at close distance and further out where charge screening is ineffective thus exhibiting asymptotic freedom. For a remote observer at far field distances ($r \gg r_C$), charges appear to cancel.

Progress is a dynamic thing, and you had to ride leaning forward a little, like a surfboard because if you stood there flat-footed you'd get drowned. – Theodore Sturgeon

When someone says it can't be done, do it anyway. – Nikola Tesla

Some quotations are greatly improved by lack of context. – John Wyndham

I am very seldom interested in applications. I am one interested in the elegance of a problem. Is it a good problem, an interesting problem. – Claude Shannon

What is the go of it? – James Clerk Maxwell

27. Bound Particle States

A baryon consists of three quarks and anti-baryon consists of three anti-quarks. A meson consists of a quark and anti-quark. The proton and neutron each have a baryon no. of +1 while the anti-proton and anti-neutron have a baryon no. of −1. Mesons have baryon no. of 0. The baryon quantum number of a composite particle made of quarks represents $1/3^{rd}$ of the difference between the number of quarks and anti-quarks. Conservation of baryon number represents conservation of the net number of quarks. The baryon number is described by SU(3) symmetry group corresponding to a winding number (color twist).

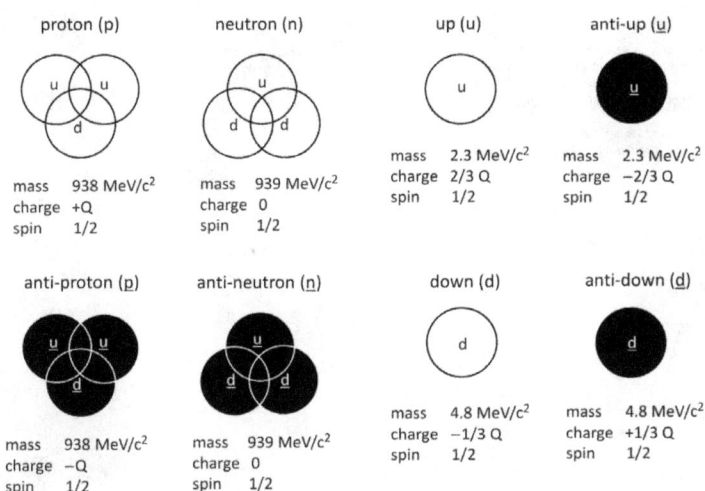

Fig. 27-9. Quantum chromodynamics (QCD) representations of protons and neutrons shown as quark composites. Each flavor of quark u, d, s, c, t and b exhibit any one of three color charges red (R), green (G) or blue (B). Antiquarks $\bar{u}, \bar{d}, \bar{s}, \bar{c}, \bar{t}$ exhibit anticolors: $\bar{R}$, $\bar{G}$ and $\bar{B}$. Fractional charge characteristics of quarks indicates quarks are themselves composites of electrons and positrons. This is apparent in observed decays such as $u + \underline{u} \rightarrow Z_0 \rightarrow e^- + e^+$ and $u + d \rightarrow W^- \rightarrow e^- + \nu_e$ and interactions such as $u + d \rightarrow W^+ \rightarrow e^+ + \nu_e$.

Gell-Mann and Ne'eman recognized that if electric charge were to be part of the SU(3) description, particles whose electric charges were integer multiples of |e| could belong only to certain families... The baryons fit such a family exactly, leading Gell-Mann to call his scheme the "Eightfold Way." – J. L. Rosner

We still don't know what's giving mass to the W and Z. We just know that SU(2) x U(1) symmetry is broken. – Howard Georgi

I saw the whole universe laid out before me, a vast shining machine of indescribable beauty and complexity. Its design was too intricate for me to understand, and I knew I could never begin to grasp more than the smallest idea of its purpose. But I sensed every part of it, from quark to quasar, was unique and – in some mysterious way – significant. – R. J. Anderson

The marriage of the electroweak theory and quantum chromodynamics produced the standard model - the basic theory of elementary particles and the laws that govern them. – Mario Livio

27. Bound Particle States

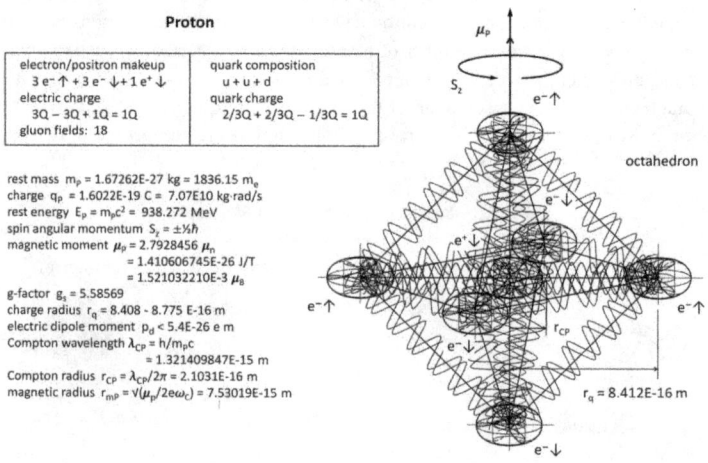

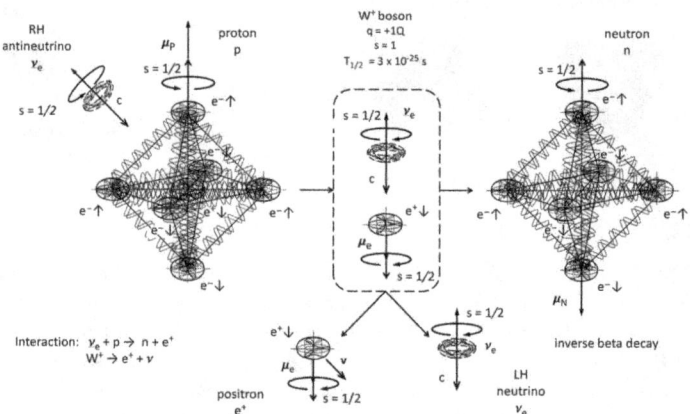

Fig. 27-10. A proton composed of electrons and positrons is equivalent to QCD quark representation: uud [↑↑↓] with +1 charge. An example interaction is the absorption of a antineutrino by a proton and transformation into a neutron with the emission of a positron in a process known as inverse beta decay. The bulk of the proton's mass is due to the kinetic energy of its component quarks. The rest mass of the proton is ~938 MeV while the up quark has a mass of ~4 MeV and the down quark has a mass of ~7 MeV.

You see, the chemists have a complicated way of counting: instead of saying "one, two, three, four, five protons", they say, "hydrogen, helium, lithium, beryllium, boron." – Richard P. Feynmann

Perhaps the equivalence of mass and energy can be checked at some future date by observation on the stability of nuclei. – W. Pauli

27. Bound Particle States

Neutron Model

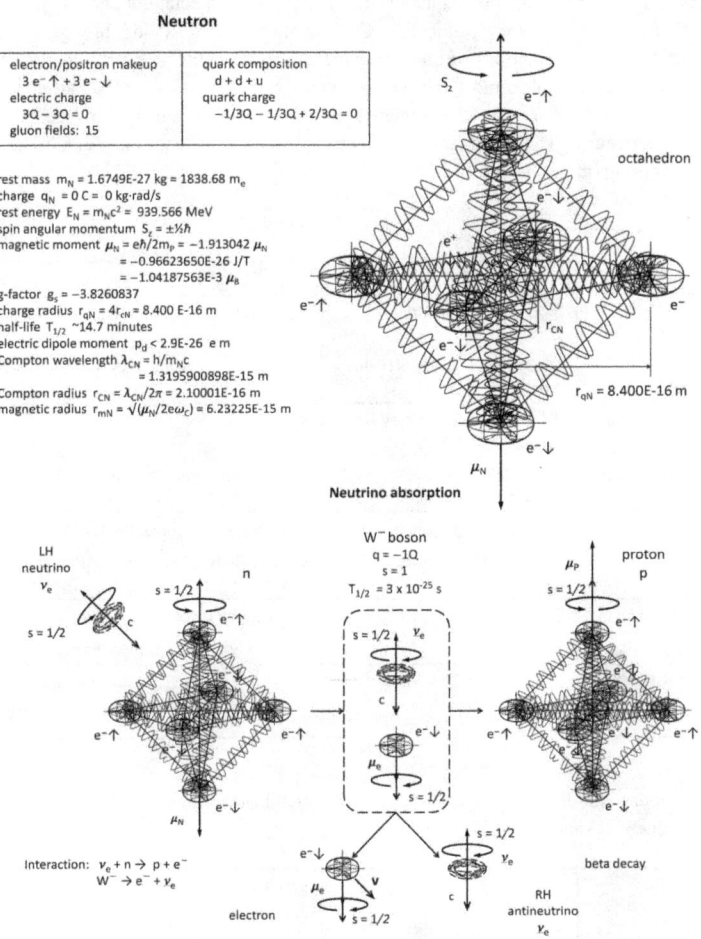

Fig. 27-11. A neutron composed of electrons is equivalent to QCD quark representation: ddu [↑↑↓] with 0 charge. An example interaction is the absorption of a neutrino by a neutron and transformation into a proton with the emission of an electron in a process known as beta decay. Protons and neutrons are isotopes in a sense as they have nearly the same mass but different charge.

I am afraid neutrons will not be of any use to any one. – James Chadwick

The fundamental point in fabricating a chain reacting machine is of course to see to it that each fission produces a certain number of neutrons and some of these neutrons will again produce fission. – Enrico Fermi

As neutrons can be packed much more closely than ordinary nuclei and electrons, the gravitational packing into a cold neutron star may become very large, and under certain conditions may far exceed the ordinary nuclear packing fractions... – Fritz Zwicky

27. Bound Particle States

The strong force involving interactions of the triplet of quark color charges are related by a group of symmetry operations under SU(3). SU(3) is a special unitary group in which two of the members are identical. Commutation relations in this group allow representation of combinations of eight possible charge combinations including antitriplet states, antiquark, anticolor and gluon states. Discrete symmetries such as an equilateral triangle are described by the S3 symmetric group. Gauge symmetries are described in terms of Lie groups. The standard model of strong, weak and electromagnetic interactions in terms of group theory is described as

$$SU(3) \times SU(2) \times U(1)$$

where

Gauge group	Order	Force field
SU(3)	3rd order special unitary group	strong force
	3 x 3 unitary matrices (color twist)	color charges; quark color; baryon number
SU(2)	2nd order special unitary group	weak force
	2 x 2 unitary matrices (up/down twist)	weak charge; spin doublets
U(1)	1st order unitary group	electromagnetic force, Planck force
	1 x 1 unitary matrices (phase twist)	electric charge, hypercharge, Planck charge
SU(2) x U(1)	2nd order and 1st order groups	electroweak force
		neutral currents, broken symmetry
SU(5)	3rd order and 2nd order groups	strong, weak & electromagnetic force

The standard model does not include gravity which instead may be described in terms of other symmetries such as

T4	space-time translational T4 Abelian subgroup of Poincaré group	gravitation force
		gravitational charge (mass)

Nuclear matter on the whole contain equal amounts of electrons and positrons bound into quarks. This would seem to account for the apparent preponderance of matter over antimatter in the universe. Anti-matter formed in the Big Bang need not be mysteriously and preferentially annihilated in charge parity violation or 'flowed backward' in time before the Big Bang. Antimatter in the form of positrons is still here in the present universe bound into quarks in quadrature phase-lock in bound states with electrons. Charge parity violation is associated with the weak force between quarks as exhibited by K, B and D meson decay. The strong nuclear force interaction is CPT invariant. The weak force strength is $\sim 10^{-6}$ of the strong force with a range of $\sim 10^{-18}$ m (0.1% of proton dia.). The strong force, electromagnetic force, weak force and gravitation force are thought to equalize at the Planck energy scale of $\sim 10^{19}$ GEV where all the force carriers become massless at extremely high temperatures. All of these forces are ultimately due to wave interactions of Planck mass dipoles that makeup the physical vacuum.

27.3 Neutrinos

Dear Radioactive Ladies and Gentlemen, ... look and judge. – Wolfgang Pauli

I suppose the neutrino can be described as the missing link between energy and matter – Exeter (This Island Earth)

At present, the mass spectrum of neutrinos is not precisely determined and it is an open question if neutrinos are Dirac or Majorana type. A massive Dirac neutrino field has four degrees of freedom (2 helicities and 2 particle-antiparticle states) whereas a Majorana neutrino field has two degrees of freedom (2 helicities, 1 particle/anti-particle state). Neutrinos are spin ½ fermions that are nearly massless and without electric charge. Only left-hand neutrinos (negative helicity) with spins anti-parallel to linear momenta are observed. Right-hand anti-neutrinos (positive helicity) exhibit spins parallel to the linear momentum vector. Neutrinos have only one helicity state and obey charge parity (CP) symmetry, i.e., a LH neutrino may be transformed into a RH anti-neutrino. Neutrinos have a very small interaction cross-sections and interact weakly with matter. It is estimated roughly 10 percent of solar radiation is in the form of neutrinos. Neutrino flux at the Earth's surface is estimated $\sim 10^{15}$ per meter^2/second with energies up to ~ 100 GeV – 2000 TeV. Representative electromagnetic properties of neutrinos are summarized in Table 27-2.

A ring-like vortex model of the neutrino has been previously proposed such as that by Winterberg[7], Sternglass[15] and Meyl[80]. In a geometrical model considered here, neutrinos are conceptualized as vortex rings composed of Planck dipoles that propagate along the spin axis. The toroidal spin is half of the poloidal spin corresponding to a two-component spinor with net spin of ½. The poloidal current induces a nonzero azimuthal magnetic field inside the torus with an intrinsic toroidal dipole (anapole) TDM moment axial-vector that reacts with an external magnetic field. The geometry is conjectured to be similar to that of a Helmholtz vortex ring, however, with a toroidal spin component. The axial velocity of propagation and the tangential spin component in free space equal the velocity of light c. A neutrino has an electric dipole moment $H = \pm \epsilon E_0$ and magnetic dipole moment $H = \pm \mu B_0$ and may exhibit electromagnetic coupling including Cherenkov radiation decay ($\nu \rightarrow \nu'$ + γ), plasma photon decay ($\gamma \rightarrow \nu + \bar{\nu}$), photon exchange scattering ($\nu + e^- \rightarrow \nu + e^-$), bremsstrahlung emission ($\nu + e^- \rightarrow \nu + e^- + \gamma$), spin precession or spin flips ($\nu_L + \gamma \rightarrow \nu_R$, $\nu_L \rightarrow \nu_R + \gamma$). Neutrinos have no electric charge suggesting, perhaps, they are composed only of an equal number of Planck dipoles and there is no spin precession.

The three known types of neutrinos, electron neutrino ν_e, muon neutrino ν_μ, and tau neutrino ν_τ have an electroweak interaction cross-section which is an indication or measure of size of quantized radii. The neutrino size is a function of internal energy and energy density of the surrounding medium. The electron neutrino ν_e, for example in the vortex model considered, is approximately 1/1000 the size of an electron with a core radius on the order of the Planck length. A neutrino may acquire electric charge in magnetized matter. If neutrinos have magnetic moments, they can precess in a transverse magnetic field. Heavier neutrinos may undergo radiative decays into lighter neutrinos $\nu_l \rightarrow \nu_k + \gamma$. In propagation through matter, resonant flavor mixing and transitions may occur such as oscillations generated by phase difference in masses in passage through varying density EM fields. This matter effect, known as the Mikheyev-Smironov-Wolfenstein (MSW) effect, is attributed to the charged current coherent scattering due to interaction with electrons. Apparent flavor change of neutrinos in propagating beams may correspond to a resonance conversion involving diametral change or change in internal relative phase shift (dipole twist).

27. Bound Particle States

If a neutrino has a vortex ring geometry, how is it generated? Comparison of the decay of the neutral pion ($\pi_0 \rightarrow \gamma + \gamma$) and the charged pion ($\pi^1_0 \rightarrow \nu + \nu$), for example, is suggestive of a possible sequence. A charged pion π^1_0 consisting of an electron e⁻↓ and positron e⁺↑ with opposite spin and magnetic moments ↓↑ undergoing annihilation as the electron and positron spiral infall and collide may be expected to produce opposed jets of Planck dipoles in the plane of rotation perpendicular to the symmetry (rotation) axis. The jets entrain Planck dipoles into a set of vortex rings $\nu\uparrow + \nu\downarrow$ ejected outward with opposite momenta. A neutral pion π_0 consisting of an electron e⁻↑ and positron e⁺↑ with aligned spins but opposite magnetic moments ↓↑ upon collision generate a pair of photons $\gamma\uparrow + \gamma\downarrow$ ejected in outward directions along the symmetry axis. The vortex core radius corresponds approximately to the interaction cross-section which is on the order of the Planck length ($l_P = 1.616 \: 10^{-35}$ m).

Observed neutrino spin, linear momentum and helicity characteristics are illustrated in Fig. 27-12. A vortex ring model is depicted in Fig. 27-13. A speculative model of the three neutrino types ν_e, μ_e, and τ_e formed from Planck dipoles is depicted in Fig. 27-14. The toroidal major diameter corresponds approximately to the particle collision impact cross-section at the time of neutrino formation. Neutrinos interact weakly with matter owing to their small size, mass, and lack of electric charge. For comparison, Storti et al, **QE3**[34] estimates neutrino radii as shown in Table 27-3 based on mass estimates relative to the electron assuming a common ω_Ω cut-off frequency.

Massive neutrinos have been proposed as a possible candidate for hot dark matter (~0.02-2 eV) in order to explain the balance of positive kinetic energy of the Hubble expansion of the universe and the negative gravitational energy of mass attraction. Apparent neutrino oscillations detected by the Super Kamiokande detector experiment have been posited to be the result of neutrino oscillations induced by passage through the Earth's interior. Neutrinos interacting with matter can result in production of electrons and protons as end products as absorbed energy is converted into matter. Given the large neutrino influx bombarding Earth, Meyl[80] hypothesizes that the conversion of neutrinos and fusion materialization of elementary particles takes place in the Earth's core and it is this process responsible for the apparent expansion of the Earth (e.g., sea floor spreading, rift valley stretch marks, continued volcanism, prehistoric species gigantism, day lengthening, increase in the acceleration gravity, etc) over the last 200 million years.

What a strange world we live in... Said Alice to the Queen of Hearts (Lewis Carrol)

There are 300 neutrinos in every cubic centimeter of the entire universe. – Stanley Wojcicki

The primary ingredient of reality contains a metric field that gives space-time rigidity and causes gravity. – Frank Wilczek

There can be no question that is the Schrödinger Equation we very nearly have the mathematical foundations for solution of the whole problem of atomic and molecular structures. – Gilbert Lewis

Young man, if I could remember the names of these particles, I would have been a botanist. – Enrico Fermi

We classify nature into a coherent system which appears to do what we say it does. – James Burke

The universe never did make much sense; I suspect it was built on government contract. – Robert A. Heinlein

27. Bound Particle States

Table 27-2 Summary of Electromagnetic Properties of Neutrinos.

Neutrino ν	Spin s	Electric Charge q	Quantum radius[1] R_q	Electroweak charge radius[2] $\langle r_\nu^2 \rangle_{EW}$	Mass Energy m	Magnetic dipole moment[3] μ_ν	Toroid dipole moment[4] a_ν
ν_e	1/2	0	< 1.43E-26 m	≈ 3.2 – 4.9E-35 m^2	< 2.2 eV/c^2	$\mu\nu_e$ < 1.5E-10 μ_B	$a_{\nu e}$ ≈ 6.8E-36 m^2
ν_μ	1/2	0	< 1.85E-28 m	≈ 1.7 – 3.2E-35 m^2	< 170 eV/c^2	$\mu\nu_\mu$ < 6.E-10 μ_B	$a_{\nu\mu}$ ≈ 4.0E-36 m^2
ν_τ	1/2	0	< 2.03E-33 m	≈ 1.0 – 2.3E-35 m^2	< 15.5 MeV/c^2	$\mu\nu_\tau$ < 3.9E-7 μ_B	$a_{\nu\tau}$ ≈ 2.5 E-36 m^2

1. $R_q = \hbar c/E$ 2. $r_{\nu e}^2 = n \times 10^{-35}$ m^2.   3. Bohr magneton $\mu_B = e\hbar/2m_e$.   4. Anapole moment $a_\nu| = 1/6(r_{\nu L}^2)$ for massless ν.

We can perhaps best describe the neutrinos as little bits of spin-energy that have got detached. I am not much impressed by neutrino theory. – Sir Arthur Stanley Eddington

The electron is for us the archtype of [an] isolated parcel of energy... – Louis de Broglie

Neutrinos ...win the minimalist contest: zero charge, zero radius, and very possibly zero mass. – Leon M. Lederman

500 billion neutrinos pass through you each second no matter how dense. – Major Carter. No matter how dense? – C ol.Jack O'Neill

Reversible processes are not, in fact, processes at all, they are sequences of states of equilibrium. The processes which we encounter in real life are always irreversible processes. – Arnold Sommerfeld

It therefore appears unavoidable that physical reality must be described in terms of continuous functions in space. – Albert Einstein

27. Bound Particle States

Neutrino characteristics

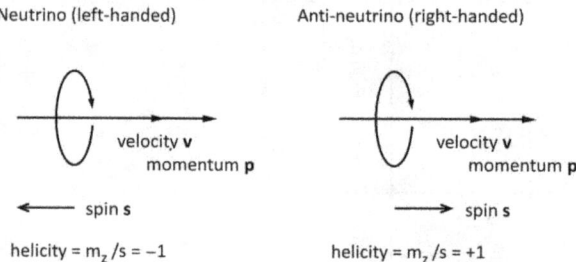

Fig. 27-12. Neutrino and antineutrino spin, linear momentum and helicity characteristics. The difference in helicity suggests a possible matter interaction effect and may account for apparent preponderance of matter over anti-matter in the observable universe.

Neutrino vortex spin model

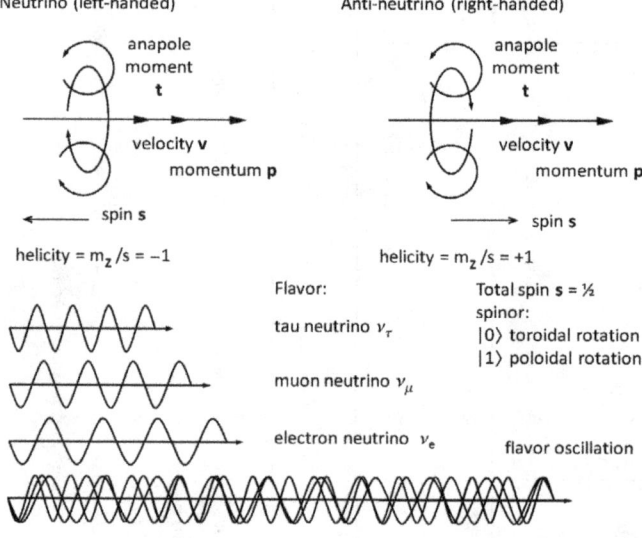

Fig. 27-13. Neutrino and antineutrino Helmholtz vortex ring spin model. As there is no precession and an even number of Planck dipole masses M_P^+, M_P^-, there is no net electrical charge. Analogous to common vortex 'smoke-rings', propagation direction is unidirectional. Vortex rings are produced by a high velocity jet. Neutrinos are produced in processes that impart high energy such that neutrinos travel at nearly the velocity of light. Topologically, a torus may be transformed into a sphere. However, like smoke rings, neutrinos evidently do not undergo such a transformation in a collision. In the limit for m = 0, chirality equals helicity. Flavor oscillations between the electron, muon and tau neutrinos may be due to torsional windup/unwind as a result of matter interaction depending on helicity yielding a nondegenerate eigenmode anomalous dispersion.

27. Bound Particle States

Table 27-3. Neutrino radii					
Description	Symbol	Mass Energy	Units	Radii	Units
Electron neutrino	ν_e	< 3	eV/c^2	2.7499E-20 - 9.5397E-20	m
Muon neutrino	ν_μ	< 0.19	MeV/c^2	6.5524E-19 - 7.6557E-19	m
Tau neutrino	ν_τ	< 18.2	MeV/c^2	1.9587E-18 – 2.8205E-18	m

Neutrino Planck dipole vortex ring

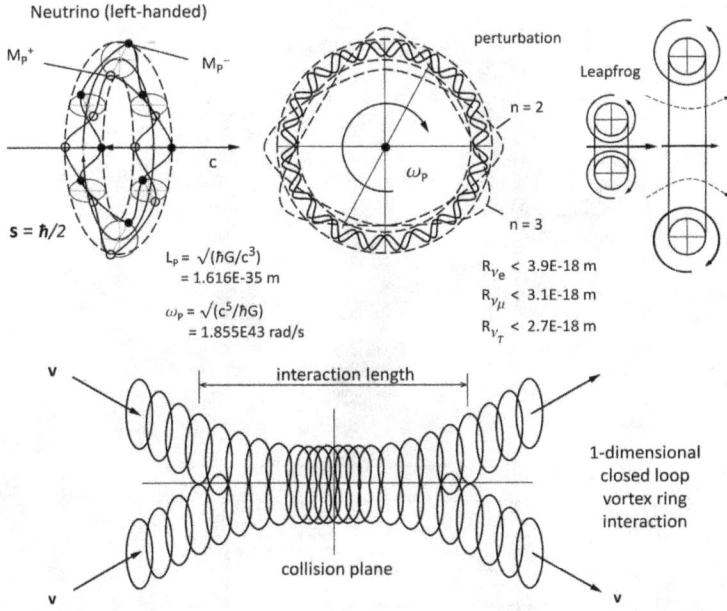

Fig. 27-14. Model of a neutrino formed by a dipole vortex ring composed of spinning Planck dipoles. A Planck mass current loop is dimensionally equivalent to an electric current. Hence, a neutrino magnetic moment is expected. For a Dirac neutrino of mass m_ν, a magnetic dipole moment μ_ν (= $(1/16\pi^2)(m_e m_\nu/M_W^2)\mu_B$) is on the order 3×10^{-19} μ_B. A magnetic moment will lead to neutrino precession in a sufficiently intense magnetic field resulting in a $\nu_L \rightarrow \nu_R$ transition. An equal number of positive and negative Planck masses averages to zero mass. The small effective mass of the neutrino may represent a torsion oscillation creating wavefunction interference → rational/irrational winding no. oscillation.

A billion neutrinos go swimming in heavy water: one gets wet. – Michael Kamakana

Hey, if you'd been listening, you'd know that Nintendos pass through everything. – Col. J. O'Neill

27. Bound Particle States

Supposing, for a moment, the neutrino is in the form of a relativistic spinning vortex ring of spin ½, how does a neutrino undergo spin reversal into an anti-neutrino in an interaction such as $v_L + \gamma \rightarrow v_R$ or $e^- + v_e \rightarrow W^- \rightarrow e^- + \bar{v}_e$. In the vortex ring model described, neutrino spin reversal sequence is as follows: On encountering the increased electromagnetic energy density field of the electron or a photon, the neutrino flattens, momentarily increasing effective mass. The constituent Planck dipoles each reorient spin along the direction of the density gradient while remaining phase-locked. On emerging into a region of decreased energy density, the ring decompresses and the Planck dipoles spin flip reversing the spin direction from left hand to right-hand. The spin flip sequence for an arbitrary number of dipoles is illustrated in Fig. 27-15. For a Dirac spinor roton, a minimum of rotating two dipoles diametrically opposed orbiting a common mass center is required. Absorption of the energy of the neutrino v_e is briefly stored in the W^- vector boson field prior to antineutrino $\bar{v}_e$ emission. The mass energy of the electron neutrino (<2 eV/c²) corresponds to the gravitational field mass of the Planck dipole spinor field.

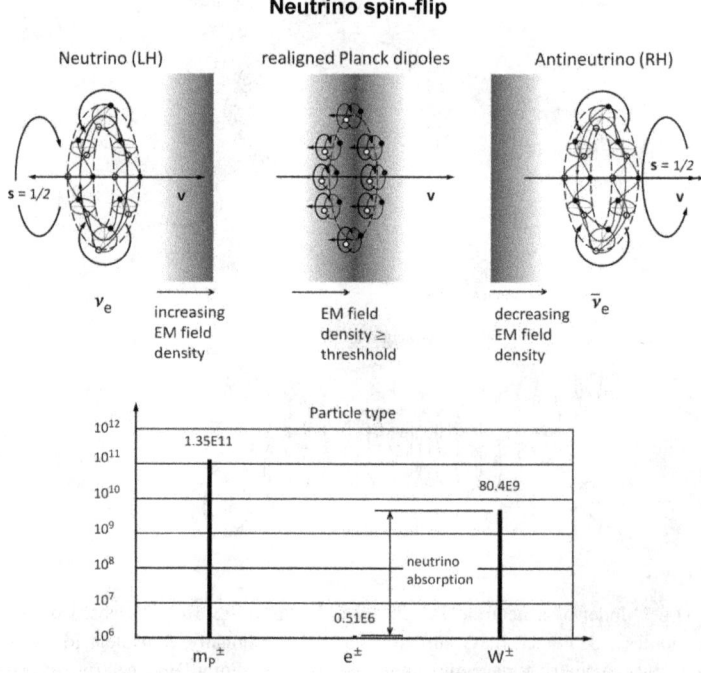

Fig. 27-15. Hypothetical electron neutrino spin-flip sequence for transformation from LH to RH spin. Chirality remains Lorentz invariant, helicity reverses.

I have done a terrible thing. I have postulated a particle that cannot be detected. – Wolfgang Pauli

Intuition is more important to discovery than logic. – Henri Poincaré

Let no one unversed in geometry enter here. – Plato's academy motto.

28. Lagrangian

The Lagrangian function (L) represents a single function description of the differential energy of motion (action) equal to the kinetic energy (T) minus the potential energy (V). The action is equal to the kinetic energy minus the potential energy summed over the history. The Lagrangian of a free particle path describes a trajectory of least action that follows a worldline geodesic independent of coordinate system. The action (S) for a system described by a Langrangian function L is represented by an integral over time

$$S = \int L\,dt = \int (T - V)\,dt \tag{28-1}$$

where:

S	Action		
	Value: Varies (≥ 0)	Units: joule·sec	Dimensions: ML^2T^{-1}
L	Lagrangian		
	Value: Varies (≥ 0)	Units: joules	Dimensions: ML^2T^{-2}

The action is invariant under a Lorentz transformation.

In a gravitational potential field, the Lagrangian of motion of a mass (m) with velocity (v) at height (h) above the Earth's surface is

$$L = T - V \tag{28-2}$$
$$= \tfrac{1}{2}mv^2 - V = \tfrac{1}{2}m((dx/dt)^2 + (dy/dt)^2 + (dz/dt)^2) - mgh$$

where:

L	Lagrangian		
	Value: Varies (≥ 0)	Units: joules	Dimensions: ML^2T^{-2}
T	kinetic energy ($= \tfrac{1}{2}mv^2$)		
	Value: Varies	Units: joules	Dimensions: ML^2T^{-2}
V	potential energy ($= mgh$)		
	Value: Varies	Units: joules	Dimensions: ML^2T^{-2}

For a relativistic single point particle of mass m with electric charge q in a potential ϕ with velocity **v**, the Lagrangian is $L = -m_0 c^2 \sqrt{(1 - v^2/c^2)} - q(\phi - \mathbf{v}\cdot\mathbf{A})$. Force $\mathbf{F} = m\mathbf{a} = \delta L/\delta x$.

Mass-energy Lagrangian for the electromagnetic field is $L = \tfrac{1}{2}m(\mathbf{E}^2 - c^2\mathbf{B}^2) = E_{s0} - E_{m0}$. Zero mass corresponds to $L = (\mathbf{E}^2 - c^2\mathbf{B}^2) = 0$ where $E = pc$. Negative mass corresponds to $L < 0$ where the magnetostatic energy E_{m0} exceeds electrostatic energy E_{s0} or equivalently $\mathbf{E}^2 < c^2\mathbf{B}^2$. The Lagrangian and Hamiltonian of a matter wave is illustrated in Fig. 28-1.

The great masters of modern analysis are Lagrange, Laplace, and Gauss, who were contemporaries. It is interesting to note the marked contrast in their styles. Lagrange is perfect in form and matter, he is careful to explain his procedure, and though his arguments are general they are easy to follow. Laplace on the other hand explains nothing, is indifferent to style, and, if satisfied his results are correct, is content to leave them either with no proof or with a faulty one. Gauss is as exact and elegant as Lagrange, but even more difficult to follow than Laplace, for he removes every trace of the analysis by which he reached his results, and studies to give a proof which while rigorous shall be as concise and synthetical as possible. – W. Rouse Ball

28. Lagrangian

Lagrangian and Hamiltonian

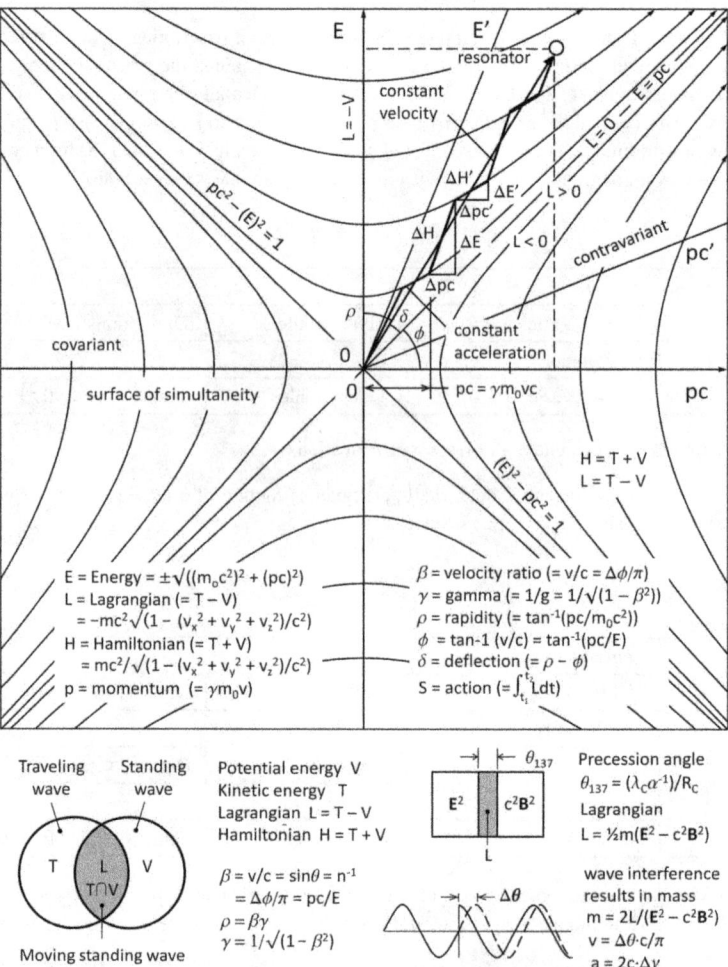

Fig. 28-1. Representation of a de Broglie matter wave illustrating the Lagrangian L and the Hamiltonian H. The energy-density scalar represents the total system energy. Only when there is overlap of potential and kinetic energy is there mass-energy of interaction. Mass is a measure of wave function dissonance. The path or trajectory followed is a geodesic for which the action is minimized (principle of least action). Action for a material particle is defined as the integral of the particle's velocity along its spacetime trajectory. Action S has dimensions of [energy]·[time] or alternatively [flux x charge]. The integrand L of the action integral is the Lagrangian with dimensions of energy. Least action of the path integral may be understood as a result of destructive interference of electromagnetic waves.

Mathematics is a game where the rules are invented by mathematicians. Physics is a game where the rules are invented by nature but they have to guess them. – Kurt Friedrich Gödel

29. Hamiltonian

In the classical mechanics phase space momentum concept formulated by William R. Hamilton in 1833, both position and velocity are taken as independent variables. In a closed system, the total energy is represented by the Hamiltonian function H as two equations in terms of position q and momentum p in phase space.

$$\dot{p}_I = -\delta H/\delta q_i \quad (29\text{-}1)$$
$$\dot{q}_I = \delta H/\delta p_i \quad (29\text{-}2)$$

Three coordinates are required to define position (x, y, z) and three dimensions are required for momentum (p_x, p_y, p_z). Each point in phase space represents the energy state of a physical system at a moment in time. Confining a particle to a small region of space, increases the momentum and, hence, pressure on the confining walls. Position and momentum are independent variables and cannot be simultaneously determined. The equations $\delta p/\delta t = -\delta H/\delta q$, $\delta q/\delta t = \delta H/\delta p$ are called Hamilton's equations of motion. Pairs of variables, like position and momentum, whose operators do not commute AB ≠ BA cannot be determined simultaneously. Given p and q at a certain instant, the instantaneous rates of change of p and q, denoted by $\dot{p}$ and $\dot{q}$, may be determined from the Hamiltonian: $\dot{p}$ = [p, H], $\dot{q}$ = [q, H]. The Heisenberg uncertainty principle states $\Delta x \cdot \Delta p_x \geq \hbar/2$ where $\hbar$ is the reduced Planck's constant (= $h/2\pi$). A similar relation for energy and time is $\Delta E \cdot \Delta t \geq \hbar/2$. Momentum p (= $2\pi\hbar/\lambda$) varies inversely with wavefunction wavelength λ. An increase in uncertainty of momentum p_x results as the position is localized over a smaller Δx.

The Heisenberg commutation relation in terms of noncommutating operators p and q is given by

$$[\mathbf{p},\mathbf{q}] = qp - pq = i\hbar \, \mathbf{I} \quad (29\text{-}3)$$

where [ ] denotes the Poisson bracket, boldface type represents matrices, and **I** denotes the identity matrix. This relation contains the quantization of the Heisenberg's uncertainty principle and shows that the position and momentum operators do not commute. The metric for Euclidean space $[|p - q|]^2 = (p - q)^*(p - q)$. The taxi-cab distance between two vectors **p**, **q** is $|p_1 - q_1| + |p_2 - q_2|$. The Hamiltonian H is a function of position, momentum and time representing the total energy of a system which is the sum of the kinetic and potential energy. The potential energy represents hidden kinetic energy.

$$H = f(\mathbf{q}, \mathbf{p}, t) = \Sigma p_i q_i = T + V \quad (29\text{-}4)$$

where:

H	Hamiltonian (= $\sqrt{(m^2c^4 + c^2(\mathbf{p} - q\mathbf{A})^2 + \phi)}$ = ½m($\mathbf{E}^2 + c^2\mathbf{B}^2$))		
	Value: Varies (≥ 0)	Units: joules	Dimensions: ML^2T^{-2}
T	Kinetic energy		
	Value: Varies	Units: joules	Dimensions: ML^2T^{-2}
V	Potential energy		
	Value: Varies	Units: joules	Dimensions: ML^2T^{-2}

One may view the world with the p-eye and one may view it with the q-eye... – Wolfgang Pauli

29. Hamiltonian

The Hamiltonian operator for a free particle of mass m in a potential U is given by

$$\hat{H} = \hat{p}^2/2m + U(\hat{x}) \tag{29-5}$$
$$\hat{H}\psi = [-(\hbar^2/2m)\nabla^2 - e^2/r]\psi = i\hbar\delta\psi/\delta t$$

where: ^ indicates that it is a Hermitian operator (termed an observable in QM where the eigenvalues are real numbers), $U(x)$ = potential, ψ = wave function, $\nabla^2 = \delta^2/\delta x^2$, $\hat{p}$ and $\hat{x}$ are momentum and position operators, respectively. Hamiltonians are a method for finding the minimum of a given equation and oft used to determine path of least action. The Hamiltonian for a harmonic oscillator in phase space is illustrated in Fig. 29-1. The Schrödinger wave equation $H\psi = E\psi$ is an eigenvalue equation where the Hamiltonian H is a differential operator, the energy E is the eigenvalue of the operator corresponding to the observable and the wavefunction ψ is the eigenfunction representing the system state.

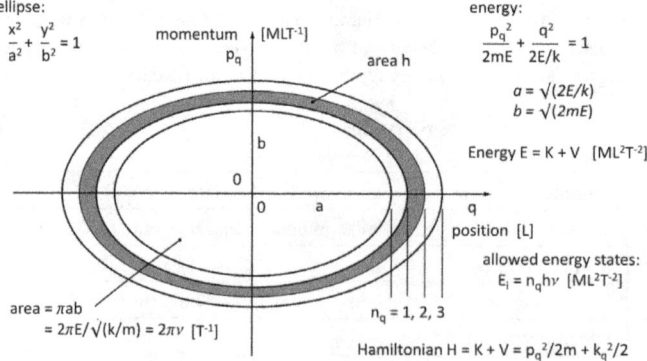

Fig. 29-1. Phase space diagram of position q vs. momentum p of a harmonic oscillator. Energy $E = H(q,p) = (\ddot{q}/\hat{p})(q^2 + p^2)$. Indeterminacy of position and momentum in accordance with the uncertainty principle limits $\Delta p\Delta q \geq \hbar/2$. The difference $pq - qp = [p,q] = \hbar/I$ remains constant. The area h remains invariant in a Lorentz boost.

30. Laplacian

30.1 Wave function curvature.

The curvature of a wave function ψ is measured by the Laplacian $\nabla^2\psi$ and represents flux density of the gradient of energy flow. Mass-energy and gravitation are associated with positive curvature (compression of spacetime volume). Curvature varies inversely with wavelength ($k = 1/\lambda$). A positive (convex) curvature $\delta f^2/\delta x^2 > 0$ corresponds to E < V where $\psi > 0$ and E > V where $\psi < 0$. A negative (concave) curvature $\delta f^2/\delta x^2 < 0$ corresponds to E > V where $\psi < 0$ and E < V where $\psi > 0$. The Laplacian may be thought of as a measure of the concavity of the wave function. The in-surface (intrinsic) curvature K on a 2D surface is related to the radius of curvature by the Gaussian curvature

$$K = k_1 k_2 = -1/(\text{rafter x swing}) = -1/\rho_1\rho_2 \qquad [L^{-2}] \qquad (30\text{-}1)$$

where k_1 and k_2 are the principal curvature and ρ_1 and ρ_2 are the principal radii of curvature, respectively.

The extrinsic curvature is given by

$$K = k_1 + k_2 = 1/\rho_1 + 1/\rho_2 \qquad [L^{-1}] \qquad (30\text{-}2)$$

The mean curvature is given by

$$M = (k_1 + k_2)/2 \qquad [L^{-1}] \qquad (30\text{-}3)$$

For n-dimensional surfaces, the generalized Gaussian curvature is

$$K_n = k_1 k_2 \ldots k_n \qquad [L^{-n}] \qquad (30\text{-}4)$$

where k_n = principal curvatures.

The del operator is called the Laplacian. The Laplacian operator ∇^2 equals the divergence of the gradient of a function: $\Delta q = \nabla^2 q = \nabla \cdot \nabla q$. For an electrostatic potential ϕ, the charge distribution is given by $q = \Delta\phi$. Similarly for a gravitational potential ϕ, the mass distribution is given by $m = \Delta\phi$. Divergence is the second derivative of a gradient. The divergence of the gradient of a scalar field is $-\nabla \cdot \nabla \phi = \nabla^2 \phi = 0$.

In Cartesian coordinates, the Laplacian operator in three dimensions is given by

$$\nabla^2 = \nabla \cdot \nabla = \delta^2/\delta x^2 + \delta^2/\delta y^2 + \delta^2/\delta z^2 \qquad (30\text{-}5)$$

The divergence of the gradient of a function f is

$$\nabla^2 f = \Delta f = \delta^2 f/\delta x^2 + \delta^2 f/\delta y^2 + \delta^2 f/\delta z^2 \qquad (30\text{-}6)$$

Laplace's equation is $\nabla^2\psi = 0$ which is a partial, second-order, homogenous differential equation. For the electric potential ϕ, $\nabla^2\phi = 0$ and vector potential **A**, $\nabla^2\mathbf{A} = 0$ where $\rho = 0$ and **J** = 0, respectively. The one-dimensional wave equation is $\nabla^2\psi = 1/v^2 \; \delta\psi^2/\delta t^2$ where ψ is position. In Schrödinger's wave equation, the Laplacian of the wave function is $\nabla^2\psi(r,t)$ where $|\psi(r,t)|^2$ is interpreted as a probability amplitude.

30. Laplacian

A generalization of the Laplacian in flat Minkowski 4D spacetime is the d'Alembertian operator:

$$\Box = 1/c^2 \delta^2/\delta t^2 - \delta^2/\delta x^2 - \delta^2/\delta y^2 - \delta^2/\delta z^2 \qquad (30\text{-}7)$$

The d'Alembertian represents a 4D gradient and is measure of the difference between the value of a function at a point and the average value of adjacent points in the region indicative of the local rate of change. The wave equation is given by $\Box A^\mu = 0$. An analog of the continuous Laplace operator is the discrete or mesh Laplace operator that finds application in loop quantum gravity.

The Laplacian is oft utilized in signal/image filter processing. A Gaussian filter is typically employed first to remove noise and then a Laplacian of the Gaussian (LoG) is performed to detect zero crossing edge detection. See Fig. 30-1. The Laplacian may be approximated with a difference of Gaussians. The Laplacian of the Gaussian may be approximated by a Mexican hat wavelet.

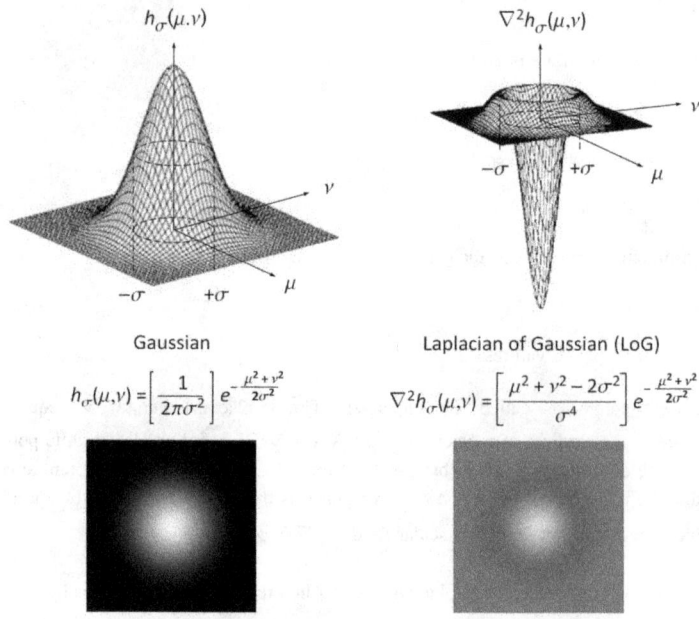

Fig. 30-1. Comparison of Gaussian and Laplacian of Gaussian profiles.

30.2. Laplace equilibrium surfaces.

Laplace, a mathematician and astronomer, in 1808 developed a theory describing the motion of the Galilean satellites of Jupiter showing the orbital plane of the satellites lies on a warped Laplace surface depending on the distance from the planet, the planet's orbital plane and quadrapole precession due to oblateness. The general equilibrium surface contour is illustrated in Fig. 30-2. A similar Fedora hat brim warping may be seen in the

30. Laplacian

galactic disc plane of Andromeda M31 attributed to gravitational perturbations induced by disc penetrations by satellite galaxies M32 and M110.

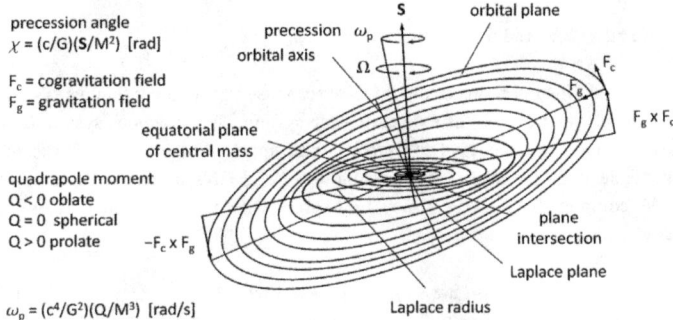

Fig. 30-2. Warped Laplace equilibrium surface of satellite orbits lying between planet equatorial and orbital planes due to oblateness-induced quadrupole precession.

30.3 Laplace transforms

A Laplace transform is a linear transformation that converts a time-domain function f(t) of a real variable t (time) into a function F(s) of a complex variable s (frequency) in the s-domain making it useful for analysis of linear, dynamical systems involving linear ordinary differential equations. The Laplace transform of a function f(t) for all real numbers t ≥0 is the function F(s) evaluated from 0 to ∞.

$$\mathcal{L}[f(t)] = F(s) = \int f(t)e^{-st}dt \qquad (30\text{-}8)$$

where s is a complex frequency ($= \sigma + i\omega$).

While the Laplace transform is used to simplify analysis of continuous differential equations, the z-transform facilitates analysis of discrete difference equations. A z-transform is a discrete time equivalent of the Laplace transform. The z-transform of a discrete sequence h(n) represented as $H(z) = H(re^{j\omega}) = P(z)/Q(z)$ a ratio of polynomials of the complex variable z is commonly applied to determine time-domain responses and stability of discrete control systems and digital filters. The characteristic equation is the denominator of the transfer function. Given the H(z) transfer function of a system, the function's pole location determine system stability. Evaluating the z-domain transfer function, the H(z) modular surface gives the system's frequency response. The poles and zero locations determine the critical frequencies to avoid instability. For a transfer function in continuous time defined as a ratio of two polynomials H(s) = N(s)/D(s), poles and zeros of a transfer function are the frequencies in which the value of the denominator and numerator become zero, respectively. Zeros are roots obtained by setting the numerator N(s) = 0 and solving for s. Poles are the roots of the denominator D(s) obtained by setting D(s) = 0 and solving for s. An example of a transfer function is impedance of a filter. A zero corresponds to a short circuit while a pole corresponds to an open circuit.

"Il est facile de voir que..." [It is therefore obvious that...] – Pierre-Simon Laplace

30. Laplacian

For the generalized transfer function

$$H(s) = [(s-z_1)\cdot(s-z_2)\ldots \cdot(s-z_n)]/[(s-p_1)\cdot(s-p_2)\ldots \cdot(s-p_n)] \tag{30-9}$$

the zeros are $z_1, z_2, \ldots z_n$ and the poles are $p_1, p_2, \ldots p_n$.

Stability can be defined in terms of the impulse response $y_\delta(t)$ of a continuous system or the Kronecker delta response $y_\delta(k)$ of a discrete-time system. A continuous system is stable if its impulse response $y_\delta(t)$ decays to zero as time approaches infinity. A discrete-time system is stable if its Kronecker delta response $y_\delta(k)$ decays to zero as time approaches infinity. A complex function $P(z)$ may be expressed (e.g., via Taylor's Theorem) as a power series

$$P(z) = \Sigma a_j z_j = a_0 + a_1 z + a_2 z^2 + a_3 z^3 + \ldots + a_n z^n = 0 \tag{30-10}$$

where z is a complex variable evaluated from 0 to ∞. The stability of a system is determined by the roots of the characteristic polynomial equation. The system stability condition is satisfied when all the roots of the characteristic equation have a magnitude less than one, i.e., within the unit circle $|z| = 1$ in the z-plane. A pole-zero plot can represent either a continuous time (CT) or discrete-time (DT) system. For a CT system, the poles and zeros are mapped in the s-plane of a Laplace transform. For a DT system, poles and zeros are mapped in the z-plane. A mapping of the Laplace s-plane onto the z-plane is illustrated in Fig. 30-3. The system impulse transient response is composed of steady-state output, exponential and sinusoidal terms which vary with location of roots of the characteristic equation.

A *first-order* system refers to a system involving only one energy storage element for which the governing equation is a first-order differential equation. First-order refers to the order of the highest derivative present. An example of a first order system is an RC circuit which exhibits an exponential current decay ($I = I_0 e^{-t/RC}$). If there are two energy storage elements such as an RLC circuit, the characteristic equation is of *second-order*. The natural response of a second order system is determined by the roots of the characteristic equation. The general solution is $I = A_1 e^{s_1 t} + A_2 e^{s_2 t}$. The values of R, L, C are all real and positive, but the roots s_1 and s_2 may be real, complex or purely imaginary. A system described by an equation with complex roots that are complex conjugates (exponentially decaying sinusoid) is said to be "oscillatory" or "underdamped" whereas when the roots are real, negative and distinct (sum of two decaying exponentials), the system response is said to "overdamped". The limiting condition when the roots are real and equal represents the "critically damped" case. In general, roots are located in the complex plane referred to as the s-plane or complex frequency plane (s = σ + jω) in units of (1/s = Hz). A zero of the complex impedance function Z(s) V(s)/I(s) indicates the possibility of an internal current without an applied voltage. A pole of the impedance function indicates the possibility of a voltage without an applied current. An example of representative H(z) pole locations and discrete time-domain impulses responses are illustrated in Fig. 30-4. The transient response of a closed-loop feedback control system can be represented in terms of the location of poles of the transfer function.

Reason is a supple nymph, and slippery as a fish. She had as leave give her kiss to an absurdity any day, as to syllogistic truth... – D.H. Lawrence

30. Laplacian

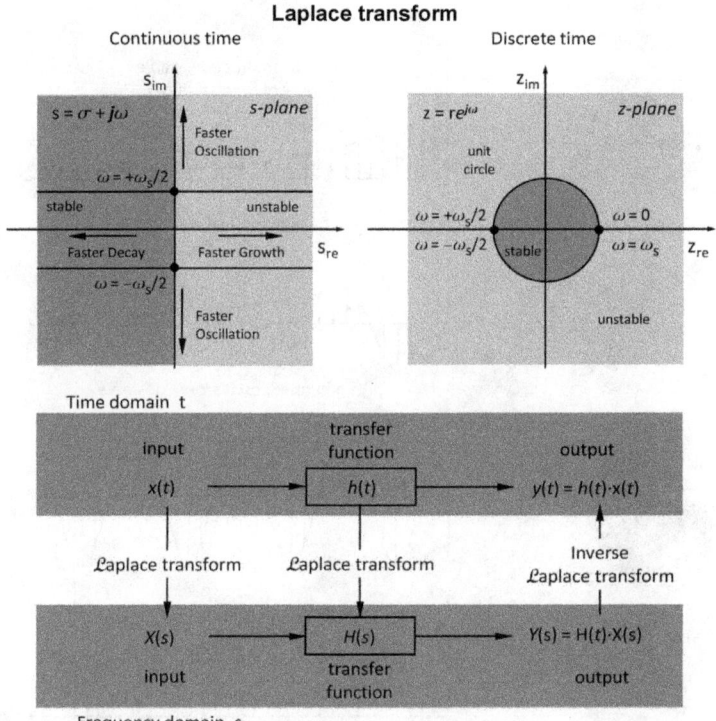

Fig. 30-3. Laplace s-plane of a complex frequency mapped to the complex z-plane.

The Laplace transform X(s) is used to simplify analysis of continuous differential equations while the z-transform X(z) facilitates analysis of discrete difference equations. Dynamic systems characterized by continuous time variables exhibit stability in the left half s-plane where the roots of the characteristic equation all have negative real components of the complex frequency $s = \sigma + j\omega$.. For systems characterized by discrete time variables, the stability region lies within the unit circle in the z-plane where the roots of the characteristic equation all have magnitudes less than unity. Poles located within the zone of convergence inside the unit circle denote region of stability whereas poles located on or outside the unit circle mark region of instability.

Such is the advantage of a well constructed language that its simplified notation often becomes the source of profound theories. – Pierre-Simon Laplace

Amplifiers are oscillators that don't and oscillators are amplifiers that do. – R.F. Anon.

Technology is anything that doesn't quite work yet. – Danny Hillis

In any physical science, we must ascend from facts to laws, by way of induction and analysis; and we must descend from laws to consequences, by the deductive and synthetical way. – W. R. Hamilton

30. Laplacian

Pole-zero map

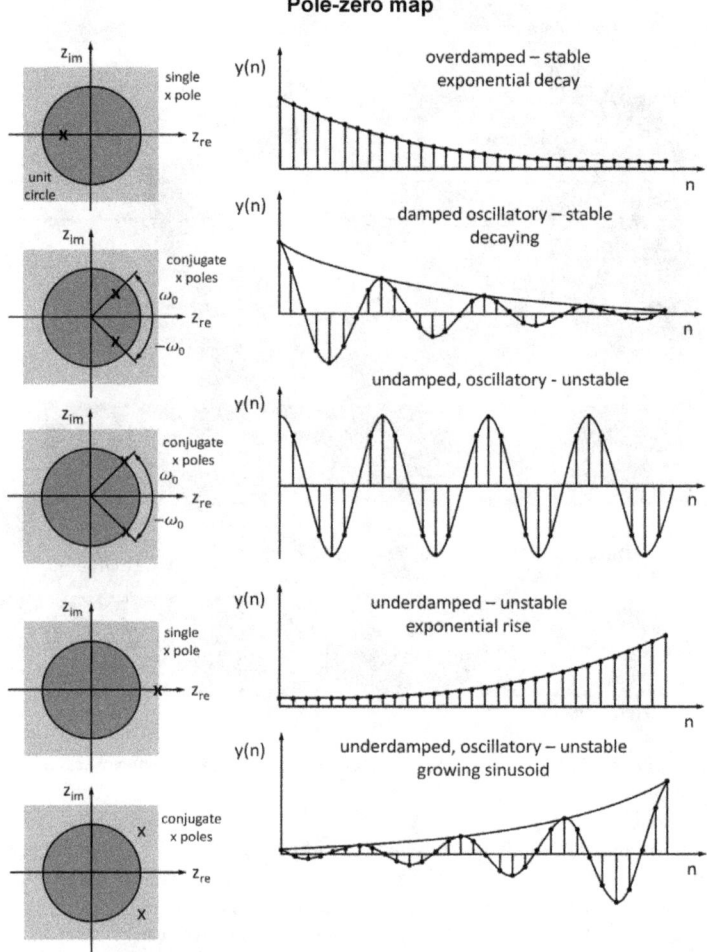

Fig. 30-4. Example pole-zero map of pole locations and discrete time domain impulse frequency responses for various stable and unstable conditions are illustrated in the z-plane. The poles and zeros represent the input-output characteristics of the system transfer function. A transfer function $H(s) = N(s)/D(s)$ is defined as the relation between the system output $N(s)$ to the system input $D(s)$. If the transfer function is known, the system response $N(s)$ can be found by taking the inverse Laplace transform of $H(s) \cdot D(s)$. The system gain K together with the transfer function define the system dynamic response. A digital resonator is a two-pole bandpass filter with a pair of complex conjugate poles near the unit circle.

> ...all these [real and virtual processes] problems of vacuum polarization, etc. could be studied by studying the real processes such as pair production to which they are related as absorption is to dispersion. The real processes represent residues at the poles of some complex function. The virtual processes give the remainder of the description of the function, which should however be determined by the character of its poles... – R. P. Feynman

31. Tensors

In geometrical and physical applications, it always turns out that a quantity is characterized not only by its tensor order, but also by symmetry. – Hermann Weyl

A tensor is a mathematical object that represents the generalization of a vector in which the magnitude and direction can vary at each point in a field independent of any selected frame of reference. A tensor **T** is a linear operator or function which given an input **v** outputs another tensor **T·v** of one lower order. A zero-order tensor is a scalar T. A tensor of order one is a linear function that maps every vector into a scalar. A first order tensor corresponds to a vector which may be either a covariant tensor T or a contravariant tensor T^*. A tensor of order two is a linear function that maps every vector into a vector. A second order tensor is a gradient of a vector function and can be represented as a matrix denoted by two subscripts or superscript indices T_{ij}, T^{ij} or $T_i{}^j$. The T_{ij} are co-variant components, T^{ij} are contravariant components, and $T_i{}^j$ are mixed components of **T**. Rank is indicated by the number of different indices. A 2nd order or rank-2 tensor has magnitude and two directions and can be described by a 3 x 3 spatial matrix such as a stress or strain tensor which combined result in a 4th rank tensor. A tensor of order three is a linear function that maps every vector into a dyadic, a product of a pair of vectors **ab**. Any order three tensor can be represented as a triadic, a linear combination of triads of three adjacent vectors **uvw**. Tensors need not be resolved along triads of base vectors that are orthonormal, but may be generalized to a general basis denoted by (**g**$_1$, **g**$_2$, **g**$_3$) that are neither orthonormal or of unit length. The reciprocal basis (**g**^{1}, **g**^{2}, **g**^{3}) is related to the original basis by the Kronecker tensor $\delta_{\text{I}}{}^j$ which is a generalization of the Kronecker delta δ_{ij}, $g_{ik}g^{kj} = \delta_{\text{I}}{}^j$. A metric tensor is used to interconvert vectors and one-forms.

Some tensors are associated with metric properties of a manifold such as the metric tensor $g_{\mu\nu}$ which relates coordinates to the invariant 4D space-time interval or separation between events. The symbol $g_{\mu\nu}$ represents the matrix components of a second rank tensor where the lowered indices u identifies the row and v identifies the column of the matrix. The sum of the diagonal elements of the matrix (metric coefficients) is the signature or trace. For a flat metric of Minkowski spacetime denoted by the symbol $\eta_{\mu\nu}$, the signature is $-1, 1, 1, 1 = 2$ corresponding to the Lorentz invariant interval ds where $ds^2 = -dt^2 + dx^2 + dy^2 + dz^2 = \eta_{\mu\nu}dx^\mu dx^\nu$.

$$x^2 + y^2 + z^2 - t^2 = \eta_{\mu\nu}x^\mu x^\nu \qquad (31\text{-}1)$$
$$\eta_{00} = -1, \ \eta_{11} = \eta_{22} = \eta_{33} = 1$$

A four vector V in spacetime is given by $V^{\alpha\beta\gamma\delta} = (V^0, V^1, V^2, V^3)$ where the Greek letter indices denote space-time. A three vector V in space is given by $V^{ijk} = (V^1, V^2, V^3)$ where the Latin indices denote spatial components. Using the Einstein summation convention, a vector V may be expressed in terms of a basis vector $V = V^a e_a$ where V is called a one-form vector represented with contravariant components and e_α is the basis vector. Covariance and contravariance refers to how the components of a vector transform under a change in basis. Basis vectors are frame dependent. The one-form and the basis vectors are duals. In Euclidean 3-space, tensors are called Cartesian and the covariant and covariant indices are equivalent. Using the metric, the covariant components of V$_a$ may be obtained from the contravariant vector v^b by lowering the indices $V_a = g_{ab}V^b$. Similarly $V^a = g^{ab}V_b$ is obtained by raising the indices. In Einstein notation, contravariant components are denoted with

315

31. Tensors

raised indices $\mathbf{v} = v^i e_i$. Coordinates, velocity vectors, acceleration vectors and physical quantities are generally represented with raised indices while derivatives and gradients are shown with lowered indices $\mathbf{v} = v_i e^i$. Some other types of tensors describe the orientation of curves such as geodesics with tangent vectors or 2D surfaces such as bivectors. Tensors are functions that map vectors or one-forms to a real number which will be the same in any reference frame. For example, the curvature of a Riemannian manifold is given by the Riemannian curvature tensor R_{ijkl}. The Riemann curvature tensor equals the sum of the Weyl and Ricci tensors ($R_{ijkl} = C_{ijkl} + R_{ij}$). For physical materials, the strain tensor describes the local distortion while the stress tensor describes the internal force distribution. Piezoelectricity is described by a third rank tensor and strain gradient effects are described by a fifth rank tensor. For covariant differentiation, the Christoffel operator ∇_μ maps a tensor field into another denoted by $\nabla_i \phi$ which describes the rate at which ϕ varies.

In General Relativity (GR), the Einstein field equation is given by

$$G_{\mu\nu} = (8\pi G/c^4)T_{\mu\nu} = \kappa T_{\mu\nu} = R_{\mu\nu} - \tfrac{1}{2} g_{\mu\nu} R + \Lambda g_{\mu\nu} \tag{31-2}$$

where

$G_{\mu\nu}$	Riemann-Einstein tensor (curvature of spacetime)
G	Newton's Gravitational constant (= $G_0/\Gamma^3 = c^2 R_{univ}/M_{univ}$)
c	speed of light (= c_0/Γ)
$T_{\mu\nu}$	stress-energy-momentum tensor (ν flux of μ momentum)
$R_{\mu\nu}$	Ricci curvature tensor (= $g_{\mu\nu} R_{\mu\nu}$ = trace component of Riemann tensor $R^\rho{}_{\sigma\mu\nu}$)
$g_{\mu\nu}$	Einstein metric tensor (= $\eta_{\mu\nu} + h_{\mu\nu}$)
R	Ricci curvature scalar (= $R(\Gamma)$) = trace of Ricci tensor (= $R_{00} + R_{11} + R_{22} + R_{33}$)
Λ	cosmological constant = vacuum energy density (= $8\pi G \rho_\Lambda$)

The stress-energy-momentum tensor $T_{\mu\nu}$ describes the distribution of mass and energy. Matter and radiation may be represented by the stress-energy tensor as a perfect fluid:

$$T_{\mu\nu} = (\rho + p)u_\mu u_\nu - p g_{\mu\nu} \tag{31-3}$$

where

$T_{\mu\nu}$	stress-energy-momentum tensor (= T_0/Γ)
u_μ	fluid velocity (= u_0/Γ)
ρ	energy density (= ρ_0/Γ)
p	fluid pressure (= p_0/Γ)
$g_{\mu\nu}$	metric tensor ($g_{00} = 1/\Gamma^2$). $g_{00} = (1 - 2GM/c^2 r)$; $h_{00} = -2GM/c^2 r$

Do not worry about your difficulties in mathematics. I can assure you mine are still greater. – Albert Einstein

If I had inherited a fortune I should probably not have cast my lot with mathematics. – Joseph-Louis Lagrange

Ist das wirklich so? – Albert Einstein

31. Tensors

In a 4 x 4 matrix of vector components of the $T_{\mu\nu}$ tensor function, T_{00} represents energy density; T_{01}, T_{02}, T_{03} denote energy flux; T_{10}, T_{20}, T_{30} denote momentum density; T_{12}, T_{13}, T_{30} denote shear stress and T_{21}, T_{31}, T_{32} denote momentum flux. The trace of the tensor T_{11}, T_{22}, T_{33} denotes a pressure with dimensions equal to energy density. The Ricci scalar curvature (volume-reducing) tensor is proportional to pressure. A false vacuum corresponds to positive energy and negative pressure. Dark energy is hypothesized as a perfect fluid with negative pressure ($p + 3\rho < 0$) attributed as a possible cause for expansion of the universe. In the EGM model, the external quantum vacuum pressure acting on an enclosed mass locally compresses the EM spectrum into higher frequencies and fewer modes. The compressed EGM spectrum within the mass provides an opposing radiation pressure. The distortion in the surrounding Bose-Einstein Condensate (BEC) vacuum composed of positive and negative Planck masses M_P^+, M_P^- induces a distortion field with accelerative curvature similar to a polaron lattice defect.

The energy tensor can be regarded only as a provisional means of representing matter. In reality, matter consists of electrically charged particles. It is only the circumstance that we have no sufficient knowledge of the electromagnetic field of concentrated charges that compels us, provisionally, to leave undetermined in presenting the theory, the true form of this tensor. – Albert Einstein

Nothing is settled. Everything can still be altered. – Claude Levi-Strauss

Ask the next question. – Theodore Sturgeon

I was sitting in a chair in the patent office in Bern when all of a sudden a thought occurred to me: "If a person falls freely he will not feel his own weight." I was startled. This simple thought made a deep impression on me. It impelled me toward a theory of gravitation. – A. Einstein

Gravity is that field which corresponds to a gauge invariance with respect to displacement transformations. – Richard Feynman

Young man, in mathematics you don't understand things. You just get used to them. – John von Neumann

It's nice to know the computer understands the situation, but I would like to understand it too. – Eugene Wigner

Mathematics has the inhuman quality of starlight, brilliant and sharp, but cold. – Hermann Weyl

The notion of parallelism in Riemannian spaces permitted one to give a geometrical meaning to the fundamental operation of covariant derivative. – Tullio Levi-Civita

Symmetry ... is one idea by which man through the ages has tried to comprehend and create order, beauty and perfection. – Hermann Weyl

What we know is not much. What we do not know is immense. – Pierre-Simon Laplace

Time and concentration allow the intellect to perceive a ray of light in the darkness of the most complex problem. – Santiago Ramòn y Cajal

If you are receptive and humble, mathematics will lead you by the hand. Again and again, when I have been at a loss how to proceed, I have just had to wait until I have felt the mathematics led me by the hand. It has led me along an unexpected path, a path where new vistas open up, a path leading to new territory, where one can set up a base of operations, from which one can survey the surroundings and plan future progress. – P. A. M. Dirac

32. Nonuniform Force Fields

32.1 Dielectrophoretic force

Nonconductive dielectric materials immersed in a nonuniform electric field undergo polarization and experience a net electric translational dielectrophoretic force acting in the direction of the maximum field gradient.

$$\mathbf{F}_{DEP} = q(\mathbf{E} + \Delta\mathbf{E}) - q(\mathbf{E} - \Delta\mathbf{E}) = q(2\Delta\mathbf{E}) \qquad (32\text{-}1)$$

Substituting p (= qΔx) for the induced electric dipole moment yields

$$\mathbf{F}_{DEP} = \mathbf{p}(2\Delta\mathbf{E}/\Delta x) \approx \mathbf{p}(d\mathbf{E}/dx)_{max} \qquad (32\text{-}2)$$

In the differential limit for a small sphere, the dielectrophoretic force is proportional to the electric field gradient. See Fig. 32-1. The dielectrophoretic force is weaker than electrophoretic force experienced between opposite charges. The translational force is directed toward region of increased field intensity regardless of the polarity of the applied field, hence, a net thrust force is produced as well for an applied AC signal. In this respect, diamagnetism is similar to gravity which is always attractive. A similar effect occurs for weakly magnetic (paramagnetic) materials with unpaired electrons

$$\mathbf{F}_m = \mu(2\Delta\mathbf{B}/\Delta x) \approx \mu(d\mathbf{B}/dx)_{max} \qquad (32\text{-}3)$$

where μ = the induced magnetic dipole moment. A paramagnetic body immersed in a non-uniform magnetic field displaces a volume of the medium of permeability μ and undergoes a change in potential energy. For a diamagnetic material, the translational force is repulsive. In a superconductor, as a result of the Meisner effect, Cooper pairs of electrons form creating a magnetic field in response to the externally applied magnetic field, the superconductor becomes perfectly diamagnetic expelling magnetic fields. The Biefeld-Brown effect associated with force effects of non-ionic nonuniform electric fields and graded, high K dielectrics in asymmetrical capacitors may be attributed in part to dielectrophoresis consistent with the comprehensive nonuniform dielectric field theory developed by Dr. H. A. Pohl. This was further confirmed in communications and visit with T. Townsend Brown at his Catalina island laboratory by the author.

The polarizability of a material medium is a function of frequency and described by a complex permittivity

$$\epsilon^* = \epsilon_0 \epsilon_r - i\sigma/\omega \qquad (32\text{-}4)$$

where
- ϵ_0 = permittivity of free space
- ϵ_r = relative permittivity
- σ = electrical conductivity
- ω = applied angular frequency of electric field (= $2\pi f$)

The dipole moment may be expressed as

$$\mathbf{P} = 4\pi r^3 \epsilon_m [(\epsilon_p - \epsilon_m)/(\epsilon_p + 2\epsilon_m)] \mathbf{E}_0 \qquad (32\text{-}5)$$

where
- ϵ_p = permittivity of particle
- ϵ_m = permittivity of medium
- r = particle radius
- E = electric field strength

32. Nonuniform Force Fields

The dielectrophoretic force is then

$$F_{DEP} = (P \cdot \nabla)E \qquad (32\text{-}6)$$

The time-averaged dielectrophoretic force on a spherical particle immersed in a dielectric fluid is given by

$$\langle F_{DEP} \rangle = 2\pi r^3 \epsilon_m [\epsilon_p - \epsilon_m / \epsilon_p + 2\epsilon_m] \nabla E^2 = 2\pi r^3 \epsilon_m Re[K(\omega)] \nabla E^2 \qquad (32\text{-}7)$$

where
- ϵ_p = permittivity of particle
- ϵ_m = permittivity of medium
- r = particle radius
- E = electric field strength
- $Re[\ ]$ = real part of bracketed term
- $K(\omega)$ = Clausius-Mossotti factor = $[\epsilon_p - \epsilon_m / \epsilon_p + 2\epsilon_m]$

The direction of the dielectrophoretic force F_{DEP} is positive towards the region of higher field strength. If the particle is more polarizable than the surrounding medium where $Re[K(\omega)]$ term is positive and oppositely directed if negative. For a nonmaterial polarizable medium such as the physical vacuum, a somewhat similar force arises as a result of a gradient in the electromagnetic energy density. As the EM field strength is very much weaker than that associated with matter, the translational forces are many magnitudes weaker. In a material medium, DEP trapping and levitation provides an alternative to optical tweezers for micro-manipulation of small scale objects.

Shulgin[82] observes that a point electric charge attracts a dielectric with a force

$$F = (\epsilon_r - 1/\epsilon_r + 1)(e/2Z_0) \qquad (32\text{-}8)$$

where
- ϵ_r = relative permittivity
- e = electrical charge
- Z_0 = distance from charge to dielectric

Nonuniform potential energy fields give rise to force fields acting on immersed objects in the direction of maximum scalar potential field gradient satisfying Laplace's equation:

$F = -m(dU/dx)_{max}$	Gravitation (Electrogravimagnetic)	$g = -\nabla \phi_g$
$F = +p(dE/dx)_{max}$	Electrostatic (Dielectrophoresis)	$E = -\nabla \phi_E$
$F = +\mu(dB/dx)_{max}$	Magnetostatic (Magnetophoresis)	$B = -\nabla \phi_B$

By analogy with a dielectric material, the Planck vacuum acts as a polarizable medium and is compressed by the action of gravitation. The negative vacuum energy density arises from the distortion of the Zero-point field (ZPF) producing a gravitationally squeezed vacuum quantum state. Compression of the Planck vacuum increases both the electric permittivity and the magnetic permeability of the vacuum thus increasing EM field energy density and reducing local light velocity $c = 1/\sqrt{(\epsilon \mu)} = 1/\sqrt{(\Gamma \epsilon_0 \Gamma \mu_0)} = c_0/\Gamma = c_0/\sqrt{K_{PV}}$.

"Dielectrophoresis" is defined as the motion of matter caused by polarization effects in a nonuniform field. – Herbert Ackland Pohl

A definition of the electrogravitic force might be "the ponderomotive force developed with a high K-dielectric under electrical strain." – Thomas Townsend Brown

32. Nonuniform Force Fields

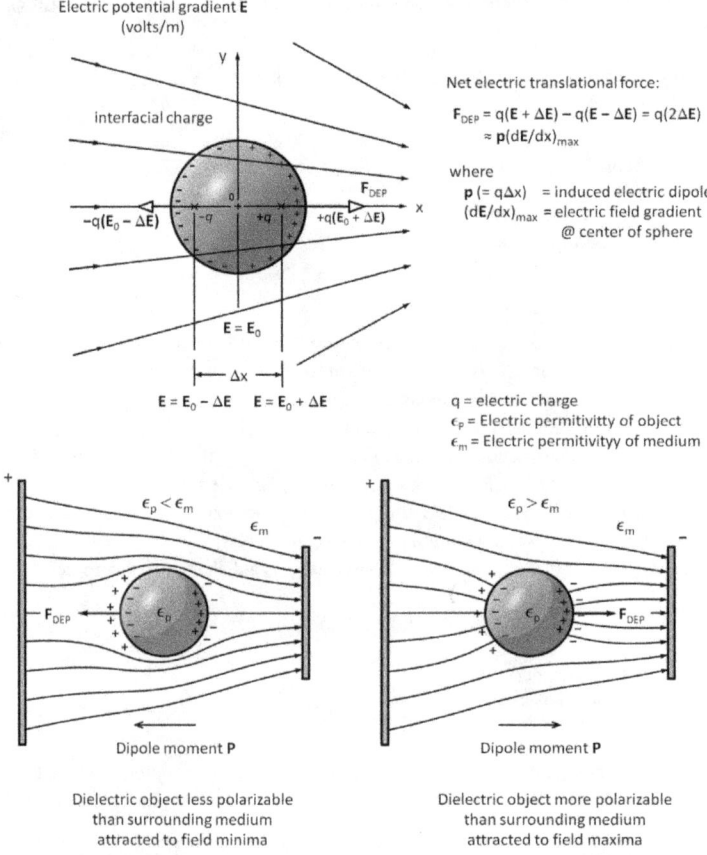

Fig. 32-1. Dielectrophoretic force, F_{DEP}, acting on a dielectric sphere in a non-uniform electric field. An external electric field induces separation of charges creating an electric dipole moment. For a particle more polarizable than the medium, the net electric translational force is directed toward the region of higher field strength regardless of the external field polarity or frequency.

The typical frequency response of a polarizable medium and dielectrophoretic force F_{DEP} on an immersed particle varies with frequency as illustrated in Fig. 32-2. Nonlinear polarization is typically expressed as a power series of electric field intensity **E** nonlinear susceptibilities $\chi^{(n)}$ as $\mathbf{P} = \epsilon_0(\chi^{(1)} + \chi^{(2)}\mathbf{E}^2 + \chi^{(3)}\mathbf{E}^3 + \ldots)$. The dielectrophoretic force propels particles toward the electric field maxima or minima depending on the sign of the Clausius-Mossotti (K_{CM}) factor and increases in strength with increasing excitation frequency difference. If a suspended particle has polarizability greater than the medium, F_{DEP} is positive and will push the particle towards the region of higher electric field strength. If a suspended particle polarizability is less than the immersion medium, F_{DEP} is negative and the particle will be driven towards the region of lower field strength. The degree of polarization is a measure of directivity and represents the difference between an isolated state and a coupled state.

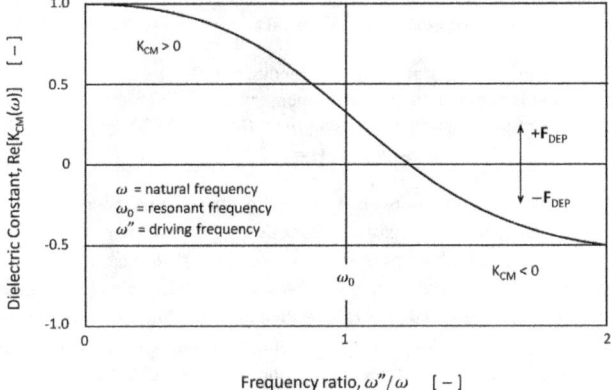

Fig. 32-2. General frequency response characteristic of a polarizable medium varies with frequency as a result of molecular, ionic and atomic effects. At resonance frequency, the dielectric constant $K_{CM} = 0$ where the phase lag between input and response is 90°. The dielectrophoretic force F_{DEP} becomes positive at frequencies greater than resonance. At high frequencies, the response approaches 180° out-of-phase with the frequency of the applied electric field as the medium effectively saturates.

Ferroelectric materials, usually ceramics such as barium titanate, exhibit dielectric hysteresis and often a piezoelectric electric effect. Certain electret waxes may also retain a weak permanent polarization. These properties are analogous to ferromagnetism.

32.2 Van der Waals forces

Van der Waal forces are short-range intermolecular and interatomic forces that are electrostatic in origin and may be attractive or repulsive. There are three types: 1) dipole-dipole electrostatic attractions between polar molecules with permanent dipole moments; 2) dipole-induced polarization (induction) and attraction of nonpolar molecules by polar molecules with a permanent dipole moment (also known as Debye force) and; 3) London dispersion attractive forces between non-polar molecules or atoms arising due to instantaneous multipole interactions (average of square of dipole moment $\langle |\mathbf{p}|^2 \rangle$ is non-zero due to charge fluctuations). The more electrons a molecule has, the greater its polarizability. The greater the mass of the molecule, and hence, more electrons, the higher the melting and boiling points.

It's a force field. Try to relax. – Dr. Who

It is an electrogravitic field acting on all the parts simultaneously. – "Towards Flight without Stress or Strain... or Weight", Interavia 11(5), May, 1956: p. 373-4.

32. Nonuniform Force Fields

32.3 Magnetophoretic force

Analogous to dielectrophoresis, magnetophoresis results in object motion with respect to a surrounding medium due to a nonuniform field gradient interaction of the induced magnetization by an external field. The magnetophoretic force varies as

$$\mathbf{F}_m = (\mu_m \mathbf{m}) \cdot \nabla \mathbf{H} = \nabla(\mathbf{m} \cdot \mathbf{B}) \qquad [\text{N}] \qquad (32\text{-}9)$$

where
- μ_m = magnetic permeability of the medium (= B/H)
- $\mathbf{m}$ = induced magnetic dipole moment (= $\pm\mu_m \mathbf{B}_0$)
- $\mathbf{H}$ = external magnetic field strength (= $\mathbf{B}/\mu_0$ = n·E/Z_0)
- $\mathbf{B}$ = magnetic flux density (= $\mu_0 \mathbf{H} = -\nabla\Phi_m$)

The magnetization of materials is classified as to type: diamagnetic, paramagnetic and ferromagnetic. Diamagnetism (negative magnetic susceptibility) is attributed to orbital motion of electrons in an atom producing a magnetic field that opposes an external magnetic field in accordance with Lenz's Law. A superconductor expels a magnetic field as a result of the Meissner effect and is perfectly diamagnetic. Paramagnetism (weak positive magnetic susceptibility) is due to unpaired electrons resulting in a magnetic moment that is capable of being aligned in the direction of the applied field. Ferromagnetism (large magnetic susceptibility) is explained by formation of domains consisting of a group of atoms with magnetic moments aligned in the same direction. When the domains are aligned, the magnetization is maximized (magnetic saturation) and insensitive to further magnetization by an external field. The magnetic moment of ferromagnetic atoms is attributed to the spin of unfilled inner shell of electrons. Other forms of magnetism include antiferromagnetism and ferrimagnetism. A ferromagnetic material is attracted to a region of higher magnetic field strength. A Halbach array of oriented magnets may be used to augment or diminish magnetic field strength. A paramagnetic material is weakly repelled by an external magnetic field and, if sufficiently diamagnetic, such as bismuth or pryrolytic graphite, may be levitated in a magnetic field. See Fig. 32-3.

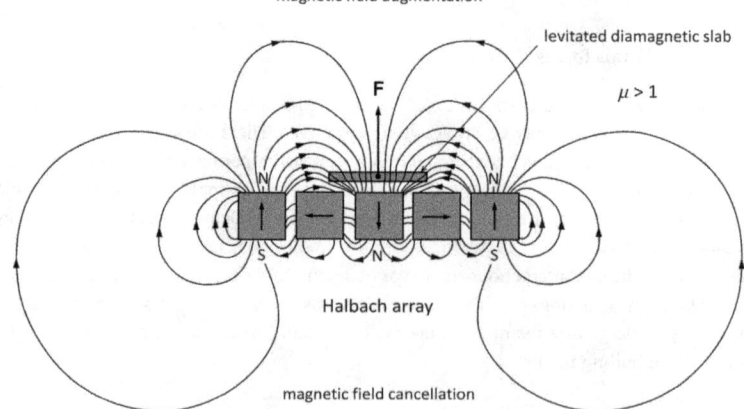

Fig. 32-3. Diamagnetic levitation effect. The magnetic flux density asymmetry is induced by oriented magnetic dipole moments (spatial phase rotation).

32. Nonuniform Force Fields

32.4 Electromagnetic Magnus effect

An electrostatic motor comprised of charging electrodes and metal disc with an intervening dielectric experiences a torque inducing rotary motion in the vicinity of a magnetic field. Moving charges in a magnetic field undergo deflection as a result of a Lorentz force $\mathbf{F} = q^+\mathbf{v}_e \times \mathbf{B}$ and a torque due to interaction of the electron magnetic dipole moment and the external magnetic field. The disc torque varies as $T = BQ\omega R^2/4$ where $\mathbf{B}$ = the magnetic flux density, Q = induced charge, ω = angular velocity and R = disc radius. The disc may be levitated above a magnet with a lift force $\mathbf{L} = BQ\omega R/2$. An electromagnetic plane incident on a rotating conducting cylinder imparts a lift force perpendicular to the rotation axis and the propagation direction of the incident wave analogous to the hydrodynamic effect. The EM field distortion is a measure of the pressure imbalance.

An optical Magnus effect discovered by Zel'dovich in 1990 consists of propagation of a wave of right and left circular polarization in an inhomogeneous medium in different directions. A comparable optical effect (spin Hall effect) occurs with light by virtue of photon spin inducing wavefront deflection. A circularly polarized EM wave propagating in an inhomogeneous medium undergoes a twist about the propagation axis.

32.5 Ponderomotive (radiation pressure) force

For a charged particle in a nonuniform alternating field, a nonlinear ponderomotive force arises given by

$$\mathbf{F}_P = (e^2/4m\omega^2)\nabla \mathbf{E}^2 \qquad [N] \qquad (32\text{-}10)$$

where
- e = electrical charge ($= F/E = \hbar k/A = E/V$)
- m = particle mass ($= F/a = W/g = \hbar k/v = \Gamma m_0 = \gamma m_0 = |V \cdot e|/c^2$)
- ω = angular frequency of the external field ($= 2\pi f = \omega_0/\Gamma$)
- E = electric field strength ($= F/q^+ = \mathbf{B} \cdot c$)

The ponderomotive force $\mathbf{F}_P$ is in the direction of the field minima regardless of the charge polarity. The charge experiences a Lorentz force $\mathbf{F}_L = e(\mathbf{E} + \mathbf{v} \times \mathbf{B})$ where the velocity $v = eE/m\omega \cos(\omega t)$. Radiation pressure has the same dimensions as EM energy density [Pa].

Ponderomotive acceleration of electrons using ultra-high intensity lasers has been experimentally demonstrated with E-fields of $10^{14} - 10^{18}$ V/m yielding accelerations of $\sim 10^7 - 10^{30}$ g's. For comparison, the vacuum breakdown field strength $E_c = 2m_e^2 c^3/\hbar e \approx 10^{18}$ V/m. Materialization of an electron-positron pair from the vacuum can occur above the Schwinger electric field intensity limit of $E_{cr} = 1.3 \times 10^{16}$ V/m ($= m_e^2 c^3/q_e \hbar$) which is ~ 137 times the electric field strength of an electron at the Compton radius.

32.6 Travelling wave electromagnetic fields

Phase-displacement coupled with space-displacement gives rise to travelling waves. A group of objects separated in space and moving cyclically and out-of-phase relative to each other produces a travelling wave. This effect can produce motion such as that of a millipede by the travelling wave action of its legs or of electrocatapults, rail guns and levitated high speed monorail vehicles propelled by a linear induction motor. Linear induction machines with suspension capability are in wide use in a variety of MagLev applications using attractive and/or repulsive levitation. The magnetic flux interaction in a simple repulsive levitation scheme is depicted in Fig. 32-4 in which a travelling current carrying coil induces

32. Nonuniform Force Fields

an opposing current in a stationary coil. In a coil-loop arrangement, on approach, increasing flux linkage generates an opposing current in the stationary conductor loop. When directly aligned, the magnetic flux no. is changing and the opposing current begins to decay due to resistance. On retreat, decreasing flux linkage induces an opposite e.m.f. cancelling the remaining current. In a coil-sheet arrangement, during relative motion, an opposing eddy current is induced in the conductive sheet (stator). With increasing velocity, most of the magnetic flux is repelled from the conductive sheet generating a pattern equivalent to interaction with a current image.

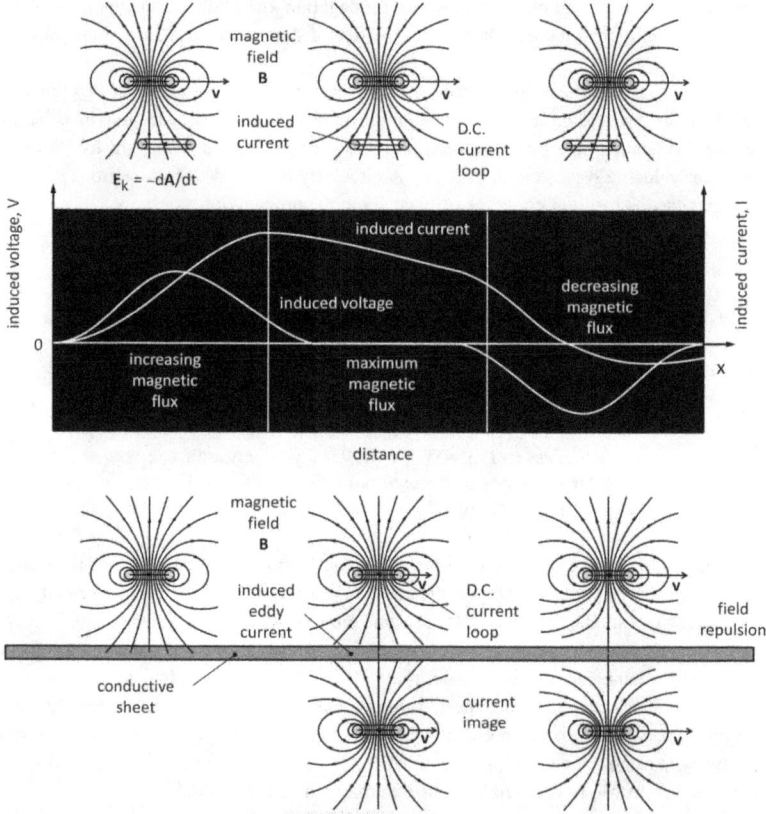

Fig. 32-4. Repulsive levitation may be demonstrated using direct current electromagnetic coils interacting with inductive loops or conducting, nonmagnetic sheets. Forces arise with changes in scale (symmetry breaking) due to phase and frequency shift.

> My curiosity in new machines was first aroused in 1948 when I learned that Westinghouse of America had launched aircraft by means of an induction motor which operated in straight line. – Eric Laithwaite

> Electrogravitics must be understood as an entirely new field of scientific investigation and technology development. – Will Cady (ONR)

32. Nonuniform Force Fields

Electromagnetic levitation effects can readily be produced by nonuniform, travelling wave electromagnetic fields by multiphase alternating current. As an example, a linear induction 3-phase A.C. motor providing magnetic levitation (Maglev) and propulsion as a "magnetic river" was developed in the late 1960s by Eastham and Laithwaite. The general concept is depicted in Fig. 32-5 illustrating a Maglev design consisting of a two pole transverse arrangement of linear induction alternating current motor magnets with stacked U-shaped ferromagnetic core plates. Lateral stability is achieved with staggered coils with transversely aligned pole pairs and a U-shaped ferromagnetic core acting as a waveguide with two eddy currents one above each magnetic pole. Each pole represents a sudden change in reluctance in the magnetic circuit. Levitation with a single electromagnet may be effected with electroptic position sensing coupled with PD current control.

In another example of electrodynamic suspension, Laithwaite[82] describes an apparent 'anti-magnetic' effect consisting of levitation of a nonmagnetic aluminum sphere above an electromagnetic coil (Bedford levitator). The effect may be understood as arising from the effects of opposed travelling magnetic fields with varying phase gradient. An alternating current applied to an inner and outer coil arrangement with an opposing inner coil current that lags in phase. The inner and outer coil currents moves in opposite directions. An upward travelling magnetic field generates a lift force that is counterbalanced by an inner downward travelling magnetic field that acts to provide radial stability. The auxiliary field generated by the inner coil produces an inward radial force component that offsets the outward travelling field. Levitation of an aluminum sphere by superposition of opposed travelling wave electromagnetic fields illustrating phase progression is depicted in Fig. 32-6. The alternating induced eddy currents are proportional to the magnitude of and in the direction of the electrokinetic field. A conductive free rotating cylinder test probe provides an indication of the wave direction.

The coil phase currents generate an electrokinetic field $\mathbf{E}_k$ $(= -d\mathbf{A}^*/dt = \mathbf{F}_k/q_0)$ in the sphere producing induced eddy currents in opposition to the direction of the coil current flow. The direction of the electrokinetic field is opposite in sign of the time derivative of the inducing current providing an explanation of Lenz's law. The induced currents are accompanied by a circumferential magnetic induction field $\mathbf{B}_i$ which varies with time. The direction of the induced magnetic field is given by the Fleming left-hand rule in which the thumb points in the direction of current. The resultant magnetic field $\mathbf{B}$ acting on the sphere at any given instant equals the superposition of the external magnetic fields $\mathbf{B}_e$ generated by the outer and inner coil windings and the induced magnetic field $\mathbf{B}_i$ generated by eddy current flow. The induced eddy current loops coincide with lines of constant flux phase ϕ of the resultant magnetic field.

I have no idea how it works. – Dr. Edward Teller

It is not knowledge, but the act of learning, not possession but the act of getting there, which grants the greatest enjoyment. – Carl Friedrich Gauss

Creativity is intelligence having fun. – Albert Einstein

The true sign of intelligence is not knowledge but imagination. – Albert Einstein

Truth like a torch, the more 'tis shook, it shines. – Sir William R. Hamilton

It is the lone worker who makes the first advance in a subject: the details may be worked out by a team, but the prime idea is due to enterprise, thought, and perception of an individual. – Alexander Fleming

32. Nonuniform Force Fields

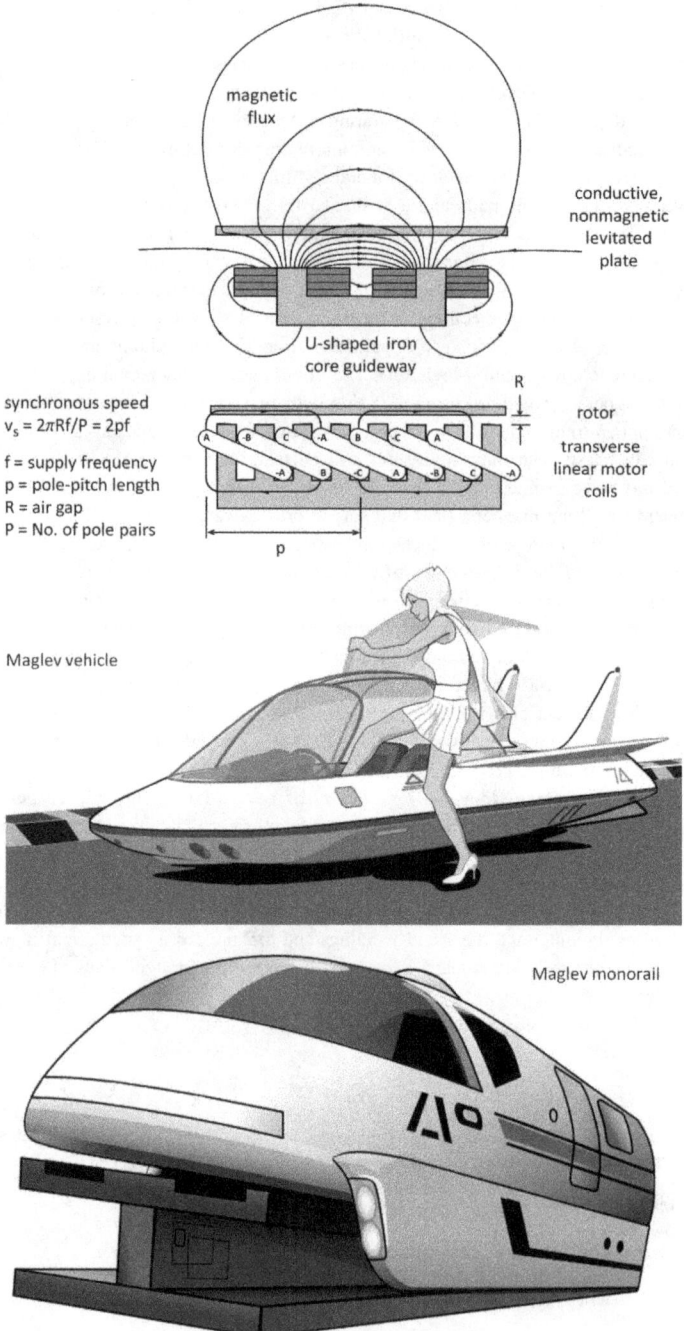

Fig. 32-5. Maglev electromagnetic travelling wave propulsion and levitation.

32. Nonuniform Force Fields

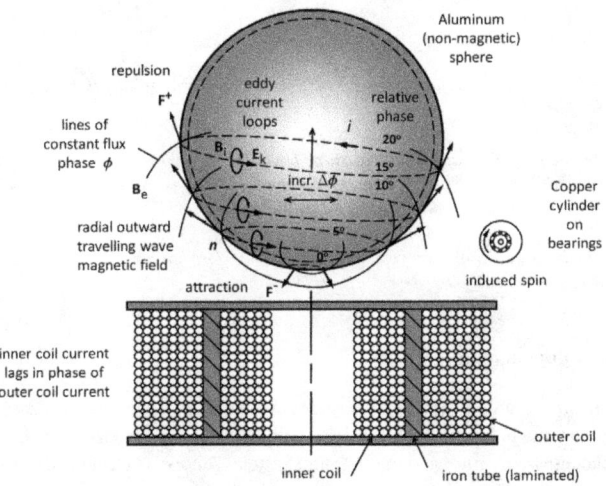

Inductance $L = N^2 \mu A/l$ [V·s/A]
Magnetic potential energy $U_B = \frac{1}{2} L i^2$ [J = N·m]]
Electrostatic potential energy $U_E = \frac{1}{2} C V^2$ [J = N·m]
Magnetic energy density $u_B = U_B/\text{vol.} = B^2/2\mu_0$ [J/m³]
Electrostatic energy density $u_E = U_E/\text{vol.} = \frac{1}{2}\mu_0 E^2$ [J/m³]
Energy density $T^{00} = (E^2 + B^2)/8\pi$ [J/m³]
Poynting flux $T^{0i} = (E \times B)/4\pi$ [V²·s/m³]
Faraday 2-form $F = dA = E_z dz \wedge dt + B_z dx \wedge dy$ [m/rad]
Poynting unit vector $\boldsymbol{n} = \tanh 2\alpha = 2|E \times B|/(E^2 + B^2) = T^{0i}/T^{j0}$

Electromagnetic field energy density ρ_{em}
 $= \epsilon_0 E^2/2 + B^2/2\mu_0$ [J/m³]
Lorentz force $\mathbf{F} = q\mathbf{E} + q\mathbf{v} \times \mathbf{B}$ [N]
Magnetic field $\mathbf{B} = \mathbf{B}_o(t) + \mathbf{B}_i(t) = \nabla \times \mathbf{A}$ [Wb/m²]
Electrokinetic field $\mathbf{E}_k = -d\mathbf{A}/dt$ [V/m]
Electrokinetic force $\mathbf{F}_k = q\mathbf{E}_k$ [N]
Acceleration of gravity $\mathbf{g} = \mathbf{W}/m$ [m/s²]
Acceleration $\mathbf{a} = \mathbf{F}/m = (F/mc^2)\cdot(E/B)^2$ [m/s²]
Velocity $v = \tanh \alpha$ [m/s]

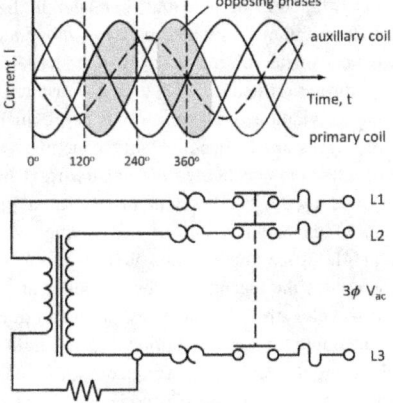

Fig. 32-6. Levitation by polyphase electromagnetic induction. [After Laithwaite]

In nature we find the oldest forms of propulsion, none of which employs wheels. – Eric Laithwaite

Badness $\propto$ Resistance x Reluctance. Goodness = 1/Badness. – E.R. Laithwaite

Art is the method of levitation, in order to separate one's self from enslavement by the earth. – Anais Nin

Think of the happiest things. It's the same as having wings. – J. M. Barrie (Peter Pan)

Section 3 - Gravity

33. Gravitation

Sir Isaac Newton when asked how he discovered the law of gravity is said to have replied "By thinking about it all the time." – Isaac Newton

I consider it quite possible that physics cannot be based on the field concept, i.e., on continuous structures. In that case, nothing remains of my entire castle in the air, gravitation theory included, [and of] the rest of modern physics. – Albert Einstein

If you have built castles in the air, your work need not be lost; there is where they should be. Now build your foundations under them. – Henry D. Thoreau

33.1 Gravity of the matter

Gravitation in a quantum gravity theory is a result of resonant electromagnetic (EM) wave interactions in a polarizable vacuum (PV) with a variable refractive index. Unlike the geometric spacetime curvature assumed in the Einstein Theory of General Relativity (GR), gravitation in the Puthoff *et al* PV model[29,30] is described instead by variation in EM wave energy and density due to local variation in the vacuum refractive index K_{PV}. The PV theory was proposed by H. Wilson[27] in 1921 and later developed by R.H. Dicke[28] in 1957, Puthoff[30] and Krogh[32]. The refractive index n is a ratio of the phase velocity v in a medium to velocity of light c in a gravity-free, zero curvature vacuum ($n = c/v$). Variable vacuum electric permittivity and magnetic permeability results in a variation in the speed of light providing an explanation of bending of light and gravitational attraction in terms of local scalar ϵ and μ fields. Einstein, in 1911, initially proposed a gravitation theory with a variable speed of light as a function of potential, but later in 1916, settled for a purely mathematical GR metric description[35] in which particles move along geodesics without a defined causal physical mechanism for means by which mass-energy curves spacetime. This metaphysical theory, further refined in 1918, uses a metric tensor field to describe both spacetime geometry and gravitation without a physical explanation. In lieu of an unexplained mechanism for assumed "bending" of spacetime (e.g., alteration in space (contraction) and time (dilation) in accelerated inertial frames), the Lorentz contraction in the Ivanov-LaFreniere wave model[36,37,38] refers instead to a physical EM wavelength contraction (compression of the nodal distance) and frequency reduction of a standing matter waves in motion. 4D spacetime remains invariant. A variation in the vacuum dielectric constant K_{PV} provides the mechanism for alteration in wavelength which occurs in quantized multiples of wavelength. The EM wavelength contraction and frequency shift in a polarizable vacuum accounts for mass in motion and gravitational effects including the energy change, deflection of light, gravitational frequency shift and clock slowing.

The gravitational attraction between masses modeled as EM oscillators, as shown by Ivanov, is the result of an arrhythmia (frequency pulling effect) due to a difference in frequencies. Mass represents a resistance to frequency change. An energy flow occurs from the oscillator with higher frequency to the lower frequency oscillator as a clock synchronization effect between coupled oscillators. The energy flow between masses is transmitted via electromagnetic waves in a standing wave interaction (mode-locking) between coupled oscillators. The acceleration of gravity is developed within each mass in response to the frequency differential to reduce the overall energy level. The energy flow is modulated by a Fresnel zone effect increasing the local EM flux density between coupled masses. The increased EM flux density (**D**, **B**) increases the local K_{PV} (r,ω,M) dielectric constant and gradient with a resultant increase in an associated gradient force of attraction

($\mathbf{F}_g \propto \nabla K_{PV}^2$) proportional to the gradient of K_{PV}^2 and, hence, is always attractive. The family of ellipsoidal Fresnel zones encapsulate the coupled masses forming in effect a graded dielectric lens with the center of masses at the ellipse focii. Photons emitted from each oscillator mass are reflected at the Fresnel zones boundaries towards the opposite foci. The augmented gradient produces a force of attraction substantially greater than the radiation pressure imbalance from a long range Casimir force. The variation in EM flux energy density produces a nonlinear variation in ϵ_0 and μ_0 of the interacting mass altering the local speed of light. The alteration of ϵ_0 and μ_0 results in local variation of the EM energy density as measured by the vacuum refractive index K_{PV} and provides a mechanism for EM wave contraction effects.

A comparison of physical constants and units of measure in a spacetime GR model in terms of the gravitational Γ with physical parameters as a function of the vacuum dielectric constant K_{PV} may be shown illustrating an exact one-for-one correspondence as does the γ factor in SR. The equivalence between gravitational mass and inertial mass has a common origin – acceleration into a region of increased EM energy density i.e., increased dielectric constant K_{PV}.

According to the Ivanov 'Rhythmodynamics' model of gravitation[37,38], matter consists of packages of standing waves of collectively synchronized elements linked by wave fields – in effect, an assembly of standing wave resonators formed of fermions. An electron/positron as a fermion constitutes an elementary standing wave resonator. In the presence of a gravitational field, the mass oscillators undergo a phase shift. Displacement of potential holes (standing wave minima) is triggered by a phase shift. Wave system velocity v_g is proportional to phase difference ($v_g = \Delta\phi \cdot c/\pi = \beta \cdot c$). Accumulation of phase displacement between moving elements makes the system self-accelerate. Gravitational acceleration **g** is proportional to the net frequency difference (= $2c \cdot \Delta \nu \cdot \mathbf{r}_u$) which is a function of the gradient in EM energy density (= $\nabla \nu$) and, hence, may be subject to control as discussed in subsequent sections.

The fall of an object under a gravitational field involves a conversion of potential energy into kinetic energy according to conservation of energy. The kinetic energy resides with the falling object. Where is the potential energy stored? In the physical aether model, the potential energy represents kinetic energy of Planck scale dipoles composed of positive and negative Planck masses which make up the vacuum and matter fields. Jefimenko[16] notes that the process by which this interchange takes place may be explained as a consequence of influx of gravitational-cogravitational field energy via a gravitational Poynting vector. The self-induced gravito-magnetic field $\mathbf{B}_g$ of a falling object induces a circumferential field surrounding the object due to a mass current similar to a magnetic field B induced by an electrical current. The net gravitational acceleration acting on a falling object is the sum of the Earth's external field **g**, the gravitational self-generated gravitational field $\mathbf{g}_c$ and the self-induced motional gravito-magnetic field $\mathbf{B}_g$. The gravito-magnetic Poynting field $\mathbf{S}_g$ (= $\mathbf{E}_g \times \mathbf{H}_g$) is radially inward directed whereas the gravito-magnetic field $\mathbf{B}_g$ is a circular field oriented perpendicular to the direction of motion in alignment with **g** and directed according to the left-hand rule relative to the mass current direction.

The introduction of the "cosmological member" into the equations of gravity, though possible from the point of view of relativity, is to be rejected from the point of view of logical economy. – Albert Einstein

The key difference between GR and EGM is that EGM explicitly describes why spacetime physically becomes refractive in the presence of matter; GR does not. – Ricardo C. Storti

33. Gravitation

33.2 Newton's law of gravitation

Halley and Wren, in 1684, speculated that the force of gravity must be inversely related to separation distance based on a study of planetary orbits as had Hooke, Bullialdus and Newton previously. At Halley's urging, Isaac Newton published the *Philosophiæ Naturalis Principia Mathematica* in 1687 which included a derivation of force of attraction in accordance with Galileo's and Kepler's laws of motion that diminished with the square of distance. The Newtonian gravitational coupling force of attraction between two idealized point masses in scalar notation is given by

$$F_g = GmM/r^2 = \mu m/r^2 \qquad (33\text{-}1)$$

where:

F_g	Force of gravity		
	Value: Varies	Units: N [= kg(m/sec^2)]	Dimensions: MLT^{-2}
G	Universal gravitation constant (= $G_0/\Gamma^3 = c^2 R_{univ}/M_{univ} = 1/4\pi\epsilon_g = c^3/Z_s$)		
	Value: 6.67384 x 10^{-11} (measured)	Units: N·m^2/kg^2 [= m^3/kg·sec^2]	Dimensions: L^3/MT^{-2}
m	Orbital (passive) mass		
	Value: Varies	Units: kg	Dimensions: M
M	Central (active) mass		
	Value: Varies	Units: kg	Dimensions: M
μ	Standard gravitational parameter = GM = $\phi \cdot r$		
	Value: Varies	Units: m^{3}s^{-2}	Dimensions: M^{3}T^{-2}
r	Radius vector r(t) = instantaneous distance between mass centers		
	Value: Varies	Units: m	Dimensions: L

In vector notation, the above may be written as $\mathbf{F}_g = -(GmM/r^2)\mathbf{r}_u$ where $\mathbf{r}_u$ is the unit vector. The Gravitational constant G in terms of Planck units becomes $G_P = 2\pi G = c^3 t_P/m_P = c^3/Z_s = l_P^3/m_P t_P^2 = 4.2005\text{E-}10$ m^3/s^2·kg where Z_s = spacetime impedance (= $m_P/t_P = c^3/G$), t_P = Planck time, m_P = Planck mass, l_P = Planck length and c = velocity of light = l_P/t_P.

The variation of force with the inverse square of distance in Newton's law of gravitation is illustrated in Fig. 33-1 for a pair of spherical masses. Newton's law of gravitation is independent of time and describes instantaneous action at a distance without description of causal mechanism. Newton observes in a letter to Bentley "Tis inconceivable that inanimate brute matter should (without the mediation of something else which is not material) operate upon & affect other matter without mutual contact....That gravity should be innate & {essential} to matter so that one body may act on another at a distance through a vacuum without the mediation of anything else by & through which their action or force {may} be conveyed from one to another is to me so great an absurdity that I believe no man who has in philosophical matters any competent faculty can ever fall into it." In an essay included in the second edition to the *Principia* (1713), Newton famously remarks "Hypotheses non fingo". In modern translation, "I have not as yet been able to discover the reason for these properties of gravity from phenomena, and I do not feign hypotheses".

Nature uses as little as possible of anything. – Johannes Kepler

33. Gravitation

The gravitational force F_1 experienced by mass m_1 equals the gravitational force F_2 experienced by mass m_2

$$F_1 = |F_2| = G(m_1 \times m_2)/r^2 \qquad (33\text{-}2)$$

The gravitational charge of active mass M acting on a passive mass m generates an acceleration $g = GM/r^2$. The force exerted on another mass m is equal to mass x acceleration

$$F = mg = (m) \times GM/r^2 = GMm/r^2 \qquad (33\text{-}3)$$

The gravitational parameter GM expresses the relationship of gravitational potential $\phi(r)$ and mass radius R: $\phi = -GM/R$ and the acceleration due to gravity: $g = GM/R^2$. The mass of a central body M and the period P of an orbiting mass m with velocity V in a circular orbit are related by the gravitational parameter: $GM = 3\pi V/P^2$. In terms of escape velocity V_e, $GM = V_e^2 R/2$. The GM parameter is also directly related to the Schwarzschild radius r_S of an event horizon: $GM = c^2 r_S/2$.

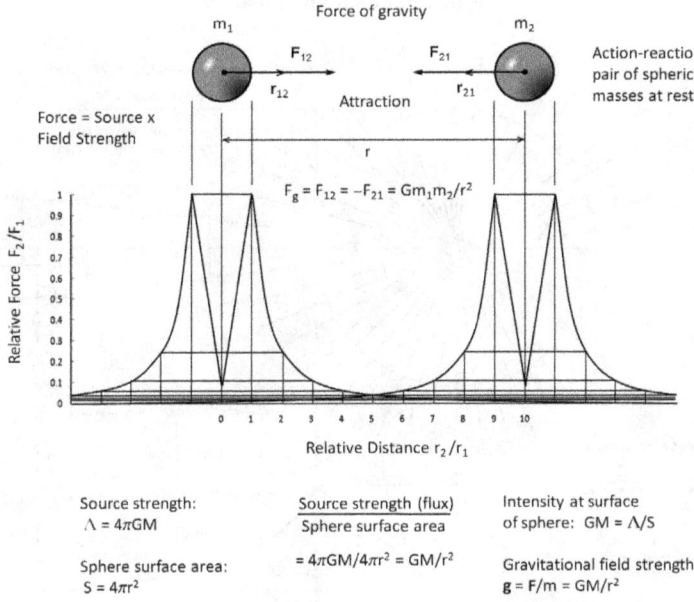

Fig. 33-1. The Newtonian gravitational force equation describes a static field configuration with instantaneous, inverse square, action-at-a-distance gravitational force between a pair of spherical mass objects represented as idealized points. The gravitational force is always attractive and varies as $1/r^2$. Velocity dependent effects are not included, hence, Newton's equation is not relativistic.

It is better to be approximately correct rather than very precisely wrong. – Ricardo C. Storti

33. Gravitation

33.3 Gravitational flux intensity

In the non-relativistic Newtonian theory of gravity, the magnitude of force depends only on the separation distance, and not on the velocity or acceleration of the masses – gravity is assumed to act instantaneously. The theory describes action at a distance without explanation for the means of force transmission or causal mechanism. The given expression for gravitational force is described in terms of the coordinates of an unaccelerated inertial frame of reference. The inverse square law relation is scale invariant and independent of time with an implicit assumption of absolute space and time with instantaneous velocity of propagation. Gravitational waves are inconsistent with the concept of instantaneous action. As such, the Newtonian relation represents a static, non-relativistic approximation. Newton's law of universal gravitation is not in accordance with Einstein's principle of covariance postulate in Special Relativity (SR) for a time dependent, causal relationship independent of reference coordinate system (Lorentz frame). The inverse square relation is illustrated in Fig. 33-2.

The gravitational flux intensity variation for a pair of point masses in the non-relativistic Newtonian theory of gravity is illustrated in Fig. 33-3. As shown, the radial flux overlap results in a dipolar Moiré fringe pattern of constructive and destructive interference.

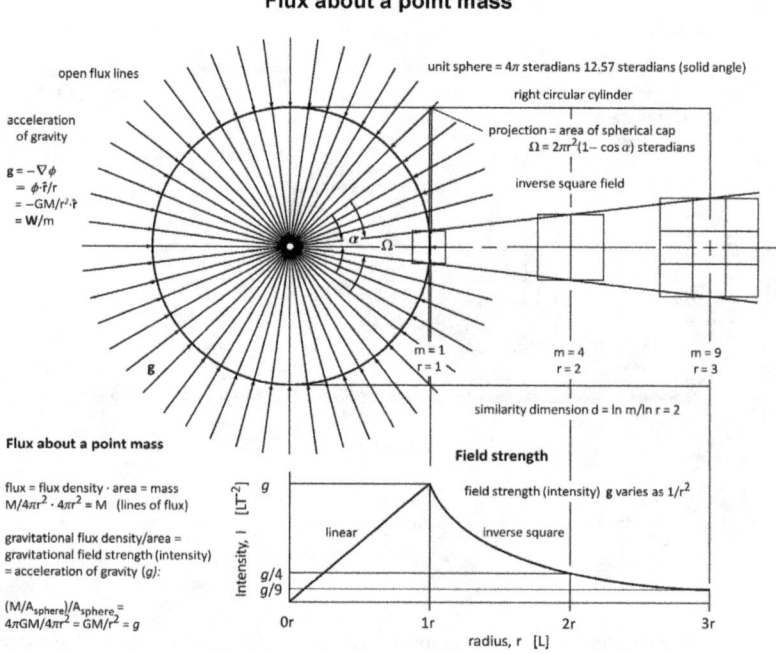

Fig. 33-2. Gravitational flux intensity from a point mass varies as $1/r^2$ in 3D space.

I had not thought of this regular decrease in gravity, namely that it is as the inverse square of the distance; this is a new and highly remarkable property of gravity. – Christiaan Huygens

33. Gravitation

Flux about coupled point masses

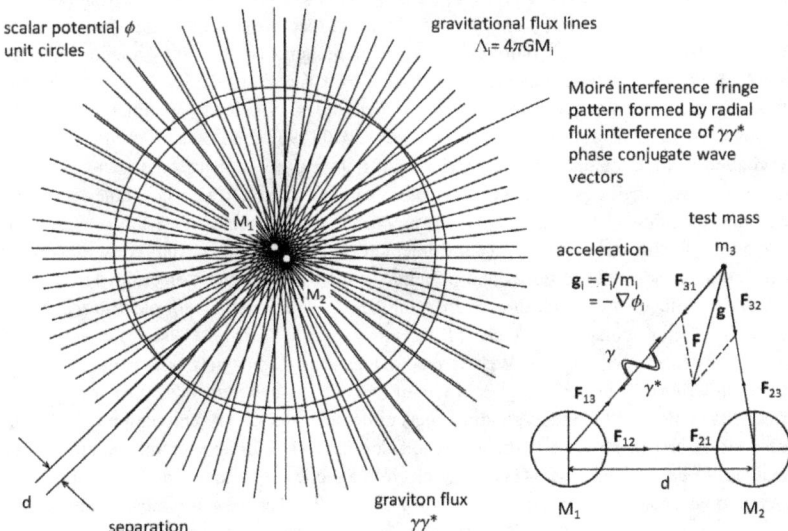

Fig. 33-3. Gravitational flux of two interacting masses produces Moiré pattern of constructive and destructive interference of narrowband polarized (PV) spectrum surrounding each mass or equivalently as broadband Electro-Gravi-Magnetic (EGM) paired conjugate wavefunctions (gravitons) with momentary overlap and zero group velocity.

33.4 Comparison of gravity and electricity

The force of gravity between static masses given by Newton's law is similar to Coulomb's law for the electrostatic force between static electric charges both varying as $1/r^2$. Based on this similarity, Maxwell speculated if gravitational attraction like electrostatic attraction was 'not also traceable to the action of the surrounding medium' A comparison of fundamental equations for each in terms of scalar potentials is summarized in Table 33-1. In a sufficiently small localized region where the field gradient is low (i.e., linear), the field is approximately uniform. The force of gravity is attractive only wheras electric force is attractive or repulsive. The force of gravity is weaker than electrostatic force by a factor of $\sim 10^{42}$ ($F_{elec}/F_{grav} = q^2/G\epsilon_0 m^2 = 4.17 \times 10^{42}$) as gravity is an acceleration field while an electrical field is a mass velocity dependent force field. The Newtonian Gravitational constant G is assumed always positive. Gravitational fields change velocity of EM waves while electromagnetic fields change EM direction. Based on Riemannian geometry of curved manifolds, Clifford suggested that the force of gravity like electromagnetic forces was the result of variation in curvature of higher-dimensional spaces. Static potentials resemble dimples; waves resemble ripples; and particles resemble knots. Analogous to electrostatic attraction, masses attract other masses via the "gravitoelectric" field – a scalar potential effect. Similarly, moving masses attract each other via the "gravitomagnetic" field – a vector potential effect. The Lense-Thirring effect represents an example of gravitational drag anaologous to the electrokinetic effect.

333

33. Gravitation

The analogy between gravitation and electromagnetism has been developed extensively in a gravitoelectro-magnetic description of Newton's law with a gravitomagnetic field due to a mass current. Theories advanced include gravitoelectromagnetism (GEM), Lorentz-invariant theory of gravitation (LITG) and covariant theory of gravity (CTG). Similar to Weber's modification of Coulomb's law, Holtzmüller (1870) and Tisserand (1872) modified Newton's law to include a force term as a function of relative velocity of the attracting masses. Gerber (1898) proposed that the gravitational potential velocity of transmission varies with relative velocity. Heaviside (1893) and Poincaré (1905) proposed gravitation has two components, a gravitational field **g** proportional to the vector distance between masses and a gravimagnetic or cogravitational field **k** which is a function of the relative velocities. Jefimenko (1991) proposed a generalized Newton's theory of gravitation to include time-dependent interactions which satisfy the conservation of momentum, principle of causality and principle of relativity. Weber's electrodynamic force between moving charges with time retardation yields wave equations for electric scalar and magnetic vector potentials. Assis (1992) proposed a non-relativistic derivation of gravitation from generalized Weber electrodynamics force between neutral dipoles including terms of fourth and higher orders of $\dot{r}_{12}/c$ in which electrostatics is a zeroth-order effect, magnetism and Faraday's induction as a second-order effect, gravitation as a fourth-order effect and inertia and perhelion precession as a sixth-order electromagnetic effect. Analogous parameters in Maxwell's electromagnetic equations in a vacuum and gravitational equations are summarized in Table 33-2. The dimensionless electron fine structure constant α (= $e^2/\hbar c \cong 0.007297$) has an analogous gravitational fine structure constant α_G (= $GH^2/\hbar c$ = $(t_P\omega_C)^2 \cong 1.7518E-45$) where H = mass of a nucleon, t_P = Planck time, ω_C = Compton angular frequency of the electron. Expressed in so-called 'natural' units ($4\pi G = c = \hbar = \epsilon_0 = 1$), the ratio α/α_G (= $(e/m_e)^2 \approx 10^{-39}$) is a measure of the relative strength of the electrostatic and gravitational forces between electrons.

The gravitomagnetic field $\mathbf{H}_g$ of mass currents is the mechanism held responsible for frame-dragging or Lense-Thirring effect. These interactions are second order in v/c and very weak as a result. Generation and spin alignment of plasma jets along the spin axis of quasars and black hole nuclei is attributed to the gravitomagnetic field. Interaction with the gravitomagnetic field results in accretion disc precession and jet alignment (Bardeen-Petterson effect). The immense gravitomagnetic rotational energy of supermassive black holes and conversion to relativistic plasma jets is known as the Blandford-Znajek mechanism.

Half of science is putting forth the right questions. – Francis Bacon

Faraday's idea was that it might be possible to convert electricity into gravity, or vice versa. The grounds for his expectation were two theories he held, the unity of force and the conservation of force. – William Berkson

In 1914, a Finnish physicist named Gunnar Nordstrom found that all you had to do to unify gravity with electromagnetism was to increase the dimension of space by one. – Lee Smolin

It is known from oberved phenomena and established physics... that gravity, electromagnetism and spacetime are interelated phenomena. These ideas have led to questioning if gravitational or inertial forces can be created or modified using electromagnetism. – Marc Millis

Behind it all is surely an idea so simple, so beautiful, that when we grasp it - in a decade, a century, or millenium - we will all say to each other, how could it have been otherwise. – J. A. Wheeler

33. Gravitation

Table 33-1. Comparison of gravity and electricity under static conditions*

Parameter	Gravitation	Units	Electrostatics	Units
charge	$\mathbf{m}$ = mass = $\mathbf{F}/\mathbf{a}$ = E/ϕ_g = $\hbar \mathbf{k}/\mathbf{v}$ = E/c^2 = $|\mathbf{V}\cdot\mathbf{q}|/c^2$ = $(\phi_E \cdot \mathbf{q})/c^2$	kg	$\mathbf{q}$ = charge = $\mathbf{F}/\mathbf{E}$ = E/ϕ_E = $\hbar \mathbf{k}/\mathbf{A}$ = E/V = mc^2/V = $m((\omega_C + \omega_{e/m})$	Coulombs
flux	$\Lambda = 4\pi GM$	$m^3 s^{-2}$	$\phi_E = q/\epsilon_0$	volt·meter
divergence of gradient	$\nabla^2 \phi = -4\pi G \rho_m$ Poisson eqn $\nabla \cdot \mathbf{g} = -4\pi G \rho_m$	sec^{-2}	$\nabla^2 \phi = -\rho_{mq}/\epsilon_0$ Poisson eqn $\nabla \cdot \mathbf{E} = -\rho_q/\epsilon_0$	$kg \cdot C^2/sec^2$
coupling constant	$G = \frac{1}{4\pi \epsilon_g} = 6.67262 \times 10^{-11}$ (Newton's gravitational constant)	$N \cdot m^2/kg^2$	$k_C = \frac{1}{4\pi \epsilon_0} = 8.987551 \times 10^9$ (Coulomb's constant)	$N \cdot m^2/C^2$
Uniform field				
Field strength	$\mathbf{g}$ = const. = $-\Delta \phi/\Delta x$ = $-\nabla \phi = -\nabla \phi_g$ (acceleration of gravity, gravitational field strength)	m/sec^2	$\mathbf{E}$ = const. = $-\Delta V/\Delta x$ = $-\nabla V = -\nabla \phi_E$ (electric field strength, voltage gradient)	volts/m
Force	$\mathbf{F} = W = m\mathbf{g} = -m\nabla \phi_g = \delta L/\delta x$ $= d\mathbf{P}/dt$ (Gravity force)	$kg \cdot m/sec^2$ = N	$\mathbf{F} = q'\mathbf{E} = qV/r = -q\nabla \phi_E$ (Coulomb force)	$kg(m/sec^2)$ = N

cont

335

33. Gravitation

Table 33-1. Comparison of gravity and electricity under static conditions* (Cont)

Parameter	Gravitation	Units	Electrostatics	Units
Uniform field (cont)				
Scalar potential	$\phi = \phi_g = gh$ (gravitational potential)	joules/kg $= m^2/sec^2$	$V = \phi_E = Ed$ (electrostatic potential – voltage)	volts $= m^2/s \cdot rad$
Potential energy	$U = m\phi = mgh$	$kg \cdot m^2/sec^2$ $= J$	$U = qV = qEd$	C·volts = Joules
Non-uniform field (point source)				
Field strength	$\mathbf{g} = -GM/r^2 \cdot \mathbf{r}_u$ $= -\text{grad } \phi_g$ $= -\nabla \phi_g$	m/sec^2	$\mathbf{E} = (1/4\pi\epsilon_0)q/r^2 \cdot \mathbf{r}_u$ $= k_e q/r^2 \cdot \mathbf{r}_u = -\text{grad } \phi$ $= -\nabla \phi$	volts/m
Force	$\mathbf{F} = -Gm_1m_2/r^2 \cdot \mathbf{r}_u$ $\mathbf{F} = -m\nabla\phi = -\Delta U/\Delta x$	$kg \cdot m/sec^2$	$\mathbf{F} = q_1q_2/(4\pi\epsilon_0 r^2) \cdot \mathbf{r}_u$ $\mathbf{F} = -q\nabla\phi = -\Delta U/\Delta x$	N = $kg(m/sec^2)$
Scalar potential	$\phi = \phi_g = -GM/r = (\Delta v/v)c^2$ $= (\mathbf{A}_g/\mathbf{A})\phi_E$	joules/kg $= m^2/sec^2$	$V = \phi_E = (1/4\pi\epsilon_0) \, q/r = E/q$ $= (\mathbf{A}/v)c^2 = (\mathbf{A}/\mathbf{A}_g)\phi_g$	volts $= m^2/s \cdot rad$
Potential energy	$U = -Gm_1m_2/r$	Joules	$U = (1/4\pi\epsilon_0T) \, q_1q_2/r$ $= k_e \, q_1q_2/r$	Joules

*Note: The above equations do not include charge velocity or acceleration dependent effects.

Table 33-2. Analogy between gravity and electromagnetism. [After Jefimenko[16], Evans[86]]

Gravitation		Electromagnetic	
Parameter	Symbol/Relation/Units	Parameter	Symbol/Relation/Units
mass	$m = \mathbf{g}/W = \mathbf{F}/\mathbf{a} = E/\phi_g = \Gamma m_0$ $= \hbar \mathbf{k}/\mathbf{v} = \mathbf{F}_g \cdot \mathbf{r}^2/GM \cdot \mathbf{r}_u$ $= E/c^2 = |\mathbf{V} \cdot \mathbf{q}|/c^2 = E/(\mathbf{E}/\mathbf{B})^2$ $= E/\epsilon_0 \mu_0 = E/v_p v_g$ $= \mathbf{F}/(\mathbf{g}_i - \mathbf{g}) = \mathbf{S}/vc^2$ $= 2L(\mathbf{E}^2 - c^2\mathbf{B}^2)$ (kg)	electric charge	$q^+ = \mathbf{E}/\mathbf{F} = E/V$ $= \hbar \mathbf{k}/A = mv/A$ $= mc^2/V = C \cdot V$ $= mc^2/\phi_E$ (C)
volume mass density	$\rho_m = dm/dV$ (kg·m^{-3})	volume charge density	$\rho = dq/dV$ (C·m^{-3})
surface mass density	$\sigma_m = dm/dS$ (kg·m^{-2})	surface charges density	$\sigma = dm/dS$ (C·m^{-2})
line mass density	$\lambda_m = dm/dL$ (kg·m^{-1})	line charge density	$\lambda = dm/dL$ (C·m^{-1})
mass current	$I_m = dm/dt = \rho_m vA$ (kg·s^{-1}) $= \mathbf{B}_g(2\pi r)\mu_g$	electric current	$I = dq/dt = \rho vA$ $= \mathbf{B}(2\pi r)/\mu_0)$ (A)

cont

Table 33-2. Analogy between gravity and electromagnetism. [After Jefimenko[16], Evans[86]] (Cont)

Gravitation		Electromagnetic	
Parameter	Symbol/Relation/Units	Parameter	Symbol/Relation/Units
mass current density	$\mathbf{J}_m = \rho_m \mathbf{v}$ $\quad$ (kg·s^{-2})	convection current density	$\mathbf{J} = \rho \mathbf{v}$ $\quad$ (A·m^{-2})
gravitational field strength (acceleration)	$\mathbf{g} = \mathbf{F}/m = \mathbf{W}/m$ $= -G(m/r^2)\mathbf{r}_u$ $= -\nabla\phi - \delta\mathbf{A}/\delta t$ $= -\nabla \times \mathbf{A}_g$ $= -G([m(1-v^2/c^2)]/(r_0^3[1-(v^2/c^2)\sin^2\theta]^{3/2})\mathbf{r}_0$ $= c^2/2r_S$ $= \mathbf{H}_g \cdot Z_0 / n$ $= c \cdot \mathbf{B}_g / \theta \quad$ (N/kg = m/s^2)	electric field strength	$\mathbf{E} = \mathbf{F}/q = q\mathbf{E}/V = k(q/r^2)\mathbf{r}_u$ $= -\nabla V - \delta\mathbf{A}/\delta t$ $= ([e(1-v^2/c^2)]/(r_0^3[1-(v^2/c^2)\sin^2\theta]^{3/2})\mathbf{r}_0$ $= \mathbf{g} \cdot \phi / c^2$ $= \mathbf{g} \cdot V / c^2$ $= \mathbf{H} \cdot Z_0 / n$ $= c \cdot \mathbf{B}$ $\quad$ (V/m)
gravitomagnetic field strength	$\mathbf{H}_g = \mathbf{B}_g / \mu_g$ $\quad$ (kg·m^{-1}·s^{-1}) $= \epsilon_g c^2 \mathbf{B}_g$ $= \mathbf{n} \cdot \mathbf{g} / Z_0 = \mathbf{J}_m / 2\pi r$	magnetic field strength	$\mathbf{H} = \mathbf{B} / \mu_0$ $\quad$ (Wb/m) $= \epsilon_0 c^2 \mathbf{B}$ $\quad$ (A·m^{-1}) $= \mathbf{n} \cdot \mathbf{E} / Z_0 = \mathbf{I} / 2\pi r$
gravitational gamma	$\Gamma = dt/d\tau = 1/\sqrt{(1 - v_c^2/c^2)}$ $= 1/\sqrt{(1 + 2\phi/c^2)} = 1/\sqrt{g_{00}}$	Lorentz factor	$\gamma = \rho/\beta = 1/\sqrt{(1-v^2/c^2)}$ $= 1/\sqrt{(1-\beta^2)}$

cont

Table 33-2. Analogy between gravity and electromagnetism. [After Jefimenko[16], Evans[86]] (Cont)

Gravitation			Electromagnetic	
Parameter	Symbol/Relation/Units		Parameter	Symbol/Relation/Units
cogravitation force, kinemsssic force	$\mathbf{F} = m(\mathbf{v} \times \mathbf{K})$ (N) $(= kg \cdot m/s^2)$		magnetic force	$\mathbf{F} = q_0^+ (\mathbf{v} \times \mathbf{B})$ (N) $(= kg \cdot m/s^2)$
gravitomagnetic flux density, torsion field, torsion flux density	$\mathbf{B}_g = \nabla \times \mathbf{A}_g$ rad·s^{-1} $= \mu_g \mathbf{H}_g$ $= \frac{1}{2}\nabla \times \mathbf{A}_{gm} = -2\Omega$ $= 4\pi G/c^2 \mathbf{H}_g$ $\mathbf{g} \cdot \theta /c = (\mathbf{v} \times \mathbf{g})/c^2$		magnetic flux density, magnetic induction	$\mathbf{B} = \nabla \times \mathbf{A}$ (T = Wb/m^2) $= \mathbf{E}(v/c)\sin\theta$ (= 1/rad) $= \mu_0 \mathbf{H} = 1/\epsilon_0 c^2 \mathbf{H}$ $= \mu_0(\mathbf{H} + \mathbf{M})$ $= \mathbf{E}/c = \mu_0 I / 2\pi$ $= (\mathbf{v} \times \mathbf{E})/c^2$
gravitational flux density	$\mathbf{D}_g = \epsilon_g \mathbf{g}$ (kg·m^{-2}) $= \mathbf{g}/8\pi G$		electric flux density	$\mathbf{D} = \epsilon_0 \mathbf{E}$ (C·m^{-2})
gravitational charge	$m = \mathbf{F}/\mathbf{g}$ (kg)		electric charge	$q = e = \mathbf{E}/E$ (C)
gravitomagnetic charge	$b_g = m/c$ (kg·s·m^{-1})		magnetic charge	$b = -e/c$ $= -2.359E\text{-}18$ (kg/m)

cont

33. Gravitation

Table 33-2. Analogy between gravity and electromagnetism. [After Jefimenko[16], Evans[86]] (Cont)

Gravitation		Electromagnetic	
Parameter	Symbol/Relation/Units	Parameter	Symbol/Relation/Units
cogravitation flux, cogravitational scalar potential, kinemassic flux	$\phi_C = \xi I_m = \int \phi_g dt = \mathbf{B}_g \cdot \mathbf{A}$ $= \oint \mathbf{A} \cdot d\mathbf{l}$ (m/s)	magnetic flux, magnetic scalar potential	$\phi_B = LI = \int V dt = n\phi_0 = \mathbf{B} \cdot \mathbf{A}$ $= \oint \mathbf{A} \cdot d\mathbf{l}$ (Weber = volt·sec)
scalar potential (gravitation)	$\phi = -G \sum_{i=1}^{n} \frac{m_i}{r_i}$ $(J \cdot kg^{-1})$ $\phi_g = (\mathbf{A}_g / \mathbf{A})\phi_E$ $(= m^2/s^2)$	scalar potential (voltage)	$\phi = V = \frac{1}{4\pi\epsilon_0} \sum_{i=1}^{n} \frac{q_i}{r_i}$ $\phi_E = (\mathbf{A}/\mathbf{A}_g)\phi_g$ $(V = J \cdot C^{-1})$
cogravitation vector potential, gravitational vector potential, kinemassic vector potential	$\mathbf{A} = -\frac{G}{c^2} \sum_{i=1}^{n} \frac{m_i \bar{v}_i}{r_i}$ $\mathbf{A}_g = v\phi_g/c^2 = \hbar k/v = \epsilon_g \mu_g v \phi_g$ $\mathbf{A}^* = -\int \mathbf{g}_k dt + \mathbf{A}_0$ (m/s) $\mathbf{A}_g = (\phi_g/\phi_E)\mathbf{A}$	vector potential (magnetic)	$\mathbf{A} = \frac{1}{4\pi\epsilon_0 c^2} \sum_{i=1}^{n} \frac{q_i \bar{v}_i}{r_i}$ $\mathbf{A} = v\phi_E/c^2 = \hbar k/q = mv/e$ $\mathbf{A}^* = -\int \mathbf{E}_k dt + \mathbf{A}_0$ $\mathbf{A} = \mathbf{A}_g(\phi_E/\phi_g)$ $(T \cdot m = Wb/m$ $= V \cdot s \cdot m^{-1} = m/rad)$
gravikinetic force	$\mathbf{F}_g = G(m_1 m_2/rc^2)\mathbf{a}$ (N)	Electrokinetic force	$\mathbf{F}_e = k(q_1 q_2/rc^2)\mathbf{a}$ (N)

cont

33. Gravitation

Table 33-2. Analogy between gravity and electromagnetism. [After Jefimenko[16], Evans[86]] (Cont)

Gravitation		Electromagnetic	
Parameter	Symbol/Relation/Units	Parameter	Symbol/Relation/Units
group velocity (free space)	$c = 1/\sqrt{(\mu_g \epsilon_g)} = \mathbf{g}/\mathbf{K} = \omega/k$ $= c_0/\Gamma$ (m/s)	group velocity (free space)	$c = 1/\sqrt{(\mu_0 \epsilon_0)} = \mathbf{E}/\mathbf{B} = \omega/k$ $= c_0/\gamma$ (m/s)
gravitational force	$\mathbf{W} = m\mathbf{g}$ (N)	electric force	$\mathbf{F} = q^+\mathbf{E}$ (N)
Newton gravitational constant	$G = 1/(4\pi \epsilon_g) = 6.67262\text{E-}11$ $(\text{kg}^{-1}\cdot\text{m}^3\cdot\text{s}^{-2})$	Coulomb's constant	$k_C = 1/(4\pi \epsilon_0) = 8.987\text{E}9$ $(\text{N}\cdot\text{m}^2/\text{C}^2)$
gravitomagnetic force	$\mathbf{F}_k = m(\mathbf{v} \times \mathbf{K})$ (N)	magnetic force	$\mathbf{F}_m = q^+(\mathbf{v} \times \mathbf{B})$ (N)
dynamic gravitation force	$\mathbf{F}_K = m(\mathbf{g} + \mathbf{v} \times \mathbf{K})$ $= m(\mathbf{E}_g + 4\mathbf{v} \times \mathbf{B}_g)$ (N)	Lorentz force	$\mathbf{F}_L = q^+(\mathbf{E} + \mathbf{v} \times \mathbf{B})$ (N)
gravikinetic field	$\mathbf{g}_k = -\mathbf{v} \times \mathbf{K} = -d\mathbf{A}*/dt$ (m/s²)	electrokinetic field	$\mathbf{E}_k = -d\mathbf{A}*/dt$ (volts/m)

cont

33. Gravitation

Table 33-2. Analogy between gravity and electromagnetism. [After Jefimenko[16], Evans[86]] (Cont)

Gravitation		Electromagnetic	
Parameter	Symbol/Relation/Units	Parameter	Symbol/Relation/Units
divergence of gravitational field	$\nabla \cdot \mathbf{g} = -4\pi G \rho_g$ (N/m·kg)	divergence of electric field	$\nabla \cdot \mathbf{E} = -\rho/\epsilon_0$ (C·F^{-1}·m^{-2}) Gauss's law (**E**-fields)
divergence of torsion field	$\nabla \cdot \mathbf{K} = 0$ (m^{-1}·s) $\nabla \cdot \mathbf{B}_g = 0$	divergence of magnetic field	$\nabla \cdot \mathbf{B} = 0$ (T/m) Gauss's law (**B**-fields)
curl of gravitational field	$\nabla \times \mathbf{g} = -\delta \mathbf{K}/\delta t$ (s^{-2}) $\nabla \times \mathbf{E}_g = -\delta \mathbf{B}_g/\delta t$	curl of electric field	$\nabla \times \mathbf{E} = -\delta \mathbf{B}/\delta t$ (T/s) Faraday's law
curl of torsion field	$\nabla \times \mathbf{K}$ $= -(4\pi G/c^2)\mathbf{J} + 1/c^2\,\delta \mathbf{g}/\delta t$ $\nabla \times \mathbf{B}_g$ $= -(4\pi G/c^2)\mathbf{J} + 1/c^2\,\delta \mathbf{E}_g/\delta t$ (m^{-1}s^{-1}) (J)	curl of magnetic field	$\nabla \times \mathbf{B}$ $= 1/\epsilon_0 c^2\,\mathbf{J} + 1/c^2\,\delta \mathbf{E}/\delta t$ (V/m·s) Ampere's law
gravitomagnetic energy Hamiltonian	$H = \tfrac{1}{2} m(\mathbf{g}^2 + c^2 \mathbf{B}_g^2)$ (J)	electromagnetic Hamiltonian	$H = \tfrac{1}{2} m(\mathbf{E}^2 + c^2 \mathbf{B}^2)$ (J)

cont

33. Gravitation

Table 33-2. Analogy between gravity and electromagnetism. [After Jefimenko[16], Evans[86]] (Cont)

Gravitation		Electromagnetic	
Parameter	Symbol/Relation/Units	Parameter	Symbol/Relation/Units
gravitoelectric permittivity	$\epsilon_g = 1/(4\pi G) = 1.1972708\text{E}9$ $= Z_s/4\pi c3$ (kg·s^{2}m^{-3})	electric permittivity	$\epsilon_0 = 1/\mu_0 c^2$ $= 8.854187\text{E-}12$ (F·m^{-1}) $= 1.7251\text{E}8$ (kg·rad^2/m^3)
gravitomagnetic permeability, inertial permeability	$\mu_g = 4\pi G/c^2 = 9.328772\text{E-}27$ $= 4\pi c/Z_s$ (m/kg)	magnetic permeability	$\mu_0 = 4\pi\text{E-}7$ (V·s/A·m) $6.4498\text{ E-}26$ (m·s^2)/(kg·rad^2)
divergence of gravitational flux density	$\nabla \cdot \mathbf{D}_g = \rho_g$ (kg·m^{-3})	divergence of electric flux density	$\nabla \cdot \mathbf{D} = \rho$ (C·m^{-3})
curl of gravitomagnetic field strength	$\nabla \times \mathbf{H}_g = \mathbf{J}_m + \delta \mathbf{D}_g/\delta t$ (kg·s^2)	curl of magnetic field strength	$\nabla \times \mathbf{H} = \mathbf{J} + \delta \mathbf{D}/\delta t$ (A·m^2)
gravitational energy density	$U_g = \frac{1}{2}(\mathbf{g} \cdot \mathbf{D}_g + \mathbf{B}_g \cdot \mathbf{H}_g)$ $= \frac{1}{2}(\epsilon_g \mathbf{g}^2 + \mu_g \mathbf{B}_g^2)$ (J/m^3)	electromagnetic energy density	$U = \frac{1}{2}(\mathbf{E} \cdot \mathbf{D} + \mathbf{B} \cdot \mathbf{H})$ $= \frac{1}{2}(\epsilon_0 \mathbf{E}^2 + \mu_0 \mathbf{B}^2)$ (J/m^3)
gravitomagnetic Lagrangian	$L = \frac{1}{2}\text{m}(\mathbf{g}^2 - c^2\mathbf{B}_g^2)$ (J)	electromagnetic Lagrangian	$L = \frac{1}{2}\text{m}(\mathbf{E}^2 - c^2\mathbf{B}^2)$ (J) $\mathbf{L} = (\mathbf{F}^2 - \mathbf{G}^2)/\mathbf{FG}$ where $\mathbf{F} = \mathbf{B} - i\mathbf{D}, \mathbf{G} = \mathbf{E} + i\mathbf{H}$

cont

33. Gravitation

Table 33-2. Analogy between gravity and electromagnetism. [After Jefimenko[16], Evans[84]] (Cont)

Gravitation		Electromagnetic	
Parameter	Symbol/Relation/Units	Parameter	Symbol/Relation/Units
gravito-Poynting vector	$\mathbf{S}_g = \mathbf{g} \times \mathbf{H}_g$ $\quad$ (kg/s^2) $= -(c^2/4\pi G)\mathbf{E}_g \times 4\mathbf{B}_g$ $= (c^2/4\pi G)\mathbf{K} \times \mathbf{g}$	Poynting vector	$\mathbf{S} = \mathbf{E} \times \mathbf{H}$ $\quad$ (W·m^{-2} = kg/s) $= c^2 \epsilon_0 \mathbf{E} \times \mathbf{B}$
wave equation	$\nabla^2 \mathbf{g} = 1/c^2 \, \delta^2 \mathbf{g}/\delta t^2 = 0$ $\quad$ (m^{-1} s^{-2}) $\nabla^2 \mathbf{K} = 1/c^2 \, \delta^2 \mathbf{K}/\delta t^2 = 0$ $\quad$ (m^{-2} s^{-1})	wave equation	$\nabla^2 \mathbf{E} = 1/c^2 \, \delta^2 \mathbf{E}/\delta t^2 = 0$ $\quad$ (m^{-1}·s·rad) $\nabla^2 \mathbf{B} = 1/c^2 \, \delta^2 \mathbf{B}/\delta t^2 = 0$ $\quad$ (m^{-2}·rad)

Gravity: Surely this force must be capable of an experimental relation to electricity, magnetism, and the other forces, so as to bind it with them in reciprocal action and equivalent effect. – Michael Faraday

The theory that ether does not exist, and that gravity is not a force but a property of space can only be described as a crazy vagary, a disgrace to our age. – Charles Lane Poor

Any opinion as to the form in which the energy of gravitation exists in space.... and probable... will make an enormous stride in physical speculation. –James C. Maxwell

E and B are slowly disappearing from the modern expression of physical laws; they are being replaced by A and ϕ – Richard P. Feynman

Both the electrokinetic field and the magnetic vector potential are simultaneously caused by the same electric current. – Oleg D. Jefimenko

33. Gravitation

33.5 Kepler's laws

Newton's laws of motion and gravitation were used to derive Johannes Kepler's laws of planetary motion published in *Astronomia Nova* in 1609 and *Harmonics Mundi* in 1618. Kepler's heuristic analysis of astronomical observations by Tycho Brahe are described by three laws as summarized in Table 33-3 below.

Table 33-3 Kepler's Laws

First Law (Law of orbits)	All planetary orbits are ellipses with the center-of-mass of gravitational sources located at the foci.
Second Law (Law of Areas)	Equal areas are swept out during equal time intervals.
Third Law (Harmonic Law)	The square of the orbital period T is proportional to the cube of the semi-major axis. Kepler's constant $K_s = T^2/r^3$ where $T^2 = 4\pi^2 a^3/GM$

A point mass M has a Keplerian potential $\phi(r) = -GM/r$ since orbits follow Kepler's orbital laws. A circular orbit at radius r has velocity $v(r) = \sqrt{(GM/r)}$. A uniform sphere of mass M and radius a has potential: $\phi(r) = -2\pi Gr \cdot (a^2 - r^2/3)$ for $r < a$ and $\phi(r) = -GM/r$ for $r > a$. A diffuse mass orbital distribution tends to exhibit torsional windup. Refer to Fig. 33-4.

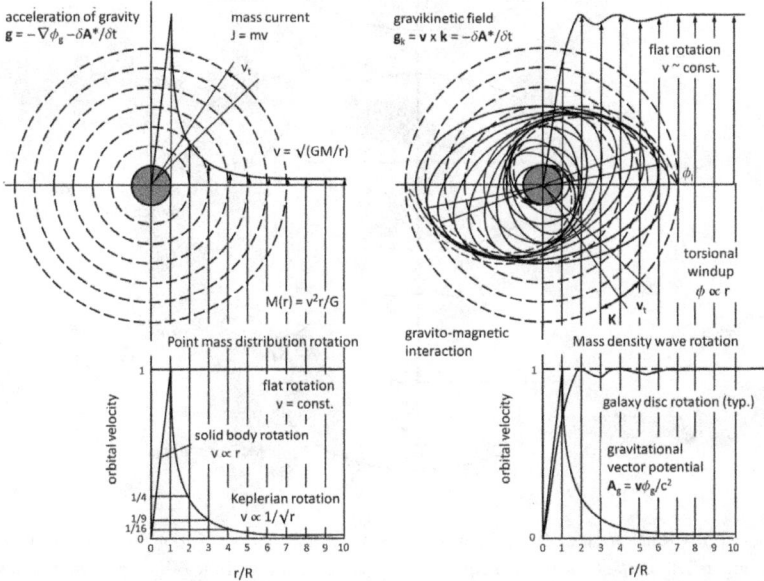

Fig. 33-4. Keplerian and non-Keplerian orbital motion exhibit different tangential velocity profile curves due to cogravitation field mass current interaction effects. Accelerating mass currents results in gravitational [frame] drag on neighboring masses and increased inertia.

33. Gravitation

Keplerian orbital motion of a mass m about a central mass M are illustrated in Figs. 33-5 and 33-6. Keplerian planetary motion orbital velocity varies as $1/\sqrt{r}$ for discrete masses orbiting about a large central mass. Spiral arm galactic discs exhibit density waves of diffuse matter with linear velocity rotation curves. The orbital speed of stars in the outskirts of spiral galaxies is found to vary linearly in a log V vs. log R plot according to the Tulley-Fischer rotation relationship. Objects in the inner spiral arms in the star forming regions travel at rotational velocities less than the density waves when the total energy is less than the wave potential energy and are confined. Gas and dust outflows occur in the outer arms where the potential energy oscillations are less and thus matter can escape with the energy imparted by the spiral arms. Electromagnetic forces acting on plasma Birekeland currents are many orders of magnitude greater than gravitational forces and fall off inversely as the first power of distance from the plasma electric current channels. Lorentz forces acting on ionized galactic plasma and mass current gravito-magnetic forces can explain the observed galactic velocities profiles and evolutionary shapes without recourse to Modified Newtonian Dynamics (MOND) theories or exotic forms of dark matter conjectures.

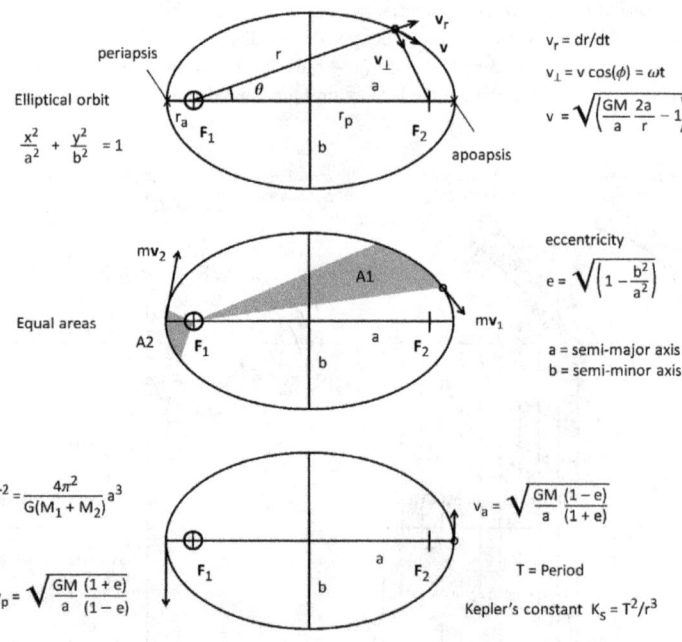

Fig. 33-5. Illustration of Kepler's three laws of orbital motion. Elliptical orbits result if the force-field is linear or inverse-square. The center-of-mass of the orbital pair constitutes an inertial frame. Kepler's laws may be derived from Newton's law of gravitation.

If the facts don't fit the theory, change the facts. – Albert Einstein

33. Gravitation

Kepler's laws are found not to accurately describe situations where the mass of the satellite is not negligible in comparison to the central mass, n-body systems of three or more bodies, or when the satellite velocity exceeds escape velocity. Spiral galaxies are observed not to obey Keplerian motion as the orbital speed of outlying stars appears anonymously high compared with predicted Newtonian values. Various hypotheses have been advised to account for this discrepancy including modification of Newton's laws, dark matter halo, and decay of gravity over time, etc., however, the velocity variation can be accounted for classically. Orbital precession $\Delta\phi$ (= $6\pi GM/a(1 - e^2)c^2$) is attributed to perturbations due to other orbiting bodies, oblateness (quadrupole moments) of the central mass, spin-orbit resonances and relativistic corrections to Newtonian gravity.

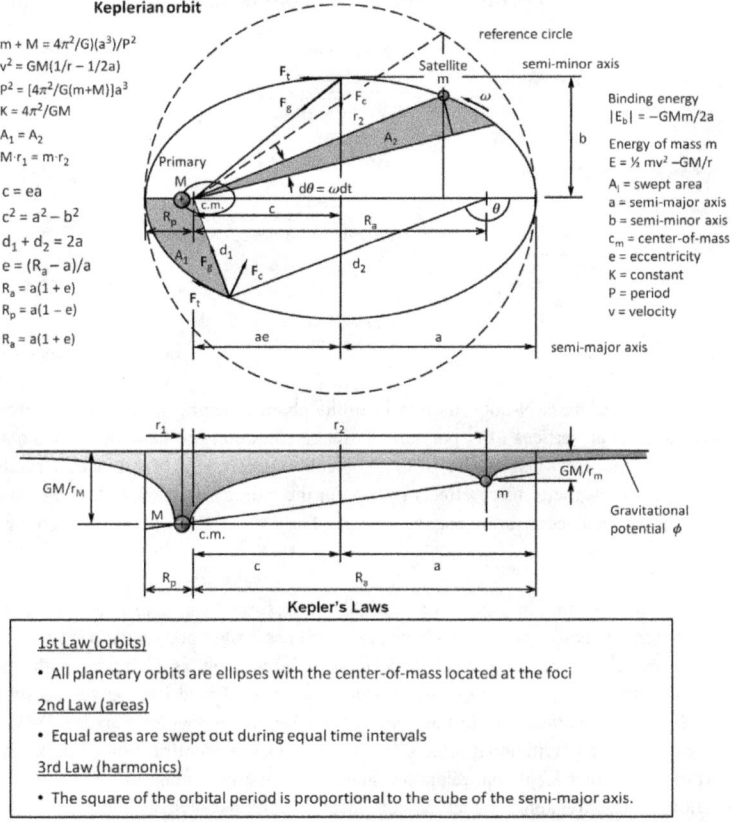

Fig. 33-6. Kepler's three laws of orbital motion are described by Newton's equations. Illustrated is the orbital motion of satellite and primary about a common mass center within an orbital plane. Orbital interactions of multiple orbiting bodies tend to stabilize with planetary periods and spacings in a power law relationship (d = ab^n) depending on orbital resonance ratios of simple whole numbers.

Where there is matter, there is geometry. – Johannes Kepler

33. Gravitation

33.6. N-body gravitation

Lagrange in 1772 discovered relative equilibrium of three bodies forming an equilateral triangle in orbital motion. Maxwell (1859) studying Saturn's rings showed small bodies of equal masses in planar orbit around a central mass formed a regular polygon that was stable rotating as if a rigid body. In 1870, Shering proposed that F_g varies as $1/\sinh^2 r$. Liebmann (1902) showed that the orbits are conic sections in the hyperbolic plane and that the potential is a function of $\coth r$. Refer to Fig. 33-7. Poincaré found most solutions on N-body problems are unstable with sensitive dependence on initial conditions resulting in chaotic orbits making prediction impossible.

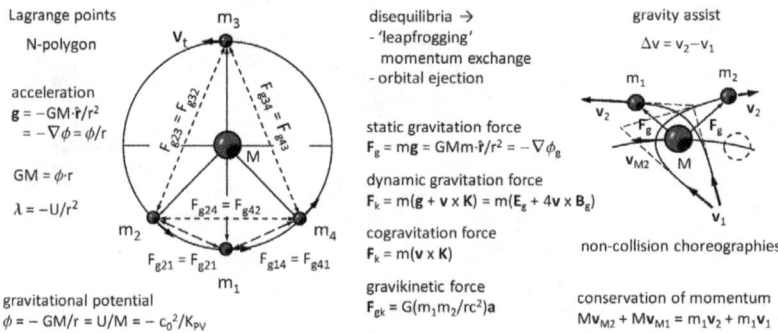

Fig. 33-7. Discrete mass N-body motion in orbits about a central mass tend to describe mass distributions at vertices of N-polygons in relative equilibrium that vary in scale λ (= $-U/r^2$) with total mass. Mass motion of large objects at relativistic velocities may result in significant gravitomagnetic force effects such as in the vicinity of black holes. Static and dynamic gravitational acceleration may be expressed as $\mathbf{g} = -G(M/r^2)\cdot\hat{\mathbf{r}} - G^2(M^2/2c^2r^2)\cdot\hat{\mathbf{r}}$.

D. Saari[84] proposes that in a system of variable masses, the mass distribution with equal angular spacing and equal masses on a given radii orbit as a rigid body with $v(r)$ and $M(r)$ a linear function of r exhibiting discrete scale invariance such as observed with spiral galaxies. Assertions of presence of dark matter have been inferred in observations of thin disc spiral galactic rotation that do not comply with Keplerian rotation curves. Dwyer[85] notes overestimating gravitational effects results in erroneous identification of dark matter as Newton proved that Keplerian relations cannot be properly applied to large scale mass aggregations and only apply under certain ideal conditions such as the Solar system dominated by a single, discrete mass (e.g., Sun contains an estimated 99.86% of total mass of Solar system). The motion equations for Newtonian N-bodies involve nonlinear systems of ordinary differential equations and, in general, difficult to solve. N-body multipole expansion algorithms involving position and velocity coordinates for each body have been developed allowing large scale numerical simulations for gravitational interactions and relative equilibria in celestial mechanics. Orbital resonances tend to occur with orbital periods occuring in small interger ratios. Destabilization arises as a result of momentum exchange between orbiting bodies producing unstable resonances and radial displacement.

34. Gravitation as a Harmonic Phenomena

Something completely new has to be found... something that is somehow based on the ideas of General Relativity, – Albert Einstein

Gravitation is, so far, not understandable in terms of other phenomena. – Richard P. Feynman

The larger central body mass denoted by the large letter 'M' in Newton's Eqn (33-1), is sometimes referred to as the 'active' mass, whereas the smaller orbiting mass denoted by the small letter 'm' is the 'passive' mass. In oscillator terms, M denotes the master oscillator and m denotes the slave oscillator. Each mass represents an ensemble of one or more autonomous, individual, periodic, self-sustained oscillators (e.g., electrons, atoms, etc) radiating a broad spectrum of EM waves. Oscillators constitute clocks which are formed of standing matter waves. In synchronicity, the coupled oscillators reset to a common base frequency. Clocks (and standing waves) contract due to Lorentz contraction, slow down when in motion and undergo a time shift due to Lorentz Doppler effect. In free fall, the frequency discordance is reduced to zero, whereas at a fixed elevation potential, the oscillators are prevented from approaching synchronicity with a resultant net acceleration (weight) equivalent to the net frequency differential.

Gravity has long been speculated to involve some sort of oscillatory effect. Pythagoras, for example, over 2000 years ago surmised that planetary motions were in accordance with harmonic proportions in a meta-physical model that later became known as *Musica Universalis* or "the Music of the Spheres". Robert Hooke in 1691, in attempting to relate gravitation to waves propagated in a medium, demonstrated that floating objects on water agitated by waves were attracted to the center of agitation. Ruggero Boscovitch in his work *Theoria Philosophiae Naturalis* in 1758 proposed a law of gravitational force as a harmonic convergent series that is scale dependent. At close range, the force between matter points is strongly repulsive and alternately becomes attractive or repulsive with distance and approaches Newton's inverse square law at large distances. This theory of forces was proposed to expand Newton's theory which is scale invariant to explain the existence of size of material structures resulting as a balance between attractive and repulsive forces as illustrated by the Boskovic curve and anticipated nuclear forces.

A hydrodynamic analogy to gravitation and electromagnetism was demonstrated by William Thomson (Lord Kelvin) circa 1870 in which solid spherical bodies immersed in an incompressible fluid induced to oscillate by electric currents or mechanical oscillations could cause attraction or repulsion depending on relative density ratio of the fluid[86]. A large sphere of radius (r) free to move was attracted to a smaller immobilized sphere depending on the density of the free globe (ρ), if its density was greater than the surrounding fluid and separation distance less than $r/\sqrt{(1-5\sqrt{((1 + 2\rho)/3)})}$. For a massless sphere ($\rho = 0$), the critical distance for a free rigid body, Thomson calculated as ~2¼ radii of the free globe. Similar effects were already well known in acoustic vibration effects of tuning forks on candle flames, induced synchronous vibrations in pendulums, etc. Thomson's investigations were said to have inspired Nikola Tesla's subsequent discoveries of resonant EM phenomena. This effect was later further investigated by Carl A. Bjerknes and in 1881 in Paris International Electrical Exposition demonstrated attraction and repulsion of oscillating masses set in rhythmic vibration immersed in fluid[86]. With a pair of submerged spheres pulsating with changes in volume with equal periods T, they will attract each other if the phase difference ϕ is < T/4; but will repel each other if the phase difference is T/4 > ϕ < 3T/4. The phased waves produced a force that varied as $1/r^2$ in

34. Gravitation as a Harmonic Phenomena

accordance with Newton's law. The Bjerknes effect demonstrated in an incompressible medium is opposite that of electromagnetism in which like poles or charges repel and unlike poles or charges attract.

The pulsatory hypothesis of gravitation was further developed by W.M. Hicks[87]. A.H. Leahy subsequently demonstrated in 1885 the opposite effect occurs in an elastic, compressible medium[88]. If the waves are in phase, the interference is constructive. If the waves are out of phase, the interference is destructive and if the amplitudes are identical the amplitude sums to zero. The standing wave force imbalance between coupled oscillators is proportional to the difference in wave energy density. The instantaneous wave energy density is in turn proportional to the square of the wave amplitude.

Wave synchronization and standing wave interference effect of two coupled oscillators is shown in Figs. 34-1 and 34-2. Interferemce node no. denote degree of entanglement.

Electrodynamic wave phase interference

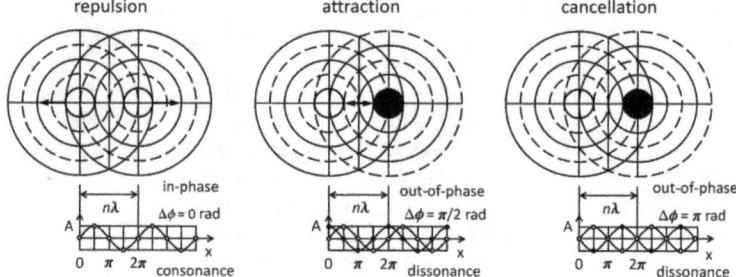

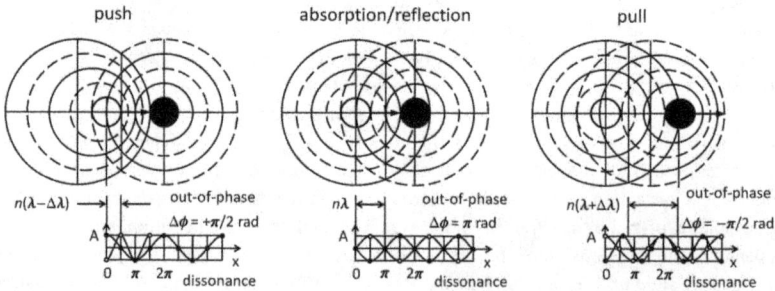

Fig. 34-1. Wave phase interference effects for a pair of interacting oscillators in an elastic, nondissipative, nonviscous perfect fluid medium. Motion of matter arises as a result of dissonance in which the wave system is attracted to wave antinodes of out-of-phase oscillator pairs in alternating push/pull fashion.

The only thing that interferes with my learning is my education. – Mark Twain

Imagination is the highest form of research. – Albert Einstein

34. Gravitation as a Harmonic Phenomena

Standing wave interference

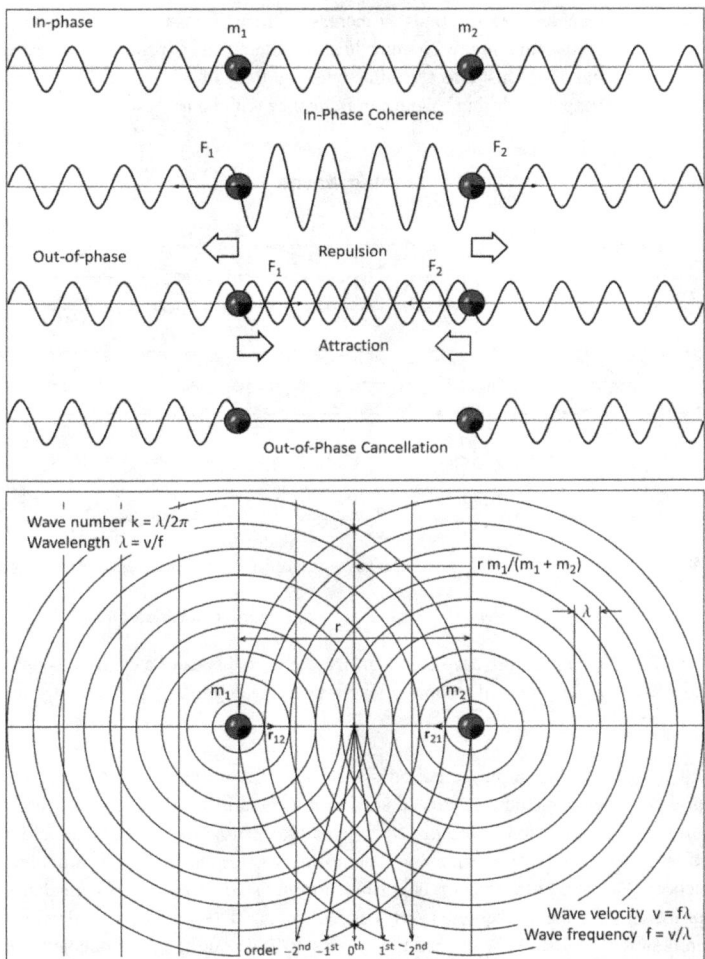

Fig. 34-2. Standing wave interaction of a pair of oscillators of equal frequency in an idealized, nondissipative elastic medium results in attraction or repulsion depending on phase synchronization. The force of attraction is equivalent to that of gravity – an effect due to frequency discordance. The force imbalance is proportional to the difference in wave energy density between oscillators and inversely to the wave velocity.

Standing wave boundary effects vary with type of end conditions, i.e., fixed or free floating as illustrated in Fig. 34-3. A fixed end reflection of a travelling wave from a denser medium induces a π radian phase shift. A free end corresponds to reflection from a less dense medium. An impedance mismatch results in a reflected wave interfering with the forward wave producing standing waves along a transmission line or waveguide. The ratio of the maximum amplitude at an antinode to the minimum amplitude at a node is

34. Gravitation as a Harmonic Phenomena

defined as the standing wave ratio (SWR). The reflection coefficient Γ is defined as the ratio of the ratio of the amplitude of the forward wave and reflected wave. Increased nodal density (scattering centers) corresponds to increased energy density and results in partial wave reflection. The standing wave envelope is altered as a result as measured by increased SWR and Γ. This effect is illustrated in Fig. 34-4. Node displacement is symmetrical occurring only during changes in frequency with no net resonator impulse.

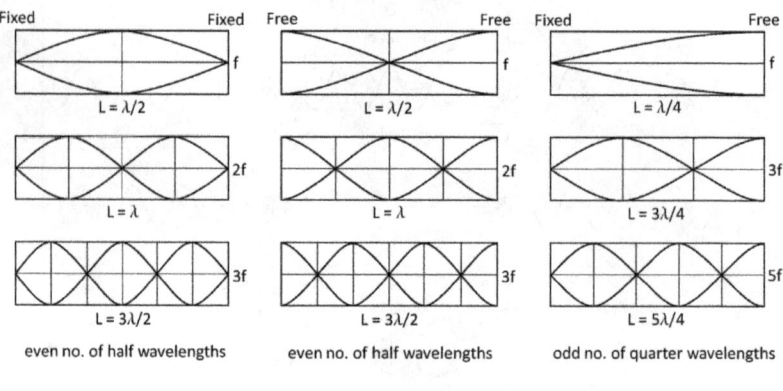

Fig. 34-3. Standing wave boundary effects of free and fixed ends of a cavity resonator for different whole number ratios of resonator length vs. wavelength.

In the Bjerknes effect, in-phase periodic harmonic motion interference in an isotropic, incompressible, dense medium results in attraction. Conversely, anharmonic or discordant motion results in repulsion. In this hydrodynamic analog to gravitational attraction, induced in-phase bubble oscillation by fluid excitation results in bubble attraction and coalescence. The force of attraction is attributed to net static pressure (p_0) imbalance as a result of increased dynamic pressure ($\frac{1}{2}\rho v^2$) due to increased average flow field velocity (v) between bubbles. Pressure waves result in expansion and contraction of bubbles inducing fluid flow. With in-phase oscillations perpendicular to the symmetry plane, fluid flow diverges outward parallel to symmetry plane and acts as if there was a repulsive force between each bubble and its anti-phase image. Conversely, out-of-phase oscillations result in alternating fluid flow between source and sink perpendicular to the mirror image symmetry plane. In-phase oscillation increase dynamic pressure and decrease static pressure while out-of-phase oscillations result in the opposite effect. The resultant fluid pressure imbalance results in the apparent force of attraction or repulsion. In this hydrodynamic analog to gravity, the velocity flow field pattern associated with attraction and repulsion is counter to that of electrostatic or magnetostatic fields in which like charges repel and opposite charges repel. For the hydrodynamic case, open flux lines are associated with attraction and closed flux lines are associated with repulsion. With electrostatic or magnetostatic fields, the opposite holds true as unlike charges attract (closed flux lines) and like charges repel (open flux lines). Gravitostatic fields attract (open flux lines) while gravitomagnetic fields (closed flux lines) repel. See Figs. 34-5 and 34-6.

34. Gravitation as a Harmonic Phenomena

Winterberg[7] theorizes that quantum fluctuations of Planck masses produce virtual longitudinal compression waves and are the cause of the attractive Newtonian gravitational potential obeying an inverse square law. In classical fluid dynamics, an attractive inverse square force law is created between two pulsating spheres immersed in an incompressible fluid. The wave equation for Planck vortices has the same form as the scalar component ϕ of the electromagnetic vector potential **A**. Gravity may be viewed as a weak interference effect of electromagnetic waves in a Planck vacuum which behaves as an elastic, nondissipative Bose-Einstein condensate (BEC) superfluid. As energy is proportional to frequency, the force of gravity increases with frequency and energy density differential. Gravitation may be viewed as a broadband interference effect of interacting quantum oscillators of predominantly extremely high frequency with EM wave velocity propagating at $v = c = \Gamma c_0$.

Mass is a wave interference effect and represents an obstruction to flow of wave energy. Energy represents the curvature of the wave function. For a collection of interacting mass oscillators, the origin of mass in terms of the Einstein relation ($E = mc^2$), the Planck relation ($E = nh\nu$) and the deBroglie equation ($\lambda_{dB} = h/p = 2\pi/k$) is described as a wave interaction. Mass energy equivalence ($m = E/c^2$) may be understood as a statement that the volumetric density of wave scattering centers (nodes) is proportional to the wave frequency (ν) divided by the square of wave propagation velocity (c^2). The greater the frequency of a wave function $\psi(r,t)$, the greater the wave function curvature. The greater the frequency, the greater the linear momentum (p) and nodal density. The magnitude of the curvature increases as the total energy E exceeds the potential energy V. The kinetic energy K is proportional to curvature and is proportional to the square of momentum (p). The greater the momentum, the greater force is exerted in a collision. A superposition of interfering wavefunctions results in a localized wave packet that has a sharply defined momentum and wavelength corresponding to a classical particle. The slope or mean gradient (first derivative) of the wave function $\psi(r,t)$ represents the momentum of the wave/particle. The mean curvature (second derivative) represents the kinetic energy of the wave/particle. Field energy density is proportional to the number nodes. Effective node width and reflection coefficient increases with nodal density. Mass varies with nodal cross-sectional area times no. of nodes times overlap period. Negative mass is equivalent to negative energy, frequency, curvature, and EM wave impedance corresponding to negative Lagrangian.

Inertial mass ($m = \gamma m_0 = F/a = W/g = \hbar k/v$) is a measure of an object's resistance to acceleration when a force is applied. Acceleration increases EM wave nodal density. Mass is an electromagnetic interference phenomena and may be expressed in terms of electric field strength **E** and magnetic flux density **B** as follows: $m = E/(E/B)^2$. In quantum mechanics, mass appears as a scalar quantity in the Schrödinger wave equation

$$i\hbar \delta/\delta t(\psi(r,t)) = -\hbar/2m \nabla^2 \psi(r,t) + V(r,t)\psi(r,t) \tag{34-1}$$

where the time rate of change of the wave function is a function of the Dirac constant $\hbar$, mass m, Laplacian of the wave function ∇^2 and potential energy $V(r,t)$. The development of de Broglie standing waves is deterministic. Given a standing wave pattern in initial state with a matter wave amplitude $\psi(t)$, the probability that it will be observed in a subsequent state $\phi(t)$ is $|\langle \psi(t) | \phi(t) \rangle|^2$. For a standing wave resonator, mass $m = 4V\rho_e/3c^2$, where V is the volume of the cavity, ρ_e the energy density at rest, and c the velocity of light.

Though the ether hypothesis passed away, the electromagnetic wave remained, for now it became possible to view the oscillating field as oscillating change in geometry of space. – Isaac Asimov.

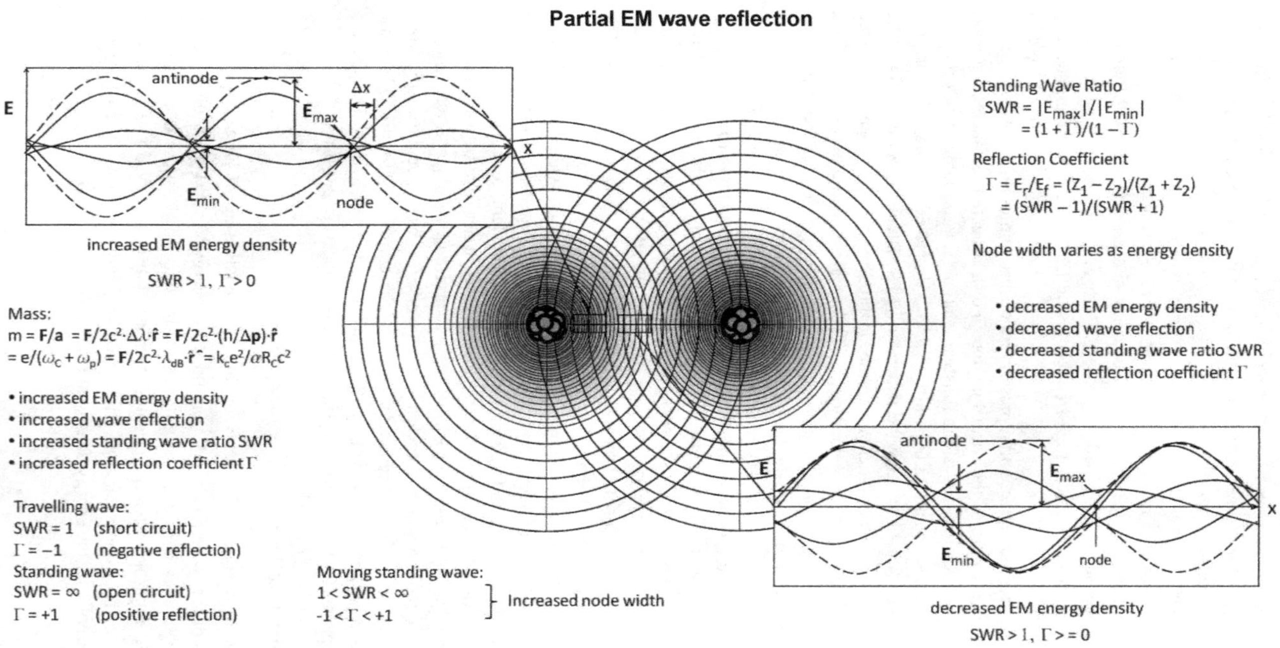

34. Gravitation as a Harmonic Phenomena

Partial EM wave reflection

Fig. 34-4. Impedance mismatch and partial reflection of EM waves radiated from a pair of interacting mass oscillators results in decreased energy density. Node width varies as energy density. For simplicity, only a single frequency is illustrated. The gravitational frequency spectrum covers the entire EM frequency spectrum. The farfield vacuum impedance for electromagnetic waves $Z_0 = |\mathbf{E}|/|\mathbf{H}| \cong 120\pi$ Ω. A black hole corresponds to a perfect absorber (impedance match). The Big Bang singularity corresponds to a white hole or perfect reflector (impedance mismatch).

354

34. Gravitation as a Harmonic Phenomena

Oscillator in-phase interference in hydrodynamic fluid

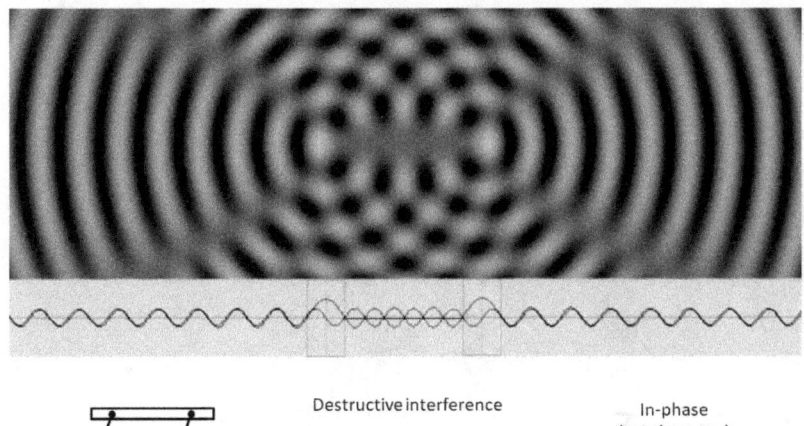

Consonance   Destructive interference   In-phase (synchronous)

$\lambda_1 = \lambda_2$ 0° phase lag

Attraction

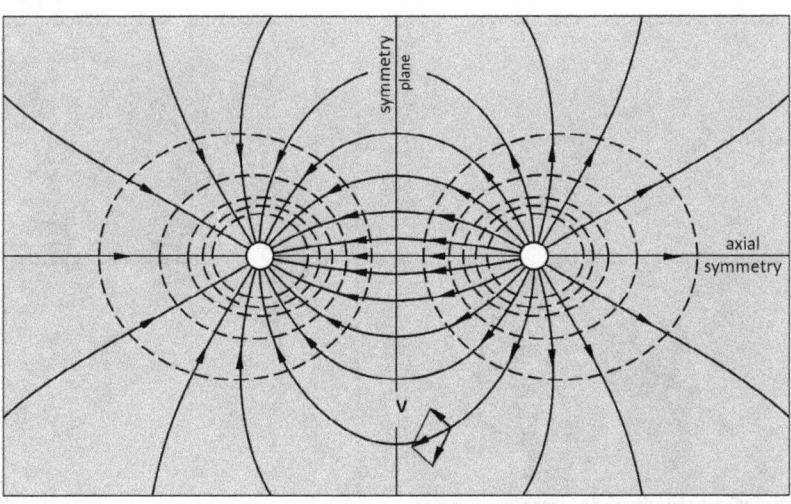

Anti-phase

Fig. 34-5. In-phase constructive wave interference by interaction of a pair of oscillators in a dense, incompressible hydrodynamic fluid. In-phase pulsations increase dynamic pressure and decrease static pressure attracting less dense bubbles. Electrostatic and magnetostatic fields where unlike charges or poles attract behave oppositely to the hydrodynamic case. At any given point, the flux direction is tangent to the intensity. The flux pattern resembles that of two unlike electrical charges. The flux lines are closed and everywhere normal to the equipotential surfaces.

34. Gravitation as a Harmonic Phenomena

Oscillator out-of-phase interference in hydrodynamic fluid

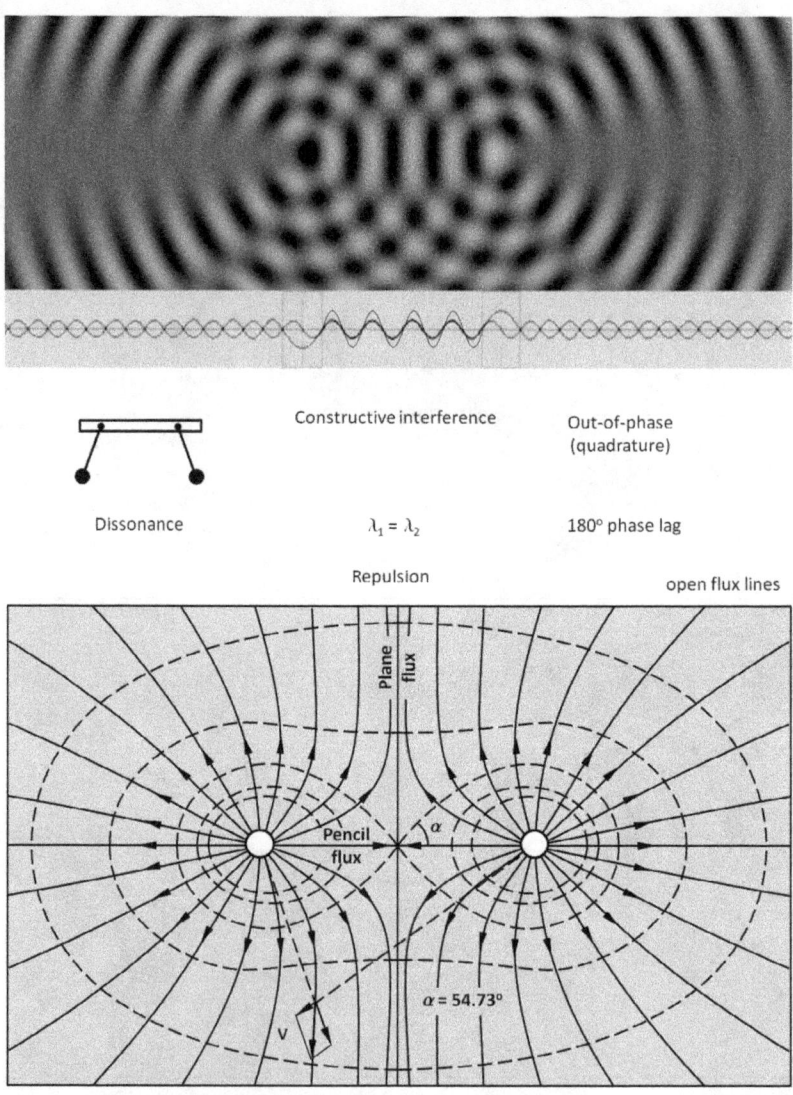

Fig. 34-6. Out-phase destructive wave interference by interaction of a pair of oscillators in a dense, incompressible hydrodynamic fluid. Out-of-phase pulsations decrease dynamic pressure and increase static pressure repelling less dense bubbles. A similar flux field is produced by two like electric charges. The equipotential contours are Cassinian curves with a two-loop lemniscate equipotential with foci at mass centers. In the far field region, the flux lines become asymptotic to radial lines though the center of gravity.

35. Gravitational Frequency Domain

There's a certain irrationality to any work in gravitation... because the dimensions are so peculiar. – Richard Feynman

The electromagnetic (EM) frequency spectrum in terms of octaves is illustrated in Fig. 35-1. At x-ray frequencies (f > 10^{16} Hz), anomalous dispersion ceases and the refractive index for dielectric materials approaches 1. All natural dielectric materials become transparent (non-absorbing) and non-refracting at γ–ray frequencies (f > 10^{19} Hz). Effects of gravitation are associated with principally the extreme high energy end of the EM spectrum. Electromagnetic frequencies associated with gravitation encompass a broad spectrum and estimated by Boutorin, Bershadsky and Akimov to lie predominately in the 10^{20} to 10^{40} Hz range[89, 90, 91]. At the Earth's surface, the bulk of the PV spectral energy is well above the "THz" range as estimated by Storti *et al*[33]. The Earth's gravitational frequency spectrum is estimated to extend into the Yottahertz (10^{24} Hz) range.

EM radiation of mass objects consist of photon emission and absorption by electrons/positrons which form the basic building block of matter including constituent atoms and molecules. The electron Compton frequency f_C is 1.236 x 10^{20} Hz and represents the upper frequency limit of electrons and positrons. Electromagnetic radiation is associated with a changing magnetic quadrupole. Gravitons ($\gamma\gamma^*$) consists of superposition states of photons (γ) and counter-propagating phase conjugate photons (γ^*) and are not directly detectable. As such, graviton frequencies are identical to the associated photon frequencies. The Planck frequency (f_P = 2.913 YHz) represents the apparent upper frequency limit supported by the vacuum. Below this limit only potentials exist rather than fields.

Large scale gravitational waves are produced by dynamic interaction of coupled masses such as merging of binary black holes. The frequency of gravitational waves depends on the nature of the source. As there are no naturally occurring macroscopic negative gravitational masses, gravitational radiation is of a quadrupole radiation pattern rather than dipolar radiation pattern of oscillating electrical charges within a Hertzian antenna. Gravitational wave frequencies of astronomical objects are estimated to lie in the lower end of the electromagnetic spectrum roughly in the range 10^{-16} to 10^{10} Hz. For pairs of supermassive black holes the gravitational frequency range is on the order of ~10^{-10} to 10^{-2} Hz while pairs of dense stars is ~10^{-3} to 10^2 Hz. The gravitational wave frequency of pulsars is ~10^{-9} to 10^{-7} Hz. Detection of gravitational waves emitted by the spiral and merger of two black holes in a signal identified as GW150914 by the ground-based Laser Interferometer Gravitation-Wave Observatory (LIGO) in 2016 was in the range of 35 to 450 Hz with a chirp duration of ~0.25 sec.

Happy is he who can trace effects to their causes. – Publius Vergilius Maro (Virgil)

To know a thing well, know its limits. Only when pushed beyond its tolerances will true nature be seen. – Frank Herbert

An ocean traveller has even more vividly the impression that the ocean is made of waves than it is made of water. – Arthur Stanley Eddington

If we wish to make an observation on a system of interacting particles, the only effective method of procedure is to subject them to a field of electromagnetic radiation. Thus the role of the field is to provide a means for making observations. The very nature of observation requires an interplay between the field and the particles. – P.A.M. Dirac.

35. Gravitational Frequency Domain

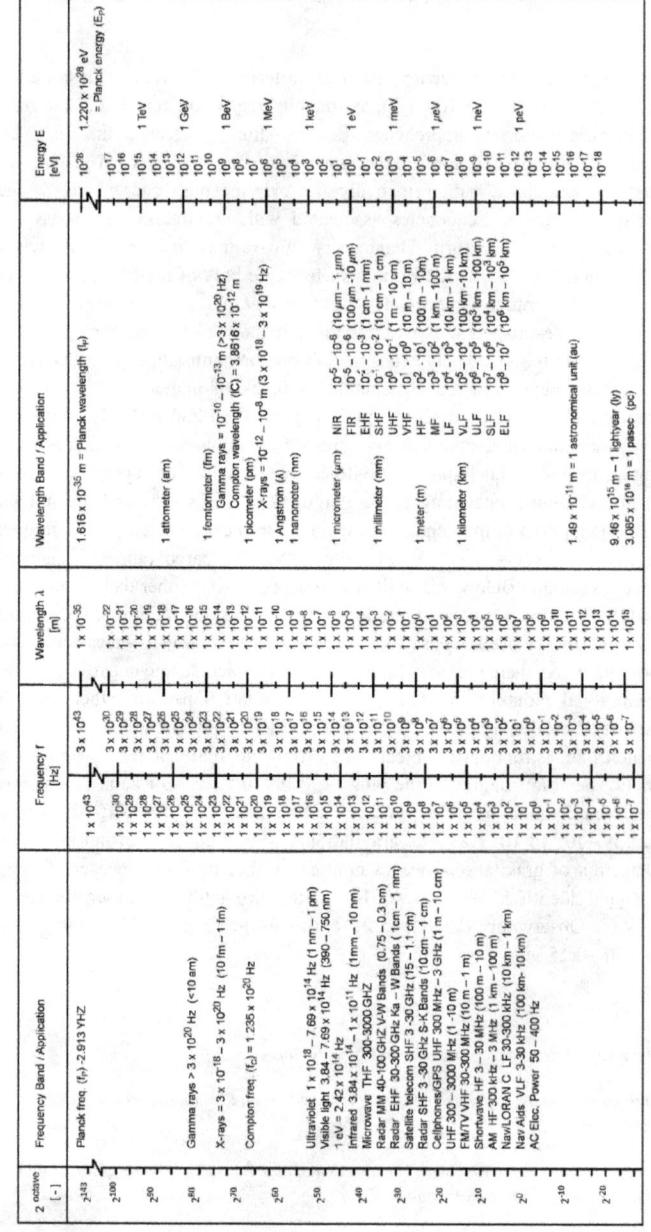

Fig. 35-1. Gravitation frequencies encompass the entire electromagnetic frequency spectrum with energy concentrated in the extreme high frequency range well above visible light frequencies. Gravitational wave frequencies lie in the lower end of the EM spectrum.

36. Mass Scaling

The mass of a body is a measure of its energy content. – Albert Einstein

All the evidence, experimental and even a little theoretical, seems to indicate that it is the energy content which is involved in gravitation, and therefore, since matter and antimatter both represent positive energies, gravitation makes no distinction. – Richard P. Feynman

The range of mass of the observable universe from the subatomic scale to the galactic scale is illustrated in Fig. 36-1 as a function of the Schwarzschild radius R_S (= $2GM/c^2$). As shown, the observed masses follow a logarithmic progression. The Schwarzschild radius represents the radius of a sphere into which matter must be compressed in order to form a black hole. The surface of a sphere with this radius corresponds to the black hole event horizon from which neither mass or energy can escape. Mass appears related to a contraction in radius such that the difference in circumference equals the Schwarzschild wavelength λ_S (= $2\pi R_S$). The vacuum refractive index K_{PV} may likewise be expressed as a function of Schwarzschild radius $K_{PV}(r,\omega,M) = e\wedge(2GM/rc^2) = e\wedge(R_S/r)$. Gravitational permeability equals Schwarzschild radius/Mass ratio $R_S/M = \mu_g = G/c^2 = 7.4604\text{E-}28$ m/kg.

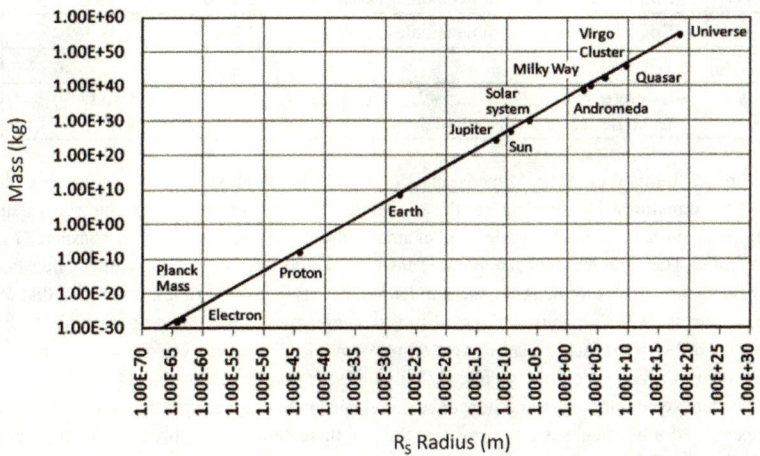

Fig. 36-1. Log plot of Mass vs. Schwarzschild radius R_S of subatomic and astrophysical objects. The gravitational permeability G/c^2 appears constant in time and distance.

In the electron-positron model proposed by Sternglass[15], the Gravitational constant, electron mass radius and Schwarzschild radius increase with time. A logarithmic plot of mass of atoms, planets and galaxies expressed in units of electron mass exhibits a linear relation with radius expressed in multiples of the Compton radius with a grouping that varies in scale multiples of 2^{10} = 1024. Just as mass exhibits a linear relationship with Schwarzschild radius, a linear relationship between hadron particle spin angular momentum J ($\propto \alpha M^2$) and the square of mass, is apparent in Regge trajectory plots of experimental

36. Mass Scaling

mass spectra suggestive of a spinning string under tension. The universe does not appear to be globally scale invariant as the vacuum state at the Planck scale is discrete (quantum mechanical) while at cosmological scales appears continuous (classical). Breaking of conformal scale symmetry may be gravitationally induced.

Oldershaw[92] proposes the universe appears to exhibit discrete scale relativity (DSR) effects governing size of aggregations of matter with distinct morphological differences in which gravity plays a central role. Oldershaw has identified discrete scale bands corresponding to subatomic, atomic, planetary, stellar and galactic size ranges that display a clustering or fractal characteristic termed Discrete Scale Relativity (DSR) corresponding to scale-free (SF) resonant coupling of oscillators. Oldershaw notes that from the smallest observable subatomic particles to the largest cosmological structures spans nearly 80 orders of magnitude. Three dominant mass ranges, i.e., Atomic, Stellar, and Galactic scales, each extending over a relatively narrow range of 5 orders of magnitude account for >99% of all mass. The mass range of these three scales are roughly as follows: Atomic (10^{-30} kg to 10^{-25} kg, Stellar (10^{25} kg to 10^{30} kg and Galactic scale (10^{35} kg to 10^{40} kg. Oldershaw theorizes mass (M), length (L) and time (t) are related by discrete self-similar transformation equations.

$$L_\psi = \Lambda \, L_{\psi-1} \quad (36\text{-}1)$$
$$T_\psi = \Lambda \, T_{\psi-1} \quad (36\text{-}2)$$
$$M_\psi = \Lambda^D M_{\psi-1} \quad (36\text{-}3)$$

where

Λ	= Empirically derived dimensionless constant (= 5.2×10^{17})
D	= Empirically derived dimensionless constant (= 3.174)
Λ^D	= 1.70×10^{56}
ψ	= Discrete scale index (= ..., −2, −1, 0, 1, 2, ...). Atomic, Stellar, and Galactic scales are assigned $\psi = -1$, $\psi = 0$, and $\psi = +1$, respectively.

According to this theory, any dimensional constant or physical parameter for a given scale may be transformed according to these scaling relations to determine the equivalent parameter on a neighboring scale. For example, the Newtonian gravitation constant G = $[\Lambda^{1-D}]\psi G$. The gravitational coupling constant in this model varies with scale. For transformation to an atomic scale, the gravitational constant is 10^{38} times larger than that on the stellar scale. The cause of the observed aggregation size clumping remains to be elucidated, but presumably may be related to harmonic dissonance effects of resonant phi damping ratios and synchronicity of coupled, nonlinear chaotic systems.

Discrete scale relativity is a renormalization theory in that it attempts to describe how the properties of a physical system change as the length scale on which observation are made are changed form the quantum to the cosmological scale mediated by scale-dependent feedback loops. The scale groupings represent renormalization groups – a mathematical formalism describing how things scale. Simplicity and beauty may be found at the endpoints of such groupings.

The joy of engineering is to find a straight line on a double logarithm diagram. – Thomas Koening

It is always pleasant to have exact solution in simple form at your disposal. – Karl Schwarzschild

The quantum theory of gravity has opened up a new possibility, in which there would be no boundary to space-time and so there would be no need to specify the behaviour at the boundary.
– Stephen Hawking

37. Generalized Newtonian Gravitational Law

A general force law, as derived by Pierre-Simon Laplace in *Méchanique Céleste (Celestial Mechanics)* published in 1799, has the form $F = Ar + B/r^2$. A generalized Newtonian gravitational force equation may be expressed as $F = \lambda mMr - GmM/r^2$. The corresponding potential becomes $\phi = -GmM/r - (\lambda mM/2)(r^2 + R^2)$. Newton noted in the *Principia* that forces of attraction varying with inverse square distance or with linear distance can both cause bodies to move in stable planetary orbits described by conic sections but dispensed with a 'harmonic' linear force law as not in accordance with experimental observations. A linear r dependency results in a circular or elliptical orbit while a $1/r^2$ dependency also includes parabolic and hyperbolic orbits. Inclusion of both terms results in orbital precession as in Einstein's General Relativity. Barrow[93,94] and Calder & Lahav[95] note that the constant in the linear term λM corresponds to Einstein's GR cosmological constant Λ ($= 3\lambda/c^2$) describing the effect of dark energy (quintessence) which may be related to the total mass of the universe. This postulated dark energy repels itself gravitationally ($F = -GM/r^2 + \Lambda/3\ r$). The cosmological constant was included in Einstein's gravitational field equation $G_{\mu\nu} - \Lambda g_{\mu\nu} = 8\pi G T_{\mu\nu}$ to offset gravitational expansion to maintain a steady-state, constant radius universe but was initially discarded for lack of observational evidence. The linear term in the generalized Newtonian equation dominates at cosmological scales. Jefimenko[17] derives a generalized gravitational force equation $\mathbf{F} = -G[mM/r^2]\cdot\mathbf{r}_u - G[mMv^2/c^2r^2]\cdot\mathbf{r}_u$ which accounts for Mercury's perihelion advance.

In the electrodynamic theory advanced by Lucas[57], the emission of radiation causes a decay in the force of gravity over time due to a decrease in energy and in the value of mass. Radiation emission of energy L ($= hf = m_0c^2(1-\sqrt{(1-(1-v^2/c^2)}))$) reduces the mass by L/c^2 where L is the Lagrangian. Atoms nearest the center of an astronomical body lose their oscillation energy more slowly than those atoms nearest the surface. The rate of decay of gravitational force varies with the ratio of volume to surface area. Lucas maintains this weakening gravity effect accounts for conjectured expansion of the Aechean Earth since its surface solidified while conserving energy and momentum. According to the expanding Earth model, the interior volume has been increasing since initial formation with rupturing and fracture of the cold, thin, brittle outer crust (Lithosphere). Images of Mars, Mercury, Moon, Ganymede, Enceladus, Miranda, Europa, Dione, Ariel, for example, likewise show large scale rift valleys, surface striations, crustal fracturing/cracks, wrinkle ridges, etc suggestive of growth stretch marks and surface tension not explained by tectonics and subduction. This model assumes essentially no mass accretion since formation. The expansion may be attributed, in whole or in part, to thermal expansion. It is generally accepted that the universe as a whole is expanding as indicated by the cosmological redshift with a conversion from a radiation-dominated to a mass-dominated universe. Rather than postulate that every point in space is expanding spherically, a decay in the force of gravity results in expansion of matter and increase in enclosed volume. The weakening of the force of gravity with time may account for relatively high velocity of stars in spiral galaxy discs and the overall expansion of the universe without recourse to assumption of dark energy.

One of the possible explanation proposed for accelerated expansion of the universe is gravitational repulsion due to dark energy with negative pressure and positive energy density. Zel'dovich[96,97] gives the mass density of the vacuum $\rho_\Lambda = c^2G/8\pi\Lambda$, energy $E_\Lambda = c^2\Lambda/8\pi G$ and pressure $P_\Lambda = -E_\Lambda$. where $\Lambda \cong G^2m^6/\hbar^4$. Estimates of dark energy density ρ_Λ is approximately the same order as mass density ρ_M of the known universe. Negative gravitational field energy equals positive mass energy density. The vacuum energy consisting of EM radiation flux acts a frictionless, perfect fluid with constant energy density

37. Generalized Newtonian Gravitational Law

$\rho_\Lambda = \Lambda/8\pi G$. A dark energy density is equivalent to a negative pressure $p = -\rho_\Lambda$. The ratio of the dark energy pressure to the dark energy density defines the dark energy equation of state $w_\Lambda = P_\Lambda/\rho_\Lambda$. This negative pressure is attributed to apparent repulsive gravity accelerating the expansion of the universe. At present, dark energy and non-baryonic dark matter are thought to account for ~74% and ~22% of the total energy and matter of the universe with the remaining ~4% ordinary baryonic matter (protons, planets, stars, galaxies, etc) of which ~0.5% is visible. See Fig. 37-1. Dark energy appears as a dissonance effect of interaction of positive and negative energy and curvature including, for example, local vacuum spectral energy density compression due to creation of Planck mass → repulsion.

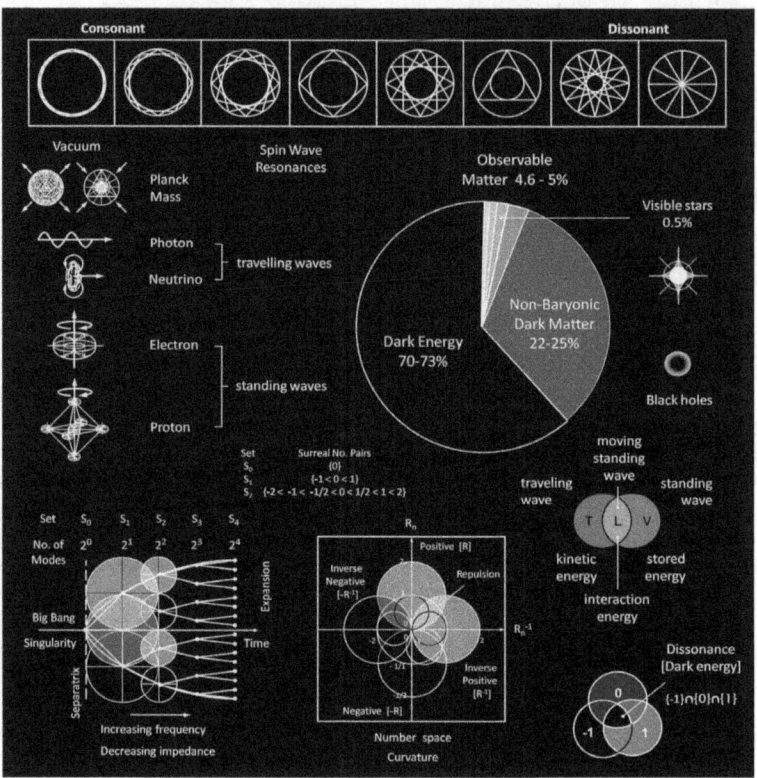

Fig. 37-1. Most of the energy density of the universe appears stored as dark energy. All observable light and elemental particles spin. Spin wave disturbances in the Planck vacuum formed of Planck dipoles of positive and negative mass result in travelling wave excitations such as light waves and neutrinos and resonant, closed-loop standing wave structures such as electrons, protons, neutrons as well as composite atomic and molecular groupings. Total energy and curvature of the universe equals zero of which half is positive and half negative.

It happens that the requirements of simplicity and beauty are the same, but where they clash, the latter must take precedence. – P.A.M. Dirac.

38. Gravitational Potential

Gravitational potential is a three dimensional scalar function ϕ (x,y,z) and may be interpreted as a measure of electromagnetic energy (EM flux) density. The gravitational potential ϕ at a distance r for a spherically symmetrical mass distribution M(r) represented as a point mass within the shell radius is given by

$$\phi(x,y,z) = U(r) = -GM/r \qquad (38\text{-}1)$$

where:

ϕ	Gravitational potential (potential energy/unit mass) = ϕ_g		
	Value: Varies (≤ 0)	Units: Joules/kg (= m^2/sec^2)	Dimensions: L^2T^{-2}
G	Universal Gravitational const. (Newtonian gravitational coupling) = $c^2(\sum r_i/m_i)$		
	Value: 6.67384(80) x 10^{-11} (2010 CODATA)	Units: $N\cdot m^2/kg^2$ (= $m^3/kg\cdot sec^2$)	Dimensions: L^3/MT^{-2}
M	active mass of central object (= $-\phi_g R/G$ = $E/(\mathbf{E}/\mathbf{B})^2$ = ($\hbar \mathbf{k} = q_0\mathbf{A})/\mathbf{v}$)		
	Value: Varies (≥ 0)	Units: kg	Dimensions: M
r	Gravitational radius (distance from center of mass to surface)		
	Value: Varies (≥ 0)	Units: m	Dimensions: L

For a sphere, the potential may be represented as a series expansion

$$\phi = -GM/r - G^2M^2/4r^2c^2 + \ldots \qquad (38\text{-}2)$$

The force of gravity $F_g = -m\Delta\phi/\Delta r = -m\nabla\phi$ (N = kg·m/s²) where m is the passive gravitational mass (kg). For a planetary or stellar mass, conversion of gravitational energy generates heat within the core driving convective thermal flows.

To represent a timelike causal event ordering relation, by analogy with retarded potential of the electromagnetic field, the retarded gravitational potential shown by Jefimenko[16] may be written as

$$\phi([\mathbf{r}],t) = U([\mathbf{r}],t') = -GM(t')/r \qquad (38\text{-}3)$$

where:

[**r**] is the retarded position vector of the moving mass point to be evaluated for retarded time t' = t – r/c, t = time for which **g** is evaluated, r = distance between field point and source and c = velocity of propagation velocity of gravitation.

Barrow and Tipler[93] add a cosmological term to the Poisson equation

$$\phi = -GM/r - \Lambda/6\, r^2 \qquad (38\text{-}4)$$

where $\Lambda = 8\pi G\rho_\Lambda$.

...natural phenomena involving gravitation and inertia...electricity and magnetism...are not independent of one another, but are intimately related.... – Hendrik A. Lorentz

38. Gravitational Potential

The gravitational potential for an idealized spherical mass is illustrated in Fig. 38-1. The potential shown assumes a uniform solid sphere. The normalized potential is expressed in terms of units of $-\mu/r$. The potential energy of an object of mass m in the vicinity of a central mass M varies as $mg\Delta r$ where g $(= -GM/r^2)$ is the gravitational field strength. In GR, the metric and the gravitational potential are described by the metric tensor while the geodesics and gravitational force field are described by the metric connection usually represented in terms of Christoffel symbols Γ^i_{jk} for I, j, k = 1,2, ... n.

In analogy with electromagnetic potential ϕ_e, Krogh[32] provides the following representation for gravitational potential ϕ_g [$q/4\pi\epsilon_0 \rightarrow -Gm_o$] for a moving particle of rest mass m_0 in rectangular coordinates moving in the x-direction

$$\phi_g(x,y,z) = -Gm_0/\sqrt{(x^2 + y^2 + z^2)(1 - v^2/c^2)} \tag{38-5}$$

where:

ϕ_g	Gravitational potential (potential energy/unit mass) $(= -GM/r = c_0^2/K_{PV} = c_0^2/\Gamma^2)$		
	Value: Varies (≤ 0)	Units: J/kg = m²/s²	Dimensions: L^2T^{-2}
G	Universal Gravitational const. $= c^2(R_{univ}/M_{univ}) = c^2(\sum r_i/m_i) = \Gamma_0/\Gamma^3 = \phi_g\mu_g/4\pi$		
	Value: 6.67384(80) x 10^{-11} (2010 CODATA)	Units: N·m²/kg² (= m³/kg·sec²)	Dimensions: L^3/MT^{-2}
m_0	particle rest mass $(= m/\Gamma = m/\sqrt{K_{PV}} = E/(E/B)^2 = 2L/(E^2 - c^2B^2) = \hbar k/v)$		
	Value: Varies (≥ 0)	Units: kg	Dimensions: M
x, y, z	rectangular coordinate (direction of motion along x-axis)		
	Value: Varies	Units: m	Dimensions: L
v	particle velocity $(= v_g = v_0/\Gamma = v_0/\sqrt{K_{PV}} = (\hbar k - qA)/m_i = c \cdot (\Delta\phi/\pi) = \beta \cdot c)$		
	Value: Varies	Units: m/sec	Dimensions: LT^{-1}
c	velocity of light $(= c_0/\Gamma = c_0/\sqrt{K_{PV}} = E/B = \phi_E/\phi_B = l_P/t_P = \sqrt{(v_g v_p)} = v/\beta)$		
	Value: ≈ 2.997924 x 10^8	Units: m/sec	Dimensions: LT^{-1}

Eq. (38-4) is comparable to the scalar electromagnetic potential ϕ_e for a particle at the origin moving in the x-direction.

$$\phi_e(x,y,z) = -q/\sqrt{(x^2 + (y^2 + z^2)(1 - v^2/c^2)} \tag{38-6}$$

where q = electric charge. Particles that are self-conjugate have zero charge.

I recently resumed the enquiry by experimenting in a most strict and searching manner, and have at last succeeded in magnetising and electrifying a ray of light and illuminating a line of magnetic force. – Michael Faraday

According to the general theory of relativity, space is endowed with physical qualities... space without ether is unthinkable. – Albert Einstein

Using de Broglie's relation between momentum and wavelength, Schrödinger was able to guess the wave equation for a quantum object travelling in a potential. – Tony Hey

38. Gravitational Potential

Normalized gravitational potential for a central spherical mass

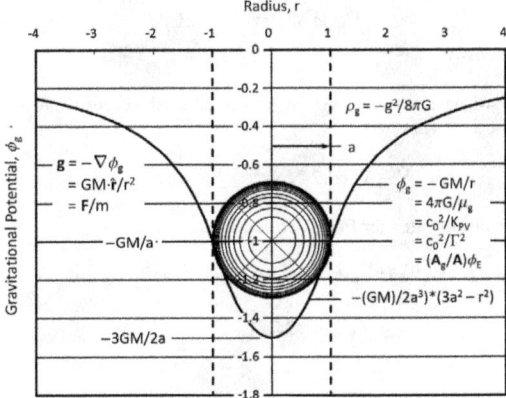

Fig. 38-1. Gravitational potential ($\phi_g = -GM/r = c_0^2/K_{PV} = c_0^2/\Gamma^2$ is a measure of electromagnetic EM energy density of mass M of volume V. In the far field region where EM energy density $\to 0$, $\phi(r) \to 0$ as $r \to \infty$. The gravitational field energy density ρ_g (= $-g^2/8\pi G$) is a measure of local spacetime impedance Z_S (= c^3/G) and local electromagnetic radiation pressure P (= ϕ_E/c). The Laplacian potential $\nabla^2\phi = -4\pi G\rho_g = 4\pi GM/V$.

Krogh observes that in Einstein's GR weak equivalence principle, the motion of a test particle in a gravitational field is independent of its mass, however, according to QM theory, the motion is a function of mass. The conflict may be resolved by relating effects of gravitational potentials as phases of de Broglie waves rather than spacetime curvature. The phase shift ΔS of de Broglie waves associated with a charged particle evaluated over the trajectory path is given by

$$\Delta S = q/h \int \phi dt - q/h \int \mathbf{A} \cdot d\mathbf{s} \tag{38-7}$$

where ϕ and $\mathbf{A}$ are the scalar and magnetic vector potentials, respectively.

The phase shift undergone by de Broglie waves of uncharged matter in a gravitational potential may be represented as

$$\Delta S = m/h \int \phi_g dt \tag{38-8}$$

Electromagnetic potentials represented by the wave equation are

$$\nabla^2 \Phi - (1/c^2)\delta^2\Phi/\delta\tau^2 = -4\pi\rho \tag{38-9}$$

$$\nabla^2 \mathbf{A} - (1/c^2)\delta^2\mathbf{A}/\delta\tau^2 = -4\pi\mathbf{j}/c \tag{38-10}$$

...*the vast majority of mass of the universe seems to be missing.* – *William J. Broad*

I think it would have been much better if Newton had contemplated how the apple got up there in the first place. – *Viktor Schauberger*

38. Gravitational Potential

Krogh develops the following equation for the gravitational potential

$$\nabla^2 \Phi g - (1/c^2)\delta^2 \Phi g/\delta\tau^2 = -4\pi G \rho_m \tag{38-11}$$

where ρ_m = mass density.

The velocity V of de Broglie waves in a gravitational potential slows in proportion to light

$$V = V_0 \, e^{\wedge}(2\Phi g/c_0^2) \tag{38-12}$$

where the 0 subscript indicates far field, zero curvature value.

The corresponding de Broglie frequency ν_{dB} and wavelength λ_{dB} are

$$\nu_{dB} = \nu_{dB0} \, e^{\wedge}(2\Phi g/c_0^2) \tag{38-13}$$

$$\lambda_{dB} = \lambda_{dB0} \, e^{\wedge}(2\Phi g/c_0^2) \tag{38-14}$$

The Einstein relation may be expressed as

$$m_0 c^2 = m_{00} c_0^2 e^{\wedge}(\Phi g/c_0^2) \tag{38-15}$$

Substituting $c = c_0 \, e^{\wedge}(2\Phi g/c_0^2)$ into Eq. (15) yields the rest mass transformation in a gravitational potential

$$m_0 = m_{00} e^{\wedge}(-3\Phi g/c_0^2) \tag{38-16}$$

Kristoffen[98] notes that transport flow media in moving flow fields that give rise to nonlinear effects as a result of repetitive coherent wave-trapping effect in a higher spatial phase ϕ-dimension (5-D spacetime). Unlike linear superposition in linear media, flow waves can refract and reflect one another producing moving Moiré patterns. Moiré waves correspond to de Broglie waves.

For a potential in a moving flow field, Kristoffen derives the following equations

$$\phi = \phi_0/\sqrt{(1 - w^2/c^2)} \tag{38-17}$$

where ϕ_0 is the stationary scalar potential, w = velocity of potential region. The relativistic equivalent of the vector potential attributed to the flow of potential is

$$\mathbf{A} = \mathbf{w}/c\phi = \mathbf{w}\phi_0/c\sqrt{(1 - w^2/c^2)} \tag{38-18}$$

For a charged particle in both scalar and potential fields, the momentum is

$$\mathbf{p} = m_0 \mathbf{v}/(\sqrt{(1 - v^2/c^2)}) + (q/c)\mathbf{A} \tag{38-19}$$

and the energy is given by

$$E = m_0 c^2/(\sqrt{(1 - v^2/c^2)}) + q\phi = (m_0 c^2/g) + q\phi \tag{38-20}$$

All the effects of Nature are only the mathematical consequences of a small number of immutable laws. – Pierre-Simon Laplace

38. Gravitational Potential

The associated gravitational vector component $\mathbf{A}_g$ in analogy with the electromagnetic four-vector $\mathbf{A}_\mu$ is

$$\mathbf{A}_g = (\mathbf{v}/c^2)\phi_g \qquad (38\text{-}21)$$

where:

$\mathbf{A}_g$	Gravitational vector potential ($= \mathbf{v}\cdot\phi_g/\varepsilon_0\mu_0 = \mathbf{v}\cdot\phi_g/c^2 = \mathbf{v}\cdot\phi_g/(\mathbf{E}/\mathbf{B})^2 = (\phi_g/\phi_E)\mathbf{A}$)		
	Value: Varies (≤ 0)	Units: m/s	Dimensions: LT^{-1}
$\mathbf{v}$	particle velocity ($= \mathbf{v}_g = \beta\cdot c = \mathbf{v}_0 + \mathbf{a}t = c\cdot(\Delta\phi/\pi) = (\mathbf{A}_g/\phi_g)c^2 = (\hbar\mathbf{k} - q\mathbf{A})/m$)		
	Value: Varies	Units: m/sec	Dimensions: LT^{-1}
ϕ_g	gravitational scalar potential ($= -GM/r = (E/m)\mathbf{A}_g/\mathbf{v} = (4\pi G/\mu_g)\mathbf{A}_g/\mathbf{v}$)		
	Value: Varies (≥ 0)	Units: Joules/kg $= m^2/s^2$	Dimensions: L^2T^{-2}

The change in gravitational potential as a function of height h and Earth's radius is

$$\Delta\Phi = [GM/R + h] - [GM/R] = -gh/[1 + h/R] \qquad (38\text{-}22)$$

Near the earth's surface, the gravitational potential is linearly approximated by

$$\Delta\Phi = \phi(R + h) - \phi(R) = -gh \quad \text{(uniform gravitational field)} \qquad (38\text{-}23)$$

The acceleration of gravity **g** represents the gravitational field strength ($= \mathbf{F}/m$) and is opposite in direction to the potential gradient, i.e., gravity is the gradient of a scalar gravitational potential ϕ_g defined as

$$\mathbf{g} = -\mathbf{grad}\ \phi_g = -\nabla\phi_g \qquad (38\text{-}24)$$

$$\mathbf{g} = \text{flux/area} = 4\pi GM/4\pi r^2 = -GM/\mathbf{r}^2 \qquad (38\text{-}25)$$

For an unaccelerated point mass moving with velocity v relative to an observer in a time-dependant system, in analogy with the electric field of a uniformly moving point charge as originally derived by Heaviside, the gravitational field vector **g** may be represented as [16]:

$$\mathbf{g} = -G([m(1 - v^2/c^2)]/(r_0^3[1 - (v^2/c^2)\sin^2\theta]^{3/2})\mathbf{r}_0 \qquad (38\text{-}26)$$

where θ is the angle between **v** and $\mathbf{r}_0$.

Jefimenko[16] derives the following relation for acceleration of gravity in terms of gravitational vector potential $\mathbf{A_g}$ ($= (\mathbf{v}/c^2)\phi_g = (\phi_g/\phi_E)\mathbf{A}$)

$$\mathbf{g} = -\nabla\phi_g - \mathbf{A}_g/dt \qquad (38\text{-}27)$$

where $\nabla\cdot\mathbf{g} = 0$ and $\phi_g = -GM/r = c^2\cdot\mathbf{A}_g/\mathbf{v} = (4\pi G/\mu_g)\mathbf{A}_g/\mathbf{v} = c^2\mathbf{A}_g/(e/m/A) = (\mathbf{A}_g/\mathbf{A})\phi_g$.

You just think lovely wonderful thoughts and they lift you up in the air. – J.M. Barrie

Architects spend an entire life with this unreasonable idea that you can fight against gravity. – Renzo Piano

I tried being reasonable once. I didn't like it. – Clint Eastwood

38. Gravitational Potential

In the Evans' electrogravitic model[5], based on tetrad formalism, acceleration **g** in terms of electric field strength **E** is given as:

$$\mathbf{g} = \mathbf{E}c^2/\phi^{(0)} \tag{38-28}$$

where $\phi^{(0)}$ = electrostatic scalar potential (volts).

Gauss's law of gravity (differential form) which is derivable from Newton's Law of gravity is

$$\nabla \cdot \mathbf{g} = -4\pi G\rho = -\Phi_g \rho \tag{38-29}$$

where Φ_g = gravitational flux and ρ = mass density.

If gravitational energy is considered itself as a source of gravitation, then a second cogravitational field similar to the magnetic field arises according to Jefimenko[16] introducing an additional non-linear term

$$\nabla \cdot \mathbf{g} = -4\pi G\rho + g^2/2c^2 = -\Phi_g \rho + g^2/2c^2 \tag{38-30}$$

The total mass density including the effect of gravitational energy on the gravitational field is then

$$\rho_t = \rho + \rho_{gm} = \nabla \cdot \mathbf{g}/4\pi G + g^2/8\pi Gc^2 \tag{38-31}$$

where ρ_{gm} is the mass density of the gravitational field and $\mathbf{g} = \sqrt{(\rho_g/8\pi G)}$.

The differential form of Gauss's law is Poisson's equation which relates the Laplacian of the potential ϕ to the mass density:

$$\nabla^2 \phi = \nabla \cdot (-\nabla \phi) = d^2\phi/dx^2 + d^2\phi/dy^2 + d^2\phi/dz^2 \tag{38-32}$$
$$= 4\pi G\rho$$

where successive operation of the gradient and divergence $\nabla \cdot (\nabla \phi)$ is represented by the Laplacian operator ∇^2.

Electromagnetic potentials ϕ, **A** in terms of the wave equation:

$$\nabla^2 \phi = -(1/c^2)\, \delta^2\phi/\delta t^2 = -\rho/\epsilon_0 \tag{38-33}$$
$$\nabla^2 \mathbf{A} = -(1/c^2)\, \delta^2 \mathbf{A}/\delta t^2 = -\mathbf{j}/\epsilon_0 c \tag{38-34}$$

To incorporate general covariance, the Laplace equation is modified to include time dependency as a weak field approximation as shown by Nyambuya[99] and Krogh[32]

$$\nabla^2 \phi_g = -(1/c^2)\, \delta^2 \phi/\delta t^2 = 4\pi G\rho_m \tag{38-35}$$
$$\nabla^2 \mathbf{A}_g = -(1/c^2)\, \delta^2 \mathbf{A}/\delta t^2 = 4\pi G \mathbf{j}_m/c \tag{38-36}$$

Adding a cosmological term to account for the expansion of the universe, Trautman[100] notes the correct form is

$$\nabla^2 \phi_g = -4\pi G\rho_m - \Lambda \tag{38-37}$$

where Λ is the cosmological constant.

38. Gravitational Potential

In terms of the vacuum refractive index K_{PV} of the polarizable vacuum, the time-independent form as derived by Desiato and Storti[101] is

$$\nabla^2 \phi = \nabla^2 c_0^2 / K_{PV}(r) = \nabla^2 c_0^2 / (1/(1 + 2\phi/c^2)) \qquad (38\text{-}38)$$

A positively curved spacetime corresponds to a converging refractive index medium ($K_{PV} > 1$) in which light slows down and material objects contract in size. The curvature is the inverse of the radius of a circular arc which approximates the curve at given point of a function. Curvature of a function f is measured by the Laplacian $\nabla^2 f$. The concavity of the refractive index, gravitational frequency or gravitational potential of a central mass corresponds to the second derivative with respect to arclength of the given function. For a gravitational potential well, the curvature in tangent space manifold is concave up while the refractive index and frequency hill is concave down. The inflection point where the 2nd derivative changes sign corresponds to the mass surface radius.

The Earth's gravitational potential well is illustrated in Fig. 38-2. The potential shown assumes a uniform solid sphere. The acceleration of gravity $g = W/m = GM/R^2 = -\nabla \phi$ where W = weight of passive mass m, M = active mass, G = gravitational constant, R = radius and ϕ = gravitational potential. Simply dividing the local acceleration of gravity g (= 9.81 m/s^2) by the speed of light c (= 2.997 x 10^8 m/s) yields a quantity with the dimensions of frequency ν = g/c = 3.273E-08 Hz which is interpreted as a beat frequency ($\Delta \nu = \Delta c/\Delta \lambda = c(1 - K_{PV})/\Delta \lambda$ produced by superposition of waves of with slightly different net frequencies as they move in and out of phase. At the Earth's surface, the acceleration of gravity in terms of EM frequency difference is approximated as $g = 2c \cdot \Delta \nu$. In a more precise representation, the geophysical International Gravity formula (IGF) for the acceleration of gravity $g(\lambda) = g_e(1 + \alpha \sin^2(\lambda) + \beta \sin^2(\lambda))$ relates the variation of g with latitude based on a reference ellipsoid giving a best fit with the equipotential mean sea level surface where $U = -GM/r + (GMJ_2 a^3/2r^3)*(3\sin^2\theta - 1)$. To account for the Eötvös effect, the gravitational acceleration $\mathbf{g}$ = acceleration due to gravity ($\mathbf{a}_g$) – centrifugal acceleration ($\mathbf{a}_c$) = $Gm/r^2 - v^2 \cos(\phi)^2/r$.

For comparison, an associated gravitational frequency ω_{PV} at the Earth's surface is also illustrated for harmonic modes for a polarizable vacuum (PV) model based on Fourier analysis by Storti and Diemer[35]. The results of this analysis indicate the fundamental frequency at the Earth's surface may be roughly estimated as $\omega_{PV} = \sqrt[3]{(\Delta \nu \cdot c^2/r^2)} = 0.33606$ Hz for an Earth radius of 6.37718E6 m and a surface acceleration of gravity $g = 2c \cdot \omega_{PV}^3 \cdot r^2/c^2 = 9.81$ m/s^2.

Einstein has put an end to this isolation; it is now well established that gravitation affects not only matter, but also light. – Hendrik Antoon Lorentz

I think the material bodies do not gravitate between each other but it is the ether that makes one material body to press to another. We wrongly call this phenomena gravitation. – Nikola Tesla

Ideas, like large rivers, never have just one source. – Willy Ley

Matter moves, but ether is strained. – Oliver Lodge

I defy gravity. – Marilyn Monroe

...that a spin-2 field has this geometrical property...is just marvelous. – Richard Feynman

The very nature of an observation requires an interplay between the fields and the particles. – P. A. M. Dirac

38. Gravitational Potential

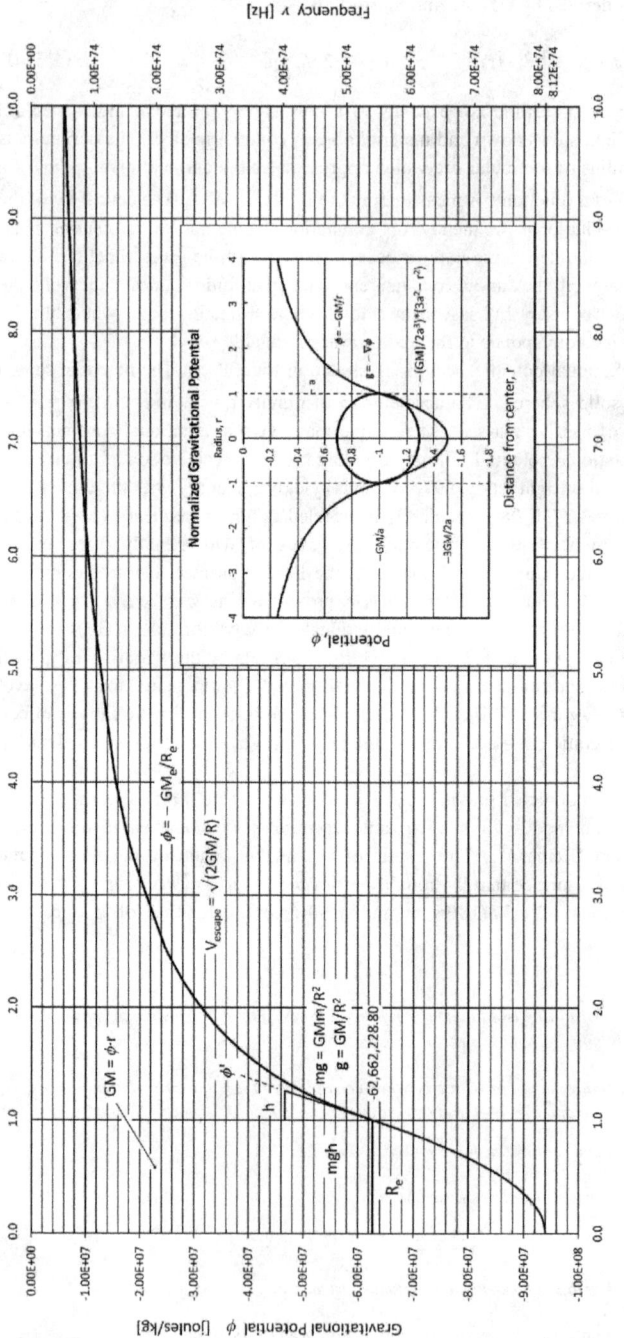

Fig. 38-2. Earth's gravitational potential well approximated as a uniform sphere with mass density $\rho = 4/3(\pi a^3)$. The acceleration of gravity **g** arises from an energy density gradient denoted by the Poisson relation ($= -\nabla \phi$) where $\phi =$ scalar energy potential. The local variation in EM field energy density results in a reduction in light velocity c ($= c_0/T = c_0\sqrt{}/K_{PV}$) and a Lorentz-Doppler frequency differential $\Delta \nu$ which is proportional to the de Broglie wave frequency ν_{dB}. Hence, acceleration of gravity for a matter object may be expressed as $\mathbf{g} = \mathbf{W}/m = (\mathbf{F}/\mathrm{E})c^2 = 2c\mathrm{E}/((\Delta \mathrm{E}/\Delta \nu))\cdot \hat{\mathbf{r}} = (\Delta \nu/\nu)(c^2/r)\cdot \hat{\mathbf{r}} = 2c\cdot\Delta \nu\cdot \hat{\mathbf{r}} = (2c^2/h)(m\nu/\Delta \lambda)\cdot \lambda_{dB}\cdot \hat{\mathbf{r}}$.

38. Gravitational Potential

A general illustration of tangent space of a gravitational potential well and gravitational frequency hill is depicted in Fig. 38-3. The gravitational potential well and gravitational frequency hill shown represent examples of parameter dimensions in tangent space manifolds. Unlike embedding diagrams associated with illustration of spacetime curvature in GR, spacetime remains Euclidean while variables such as EM nodal distances and frequencies represented in tangent space exhibit dilation and contraction and curvature. In contrast to GR (with unexplained mechanism for assumed spacetime distortion), gravitational effects in a polarizable vacuum (including length contraction, time dilation, frequency shift, alteration in the speed of light, etc) are EM wave interaction effects due to local variation in the vacuum dielectric constant $K_{PV}(r,\omega,M)$.

Coordinate length and time are represented in flat (gravity-free) Minkowski spacetime. Proper length and time in curved spacetime are represented in tangent space. Proper length is a measure of increased EM wave-length (redshift). Proper time corresponds to decreased EM frequency and increased period (time dilation). Proper length and time transformations are

$$L = L_0\sqrt{(1 - 2GM/Rc^2)} = L_0\sqrt{(1 + 2\phi/c^2)}$$
$$= L_0\sqrt{(1 - v_e^2/c^2)} = L_0/\Gamma \quad (38\text{-}39)$$

$$T = T_0/\sqrt{(1 - 2GM/Rc^2)} = T_0/\sqrt{(1 + 2\phi/c^2)}$$
$$= T_0/\sqrt{(1 - v_e^2/c^2)} = T_0\Gamma \quad (38\text{-}40)$$

where G = gravitational constant, M = active mass, ϕ = gravitational potential, R = radius, v_e = escape velocity, c = speed of light. Escape velocity $v_e = \sqrt{(2GM/R)}$. The invariant interval $ds^2 = -dt^2 + dx^2 + dy^2 + dz^2 = g_{\mu\nu}dx^\mu dx^\nu$. The metric is Lorentz invariant.

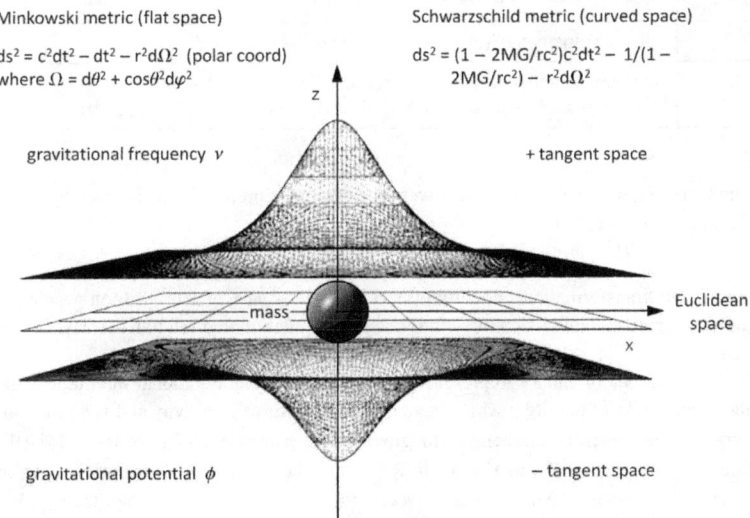

Fig. 38-3. Tangent space representation of EM frequency hill and gravitational potential well in a flat 4D spacetime. Proper length and time are represented in tangent spacetime while coordinate length and time are represented in flat Minkowski spacetime.

39. Gravitational Gamma

The gravitational gamma Γ conversion constant (nondimensional) relates transformations of physical units and constants in a gravitational field in terms of proper length and time. The gravitational gamma transformation operator, adopting the Macken definition[20], is expressed as

$$\Gamma = dt/d\tau = dr/dR = 1/\sqrt{(1 - (2GM/c^2R))} \tag{39-1}$$
$$= 1/\sqrt{(1 - (v_e^2/c^2))} = 1/\sqrt{(1 - \beta^2)} \tag{39-2}$$
$$= 1/\sqrt{(1 + (2\phi_g/c^2))} = \sqrt{(c_0^2/\phi_g)} \tag{39-3}$$
$$= 1/\sqrt{(g_{00})} = \sqrt{(1/K_{PV})} \tag{39-4}$$
$$\approx 1 + GM/Rc^2 \approx 1 + \beta \quad \text{(weak field approx.)} \tag{39-5}$$

where:

dt	coordinate time (remote observer – far field). Stationary metric.	
$d\tau$	proper time (local observer) ($= dt/\Gamma = dt/\sqrt{K_{PV}} = dt/n$)	
G	Newtonian Gravitation constant ($= G_0/\Gamma^3 = c^2(R_{univ}/M_{univ}) = \hbar c/m_P^2 = \mu_g c^2/4\pi$)	
M	mass ($= \Gamma M_0 = R_S c^2/2G = E/c^2 = E/(\mathbf{E}/\mathbf{B})^2 = 2L/(\mathbf{E}^2 - c^2\mathbf{B}^2) = E/\phi_g = \hbar k/v$)	
c	speed of light ($= c_0/\Gamma = c_0/\sqrt{(\epsilon_0\mu_0)} = |\mathbf{E}|/|\mathbf{B}| = \sqrt{(v_g v_p)} = c_0/\sqrt{K_{PV}} = c_0/n$)	
n	index of refraction ($= c_0/c = 1/\sqrt{g_{00}} = c_0/v_p = c_0/(c^2/v_g) = \Gamma = \sqrt{K_{PV}} = \sqrt{(\epsilon_r\mu_r)}$)	
R	radius = coordinate distance; r = proper distance	
ϕ_g	gravitational potential ($= -GM/R = c_0^2/K_{PV} = c_0^2/\Gamma^2 = c^2 \mathbf{A}_g/\mathbf{v}$)	
g_{00}	metric coefficient in General Relativity ($= 1 + 2\phi/c^2 = 1/n^2 = 1/K_{PV} = 1/\Gamma^2$)	
v_e	escape velocity $= \sqrt{(2GM/r)} = \sqrt[4]{(c^4(1/\Gamma - 1))} = \sqrt[4]{(c^4 \beta_g)}$	
K_{PV}	vacuum refractive index ($= c_0^2/\phi_g = (1 - GM/rc_0^2) = e^{\wedge}(2GM/rc^2)$ $= e^{\wedge}r_S/r = g_{11} = g_{22} = g_{33} = 1/g_{00} = \Gamma^2 = 1 - (v_e^2/c^2) = 1/(1 + (2\phi_g/c^2))$)	

For comparison, in special relativity, lower case gamma (Lorentz factor) is given by:

$$\gamma = 1/\sqrt{(1 - v^2/c^2)} \approx 1 + \tfrac{1}{2}(v^2/c^2) + 3/8(v^4/c^4) + 5/16(v^6/c^6) \tag{39-6}$$

which relates transformation of relative motion in flat space of inertial (non-accelerated) reference frames including Lorentz contraction, time dilation and relativistic mass increase effects.

The gravitational gamma Γ represents the strength of the gravitational field independent of the acceleration of gravity **g** which is an effect of the force of gravity and is a function of the gravitational potential gradient. The gravitational gamma varies from $\Gamma = \infty$ (@ Black hole event horizon $R = R_s$) to $\Gamma = 1$ (@ $R = \infty$) at the gravity free asymptotic limit in a Schwarzschild universe. Macken defines a corresponding gravitational magnitude

$$\beta_g = \sqrt{(1 - 1/\Gamma^2)} = \sqrt{(1 - (d\tau/dt)^2)} \tag{39-7}$$

The gravitational magnitude varies from $\beta_g = 1$ (@ Black hole event horizon ($R = R_s$) to $\beta_g = 0$ (@ $R = \infty$).

39. Gravitational Gamma

The acceleration of gravity in terms of the gradient in gravitational magnitude for small vertical elevation height is given by

$$g = c^2(d\beta_g/dh) = -c^2 d((\tau - \tau_0)/(t - t_0))/dh \qquad (39\text{-}8)$$

where
- $d\tau$ = proper time @ detector
- $d\tau_0$ = proper time @ source
- g = acceleration of gravity
- h = proper length in vertical direction

An example illustration of gravitational gamma for the Earth is shown in Fig. 39-1. The gravitational gamma Γ is equivalent to the vacuum refractive index $K_{PV}(r,\omega,M)$ in the Polarizable Vacuum (PV) model where $\Gamma = \sqrt{K_{PV}}$ (weak field) = $\sqrt[4]{K_{PV}}$ (strong field).

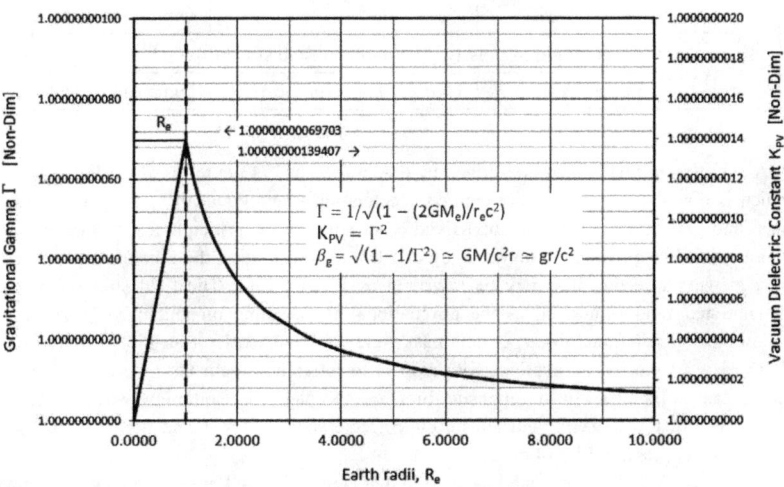

Fig. 39-1. Calculated gravitational gamma Γ, gravitational magnitude β_g and vacuum refractive index $K_{PV}(r, \omega, M)$ for the Earth as a function of planet radius.

Gravitation affects not only matter, but also light. Thus strengthened in the faith that his theory already has inspired, we may assume with him that there is not a single physical or chemical phenomenon – which does not feel, although very probably in an unnoticeable degree, the influence of gravitation. – Hendrik Antoon Lorentz

Difform motion will in every case produce the same effects as gravitation, – Hendrik Antoon Lorentz

...his [Einstein's] ideas were more radical [than Bohm's], but 'music of the future'. – Max Born

There is no instrument for measuring the pressure of the Ether, which is probably millions of times greater: it is altogether too uniform for direct apprehension. A deep-sea fish has probably no means of apprehending the existence of water, it is too uniformly immersed in it; and that is our condition in regard to the Ether. – Oliver Lodge

40. Newton's Second Law

The Vis inertiae is a passive Principle by which bodies persist in their Motion or Rest, receive Motion in proportion to the force impressing it, and resist as much as they are resisted. – Isaac Newton

The inertial force due to gravity in terms of the gravitational gradient, in accordance with the modern day interpretation of Newton's second law of motion, is defined as

$$F = m_i \, a = W = m_i \, g \qquad (40\text{-}1)$$

where:

$F \ (=W)$	Force (or weight)		
	Value: Varies	Units: N $(= kg(m/sec^2))$	Dimensions: MLT^{-2}
m	mass (m_i denotes inertial mass) = $m_0\Gamma = m_0\sqrt{K_{PV}} = \hbar k/v = q_0 A/v = E/c^2$		
	Value: Varies	Units: kg	Dimensions: M
$g \ (=a)$	Acceleration of gravity (gravitational field strength) = $-\nabla \phi_g$		
	Value: Varies	Units: N/kg $(= m/sec^2)$	Dimensions: LT^{-2}

A variation of this law was described by d'Alembert in 1743 in terms of inertial Force F_i which is directed opposite to the applied acceleration **a** and equal to $F_i = -ma$. Inertial mass and gravitational mass in Eötvös and other equivalence principle tests[102] have been measured to be equivalent to an accuracy of $>10^{-12}$. Substituting the Einstein relation E = mc^2, Newton's second law may be rewritten as $F = (E/c^2)a$. The field strength of the gravitational field is defined as the gravitational force acting on unit mass. Hence, it follows that gravitational force per unit mass = gravitational acceleration g.

The law of motion is invariant with respect to reflection about the time axis, $t \rightarrow -t$. Orbits can be traversed in the opposite direction. Kepler's harmonic law was derived by setting $F = ma = GmM/r^2$ and substituting acceleration $a = v^2/r$ and period $P = 2\pi r/v$ yielding an expression $P^2 = (4\pi^2/G(M + m)r^3$.

George Green[103] demonstrated circa 1836 that Newton's law $F = ma$ should be modified for an object moving through an incompressible fluid. The inertial mass m is replaced by the renormalized mass m + M/2 where M is the mass of the fluid such that $F = (m + M/2)a$. Jefimenko[16], in an extended Newtonian theory of gravity, showed that the Newtonian gravitational mass m should be replaced by $m_t = m + m_g = m(1 + G^3 m/5c^2 r)$ to include the equivalent mass of the gravitational field.

Spears[25] observes just as the Newton expression for a gravity force between two masses has many of the attributes of the Coulomb expression for a charge force, the $F = ma$ expression contains an acceleration parameter that acts on mass to create an opposing force similar the effect of inductance on electric current to create an opposing force (back e.m.f.). Likewise a change in vacuum permeability may be anticipated to oppose changes in charge velocity which constitutes an electric current suggesting there may be an electromagnetic interpretation of $F = ma$ in terms of electrodynamic momentum (eg., $m_i = k\hbar/v = q_0 A/v$).

Every body perseveres in its state of being at rest or moving uniformly forward, except as it compelled to change its state by forces impressed. – Isaac Newton

40. Newton's Second Law

Jennison and Drinkwater[39] in 1977 derived Newton's second law for a phase-locked cavity model of a wave system representing a fundamental particle. A trapped standing wave exhibits not only rest mass but also intrinsic inertia. Under acceleration, forward and backward propagating waves interact undergoing Doppler shifts resulting in an imbalance of radiation pressure. The total energy E_T of the system consists of the potential energy E_P required to hold the system together plus the wave energy $E_{WF} + E_{WB}$. At rest, the wave energy equals the binding energy. Application of an external force **F** results in an acceleration $\mathbf{F} = (2/c^2)(E_T - E_P)\mathbf{a} = E_t \mathbf{a}/c^2 = m_0 \mathbf{a}$ provided that the force is applied for a duration δt greater than or equal to the time to complete a feedback loop otherwise, the excess incident radiation is re-radiated back into space. The inertial force is a function of only half the total energy of the system as the potential energy makes no contribution.

Ivanov[36] derives the following expression for acceleration of a wave system of coupled oscillators in terms of phase displacement in time $\Delta\phi$, frequency differential $\Delta\nu$ and the velocity of light c

$$a = c/\pi \cdot (\Delta\phi_2 - \Delta\phi_1)/\Delta t = 2c \cdot \Delta\nu \qquad (40\text{-}2)$$

For an acceleration of gravity **g** = 9.81 m/s² at the Earth's surface, the corresponding frequency shift of a standing wave system restrained from free fall is

$$\Delta\nu = a/2c = (9.81 \text{ m/s}^2)/2(2.99 \times 10^8 \text{ m/s}) \qquad (40\text{-}3)$$
$$= 1.636 \times 10^{-8} \text{ Hz}$$

As an example, the acceleration of gravity profile for the Earth is depicted in Fig. 40-1 assuming an idealized solid sphere with linear subsurface variation in density. As shown, acceleration varies with elevation increasing to a maximum at the Earth's surface. The below depth variation of gravitational field strength is shown with the simplifying assumption of a linear variation in density with depth. The dashed line corresponds to a uniform average density for a sphere of radius R_E. Also shown is a corresponding gravitational frequency differential at the Earth's surface based on the local **g** value. The area under the curve **g** vs. r is equal to the gravitational potential ϕ.

The relation of the gravitational potential ϕ, frequency differential $\Delta\nu$ and acceleration of gravity **g** is illustrated in Fig. 40-2. The nonuniform field of a spherical mass is approximated over a small region by a uniform field close to the surface. The nonuniform fields associated with a central spherical mass may be approximated by a uniform field over a sufficiently small local region close to the surface where the height h above the surface is much less than the mass radius (h << R). Work (= force·distance) must be performed on the gravitational field to move a test mass away from the central mass. Conversely, work is performed by the gravitational field to move a test mass toward the central mass. According to the virial theorem, half the binding energy is released as radiation during contraction while the remainder is converted to heat within the consolidated mass.

Is spacetime really curved? Isn't it conceivable that spacetime is actually flat...? – Kip Thorne

Delivered from gravity and buoyancy, I flew around in space. – Jacques Ives Cousteau

To find the new truth, we have to go far out on the knowledge spectrum to the edgy part... the shady and speculative part... where the kooks, geniuses, and gurus are. – Bill Bonner

If we succeed by operational means to make C → ∞, we would have zero mass for the particle and thus no inertia. – Martins and Pinheiro

40. Newton's Second Law

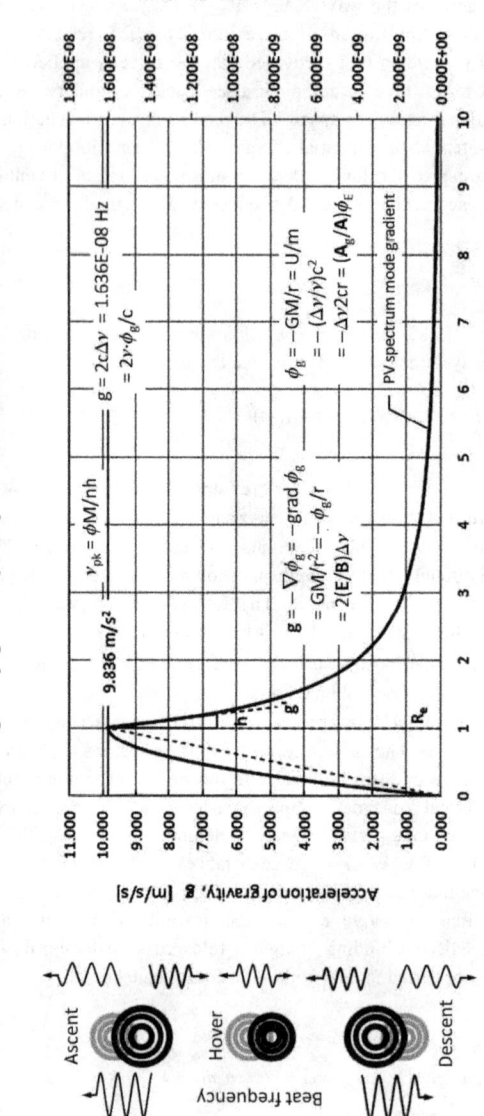

Fig. 40-1. Variation of acceleration of gravity **g** in Earth's gravitational field as a function of distance from center of mass. Also shown is the associated frequency shift of a standing wave system restrained from free fall. The frequency difference is the result of a Lorentz-Doppler shift of counter-propagating components of a contracted moving standing wave. The acceleration of gravity **g** is proportional to the Lorentz-Doppler beat frequency $\Delta\nu$ ($= (\nu_{blue} - \nu_{red}) = g/2c = v \cdot \phi_g/c^2$) which is a measure of the EM energy density gradient. The gravitational scalar potential ϕ_g ($= -GM/r = c_{ro}^2/T^2 = (\mathbf{A}_g/\mathbf{A})\phi_E = (\mathbf{A}_g/\mathbf{A})(E\phi_B/h)$) is a measure of the local EM field energy density U ($= \frac{1}{2}(\phi_x^2 + \phi_y^2 + \phi_z^2) = \frac{1}{2}(\epsilon_0 \mathbf{E}^2 + \mathbf{B}^2/\mu_0)$). The gravitational potential ϕ_g is related to the electrostatic potential ϕ_E and the magnetostatic potential ϕ_B by the ratio of the magnetic vector potential **A** and gravitomagnetic vector potential $\mathbf{A}_g$. The de Broglie matter wave frequency ν_{dB} ($= c/\lambda_{dB} = mvc/h$) is proportional to momentum **p** ($= m\mathbf{v}$).

40. Newton's Second Law

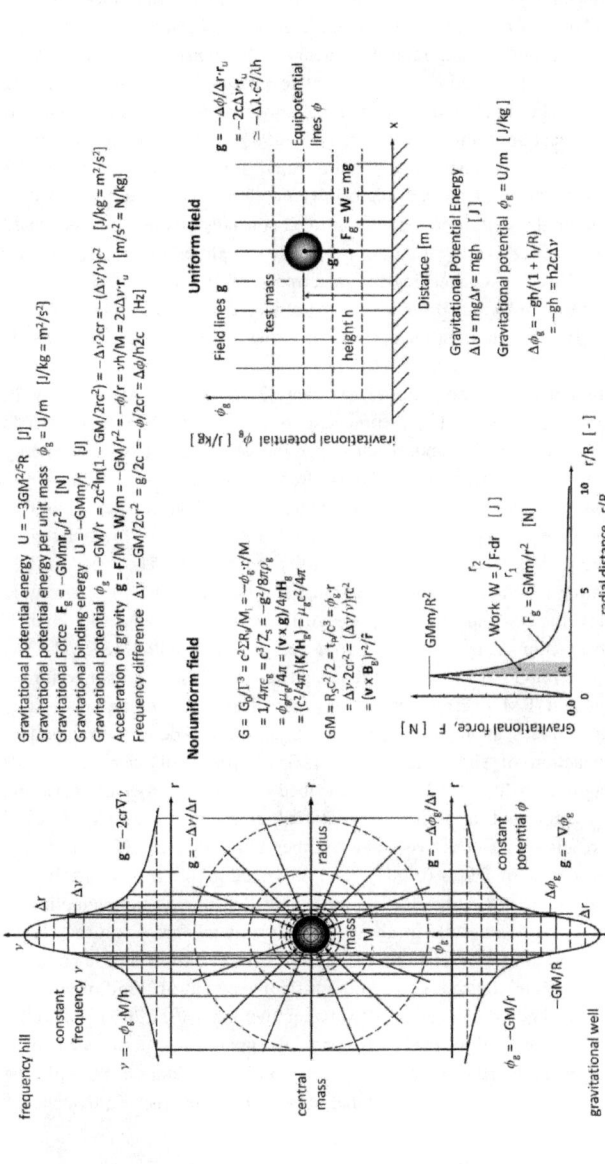

Fig. 40-2. Nonuniform gravitational potential well and frequency hill of a spherical mass. A change in gravitational potential ($\Delta\phi = \Delta\nu/Ehc^2$) is associated with a matter wave frequency shift within a test mass resulting in an internal radiation pressure imbalance and induced motion. In a free fall weightless condition (W = 0), this internal matter wave frequency difference $\Delta\nu$ is reduced to zero. Weight W (= mg) equals the spatial rate-of-change of the Lagrangian interaction energy $\delta L/\delta z = \delta(T-V)/\delta z = \delta(\tfrac{1}{2}mv^2 - mgh)/\delta z$. A free-falling object follows a trajectory described by a geodesic curve with zero acceleration. Change in phase is a measure of displacement from potential hole minima (standing wave node).

41. Gravitational Constant

41.1 Newtonian gravitation constant

The gravitational constant G in Newton's gravitational formula, first measured by Cavendish in 1798, is taken as a positive number as mass density is positive and gravitation appears only attractive. In the general relativity theory, the constant of gravity G is arbitrary with the curvature of spacetime proportional to G; gravitation and inertia are referenced to the speed of light. In the current epoch, the universal gravitation constant G is taken as an empirical physical constant (G ≈ 6.67 x 10^{-11} m^3/kg·sec^2). Experimental limits on large scale variation of G in space, spatial anisotropy, from analysis of LAGEOS satellite and laser ranging data indicate a limit of $|\delta G|/G$ ~2 x 10^{-12} on a solar system scale[102]. A possible temporal variation in the gravitational constant G(t) may be associated with a time dependent potential $\phi(t)$ and variation in velocity of light c. In a Large Number Hypothesis advanced by Dirac[104] concerning the occurrence of 10^{39} ratio of electrical to gravitational force ($e^2/Gm_p m_e$ ≈ 10^{39}) and the age of the universe in atomic units ($m_e c^3/e^2 \hbar$ ≈ 10^{39}) such that t/t_0 ≈ 10^{39}, the gravitational constant varies inversely with time as the Universe expands (G ≈ $c^2 R_{univ}/M_{univ}$ ~ 1/t).

Sakharov[105], hypothesized the Gravitational constant G is measure of elasticity of spacetime with G inversely proportional to a truncated spectra integral of virtual particle momenta in a quantum vacuum and proposed an upper frequency cutoff to limit energy densities. Gravitation is thus an emergent phenomena from change in action of zero-point energy vacuum fluctuations due to presence of matter-induced spacetime curvature. Sakarov obtains an expression for the gravitational constant G = $c^3 L^2/\hbar$ = $\hbar c^3/p_0^2$ where L is an elementary length and $p_0 = \hbar k$ is a corresponding angular momentum. The speed of light is relatively high indicating the vacuum is of large, but finite bulk modulus of elasticity K (= $\Delta P/\Delta V/V$) which has dimensional units of pressure (kg/m·s^2). Likewise, the characteristic impedance of the vacuum Z_s (= m_P/t_P = $c^3/|G|$) is very high (= 4.0302E35 kg/s)[106,20]. Spacetime stiffness k_{vac} = $c^4/8\pi G$ = 5.998E6 N/m represents energy density/unit curvature.

A local shift in vacuum EM energy density is equated with a gravitational field. Localized mass energy induces an equilibrium shift in the action density of quantum fluctuations where the action of spacetime (S = $-[1/16\pi G]\int \sqrt{[-g]}R$) depends on its curvature R and the Lagrangian function L(R) is described in a power series of curvature. In particular, Sakharov proposed that the Gravitational constant (G = $2Ac^5/\hbar \omega c^2$ = $-1/(16\pi A\int k dk)$ is the result of an effective wave number ultraviolet cutoff limit of the action integral at a Planck scale of $1/L_p$ ≈ $(c^3/\hbar G)^{1/2}$ = 1/1.616252 x 10^{-35} m^{-1} or k_0 ≈ 10^{28}eV ≈ 10^{33} cm^{-1}. Puthoff developed the idea further proposing that interparticle coupling to zero-point Zitterbewegung motion results in attractive gravitational force akin to a long-range retarded van der Waals force[29]. The relative weakness of gravity is attributed to the inverse variation of the coupling constant G to the high-frequency cut-off limit of the ZPE spectrum. Puthoff has calculated the upper cutoff wavenumber k_c^* = $\sqrt{(\pi c^3/\hbar G)}$ [=$\sqrt{\pi/r^*}$] = 1.09577E35. Gravitation is the result of the electromagnetic quantum vacuum acceleration past a mass object whereas inertia is the result of object acceleration through the electromagnetic quantum vacuum, hence gravitational mass and inertial mass equivalence.

This is just the tip of the iceberg. I will, of course, not go beyond that. – Prof. R. Mohapatra

Only black holes of very low mass would emit a significant amount of radiation. – S. W. Hawking

41. Gravitational Constant

Cahill[107] described the gravitational constant $G = \hbar c/m_P^2 = 6.68541E-11$ [m³/kg·s²], where m_P = Planck mass, arguing the expectation value of the distribution of energy in spacetime is on the order of the Planck mass and G is then a measure of the large-scale structure of the universe. In a wave-based model of spacetime filled with ZPF vacuum energy, Macken[20] shows that $G = c^4/F_P$ where the Planck force $F_p = c^4/G \approx 1.2 \times 10^{44}$ N. Setting the Newtonian $G = l_P^3/(m_P t_P^2) = 6.68541E11$ m³/kg·s² and the Planck gravitational constant $G_P = t_P c^3/m_P = 4.2005E-10$ m³/kg·s², the ratio of $G_P/G \cong 2\pi = 6.2831...$

Based upon an electron formed of dipolar quasiparticles forming a vortex resonance, Winterberg[7] derives an expression for $G = r_P c^2/m_P = (2m)^2 \hbar c/R_e^6$ where Reynolds no. $R_e = (2m_P/m)^{1/3} = 3.6 \times 10^7$ and m is the mass of the gravitational spinor field. The small value of G is a consequence of the large quantum Reynolds number. In terms of Planck mass, a dimensionless Gravitational constant $G = l_P^3/m_P t^2 = 1$ where Planck quantities are set equal to unity. In dimensioned Planck units, $G = c^2(r_P/m_P) = 6.67696E-11$ m³/kg·sec².

Daywitt[12] derives an expression for the Newtonian gravitational constant $G = e^{*2}/m^{*2}$ where e^* is the charge and m^* the mass of a Planck particle, respectively. The Planck particles comprise the Planck vacuum (PV) and are somewhat analogous to the Dirac 'sea' in quantum mechanics in which all the negative energy states are filled. The gravitational force between two particles of mass m is expressed as $F_g = -m^2 G/r^2 = (m/m_*) \cdot 2(r_*/r)^2 (m_*^2 G/r_*^2) = (mc^2/r)^2/(c^4/G)$. Yet another relation for gravitation coupling may be derived by considering two masses in orbital relationship with orbital velocity at the speed of light. Setting $F_g = Mm/r^2 = F_c = mv^2/r$ with $r = ct$ and $v = c$ yields $G = tc^3/M$. For an electron, $G_e = t_C c^3/m_e = 2.3938E35$ m³/s²·kg where t_C is the Compton time and m_e is the electron rest mass. Similarly, for a Planck mass particle, $G_P = t_P c^3/m_P = \hbar c/m_P^2 = 4.2005E-10$ m³/kg·s² where t_P is the Planck time and m_P is the Planck mass.

Roxburgh[108] in considering the large number coincidences with G, m and c finds $G = 1/6\pi r t^2$ for an Einstein de Sitter cosmology. The constant G is thus determined by the cosmological distribution of matter. In GR, the constant G is a conversion factor relating space-time curvature and energy-momentum density. The present observational limit on the variation of G with time in Lunar Laser Ranging Tests[102] appear to demonstrate an upper bound of $|\dot{G}/G| \leq \approx 10^{-13}$/yr. Other measurements, such as timing of the binary pulsar PSR1913+16, 21,000 ly distant discovered in 1974, indicates a rate of change of $G < \approx 10^{-11}$/yr. Assuming an age of the universe of 13.8×10^9 yr and constant rate of change of G indicates a corresponding decrease in G of < 0.13% to present. Time variation in G reduces the force of gravity and may explain the early flatness of the early universe following the Big Bang without need for exponential inflation.

In a hypothesis by John Hunter[109], the gravitation constant is related to the total mass and size of the universe: $E = Mc^2 = M^2G/R$ and the black hole event horizon radius $R = ct = MG/c^2$. For a galactic mass, $G = GMm/R + Gm^2/r$ where M = mass of universe, R = horizon radius, r = galaxy radius and m = galaxy mass within r. For a star orbiting a galaxy, the tangential velocity $v = \sqrt{(Gm/r)}$. A reduction in the effective value of G at the center of galaxies is said to explain the constant velocity profile for stars in a spiral galaxy and emergence of jets. Krizek and Somer[110] notes that the expansion of the universe as measured by the Hubble constant H_0 is also observable on scales smaller than cosmological scales such as Earth-Sun and Earth-Moon distance scales. For example, the Earth-Sun distance increases at a rate by various estimates of ~4-6 m/yr (~0.5 H_0) and the Earth-Moon distance increases by ~0.038 m/yr (~0.66 H_0). Krizek and Somer advance a hypothesis to explain the apparent expansion and increase in energy due to gravitational aberration resulting in the apparent repulsion (anti-gravity) effect of dark energy. This effect is predicted to arise as a result of finite velocity of gravitational interaction producing disequilibria. For example, two orbiting gravitational bound masses are predicted to slowly

41. Gravitational Constant

increase in angular momentum slowing increasing separation distance in a double Archimedean spiral. Since the mutual gravitation interaction velocity is finite, two orbiting masses are attracted in the direction of the retarded positions of interacting mass. In the case of the Earth-Moon system, the dark energy effect is comparable in magnitude to tidal interaction and accounts for ~55% of the average increase in separation (~ 0.017 m/yr).

In a scale invariant, cosmological, bimetric model with variable light velocity, Petit and Midy[111, 112] examined implications of universe evolution with variable c, h, G where the Einstein constant $\chi = -8\pi G/c^4$ in the field equation remains constant, i.e., the ratio G/c^2 ($\approx$ R/M) remains absolute with secular variation G and c. In this model with R(t) as a characteristic length scale factor (gauge parameter), Planck's constant h and Planck time t_P [$= (hG/c^5)^{-1/2}$] varies with time $t \approx R^{3/2}$; gravitational constant $G \approx c^2 \approx 1/R$; Planck length l_P [$= (hG/c^3)^{-1/2}$], mass M and Schwarzschild length r_s [$= GM/c^2$] $\approx R$; mass density $\rho \approx 1/R^2$; gravitational forces $\approx 1/R$; and light velocity $c \approx R^{-1/2}$. R(t) and c(t) are related such that $\phi(R,c) = K$ (a constant). The fine structure constant α remains constant as a function of time. Gravitational redshift and cosmological redshift are associated with spatial and temporal variation of metric coefficient g_{00}, respectively. Cosmological redshift results from variation in c, G and $\hbar$. Universe expansion does not require inflation, dark energy produces the late acceleration – in physical time the origin of the universe is shifted to $-\infty$. Spacetime expands in the absence of gravity, gravity locally contracts spacetime breaking O_3 symmetry in the radiation-dominated era.

Sternglass[15] observes that a time variation in G is implicit in the Planck energy density relation $U_P = \hbar c^7/G^2$ and is consistent with a rigidly rotating universe of successively dividing electron-pair model. This fire-works model indicates the observable universe is an oblate spheroid and the expansion of the universe may be expressed in terms of the Hubble constant H relating the radial velocities measured by an observer $v_r = H \cdot R$. The local gravitational constant G increases with time with each pair division in proportion to the square-root of mass. The Schwarzschild radius R_S (= $2GM_u/c^2$) of the universe of mass M_u likewise increases with time rapidly increasing during the initial expansion following the Big Bang up to an inflection point corresponding to the formation of protons and neutrons and asymptotically approaches a stable maximum limit of 2.4E29 m in the far distant future.

> But when we face the great questions about gravitation, i.e., Does it require time? Is it polar to the 'outside of the universe' or to anything? Has it any reference to electricity? or does it stand on the very foundation of matter-mass or inertia? then we feel the need of tests, whether they may be comets or nebulae or laboratory experiments or bold questions as to the truth of received opinions. – James Clerk Maxwell

> We could conceive the beginning of the universe in the form of a unique atom, the atomic weight of which is the total mass of the universe. This highly unstable atom would divide into smaller and smaller atoms by a kind of superradioactive process. – Georges Lemaître

> I have felt a homesickness for the paths of physical science where there were more or less discernible handrails to keep us from the worst morass of foolishness. – Arthur Stanley Eddington

> It is better to debate a question without settling it than to settle a question without debating it. – Joseph Joubert

> Why should free space conduct magnetism? Einstein tried for many years to relate μ_0 to the gravitational constant, a relationship which, if established would be, as he put it, "the key to the cosmos", but he failed to establish it. – E. R. Laithwaite

41. Gravitational Constant

41.2 Gravitational G as a function of Γ

Gravitational constant G varies with spacetime curvature as Macken[20] derives in the following relation for the normalized Newtonian gravitational constant referenced to a gravity free, zero curvature vacuum in a hybrid coordinate system with constant proper length

$$G_0 = L_0^3/M_0T_0^2 = \Gamma^3 G_g = G_0/\sqrt{(g_{00})^3} \qquad (41\text{-}1)$$

where:

G_0	Normalized Gravitational constant (Newtonian gravitational coupling) [gravity free where $\Gamma = 1$]		
	Value: $6.67384(80) \times 10^{-11}$ (2010 CODATA)	Units: N·m²/kg² (= m³/kg·sec²)	Dimensions: $L^3 M^{-1} T^{-2}$
L_0	unit of length transformation (invariant length) (= L_g/Γ where L_g = proper length in gravity field)		
	Value: Varies (≥ 0)	Units: m	Dimensions: L
M_0	normalized mass (rest mass, invariant or intrinsic mass) (= M_g/Γ where M_g = proper mass in gravity)		
	Value: Varies (≥ 0)	Units: kg	Dimensions: M
T_0	normalized time (invariant time) [= T_g/Γ where T_g = proper time τ in gravity field]		
	Value: Varies (≥ 0)	Units: s	Dimensions: T
Γ	Gravitational Gamma (= $dt/d\tau = 1/\sqrt{(1 - 2GM/c^2R)} = 1/\sqrt{(1 + 2\phi_g/c^2)} = 1/\sqrt{g_{00}} = 1/\sqrt{1 - \beta^2}) = \sqrt{K_{PV}} = 1/\sqrt{(1 - v_e^2/c^2)})$ where $v_e = \sqrt{(2GM/r)}$. $\Gamma \simeq 1 + \beta$		
	Value: Varies (≥ 1)	Units: -	Dimensions: -
G_g	Gravitational constant in curved spacetime (= $G_0/\Gamma^3 \simeq R_{univ} c^2/M_{univ}$) [$\Gamma > 1$]		
	Value: Varies	Units: N·m²/kg² (= m³/kg·sec²)	Dimensions: $L^3 M^{-1} T^{-2}$
g_{00}	GR metric coefficient (= $1/\Gamma^2 = 1/K_{PV} = (1 - 2GM/c^2r)) = 1/(1 + 2\phi_g/c^2))$		
	Value: Varies	Units: -	Dimensions: -

The acceleration of gravity g is related to the gradient of gravitational magnitude β_g: $g = c^2(d\beta_g/dh) = -GM/r^2$. Gravity may be detected as an acceleration difference between locations or equivalently as a difference in vacuum refractive index, spectral energy density or frequency gradient.

Substituting $\Gamma = \sqrt{K_{PV}}$ (weak field), Eqn (41-1) may be written in terms of the vacuum refractive index K_{PV} as

$$G_0 = G_g/\sqrt{(1/\Gamma)^3} = (1/\sqrt{K_{PV}})^3 G_g. \qquad (41\text{-}2)$$

Macken shows that $\Gamma_u(t)$ of the universe increases with time resulting in the observed stellar redshift of distant galaxies.

41. Gravitational Constant

The relation of gravitational gamma Γ ($= 1/\sqrt{(1-1/\beta^2)} \simeq 1+\beta$) and gravitational magnitude β_g ($= \sqrt{(1-1/\Gamma^2)} \sim GMc^2r \sim R_S/r$) to other gravitational parameters are illustrated below in Fig. 41-1. Shown is a comparison of normalized gravitational parameter variation from a reference far field, zero curvature (flat) vacuum state. In the near field region, the relation between the diameter and circumference in tangent space is no longer Euclidean due to the variation in energy density resulting in an apparent spacetime distortion.

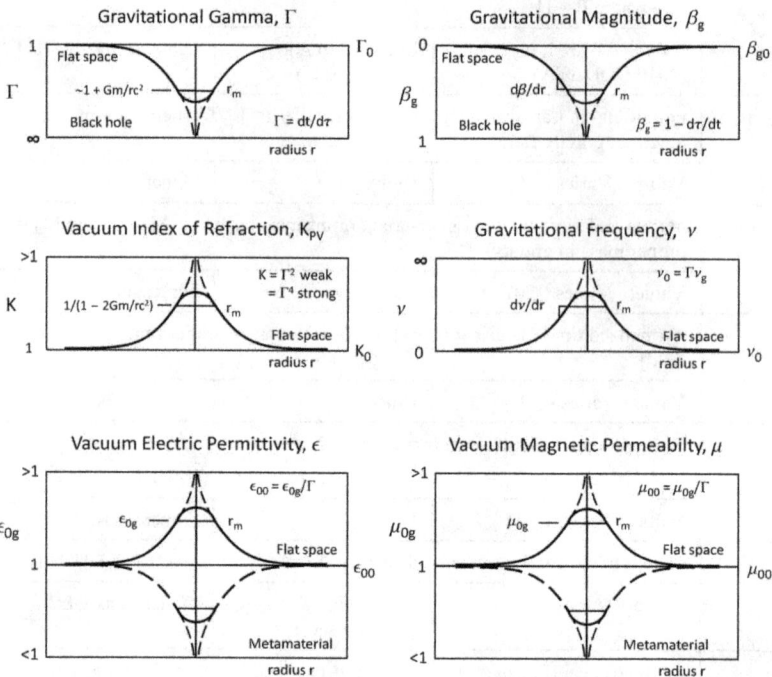

Fig. 41-1. Normalized gravitational parameter comparison illustrating magnitude range limits as a function of radius from a central mass. Dimensionally, mass m corresponds to a change in local electric permittivity, i.e., $m = (Rc^2/4\pi G)(\Delta\epsilon/\epsilon_0) = \Gamma m_0 = E/\phi_g$. Mass resides in the energy of the surrounding electric field E with a mass density $\rho_m = (\epsilon_0/c^2)E^2$. For a sphere of uniform mass density, the change in permittivity $\Delta\epsilon = 4\pi G\epsilon_0 m/Rc^2$.

We do not describe the world we see, we see the world we can describe. – René Descartes

There are two ways to be fooled. One is to believe what isn't true; the other is to refuse to believe what is true. – Soren Kierkegard

In science, mistakes always precede truth. – Horace Walpole

What we observe is not nature itself but nature exposed to our method of questioning. – Werner Heisenberg

41. Gravitational Constant

41.3 Gravitational G as a function of permittivity

Spears[25] derives an expression for the Gravitational constant G and gravitational force F_g in an electrostatic approach to gravity. By equating the electrostatic force F_e between two separated electrons to Newton's empirical gravity force $F_{gN} = GM_eM_e/r^2$, a value of G_e in MKS electrostatic units is defined as

$$G_e = G = F_g r^2/M_e^2 = G_0 \epsilon_{12}^2/\epsilon_0^2 \qquad (41\text{-}3)$$

where:

G_e	Gravitational constant (= G Newtonian gravitational coupling) (= G_0/Γ^3)		
	Value: -6.68541×10^{-11} (Calculated)	Units: (Coulomb-volt·meters)/kg^2	Dimensions: QVL/M^2
F_e	electrostatic force between electrons ($F_e = (1/4\pi\epsilon_0)Q_eQ_e/r^2 = F_0/\Gamma$)		
	Value: Varies	Units: N (= kg·(m/s^2))	Dimensions: MLT^{-2}
ϵ_0	Vacuum electric permittivity in free space (= $\epsilon_{00}/\Gamma = 1/\mu_0 c^2 = 1/\mu_0 v_p v_g$)		
	Value: 8.85419×10^{-12}	Units: farads/m	Dimensions: M^{-1}L^{-3}T^{2}Q^2
ϵ_{12}	effective electric permittivity between masses m_1 and m_2		
	Value: varies	Units: farads/m	Dimensions: M^{-1}L^{-3}T^{2}Q^2
M_e	electron mass (**F/a** = $\Gamma M_0 = \gamma M_0 = E/c^2 = E/(\mathbf{E/B})^2$ = |volts·charge|/c^2 = $\hbar\mathbf{k}/\mathbf{v}$)		
	Value: 9.10953×10^{-31}	Units: kg	Dimensions: M
Q_e	electron charge (= F/E = mc^2/volts = volts·capacitance = $\hbar\mathbf{k}/\mathbf{A}$)		
	Value: -1.60219×10^{-19}	Units: Coulomb	Dimensions: Q

The calculated numerical value of $G = G_e = -6.68541 \times 10^{-11}$ coulomb-volt-m/kg^2 is within 0.2% of a widely accepted value of $G = -6.67259(85) \times 10^{-11}$ m^3/kg·s^2. Variations in empirical measurements of G described in the literature (e.g., $6.6656(6) \times 10^{-11}$ [*Fitzgerald et al.*] to 6.71540×10^{-11} m^3/kg·s^2 [*Michaelis et al.*]) may be ascribed in part to stray capacitances (electrostatic linkages) in the individual test measurement physical layout.

Coulomb's law is similar in form to Newton's law, however, the electrostatic force is $\sim 10^{43}$ times as strong as the gravitational force. For example, for two electrons separated by a distance of 1 mm:

$$F_{elec} = k_C(q_e^2)/r^2 = (9 \times 10^9)((1.6 \times 10^{-19})^2)/(10^{-3})^2 = 2.3 \times 10^{-22} \text{ N}$$

$$F_{grav} = G(m_e^2)/r^2 = (6.67 \times 10^{-11})((9.1 \times 10^{-31})^2)/(10^{-3})^2 = 5.5 \times 10^{-65} \text{ N}$$

Noting the similarity in equations, Spears derived an expression for the gravitational constant in terms of relative electric permittivity for two coupled charges. An electrostatic circuit model relating the electrostatic force between electrons in terms of capacitance is illustrated in Fig. 41-2. The local permittivities ϵ_1, ϵ_2 and path permittivity ϵ_{12} are increased over the far field permittivity ϵ_0 as a result of increased density of hypothetical vacuum quantum particles which increase the capacitance similar to the effect of fine metallic

41. Gravitational Constant

particles distributed between poles of an air-dielectric capacitor. This alters the voltage/capacitor length (V/S) gradient. For the capacitance between electrons in free space separated by one meter, for example, the capacitor length $S_{12} = r/4\pi\epsilon_0 R_e^2 = 1.13181 \times 10^{39}$ darafs (inverse farad units of capacitance). Each daraf then corresponds to 8.83538×10^{-40} meters.

Electrostatic circuit model of hydrogen atom and electron

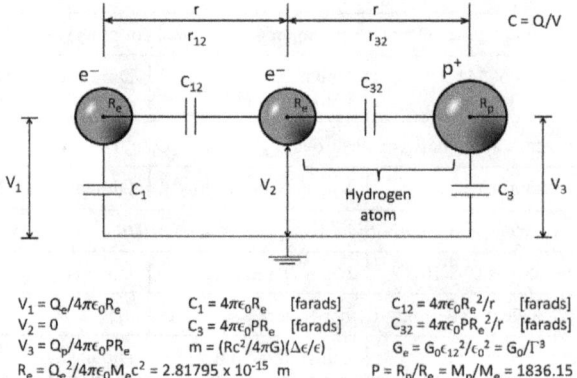

$V_1 = Q_e/4\pi\epsilon_0 R_e$
$V_2 = 0$
$V_3 = Q_p/4\pi\epsilon_0 PR_e$
$R_e = Q_e^2/4\pi\epsilon_0 M_e c^2 = 2.81795 \times 10^{-15}$ m

$C_1 = 4\pi\epsilon_0 R_e$ [farads]
$C_3 = 4\pi\epsilon_0 PR_e$ [farads]
$m = (Rc^2/4\pi G)(\Delta\epsilon/\epsilon)$

$C_{12} = 4\pi\epsilon_0 R_e^2/r$ [farads]
$C_{32} = 4\pi\epsilon_0 PR_e^2/r$ [farads]
$G_e = G_0 c_{12}^2/c_0^2 = G_0/\Gamma^3$
$P = R_p/R_e = M_p/M_e = 1836.15$

Fig. 41-2. Simplified schematic diagram of electrostatic circuit model (nonrelativistic) to derive force between an isolated electron e⁻ and hydrogen atom consisting of a proton P⁺ and an electron e⁻. The capacitor circuit elements represent a polarizable vacuum ($K_{PV}(r,\omega,M) > 1$) with current consisting of motion of Planck charges. Mass may be expressed as a function of voltage and charge, i.e., $m = |\text{volts} \times \text{charge}|/c^2 = E/c^2$.

We can see that, the constant in the law of gravitation being fixed, there may be some upper limit to the amount of matter possible... – Sir Arthur Stanley Eddington

I cannot conceive curved lines of force without the condition of a physical existence in that intermediate space. – Michael Faraday

We call that fire of the black thunder-cloud "electricity", and lecture learnedly about it, and grind the like of it out of glass and silk: but what is it? What made it? Whence comes of it? Whither goes it? – Thomas Carlyle

The gravitational force is the oldest force known to man and the least understood. – Peter van Nieuwenhuizen

The present theory of relativity is based on a division of physical reality into a metric field (gravitation) on the one hand and into an electromagnetic field and matter on the other hand . – Albert Einstein

It (proof by contradiction) is a far finer gambit than any chess gambit. – G. H. Hardy

Think like an outsider. – Robert Green

What we would like is some nice, clean way, probably electrical or atomic, of abolishing gravity at the throw of a switch. – Arthur C. Clarke [Profiles of the Future]

41. Gravitational Constant

Empirical measurements of Newtonian gravitational 'constant' G show notable variation of up to ~0.7% (e.g., 6.57 – 6.7164 x 10^{-11} $m^3s^{-2}kg^{-1}$) with average $\langle G \rangle \sim 6.6666 \times 10^{-11}$ $m^3s^{-2}kg^{-1}$. For example, measured variation is summarized in Table 41-1.

Table 41-1. Variation in measured Newtonian gravitational constant G

Investigator	G value	Method
Fitzgerald et al. (1995)	6.6656(6) x 10^{-11} $m^3/kg \cdot s^2$	Torsion balance (null deflection)
Michaelis et al. (1995)	6.71540 x 10^{-11} $m^3/kg \cdot s^2$	Flotation (null deflection)
Meyer et al. (1999)	6.6685 x 10^{-11} $m^3/kg \cdot s^2$	RF cavity resonance
Quinn et al. (2000)	6.67565 x 10^{-11} m^3 kg^{-1} s^{-2}	Torsion balance (deflection)
NIST/CODATA (2010)	6.67384(80) x 10^{-11} m^3 kg^{-1} s^{-2}	Averaged

Ref: 1) U.S. National Institute of Standards and Technology (NIST).

If the standard gravitational parameter μ = GM = (G + ΔG)*(M + ΔM) is constant, G remains constant only if M is constant. Substituting the relativistic mass M = γM_0 in terms of the Lorentz factor $\gamma (= 1/\sqrt{1 - v^2/c^2})$ yields μ = GM = $G\gamma M_0$. Hence, G may be expected to vary inversely with mass velocity which may account for some of the observed variation in measured values over time and velocity vector direction (G anisotropy), in addition to experimental errors. The gravitation permeability $\mu_g = G/c^2$ appears constant.

The force of gravity in electrostatic terms is shown equivalent to the Newtonian description. A comparison summary is shown in Fig. 41-3 illustrating determination of an electrostatic gravitational constant G_e as a function of electric charge vs. the Newtonian gravitation constant G_N as a function of mass. In terms of Planck masses, the Gravitational coupling constant $G_N = l_P^3/(m_P t_P^2) = t_P c^3/m_P = c^3/Z_s = 6.67262E-11$ $m^3/kg \cdot s^2$ where $\Gamma = 1$. The corresponding Coulomb constant $k_e (= 1/4\pi\epsilon_0 = \mu_0 c_0^2/4\pi) = 8.987551E09$.

An equivalent capacitor circuit of a Coulomb electrostatic field of a point charge is shown in Fig. 41-4. A capacitive network stores energy as does the electrostatic field. A potential field between charges can be simulated by a combination of such networks. In the Planck vacuum model, each capacitor element corresponds to a rotating Planck dipole aligned in synchronous phase with neighboring dipoles.

Along a series of lines running from longer to shorter wavelengths, the effect of the electric field becomes greater as the serial numbers increase... – Johannes Stark

For as long as one has no further point of reference, apart from the position of the maximum, the wavelength thus remains uncertain by an integral factor. – Max von Laue

The expansion of the whole cosmos was but the shrinkage of all the physical units and of the wavelength of light. – Olaf Stapledon

Gravitational and electromagnetic interactions are long-range interactions meaning they act on objects no matter how far they are separated from each other. – Francois Englert

EGM interference patterns form when two or more gravitation fields interact. – Ricardo C. Storti

General relativity predicts that time ends inside black holes because the gravitational collapse squeezes matter to infinite density. – Lee Smolin

41. Gravitational Constant

Gravitational force between electrons

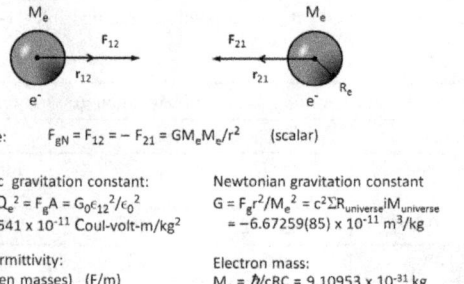

Newtonian gravity force: $F_{gN} = F_{12} = -F_{21} = GM_e M_e / r^2$ (scalar)

Electrostatic gravitation constant:
$G_e = F_{ge} r^2 / Q_e^2 = F_g A = G_0 \epsilon_{12}^2 / \epsilon_0^2$
$= -6.68541 \times 10^{-11}$ Coul-volt-m/kg^2

Newtonian gravitation constant
$G = F_g r^2 / M_e^2 = c^2 \Sigma R_{universe} i M_{universe}$
$= -6.67259(85) \times 10^{-11}$ m^3/kg

Effective permittivity:
ϵ_{12} (between masses) (F/m)

Electron mass:
$M_e = \hbar / cRC = 9.10953 \times 10^{-31}$ kg

Gravitational Gamma:
$\Gamma = \sqrt[3]{(\epsilon_{12}^2 / \epsilon_0^2)} = 1/\sqrt{(1 - 2GM/c^2R)}$

Effective radius of electron:
$R_e = Q_e^2 / 4\pi\epsilon_0 M_e c^2$
$= R_e C_1 / C_e = R_e C_2 / C_e = R_e M_1 / M_e = R_e M_2 / M_e$
$= 2.81795 \times 10^{-15}$ m

Electron capacitance
$C = Q/V = 3.12812E-25$ Coul = rad$^2 \cdot$kg/m^2

Electrostatic force between electrons

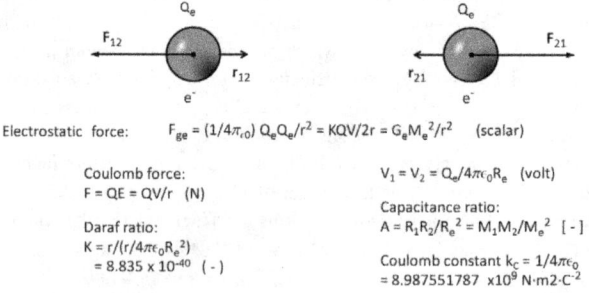

Electrostatic force: $F_{ge} = (1/4\pi r_0) Q_e Q_e / r^2 = KQV/2r = G_e M_e^2 / r^2$ (scalar)

Coulomb force:
$F = QE = QV/r$ (N)

Daraf ratio:
$K = r/(r/4\pi\epsilon_0 R_e^2)$
$= 8.835 \times 10^{-40}$ (-)

$V_1 = V_2 = Q_e / 4\pi c_0 R_e$ (volt)

Capacitance ratio:
$A = R_1 R_2 / R_e^2 = M_1 M_2 / M_e^2$ [-]

Coulomb constant $k_c = 1/4\pi\epsilon_0$
$= 8.987551787 \times 10^9$ N·m2·C^{-2}

Fig. 41-3. Newtonian gravitational constant G (= $G_0 \epsilon_{12}^2 / \epsilon_0$) expressed as a function of effective electric permittivity ϵ_0 (= $\epsilon_{00}/\Gamma = 1/\mu_0 c^2 = 1/\mu_0 v_p v_g$) according to Spear's model.

Intervals betwixt the Stars and Planets... are perfectly fill'd, but by a Matter far subtiler than Air, which some call Celestial, and others Aether... Robert Boyle

There is much reason to be attracted to a theory with no space, no time. But nobody has any idea how to build it. – Albert Einstein

You have not understood something unless you can explain it in terms that can be understood by an English barmaid. – Ernest Rutherford

To change the speed of an object, you have.to exert an external force. But not many know that the speed can be influenced, for example, by a phase shift from inside the system. – Yuri Ivanov

To achieve anti-gravity, it is only necessary to equalize frequencies. – Yuri. N. Ivanov

We have turned a corner in the path of progrss and our ignorance stands before us, appalling and insistent. There is something wrong with the present fundamental conception of physics and we do not know how to set it right. – Sir Arthur Eddington.

41. Gravitational Constant

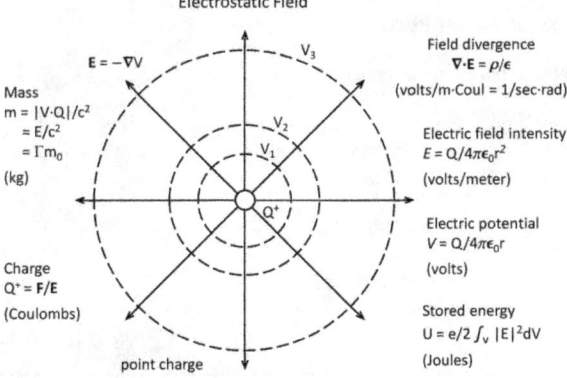

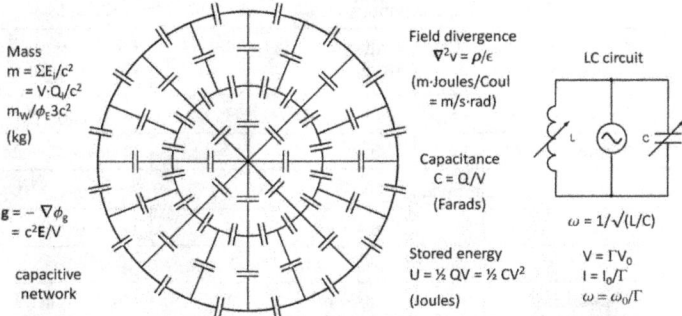

Fig. 41-4. Representation of a capacitive network equivalent to an electrostatic field. An analogous gravitational equivalent may be represented as an inductively-coupled L-network of graded concentric n-gon shells excited by high voltage, high frequency oscillators with radial frequency gradient. Mass elements are equivalent to LC standing wave resonators.

The important thing in science is not so much to obtain new facts as to discover new ways of thinking about them. – Sir Lawrence Bragg

I believe it is the first feeble ray of light on this worst of our physical enigmas. – Albert Einstein (comments on de Broglie thesis)

Bohr's standpoint, that a space-time description is impossible, I reject a limine. – Erwin Schrödinger

My object has been, first to discover correct principles and then to suggest their practical development. – James Prescott Joule

I work on hunches rather than mere facts, and the result is sometimes open to criticism by purists. – Robert Benchley

I have ever been prone to seek adventure and to investigate and experiment where wiser men would have left well enough alone. – Edgar Rice Burroughs

387

42. Newtonian Gravitational Force

42.1 Gravitational force between electrons

Based on a rotating dipole wave (rotor) in a spacetime model, Macken[20] derived the Newtonian gravitational force equation from gravitational force between two electrons separated by distance r.

$$F_g = \Delta\beta_g F_m \approx (Gm_2 R_C/c^2 r^2)(m_1^2 c^3/\hbar) = (Gm_2/c^2 r^2)(\hbar/m_1 c)(m_1^2 c^3/\hbar) \qquad (42\text{-}1)$$

cancelling terms

$$F_g \approx G m_1 m_2 / r^2$$

where:

F_g	Force of gravity between quantum rotars		
	Value: varies	Units: N [= kg(m/sec^2)]	Dimensions: MLT^{-2}
G	Universal gravitation constant (= $G_0/T^3 = c^2 \cdot R_{univ}/M_{univ}) = c^2 K/4\pi H_g$		
	Value: 6.67262 x 10^{-11} (calculated)	Units: N·m^2/kg^2 [= m^3/kg·sec^2]	Dimensions: L^{3}M^{-1}T^{-2}
m_1, m_2	Mass of electron (= **F/a** = |V·q|/c^2 = m_e/T = E/(**E/B**)^2 = e/($\omega_C + \omega_p$))		
	Value: 9.10938E-31	Units: kg	Dimensions: M
β_g	Gravitational magnitude ($\Delta\beta_g$ = Gm/c^{2}r) − (Gm/c^2(r + R$_q$)) ≈ GmR$_q$/c^{2}r^2		
	Value: Varies	Units: -	Dimensions: -
R_C	Electron Compton radius (= c/ω_c = $\hbar$/mc = $\hbar$c/E = $\lambda_C/2\pi = \lambdabar_C$)		
	Value: 3.8616E-13	Units: m	Dimensions: L
r	Separation distance		
	Value: Varies	Units: m	Dimensions: L
F_m	Maximum repulsive force at distance R_C (= m^{2}c^3/$\hbar$ = P$_c$/c)		
	Value: 0.21201	Units: N [= kg(m/sec^2)]	Dimensions: MLT^{-2}
$\hbar$	Dirac $\hbar$ = reduced Planck's unit of action constant h/2π = $\Delta p/\Delta mv$ = $\Delta E/\Delta t$ = 6.6260935E-34/2π = 1.0545E-34)		
	Value: 1.0545E-34	Units: J·s [= rad·m^2·kg·s^{-1}]	Dimensions: ΘML^{2}T^{-1}
ω_C	Compton angular frequency (= 2πf$_C$ = c/R$_C$ = mc^2/$\hbar$ = e/m = 1/$\sqrt{(\epsilon_0\mu_0)}$)		
	Value: 7.7634E20	Units: rad/s	Dimensions: T^{-1}
c	Velocity of light (= f$_C\lambda_C$ = **E/B** = ϕ_E/ϕ_B = c$_0$/T = 1/$\sqrt{(\epsilon_0\mu_0)}$ = $\sqrt{(v_g v_p)}$)		
	Value: 2.997924E8	Units: m/sec	Dimensions: LT^{-1}

It's of no practical use whatsoever... this is just an experiment to prove Maxwell was right. – H. Hertz

Not believing in force is the same as not believing in gravitation. – Thomas Hobbes.

Dubrium sapientiae initium. (Doubt is the origin of wisdom) – René Descartes

"So you're not teaching high-energy physics?", I asked. "No," Feynman replied, "low-energy mathematics". – Carver Mead

42. Newtonian Gravitational Force

The attractive force of gravity between electrons is the result of an imbalance of the force of repulsion exerted on each electron by vacuum energy. The imbalance force is the result of a difference in gravitational gamma Γ ($=\sqrt{K_{PV}}$) which is equivalent to a difference in EM energy density. In the Macken derivation, the imbalance force ΔF (= $\Delta\beta F_m$) is shown equal to the gravitational force F_g where $\Delta\beta$ is the gravitational gradient and F_m is the repulsive force. A representation of the gravitational force in terms of the ansatz electron phase-locked toroidal spin wave model in the Earth's gravitational field is shown in Fig. 42-1. The gravitational force acting on an electron in Earth's gravitational field is proportional to the gravitational field energy density difference. The flux difference acting on an electron results in an unbalance internal radiation pressure force when the metric is no longer uniform.

Macken has proposed that the vacuum must be a nonlinear medium for waves in spacetime and that gravity is the result of this nonlinearity. This argument is consistent with a polarized physical vacuum model in which the degree of polarization (separation of Planck charges) corresponds to a strain providing an explanation for the nonlinearity. A spacetime strain equation introduced by Macken for a point lying on the electron Compton radius R_C (= length L) with oscillation amplitude ΔL = Planck length L_P has the form

$$\epsilon_P(t) = \Delta L/L = \epsilon_P \sin\omega_C t + ((\epsilon_P \sin\omega_C t)^2 + \ldots \quad (42\text{-}2)$$

where ϵ_P = max. strain amplitude (= $L_P/R_C = T_P \omega_C = \omega_C/\omega_P$), ω_C = Compton angular frequency, t = time, T_P = Planck time.

The second order nonlinear term corresponds to the gravitational magnitude β_C of the electron at the Compton radius R_C and may be expanded as

$$\beta_C = (\epsilon_P \sin\omega_C t)^2 = \epsilon_P^2 \sin^2\omega_C t = \tfrac{1}{2}(\epsilon_P^2 - \epsilon_P^2 \cos^2\omega_C t) \quad (42\text{-}3)$$

As defined by Macken, the strain amplitude ϵ_P (= H_β) is a dimensionless ratio representing an energy resonance condition relating the Compton scale and the Planck scale of the Planck vacuum ($\epsilon_P = L_P/R_C = E_i/E_P = m_e/m_P = \omega_C/\omega_P = \sqrt{(Gm^2/\hbar c)}$ = 4.185 x 10^{-23}). An assumed form of the vacuum nonlinear frequency response is shown in Fig. 42-2. Amplitude is proportional to frequency squared. The weak force of gravity is attributed to the second order term and is always positive. The relative strength of gravity to that of the electromagnetic interaction F_G/F_E = 4 x 10^{-40}. The anharmonic characteristic has the same form as a nonlinear polarized medium.

Gravity is a mutual affection between cognate bodies towards union or conjunction. – Johannes Kepler

Simple as the law of gravity now appears, and beautifully in accordance with all the observations of past and of present times, consider what it has cost of intellectual study... All that the human mind has produced – the brightest in genius, the most perservering in application, has been lavished on the details of the law of gravity. – Charles Babbage

There are still much major sway in the thing that my confidence in the admissibility of the theory is still shaky. – Albert Einstein

The real theory which we would like to get to should include gravity with all those theories in such a way that gravity is seen to be a consequence of the octonions and the exceptional group. – Michael Atiyah

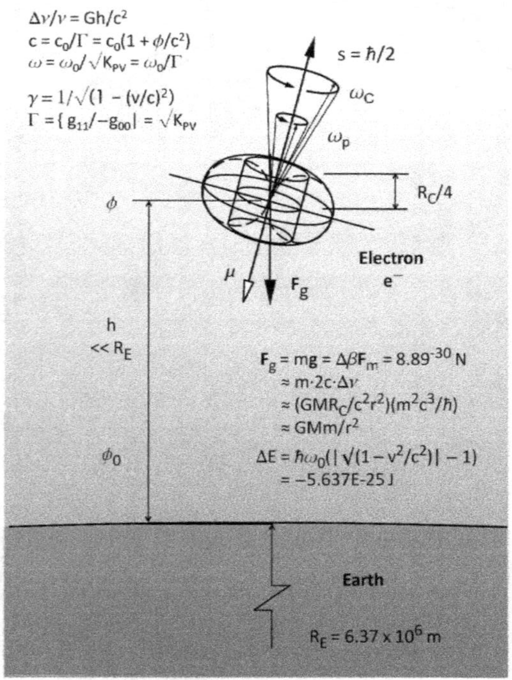

42. Newtonian Gravitational Force

Parameter	Symbol	Relation	Value	Units
gravitational gamma	Γ	$= dt/d\tau = 1/\sqrt{(1 - v_e^2/c^2)} = \sqrt{c_0^2/\phi} = \sqrt{K_{PV}}$	$1.0 + 0.692\text{E-}9$	-
gravitational magnitude	β_g	$= \sqrt{(1 - (1/\Gamma^2))}$	6.95E-10	-
max. repulsive force	F_m	$= m^2 c^3/\hbar = \hbar c/R_C^2$	0.2119	N
Compton radius	R_C	$= \hbar/mc = \lambda_C/2\pi$	3.8616E-13	m
gravitational gradient	$d\beta/dr$	$= \beta/R_E$	1.091E-16	m^{-1}
gravitational delta	$\Delta\beta$	$= (Gm/c^2 r) - (Gm/c^2(r + R_C)$	4.213E-29	-
electron mass	m_e	$= E/c^2 = \hbar/cR_C = \omega_C \hbar/c^2 = \hbar k/v = \gamma m_0$ $= E/(E/B)^2 = k_e e^2/\alpha R_c c^2 = e/(\omega_c + \omega_p)$	9.1094E-31	kg
acceleration of gravity	**g**	$\mathbf{W}/m = -GM/(R + h)^2 = -\nabla\phi$ $\equiv c^2(d\beta/dh) = (\Delta v/\gamma)\cdot(c^2/\hbar)$ $\cong 2c\cdot\Delta v = -c^2(v - v_0)/v\cdot h$ $\cong c^2\{(d\tau - d\tau_0)/d\tau\cdot dh\} = -\phi/h$	9.80	m/s^2
frequency difference	Δv	$= -GM/(2c^2) = g/2c = \Delta\phi/h2c$	1.64E-08	Hz
gravitational potential	ϕ	$= -GM/r = c_0^2/K_{PV} = c_0^2/\Gamma^2$	6.255044E07	J/kg
potential energy	U	$= -GMm/r \cong mgh = m\Delta\phi$	8.92719E-30	J
gravitational redshift	z	$= \Delta v/v = 2\Delta v\Delta h/c$ $= -\Delta\phi/c^2 = -gh/c^2 = (\lambda - \lambda_0)/\lambda_0$ $= 1/\sqrt{(1 - 2GM/rc^2)}$	2.46E-15	-
height above surface	h	$= r - R_E$	1.0	m
gravitational field energy density	u_g	$= -g^2/8\pi G = GM^2/(8\pi r^4)$	5.7368E10	J/m^3
gravitational field energy density difference	$\Delta\rho_E$	$= F_g/(\pi R_C^2)$	1.8977E-14	Pa

$\Delta v/v = Gh/c^2$
$c = c_0/\Gamma = c_0(1 + \phi/c^2)$
$\omega = \omega_0\sqrt{K_{PV}} = \omega_0\Gamma$
$\gamma = 1/\sqrt{(1 - (v/c)^2}$
$\Gamma = \sqrt{g_{11}/-g_{00}} = \sqrt{K_{PV}}$

$s = \hbar/2$
ω_C
ω_p

$R_C/4$

Electron
e$^-$

$\mathbf{F_g} = mg = \Delta\beta F_m = 8.89\text{E-}30$ N
$\approx m\cdot 2c\cdot\Delta v$
$\approx (GMR_C/c^2 r^2)(m^2 c^3/\hbar)$
$= GMm/r^2$

$\Delta E = \hbar\omega_0(|\sqrt{(1 - v^2/c^2)}| - 1)$
$= -5.637\text{E-}25$ J

μ
ϕ
h
$<< R_E$
ϕ_0

Earth
$R_E = 6.37 \times 10^6$ m

Fig. 42-1. Gravitational force acting on an electron suspended in Earth's gravitational field one meter above ground surface. A falling electron undergoes a reduction in frequency (redshift) due to energy loss from damping in regions of increased EM energy density and conservation of spin angular momentum. In contrast, a photon loses energy and undergoes a resultant frequency redshift during ascension in a gravitational field.

42. Newtonian Gravitational Force

Since energy gravitates, substantial variation in the force of gravity at small distances is conceivable with extremely strong gravitational coupling between point-like particles at the Planck scale $E_{Pl} \sim 10^{19}$ GeV. Substituting $m = E/c^2$ into Eqn (42-1) yields

$$F_g = GmM/r^2 = GE_1E_2/c^4r^2 \qquad (42\text{-}4)$$

As radius r decreases, the uncertainty in momentum **p** ($= \hbar/\lambda$) increases with momentum such that $p^2c^2 \gg$ rest mass energy m_oc^2

$$F_g = (G\hbar c^2/r^2)/c^4r^2 = G\hbar^2/c^2r^4 \qquad (42\text{-}5)$$

Substituting the electron Compton radius ($R_C = 3.816\text{E-}13$ m) into the above yields a gravitational force of 3.7132E-46 N. For comparison, the maximum repulsive force F_m between electrons at the minimum separation distance equal to the Compton radius is 0.212 N while the Planck force between Planck charges F_P is 1.2103E44 N.

The electron Compton and Planck dimensionless strain ratio ϵ_P ($= \Delta L/L = l_P/R_C = 4.18509\text{E-}23$) may be interpreted as an interaction between the electron and a physical vacuum of Planck dipoles. For an electron represented as a spin wave of Planck dipoles, the resonance condition corresponds to an in-phase spatial and temporal strain of the electron and the Planck vacuum. See Fig. 42-3. The electron Compton and Planck resonance ratio r_{PC} reflects the large difference in the Compton and Planck scales ($\epsilon_P^{-1} = 2.3893\text{E}22$). The scale is greatly exaggerated in the diagram shown to accommodate the difference in magnitudes. In this model, the non-propagating torsion field associated with the vector potential $\mathbf{A}_i$ is confined to a volume that extends beyond the Compton radius by no more than the Planck length. The eccentric rotary motion suggests an asymmetry in internal energy density due to an imbalance of electrostatic and magnetostatic energy. The variation in proper time as the electron spins results in a phase accumulation. As shown by Macken, the difference between the proper time $d\tau$ as measured by a clock at the Compton radius and the coordinate time dt as measured by a remote laboratory clock $dt - d\tau/dt > 0$ corresponds to a time loss per rotation of $\epsilon_P^2 = 1.75\text{E-}45$ or 3.59E-67 sec/rad. This time deficit may be manifest as a very small retrograde precession. To reproduce the ϵ_P^2 term proposed by Macken in the non-linear vacuum response suggests the electron is not perfectly round, but has a slight eccentricity of twice the Planck length due to alternating in-phase alignments of Planck dipoles creating an oscillating vacuum polarization in the asynchronous region just outside the Compton radius. The extreme difference in the Compton and Planck scales (i.e., $\epsilon_P^{-1} = 2.3893\text{E}22$) appears to the result of maximum radial limit for angular mode locking set by the speed of light circle.

Which was first, Matter or Force? If we think on this question, we shall find that we are unable to conceive of matter without force, or force without matter. – William Crookes

Can there be anything more beautiful than this that the necessary specialization follows from the conservation laws? – Albert Einstein

Nature is in austere mood, even terrifying, withal majestically beautiful. – Frederick Soddy

Every day sees humanity more victorious in the struggle with space and time. – Guglielmo Marconi

Good experimental physicists I have known have had an intense curosity that no 'Keep Out' sign could mute. – Luis Alvarez

42. Newtonian Gravitational Force

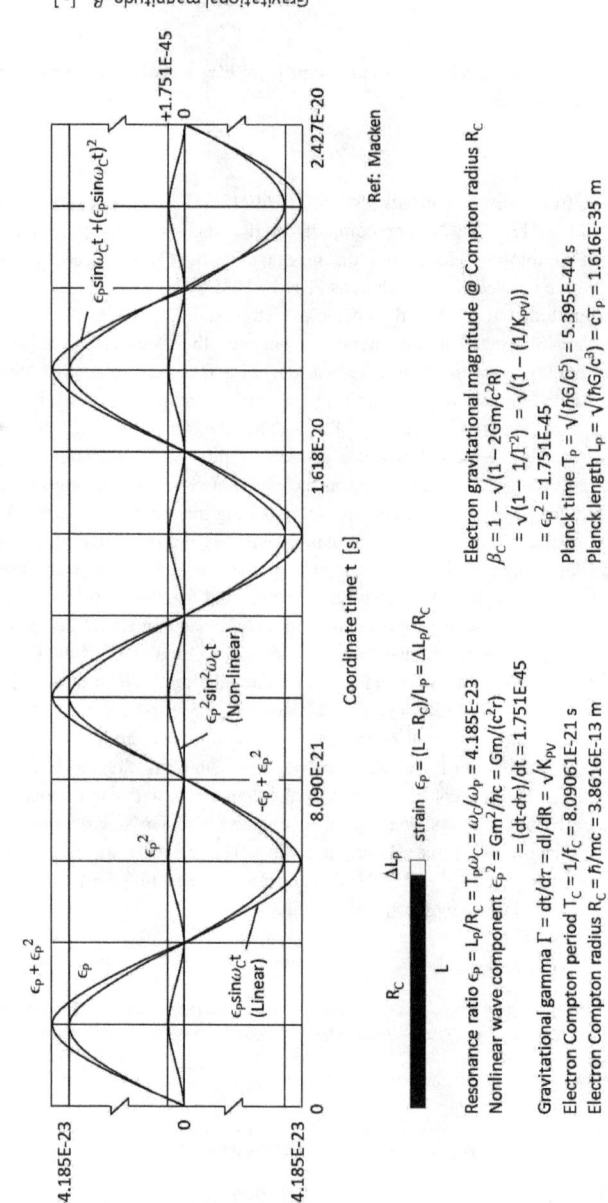

Fig. 42-2. Assumed form of non-linearity of Planck frequency response of the vacuum at Compton frequency. Electron Compton and Planck vacuum resonance is characterized by a dimensionless ratio ϵ_P (= 4.185E-23).

42. Newtonian Gravitational Force

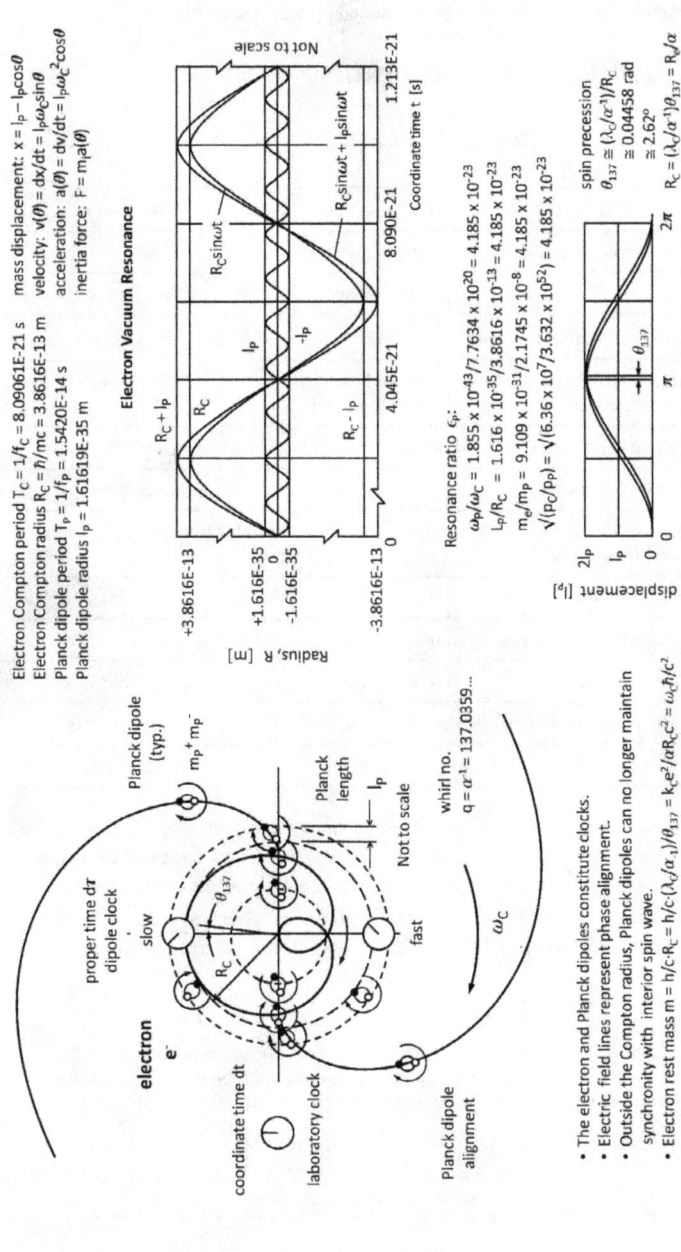

Fig. 42-3. Electron charge results from a slight electron spin precession with whirl no. q = inverse fine structure constant α^{-1}. Electron rest mass arises due to wave function interference (spin precession effect). Precession results from an imbalance of electro- and magnetostatic field energy.

393

42. Newtonian Gravitational Force

42.2. Relation between electrostatic force and gravitational force

Macken[20] derives a dimensionless relation between the electrostatic force $F_{e/P}$ and gravitational force $F_{g/P}$ expressed in Planck units between two equal mass particles with hypothetical Planck charge ($q_P = \sqrt{(4\pi\epsilon_0 \hbar c)}$) separated by a distance r given here as

$$F_{g/P} = F_g/F_P = (Gm^2/r^2)/(c^4/G) = F_{e/P}^2 N^2 = G^2/r^2c^4 \qquad (42\text{-}6)$$

where

$F_{g/P}$	Gravitational force ($= Gm^2/R_C^2)/(c^4/G)$)		
	Value: 3.07×10^{-90}	Units: None	Dimensions: -
$F_{e/P}$	Electromagnetic force ($= e^2/(4\pi\epsilon_0 R_C^2)/(c^4/G)$)		
	Value: 1.28×10^{-47}	Units: None	Dimensions: -
G	Universal gravitation constant ($= G_0/\Gamma^3 = c^3(t_P/m_P) = \hbar c/m_P^2 = G_P/2\pi = c^3/Z_s$)		
	Value: 6.672621×10^{-11} (calculated)	Units: N·m²/kg² [= m³/kg·sec²]	Dimensions: $L^3 M^{-1} T^{-2}$
N	Ratio of radius to Compton radius ($= r/R_q = r/\lambda_C = r(mc/\hbar) = r\omega_C/c$)		
	Value: Varies	Units: None	Dimensions: -
$\omega_{C/P}$	Ratio of Compton angular frequency to Planck angular frequency ($= \omega_C/\omega_P = (m^2/\hbar)/\sqrt{(c^4/\hbar G)}$)		
	Value: 8.5047×10^{-39}	Units: None	Dimensions: -
r	Separation distance between masses ($= R_q = \lambda_C = c/\omega_C$)		
	Value: Varies	Units: m	Dimensions: L
c	Velocity of light ($= l_P/t_P = \tfrac{1}{4\pi\epsilon_0} R_P = \phi_E/\phi_B = E/B = 1/\sqrt{(\epsilon_0\mu_0)} = \sqrt{(v_g v_p)}$)		
	Value: 2.997924×10^8	Units: m/sec	Dimensions: LT^{-1}

The force of gravity $F_{g/P}$ equals the square of the electromagnetic force $F_{e/P}^2 = (F_{e/P}/\alpha)^2$ where the fine structure constant $\alpha = e^2/4\pi\epsilon_0\hbar c$. For Planck charges separated by a distance equal to the Compton radius R_C, $F_{g/P} = F_{e/P}^2$. On a log scale, the electromagnetic force, $F_e = F_{e/P} \cdot F_P = 0.212$ N ($= \tfrac{1}{2}(F_P + F_G)$). The ratio of forces is given by $F_g/F_e = F_e/F_P$. The comparative strength of forces of Planck charges and electric charges is summarized in Table 42-1.

> ...it may be that when the structure of an electron is understood, we shall see that an 'ever-powered' stress in the surrounding aether is necessarily involved...for all these speculations regard the atom as a foreign substance – a sort of 'grit' in the aether.... – Oliver Lodge

> The key to understanding complicated things is knowing what not to look at. – Gerald Jay Sussman

> Timetable for the future: Gravity control 2050. – Arthur C. Clarke (Profiles of the Future)

> One had to be a Newton to notice that the moon is falling, when everyone sees that it doesn't fall. – Paul Valery

42. Newtonian Gravitational Force

42.3 Gravitational force as a function of capacitance

As shown by Spears[25], a general gravitational force equation between any two masses F_g is converted from the gravity force between electrons F_{ge} by a multiplier term A equal to the ratio of capacitance between masses m_1 and m_2 to the capacitance between two electrons. Refer to Fig. 41-3.

$$F_g = F_{ge}A = [KQV/2r][R_1R_2/R_e^2] = F_{ge}M_1M_2/M_e^2 \quad (42\text{-}7)$$
$$= [KQVA/2r][\epsilon_{12}^2/\epsilon_1\epsilon_3]$$

where:

F_{ge}	gravitational force between electrons (= ½QV/r* = (KQV)/2r = KF_e) where r = r* K and F_e = QV/2R		
	Value: −5.54779 x 10^{-71} (Calculated)	Units: N = (Coulomb-volts)/m	Dimensions: QV/M = N
A	ratio of capacitance between masses m_1 and m_2 to the capacitance between two electrons A = $(4\pi\epsilon_{12}R_1R_2/r)(4\pi\epsilon_{12}R_e^2/r) = R_1R_2/R_e^2 = M_1M_2/M_e^2$		
	Value: Varies	Units: None	Dimensions: -
K	distance/no. of darafs = $r/(r/4\pi\epsilon_0 R_e^2) = 4\pi\epsilon_0 R_e^2$ (m/(farad-meters)$^{-1}$)		
	Value: 8.83538 x 10^{-40}	Units: None	Dimensions: -
Q	capacitor charge (F/E = E/V)		
	Value: Varies	Units: Coulomb (= volts/daraf)	Dimensions: Q (= kg·rad/s)
V	voltage (= $F_e/Q^2 r$ = Q/C = E/Q = mc^2/Q)		
	Value: Varies	Units: None	Dimensions: -
R_1, R_2	effective radii between m_1 and m_2 ($R_1 = R_eC_1/C_e = R_e(M_1/M_e)$, $R_2 = R_eC_2/C_e = R_e(M_2/M_e)$		
	Value: Varies	Units: m	Dimensions: L
R_e	effective radius of electron ($R_e = Qe^2/4\pi\epsilon_0 M_e c^2$)		
	Value: 2.81795 x 10^{-15}	Units: m	Dimensions: L
M_1, M_2	mass of objects 1 and 2		
	Value: Varies	Units: kg	Dimensions: M
M_e	electron mass (F/a = M_0/Γ = |volts·charge|/c^2 = $(\hbar k - q_0 A)/v$ = $e(\omega_C + \omega_p)$)		
	Value: 9.10953 x 10^{-31}	Units: kg	Dimensions: M

[As a youth, fiddling in my home laboratory] I discovered a formula for the frequency of a resonant circuit which was $2\pi x$ sqrt(LC) where L is the inductance and C the capacitance of the circuit. And there was π, and where was the circle? ... I still don't quite know where that circle is, where that π comes from. – Richard P. Feynman

But still try, for who knows what is possible. – Michael Faraday

Nothing brings fear to my heart more than a floating point number. – G. J. Sussman

Only three constants are significant for star formation: the gravitational constant, the fine structure constant, and a constant that governs nuclear reaction rates. – Ian Stewart

42. Newtonian Gravitational Force

42.4 Force of gravity as a gradient of EM spectral energy density

Matter consists of fermions (e.g., electrons, protons and neutrons) which are, in effect, self-excited standing wave mass oscillator resonators continuously emitting and absorbing EM waves. Matter in motion undergoes a Lorentz-Doppler frequency shift and generates internal moving standing waves as a result of interference between internal blue-shifted EM waves and counter-propagating red-shifted travelling waves. Matter undergoes acceleration in response to an energy density gradient converting potential energy into kinetic energy of motion. Falling matter loses energy undergoing red-shift during descent into increasing EM field energy density. Conversely, photons lose energy undergoing red-shift during ascent. Acceleration is proportional to the gravitational gradient which, in turn, is proportional to the EM frequency difference ($g = W/m = -\nabla\phi = 2c\cdot\Delta\nu$). See Fig. 42-4. The gravitational scalar potential ϕ is a measure of the local EM field energy density which is represented by the local vacuum refractive index K_{PV} (= $\Gamma^2 = c_0^2/\phi = (1 - GM/rc_0^2)$. The spectral energy density is the energy density per frequency mode. The deBroglie matter wave frequency ν_{dB} is proportional to the moving standing wave velocity (ν_{dB} = c/h/mv) which is proportional to phase displacement $\Delta\phi = \pi\cdot(2gh)/c = v_g\cdot\pi/c = (v_p/\epsilon_0\mu_0)(\pi/c) = \pi\cdot\beta$.

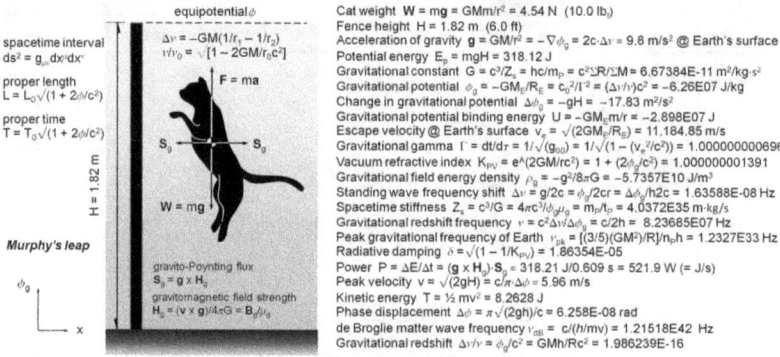

Ref: "Spacetime for Springers" by Fritz Leiber, *Star Science Fiction Stories*, 1958

Fig. 42-4. Example potential and kinetic energy conversion in a gravitational field. Conversion involves matter wave contraction with associated phase and frequency shift.

Mass represents the resistance to acceleration (m = **W**/g = F/2c·$\Delta\nu$), i.e., change in frequency. The degree of synchronization between mass oscillations is proportional to the frequency difference $\Delta\nu$. In free fall, $\Delta\nu$ vanishes. Hence, for stationary levitation to occur, this frequency difference must be nullified as suggested in the doggerel shown in Fig 42-5.

I do not define time, space, place, and motion as being well known to all. – Isaac Newton

The best that most of us can hope to achieve in physics is simply to misunderstand at a deeper level. – Wolfgang Pauli

Even if we consider the filled vacuum as a clumsy description of reality, the existence of virtual pairs and of pair fluctuations shows that the days of fixed particle numbers are over. – Victor F. Weisskopf

42. Newtonian Gravitational Force

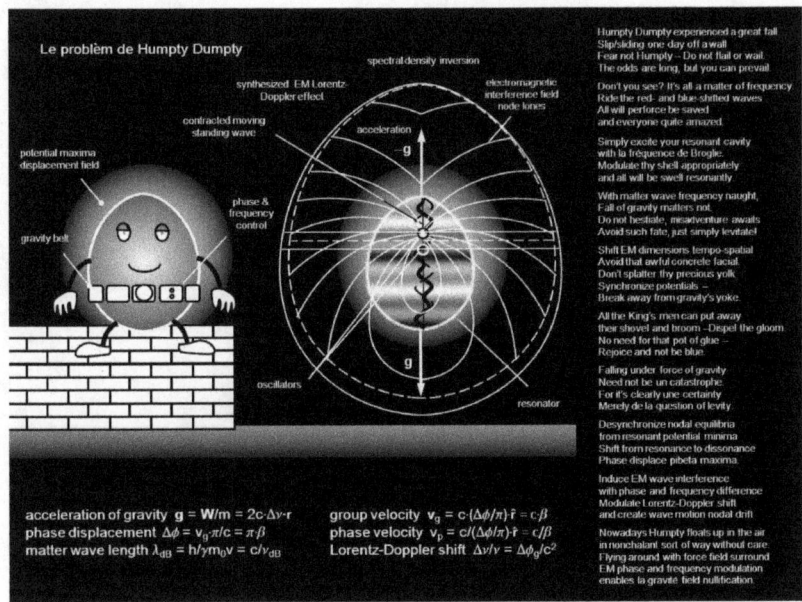

Fig. 42-5. Nursery rhyme cartoon illustrating gravity neutralization of a floating standing wave resonator by phase displacement due to internally generated moving standing waves.

You can't really understand anything unless you can calculate it. – Freeman Dyson

All motion consists of two components. One component serves inwardness (internalization) and the other outwardness (dispersion). Both preconditions of motion regulate the external flow of metamorphosis (panta Rhei). – Viktor Schauberger

He saw no prospect of engaging in the physics of other particles until the electron was well understood. – Helge Kragh

Nothing happens until something moves. – Army Transportation Corps motto

In every case do the opposite to whatever technology does today. Then you will always be on the right track. – Viktor Schauberger

If I were to bequeath to every young man one virtue, I would give the spirit of diverse discontent. – Charles Proteus Steinmetz

The fine structure constant isn't dimensional, so one can have hope that we can calculate that number. – Clive Kilmister

And now I see with eyes serene the very pulse of machine (of reason). – William Shalespeare

Music gives a soul to the universe, wings to the mind, flight to the imagination and life to everything. Geometry will draw the soul to the truth and create the spirit of philosophy. – Plato

I expect in the long run, when we get right down to the fundamental stuff of the universe, we'll find that there's nothing there at all – just nothings moving no-place through no-time. – James Blish

All beginnings are obscure. – Herman Weyl

The most dangerous person in the world is the one who listens, thinks and observes. – Bruce Lee

42. Newtonian Gravitational Force

42.5 Imbalance force between dipoles

An electrodynamic origin of the force of gravity has been proposed by Lucas[57] as a residual EM force between neutral electric dipoles. The residual electrodynamic force is said to arise due to $(v/c)^4$ terms. The non-radial term defined in terms of electrodynamic factors results in quantization of orbits similar to Bode's Law. The conjectured vibrational mechanism responsible for gravitation decays over time and produces such effects as expansion in sizes planets/moons, cosmic background radiation, Hubble's red shifts, Tift's redshifts, Tulley-Fischer luminosity relationship for spiral galaxies, unexpected high velocity of stars in spiral galaxy discs, etc.

The force of gravity as derived by Lucas is given by

$$F_g = Gm_1m_2/r^2 = (1/4\pi\epsilon_0(2e^2/5\pi)A_1^2\omega_1^2/c^2)(A^2\omega^2/c^2))/r^2 \qquad (42\text{-}9)$$

where:

G	Universal gravitation constant $(= G_0/\Gamma^3 = c^3(t_P/m_P) = \hbar c/m_P^2 = G_P/2\pi = c^3/Z_s)$		
	Value: $6.67384(80) \times 10^{-11}$ (2010 CODATA)	Units: $N \cdot m^2/kg^2$ $[= m^3/kg \cdot sec^2]$	Dimensions: $L^3M^{-1}T^{-2}$
m_i	mass $(= \mathbf{W}/g = \Gamma m_0 = (\hbar \mathbf{k} - q_0\mathbf{A})/\mathbf{v} = E/(\mathbf{E}/\mathbf{B})^2 = 2L/(\mathbf{E}^2 - c^2\mathbf{B}^2) = |eV|/c^2)$		
	Value: Varies	Units: kg	Dimensions: M
ϵ_0	Vacuum electric permittivity in free space $(= \epsilon_{0g}/\Gamma = 1/\mu_0c^2 = 1/v_gv_p\mu_0)$		
	Value: 8.85419×10^{-12}	Units: farads/m	Dimensions: $M^{-1}L^{-3}T^2Q^2$
e	electron charge $(= \mathbf{F}/\mathbf{E} = \hbar \mathbf{k}/\mathbf{A} = E/V = m(\omega_C + \omega_p) = V \cdot C)$		
	Value: -1.60219×10^{-19}	Units: Coulomb	Dimensions: Q
A_i	vibration amplitude		
	Value: varies	Units: m	Dimensions: L
ω_i	frequency $(= 2\pi f_i = \omega_0/\Gamma)$		
	Value: varies	Units: rad/sec	Dimensions: T^{-1}
c	velocity of light $(= \lambda\nu = \mathbf{E}/\mathbf{B} = \phi_E/\phi_B = \omega/k = c_0/n = 1/\sqrt{\epsilon_0\mu_0}) = \sqrt{(v_gv_p)} = c_0/\Gamma)$		
	Value: 2.997924×10^8	Units: m/sec	Dimensions: LT^{-1}
r	separation distance between masses $(= \Gamma r_0)$		
	Value: Varies	Units: m	Dimensions: L

For two hydrogen atoms separated by a distance of 1.5 Å, Eqn (42-9) reduces to

$$\begin{aligned}F_g &= Gm^2/r^2 = [(1/4\pi\epsilon_0)(2e^2/5\pi)(A^4\omega^4/c^4)]/r^2 \\ &= [(1/4\pi\epsilon_0)(2e^2/5\pi)(16\pi^4 A^4)/\lambda^4]/r^2 \\ &= 1.658 \times 10^{-43} \text{ N}\end{aligned} \qquad (42\text{-}10)$$

At a separation of 1 meter, the gravitational force $F_g = 1.868803 \times 10^{-64}$ N.

Gravitation is demonstrable by leaving a body unsupported. – Sir William Gull

... for gravitation is nothing more than a name for a general fact, the why of which we know not.
– Sir William Rowan Hamilton

42. Newtonian Gravitational Force

42.6 Gravitational vector potential

Both light and matter are subject to gravitational scalar and vector potential effects. The non-relativistic gravitational scalar and vector potentials are given by

$$\phi_g = -G \Sigma(m_i/r_i) \qquad [J/kg = m^2/s^2] \qquad (42\text{-}11)$$

$$\mathbf{A}_g = -G/c^2 \Sigma(m_i v_i/r_i)\cdot\hat{\mathbf{r}} \qquad [J/kg = m^2/s^2] \qquad (42\text{-}12)$$

where ϕ_g = gravitational scalar potential, $\mathbf{A}_g$ = gravitational vector potential and G = Newtonian gravitational constant ($= c^2/\Sigma(r_i/m_i)$). The generalized Newtonian gravitational force equation advanced by Jefimenko includes a scalar and vector potential component:

$$\mathbf{F}_g = -GmM\cdot\hat{\mathbf{r}}/r^2 - GmM\cdot\hat{\mathbf{r}}/2c^2 r^2 \qquad (42\text{-}13)$$

Light decelerates in a gravitational field exhibiting an effective mass m ($= 2c\Delta v\cdot\hat{\mathbf{r}}/F$) where $c = c_0(1 - GM/rc_0^2) = c/\Gamma$. The magnitude of the wave vector **k** increases as the EM wave encounters a region of increased EM field intensity of a gravitational field.

$$k \simeq (\omega/\omega_0)[1 + (GM/rc_0^2)]/[1 - (GM/rc_0^2)] \simeq [1 + (2GM/rc_0^2)] \qquad (42\text{-}14)$$

Similarly, matter undergoes a reduction in velocity v ($= \beta\cdot c = c/n$) in a gravitational field and a corresponding change in momentum p ($= v\cdot E/c^2 = \beta\cdot E/c = mv$). The gravitational scalar and vector potential effects are additive. Inertial mass increases when masses move relative to each other as a result of the vector potential coupling. The retarded vector potential $\mathbf{A}_g^*$ as a function of time is

$$\mathbf{A}_g^*(t) = -G/c^2 \Sigma [m_i v_i(t - r_i/c)/r_i]\cdot\hat{\mathbf{r}} \qquad (42\text{-}15)$$

where $G = c^2 \Sigma(r_i/m_i) = l_P^3/m_P t_P = \hbar c/m_P^2 = c^2 \mathbf{K}/4\pi\mathbf{H}_g = (\mathbf{v} \times \mathbf{g})\mu_g/4\pi\mathbf{B}_g = c^3/Z_s$

The cogravitation force $\mathbf{F}_k$ ($= m(\mathbf{v} \times \mathbf{K}) = m\mathbf{g}_k$) exerted by a moving mass current with velocity **v** on adjacent masses is analogous to the magnetic force $\mathbf{F}_m$ ($= q^+(\mathbf{v} \times \mathbf{B})$) exerted by a moving charge current with velocity **v** on adjacent charges. The circumferential cogravitation field **K** surrounding a mass current is analogous to the circumferential magnetic field **B** surrounding an electric current. The cogravitation field of a moving point mass acts only on moving masses and is given by

$$\mathbf{K} = -G [m(\mathbf{v} \times \hat{\mathbf{r}})/c^2 r^2 = (\mathbf{v} \times \mathbf{g})/c^2 \qquad (42\text{-}16)$$

The gravitational vector potential $\mathbf{A}_g = (\mathbf{v}/c^2)\phi_g$ is proportional to the velocity of the scalar potential ϕ_g. The gravikinetic field $\mathbf{g}_k$ ($= \mathbf{v} \times \mathbf{K} = -d\mathbf{A}_g^*/dt$) opposes change in mass motion and is analogous to the electrokinetic field $\mathbf{E}_k$ ($= -d\mathbf{A}^*/dt$) due to retarded changing magnetic field which opposes change in charge motion. A moving mass generates an acceleration of gravity $\mathbf{g} = -\nabla\phi_g - d\mathbf{A}_g/dt = \nabla\mathbf{A}_g$. The gravito-magnetic flux density $\mathbf{B}_g$ ($= \mu_g \mathbf{H}_g$) is equal to the curl of the gravitomagnetic vector potential ($\mathbf{B}_g = \frac{1}{2}\nabla \times \mathbf{A}_{gm} = -2\Omega$).

Wisdom must be earned not given. – Zen (Blake's 7)

The speed of movement of wave systems depends on the phase relationship between active elements. – Yuri N. Ivanov

42. Newtonian Gravitational Force

As previously noted, rest mass m_0 is the result of wave function interference due a slight spin precession of the electron with a whirl no. equal to the inverse fine structure constant α^{-1} where $m_0 = E/(\mathbf{E}/\mathbf{B})^2 = h/c \cdot R_C = h/c \cdot (\lambda_C/\alpha^{-1})/\theta_{137}$. Relativistic mass $m (= \gamma m_0 = \Gamma m_0 = m_0\sqrt{K_{PV}})$ increase is a result of acceleration into a region of increased EM field intensity which likewise is a wave interference effect. Inertial and gravitational force $\mathbf{F} (= d(m\mathbf{v})/dt = d\mathbf{p}/dt = m\mathbf{g} = m2c \cdot \Delta v \cdot \hat{\mathbf{r}})$ is proportional to the Lorentz-Doppler frequency shift $\Delta v (= g/2c)$. For a light wave (traveling wave) momentum $\mathbf{p} = \hbar\mathbf{k} = E/c$ whereas for a matter wave (moving standing wave) momentum $\mathbf{p} = m\mathbf{v} = \hbar\mathbf{k}/v = \beta \cdot E/c$. For electrically charged matter, momentum $\mathbf{p} = m\mathbf{v} - q\mathbf{A}$. Acceleration $\mathbf{a} (= d\mathbf{v}/dt)$ is proportional to a velocity difference which corresponds to a change in phase ϕ ($\Delta\phi = (v_g/c) \cdot \pi$). See Fig 42-6. These effects may be utilized to counter gravitational acceleration, for matter wave propulsion or the inverse effect of matter wave interferometry for precise detection of motion.

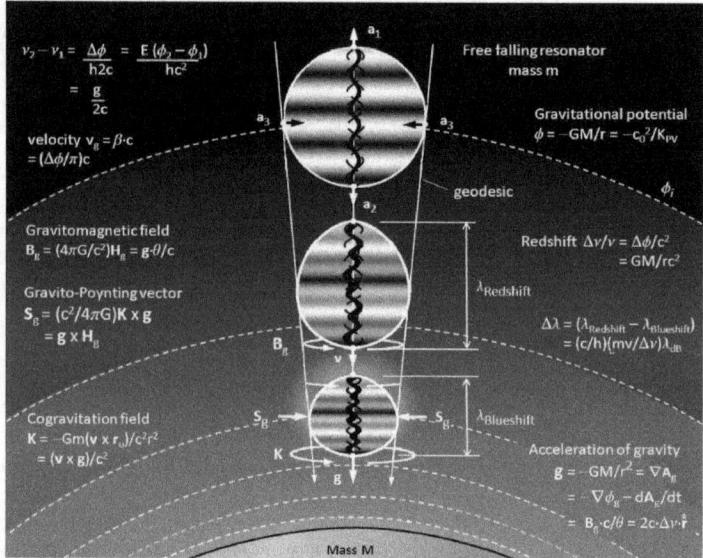

Fig. 42-6. Wave function interference produces an effective mass $m (= \Delta h v/c^2 = \Delta E/(\mathbf{E}/\mathbf{B})^2)$ of a photon traveling wave due to change in momentum. Similarly, wave function interference is responsible for the relativistic mass $m (= \gamma m_0 = \sqrt{((E/c^2)^2 - (p/c)^2)}$ increase of a matter moving standing wave due to change in momentum $p (= \beta \cdot E/c = \gamma mv)$.

Controlling phase and frequency parameters ... will lead to fundamentally new technologies both in terms of movement in space and obtaining energy.... – Yuri N. Ivanov

You will note that elapsed time appears in your answer as an undetermined unknown. Yes... yes, of course. That's the chance we have to take. Let's get busy. – Isaac Asimov

The effort to understand the universe is one of the few things that lifts human life a little above the level of farce and gives it some of the grace of tragedy. – Steven Weinberg

As the sun eclipses the stars by its brillancy, so the man of knowledge will eclipse the fame of others in the assemblies of the peoples if he proposes algebraic problems, and still more if he solves them. – Brahmagupta

43. Einstein Field Equation

Compared with understanding gravity, the special theory of relativity was mere child's play. – Albert Einstein

All the laws of physics can be contained in one equation: U = 0. – Richard Feynman

Anyone who has never made a mistake has never tried anything new. – Albert Einstein

In Einstein's general relativity, the stress-energy tensor **T** is identified as the source of the gravitational field rather than mass[35]. The Einstein GR gravitational field equation equates curvature to sources of stress-energy momentum

$$G_{\mu\nu} = R_{\mu\nu} - \tfrac{1}{2}Rg_{\mu\nu} = -(8\pi G/c^4)T_{\mu\nu} = -kT_{\mu\nu} \tag{43-1}$$

where:

$G_{\mu\nu}$	Einstein tensor [Curvature = 1/Area = Length/Volume]		
	Value: Varies	Units: m^{-2}	Dimensions: L^{-2}
$R_{\mu\nu}$	Ricci curvature tensor (contracted from Riemann curvature (= $R_{abc}^{\ b}$)		
	Value: Varies (≥ 0)	Units: m^{-2}	Dimensions: L^{-2}
R	Scalar curvature defined as trace of Ricci tensor (= $R_\mu^{\ \mu}$)		
	Value: Varies	Units: m^{-2}	Dimensions: L^{-2}
$g_{\mu\nu}$	Lorentz spacetime metric tensor (= $n_{\mu\nu} + h_{\mu\nu}$) of spacetime manifold M ($n_{\mu\nu}$ flat metric in R^4)		
	Value: Varies	Units: -	Dimensions: -
G	Newtonian Gravitational constant (= $c^2 R_{univ}/M_{univ} = c^3(t_P/m_P) = \hbar c/m_P^2 = c^3/Z_s$)		
	Value: 6.67428 ±0.00067 x 10^{-11} [2010 CODATA]	Units: N·m^2/kg^2 [= m^3/kg·sec^2]	Dimensions: L^{3}M^{-1}T^{-2}
c	velocity of light (= $c_0/\Gamma = 1/\sqrt{(\epsilon\mu)} = $ **E/B** $= l_P/t_P = \sqrt{(v_g v_p)} = c_0/\sqrt{K_{PV}}$)		
	Value: 2.997924 x 10^8	Units: m/sec	Dimensions: LT^{-1}
k	Einstein constant (= $-8\pi G/c^4$). Gravitation permeability $\mu_g = G/c^2 = \Sigma(r_i/m_i)$.		
	Value: -2.0765 x 10^{-43}	Units: 1.N	Dimensions: T^{2}M^{-1}L^{-1}
$T_{\mu\nu}$	Stress energy-momentum tensor [Energy density = Force/Area = Pressure]		
	Value: Varies	Units: kg/m^3	Dimensions: M/L^3

In contracted form Equation (43-1) becomes

$$R = -kT \quad \text{where} \quad T = T_i^i \tag{43-2}$$

In the weak field limit $R_{00} = (1/c^2)\Delta\phi = \tfrac{1}{2}\Delta g_{00} = (8\pi G/c^4)T_{00}$ where ϕ is the gravitational potential.

The Evans field equation[5,6] in tetrad formulation allowing both gravitation and torsion is expressed as

$$G^a_{\ \mu} = R^a_{\ \mu} - \tfrac{1}{2}Rq^a_{\ \mu} = k\,T^a_{\ \mu} \tag{43-3}$$

43. Einstein Field Equation

where $q^a{}_\mu$ is the tetrad eigenfunction, a = Euclidean spacetime index, μ = non-Euclidean base manifold. With the addition of a non-zero cosmological constant $+g_{\mu\nu}\Lambda$ where the vacuum energy of the universe is defined as $\rho_V = \Lambda/8\pi G$, the Einstein field equation becomes

$$G_{\mu\nu} = R_{\nu\nu} - \tfrac{1}{2} g_{\mu\nu} R + \Lambda\, g_{\mu\nu} = -k T_{\mu\nu} \qquad (43\text{-}4)$$

where Λ is a constant.

Daywitt[12] derives the following expression for the Einstein tensor $G_{\mu\nu}$ in terms of the Planck vacuum (PV) model

$$G_{\mu\nu} = 8\pi T_{\mu\nu}/(c^4/G) = 8\pi T_{\mu\nu}/(mc^2/r^*) \qquad (43\text{-}5)$$

where: m* = mass of Planck particle, r* = Compton radius of Planck particle and c^4/G (= m^*c^2/r^*) is spacetime curvature force or Planck force F_P).

The stress-energy-momentum tensor $T_{\mu\nu}$ may be equivalently written in terms of $\rho_0 c^2$ [mass-energy density] rather than ρ_0 [mass density] yielding a gravitation coupling constant $k = -8\pi G/c^4$. Gravity couples to the T_{44} energy density. The Einstein equation is equivalent to a statement that energy density (= energy/volume = force/area) equals k pressure, hence, gravitation is related to vacuum energy/pressure.

$$T_{\mu\nu} = (c^4/8\pi G) G_{\mu\nu} = k \cdot F_P\, G_{\mu\nu} \qquad (43\text{-}6)$$

where:

$T_{\mu\nu}$	stress-energy-momentum tensor $(T_{\mu\nu} = T_{\nu\mu})$		[Energy density = Pressure]
	Value: Varies	Units: N/m^{-2}	Dimensions: ML^{-1}T^{-2}
$G_{\mu\nu}$	Einstein tensor [Curvature = 1/Area = Length/Volume]		
	Value: Varies	Units: m^{-2}	Dimensions: L^{-2}
k	Einstein constant (= $-8\pi G/c^4$). Gravitation permeability $\mu_g = G/c^2$.		
	Value: -2.0765×10^{-43}	Units: s^2/kg·m = 1/N	Dimensions: T^{2}M^{-1}L^{-1}
F_P	Planck force (= $c^4/G = m_P l_P/t_P^2 = G m_P^2/l_P^2 = \tfrac{1}{4\pi\epsilon_0}(q_P^2/l_P^2)$)		
	Value: 1.21034×10^{44}	Units: N (= kg·m/s^2)	Dimensions: MLT^{-2}

In the limit of maximum curvature ($\kappa = 1/l_P$), the maximum force possible is the Planck force F_P (= 1.21034×10^{44} N).

The Einstein tensor $G_{\mu\nu}$ is symmetric with ten independent components (mass-energy density, energy flux, momentum density and momentum flux). The spacetime curvature represented by a Riemann curvature tensor $R_{\alpha\beta\mu\nu}$ has 20 independent components including a sum of background curvature and wave curvature. The spacetime metric is described by $g_{\alpha\beta} = \eta_{\alpha\beta} + h_{\alpha\beta}$ where $\eta_{\alpha\beta}$ is the Minkowski metric and $h_{\alpha\beta}$ is the perturbation term corresponding to a gravitational wave. Components the metric are given by the coefficient functions of the differentials in the line element ds^2 (interval). The Minkowski metric is $g_{00} = -1$, $g_{11} = g_{22} = g_{33} = 1$. Spacetime metric and stress-energy-momentum tensor components are illustrated in Fig. 43-1. As shown each tensor has 16 components. In a full expression of the curvature of 4D spacetime, a 20 component Riemann curvature tensor R_{ijkl} is required to include tidal effects. See Fig. 26-8. EM radiation acts a perfect fluid, i.e., with mass-density and pressure without heat conduction or viscosity. The equation of

43. Einstein Field Equation

state relating mass-energy density ρ and pressure P for EM radiation is given by $\rho = T^{tt} = 3P$. Electromagnetic field energy density $\rho_{em} = \frac{1}{2}(\epsilon_0 \mathbf{E}^2 + \mathbf{B}^2/\mu_0)$ has units J/m^3 = Pa.

Mass-energy fields (including strong, weak, and electromagnetic) produce space-time curvature in conventional GR with the exception of gravitational energy. Gravitational energy is negative and is already included in associated mass term. In the Yilmaz version of GR, a gravitational stress-energy tensor term is added to Einstein's equations yielding predictions that there are no black hole singularities or event horizons.

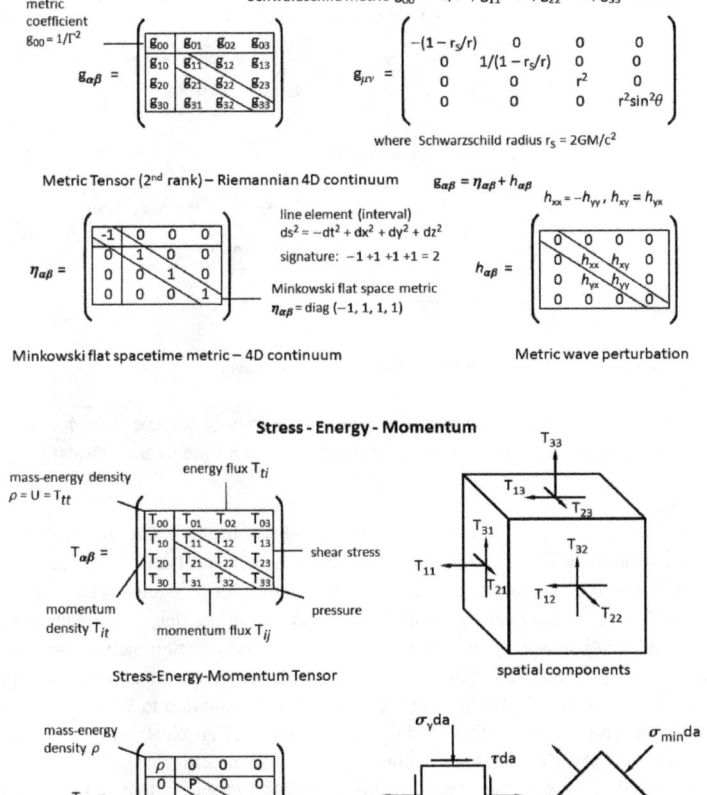

Fig. 43-1. Spacetime metric $g_{\alpha\beta}$ and stress-energy tensor $T_{\alpha\beta}$ components in 4D.

The ground of physics is littered with the corpses of unified theories. – Freeman Dyson

Ground is where electrons go to die. – David Casler

43. Einstein Field Equation

The components of the Maxwell or electromagnetic field tensor $F_{\mu\nu}$ is illustrated in Fig. 43-2.

Electromagnetic Field Tensors

Exterior derivative of EM four-potential A

$$A^\alpha = (\phi/c, \mathbf{A})$$

$$F_{\mu\nu} = \delta_\mu A_\nu - \delta_\nu A_\mu = \delta A_\nu/\delta x_\mu - \delta A_\mu/\delta x_\nu$$

$$-1/2\, F_{\mu\nu}F^{\mu\nu} = E^2 - c^2 B^2$$

Antisymmetric, traceless, rank-2, field strength tensor – Minkowki 4D spacetime

EM field components

$$F_{\mu\nu} = \begin{pmatrix} 0 & E_x/c & E_y/c & E_z/c \\ -E_x/c & 0 & -B_z & B_y \\ -E_y/c & B_z & 0 & -B_x \\ -E_z/c & -B_y & B_x & 0 \end{pmatrix}$$

$F_{\mu\nu} = -F_{\nu\mu}$ [volts/meter]

E = Electric field strength
B = Magnetic flux density

1/c factor reflects usage of SI units
+ – – – signature $x_0 = -x^0 = ct$

Covariant EM Field Tensor (2nd rank) – Riemannian 4D continuum

$$F^{\mu\nu} = \begin{pmatrix} 0 & -E_x/c & -E_y/c & -E_z/c \\ E_x/c & 0 & -B_z & B_y \\ E_y/c & B_z & 0 & -B_x \\ E_z/c & -B_y & B_x & 0 \end{pmatrix}$$

$F^{\mu\nu} = -F^{\nu\mu}$ [volts/meter]

$E = (E^x, E^y, E^z) = (F^{01}, F^{02}, F^{03})$
$B = (B^x, B^y, B^z) = (F^{23}, F^{31}, F^{12})$

$E_{\mu\nu} = \phi_{\nu,\mu} - \phi_{\mu,\nu}$

Contravariant EM Field Tensor (2nd rank) – Riemannian 4D continuum

Fig. 43-2. Electromagnetic field tensor $F_{\mu\nu}$ components in a spacetime 4 x 4 matrix representation. Covariance denoted by lowered indices refers to a Lorentz transformation.

In a zero curvature vacuum ($R_{\mu\nu} = 0$, $T_{\mu\nu} = 0$), GR reduces to the theory of a massless spin-2 field propagating in flat spacetime R^4 formulated by Fierz and Pauli[114]. A graviton $\gamma\gamma^*$ has magnitude and a superposition of two spin states and, hence, is described by a 2nd rank tensor. Jefimenko notes that the Einstein field equation in the linear approximation, the pseudo stress-energy tensor $T_{\mu\nu}$ on the right-hand side of the equation includes sources of gravitation (mass, pressure, self-gravitation and/or 4-velocity) except the energy density of the gravitational field itself. The Reissner-Nordstrom static solution to Einstein's field equations give the gravitational effect of the energy density of an electromagnetic field density corresponding to a charged, non-rotating, spherically symmetric mass.

Kaluza proposed a five dimensional modification of Einstein's four dimensional spacetime metric to incorporate the electromagnetic field of Maxwell[115]. The electromagnetic potential is described by metric components of the fifth dimension and momentum in the fifth dimension is a multiple of electric charge. See Fig. 43-3. Kaluza-Klein posited that the unseen fifth dimension is curled up in a microscopically small dimension of less than 10^{-32} m radius. In the Evans-Cartan model, gravitation corresponds to curvature whereas electromagnetism corresponds to torsion. The EM potential 4-vector A^a_μ consists of a scalar potential component (A_0) and a three dimensional magnetic vector potential (A_1, A_2, A_3) defined by a dreibein referenced to a base manifold (μ). The scalar potential represents the quantum dipole volumetric density of the polarized vacuum responsible for the electric field whereas the magnetic vector potential is a measure of the torsional spin of the quantum dipole vortices responsible for the magnetic field.

43. Einstein Field Equation

Kaluza-Klein metric

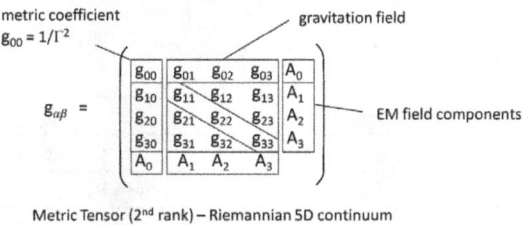

Fig. 43-3. In the 5D Kaluza-Klein geometry incorporating Einstein's 4D gravitation metric and Maxwell's electromagnetic field, the Riemann metric in five dimensions becomes a 5 x 5 matrix. The fifth dimension accounts for electric charge, a spin precession effect. Electric charge corresponds to a torsion loop closure defect.

Heim proposed a six dimensional quantized spacetime geometry that included Einstein's gravitation metric, Maxwell's electromagnetism with real and imaginary spin components to account for bosons and fermions. See Fig. 43-4.

Heim R_6 Energy Density Action Tensor

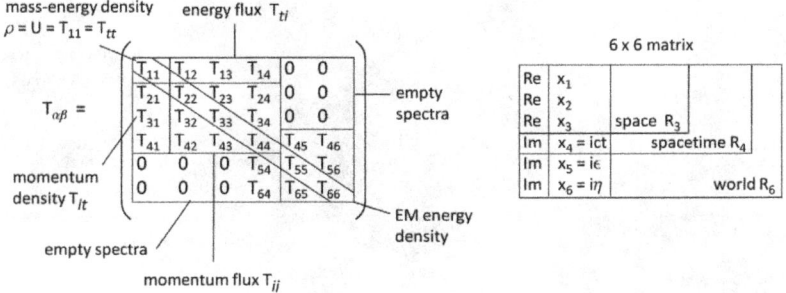

Fig. 43-4. A 6D spacetime geometry is represented in the Heim 6 x 6 matrix which includes particle spin components.

The immense unification effected by electromagnetism apparently left gravitation out of its scope.
– Hendrik Antoon Lorentz

In our work we are always between Scylla and Charybdis; we may fail to abstract enough, and miss important physics, or we may abstract too much and end up with fictitious objects in our models turning into real monsters that devour us. – Murray Gell-Man

Gravity arises from a change in energy density and physically manifests in terms of an EGM Poynting vector. – Ricardo C. Storti

405

43. Einstein Field Equation

The Evans-Cartan field equation[5,6] in terms of the metric and tetrads which includes the effects of curvature and torsion is given by

$$R^a_\mu - \tfrac{1}{2} R g^a_\mu = (8\pi G/c^2) T^a_\mu = k T^a_\mu \tag{43-7}$$

where

R^a_μ	Riemann curvature tetrad					
	Value: Varies (≥ 0)		Units: m^{-1}		Dimensions:	L^{-1}
R	Scalar curvature defined as trace of Ricci tensor (= R^a_a)					
	Value: Varies		Units: m^{-1}		Dimensions:	L^{-1}
q^a_μ	tetrad represents potential in non-Euclidean spacetime (q symbolizes gravitation, electromagnetism, gluon strong force or weak force). Index of tangent spacetime (= a). Index of base manifold (= μ).					
	Value: Varies		Units: -		Dimensions:	-
k	Einstein constant (= $8\pi G/c^4 = 8\pi\mu_g/c^2$). Gravitation permeability $\mu_g = G/c^2$.					
	Value: Varies		Units: m/kg		Dimensions:	LM^{-1}
T^a_μ	Stress energy-momentum tetrad					
	Value: Varies		Units: m^{-1}		Dimensions:	L^{-1}

This generalized equation includes a moving frame tetrad which describes curvature associated with gravitation and a rotating frame tetrad which describes torsion associated with electromagnetism. Gravity field tangent space curvature is illustrated in Fig 43-5.

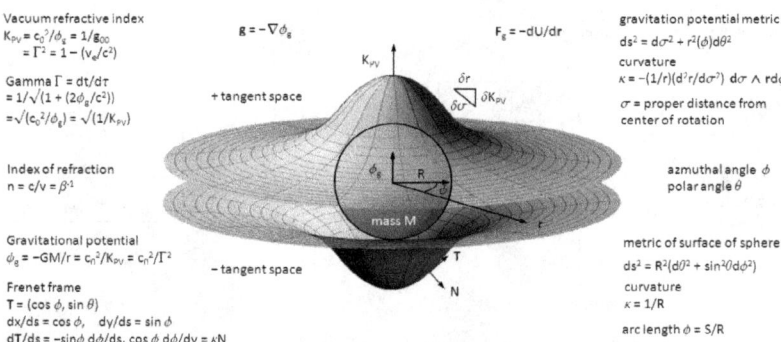

Fig, 43-5. Curvature 2-form representation of a gravitational potential of a central mass. The gravitational potential metric is a measure of volumetric EM energy nodal density.

There is a fifth dimension beyond which is known to man. It is a dimension as vast as space and as timeless as infinity... – Rod Serling

I still remember one time after those discussions with Klein in which he had told me about his five-dimensional relativity and out of that the quantum conditions would come. You see, from the periodicity in the fifth dimension you got the quantum conditions. – George Uhlenbeck

44. Quantum Gravity

General Theory of Relativity is as yet incomplete insofar as it has been able to apply the general principles of relativity satisfactorily only to gravitational fields. – A. Einstein

The quest for a quantum gravity is one of the greatest unsolved problems in all of science – Michio Kaku

Space is imagined as a lattice made of nodes connected by edges... A physicist working without a lattice is something like a trapeze artist working without a net. – Lee Smolin

44.1 Introduction

Quantum gravity theory purports to reconcile quantum theory describing elementary particles and atoms with Einstein's general theory of relativity describing events in curved spacetime under the influence of gravity. Gravitation in an optical gravity theory is a result of resonant electromagnetic (EM) wave interactions in a polarizable vacuum (PV) with a variable refractive index. Gravitation is described by variation in EM wave energy and density due to local variation in the vacuum refractive index $K_{PV}(r,\omega,M)$. Variation of the vacuum refractive index and corresponding variation in the speed of light and provides a mechanism for EM wave contraction effects. The electromagnetic field and photon/electron interactions in quantum electrodynamics theory (QED) are quantized and has provided a template for subsequent quantum field theories.

Two currently popular theoretical approaches to quantizing gravity are String/M-theory and Loop Quantum Gravity (LQG). In M-Theory, there are 10-dimensional strings in a membrane in an 11-dimensional supergravity hyperspace. A complex 6-dimensional Calibi-Yau manifold compactifies 6 of the 10 dimensions to a 4-D spacetime. 4D spacetime is regarded as a 'D-Brane' hypersurface in a 11-dimensional superspace. Particles represent vibrations of 1 dimensional Planck scale string objects in the Calibi-Yau manifold including the graviton. Quantum gravity is described in terms of dynamic intersecting, linking or knotting of loops. Particles of spin 1 and 2, i.e., the photon and graviton, are massless and correspond to zero frequencies of the string. Higher frequencies correspond to particles with mass in multiples of the Planck mass ($m_{Planck} = 10^{19}$ GeV). Electric charge is associated with open strings. Quarks are connected by open strings. String loops in 4-D may include twists or rotations in the extra six internal dimensions or degrees of freedom. In analogy with guitar strings, mass m may be associated with string tension T ($= m \cdot 4L^2 f^2$) although what the source of the tension in string theory is undefined. The amplitude envelopes of closed strings form loops moving through spacetime. Vibrations of the closed loops allow different energy levels corresponding to different particles including boson/fermion supersymmetric partners. Despite intensive search, supersymmetric particles have not been observed. String theory proponents argue string theory is the only known way that includes gravity with other forces described by quantum theory. A graviton in string theory is described by a quadrapole excitation of a closed loop string. String theory comes in many different versions with various assumptions. What is the composition of a string or how formed is undefined. With an odd number of spatial dimensions and one temporal dimension, spacetime must be an even number. However, chirality arises only with an odd number of spatial dimensions, which appears to preclude a 11-D supergravity manifold. String theory, like brane theory, appears as a fanciful, metaphysical abstraction with little substance, an idea full of holes – a doily for the mind.

44. Quantum Gravity

Gravitons in quantum field theory (QFT) are usually regarded as massless, spin-2 bosons propagating as perturbative quantum fluctuations (excitations) superimposed on a flat, Minkowski spacetime metric

$$g_{\mu\nu} = \eta_{\mu\nu} + h_{\mu\nu} \tag{44-1}$$

where $g_{\mu\nu}$ is the resultant field metric, $\eta_{\mu\nu}$ is the background metric and $h_{\mu\nu}$ is the graviton field which is a measure of deviation from the flat Minkowski metric. However, there is no explanation for causal mechanism for variation in background metric such as gravity waves, how the graviton field is generated or how exactly does gravitational attraction occur via graviton exchange. In loop quantum gravity (LQG), space-time is quantized as a network in which quantum numbers are assigned to nodes and linkages. GR is expressed in terms of Ashtekar variables in which quantum areas are quantized.

In LQG, unlike M-Theory, higher dimensions are not required. Holonomy of curved space is described in terms of rotational transformations around a loop. LQG does not describe objects within spacetime unlike string theory and appears incompatible with requirements of SR absent the effects of gravity and the principal of general covariance. LQG as presently formulated does not include the graviton of QFT nor does it describe other forces. QFT was formulated for flat spacetime and does not incorporate curved spacetime of GR. Spin networks represent quantum states of the geometry of space. The edges of spin networks correspond to discrete units of area while network nodes correspond to quantized volumes in Planck units. The spin networks are relational objects said to generate space. The geometry of spin networks may be adjusted to match a given metric. A spin network represents causal linkages, however, the causal mechanism of gravity is unclear as is just what is it that is spinning or the cause of spin.

Similar to the GR mathematical theory, both approaches attempt to quantize space and time, ascribing physical properties to geometry rather than objects in spacetime. Neither of these approaches address the fundamental nature of the physical vacuum. There is neither explanation of the origin of electric charge, mass or spin nor are specific geometric models proposed with characteristics matching particle observables. The mathematics of the theories is complex and few verifiable predictions are advanced. Laboratory experimental verification of string theory or loop quantum gravity appears at best remote if at all even possible.

44.2 Quantum diagrams

A Feynman tree level vertex diagram representing the scattering of a pair of photons via graviton resonance interaction ($\gamma + \gamma^* \rightarrow \gamma\gamma^* \rightarrow \gamma^* + \gamma$) is shown in Fig. 44-1. Graviton $\gamma\gamma^*$ corresponds to the propagator. The 4-momentum conservation is maintained across each vertex and for the sum of inputs and outputs overall:

$$\mathbf{p}_g - \mathbf{p}_g^* = -\mathbf{p}_g^* + \mathbf{p}_g \tag{44-2}$$

The momentum for s-, t-, and u-channel scattering is described by Mandelstam variables s, t, and u.

$$s = (p_1 + p_2)^2 = (p_3 + p_4)^2 \tag{44-3}$$
$$t = (p_1 - p_3)^2 = (p_2 - p_4)^2 \tag{44-4}$$
$$u = (p_1 - p_4)^2 = (p_2 - p_3)^2 \tag{44-5}$$
$$s + t + u = m_1^2 + m_2^2 + m_3^2 + m_4^2 \tag{44-6}$$

44. Quantum Gravity

The virtual graviton corresponding to the internal propagator is not directly observable. There are two different physical interpretations in the Veneziano's dual resonance model with equivalent results. In t-channel scattering, the graviton is an unstable resonance whereas in s-channel scattering, the graviton is viewed as an intermediate exchange particle. Energy and momentum are transferred to each photon. The massless graviton exists as a momentary resonance of phase conjugate photonic fields. The propagator is represented as the product of the amplitudes. The spin-2 field results in wavefront curvature and net acceleration transverse to the wavefront (spontaneous symmetry breaking) with eikonal scattering of interacting photons away from the interaction region center of rotation.

Graviton interactions

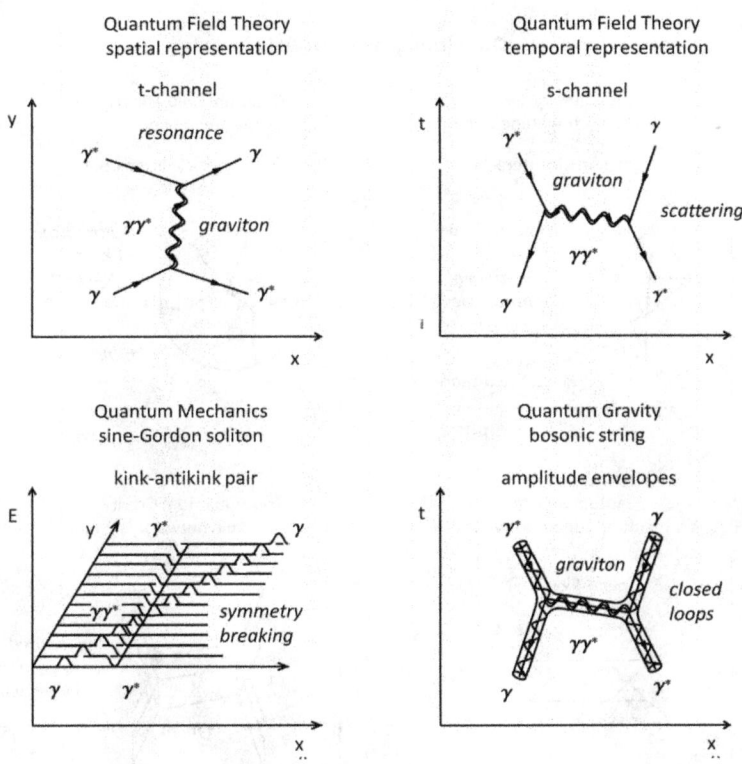

Fig. 44-1. Quantum vertex diagram illustrations of a massless spin-2 graviton $\gamma\gamma^*$ formed by interference of a pair of phase conjugate counter-propagating photons γ and γ^*. Spontaneous symmetry breaking from a false vacuum state results in resultant photon scattering.

What do you think about this? (Was meinen sie dazu)? – Wolfgang Pauli

Quantum Field Theory... is a beautiful but not very robust child. – Stephen Weinberg

44. Quantum Gravity

Gravitons are represented in string theory as the first excited mode of a Planck scale one dimensional closed loop. In loop quantum gravity, spacetime is quantized in terms of spin networks embedded in a topological curved manifold. The spin network is formed from spinors with connecting lines representing allowable angular momentum ($n\hbar/2$). The spin network describes spin angular momentum exchange between particles independent of a specific coordinate system. The combination of loop variables in Planck units allows embedding in a curved manifold such as described by GR with QM quantization. Loop quantum gravity is nonperturbative and background independent. Supergravity attempts to relate interacting fields of bosons and fermions via supersymmetry. Representations of various quantum gravity approaches are illustrated in Fig. 44-2. The Barbero-Immirzi parameter $\gamma_0 = \ln(2)/\sqrt{3}\pi$ has been interpreted in loop quantum gravity as a normalized Newton's gravitational constant G although a geometrical derivation has not yet been achieved.

Quantum gravity models

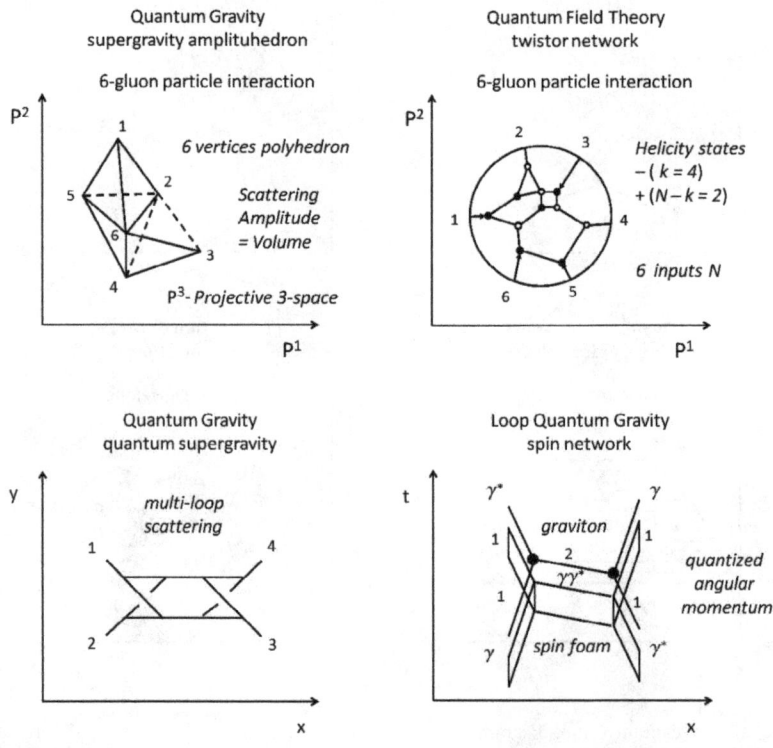

Fig. 44-2. Quantum diagram representations of symplectic structure spin loop networks. A curvilinear finite-element mesh model of a spin network with second-order triangular or tetrahedral elements requires six degrees of freedom (DOF) for each node with three DOF to describe translation along the x, y, and z axes and three DOF to describe rotation about the axes. A quantum scale representation corresponds to a reductionist viewpoint whereas thermodynamics and general relativity corresponds to a description of emergent phenomena as gravity becomes much weaker at length scales >> Planck scale ($l_P = t_P c = \sqrt{\hbar G/c^3}$).

44. Quantum Gravity

The origin of mass in quantum field theory is represented as symmetry breaking from a symmetrical unstable vacuum state to an asymmetrical lower energy state. Various representations involving transitions from unstable symmetrical potential energy states into lower energy stable unsymmetrical states are illustrated in Fig. 44-3.

Quantum Field Symmetry Breaking

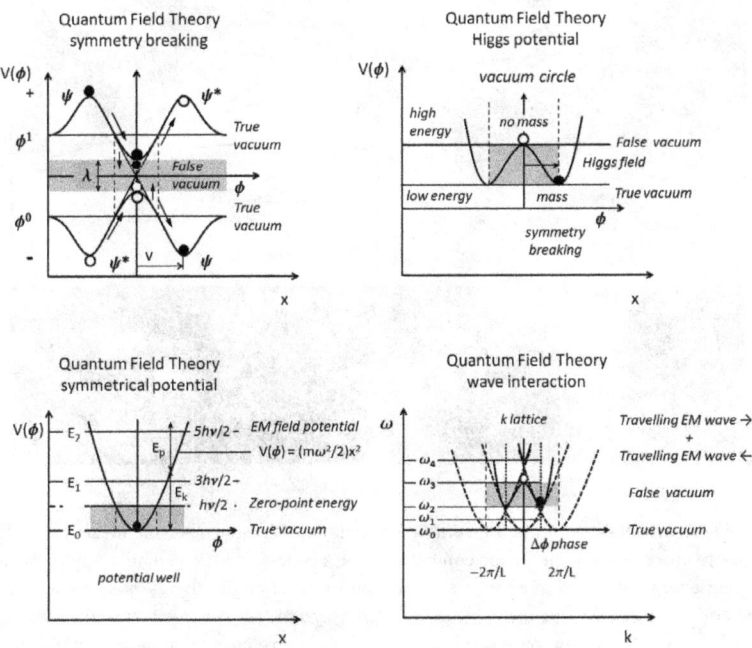

Fig. 44-3. Spontaneous gauge symmetry breaking from an unstable false vacuum state to a lower energy state. Eg., Origin of Planck mass $\pm m_P$: $S_0 = \{0\} \rightarrow S_1 = \{-m_P < 0 < +m_P\}$.

A soliton-soliton interaction in a spin network is illustrated in Fig. 44-4 representing a particle interaction. A plot of the harmonic mean of two independent variables produces a similar plot. Wave coupling occurs if the counter-propagating waves are close in frequency with the interaction zone corresponding to the coherence range. The line of impact corresponds to a decaying eigensolution (stable manifold) whereas the line of suppression corresponds to an increasing eigensolution (unstable manifold). A typical trajectory approaches the line of suppression as $t \rightarrow -\infty$ and approaches the line of impact as $t \rightarrow \infty$. A similar effect occurs in coupling of electromagnetic waves with polaritons exhibiting the quantum phenomenon of level repulsion obeying a non-crossing rule. In a system of two coupled oscillators of different natural frequencies, as the coupling strength increases the lower frequency decreases while the higher frequency increases – an apparent frequency repulsion effect.

Even the most sensible arguments for the existence of quantum gravity lack the gravitas of experimental facts. – Natalie Wolchova

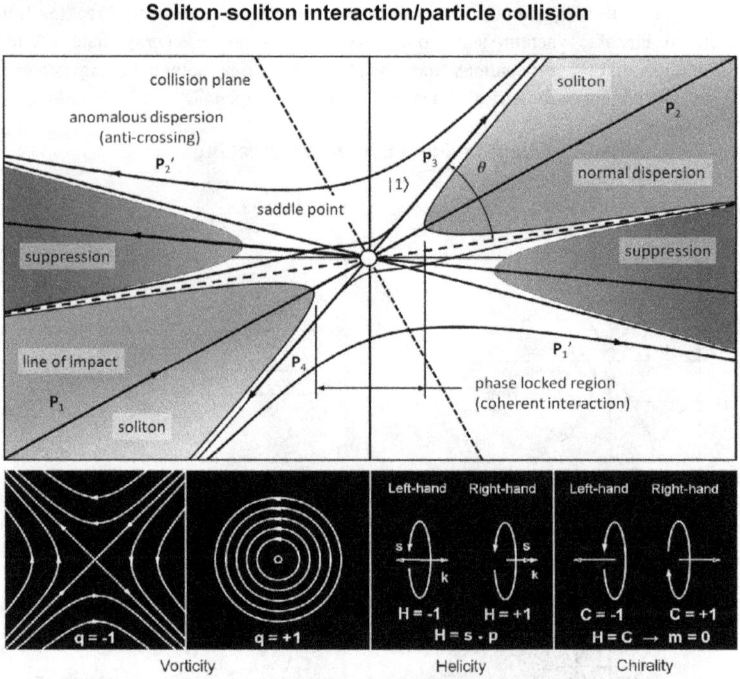

44. Quantum Gravity

Soliton-soliton interaction/particle collision

Fig. 44-4. Illustration of soliton-soliton interaction with time reversal invariance. The saddle point represents the linear combination of a 3-edge vertex pair in a spin network. Varying energy states such as e^- - e^- repulsion or bond angle obey a non-crossing rule. Converging states do not cross if of the same symmetry. Avoided crossings such as adiabatic transitions represent level repulsion as a consequence of nonpenetrability of differing energy levels. Wavefront tangent vectors correspond to different symmetry states. If counter-propagating waves are close in frequency, the waves couple to generate a single frequency in the phase-locked region. Mandelstam variables relating energy and momentum for elastic scattering of identical particles in s-, t-, and u-channels: $s = (p_1 + p_2)^2$, $t = (p_1 - p_3)^2$, $u = (p_1 - p_4)^2$. The particle trajectories resemble shear flow stream functions contours in three dimensional fluid flow such as generated by two co-rotating rollers. For n = 2, an example saddle function is $f(x,y) = x^2 - y^2$ which exhibits one saddle point. $v = (d\phi/dy, -d\phi/dx)$. The particle trajectories correspond to asymptotic lines of a hyperbola in a stereographic projection on a Riemann sphere onto a tangent plane.

Primary causes are unknown to us; but are subject to simple and constant laws, which may be discovered by observation, the study of them being the object of natural philosophy. – Joseph Fourier

We're unaware of the causes of most of the events we witness. – Ron Reed

Poincaré observed this phenomenon mathematically among colliding particles, which impart some of resonances to each other leading to a degree of synchronized resonance. – R. Marion and J. Bacon

45. Graviton Model

And so you will excuse me, Monsieur [Le Sage], if I still feel a very great repugnance for your ultramundane corpuscles; and I would rather admit my ignorance of the cause of gravity than to have recourse to hypotheses so strange. – Leonard Euler

They never called Einstein crazy. Well, they would have if he carried on like this. – Dr. Noah (Casino Royale)

The hypothetical graviton is described as a spin 2 boson posited as the carrier of the gravitational force. Photons are spin 1 bosons that are the carrier of the electromagnetic force. While photons are readily detectable, the elusive gravitons have not been detected in particle physics experiments and have remained mysterious. However, in terms of the geometrical model of the photon previously defined, the geometry, mechanism for creation, and inability for detection of the graviton may be understood. In the model described here, the graviton $\gamma\gamma^*$ is averred to be a resonance interaction of a photon γ and its phase conjugate γ^*. The phase conjugate photon γ^* is generated upon reflection from an EM wave interference pattern with nodal spacing comparable to the incident photon γ. A gravitating mass consists of a collection of EM oscillators (e.g., electrons, atoms, molecules, etc) with a frequency range over the entire EM spectrum up to the Planck frequency. The incident photon has phase and group velocities parallel and in the direction of propagation. As previously described, a phase conjugate wave has phase and group velocities anti-parallel with the phase velocity opposite to the direction of propagation. The phase conjugate is not an anti-particle as the spins are in the same direction. As a result, upon reflection, the phase conjugate photon spin adds to that of the incident photon (1s + 1s = 2s). The phase conjugate photon has an identical frequency and wavelength of the incident photon. The graviton wave train length at any given instant is the momentary length of the overlap region. The graviton resonance interaction is not directly observable, however, the gravitational effects may be observable. The effect of the resonance interaction is to locally increase the energy density of the Planck vacuum increasing the refractive index K_{PV} and gravitational gamma Γ.

Graviton formation from a counter-propagating photon and phase-conjugate photons is illustrated in Fig. 45-1 in terms of Whittaker[120,121] scalar potentials. The wave equation describes the evolution of the wave function with time. In addition, the wave function also describes the evolution of the complex conjugate wave function in time. The complex conjugate wave is a time-reversed replica of the wave equivalent to wave propagating backward in time. Unlike the Wheeler – Feynman absorber theory of radiation proposed in 1945, there no advanced wave propagating backward in time to interact with the retarded wave propagating forward in time.

Schematic diagrams of resonance interaction of a pair of phase-conjugate photons is shown in Figs. 45-2 and 45-3 illustrating graviton formation. As shown, a phase-conjugate pair of counter-propagating spin 1 photons (RH helix, LH helix) interfere to give rise to a spin 2 graviton (helicoid). The graviton has two helicity states with a spin projection on the propagation direction equal to ±2, hence, the gravitational field has two degrees of freedom. Unlike electromagnetic charge, vortices of opposite spin repel resulting in deflection of the photon wave vectors. Fields carried by exchange particles with odd integer spins with aligned spin vectors are repulsive whereas those with opposed spins are attractive. Even spins lead to attractive forces proportional energy content and varies as $1/r^2$. For a spin-2 graviton, the polarization rotational phase change corresponds to $e^{2i\theta}$ and $e^{-2i\theta}$.

45. Graviton Model

In the graviton model illustrated, the linear momentum is zero. Hence, a freely-propagating graviton as such is not predicted to exist. In addition, the theoretic Ivanenko process $(2\gamma \to e^- + e^+)$ in which two gravitons collide head-on to create matter, an electron and positron, does not occur as the relative velocity is zero. The effect of gravity is not due to the momentum exchange of gravitons in billiard ball fashion, but rather is a harmonic resonance phenomena of quantum oscillators mediated by photons. The photon and graviton flux alters the local vacuum energy density of aggregated concentrations of matter. As in quantum field theory, the massless spin-2 graviton model illustrated has just two helicity states. In string theory, bosons, including gravitons, form closed loops. Similar to string theory, the periphery of the graviton represented here forms a closed loop. As the graviton ($h_{\mu\nu}$) carries energy, it contributes to the gravitational field energy density and the energy-momentum tensor. A gravity wave as a function of time may be represented as $h_{\mu\nu} = A_{\mu\nu}\sin(t - \chi/c)$. Gravitational waves $g_{\mu\nu}(x)$ as a coherent states of many gravitons may be considered as excitation of the background metric where the metric is not the Minkowski spacetime $\eta_{\alpha\beta}$, but the electromagnetic field interference pattern metric $\eta_{\mu\nu}$: $g_{\mu\nu}(x) = \eta_{\mu\nu} + h_{\mu\nu}$. The gravitational wave amplitude is the sum of two polarizations: $g_{\mu\nu} = h_+ e_{\mu\nu}^+ + h_\times e_{\mu\nu}^\times$ where " + " and "x" denote polarization directions.

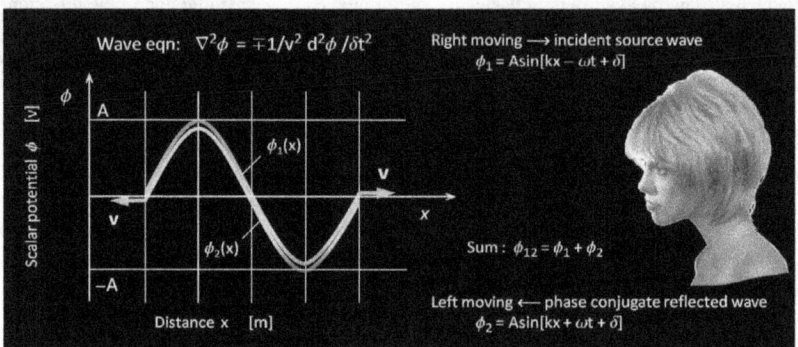

Fig. 45-1. Illustration of resonance interaction of counter-propagating phase-conjugate photons represented as counter-propagating scalar potential waves. Continuous monochromatic waves are spatially and temporally coherent with infinite coherence length $l_c = \lambda^2/2\Delta\lambda$ as the waves remain in phase for an infinite distance. Pulsed wave coherence duration is equal to the wave train duration. Linear momentum $\mathbf{p}$ ($= \hbar\mathbf{k}$) cancels to zero.

It is by logic we prove, but it is by intuition we discover. To know how to criticize is good, to know how to create is better. – Henry Poincaré

We all try. You succeed. – Humpfrey Bogart (Casablanca)

There are very few things which we know, which are not capable of being reduced to a Mathematical Reasoning; and when they cannot it's a sign our knowledge of them is very small or confus'd. – Kurt Gödel

Gravitons are the avatars of general covariance. – Frank Wilczek

45. Graviton Model

A graviton corresponds to transient standing wave produced as incident and phase conjugate reflected travelling photon waves cross during an overlap period upon reflection. The bi-directional traveling harmonic wave pattern is similar to the diatonic interference pattern energy flow model in music theory correlated to the exchange and energy flow in a standing wave for example, such as represented as y = sin(2x) and y = −sin(2x). Harmonic interference theory suggests geometric patterns of a graviton field arise as a result of interference of reflected harmonic waves with resonance and amplification at harmonic frequency multiples.

The propagation velocity and frequency of photon wave quanta are dependent on the gravitation potential which is influenced by the action of graviton flux. Gravitons are associated with gravitational waves generated from mass oscillations such as binary stars, exploding or collapsing stars, etc. Gravity is far weaker than other forces; coupling of gravitational waves/gravitons to matter fields is weak. Macken has shown for a pair of electrons in terms of Planck units, the electromagnetic force $F_{e/p} = (\omega_c)^2/N^2$ while the gravitational force $F_{g/P} = (\omega_C)^4/N^2$ where ω_c is the Compton frequency and N is a dimensionless ratio of separation distance r divided by the Compton length R_C. That $F_{g/P} = F_{e/P}^2$ is a consequence that EM forces are proportional to e^2 mediated by exchange of a pair of virtual photons ($\omega_{c1}\omega_{c2}$) whereas gravitational forces are proportional to m^2 mediated by exchange of a pair of virtual photons and a pair of phase conjugate photons $(\omega_{c1}\omega_{c2})^2$.

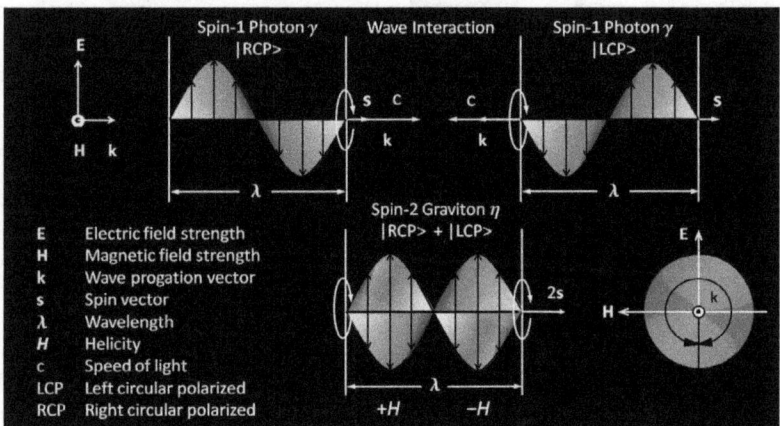

Fig. 45-2. Diagram of graviton formation from coupling of a pair of counter-propagating phase-conjugate photons. The phase conjugate photon γ^* is a time-reversed replica of the incident photon γ. The graviton has spirality with spin projection s on the propagation direction k equal to ±2. Spin angular momentum is additive (s = 2$\hbar$). Helicity H cancels..

The energy of gravitons is proportional to frequency (E = 2hν). Emitted gravitational waves are of low energy and notoriously difficult to detect. Similar to Delbrück scattering of photons by photons, gravitons should likewise undergo scattering interactions. The strength of graviton scattering interactions vary with frequency, the probability of interactions described by the vertex in terms of the fine structure constant. The higher the frequency of gravitons, the greater the strength of interaction.

45. Graviton Model

Frolov[118] has proposed a gravitation concept as a paired process of radiation and gravitation. In terms of duality of action: Radiation is emission of photons from a mass object; gravitation is the absorption of photons by a mass object. Material mass, in other words, represents the absorption and transformation of electromagnetic waves. Radiation emission is associated with a contraction of mass and absorption with an increase in mass. The instability of the interaction of phase-conjugate photons arises as the result of the phase-conjugate wave has negative action (A = $-h$) producing a negative energy state (E = $-h\nu$) in which the group velocity is opposite to the spin vector. This false vacuum instability (E = $h\nu - h\nu = 0$) results in spontaneous symmetry breaking as represented by the Higgs potential and resultant photon scattering. Storti and Diemer[33] maintain that dark energy is in the form of conjugate photons (i.e., gravitons) which would account for the observational difficulty for direct detection. Conjugate photons are a feature of the Electro-Gravi-Magnetic (EGM) quantized theory of gravity.

A vector diagram of graviton formation from phase conjugate beams is depicted in Fig. 45-3. The graviton arises from a photon gauge field and its phase conjugate emergent from an interference network lattice containing quantum spins. The graviton has only two polarizations: parallel or anti-parallel to the wave vector k. The rotational phase change varies as $e^{2i\theta}$ and $e^{-2i\theta}$. A spin 1 photon γ and its counter-propagating phase conjugate γ^* are both described as a helix. The spin 2 graviton $\gamma\gamma^*$ conforms to a helicoid geometry. Each represent minimal ruled surfaces. The curvature and torsion characteristics of each are illustrated in Fig. 45-4. Coupling of counter-propagating photon (right hand helicity) and phase-conjugate photon (left hand helicity) results in creation of false vacuum instability and photon scattering. Gravitation phenomena appears associated with photon wave-function pairing in a phase conjugate mixing interaction in which energy density is momentarily reduced to a false vacuum state as a result of vector addition of conjugate phase amplitudes. The massless spin 2 graviton $\gamma\gamma^*$ has just two helicity states. Spins are additive as chirality equals helicity resulting in a spin (torsion) field with no net longitudinal momentum transfer ($\langle S \rangle = 0$). Hence, gravitons have negligible interaction cross-section with matter.

Some years ago, Pauli and Fierz considered the question, what relativistic wave equations would be appropriate for particles of zero rest mass and spin two. Now in a relativistic theory for spin S, 2(2S + 1) components are needed, so a second-rank tensor is required... Since the only available direction is the direction of motion, it follows that the spin angular momentum must be oriented in the direction of motion. Since the gravitational forces have infinite range, it follows that the rest mass of the graviton must be zero. – Joseph Weber

Provando e riprovando. (trying and trying again) – Academia del Cimento motto

Everthing flows and nothing persists, or nothing endures. – Heraclitis

In Terra inest virtus, quae Lunam dei. (There is a force in the earth which causes the moon to move) – Johannes Kepler

Real gravitons make up what classical physicists would call gravitational waves, which are very weak – and so difficult to detect that they have not yet been observed. – Stephen Hawking

The new mathematics, which is responsible for the merger of these two theories [Riemann geometry of Einstein's theory and Lie Groups coming from quantum theory] is topology, and it is responsible for accomplishing the seemingly impossible task of abolishing the infinities of a quantum theory of gravity. – Michio Kaku

Terrans are known for their stupidity. – Philip K. Dick

45. Graviton Model

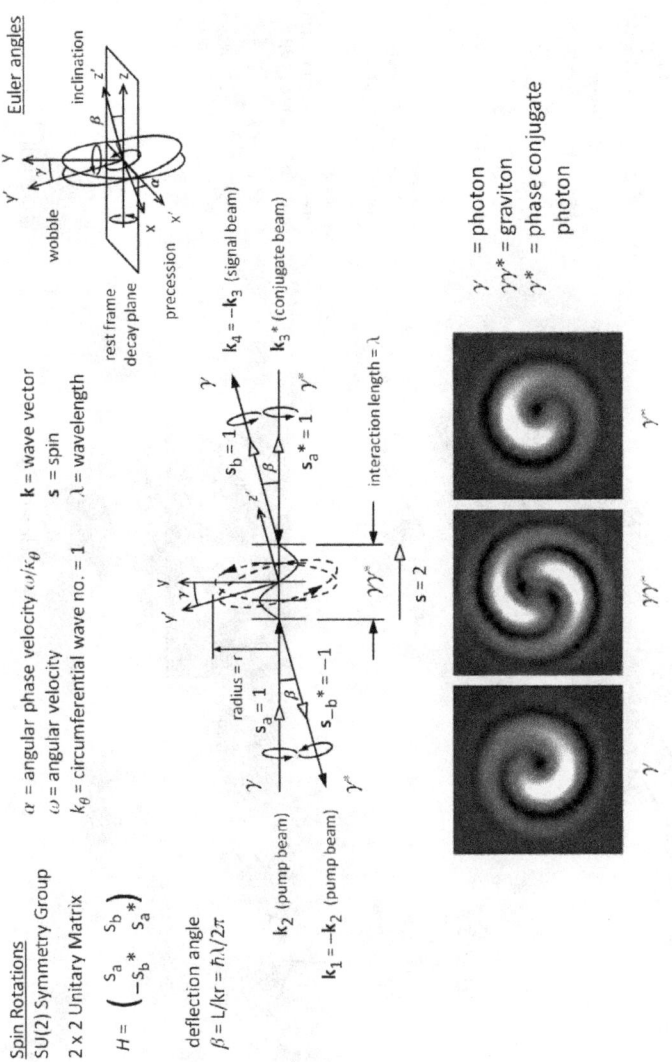

Fig. 45-3. Graviton formation by phase conjugate coupling of counter-propagating photons. Spins are additive (s = 1 + 1 = 2).

45. Graviton Model

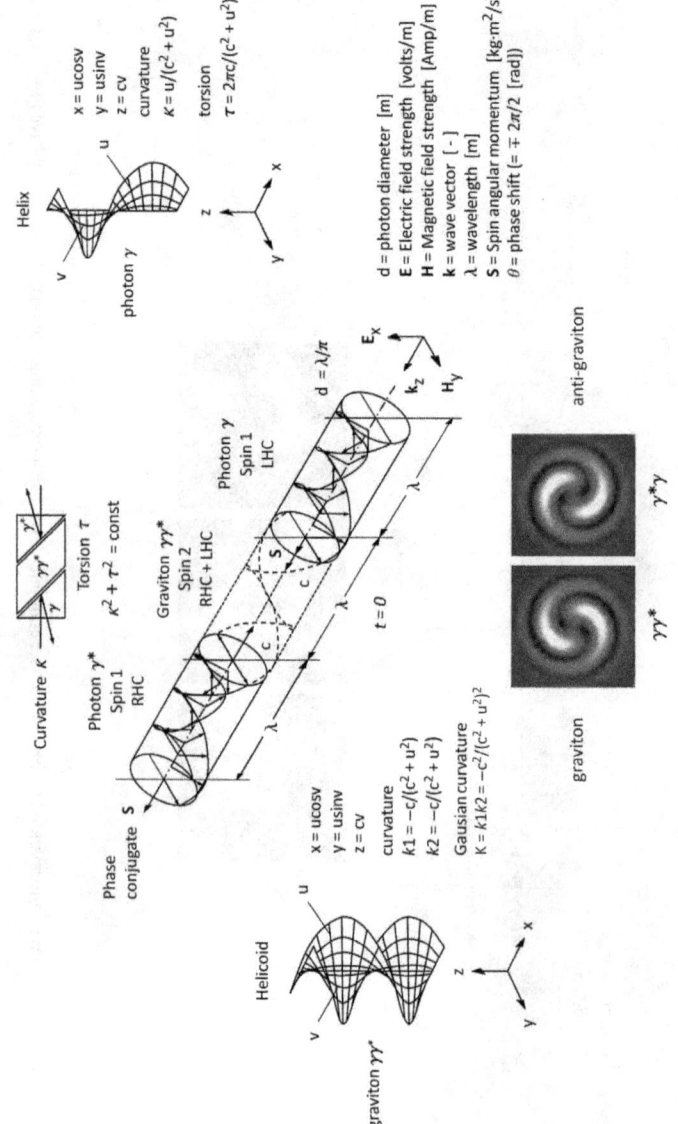

Fig. 45-4. Graviton $\gamma\gamma^*$ and photon γ curvature κ and torsion τ characteristics represented as a helicoid and helix, respectively.

46. Nonlinear Gravitational Field

Mass is associated with energy (i.e., wave function curvature) according to the Einstein mass-energy relation (E = mc^2) including gravitational energy. Consequently, both mass and gravitational energy are sources of gravitation. For mass concentrated at a point, the associated gravitational field has negative energy (–mc^2) equal to the positive mass-energy of matter (+mc^2). Rest mass produces a distortion field by transformation of energy. Condensation of matter produces an equal and opposite reaction: extraction of mass-energy from the vacuum: positive mass material and negative mass vacuum. Negative mass vacuum states may be described by contraction of dimensional spacetime coordinates. See Fig. 46-1. According to GR, gravitational energy becomes more negative as the mass density of an object increases as the volume contracts under the action of gravity. In the PV model, spacetime 4D mathematical construct remains Euclidean and the apparent contraction represents instead the contraction of electromagnetic wavelengths.

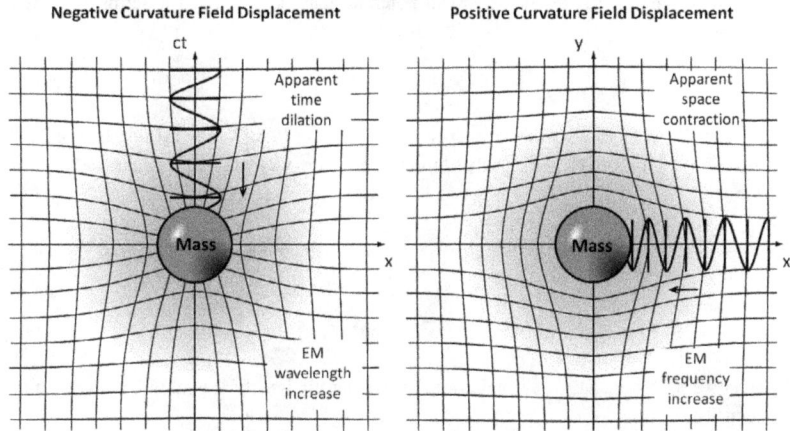

Fig. 46-1. EM wavelength increase corresponds to apparent time dilation and EM frequency increase corresponds with apparent space contraction of GR theory which posits curvature of spacetime. Change in wavelength $\Delta\lambda$ varies with change in light velocity c (= $\Delta c/\Delta \nu$). The orthogonal harmonic grid lattice of standing waves illustrated represents a metric connecting nodal points of EM waves. For a curved metric, the line element $ds^2 = g_{\mu\nu}dx^\mu dx^\nu$. Gravitational redshift frequency $\nu_\infty = \nu_{em}\sqrt{(1 - 2GM/c^2r)}$ where ν_∞ = frequency measured by a distant observer and ν_{em} = source emission frequency. Doppler frequency shift (decoherence) is given by $\nu_0 = \nu_{em}(1 - v/c) = \nu_{em}(1 - gh/c^2) = \nu_{em}(1 - 2\Delta\nu h/c)$ where ν_0 = measured velocity by an observer moving with relative velocity v, g = acceleration of gravity and h = height (vertical distance between source emitter and observer).

Winterberg[7] observes that negative gravitational field energy density implies negative mass which is one of the principal assumptions of the Planck aether hypothesis in which the physical vacuum is composed of an equal number of positive and negative Planck mass particles. See Fig. 46-2. The negative energy density of a gravitational field and positive energy density of mass may be understood as polarization effects of Planck dipoles in a

46. Nonlinear Gravitational Field

physical vacuum, i.e., gravitization analogous to magnetization. The gravitational negative energy density field surrounding the mass positive energy density field implies a shielding effect similar to Debye shielding of positive ions by surrounding electrons in a plasma. A similar effect arises in an ionic crystal in which a localized excess of charge polarizes the lattice in its vicinity forming a distortion defect known as a polaron.

As shown, the negative energy density of the surrounding gravitational field is indicated by negative curvature geodesics displaced inward toward the center of mass, while the positive energy density of the central mass is represented by a positive curvature field with geodesics displaced outward. The induced displacement distortion of the Planck dipoles represents a polarization of the Planck aether (physical vacuum). Acceleration of gravity is coincident with the positive mass displacement vector. The self-created polarization modifies the local refractive index of the vacuum. In a far-field, zero curvature vacuum remote from the gravitating mass, the population of positive energy and negative energy Planck particles is in equilibrium. A positive energy particle moving through the negative energy gravitational field induces a distortion in the dipole field increasing the effective mass of the particle by dragging of the accompanying distortion with it.

Jefimenko[16] notes that the gravitational potential may be expressed in terms of energy density [J/m^3] as

$$U_v = -g^2/8\pi G \qquad (46\text{-}1)$$

At the Earth's surface, volumetric energy density $U_V = (9.807 \text{ m/s}^{-2})^2/(8\pi \cdot 6.672 \times 10^{-11} \text{ N·m}^2/\text{kg}^2) = -5.736 \times 10^{10}$ J/m^3. Substituting $g = GM/r^2$ yields an equivalent scalar $U_v = -GM^2/8\pi r^4$.

Equating $M = U/c^2$ yields a relation for the corresponding mass density [kg/m^3] of gravitational energy

$$\rho_g = -g^2/8\pi Gc^2 = -GM^2/(8\pi r^4 c^2) \qquad (46\text{-}2)$$

The negative sign indicates the gravitational field energy density is negative with negative mass density.

In terms the scalar gravitational potential ϕ for a static field, Winterberg[7] shows the field energy density may be expressed as

$$\rho_g = -(\nabla \phi^2)/4\pi Gc^2 \qquad (46\text{-}3)$$

According to Jefimenko, the Newtonian divergence relation for a time-independent gravitational field (Eq. 38-29) should therefore modified as follows

$$\nabla \cdot \mathbf{g} = -4\pi G\rho + g^2/2c^2 = -\nabla^2 \phi + g^2/2c^2 \qquad (46\text{-}4)$$

The total mass density of a mass and accompanying nonlinear gravitational field energy density is then

$$\rho_t = \rho + \rho_g = -\nabla \cdot \mathbf{g}/4\pi G\rho + g^2/8\pi Gc^2 \qquad (46\text{-}5)$$

where the equivalent mass of the gravitational field is $m_{equiv} = Gm^2/2c^2 r$.

Visualization is more important than knowledge. – Albert Einstein

46. Nonlinear Gravitational Field

Planck vacuum polarization in a gravitational field

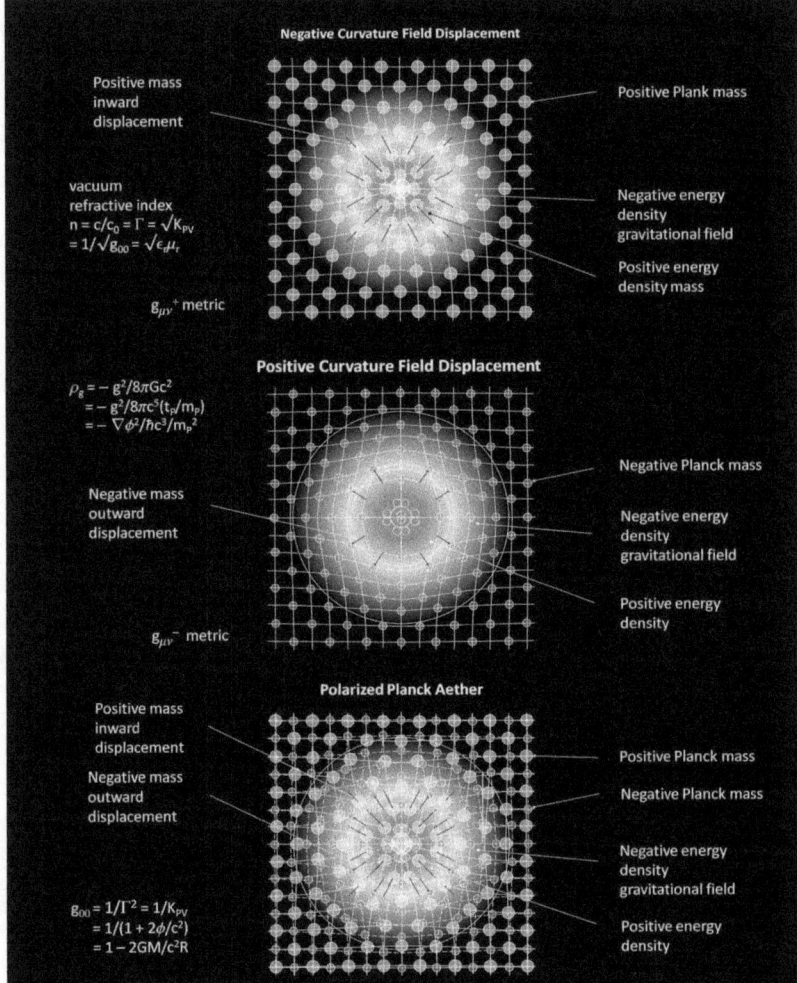

Fig. 46-2. Representation of Planck vacuum polarization in a gravitational field of a central mass. Polarization is depicted as radial displacement of Planck dipoles composed of positive and negative mass. The vacuum polarization alters the local refractive index of the vacuum $K_{PV}(r,m)$ which is a measure of electromagnetic field energy density.

It might be in the end that the field laws may be equally described in terms of a force field or a curved space theory. – William Berkson

... those who believed in Truth clung to Geometry as to a rock, and it was the highest ideal of every scientist to carry on his science "more geometrico." – Herman Weyl

46. Nonlinear Gravitational Field

In the weak field limit where Newtonian theory of gravity is valid, i.e., relative motion of sources are slow compared to the speed of light (slowly varying momentum), and material stresses are much smaller than mass-energy, the density equals mass/volume (T = m/v). In the generalized nonlinear gravitation theory advanced by Jefimenko, the Newtonian theory is extended to include the mass-energy density of the gravitational field. The total mass is

$$m_t = m + m_g = m(1 + G^3m/5c^2r) \qquad (46\text{-}6)$$

Interior to radius r, the Newtonian mass m is reduced by the equivalent mass of the self-field by a factor of $(1 - 2Gm/c^2r)$.

In 3-D space, pressure P equals the 1/3 of the vacuum (volumetric) energy density u

$$u = 3P = \rho c^2 \qquad (46\text{-}7)$$

where ρ = equivalent mass density (= kg/ms^2).

A comparison of the gravitational field for the Earth modeled as a spherical mass of uniform density (ρ) and that of a corresponding nonuniform gravitational field mass density (ρ_g) is illustrated in Figs. 46-3 and 46-4. The peak gravitational field mass energy density ρ_g is estimated as 6.372×10^{-7} kg/ms^3 which is far less than mass energy density of the Earth (ρ = 1.4274E24 kg/m^3). The mass density ratio $\rho_g/\rho = 2.223 \times 10^{30}$ at the Earth's surface. The field quantum pressure P_q (= $k/4\pi r = -\rho_g/3$) is equal to -2.1406×10^{-07} Pa (= kg/ms^2).

You sometimes speak of gravity as essential and inherent to matter... the cause of gravity is what I do not pretend to know, and therefore would take more time to consider of it. – Isaac Newton

Any opinion as to the form in which energy of gravitation exists in space is of great importance, and whoever can make his opinion probable will have, made an enormous stride in physical speculation... We cannot conceive of matter with negative inertia or mass; but we see no way of accounting for the proportionality of gravitation to mass by any legitimate method of demonstration... But when we face the great questions about gravitation: Does it require time? Is it polar to the 'outside of the universe' or to anything? Has it any reference to electricity? or does it stand on the very foundation of matter-mass or inertia? then we feel the need of tests whether they be comets or nebulae or laboratory experiments or bold questions as to the truth of received opinions. – James Clerk Maxwell

The nature of light as dependent on a medium is now very largely accepted. The presence of a medium in the phenomena of electricity and magnetism becomes more probable daily... Then what is there about gravitation that should exclude it from consideration also? – James Clerk Maxwell

As soon as matter took over, the force of Newtonian gravity, which represents one of the most important characteristics of "ponderable" matter, came into play. – George Gamow

When forced to summarize the general theory of relativity in one sentence: Time and space and gravitation have no separate existence from matter. – Albert Einstein

Physics is self explanatory except at points where symmetry breaks... – Vipal Tewari

Space music'd be really something... but they don't have no gravity up there. You couldn't have no downbeat! – Miles Davis

Die Natur kapiren und kopiren. Comprehend and copy nature. – Viktor Schauberger

46. Nonlinear Gravitational Field

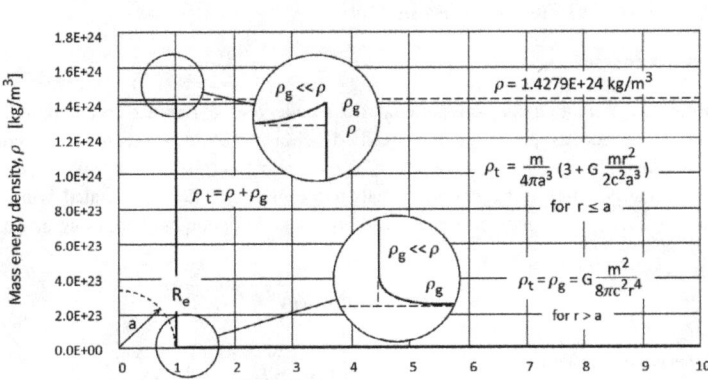

Fig. 46-3. Plot of mass energy density ρ for the Earth modeled as a sphere of uniform density. [Adapted from Jefimenko].

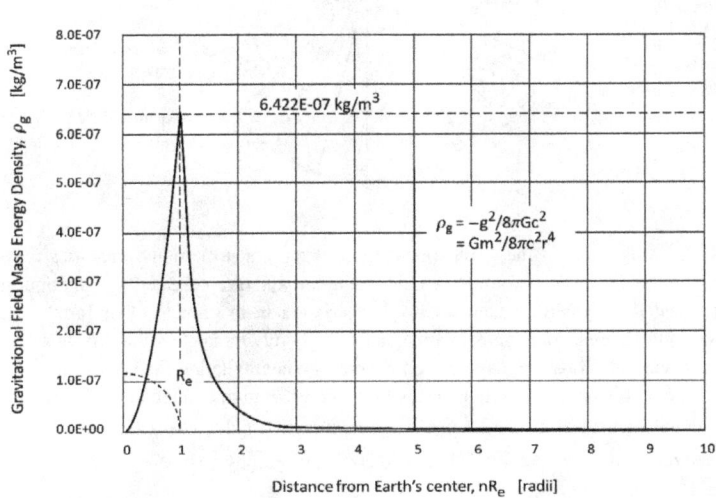

Fig. 46-4. Plot of mass energy density ρ for the Earth modeled as a sphere of uniform density and equivalent gravitational field mass energy density ρ_g. The gravitational field mass energy density varies as $1/r^4$. [Adapted from Jefimenko]

> It is noteworthy, however, that Newton's gravitational theory generalized to time-dependent systems (Heaviside's theory was essentially just that) yields several results which heretofore were believed to be the exclusive consequence of the general theory of relativity theory. – Oleg. D. Jefimenko

423

47. Gravitational Field of Mass in Motion

47.1 Mass current

For masses in motion the gravitational field undergoes a Lorentz contraction in the direction of motion as illustrated in Fig. 47-1. Mass motion represents mass current analogous to electric current. As there is a magnetic field associated with an electric current, similarly, there is a circumferential gravitomagnetic field associated with mass current. An accelerating mass generates in addition a gravikinetic field opposing motion.

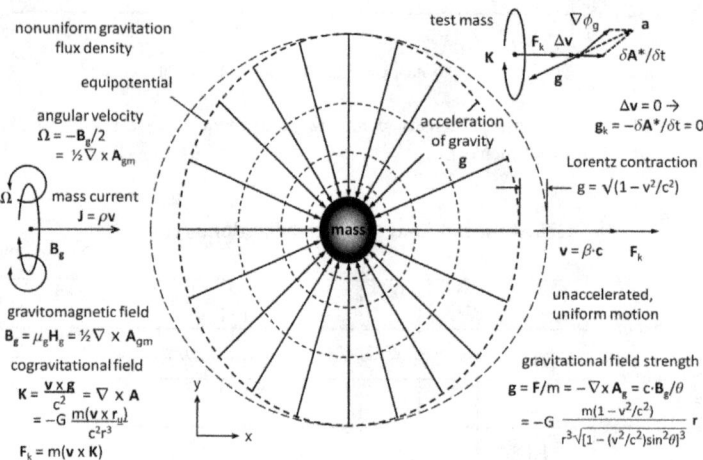

Fig. 47-1. Magnitude of the gravitational field of a mass in motion varies as a result of Lorentz contraction. The variation in field strength may be represented by variation in the gravitational flux density or equivalently by variation in magnitude of uniformly spaced field strength vectors. Analogous to a magnetic field, gravitomagnetic field B_g is generated by a mass current. Relative mass motion generates a cogravitation field K exerting a force F_k (= $m(v \times K)$) on a mass moving with velocity v in the direction of motion. The gravitational vector potential $A_g = v \cdot \phi_g/c^2$. Gravikinetic field $g_k = v \times K = -dA^*/dt$ where A^* is the retarded vector potential. The acceleration of gravity $g = -\nabla \phi_g - \delta A/\delta t = \nabla \times A_g$.

Analogous to a moving charge through an electric field, a force applied to an object induces a counter-potential that lowers the energy flow though an object and causes repositioning of the object. When the energy flow change induced by the applied force exceeds that due to mass, the object moves through the field. The resistance of mass to motion may be imagined in analogy to a resistor in a direct current loop with one end connected at a point in a prior instant in time at a higher potential and the other end connected to a displaced point in the present at a lower potential. Mass current flow corresponds to electric current flow. A gravikinetic field g_k is analogous to an electrokinetic field E_k opposing motion.

47. Gravitational Field of Mass in Motion

For a orbiting mass m around a central mass M, there are five orbital positions known as Lagrange points L1–L5 where a small satellite can, in some cases, maintain a stable relative position. On a equipotential contour plot, these positions correspond to locations where the potential gradient is zero. At these locations, the gravitational force and centrifugal force are in balance. The distance r from mass m to L1 corresponds to the Hill sphere radius ($\simeq R^3\sqrt{(M_2/3M_1)}$) where R is the distance between M and m. See Fig. 47-2.

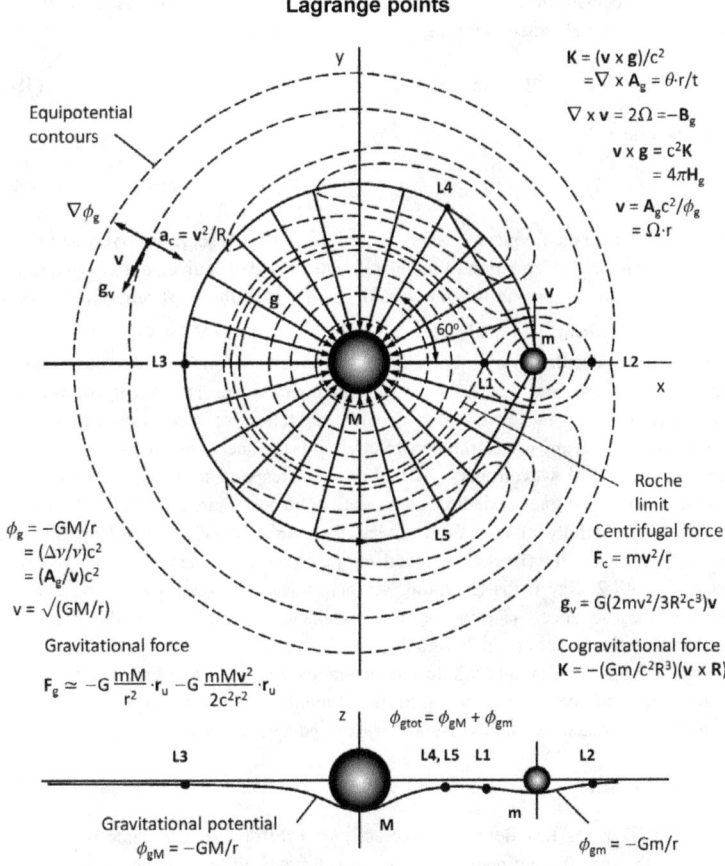

Fig. 47-2. Lagrange (or libration) points are orbital positions in a rotating coordinate frame of a pair of masses where the gravitational attraction force is equal to the centrifugal force. The Lagrange points L4 and L5 represent stable locations for a satellite to maintain position relative to central pair in a three body system.

By the labours of LaGrange, the motions of a disturbed planet are reduced with all their complication and variety to a purely mathematical question... – George Boole

Il faut que j'y songe encore. – Joseph Louis Lagrange

425

48. Wavefront (Moiré) Interference Patterns

Light brings us the news of the Universe. – Sir William Bragg

48.1 Moiré fringes

Moiré fringes are a family of cardioids such as formed by circular gratings and represent the spatial displacement field of interfering spherical wavefronts. Moiré patterns are a type of aliasing. The general equation for a spherical wavefront is

$$E_t = A\cos(\omega t - kr + \phi_0) \qquad (48\text{-}1)$$

And Stokes reflected wave by

$$E_s = B\sin(\omega_s t - k_s r + \phi_s) \qquad (48\text{-}2)$$

where ω, k, and ϕ are the frequency, wave vector, and initial phase, respectively and $\omega t - kr$ represents the phase. Deformation of the displacement field is equivalent to local strain $\epsilon = \Delta L/L_0$. The wavefront spacing is a measure of the effect of the local refractive index and refractive index gradient. The field gradient may be expressed as $\nabla F = \nabla \cdot F + \nabla \wedge F$.

Two proximate oscillators of equal frequency are depicted in Fig. 48-1 illustrating the effects of constructive and destructive wavefront interference. The density of wavefront spacing corresponds to the energy flux density and refractive index gradient which the highest density occurring where the constructive interference is maximum. The circular patterns represent EM wavefronts. The ellipses correspond to Fresnel zones and are representative of the refractive index $K_{PV}(r,\omega,M)$. Outside a circular boundary enclosing the centers of the family of wavefront circles, the Fresnel zones (Moiré fringes) form a family of hyperbolas. Interference of radial ray paths results in standing wave patterns as shown in Fig. 48-2. Ray paths (Poynting vector) indicate direction of energy flow from dissonance to consonance. Harmonics of the fundamental frequency produces overtones and undertones that combine in harmonic series. Intervals consisting of differences in frequencies occurring in simple whole number ratios correspond to consonant resonance while noninteger number ratios forming irrational numbers result in dissonance. Fibonacci ratios indicate dissonance while Pythagorean ratios indicate resonance.

48.2 Fresnel zones

La Freniere[38] notes EM fields of force act like a diffractive lens since waves cannot travel freely through the antinodes. Waves travel freely through nodes but are weakly scattered at each antinode. Amplification of EM field strength occurs in a diffractive lens with a converging effect focusing a significant part of energy towards both sources. The concentric node and antinode pattern produces a phase rotation θ as measured from the symmetry midline. Gravitation is fundamentally an EM phenomena of extreme broadband frequency spectrum with weakly interacting, diffuse Fresnel zones and converging focusing effect. The long range Casimir effect alone is insufficient to account for gravitational attraction. In the PV model, EM focusing and amplification is readily interpreted in terms of Fresnel zones with variable dielectric constant $K_{PV}(r,\omega,M)$. This effect is exploited, for example, in Gradient Index (GRIN) lenses and Luneberg lens antennas with graded dielectric constant. The gradient ∇F equals sum of scalar dot product and bivector product.

48. Wavefront (Moiré) Interference Patterns

Fresnel zones are loci of points of constant path length difference of $\lambda/2$ (180° phase difference). The dielectric medium phase discontinuity boundary provides a phase shift in the transmitted wave reflecting some of the energy back. Reflected signals off surfaces within an even Fresnel zone will be delayed 180 degrees (out-of-phase) from line of sight signal. Reflected signals within an odd Fresnel zone exhibit a 180 degree phase shift plus a 180 degree phase delay or 360 degrees phase difference (in-phase) that results in a combined signal that is at a higher level than the line of sight signal. Size of Fresnel zones and Fresnel circles (cross-sections) are frequency dependent (higher the frequency, the smaller the ellipsoid). Higher order Fresnel zones occur at delays that are integer multiples of 180 degrees plus the first Fresnel delay. Refractive index varies with EM node density.

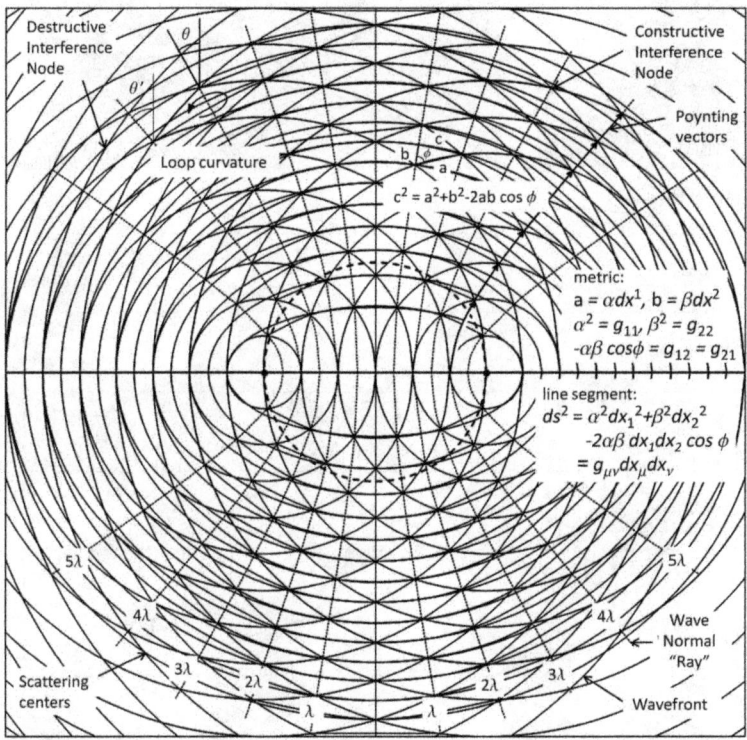

Fig. 48-1. Wavefront interference of two oscillators sources of equal wavelength results in a quantized field metric of interlocked nodes. Destructive interference occurs at nodes while constructive interference occurs at antinodes (wave crests). Two source interference: Path difference for destructive interference = $(n + \frac{1}{2})\lambda \rightarrow \frac{1}{2}\lambda$, $1 - \frac{1}{2}\lambda$, $2 - \frac{1}{2}\lambda$. Path difference for constructive interference = $n\lambda \rightarrow 1\lambda, 2\lambda, 3\lambda$. The metric is an emergent property of wave interference. Line element $ds^2 = g_{\mu\nu}dx_\mu dx_\nu$ where $g_{\mu\nu}$ is the metric. Distance (between rays) is $ds = \sqrt{g_{\mu\nu}dx_\mu dx_\nu}$. The curvilinear metric may be approximated with a second order finite-element model with triangular and tetrahedral elements with 3 degrees of freedom (DOF) at each node in terms of geometric algebra.wedge products.

427

48. Wavefront (Moiré) Interference Patterns

Standing wave interference

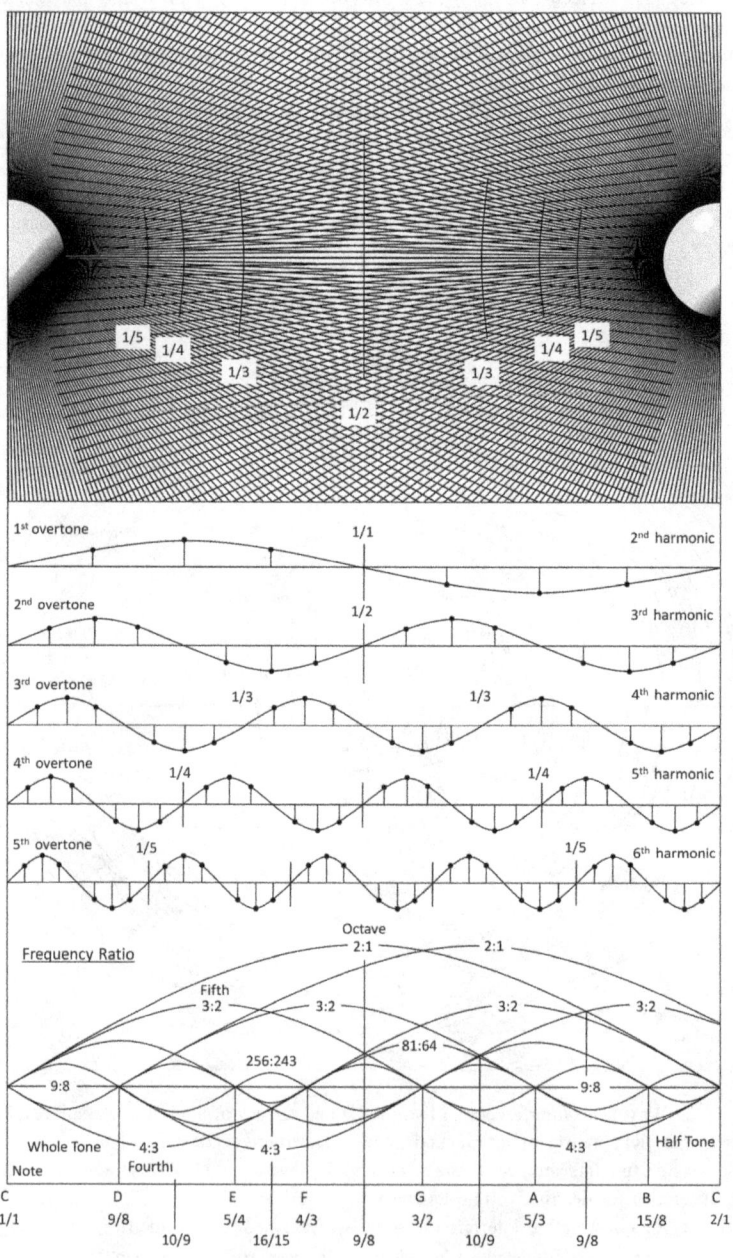

Fig. 48-2. Illustration of moiré pattern produced by radial lines simulating EM ray paths for two oscillators of equal flux density. Constructive and destructive interference of Poynting energy flow results in harmonic standing wave formation.

48. Wavefront (Moiré) Interference Patterns

The Fresnel zone boundary radius at any point P between foci is given by the Fresnel equation:

$$r_n = \sqrt{(n\lambda d_1 d_2 / d_1 + d_2)} \tag{48-3}$$

where:
- r_n = nth Fresnel zone radius
- n = no. of Fresnel zone contributing to point P
- λ = wavelength
- d_1 = axial distance from 1st foci to point P
- d_2 = axial distance from 2nd foci to point P

Light slows down in a medium due to photon's EM field interaction with the EM energy density of the medium, gravity being mediated by radiation (photons, gravitons). Gravity in a polarizable vacuum may be viewed as a dielectric gradient force effect in a Fresnel zone nonuniform field. In the standard model of particle physics, a graviton (spin-2) is described as a pair of phase-conjugate photons (spin-1). The spin-2 characteristic arises as a result (180 degree rotational symmetry). The phase conjugate photons propagating in a nonlinear medium from opposite directions interfere to produce a nonlinear standing wave that oscillates without propagation.

In the Fresnel zone PV model illustrated, the conjugation arises naturally due to the well known self-focusing action of Fresnel zone ellipsoids where the EM radiation source/absorber (co-incident with center of each mass) lies at the focii. A photon emitted from one mass follows the same path as a counter-propagating photon emitted from a mass at the opposite foci. The Fresnel zone boundaries form, in effect, phase conjugate Brillouin mirrors. The nested ellipsoidal Fresnel zones encapsulating the mass pair are confocal thereby augmenting the focusing effect.

A pair of counter-propagating conjugate EM waves corresponds to the scalar wave model proposed by Whittaker[116,117]. The conjectured spin 2 gravitons are entangled pair states of spin 1 quanta of the photonic field. In the Storti, Diemer model[33,34], the graviton consists of a photon coupled to its virtual photon zero-point fluctuation (ZPF) reactive space-time manifold response to the EM wave-induced stress forming a phase conjugate pair. The conjugate EM wave response propagates 180° out-of-phase with respect to the EM forcing function. A wave with a reversed wavefront propagating in an inhomogeneous nonabsorbing medium follows in reverse in the same path as the original wave. Phase conjugate reversal may be achieved by reflecting a wave off a mirror whose surface coincides with its wavefront. The wavefront of a conjugate back-scattered lightwave has the same shape but opposite sign as the excitation wave.

A refractive index ellipsoid (multi-Fresnel zone, graded index) representing anisotropic polarizable vacuum medium is illustrated in Fig. 48-3. For clarity, only a few Fresnel zones are shown whereas physically for a given EM spectrum, the multiplicity of zones form a near continuous field. The refractive one-form is a covariant vector in cotangent space representing the gradient of a scalar field. A contravariant vector V_α is a tangent vector on a parametized curve and acts on a one-form V_β to give a gradient of a coordinate surface as a real number scalar. In the Fresnel zone PV model illustrated, the conjugation arises naturally due to the well known self-focusing action of Fresnel zone ellipsoids where the EM radiation source/absorber (co-incident with center of each mass) lies at the focii. A photon emitted from one mass follows the same path as a counter-propagating photon emitted from a mass at the opposite foci. The Fresnel zone boundaries form, in effect, phase conjugate Brillouin mirrors. The nested ellipsoidal Fresnel zones encapsulating the mass pair are confocal eo ipso augmenting the focusing effect.

48. Wavefront (Moiré) Interference Patterns

Freely propagating radiation **E** and **H** fields in an EM traveling wave are in-phase. When EM radiation confined to a waveguide with width greater than ½ wavelength, the radiation forms standing waves with the **E** and **H** fields 90° out of phase. The Fresnel zones in a graded dielectric form the equivalent of waveguides which deflect and confine the EM radiation. The wave interference Fresnel zone description includes features in common with string, loop and lattice quantum gravity models. The **E** and **H** components of EM waves in the nearfield region of the source emitter are out-of-phase and gradually become in-phase in the far-field region. In optics and antenna theory, the near-field region is formed of evanescent waves of extremely high frequency. In part, due to the virtual, transitory nature, the elusive gravitons $\gamma\gamma^*$ have gone undetected in direct physical measurements in laboratory experiments as these are momentary superpositions of photons γ and counter-propagating phase congugate photons γ^*.

...and I think we now have strong reason to believe, whether my theory is fact or not, that the luminferous and the electromagnetic medium are one. – James Clerk Maxwell

I like relativity and quantum theories because I don't understand them and they make me feel as if space shifted about like a swan that can't settle, refusing to sit still and be measured; and as if the atom were an impulsive thing always changing its mind. – D. H. Lawrence

Einstein's relativity work is a magnificent garb which fascinates, dazzles and makes people blind to the underlying errors. The theory is like a beggar clothed in purple whom ignorant people take for a king... its exponents are brillant men but they are metaphysicists rather than scientists. – Nikola Tesla

The light-quantum has the peculiarity that it apparently ceases to exist when it is in one of its stationary states, namely the zero state, in which its momentum, and therefore also its energy, are zero. When a light-quantum is absorbed it can be considered to jump into this zero state, and when one is emitted it can be considered to jump from the zero state to one in which is physically in evidence, so that it appears to have been created. Since there is no limit to the number of light-quanta that may be created in this way, we must suppose that there are an infinite number of light-quanta in the zero state, so that the N_0 of the Hamiltonian is infinite. – P.A.M. Dirac

I am very astonished that the scientific picture of the real world around me is deficient. – Erwin Schrödinger

These, Gentlemen, are the opinions upon which I base my facts. – Winston Churchill

Most of the arguments to which I am party fall somewhat short of being impressive, owing to the fact that neither I nor my opponent knows what we are talking about. – Robert Benchley

What is the most value science can offer?... Truth. – Les Tremayne

It is always wise to look ahead, but difficult to look further than you can see. – Winston Churchill

We can only see a short distance ahead, but we can see plenty that needs to be done. – Alan Turing

Go as far as you can see; when you get there, you'll be able to see further. – Thomas Caryle

The ingenious but nevertheless somewhat artificial assumptions of [Bohr's model of the atom], ... are replaced by a much more natural assumption in de Broglie's wave phenomenon. – Erwin Schrödinger

A mathematical truth is timeless, it does not come into being when we discover it. – Erwin Schrödinger

48. Wavefront (Moiré) Interference Patterns

Refractive Index Ellipsoid and One-Form

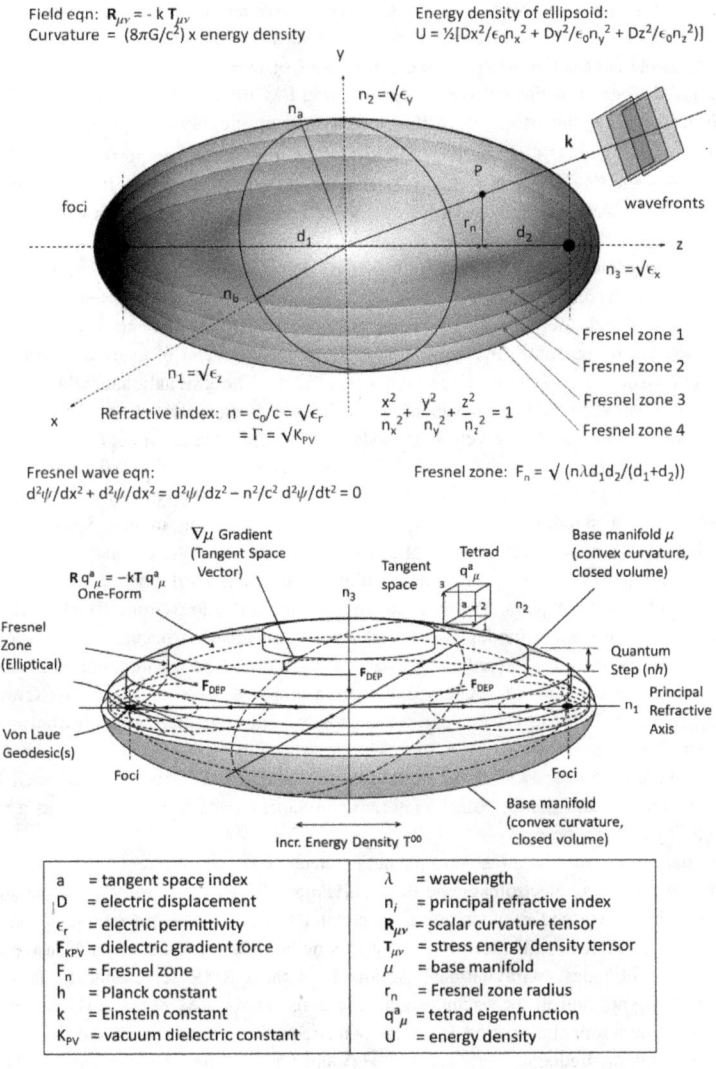

Fig. 48-3. Refractive index ellipsoid (multi-Fresnel zone, graded index) representing anisotropic polarizable vacuum of varying EM energy density.

In the polarizable vacuum PV model, gravitation is a result of variation in electromagnetic energy density equivalent to a gradient in the vacuum dielectric constant $K_{PV}(r,\omega,M)$. The PV vacuum may be characterized by the scalar potential ϕ and electromagnetic vector potential $\mathbf{A}^\alpha$ representing the density and motion, respectively of quantum Planck scale rotating dipoles with gravitation due to gradient in scalar potential and EM phenomena due to spin wave interaction. The constancy of the speed of light is a

48. Wavefront (Moiré) Interference Patterns

measure of the uniform density of quantum dipoles of the physical vacuum. Unlike General Relativity, there is no assumption of the curvature ("bending") of space-time or unexplained mechanism by which mass "distorts" spacetime. The observed Lorentz contraction is a contraction of EM wavelength nodal distance in regions of increased K_{PV} rather than the contraction of spacetime. In regions of increased EM energy density, EM waves are diffracted in Fresnel zones concentrating EM flux. EM wavefronts are slowed and deflected in regions of increased K_{PV} with the wavefront curvature associated with the 'gravitational' field. Gravitational force is interpreted as proportional to ∇K_{PV}^2 dielectrophoretic-like effect in a nonuniform field rather than Casimir effect and may be represented as a summation of Poynting vectors of EM wave interactions in nested Fresnel zones over the entire EM spectrum of interacting masses.

The gravitational flux of a gravitating mass in a polarized vacuum representation corresponds to a standing wave pattern of electromagnetic waves emitted by quantum oscillators of which the mass is comprised. A mass composed of many such interacting oscillators (electrons, atoms, molecules, etc) exhibit entanglement of many quantum body-states and associated collective electrodynamic effects. The gravitational field represents the total EM wave emission over the entire EM frequency spectrum. As a result, the wave interference pattern exhibits a very high nodal density approaching a near continuum. A simplified schematic of gravitation flux in a polarized vacuum for a limited range of frequencies and Fresnel zones is illustrated in Fig. 48-4. The gravity field is shown represented as a standing EM wave pattern for a non-rotating mass. Standing wave formation occurs as a result of equilibrium between EM gravity field and EM radiation field. Gravitons ($\gamma\gamma^*$) are formed as a result of interaction of photon (γ) and phase conjugate photon (γ^*) pairs. Phase conjugate photon reflection (from Bragg scattering centers of EM wave anti-nodes with comparable wavelengths) occurs at multiple EM Fresnel zone boundaries formed in a polarized vacuum. The vacuum refractive index $K_{PV}(r,\omega, M)$ is a measure of vacuum polarization and EM wave node density. Graviton density correlates with EM wave anti-node density. Interference nodes obstruct energy flow and represent radiation damping. The least action path is one in which wave constructive interference is maximal. Much of the internal EM radiation is reflected from wave interference anti-nodes. Blackbody radiation corresponds to leakage flux as a result of scattering at nodal points.

Gravitation is entirely an electromagnetic phenomena; spacetime is a mathematical abstraction describing electromagnetic wave curvature. Gravitons represented as the quanta of gravity are illustrated as interactions of counter-propagating phase conjugate photons produced as a result of reflection from Fresnel zone boundaries formed by EM wave front interference antinodes. Gravitational gamma Γ in the GR/SR representation is shown equivalent to variation in the vacuum dielectric constant K_{PV}. As $K_{PV}(r,\omega,M)$ varies with frequency, gravitational attraction increases with frequency extending beyond the resonant electron Compton frequency $\sim 1.24 \times 10^{20}$ Hz) and above yottahertz range ($>10^{24}$ Hz) in case of the Earth and ultimately up to the Planck frequency ($\sim 1.85 \times 10^{43}$ Hz) which may represent an upper harmonic cutoff limit of the vacuum.

The inner properties of surfaces are "most worthy of being diligently exploited by geometers". – Carl Friedrich Gauss

In the wave lies the secret of creation. – Walter Russell

I try to visualize geometric shapes and patterns slowly spinning and expanding into infinity. – Martina McBride

48. Wavefront (Moiré) Interference Patterns

Gravitational field standing wave pattern

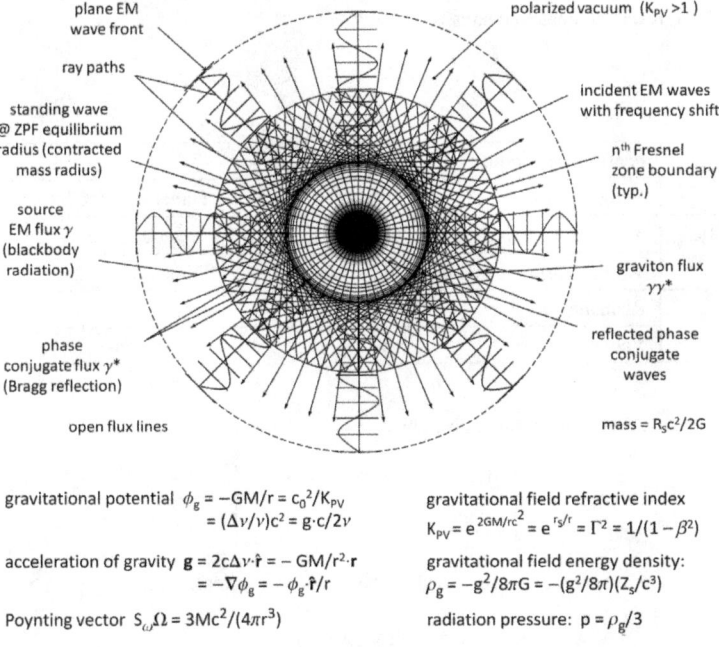

gravitational potential $\phi_g = -GM/r = c_0^2/K_{pv}$
$= (\Delta\nu/\nu)c^2 = g \cdot c/2\nu$

acceleration of gravity $\mathbf{g} = 2c\Delta\nu \cdot \hat{\mathbf{r}} = -GM/r^2 \cdot \mathbf{r}$
$= -\nabla\phi_g = -\phi_g \cdot \hat{\mathbf{r}}/r$

Poynting vector $S_\omega \Omega = 3Mc^2/(4\pi r^3)$

gravitational field refractive index
$K_{pv} = e^{2GM/rc^2} = e^{r_s/r} = \Gamma^2 = 1/(1-\beta^2)$

gravitational field energy density:
$\rho_g = -g^2/8\pi G = -(g^2/8\pi)(Z_s/c^3)$

radiation pressure: $p = \rho_g/3$

Fig. 48-4. Gravity field representation as a standing EM wave pattern for a non-rotating spherical mass. Incident EM waves undergo a relative frequency shift while propagating in a region of higher refractive index. The incident waves appear compressed in the 2D view illustrated but are stretched in tangent space. The interference of incident and reflected radiation creates nested Fresnel zone boundaries. Photons reflected from the interference field antinodes are phase conjugates of the incident photons. Superposition of incident photons and phase conjugates generate local non-propagating graviton flux with additive spins. EM radiation reflected between Fresnel zones boundaries generate standing waves created by superposition of incident and counter-propagating phase conjugate travelling waves. Blackbody radiation corresponds to leakage flux of rays propagating through interference field nodes. A rotating mass creates, in addition, a frame dragging effect due to a superimposed torsion (spin) field component. A radial effect arises during mass inflow.

If you want to find the secrets of the universe, think in terms of energy, frequency and vibration. – Nikola Tesla

If all twenty components of Riemann-Christoffel's curvature tensor were to be required to vanish, then the space-time continuum would be flat, and there would be no possibility of a gravitational field. – Peter G. Bergmann

It may not be an unattainable hope that some day a clearer knowledge of the processes of gravitation may be reached; and the extreme generality and detachment of the relativity theory may be illuminated. – Arthur Eddington

48. Wavefront (Moiré) Interference Patterns

The acceleration of gravity due to a gravitational mass may be represented in terms of relative acceleration of geodesics in curved spacetime. See Fig. 48-5. The relative acceleration in the horizontal x, y-direction and vertical z-direction of two test particles initially at rest in geometrized units is

$$R_{0x0} = m_{gu}/R_e^3 \qquad R_{0y0} = 2m_{gu}/R_e^3 \qquad R_{0z0} = -2m_{gu}/R_e^3 \qquad (48\text{-}4)$$

where:

R_{0n0}	relative acceleration		
	Value: Varies	Units: m/sec² (= N/kg)	Dimensions: LT⁻²
m_{gu}	mass (geometrized units) (= GM/c^2 = $R_S/2$ = t_P/c^5 = $-\phi_g \cdot r/c^2$ = $-\Delta v/v \cdot r$)		
	Value: Varies	Units: m	Dimensions: L
R_{abc}	Riemann curvature		
	Value: Varies	Units: m⁻²	Dimensions: L⁻²

The Riemann curvature tensor is given by

$$R_{abc} = \begin{vmatrix} R_{0x0} & 0 & 0 \\ 0 & R_{0y0} & 0 \\ 0 & 0 & R_{0z0} \end{vmatrix}$$

The curvature in spacetime is described by the Riemann-Christoffel tensor

$$R_{abcd} = C_{abcd} + R_{ab} \qquad (48\text{-}5)$$

where C_{abcd} = Weyl curvature tensor, R_{ab} = Ricci curvature tensor.

The Riemann curvature tensor in four dimensions R_{abcd} has 256 components has 20 algebraic degrees of freedom. The Weyl and Ricci tensors each have 10 components. For n dimensions, the number of independent components is given by $1/12(n^2(n^2 - 1))$. In a two-dimensional case (n = 2), the Riemann curvature tensor R_{abcd} can be replaced by a single scalar R, the Gaussian curvature.

The Ricci tensor may be determined from the Riemann tensor by contraction of indices

$$R_{ab} = R^c{}_{acb} \qquad (48\text{-}6)$$

Further contraction yields the Ricci scalar

$$R = g^{ab}R_{ab} = R^a{}_a \qquad (48\text{-}7)$$

A positive Ricci scalar indicates positive curvature (e.g., a sphere). A negative Ricci scalar indicates negative curvature (e.g., a saddle shape). A zero Ricci scalar indicates a flat geometry. The Einstein tensor is given by

$$G_{ab} = R_{ab} - \tfrac{1}{2}g_{ab}R \qquad (48\text{-}8)$$

The metric g_{ab} corresponds to the gravitational potential ϕ in Poisson's equation $\nabla^2\phi = 4\pi G\rho = R_{00}$ where $\nabla^2\phi$ represent curvature and the source of curvature is ρ = mass density)

48. Wavefront (Moiré) Interference Patterns

in which the line element $ds^2 = g_{ab}dx_a dx_b$. The corresponding field equation in terms of the stress-energy tensor T_{ab} which acts as the source of the curvature G_{ab} is

$$G_{ab} = 8\pi G T_{ab} \tag{48-9}$$

For a zero curvature vacuum, $R_{ab} = 0$ and $T_{ab} = 0$.

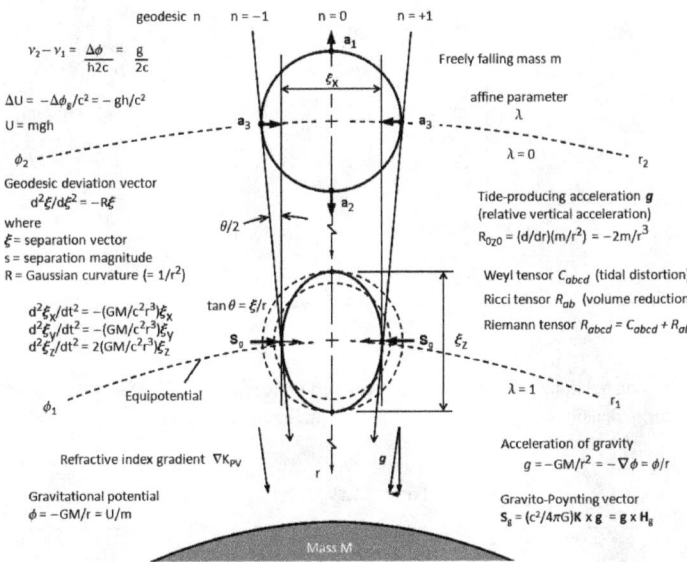

Fig. 48-5. Relative acceleration of freely-falling test particles in a local Lorentz frame in a gravity field described by acceleration of geodesics. The affine parameter λ is a proportionality factor. Inertial mass represents the resistance of particle motion to deviation from the geodesic. A comparable description of relative acceleration of geodesics in terms of an optical gravity model is expressed as a variation in EM energy density. The total tidal deformation is described by the Riemann curvature tensor which is the sum of the volume-conserving Weyl curvature and volume-reducing Ricci curvature tensors. Tangent vectors on the Riemannian manifold $\mathcal{M}$ generate a vector field. Geodesics lie in the direction of the Killing vector field $\xi(r)$. The metric product of the Killing vector and the geodesic tangent vector represents conservation of total energy. Increase in kinetic energy of a falling mass results from influx of gravitational potential energy via the gravitational Poynting vector $\mathbf{S}_g$.

There is no model of the theory of gravitation today, other than the mathematical form.
– Richard Feynman

Don't fight forces, use them. – Buckminster Fuller

In architecture, the pride of man, his triumph over gravitation, his will to power, assume a visible form. – Friedrich Nietzsche

48. Wavefront (Moiré) Interference Patterns

For a particle of density ρ_p immersed in a fluid of density ρ_f, the buoyant force is given by $\mathbf{F}_b = v(\rho_f - \rho_p)\mathbf{g}$ where v = volume of the particle and **g** = acceleration of gravity. Similarly, a particle mass immersed in a EM field experiences a buoyant force $\mathbf{F}_b = \Delta p_{rad}/A - \mathbf{F}_g$ where the radiation pressure $p_{rad} = -\rho_g$, $\mathbf{F}_g$ = particle weight, A = cross-section area, and ρ_g = EM energy density. The effects of tidal distortion, volume contraction on a system of bound masses forming a Fresnel zone are illustrated in Figs. 48-6 and 48-7.

Weyl curvature

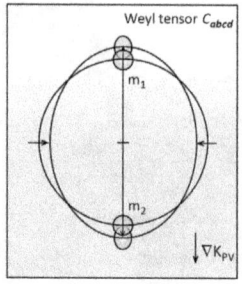 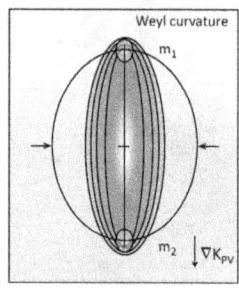

Tidal distortion of 2 equal masses due to Weyl curvature in polarized vacuum | Fresnel zone of 2 equal masses in zero gravity external field | Fresnel zone tidal distortion due to Weyl curvature

Fig. 48-6. Fresnel zone tidal-distortion (Weyl curvature) of free-fall masses due to gradient in EM energy density in a polarized vacuum producing bi-refringence.

Ricci curvature

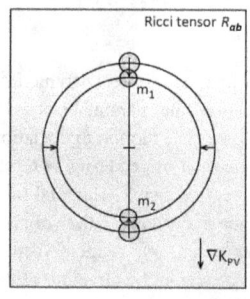 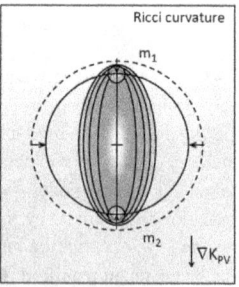

Volume distortion of 2 free-falling masses due to Ricci curvature in polarized vacuum | Fresnel zone of 2 equal masses in zero gravity external field | Fresnel zone volume distortion of 2 free-falling masses due to Ricci curvature

Fig. 48-7. Fresnel zone contraction (Ricci curvature volume reduction) due to negative pressure from positive external EM energy density in a polarized vacuum.

Gravity is solely concerned with mass, and is measured strictly by movement. – Agnes Mary Clerke

I am more in touch with positive valences. – George Saunders

48. Wavefront (Moiré) Interference Patterns

Lorentz contraction effects under rotation

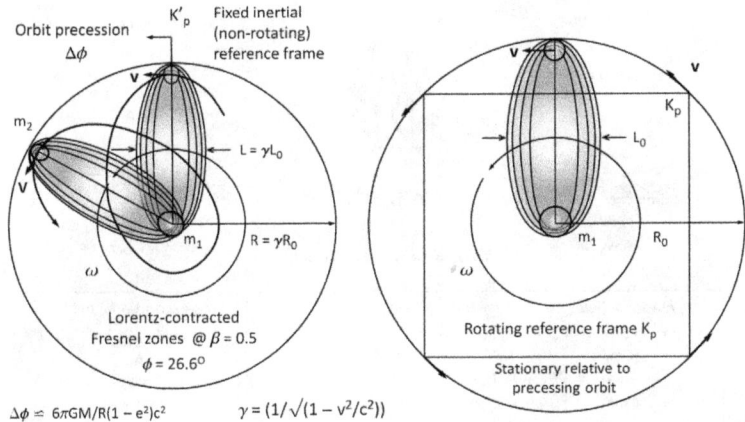

Fig. 48-8. Schematic diagram of orbital precession of two gravitationally interacting masses at relativistic velocity (v = 0.5c) illustrating Lorentz contraction of associated polarizable vacuum Fresnel zones. The relative distance between masses likewise undergoes a Lorentz contraction such that in a rotating reference frame which is stationary to the precessing orbit rosette, the Fresnel zones remain undistorted.

A time series evolution of a gravity field of mass m in relative motion to a central mass M is shown in Fig. 48-9. In addition to the gravitational attraction force F_g, a co-gravitational field F_k arises as a result of a mass current analogous to a magnetic field associated with a moving electrical current. Yet another force, a gravi-kinetic drag force F_{kd} is generated analogous to the opposing electrokinetic force associated with time rate of change of current[16]. Mass experiences Lorentz contraction plus Weyl tidal deformation. For a mass moving at relativistic velocities as shown, the gravitational field is strongly affected by velocity producing momentarily large tidal forces.

No more fiction, for now we calculate; but that we may calculate, we had to make fiction first. – Friedrich Nietzsche

Without dreams there is no art, no mathematics, no life. – Michael Atiyahs

When a screen misaligns plied with another behind, that's a Moiré! – Anon.

Concerning matter, we have all been wrong! What we have called matter is energy, whose vibration has been lowered as to be perceptible to the senses. – Albert Einstein

There is no more dangerous error than that of mistaking the consequence for the cause. – Friedrich Nietzsche

Invisible threads are the strongest ties. – Friedrich Nietzsch

I am more exempt and more distant than any man in the world. – Pierre de Fermat

The wreckage of stars – I built a world from this wreckage. – Friedrich Nietzsche

437

48. Wavefront (Moiré) Interference Patterns

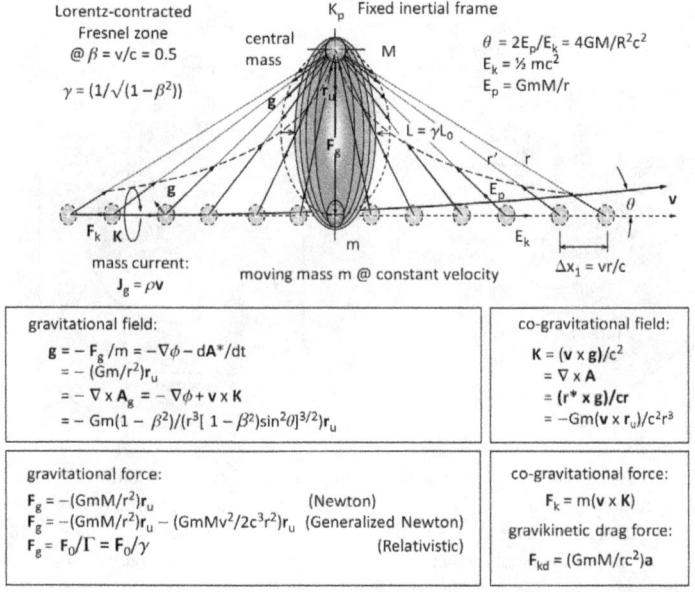

Fig. 48-9. Gravitational coupling of two proximate masses with the smaller mass moving with constant velocity (v ≈ 0.5c) relative to stationary mass at point P illustrating Lorentz contraction of Fresnel zone and gravitational force components. The gravitational force vector and acceleration is illustrated for sequential positions as the mass moves left to right. The cogravitational field **K** (= (**r*** × **g**)/cr) for accelerated motion is circumferential following a left-hand rule where **r*** is the retarded position vector. The cogravitational force F_k is in the direction of motion. An example of cogravitational force is the Lense-Thirring 'frame-dragging' effect. Both the gravitational force and cogravitational force increase at relativistic velocities and at reduced separation distance.

Discover the force of the skies O Men: once recognized it can be put to use. – Johannes Kepler

That gravitation is a phenomenon of the all-pervading aether is beyond reasonable doubt.. – Charles Francis Brush

Nothing is too wonderful to be true, if it be consistent with the laws of nature, and in such things as these, experiment is the best test of such consistency. – Michael Faraday (Diary, 19 March 1849)

We inhabit two universes, then, One is the universe inside our skulls – our viewpoint universe, as it were. – James Blish

It is by will alone I set my mind in motion. – Frank Herbert, Mentat mantra (Dune)

It is not worth an intelligent man's time to be in the majority. By definition, there are already enough people to do that. – Niels Bohr

One must never be satisfied doing what one can; rather, one must always do what one really cannot. – Niels Bohr

48. Wavefront (Moiré) Interference Patterns

Effects of harmonic distortion of Fresnel zones and gravitational wave formation of interacting mass couples is shown in Figs. 48-10 and 48-11.

Harmonic deformation effects

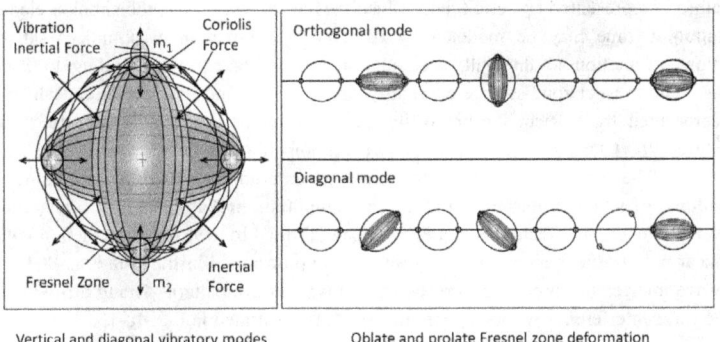

Fig. 48-10. Vibration oscillation of coupled masses results in harmonic Fresnel zone deformation.

Gravitational Wave Effects

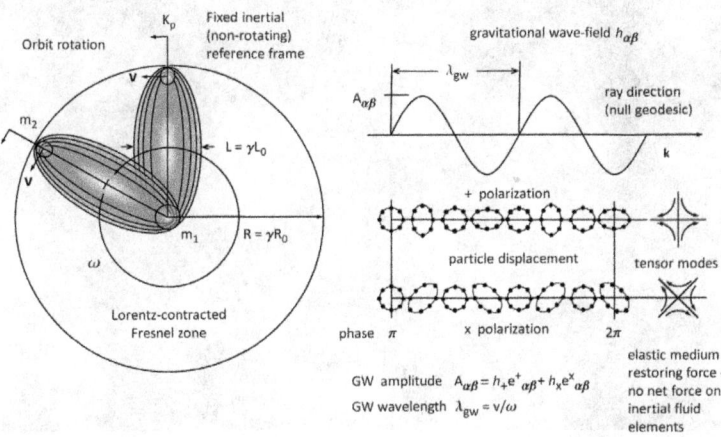

Fig. 48-11. Gravitational wave field generation by rotational oscillation motion of two proximate masses exhibit quadrapole polarization. EM waves propagating through a gravity-wave distorted region of space may be expected to exhibit a small oscillatory variation in wavelength allowing a potential indirect means of gravity wave detection.

What is now proved was once only imagin'd. – William Blake

I am constantly on the 'lookout' for something that might represent an 'electrodynamic-gravitational' coupling. – Thomas Townsend Brown

439

48. Wavefront (Moiré) Interference Patterns

The Fresnel zone wave interference pattern may be numerically evaluated utilizing Regge calculus where the nodal points are used to construct a 4-dimensional polyhedron of triangular mesh simplex elements representing each Fresnel zone for discrete quantum gravity simulation. The lengths of the skeletonized finite-element framework corresponds to ds line elements (separation interval in spacetime). The equivalent spacetime manifold curvature is represented as deficit angles localized at the vertices of the simplex elements. Evolution in time may be modeled by addition or subtraction of simplices satisfying equations of motion to the bulk triangulation hypersurfaces corresponding to Pachner moves. The Fresnel zone surface curvature corresponds to intrinsic curvature (which may be represented by a Ricci tensor) while the variation in K_{PV} corresponds to extrinsic curvature (Weyl tensor providing a discrete version of Gaussian curvature of a smooth manifold). The simplicial lattice corresponds to an amplitude polyhedron (amplituhedron), the volume of which represents the scattering amplitude probabilities. An illustration of EM wave interaction effects within a Fresnel zone formed by EM wave interaction within a nonlinear polarizable medium in a four-way mixing process is illustrated in Fig. 48-12. The EM wave interference produces a crystal-like Bravais lattice pattern. Bragg diffraction due to EM wave interference within a photonic crystal is illustrated in Fig. 48-13.

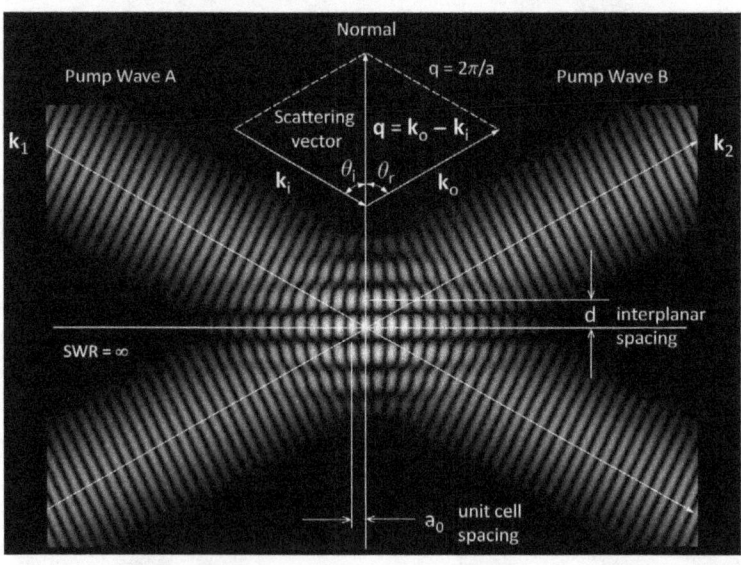

Fig. 48-12. Bravais lattice formed by EM wave interference antinode points which act as scattering centers similar to atomic lattice planes in a photonic crystal. If incident EM waves are of a wavelength not comparable to the wave interference lattice spacing, waves are not diffracted and pass through the interference zone unimpeded.

It requires a very unusual mind to undertake the analysis of the obvious. – Alfred Norse Whitehead

48. Wavefront (Moiré) Interference Patterns

The variation in the speed of light, deflection of light in a gravitational field, length contraction and time dilation effects have been described in terms of local variations in the electric permittivity and magnetic permeability of the vacuum in the presence of mass in a variable polarizable vacuum (PV) model by Wilson, Dicke, Puthoff, Storti *et al.* This model was shown to be compatible with General Relativity without recourse to the assumption of spacetime curvature. The permittivity and permeability appear as scalar fields which, in turn, influence electric field intensity **E** and magnetic flux density **B** in accordance with governing equations of electromagnetism. According to the Coulomb equation $F = Q_1Q_2/4\pi\epsilon r^2$, the Coulomb force between two charges is lower in any dielectric medium than in free space in proportion to the value of K for the medium. The cause of this decrease in force action is found in the polarization or stretching of the molecules of the dielectric material. In the PV model a variable refractive index K_{PV} is attributed to the quantum vacuum and is proportional to the EM energy density associated with the gravitational field. The variable K_{PV} describes the gradient in the refractive index acting to deflect light. The vacuum permittivity and permeability are not uniformly distributed in space but form scalar fields. The polarizability of the vacuum in the vicinity of mass differs from the asymptotic far-field value as a result of increased density of EM wave interference anti-nodes which act as scattering centers slowing and deflecting propagating EM wavefronts.

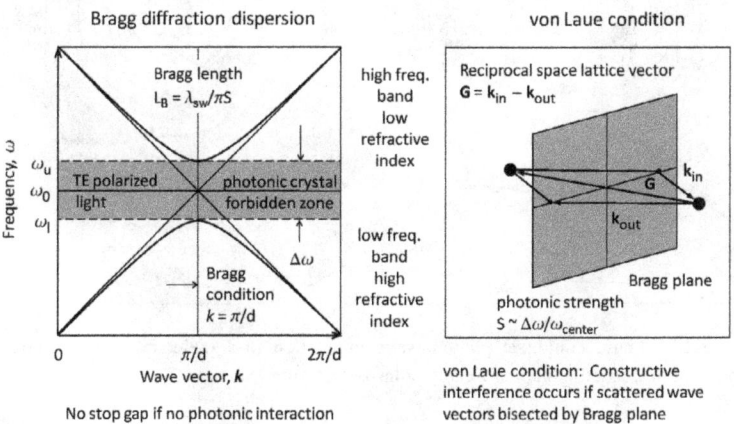

Fig. 48-13. Bragg diffraction is commonly observed in crystal lattices due to EM wave interference of waves scattered off atomic lattice planes in a photonic crystal. Forbidden zones or stop gaps in crystalline structures represent conditions where no light transmission occurs. Energy states in which single photons are neither absorbed or emitted in 2D semiconductors represent excitonic dark energy states formed by bound pairs of electrons and holes (excitons). Separation of photoexcited electrons and holes enables flow of electrical current in photovoltaic devices. Charge separation and conductance may be controlled by varying the energy band-gaps.

To understand hydrogen is to understand all of physics. – Victor Weisskopf

441

48. Wavefront (Moiré) Interference Patterns

EM wave reflection and diffraction of phase conjugate EM waves emitted from a pair mass oscillators interacting in a weakly polarized vacuum is depicted in Fig. 48-14. For simplicity, only two emission wavelengths (λ_1, λ_2) are shown whereas the EM frequency spectrum, in general, extends well into the 10^{24} Hz (yottahertz) range up to ω_{PV} cutoff frequency where Kerr-like nonlinear effects are expected to dominate. The incident photons γ and reflected phase conjugate photons γ^* are reflected from the Fresnel zone boundary. Gravitons $\gamma\gamma^*$ are formed as a superposition of counter-propagating photons and phase conjugate photons in the overlap region with net spin of 2. Spin 2 fields result in attractive forces proportional to energy density. Each Fresnel zone corresponds to an ellipsoidal whispering gallery.

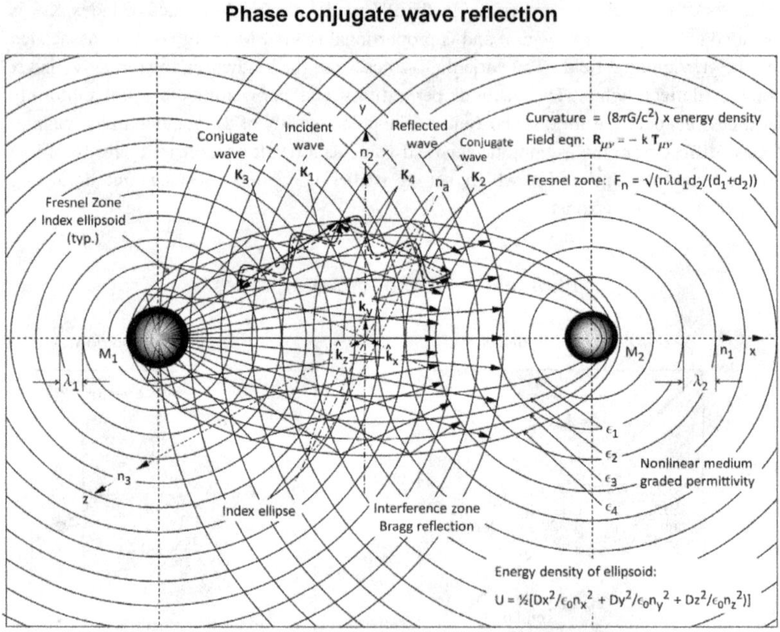

Phase conjugate wave reflection

Fig. 48-14. Phase conjugate photon wave interaction in a dielectric Fresnel zones of interacting mass oscillators in a weakly polarized quantum vacuum.

The Indian Bose has given a beautiful derivation of Planck's law, including the constant $[8\pi v^2 dv/c^3]$. – Albert Einstein

Discovery consists of looking at the same thing as everyone else and thinking something different. – Albert Szent-Györgyi

Gravity. Surely this force must be capable of an experimental relation to electricity, magnetism and other forces ... in reciprocal action and equivalent effect. – Michael Faraday

We're free out here, really free for the first time. We're floating, literally. Gravity can't bow our backs or break our arches or tame our ideas. – Fritz Leiber

We can lift ourselves out of ignorance, we can find ourselves as creatures of excellence, intelligence and skill. We can learn to be free! We can learn to fly! – Richard Bach

48. Wavefront (Moiré) Interference Patterns

Bragg reflection and diffraction of an incident EM from on interference of EM of comparable wavelength emitted from a pair of oscillators is illustrated in Fig. 48-15. For clarity, only a single frequency common to both emitters is shown.

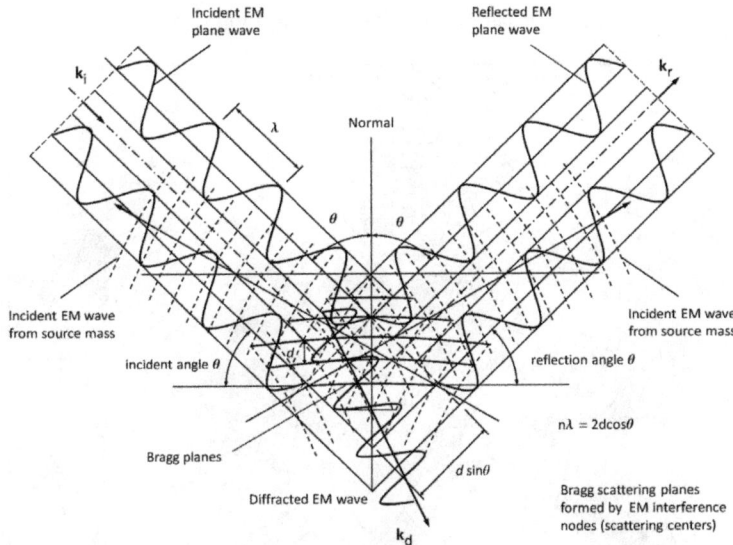

Fig. 48-15. EM wave reflection/diffraction from Bragg planes formed by EM waves emitted from a gravitational mass. Wave interference nodes act as scattering centers of a holographic amplitude grating for incident EM waves to form a reflected phase conjugate beam. Reflection occurs when the incident wavelength is comparable to the Bragg plane spacing. Effect is similar to the much stronger conventional Bragg x-ray diffraction from crystals where atomic lattices form periodic scattering centers. Irradiation of a BEC condensate by counter-propagating laser beams likewise creates an optical lattice.

[Mathematics consists of] true facts about imaginary objects. – P. Davis/R. Hersh

The gravitational field is like the electric field but requires three sets of field lines to describe it. – Lee Smolin

It is hypothesized that coupling exists between Electromagnetic (EM) fields and the magnitude of the local value of gravitational acceleration 'g'. – Ricardo C. Storti

Ask the right questions, and nature will open the doors to her secrets. – C.V. Raman

The questions are diamonds you hold in the light. – Richard Bach

Success can come to you by courageous devotion to the task lying in front of you. – C.V. Raman

It is the business of the future to be dangerous. – Alfred Norse Whitehead

One must explore deep and believe the incredible to find the new particles of truth floating in an ocean of indifference. – Joseph Conrad

443

48. Wavefront (Moiré) Interference Patterns

Polarized reflection and scattering of EM waves from a Fresnel zone boundary is depicted in Fig. 48-16. The Fresnel zone is formed by the interference pattern of EM waves emitted from a pair of interacting masses. As shown, multiple Fresnel zones are generated for each pair of source frequencies. An incident EM wave reflected from the Fresnel zone boundary undergoes polarization similar to Raman scattering.

Fresnel zone Raman scattering

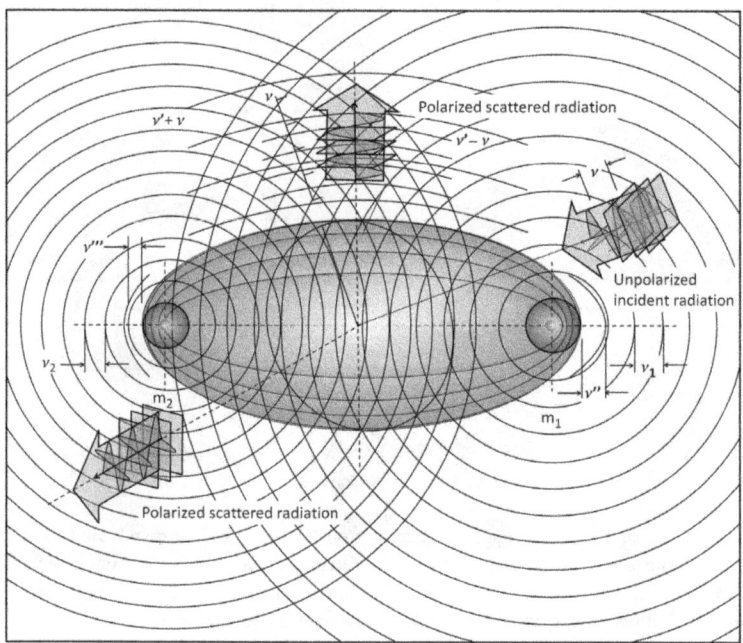

Fig. 48-16. EM wave reflection/scattering from a Fresnel zones resulting in polarized scattering with radiation components that are the sum and difference of the incident and internal frequencies similar to Raman scattering. For simplicity, only a single frequency emission from each oscillator is shown. For a given mass oscillator, the EM radiation frequency range is extremely large forming a near continuum of zone layers. Effect varies with vacuum refractive index $K_{PV}(r,M)$ and apparent under very strong gravitational fields.

The senses delight in things duly proportioned. – St. Thomas Aquinas

The vitality of thought is in adventure. Ideas won't keep. Something must be done about them. – Alfred Norse Whitehead

In the history of science, we often find that the study of some natural phenomenon has been the starting point in the development of a new branch of science. – C.V. Raman

Most of the complex structures found in the world are enormously redundant, and we can use this redundancy to simplify their description. – Herbert Simon

49. Gravitational Frequency Redshift

49.1 Photon frequency

The ratio of gravitational redshift frequencies measured by a pair of clocks in a gravitational field may be expressed as

$$\nu_1/\nu_2 = \sqrt{(1 - 2GM/r_2)}/\sqrt{(1 - 2GM/r_1)} \tag{49-1}$$

where r_1 and r_2 are radial distances from a central mass M.

The General Relativity (GR) redshift formula for the acceleration of gravity in terms of frequency gradient is

$$\Delta U = \Delta \nu/\nu = (\nu_r - \nu_e)/\nu_e = -\Delta \phi_g/c^2 = -gh/c^2 \tag{49-2}$$

where ν_e = frequency of emitter and ν_r = frequency of signal received both measured locally. Rewriting

$$g = c^2(d\tau - d\tau_0)/(d\tau dh) = c^2(d\beta/dh) \tag{49-3}$$

where:

g	Acceleration of gravity ($= GM/(R + h)^2 = F/m = W/m = -\nabla \phi_g$)		
	Value: Varies	Units: N/kg (= m/sec^2)	Dimensions: LT^{-2}
$d\tau$	proper time measured at detector ($d\tau = 1/\nu$)		
	Value: Varies	Units: s	Dimensions: T
$d\tau_0$	proper time measured at source ($d\tau_0 = 1/\nu_0$)		
	Value: Varies	Units: sec	Dimensions: T
dh	Vertical height of detector above source (proper length)		
	Value: Varies	Units: m	Dimensions: L
c	speed of light ($= c_0/\Gamma = c_0/\sqrt{K_{PV}}$ (weak field) $= c_0/\sqrt[4]{K_{PV}}$ (strong field))		
	Value: 2.997924 x 10^8	Units: m/sec	Dimensions: T
$d\beta$	gravitational magnitude ($\beta_g = \sqrt{(1 - 1/\Gamma^2)} = \sqrt{(1 - 1/K_{PV})}$)		
	Value: Varies (0 to 1)	Units: -	Dimensions: -

The heavens are full of floating mysteries. – Thomas Buchanan Read

Information is not knowledge. – Albert Einstein

In an honest search for knowledge, you quite often have to abide by ignorance for an indefinite period. – Erwin Schrödinger

Big results require big ambition. – Heraclitus

The direction of the mind is more important than its progress. – Joseph Joubert

The longer you look back, the further you can look forward. – Winston Churchill

The future influences the present just as much as the past. – Friedrich Nietzsche

49. Gravitational Frequency Redshift

The acceleration of gravity of 9.8 m/s² at the earth's surface corresponds to a frequency gradient (rate of time gradient) of ~10^{-15} s/s/m. In the Pound, Rebka, Snider experiment[119] using the Mössbauer effect, the photon frequency shift of a gamma ray source for a height (h = Δr) of 22.5 m was measured as

$$\Delta\nu_g/\nu = (\nu_0 - \nu_1)/\nu_0 = (U_0 - U_1)/c^2 = g\Delta r/c^2 \qquad (49\text{-}4)$$
$$= 2.5 \times 10^{-15}$$

The frequency ν of a photon emitted or absorbed by an electron is a function of the gravitational potential and in a uniform static gravitational field the redshift frequency is given by

$$\nu = \nu_0[1 - gh/c^2] = \nu_0[1 - gh/Kc_0^2] = \nu_0[1 - gh\Gamma/c_0^2] = \nu_0[1 - \Phi/c^2] \qquad (49\text{-}5)$$

where:

ν	frequency of source emitter as measured at source
ν_0	frequency of detector as measured at detector (proper time)
g	local acceleration of gravity at source elevation ($GM/r^2 = -\phi_g/r = -\nabla\phi_g$)
h	vertical height (proper length) of detector above source
c	speed of light (= $c_0/\Gamma = c_0/\sqrt{K_{PV}}$ (weak field) = $c_0/\sqrt[4]{K_{PV}}$ (strong field))
K	vacuum index of refraction (= $K_{PV}(r,\omega,M) = \Gamma^2 = 1/g_{00} = g_{11} = g_{22} = g_{33}$)
Γ	gravitational gamma (= $dt/d\tau = 1/\sqrt{(1-\beta_g^2)} = \sqrt{K_{PV}}$ [weak field])
Φ	gravitational potential (= $\phi_g = -GM/r = -\Delta\nu 2cr = c_0^2/K_{PV} = c_0^2/\Gamma^2 = (\Delta\nu/\nu)c^2$

Krogh[32] has derived the following expression for the effect of a strong gravitational potential representing a infinite series of mass shells each producing a weak contribution for the total potential

$$\nu/\nu_0 = \lambda/\lambda_0 = \lim_{n\to\infty}(1 + 1/n(\Phi_g/c_0^2))^n = e^{\wedge}\Phi_g/c_0^2 \qquad (49\text{-}6)$$

where:

ν	frequency of source emitter as measured at source
ν_0	frequency of detector as measured at detector (proper time)
λ	wavelength of source emitter as measured at source
λ_0	wavelength of detector as measured at detector (proper time)
Φ_g	local gravitational potential (= $-GM/r = -\Delta\nu 2cr = c_0^2/K_{PV} = c_0^2/\Gamma^2 = (\Delta\nu/\nu)c^2$)
e	Euler's constant (= $1/0! + 1/1! + 1/2! + \ldots + 1/n! = 2.718281828\ldots$)
n	number of included mass shells
c_0	velocity of light in a zero curvature, gravity free vacuum (= $g\cdot c = 1/\sqrt{(\epsilon_0\mu_0)}$)

Everyone experiences far more than he understands. Yet it is the experience, rather than understanding, that influences behavior. – John Brockman

We are as gods and might as well get good at it. – Stewart Brand

49. Gravitational Frequency Redshift

49.2 Gravitational time dilation

A clock may be defined as a device which records the frequency or number of periods of an oscillator. The frequencies of two clocks in a gravitational field separated by a height h = $r_1 - r_2$ vary as

$$\nu_1/\nu_2 = d\tau_2/d\tau_1 = \nu_e/\nu_r = 1 - GM/r_2 + GM/r_1 \qquad (49\text{-}7)$$

where frequencies $\nu_1 = 1/d\tau_1$ and $\nu_2 = 1/d\tau_2$ are related to the proper time intervals measured by local clocks at positions 1 (emitter) and 2 (receiver).

In flat spacetime as $r \rightarrow \infty$, the metric tensor component $g_{00} = 1$, but in a gravitational field of a mass object $g_{00} = 1 - 2GM/c^2r$. Proper time intervals $d\tau$ indicated by a local clock in relation to coordinate time intervals dt varies as $d\tau = \sqrt{g_{00}}dt$. Observers at different gravitational potentials measure disparate clock rates. The time $\tau_2 = \tau_1(1/(1 \pm \phi/c^2))$ where τ_2 = time at r_2 (detector) and τ_1 = time at r_1 (source).

The rate of a clock near a massive spherical body in the form of a series expansion of the exponential is given by Krogh[32] as

$$\nu = \nu_0 e^{-\mu/r} = \nu_0[1 - GM/c_0^2 r + \tfrac{1}{2}(GM/c_0^2 r)^2 + \ldots] \qquad (49\text{-}8)$$

where $\mu = GM/c_0^2$ The dimensionless potential is represented by

$$\Phi_g/c_0^2 = -GM/c_0^2 r = \mu/r \qquad (49\text{-}9)$$

49.3 Frequency shift differential

Ivanov[36,37] has derived from analysis of the photon frequency shift effect due to a difference of gravitation potential the following relationship between the force of gravity and the frequency shift of coupled masses (oscillators) in a uniform gravitational field:

$$\mathbf{F}_g = 2mc\Delta\nu\cdot\mathbf{r}_u = mg \qquad (49\text{-}10)$$
$$= 2mc_0\Delta\nu\cdot\mathbf{r}_u/\Gamma = 2mc_0\Delta\nu\cdot\mathbf{r}_u/\sqrt{K_{PV}}$$

where:

$\mathbf{F}_g$	Gravitational force (= weight $\mathbf{W} = \mathbf{F}_0/\Gamma$). Magnitude $F_g = |\mathbf{F}_g|$		
	Value: Varies	Units: N = kg(m/sec^2)	Dimensions: MLT^{-2}
m	Mass (= $\mathbf{F}/\mathbf{a} = \mathbf{W}/\mathbf{g} = E/c^2 = E/(\mathbf{E}/\mathbf{B})^2 = -2\Delta\nu cr^2/G = \Gamma m_0 = \gamma m_0$)		
	Value: Varies	Units: kg	Dimensions: M
c	Velocity of light (= $\lambda\nu = c_0/\Gamma = c_0/\sqrt{K_{PV}} = \sqrt{(v_g v_p)} = \mathbf{E}/\mathbf{B} = \phi_E/\phi_B = \omega/k$)		
	Value: 2.9974 x 10^8	Units: m/sec	Dimensions: LT^{-1}
$\Delta\nu$	Frequency differential (= $-GM/2cr^2 = g/2c = -\phi_g/2cr = \phi_g\cdot\nu/c^2$)		
	Value: Varies	Units: Hz	Dimensions: T^{-1}

If indeed we are able to alter the vacuum, then we may encounter new phenomena, totally unexpected. – T. D. Lee

By studying the type of gravitational field that was produced by a ring laser..could lead to the possibility of a time machine based on a circulating beam of light. – Ronald Mallett

49. Gravitational Frequency Redshift

As shown, the acceleration of gravity $g = 2c\Delta\nu$ over a sufficiently small region where the field is approximately uniform. Substituting a gravitational acceleration of 9.8 m/s² at the earth's surface corresponds to a frequency differential $\Delta\nu = 1.636 \times 10^{-8}$ Hz. See Fig. 40-1. In the Ivanov model, the phase and frequency relationships of standing EM waves between oscillators were shown to result in an attractive force indistinguishable from gravity. The gravitational force is attributed to an EM frequency arrhythmia between EM coupled masses.

49.4 Phase shift differential

Ivanov[36,37] also derives the following relation between wave velocity and phase displacement for a system of two coherent oscillators located at nodes of a standing wave.

$$v = c \cdot \Delta\phi/\pi = k_c \cdot \Delta\phi \qquad (49\text{-}11)$$

where:

v	System velocity ($v_g = v_0/\Gamma = v_0/\gamma = v_0/\sqrt{K_{PV}} = (\hbar k - q_0 A)/m = \beta \cdot c$)		
	Value: Varies	Units: m/sec	Dimensions: LT^{-1}
$\Delta\phi$	Phase displacement ($= (v_p/c)\pi = \beta \cdot \pi = \pi \cdot \cos\theta = \pi \cdot (\tanh\rho) = \alpha\pi/g\gamma$)		
	Value: Varies	Units: radians	Dimensions: T^{-1}
c	Velocity of light ($= c_0/\Gamma = c_0/\sqrt{K_{PV}} = \sqrt{(v_p v_g)} = 1/\sqrt{(\epsilon_0\mu_0)} = E/B = \omega/k = l_P \cdot t_P$)		
	Value: 2.9979×10^8	Units: m/sec	Dimensions: LT^{-1}

For two oscillators in phase ($\Delta\phi = 0$ rad $= 0°$), the oscillators are synchronized simultaneously emitting waves of equal amplitude. For out of phase oscillators ($\Delta\phi = \pi$ rad $= 180°$), one oscillator lags or leads the other. Significantly, phase displacement causes the system to move in the wave medium while reducing the standing wavelength. The distance between the standing wave nodes contracts $\Delta l = \lambda_{st} - \lambda'_{st}$ according to the rule

$$\lambda'_{st} = \lambda_{st} \cdot (1 - \Delta\phi^2/\pi^2) = \lambda_{st} \cdot (1 - v^2/c^2) \qquad (49\text{-}12)$$

where:

λ_{st}	Standing wavelength (system velocity = 0)		
	Value: Varies	Units: m	Dimensions: L
λ_{st}'	Contracted standing wavelength (system velocity > 0)		
	Value: Varies	Units: m	Dimensions: L
$\Delta\phi$	Phase displacement ($= (v_p/c)\pi = \beta \cdot \pi = \pi \cdot \cos\theta = \pi \cdot (\tanh\rho) = \alpha\pi/g\gamma$)		
	Value: Varies	Units: radians	Dimensions: T^{-1}
c	Velocity of light ($= c_0/\Gamma = c_0/\sqrt{K_{PV}} = \sqrt{(v_p v_g)} = 1/\sqrt{(\epsilon_0\mu_0)} = E/B = \omega/k = l_P \cdot t_P$)		
	Value: 2.9979×10^8	Units: m/sec	Dimensions: LT^{-1}

We haven't the money, so we have to think. – Ernest Rutherford

If you'e going to be a prisoner of your own mind, the least you can do is make sure it's well furnished. – Peter Ustinov

The science of today is the technology of tomorrow. – Edward Teller

49. Gravitational Frequency Redshift

The resulting acceleration, **a**, due to a changing phase shift is given by

$$\mathbf{a} = dv/dt = (v_2 - v_1)/(t_2 - t_1) = c/\pi \cdot (\Delta\phi_2 - \Delta\phi_1)/\Delta t \quad (49\text{-}13)$$
$$= c(\beta_2 - \beta_1)/(t_2 - t_1) = 2c \cdot \Delta\nu \cdot \mathbf{r}_u = (2c^2/h)(mv/\Delta\lambda)\lambda_{dB} \cdot \mathbf{r}_u.$$

where: V = velocity, t = time, $\Delta\phi$ = phase shift, $\Delta\nu$ = frequency differential, λ_{dB} = de Broglie wave length and $\mathbf{r}_u$ = the unit vector in direction of the frequency gradient.

A Lissajous figure illustration of phase shift differential is shown in Fig. 49-1. For a moving standing wave, the phase shift $\Delta\phi = (v_p/c) \cdot \pi = \beta \cdot \pi = (\sin\theta) \cdot \pi = (\alpha/g\gamma) \cdot \pi$ which is a function of the aberration angle θ and phase velocity v_p. The Lorentz Doppler frequency shift $\Delta\nu = c \cdot g/\lambda 2\beta$ where g is the Lorentz contraction (= $\sqrt{(1 - \beta^2)}$) and β is the velocity ratio (= v_p/c = $\sin\theta$). The frequency shift $\Delta\nu$ is proportional to the de Broglie matter wave frequency ν_{dB}. The frequency ratio ν_1/ν_2 equals the ratio of initial wave frequency at rest and the Lorentz Doppler frequency.

Lissajous figures

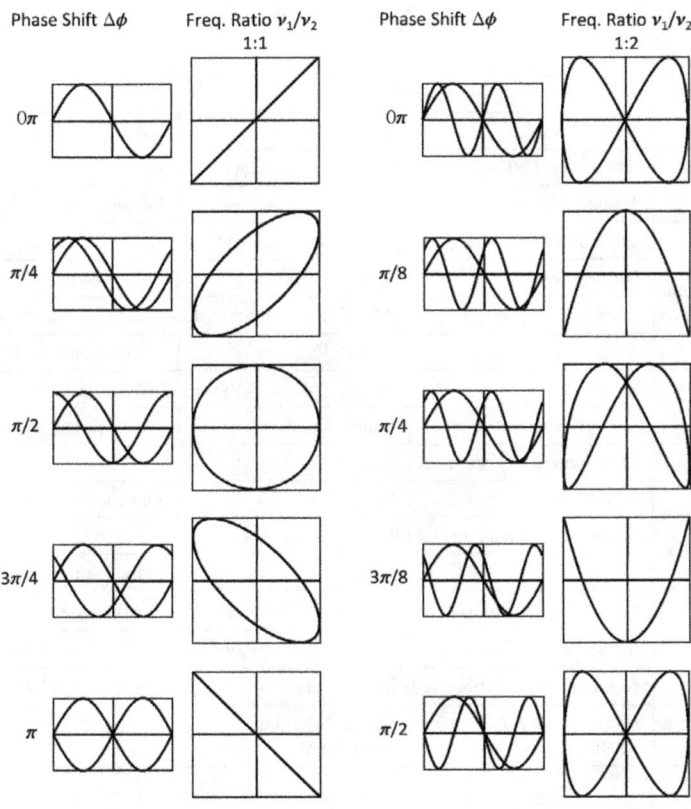

Fig. 49-1. Example of Lissajous phase shift differential vs. frequency ratio for a one octave frequency interval. A moving wave system undergoes a Lorentz contraction g and Lorentz Doppler frequency shift $\Delta\nu$ in the direction of motion. The group velocity $v_g = c \cdot \Delta\phi/\pi = c^2/v_p = 1/(\epsilon_0\mu_0 v_p) = c \cdot \beta$.

50. Gravitational frequency spectrum

In general, the objects in the universe that are very high-energy objects, or the processes that are high energy processes, will radiate more in the short wavelength towards the gamma rays or the x-rays. – Claude Nicollier

50.1 Gravitational frequency range

The range of frequencies of oscillation peculiar to gravity have been estimated by Boutorin, Bershadsky, Akimov *et al* to be 10^9 to 10^{40} Hz with 10^{20} to 10^{40} Hz more likely[89, 90, 91]. The minimum peak gravitational frequency ω_{pk} of a given mass m may be estimated in terms of the number of equivalent Planck mass oscillators from the Planck relation $\omega_{pk} = E_n/(n_P \cdot h)$ where E_n = rest mass energy, n_P = no. of equivalent Planck mass oscillators and h = Planck's constant. For a solid spherical mass, the peak gravitational frequency ω_{pk} may be estimated

$$\omega_{pk}(n_{mP}, r, m) \cong \phi \cdot M/h \cong (E_0/\Gamma)/(n_{mP} \cdot h) \qquad (50\text{-}1)$$

where

ω_{pk}	Effective cut-off frequency		
	Value: Varies	Units: Hz	Dimensions: T^{-1}
E	energy $(= E_0/\Gamma = m_0c^2/\Gamma = m_0c^2/\sqrt{K_{PV}} = \frac{1}{2}m(\mathbf{E}^2 + \mathbf{B}^2) = pc/\beta$		
	Value: Varies	Units: J = kg·m²/sec²	Dimensions: $M L^2 T^{-2}$
Γ	gravitational gamma $(= 1/\sqrt{(1 + (2\phi/c^2))} = 1/\sqrt{(1 - (v_e^2/c^2))} = 1/\sqrt{(1 - \beta_g^2)})$		
	Value: Varies	Units: None	Dimensions: -
ϕ	gravitational potential $(= -GM/r = \Delta v2cr = g \cdot r = v \cdot h/M = c_0^2/K_{PV} = c_0^2/\Gamma^2)$		
	Value: Varies	Units: J/kg = m²/s²	Dimensions: L^2T^{-2}
m	mass $(= \Gamma m_0 = \gamma m_0 = E/c^2 = \mathbf{W}/\mathbf{g} = (\hbar\mathbf{k} - q_0\mathbf{A})/v = E/(\mathbf{E}/\mathbf{B})^2 = |eV|/c^2)$		
	Value: Varies	Units: kg	Dimensions: M
m_P	Planck mass $(= \sqrt{(\hbar c/G)} = E_P/c^2 = \hbar\omega_P/c^2)$		
	Value: 2.17651E-8	Units: kg	Dimensions: M
n_{mP}	No. of Planck masses $(= m/m_P)$		
	Value: Varies	Units: -	Dimensions: -
h	Planck's quantum of action constant $(= 2\pi\hbar = \Delta E/\Delta v = 2\pi\hbar = 2\pi c^2 t_P^2 Z_s)$		
	Value: 6.62607004E-34	Units: J·s = rad·kg·m²/s	Dimensions: $\Theta M L^2 T^{-1}$
G	Newton's Gravitational constant $(= G_0/\Gamma^3 = c^3/Z_s = c^3/(m_P/t_P) = R_S c^2/2M)$		
	Value: 6.67428E-11 [2010 CODATA]	Units: N·m²/kg² $(= m^3/kg \cdot sec^2)$	Dimensions: $L^3 M^{-1} T^{-2}$

...all the stars directly behind me were now deep red, while those directly ahead were violet. Rubies lay behind me, amethysts ahead of me. – Olaf Stapledon

Wonder is the seed of knowledge. – Sir Francis Bacon

50. Gravitational frequency spectrum

For an observer at rest, the calculated peak gravitational frequency ω_{pk} at the Earth's surface is 2.95×10^{42} Hz (= 2.95×10^{18} YHz) which corresponds to a wavelength $\lambda = c/\omega_{pk} = 1.015 \times 10^{-34}$ m (= 1.015×10^{-25} nm = 1.015×10^{-28} microns). The above gravitational frequency estimate is much greater, for example, than the far infrared frequency of $\sim 0.5 \times 10^{13}$ Hz associated with the Earth's estimated blackbody radiation temperature of ~ 255 °K evaluated according to the Planck radiation formula. A plot of ω_{Pk} for the Earth's gravitational field is illustrated in Fig. 50-1. At x-ray frequencies (f > 10^{16} Hz), anomalous dispersion ceases and the refractive index for dielectric materials approaches 1. All natural dielectric materials become transparent (non-absorbing) and non-refracting at γ frequencies (f > 10^{19} Hz). For comparison, the electron Compton frequency ν_c is $\sim 1.2356 \times 10^{20}$ Hz.

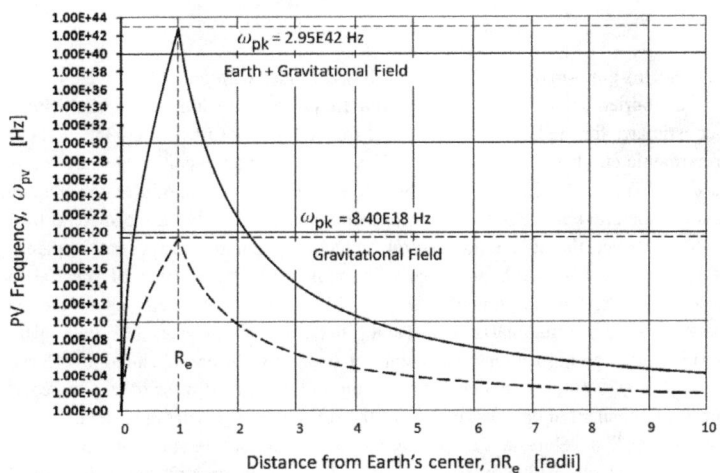

Fig. 50-1. Plot of calculated gravitational frequency ω_{PV} of Earth's gravitational field based on number of equivalent Planck mass oscillators. The peak frequency ω_{pk} lies in the far gamma-ray region of the electromagnetic spectrum beyond range of direct measurement.

50.2 Fourier spectral analysis of Earth's gravitational spectrum

For comparison, reference is made to calculation by Storti *et al*[34] of the estimated gravitational frequency of Earth based on Fourier analysis of spectral bandwidth using the Buckingham Π theorem (BPT). In BPT, an equation of n variables may be represented equivalently as an equation of n-m dimensionless parameters where n = no. of dimensionless Π groups and m = number of dimensions used. In this spectral analysis, the acceleration of gravity $a_{PV}(r,M,t)$ is modeled as a rectified square wave summation of odd-numbered wave-functions over a range of modes (n_{PV}) from the fundamental to an upper harmonic (n_Ω) encompassing frequencies of $\omega_{PV}(1,r,m)$ to $\omega_\Omega(r,m)$.

All mass radiates a spectrum of wavefunctions $\omega_{PV}(1,r,M) \leq \omega \leq \omega_\Omega(r,M)$... At each frequency wavefunctions are propagated with positive and negative amplitudes. – Ricardo C. Storti

50. Gravitational Frequency Spectrum

Using an approximate rest mass-energy density U_m of a solid spherical mass, Storti *et al* derives the following

$$U_m(r,M) = 3Mc^2/4\pi r^3 \qquad (50\text{-}2)$$

Harmonic cut-off function $\Omega_m(r,M)$

$$\Omega_m(r,M) = (108(U_m/U_\omega) + 12(768 + 81(U_m/U_\omega)^2)^{1/2})^{1/3} \qquad (50\text{-}3)$$

Harmonic cut-off mode $n_\Omega(r,M)$

$$n_\Omega(r,M) = (\Omega/12) - (4/\Omega) + 1 \qquad (50\text{-}4)$$

where Ω = harmonic cut-off function of PV = $f(n_\Omega, \omega_\Omega)$

Harmonic cut-off frequency $\omega_\Omega(r,M)$

$$\omega_\Omega(r,M) = n_\Omega(r,M)\omega_{PV}(1,\rho,M) \qquad (50\text{-}5)$$

In an electro-gravi-magnetic (EGM) PV-based model with an EM frequency spectrum obeying a Fourier polarized ZPF (zero-point frequency) spectral distribution, Storti and Diemer estimates for the Earth a fundamental frequency $\omega_{PV}(1,r,M_E) = 0.035879$ Hz and an upper harmonic cutoff $\omega_\Omega(R_E,M_E) = 519.573 \times 10^{24}$ Hz for an overall frequency bandwidth of $\Delta\omega_{PV} \sim 520$ YHz. The frequency bandwidth represents the range of graviton frequencies. The bulk of the gravitational energy is concentrated above 10^{23} Hz at the spectral limit. In the EGM construct, the maximum spectral frequency, fundamental spectral frequency and spectral frequency bandwidth increases with mass while the number of spectral modes decreases. The estimated gravitational field PV spectral characteristics of the Earth's gravitational field are summarized in Table 50-1. For frequencies less than $\sim 10^{19}$ Hz, dielectrics exhibit an absorption coefficient > 1 suggestive of partial shielding effects. The cutoff frequency $\omega_\Omega(R_E,M_E)$ is derived assuming the equivalent mass energy stored within an object is equivalent to the mass energy of the surrounding gravitational field.

Storti et al[34] develops a Fourier representation of acceleration of gravity **g** as a summation of electromagnetic frequencies. The lower spectral limit is termed the fundamental frequency $\omega_{PV}(1,r.M)$ while the upper spectral limit is termed the harmonic cutoff frequency ω_Ω. Illustrated in Fig. 50-2 is a square wave representation of the first 5 harmonics of the polarized vacuum PV spectrum. As an approximation as a square wave pulse input signal, the Fourier integral corresponds to the output to an ideal low-pass filter, cutting off all frequencies above ω_Ω. The Fourier summation with a finite number of harmonics approximates a square wave with transient overshoots at the corners with some ringing in the time domain and ripple in the frequency domain. Due to symmetry, the constant value of **g** in the time domain at a specific location may be represented as half the period of a rectified square wave composed of odd number of Fourier harmonics. In the case of a constant function, the even modes sum to zero, hence, only odd mode components need be considered. The bulk of the energy lies in the higher frequency components.

When you cannot express it in numbers, your knowledge is of a meager and unsatisfactory kind. – Lord Kelvin (William Thomson)

The ability to perceive or think differently is more important than the knowledge gained. – David Bohm

50. Gravitational frequency spectrum

Given the fundamental frequency $\omega_{PV}(1,r,M)$, the acceleration of gravity $g \cong 2c\cdot\omega_{PV}^3 r^2/c$. For a surface acceleration at the Earth's surface of g = 9.81 m/s², an Earth radius of ~6.37718E6 m and a delta frequency $\Delta v = -GM/2c\cdot r^2$, the estimated fundamental frequency $\omega_{PV}(1,r,M) \cong \sqrt[3]{(\Delta v c^2/r^2)} = 0.033067$ Hz with a period $T_{PV}(1,r,M) = 30.2412$ sec and wavelength $\lambda_{PV}(1,r,M) = 9.066E8$ m. For the Earth with a Schwarzschild radius r_S = 8.87E-3 m, the corresponding minimum ZPF mode wavelength is 0.223 m.

Fourier representation of gravitational frequency spectrum

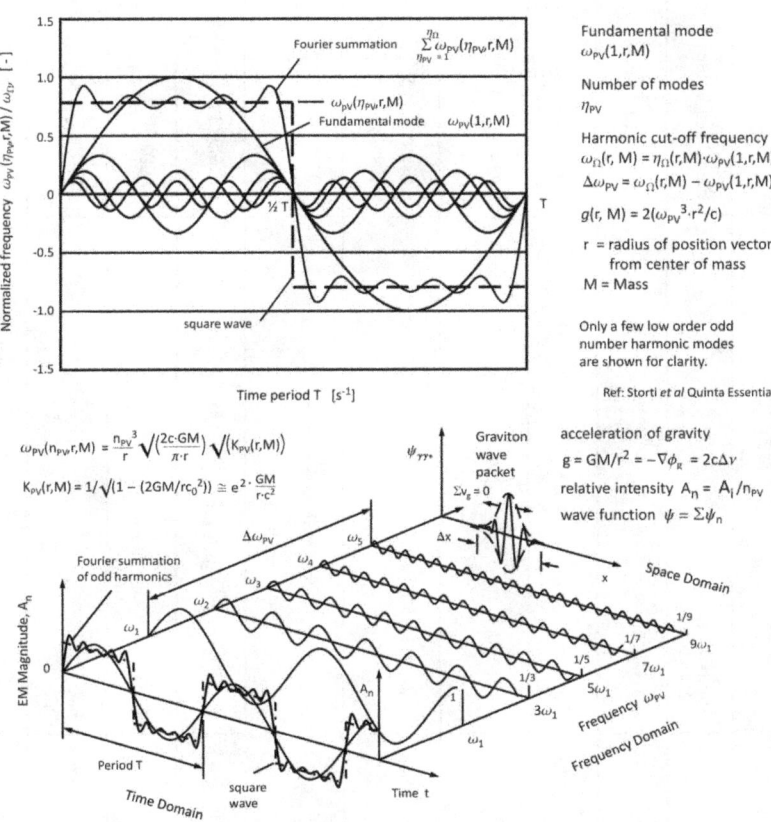

Fig. 50-2. Generalized Fourier harmonic representation of gravitational frequency spectrum ω_{PV} of a polarized vacuum is shown. The PV frequency spectrum is described as a function $\omega_{PV}(n_{PV},r,M)$ where n_{PV} is the number of harmonic frequency modes, r is the radius from the mass center, and M is the gravitational mass consisting of a collection oscillators. Only the odd modes contribute to the gravitational acceleration g. The dashed line represents a constant function generated as a non-zero summation of sinusoids (real terms) corresponding to a constant acceleration of gravity. [Adapted from Storti *et al*].

453

50. Gravitational Frequency Spectrum

Table 50-1. Fourier spectral analysis of Earth's gravitational spectrum [Ref: QE3, Storti *et al*].

Parameter	Symbol/Relation	Calculated Value	Units
Mass Earth	M_E	5.977×10^{24}	kg
Mean radius Earth	r_E	6.37718×10^6	m
mass density uniform density sphere	$\rho = 3M/4\pi r^3$	3.48409×10^{-5}	kg/m^3
Rest mass-energy density	$U_m(r,M) = 3Mc^2/4\pi r^3$	4.944814×10^{20} = 494.481475	kg/ms^2 (= Pa) EPa
Fundamental spectral frequency	$\omega_{PV}(1,r,m) = (1/r)[(2cGM/\pi r)^{1/3}]\sqrt{K_{PV}(r,M)}$	0.0358 $-\omega_\Omega < \omega_{PV} < \omega_\Omega$	Hz
Frequency bandwidth	$\Delta\omega_{PV}(1,r,m) = \omega_\Omega(r,M) - \omega_{PV}(1,r,M)$	≈ 519.573	YHz (=10^{24} Hz)
Fundamental period	$T_{PV}(1,r,M) = 1/\omega_{PV}(n_{PV},r,M)$	27.902	s
Harmonic frequency modes	n_{PV}	$-\infty < n_{PV} < \infty$ (free space)	None

cont

454

50. Gravitational frequency spectrum

Table 50-1. Fourier spectral analysis of Earth's gravitational spectrum [Ref: QE3, Storti et al] (Cont).

Parameter	Symbol/Relation	Calculated Value	Units
Harmonic cut-off function	$\Omega_m(r,M) = \sqrt[3]{(108(U_m/U_\omega) + 12\sqrt{(768 + 81(U_m/U_\omega)^2)})}$	1.73968910^{29}	None
Harmonic cut-off mode	$n_\Omega(r,M) = (\Omega/12) - (4/\Omega) + 1 = \omega_\Omega/\omega_{PV}$	1.449741×10^{28}	None
ZPF beat cut-off frequency	$\omega_\Omega ZPF(r_E, \Delta r, M_E)$	371	$PHz (= 10^{15} Hz)$
Harmonic cut-off frequency	$\omega_\Omega(r_E, M_E) = n_\Omega(r,M)\cdot\omega_{PV}(1, \rho, M)$	519.573	$YHz (= 10^{24} Hz)$
Harmonic cut-off wavelength	$\lambda_\Omega(r_E, M_E)$	5.776E-19	m
Harmonic frequencies	$\omega_{PV}(r,m) \cdot n_{PV}(r,m) = (n_{PV}^3/r)\sqrt{(2c\cdot GM/\pi r)}\sqrt{(K_{PV}(r,M)}$	varies	Hz, $-\infty < \omega_{PV} < \infty$
PV wavelength	$\lambda_{PV}(1, R_E, M_E)$	8.36497210^6	km
Euler-Masceroni const.	$\gamma = 1/1 + 1/2 + 1/3 + ... 1/n - \ln(n + \alpha)$	$\approx 0.57721566... \sim 1/\sqrt{3}$	--

cont

50. Gravitational Frequency Spectrum

Table 50-1. Fourier spectral analysis of Earth's gravitational spectrum [Ref: QE3, Storti *et al*] (Cont).

Parameter	Symbol/Relation	Calculated Value	Units
PV field energy density	$U_\omega(r,M) = (h/2c^3)\cdot\omega_{PV}(1,r,M)^4$ $U_\omega(n_{PV},r,M) = U_\omega(r,M)\cdot[(\lvert n_{PV}\rvert+2)^4 - n_{PV}^4]$	1.4727×10^{-14}	$kg/ms^2 \ (= Pa)$
PV field quantum pressure	$P_\omega = -U_\omega/3$	0.4909×10^{-14}	N/m^2
Gravitational Poynting vector magnitude	$S_m(e,M) = c\cdot U_m(r,M) = 3Mc^3/4\pi r^3$	1.48242×10^{29}	$kg/s^3 \ (= W/m^2)$
PV refractive index	$K_{PV}(r,M) = e^{\wedge}2(GM/rc^2)$	≥ 1	None
Acceleration of gravity	$g(r,M) = (GM/r^2)\sum(-2i/\pi n_{PV})e^{\wedge}(\pi n_{PV}\omega_{PV}(1,r,M)t)i$ $= GM/rc^2 = -\phi_g/r = -2cr\nabla\nu = -2c\Delta\nu = -\nabla\phi_g$	9.81 (@ surface)	m/s^2

Historically, variations in energy density are known to result in gravitation from the solutions of Poisson's equation in Newtonian gravity. – R.C. Storti

Since the Energy Density of the gravitational field is less, farther away from the centre of mass than closer to it, the change in Energy Density always acts toward the centre of mass of the object ... the change in Energy Density is always negative. – R.C. Storti

The presence of a planetary mass superimposed on the ZPF [Zero Point Field] alters the free space mode spectrum to "$-n_\Omega(r,M) \leq +n_\Omega(r,M)$". – R.C. Storti

The spacetime metric may be engineered utilizing Electro-Gravi-Magnetics (EGM), where EM fields may be applied to affect the state of the PV and thereby facilitate interactions with the local gravitational environment. – R.C. Storti

51. Coupled Oscillators

I am always keen to head where the greatest gravitation pull is tugging me. – Grant-Lee Phillips

Machines take me by surprise with great frequency. – Alan Turing

51.1 Oscillator synchronization

Mutual synchronization of coupled oscillators was described in 1657 by Huygens in development of synchronized pendulum clocks. An oscillator constitutes a clock which is ultimately formed from standing matter waves. Coupled oscillators tend to synchronize or 'mode lock' when the ratio of observed frequencies $\Delta f_1/\Delta f_2$ occur in simple integer whole number p/q ratios (e.g., 1:1 (unison), 1:2 (octave), 2:3 (perfect fifth), 3:4 (perfect fourth), etc) as represented by a tetrachord base when close to the ratio of the oscillator's intrinsic frequencies $\Omega = \omega_1/\omega_2$. In a plot of observed frequencies vs. forcing frequency, the set of mode-locked states with a fixed non-linear coupling forms a monotonic, increasing fractal 'Devil's staircase' with the width of the phase-locking plateau region $\Delta\Omega$ corresponding to a phase-locking frequency p/q decreases as the frequencies p and q increase. Between phase-locked stability plateaus, the slave oscillator unlocks from the master oscillator and drifts resynchronizing at the next integer ratio Ω/ω = p/q.

For two weakly coupled self-sustained oscillators, the synchronization between oscillators is approximated by the Adler equation[120]:

$$d\theta/dt = \Delta\omega - \epsilon\sin(\theta) \tag{51-1}$$

where

- $d\theta/dt$ = time rate of change in phase difference
- $\Delta\omega$ = $\omega_1 - \omega_2$ = frequency difference (detuning)
- ϵ = oscillator coupling constant (forcing strength)
- θ = phase difference between oscillators

Attraction is the result of in-phase synchronization whereas repulsion is the result of out-of-phase synchronization. As the coupling strength increases, the phase-locked region converges to a common frequency ω_0. The magnitude of the frequency difference between the two eigenmodes is equal to the rate of energy exchange between the two states. Mutual synchronization of weakly coupled nonlinear oscillators is described by the Kuramoto model of a system of N limit cycle oscillators.

$$\dot{\phi}_k = \omega_k + \frac{\epsilon}{N}\sum_{j=1}^{N} Sin(\phi_j - \phi_k), \qquad k = 1, 2, ...N \tag{51-2}$$

where

- $\dot{\phi}_k$ = $d\phi_k/dt$ = rate of change in phase
- θ = phase difference between oscillators = $\phi = (\theta_0 - \theta)$
- ω_k = natural oscillator frequency of kth oscillator
- ϵ = oscillator coupling strength (gain)
- ϕ = oscillator frequency

To predict the future of a curve is to carry out a certain operation on its past. – Norbert Wiener

51. Coupled Oscillators

At a critical frequency threshold, a collective synchronicity occurs resulting in entrainment of more of the N population ensemble of n-coupled oscillators. Two oscillators are synchronized if phase locked to the same mean frequency Ω. Once synchronized (n = N), the frequency differential is reduced to zero. This may be seen by depicting the center point of the phases of the oscillators plotted on a unit circle in the complex plane. Refer to Fig. 51-1.

The vector z is the average on n point vectors referenced to the center point is defined as the complex order parameter

$$re^{i\theta} = \frac{1}{N}\sum_{j=1}^{N} e^{i\theta_j} \qquad (51\text{-}3)$$

where

ϕ = argument of order parameter
$e^{i\theta}$ = $\cos\theta + i\sin\theta$
r = $|z|$ = modulus of the order parameter (phase coherence)
z = $z(r, \phi) = re^{i\theta}$ = mean field circle (or marker) radius
θ = average phase (argument of z)

Oscillator synchronization order parameter

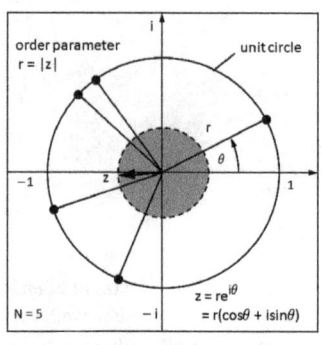
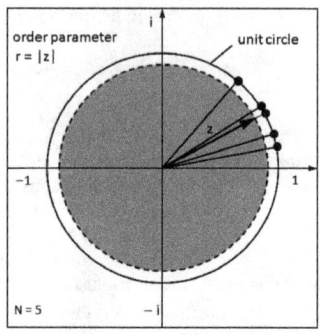

unsynchronized oscillators (r < 1)   nearly synchronized oscillators (r $\tilde{<}$ 1)
$K < K_c$ incoherent state $K > K_c$ partially synchronized state

Fig. 51-1. Order parameter plot illustrating oscillator phase synchronization. Below a critical threshold K_c, the magnitude of the order parameter (r), which is a measure of phase coherence, is small compared to unit circle radius and oscillator phases represented as points are randomly distributed on the unit circle. Above a critical coupling strength K_c, the number of entrained phase-locked oscillators (n) increases and the degree of synchronization as measured by order parameter approaches unity. Once fully synchronized (i.e., n = N and r = 1), all the oscillators are phase-locked to the same mean frequency.

Riemann hypothesis states the primes have music in them... a mathematical drum whose natural frequencies line up with the zeros of the zeta function... The answer lies in quantum mechanics... a more varied source of vibrating systems... – Barry Cipra

51. Coupled Oscillators

Uncoupled oscillators represent an isolation state. Fully coupled oscillators represents a synchronized state. The difference between an isolation and synchronization state represents the mean-field directivity. Individual oscillators in a system of N-oscillators interact with the average (mean-field) of the other oscillators partially exchanging resonances leading to a degree of synchronized resonance. A bi-stable system flips from one state to another as a consequence of the previous value providing a measure of passage of time. Multiple synchronized states may occur with synchronization occurring at rational, integer frequency ratios. For identical, strongly-coupled oscillators, astable synchronization state is realized with in-phase synchronization with out-of phase synchronization occurring for weakly-coupled oscillators with negligible frequency differences. For a finite-N oscillators, the probability of phase-locking is zero in the Kuramoto model if below a critical coupling K_c and equal one if above K_c. Refer to Figs. 51-2 and 51-3.

Phase lock synchronization

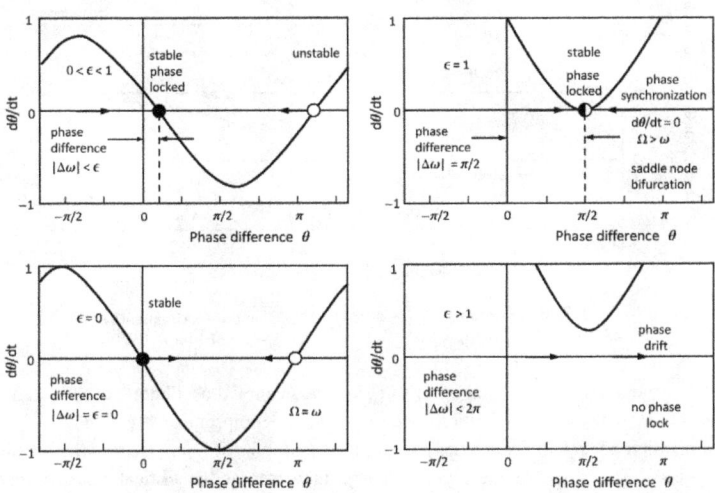

Fig. 51-2. Phase response curve $d\theta/dt$ vs. θ for a nonuniform oscillator characterized by $d\theta/dt = \omega - \epsilon\sin\theta$.

Phase-locking of interacting oscillators associated with gravitation may be expected to exhibit a Tracy-Wisdom statistical cross-over distribution of phase transitions of collective behavior of N number of correlated variables. Energy curves of strongly correlated systems exhibit a $\sqrt{(2N)}$ peak associated with a third-order phase transition separating a weak coupling phase and a strong coupling phase. Coupled mass oscillators may represent a form of the Tracy-Wisdom distribution describing the phase transition from an uncoupled state to a strongly coupled state as the oscillators synchronize on approach.

If we are going to stick to this damned quantum jumping, then I regret that I ever had anything to do with quantum theory. – Erwin Schrödinger

Mathematicians call it "the arithmetic of congruences". You can think of it as clock arithmetic. – John Derbyshire

51. Coupled Oscillators

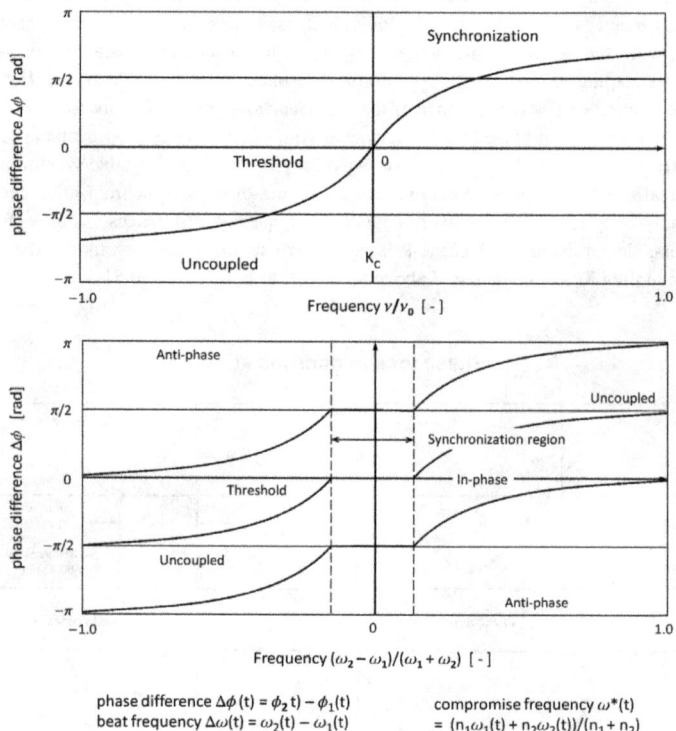

Fig. 51-3. In mutual synchronization of self-sustained coupled oscillators, each system acts on the other with synchronization occurring at a compromise frequency. Mutual synchronization occurs at frequencies that are near multiples of $\omega_1/\omega_2 = n/m$ where n and m are integers. The slope of the sigmoid curve increases at the critical phase transition occurring at the critical coupling strength K_c and depends on the number of nodes that become entrained (cross-connected).

With gravity, there is negligible damping and at large separations, the oscillators are free-running at each of their respective natural oscillator frequencies. For example, consider two coupled masses m and M represented by

$$d\theta_1/dt = \nu_1 + K_1\sin(\theta_2 - \theta_1)$$
$$d\theta_2/dt = \nu_2 + K_2\sin(\theta_1 - \theta_2) \tag{51-4}$$

where
- θ_1, θ_2 = phase of each oscillator
- $d\theta/dt$ = time rate of change in phase difference
- ν_1, ν_2 = natural frequency of each oscillator (> 0)
- K_1, K_2 = coupling constant (≥ 0)

We presuppose two things: That there is yet to be learned infinitely more than is now known, and that men can learn it. – John W. Campbell

51. Coupled Oscillators

When phase locked, the oscillators are separated by a constant phase difference and become synchronized at a phase-lock or compromised frequency

$$\nu^* = (K_1\nu_2 + K_2\nu_1)/(K_1 + K_2) \qquad (51\text{-}5)$$

which lies between the two natural oscillator frequencies. See Fig. 51-4. Phase locking ($\phi_1 = \phi_2$) with matched zero-crossings guarantees frequency locking ($\omega_1 = \omega_2$), however, two frequency-locked signals with uncontrolled relative phase ($\Delta\phi \neq$ const) varying from in-phase to 180° out-of-phase may occasionally unlock and drift over time.

Phase–lock frequency of coupled oscillators

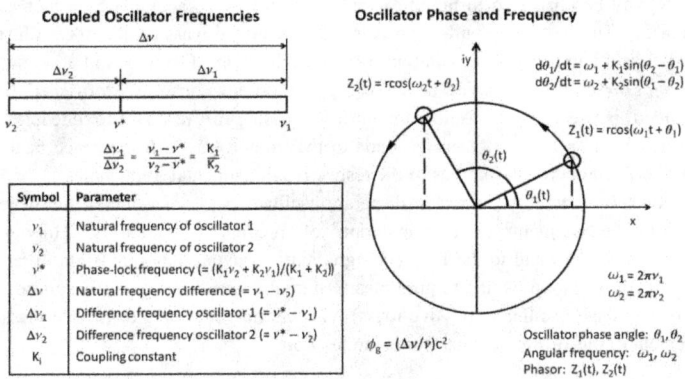

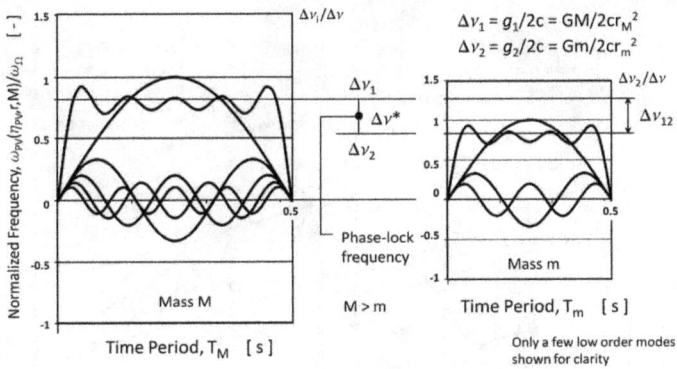

Fig. 51-4. Phase-lock frequency of a system of two coupled mass oscillators M and m.

To ask the right question is harder than to answer it. – Georg Cantor

If you formulate the problem correctly, that is, if you ask the right question, the answer emerges spontaneously. – Sin-itiro Tomonaga

51. Coupled Oscillators

Due to nonlinear vacuum polarization effects, the interaction between coupled oscillators exhibits anharmonic modulation. As a result, the net effective gravitational frequency ω_2 of the passive mass is pulled up towards the phase-locking frequency ω^* while the net effective gravitational frequency ω_1 of the active mass is pulled down towards the phase-locking frequency ω^*. In the unlocked state, the frequency modulation is anharmonic. The difference in half period duration between positive and negative modulation results in a regenerative effect tending to pull in the respective frequencies towards a compromise frequency. The period of time for near synchronous phase lock corresponds to the time required for the masses to approach and physically contact. The $\Delta\omega$ frequency differential is thus proportional to the net acceleration of gravity between the objects. Gravitational acceleration $g = 2c \cdot \Delta v \cdot r_u = -2v \cdot \phi_g/c$ where c = velocity of light, ϕ_g = gravitational potential and r_u = unit vector along a line in the direction of the frequency gradient. The waveform asymmetry may be attributed to nonlinear polarization of the vacuum which increases as $\Delta\omega$ decreases. The frequency pull-in process of two coupled mass oscillators is illustrated in Fig. 51-5 as a harmonic representation of the acceleration of gravity and a single wave function ω_Ω Fourier approximation based on the method developed by Storti *et al*. Energy is exchanged between mass oscillators during crossing of resonant frequencies. The resultant acceleration of gravity corresponds to the time-averaged interference beat wave function of odd numbered harmonics of the respective fundamental frequencies.

The frequency differential between disparate oscillators is a measure of the degree of mode-locking entrainment of a population of n-coupled oscillators. Gravitational acceleration is proportional to the frequency gradient. Gravitational repulsion, in terms of oscillator synchronization, would require means of continuous disruption of in-phase mode-locking of the slave oscillator (passive mass) with the master oscillator (active mass) and reinforcement of out-of-phase anharmonic modulation.

Interaction of two coupled mass oscillators M and m

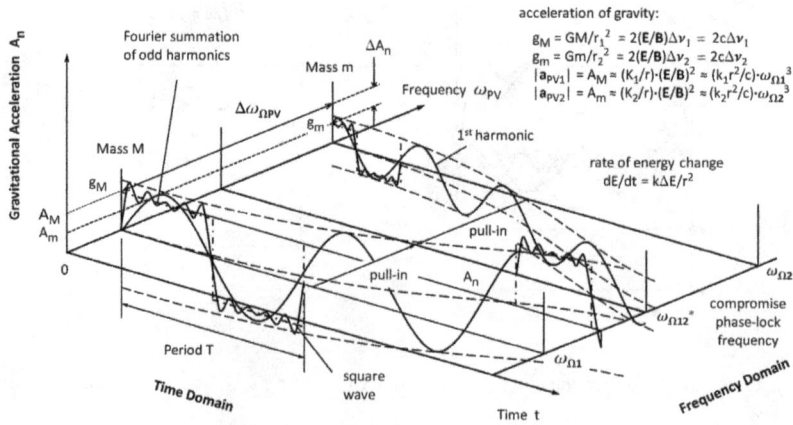

Fig. 51-5. Oscillator frequency synchronization of a coupled mass pair. Coupled mass oscillators attempt reset to an intermediate compromise phase-lock frequency on approach increasing degree of synchronization. Full synchronization occurs on merger intergration.

51. Coupled Oscillators

51.2 Frequency arrhythmia

Gravitational acceleration, as Ivanov[36,37] has shown is attributed to frequency discordance or arrhythmia between interacting oscillators resulting in asymmetry of the standing wave interference pattern. This result forms a part of Rhythmodynamic theory developed by Ivanov. The greater the frequency difference induced by acceleration or greater the phase shift induced by velocity, the greater the deformation of the interference field. The strength of coupling increases as the frequency difference decreases. A standing wave system in motion undergoes a compression of standing waves as well as an internal phase shift that varies with velocity. Motion of a standing wave system is associated with a flow of internal wave energy that opposes acceleration reducing the frequency discordance and energy flow. In a gravitational field, the natural frequency of mass oscillators is slightly reduced such as demonstrated by the Mössbauer effect. Ivanov notes there are three states of equilibrium of a standing wave system: 1) object at rest with respect to a second object, (2) object in uniform motion at constant velocity with respect to a second object, and 3) an object in free fall at constant acceleration towards a second object. A free-falling mass undergoes self-acceleration towards the higher density of potential caused by the frequency gradient between the falling mass m and the central mass M that reduces internal deformation. Motion represents a continuous symmetry transformation through translation of wave function to minimize the frequency difference. The standing wave interference pattern for coherent oscillators of equal frequency shows a symmetrical Moiré pattern. A standing wave system that is restrained from free movement produces an internal asymmetric interference pattern. For unequal frequencies, the standing wave Moiré pattern is asymmetric due to phase shift of nodes and anti-nodes. See Figs. 51-6 and -7.

51.3 Constant velocity (inertial frame)

Velocity of center of mass of a system of standing waves and phase displacement are related according to a relation derived by Ivanov

$$v = s/t = (c/\pi) \cdot \Delta\phi = c \cdot (v_1 - v_2)/(v_1 + v_2) \qquad (51\text{-}6)$$

where:

v	speed of wave propagation in a medium ($= $ distance/time $= s/t = v_0/T$)		
	Value: Varies	Units: m/sec	Dimensions: LT^{-2}
$\Delta\phi$	Phase shift between oscillators		
	Value: Varies	Units: radians	Dimensions: Θ
π	Circular measure Pi (2π radians $= 360$ degrees)		
	Value: 3.14159...	Units: radians	Dimensions: Θ
c	velocity of light ($= c_0/T = c_0/\sqrt{K_{PV}}$ (weak field) $= c_0/\sqrt[4]{K_{PV}}$ (strong field))		
	Value: 2.997×10^8	Units: m/sec	Dimensions: LT^{-1}

Enthusiasm is followed by disappointment and even depression, and then renewed by enthusiasm.
– Murray Gell-Mann

The best thing a human being can do is to help another human being know more. *– Charles Munger*

51. Coupled Oscillators

Wave interference of oscillator pair under uniform motion

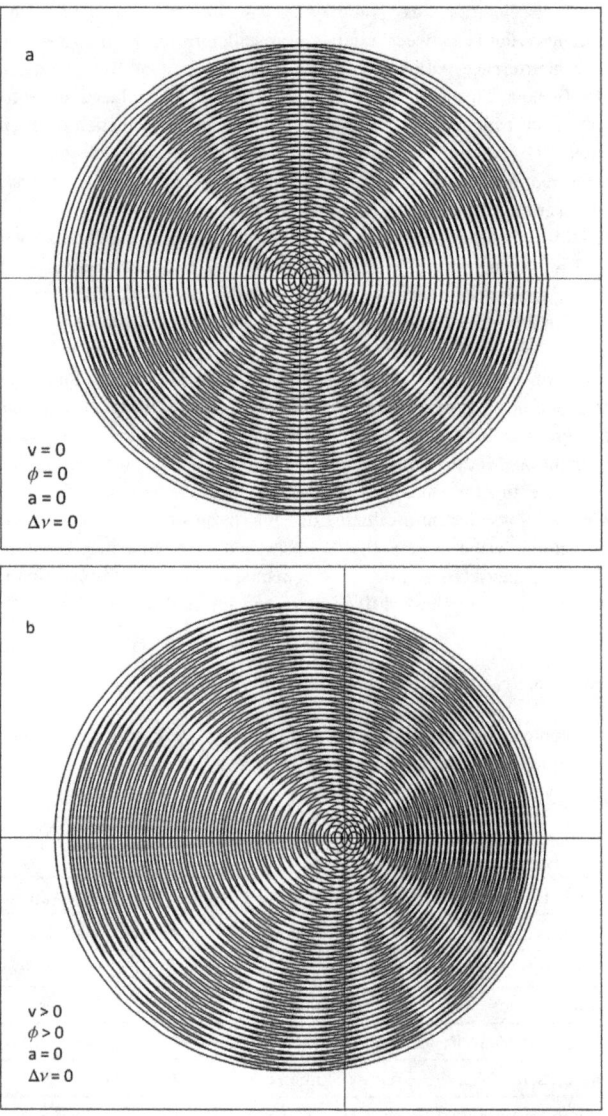

Fig. 51-6. A state of rest or uniform motion results in a symmetrical interference field pattern shown in a) as interaction of two proximate oscillators of equal frequency. In b), the interaction of two proximate oscillators of equal frequency is shown in which the oscillator at left is approaching at constant velocity a stationary oscillator on the right. The Doppler shift generates an asymmetry of the field interference pattern.

I enjoy keeping up with the front wave of technology. – Syd Mead

51. Coupled Oscillators

Nodal line symmetry breaking

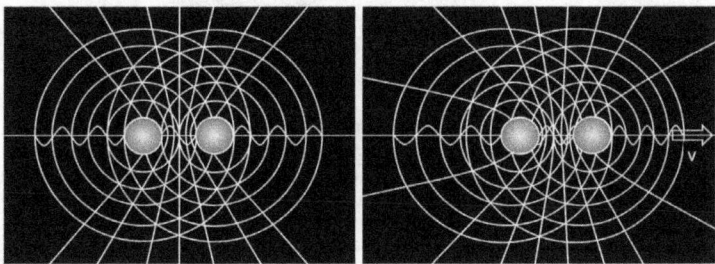

Fig. 51-7. Nodal line asymmetry for a pair of coupled oscillators at rest and in translation at positive velocity (v << c). As shown, the nodal lines are displaced to the left opposite to the direction of motion. Wave phase displacements shift potential minima. Nodal line curvature may be determined from vertex deficit angles of skeletonized simplex nodal line surfaces using Regge calculus. Forces arise with changes in scale (symmetry breaking).

The phase shift $\Delta\phi$ of a system of oscillators results in motion at constant velocity with constant phase shift. A free-floating wave system of phase-shifted standing waves will tend to move from the induced dissonant state to a consonant state and come to rest with respect to the anti-node potential holes thereby minimizing the phase difference. This effect is illustrated in Figs. 51-8 and 51-9. Accumulation of phase displacement with distance l makes the wave system self-accelerate with a propagation vector k along the path in the direction dl. The rate at which phase accumulates with time equals the difference in frequency $\Delta\nu$ of the contracted moving standing wave between the ends of the resonator. Displacement of phase triggers shifting of standing wave nodes. Total phase accumulation over a distance l is

$$\phi = \int(\nu_1 - \nu_2)\,dt = \int k \cdot dl \tag{51-7}$$

Mass transport is associated with node displacement of contracted moving standing waves. A tabulation of system velocity as a function of time phase difference between constituent oscillators is shown in Table 51-1 for a phase difference of 0 to π radians.

Nature is always speaking to us. – Morihei Ueshiba

There is no excellent beauty that hath not some strangeness in the proportion. – Francis Bacon

Regardless what form the theory of electromagnetic processes will take, the Doppler principle will remain. – Albert Einstein

Scientific progress is measured in units of courage, not intelligence. – Paul Dirac

Always take the high road, it's far less crowded. – Charles Munger

The career of a young theoretical physicist consists of treating the harmonic oscillator in ever-increasing levels of abstraction. – Sidney Coleman

Heisenberg has discussed the coupled double harmonic oscillator, and has shown that the ordinary rules of quantization lead to two non-combining sets of states in which one of the electrons are in phase and out of phase. The energy of the system is successively transferred from one to the other. – resonance! – Linus Pauling

51. Coupled Oscillators

Wave system resonator motion at constant velocity

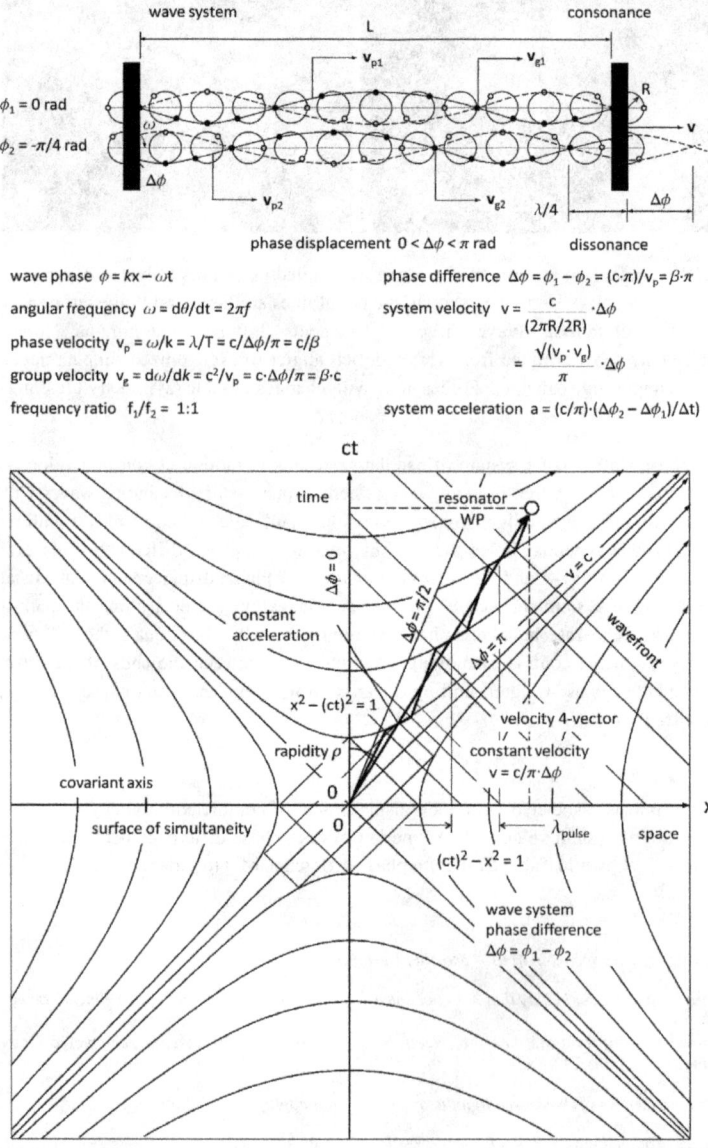

Fig. 51-8. Wave system resonator at constant velocity $v = c \cdot \Delta\phi/\pi$ corresponds to constant phase difference. Phase displacement $\Delta\phi = \pi \cdot v_g/c = \pi \cdot \beta$. For in-phase oscillation: $\Delta\phi = 0$ radians, for out-of-phase oscillation: $\Delta\phi = \pi$ radians. Proper velocity $w = dx/d\tau = pv$.

The flow of energy through a system acts to organize that system. – Harold Morowitz

51. Coupled Oscillators

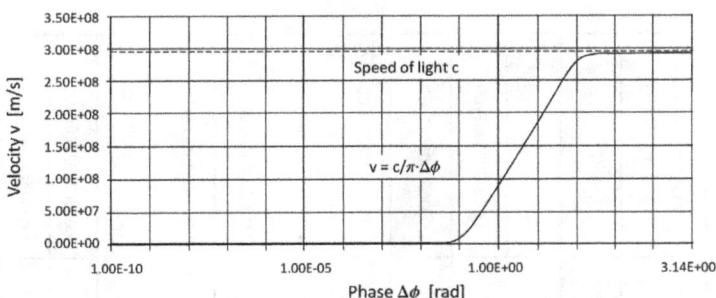

Fig. 51-9. Velocity of a standing wave system of multiple oscillators in motion in an inertial frame varies with phase difference $\Delta\phi$ independent of amplitude or frequency. Wave system group velocity $v_g = c^2/v_p = c\cdot\Delta\phi/\pi = \beta\cdot c$. Phase velocity $v_p = c/\Delta\phi\cdot\pi = \beta/c$.

The standing wave velocity v is a function of the phase difference $\Delta\phi$ and local velocity of light c and independent of the carrier frequency, pulse repetition frequency, pulse width, etc. The local velocity of light c (= $\sqrt{(v_p v_g)} = c_0/\Gamma = c_0/\sqrt{K_{PV}} = \nu\lambda = \mathbf{E}/\mathbf{B} = \phi_E/\phi_B$) varies with energy density. Ex., For an oscillator phase difference $\Delta\phi$ = 2.3578E-6 rad with c = c_0, the standing wave velocity v = 225 m/s (~500 miles/hr).

The effects of phase and frequency difference of counter-propagating travelling waves correspond to the velocity and acceleration of a moving standing wave resonator. Wave phase and frequency differences may be represented as Lissajous figures. See Fig. 51-10. A waveguide with counter-propagating travelling waves of wavelength λ in phase quadrature with a resonator length $n\lambda + \lambda/4$ corresponds to a dissonant wave system condition.

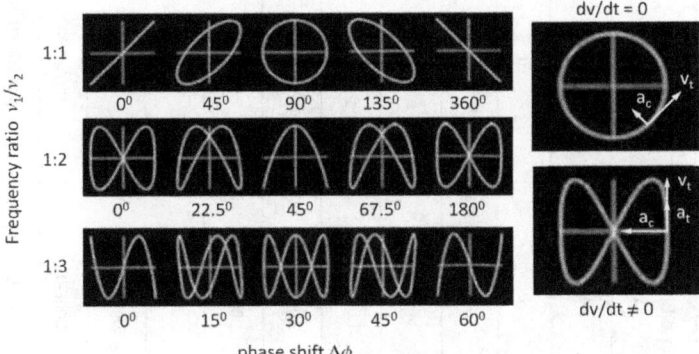

Fig. 51-10. Lissajous figures produced by intersection of two orthogonal sinusoids. Phase and frequency differences result in wave system velocity and acceleration, respectively.

Always with the negative waves Moriarty, always with the negative waves. – Oddball (Kelly's Heroes)

You can't stop the waves, but you can learn to surf. – Jon Kbat-Zinn

51. Coupled Oscillators

Table 51-1. Oscillator Time Phase Difference vs. Velocity

Phase Difference $\Delta\phi$	Velocity $v = (c/\pi)\cdot\Delta\phi$	Velocity ratio $\beta = v/c = \Delta\phi/\pi$	Lorentz factor $\gamma = 1/\sqrt{(1-\beta^2)}$	Wavelength Ratio $R = (1+\beta)/(1-\beta)$
[rad]	[m/s]	[-]	[-]	[-]
0	0	0	1	1
1.00E-10	0.00954	3.18E-11	1.0000000000000E+00	1.000000000006E+00
1.00E-09	0.09543	3.18E-10	1.0000000000000E+00	1.000000000064E+00
1.00E-08	0.95427	3.18E-09	1.0000000000000E+00	1.000000000637E+00
1.00E-07	9.54269	3.18E-08	1.0000000000000E+00	1.000000006366E+00
1.00E-06	95.4269	3.18E-07	1.0000000000000E+00	1.000000063662E+00
1.00E-05	954.27	3.18E-06	1.0000000000001E+00	1.000000636622E+00
1.00E-04	9,542.69	3.18E-05	1.0000000000051E+00	1.000006366400E+00
1.00E-03	95,426.90	3.18E-04	1.0000000005066E+00	1.000063682248E+00
1.00E-02	954,269.03	3.18E-03	1.0000000506610E+00	1.000638652667E+00

cont

51. Coupled Oscillators

Table 51-1. Oscillator Time Phase Difference vs. Velocity (Cont)

Phase Difference $\Delta\phi$	Velocity $v = (c/\pi)\cdot\Delta\phi$	Velocity ratio $\beta = v/c = \Delta\phi/\pi$	Lorentz factor $\gamma = 1/\sqrt{(1-\beta^2)}$	Wavelength Ratio $R = (1 + \beta)/(1 - \beta)$
[rad]	[m/s]	[-]	[-]	[-]
1.00E-01	9,542,690.32	3.18E-02	1.00050699122E+00	1.06575502468E+00
1.00E+00	95,426,903.18	3.18E-01	1.05486710061E+00	1.93388441385E+00
1.57E+00	149,896,229.00	0.50E+00	1.15470053837E+00	3.00000000000E+00
3.14E+00	299,792,458.00	1.00E+00	3.36890689053E+00	4.33750798356E+01

If the wave is getting bigger, it causes time to grow a little bit. If the wave is trying to contract, it reduces a little bit. So, you can see this oscillation in time on the clock. – Rainer Weiss

At first glance it might seem that the whole of the magnetic side of electromagnetism was absent from the gravitation analogy. But this was not true. – Oliver Heaviside

The science of mathematics treats its object as though it were something abstracted mentally, whereas it is not abstract is reality. – St. Thomas Acquinas

I know numbers are beautiful. If they are not beautiful, nothing is. – Paul Erdös

Progress isn't made by early risers. It's made by lazy men trying to find easier to do something. – Robert A. Heinlein

I much prefer the sharpest criticism of a single individual man to the thoughtless approval of the masses. –Johannes Kepler

I know what I don't know. – Aristotle

51. Coupled Oscillators

51.4 Constant acceleration (Rindler frame)

Acceleration of a system of oscillators results from a difference in oscillator frequency. Ivanov[36,37] has derived the following relation between acceleration and frequency difference for a system of oscillators

$$\mathbf{a} = d\mathbf{v}/dt = 2c \cdot \Delta v \cdot \mathbf{r}_u \qquad (51\text{-}8)$$

where:

a	acceleration of standing wave system (= $\mathbf{F}/m$ = $(dL/dx)/m$ = $d\mathbf{v}/dt$ = $(c^2/\phi_E)\mathbf{E}$ = $(\mathbf{F}/E) \cdot c^2$ = $(\mathbf{F}/E) \cdot (\mathbf{E}/\mathbf{B})^2$ = $\mathbf{v} \times \mathbf{B}_g$		
	Value: Varies	Units: m/sec^2	Dimensions: MLT^{-2}
Δv	Lorentz-Doppler beat frequency difference (= $|v_2 - v_1|$ = $a/2c$ = $v \cdot \phi_g/c^2$)		
	Value: Varies	Units: sec^{-1}	Dimensions: T^{-1}
c	velocity of light (= λv = $\mathbf{E}/\mathbf{B}$ = ω/k = c_0/Γ = c_0/γ = $c_0/\sqrt{K_{PV}}$ (weak field))		
	Value: 2.9979 x 10^8	Units: m/sec	Dimensions: LT^{-1}

The direction of the acceleration vector $\mathbf{a}$ is in the direction of the frequency gradient ∇v along the unit vector $\mathbf{r}_u$ between oscillator mass centers. An illustration of gravitational effect on frequency of a vertically oriented standing wave system is shown in Fig. 51-11. A frequency difference was detected on a vertically aligned oscillating gamma ray source in a gravitational field utilizing the Mössbauer effect in the Pound-Rebka experiment. Such an effect may be modelled in terms of a standing wave. For a height h, the gravitational potential energy $U = -(GMh/rc^2)v_0$. A photon emitted in a gravitational field undergoes a frequency reduction (redshift) $\Delta v = -(GM/rc^2)v_0$ where v_0 is the gravity-free frequency.

For an acceleration of 1g (= 9.8 m/s^2) at the Earth's surface, the frequency difference $\Delta v \approx 1.638 \times 10^{-8}$ Hz with corresponding $\Delta \lambda \approx 1.824E12$ m. The force of gravity (reaction force) is related to the frequency differential

$$\mathbf{F}_g = m\mathbf{g} = \mathbf{W} = 2mc \cdot \Delta v \cdot \mathbf{r}_u \qquad (51\text{-}9)$$

where

$\mathbf{F}_g$ = Force of gravity (= $-m\nabla \phi_g$ = $-GMm/r^2 \cdot \mathbf{r}_u$ = dL/dx)
m = mass (= Γm_0 = γm_0 = $\sqrt{K_{PV}}$ = $E/(\mathbf{E}/\mathbf{B})^2$ = $2L/(\mathbf{E}^2 - c^2\mathbf{B}^2)$ = $(\hbar \mathbf{k} - q_0\mathbf{A})/\mathbf{v}$)
g = acceleration of gravity (= $GM/r^2 \cdot \mathbf{r}_u$ = $2c\Delta v \cdot \mathbf{r}_u$ = $-\nabla \phi_g$ = $-d\mathbf{A}/dt$ = $\nabla \times \mathbf{A}_g$)
$\mathbf{r}_u$ = unit vector between oscillator mass centers (= $\mathbf{r}_{21}$ = $\hat{\mathbf{r}}$)

Equating Eqn (51-8) with the Newton's law Eqn (51-9): $\mathbf{F}_g = 2mc \cdot \Delta v \cdot \mathbf{r}_u = -(GmM/r^2)\mathbf{r}_u$ yields

$$\Delta v = -GM/2cr^2 \qquad (51\text{-}10)$$

Constant phase velocity corresponds to constant phase displacement. Changing phase velocity $\Delta\phi/\Delta t$ correlates with changing velocity $\Delta v/\Delta t$, i.e., acceleration. A plot of acceleration as a function of frequency difference of a pair of interacting oscillators is shown in Fig. 51-12. The delta frequency range Δv encompasses the phase-locking or compromise frequency v^*. As shown, acceleration is associated with asymmetry of wave interference patterns caused by a difference in frequency of coupled oscillators. Ivanov refers to the deformation of nodal lines connecting wave interference nodes, which resembles a spider configuration, a "spider-effect". The number of nodal lines between

51. Coupled Oscillators

oscillators is equal to twice the number of wave crests enclosed. The acceleration of gravity is attributed to the difference in frequency of oscillations of the interacting masses. The frequency arrhythmia or discordance decreases as the distance between oscillators decreases as the coupling strength increases. This effect is illustrated in Figs. 51-13 through -17.

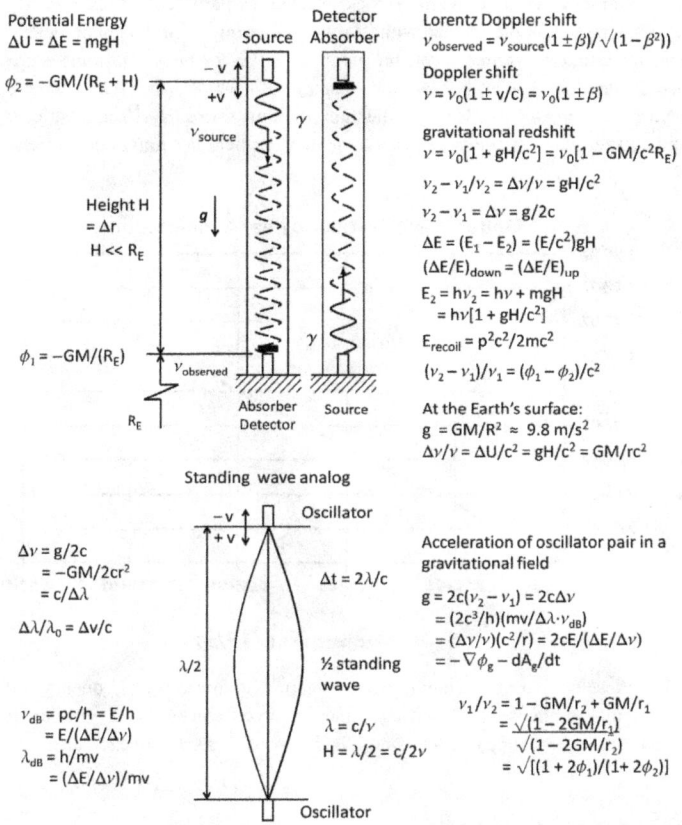

Fig. 51-11. Acceleration of a system of oscillators derived from gravitational redshift. In the Pound-Rebka experiment utilizing the Mössbauer effect, the gravitational shift of emitted gamma rays is offset by the relativistic Doppler shift of the source resulting in a frequency ratio $\Delta v/v = gH/c^2 = 2.46 \times 10^{-15}$. Substituting $H = \lambda/2$ yields $\Delta v = g/2c$, hence, local acceleration of gravity $g = 2c\Delta v = -2v \cdot \phi_g/c = -(\Delta v/v)c^2/r = -(2v/c) \cdot (GM/r)$.

> We shall therefore assume the complete physical equivalence of a gravitational field and a corresponding acceleration of the reference system. – Albert Einstein

> ...the concept of a matter wave actually existing in space is just as valid as a light wave existing in space. – Sin-itiro Tomonaga

471

51. Coupled Oscillators

The deformation pattern of nodal scattering centers represents storage of stress energy potential similar to that of two oscillating masses connected by a massless spring. Kinetic energy of motion results from conversion of this potential energy temporarily relieving the stress. Each concentric layer of anti-nodes constitutes a phase conjugate arcuate mirror. The volumetric nodal density is a measure of mass density. Creation of interference patterns (constructive or destructive) alters mass energy. With a localized wave packet in space, the momentum spread is wide corresponding to a particle with narrow interference patterns with the sum of the waves concentrated at the center of the wave packet and wave cancellation further away from the monadock. A charged particle acquires mass as a result self-energy interaction – interference with its own precessing electromagnetic field.

Ivanov notes that an atomic level, the minimal distances between atoms represents a fundamental standing wave. For atomic spacing of $\sim 1 - 10$ Å, the corresponding vibrational frequency is $\sim 10^{18}$ Hz. A collection of atoms forms, in effect, a lattice network of standing waves which will experience a frequency gradient in a gravitational field.

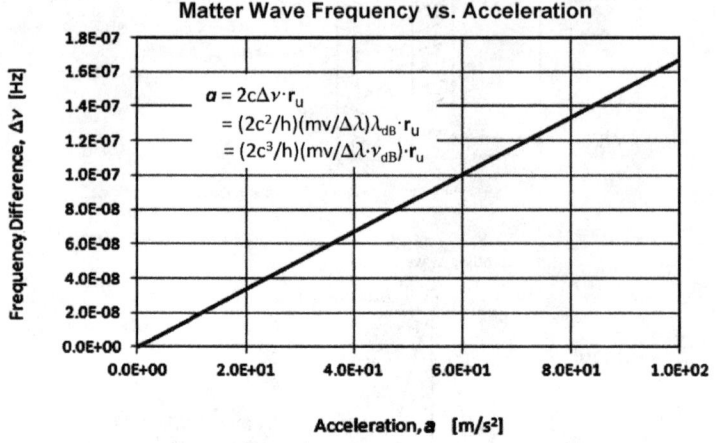

Fig. 51-12. Acceleration of a standing wave system as a function of frequency difference Δv of interacting oscillators (i.e., de Broglie matter wave frequency v_{dB}). Acceleration is in the direction of the frequency gradient and falls to zero with zero frequency difference.

Uniform acceleration of an observer in Minkowski spacetime results in thermal radiation with an Unruh temperature $T = \hbar a/(2\pi c k_B)$ where a = proper acceleration and k_B = Boltzmann's constant (= E/T = 1.381E-23 J/°K). Hence, acceleration, frequency delta and temperature are related as follows:

$$a = T \cdot (2\pi c k_B)/\hbar \qquad (51\text{-}11)$$

As an example, the Earth's surface gravitational acceleration of 9.8 m/s² is equivalent to a temperature $T \approx 4\text{E-}20$ °K, a frequency differential $\Delta v \approx 1.63\text{E-}8$ Hz and a time gradient $\beta = d\tau/dt$ of 1.09E17 s/s/m. Vacuum temperature is reduced in a negative energy density field.

Vacuum voco locum in quo corpora sine resistia movetur. – Sir Isaac Newton

51. Coupled Oscillators

Oscillator pair of equal frequencies

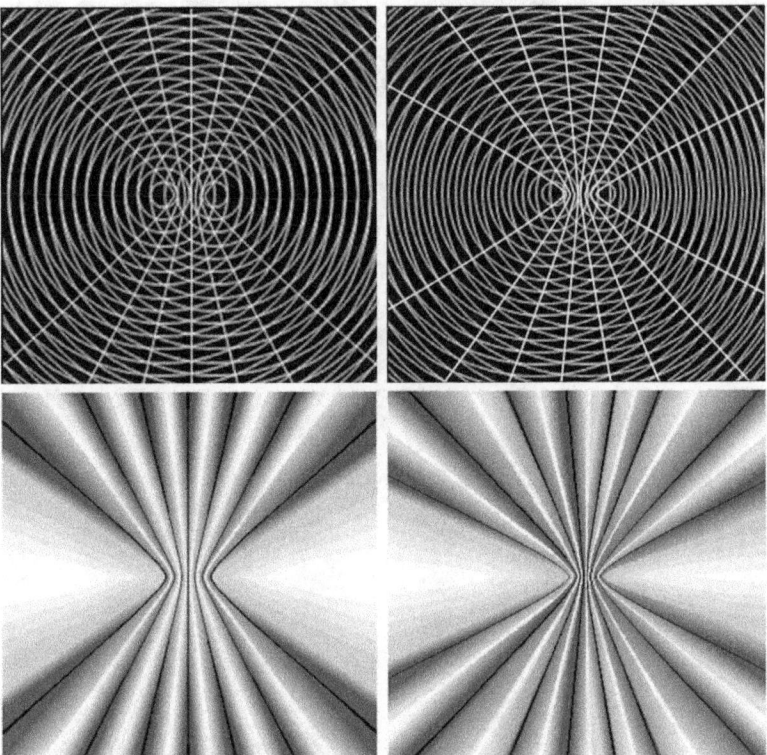

Fig. 51-13. Wave interference field pattern for a pair of proximate oscillators of equal frequency. On the left, the nodal (pinch fold) pattern is symmetrical for oscillators at rest and, on the right, the pattern is asymmetrical for oscillators in uniform motion (constant velocity). The nodal lines are displaced forward in the direction of motion. The wave system velocity is proportional to the phase displacement of potential minima relative to their initial position. The relative phase of pulsation with respect to each oscillator may be plotted as two orthogonal sinusoids as a Lissajous or Bowditch curve. Similarly, the relative phase displacement of each oscillator in uniform motion with respect to its rest state may be compared on Lissajous figures of constant phase difference. The interference pattern shown is identical to a double-slit interference pattern of a plane wave. In a sinusoidal gray scale representation shown, the nodes appear in white and anti-nodes in black. The graded interference pattern is equivalent to a holographic Gabor zone plate.

It's not what you look at that matters, it's what you see. – Henry David Thoreau

The vector equilibrium is the true zero reference of the energetic mathematics. – R. Buckminster Fuller

The wave-particle duality afflicting modern physics is best resolved in favor of waves... but there is no clear picture of matter on which physicists agree. – Erwin Schrödinger

51. Coupled Oscillators

Oscillator pair of unequal frequencies

Fig. 51-14. Effect of oscillator spacing on interference field pattern is shown for two oscillators with different natural frequencies. On the left, there is zero dipole spacing and, on the right, there is finite spacing corresponding to two wavelengths. In the zero spacing case, the two oscillators are in contact, but do not fully merge retaining individual identities and natural frequencies. Hence, a beat frequency is apparent and the oscillators do not become coherent. For finite spacing, the field asymmetry is a measure of frequency discordance. Acceleration is proportional to the difference in oscillator frequencies. The relative frequency of pulsation with respect to each oscillator may be plotted as two orthogonal sinusoids for a given frequency ratio as a Lissajous or Bowditch curve. Similarly, the relative Doppler frequency difference of each oscillator with respect to its rest state in constant accelerated motion may be compared on Lissajous figures of constant frequency difference. The direction of motion lies in the direction of the oscillator synchronization order parameter. The harmonic cut-off frequency ω_Ω is proportional to the number of enclosed volume nodal scattering centers.

Mathematics is the science of patterns. – Lynn Steen

The movement of the emitters of the spectral lines may be deduced on the basis of the Doppler principle. – Johannes Stark

51. Coupled Oscillators

Oscillator pair under acceleration

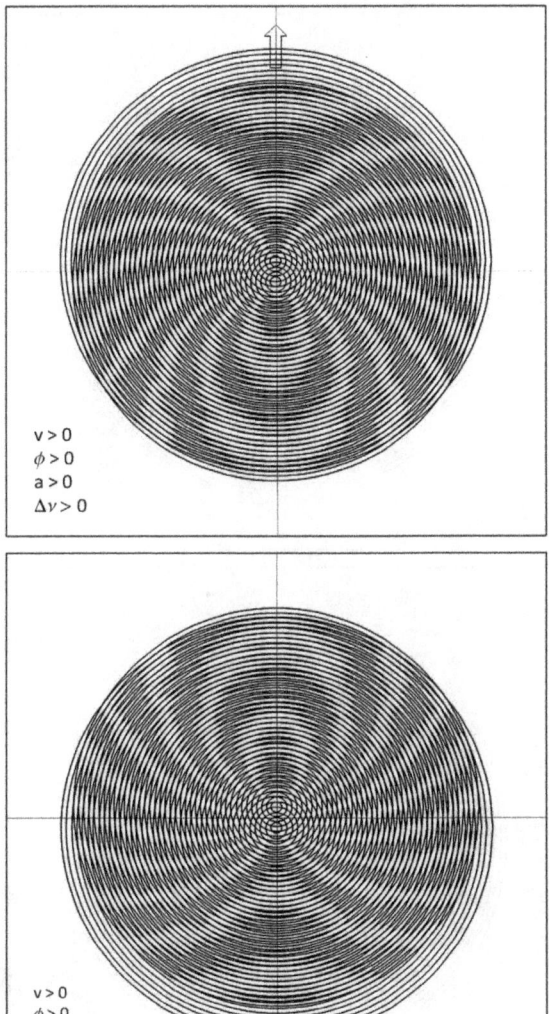

Fig 51-15. Acceleration of a high frequency oscillator is towards a low frequency oscillator. In the example illustrated, the frequency difference between oscillators is 3%. Shown in the top frame is upward acceleration and in the bottom frame downward acceleration. Energy equals force times displacement. Wave energy flows in the direction of the frequency gradient. Frequency mismatch is indicated by the asymmetry of the field interference pattern.

Whenever you can, count. – Francis Galton

475

51. Coupled Oscillators

Oscillator under zero acceleration

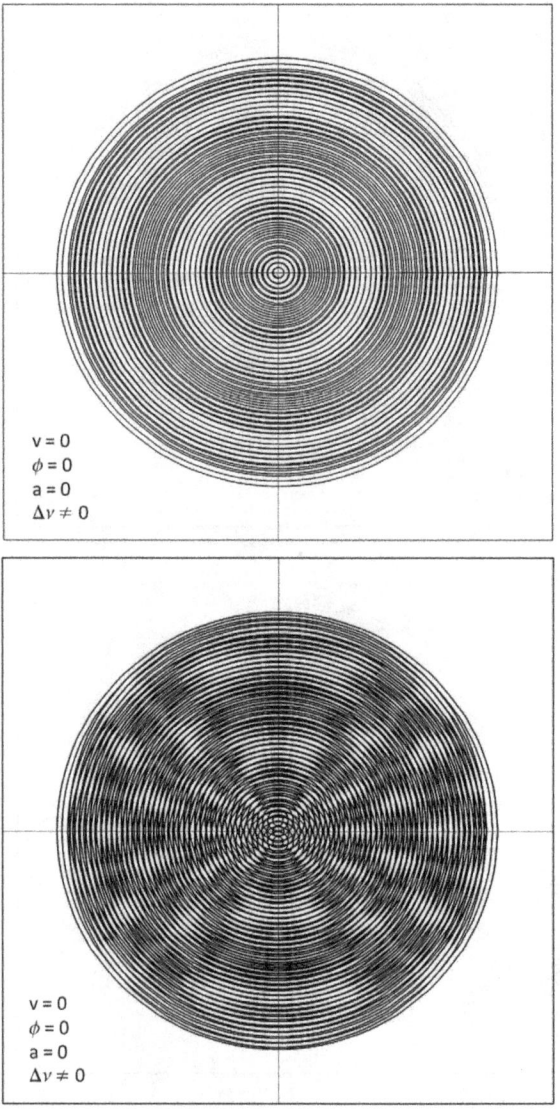

Fig. 51-16. Frequency discordance is reduced to zero when both oscillators are fully synchronized in a coherent state. Shown in the upper frame is a case of zero acceleration indicated by a symmetrical interference pattern. In the lower frame, a low frequency oscillator lies between two opposed high frequency oscillators resulting in a symmetrical field interference pattern with net zero acceleration of source oscillators.

Zero G and I feel fine. – John Glenn

51. Coupled Oscillators

Wave interference pattern of oscillator pair under acceleration

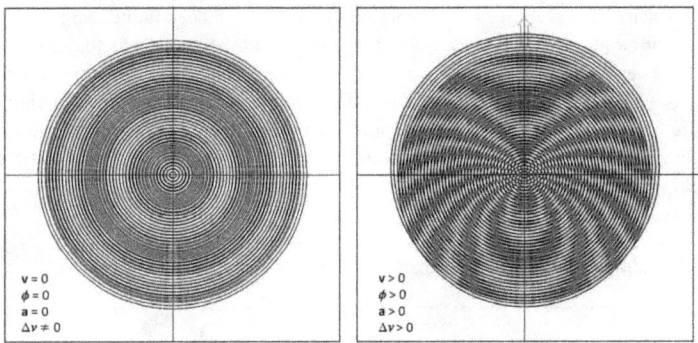

Oscillator Pair Standing Wave Interference Pattern

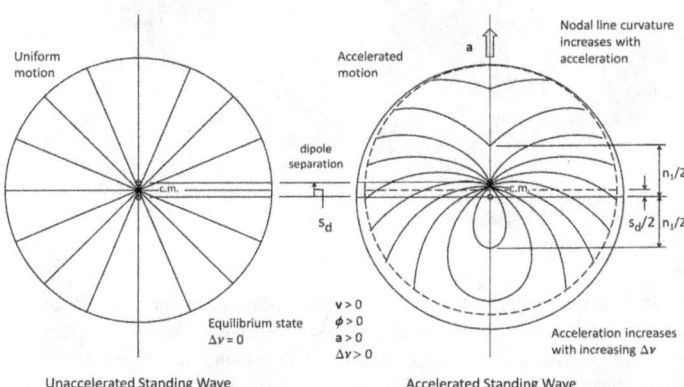

Fig. 51-17. Nodal line curvature increases with acceleration of source oscillators. Under acceleration, there is a shift in the center-of-mass in the direction of motion. For constant acceleration, the displacement $s = v_0 t + \frac{1}{2}at^2$. The number of nodal lines corresponds to the number of wave crests between the two oscillators. Potential minima corresponds to moving standing wave nodes. Interference field remains symmetrical in an unaccelerated (inertial) reference frame. The relativistic increase in mass ($m = \gamma m_0 = m_0/\sqrt{(1 - v^2/c^2)}$) is associated with breaking of symmetry of the wave interference field pattern.

One passive means to reduce interference field asymmetry is to isolate the discordant oscillators from one another with a phase conjugate reflector. Isolation of the two oscillators prevents frequency locking in which the oscillators pull each other to a common frequency. The wave interference field pattern for a dipole oscillator coupled with a phase conjugate reflector is illustrated in Fig. 51-18. As shown, the wave front propagating from each towards the adjacent oscillator is reflected back towards the source thereby preventing wave interference in the region spanned by the phase conjugate reflector. Isolation of the oscillators breaks the action-reaction symmetry. The reduction in field asymmetry impedes field energy flow and net acceleration. Phase conjugate reflection is limited to the range of EM source frequencies used. For high end gravitational frequencies in the YHZ range as in the Earth's gravitational field, the corresponding peak wavelength is on the order of 5.77E-

477

51. Coupled Oscillators

19 meter (= 5.77E-09 Å). These frequencies are well beyond manmade, controllable EM transmitters and present day metamaterial capabilities. Photonic metamaterials, for example, which rely on repeating nanoscale substructures have been limited up to the THZ range (far infrared) with a wavelength of ≈ 1.0E-04 m. Superconducting Josephson junctions are capable of 10^{10}-10^{11} Hz typically. For comparison, the carbon-carbon bond in graphene is ~1.42 Å with interplanar spacing of ~3.35 Å. The Compton wavelength of the electron is 2.426E-12 m (= 0.024 Å). Controlled production and detection of individual photons have been demonstrated recently in the low energy microwave region ~2 GHz with a microchip and light pulse durations as short as 65 femtosec.

Dipole oscillator separated by phase conjugate reflector

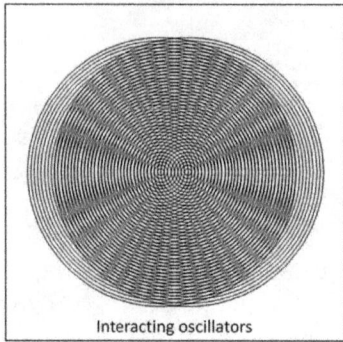
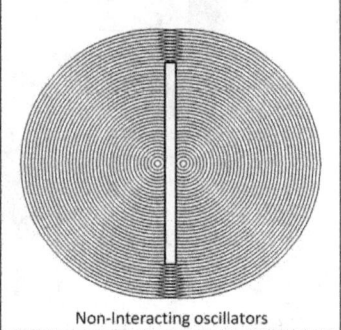

Interacting oscillators | Non-Interacting oscillators

Fig. 51-18. Frequency discordance is reduced by a phase conjugate reflector or holographic layer interposed between oscillators of disparate frequency ($v_1 \neq v_2$). In effect, elimination of anti-node interference scattering centers reduces the local graviton $\gamma\gamma^*$ density. Frequency response of the phase conjugate reflector must be compatible for oscillator frequency range for effective isolation.

An active means to reduce interference effects is to incorporate a local beat frequency oscillator (BFO) with a frequency twice the Δv delta frequency of the two oscillators. The bulk of energy of the Earth's gravitational field lies in the YHZ range well above current technology for direct EM frequency measurement, however, detection of the difference in frequency is well within existing technology limits. For example, a variable-frequency local oscillator may be adjusted maintain the difference in frequency based on an accelerometer output in single-mode heterodyne detection based on Eqn (51-7). This effect is illustrated in Fig. 51-19 for two oscillators of different natural frequency v_1 and v_2 and an interposed BFO with frequency $v_3 = (v_1 - v_2)$. The induced acceleration $a = -K_R g = -\Delta U_g/U_g$ where K_R is the critical ratio, ΔU_g is the change in gravitational potential and U_g is the gravitational potential energy. As shown, in accordance with the Ivanov model, addition of a beat frequency of $2c\Delta v$ results in a symmetrical field pattern as the induced acceleration is in opposition to the natural acceleration of the dipole oscillator.

If a [electrodynamics-gravitational] coupling did exist, what (physical) instrumentality might it resemble? – Thomas Townsend Brown

51. Coupled Oscillators

Counter acceleration

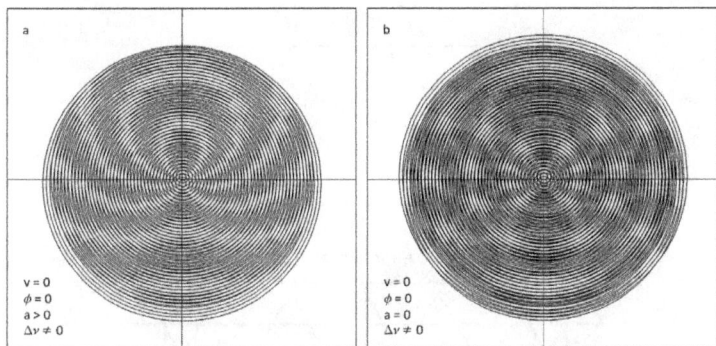

Fig. 51-19. Acceleration of oscillator dipole due to frequency difference is negated by superposition of a third oscillator with beat frequency of $v_1 - v_2$. In frame (a), the asymmetrical wave interference field pattern is illustrated for two coupled oscillators with different frequencies ($v_1 \neq v_2$). In frame (b), a third oscillator with a frequency v_3 (= Δv) is added on a line in between the paired oscillators resulting in counter acceleration with an overall wave field symmetry and oscillators stabilized at standing wave minima.

Examples of superposition effects of a pair of waves in a waveguide and a standing wave resonator are illustrated in Fig. 51-20. A waveguide may be considered as a special class of a transmission line with continuous shorting stubs providing a conduit for travelling electromagnetic waves (TE/TM modes). A standing wave cavity resonator represents a closed-end waveguide. Effects of a step change in phase and frequency in a standing wave resonator are depicted in Fig. 51-21 illustrating a nodal contraction equivalent to motion-induced Lorentz contraction due to uniform velocity and acceleration, respectively. Motion of a wave pumped single or double-ended phase conjugate resonator may be induced by amplified simulated counter-propagating Doppler shifted waves as an inverse effect of induced Doppler shifted matter waves created by motion of matter. This action may be understood as consistent with Schrödinger's 'wave mechanics' in which matter is viewed as a collection of standing waves which constructively or destructively interfere. Physically, the superposition of waves manifests itself as a wave packet which may be represented by a wave function $\psi(r,t) = \langle \psi | \hat{H} | \psi \rangle$. Matter in motion exhibits a Doppler shift in the associated matter waves such as may be produced by oscillation of a crystal diffraction grating exposed to a stream of particles which generates a shift in the interference pattern and alteration in the probability density. Matter wavelength λ_{dB} varies with momentum (= h/p).

It is not known how the energy of the electron in the X-ray bulb is transferred by wave motion to an electron in the photographic plate or in any other substance on which X-rays fall... How does energy get from one place to another? – Sir William Bragg

Equations representing causal relations must be equations involving 'retarded' (previous-time) quantities. – Oleg D. Jefimenko

I fly in dreams... To execute every sort of curve and angle with a light impulse, a flying mathematics. – Friedrich Nietzsche

51. Coupled Oscillators

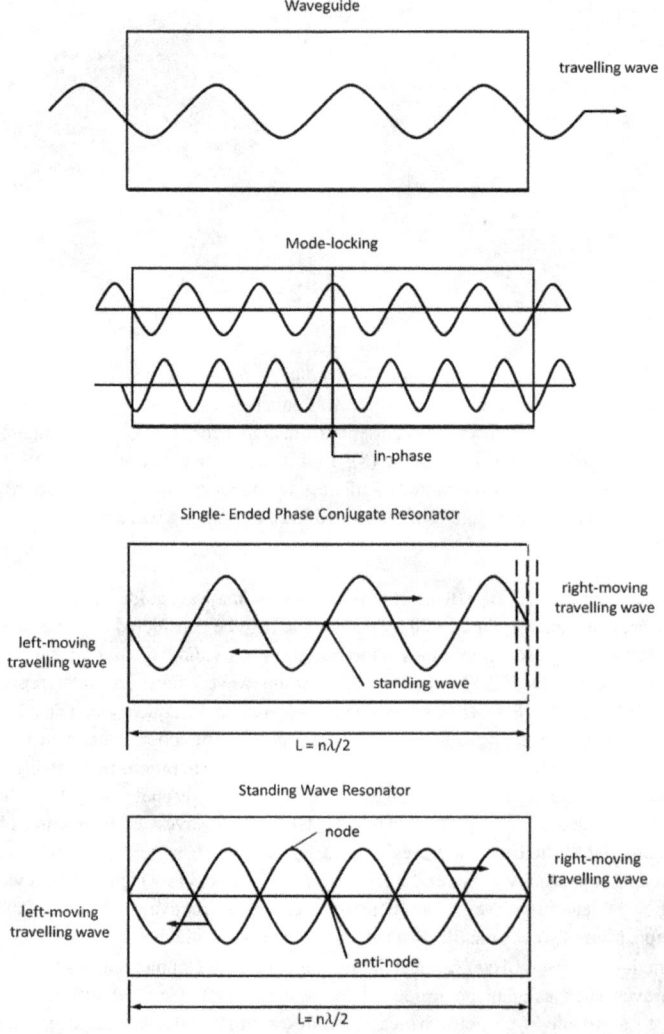

Fig. 51-20. Superposition of waves in an open-end vs. closed-end waveguides. Energy extraction or addition to the phase conjugate reflector enables modulation of the reflection coefficient Γ (= E_r/E_f) and standing wave ratio SWR (= E_{max}/E_{min}). Modulation of phase displaces SW nodal points while modulation of frequency alters density of SW nodal points. Pulse compression decreases pulse interval τ (= $1/\Delta\nu$) and increases peak power P (= E/τ). The spread in wavelength is proportional to momentum. With a narrow wavepacket localized in space, the momentum spread is wide yielding wave interference effects.

Abstract knowledge is always useful, sooner or later. – Robert A. Heinlein

51. Coupled Oscillators

Contraction of a standing wave resonator in motion

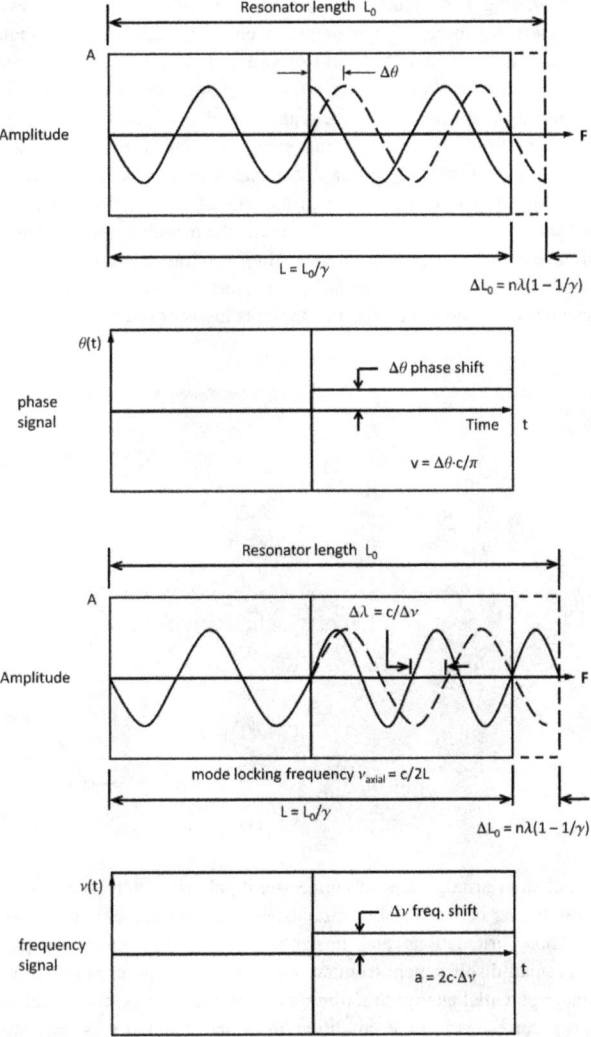

Fig. 51-21. Change in phase or frequency results in a change in wavelength. Motion of a standing wave resonator results in a Lorentz contraction in the direction of motion. Velocity is associated with a change in phase $\Delta\phi$ while acceleration is associated with a change in frequency Δv. A train of resonators with uniformly applied force may be accelerated or decelerated without structural failure or buckling on impact by a progressively applied force $\mathbf{F}$ ($= m \cdot 2c\Delta v \cdot \hat{\mathbf{r}}$) so as to minimize jerk j ($= \Delta a/\Delta t$).

Einstein's three rules of work: 1) Out of clutter find simplicity; 2) From discord find harmony; 3) In the midst of difficulty lies opportunity. – Albert Einstein

51. Coupled Oscillators

51.5 Oscillator arrays

Arrays of oscillating EM dipoles may be expected to produce cyclic variations in a gravity-like accelerative force. For example, a circular arrangement of radially aligned oscillator dipoles produce a frequency hill or valley depending on the orientation with a corresponding acceleration gradient. A circular arrangement of oscillator dipoles with tangential alignment corresponds to a circulator and can generate negative frequency rotational Doppler shifts. A linear arrangement of aligned parallel oscillator dipoles corresponds to a 2-port waveguide isolator. An analog of a torsional magnetic spring may be constructed of inner and outer concentric rings of spaced dipole oscillators. With one of the rings fixed and the other allowed to freely rotate, the movable ring will tend to latch into a stable configuration when opposing dipoles align resulting in zero torque. When offset into an unstable equilibrium, the potential energy stored in the EM field is converted into kinetic energy and rotary motion as the ring attempts to restore equilibrium. See Fig. 51-22.

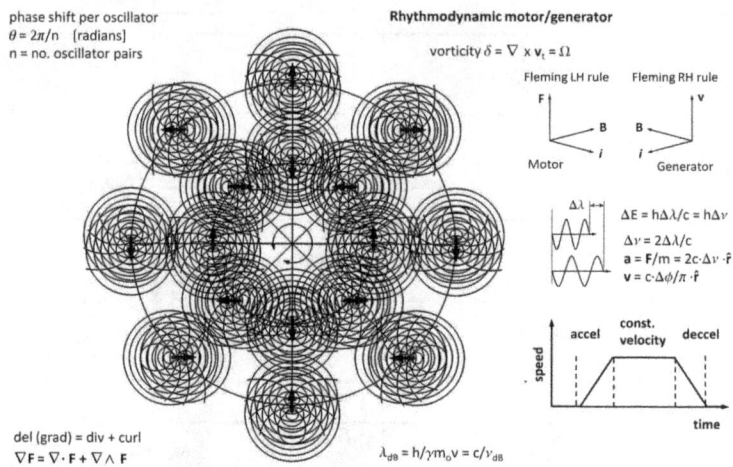

Fig. 51-22. A circular arrangement of concentric dipole rings acts as a torsional spring or steppermoto oscillating between stable and unstable equilibria. Stable positions correspond to opposing dipole orientations and unstable positions correspond to parallel dipole orientations. A small displacement from an unstable balance point results in rotary motion from a maximum potential energy to a minimum potential energy configuration. Minimum potential energy corresponds to a condition in which symmetry is maximized. Phase displacement of overlapping waves leads to shift of potential minima.

A similar effect may be realized in a linear array of oscillating dipoles. A linear or circular array can translate, oscillate, rotate, expand or contract. A linear arrangement of stacked oriented dipoles represents a frequency wiggler with a wavelength that is a function of the number and spacing of dipoles. A particle with comparable resonance wavelength injected in the gap will experience a variation in acceleration similar to a Halbach array. Oscillator dipole rotation sets up a travelling wave which may augment or impede particle motion depending on the dipole spin direction resulting in particle bunching at the antinodes. Such an arrangement is similar to Halbach magnetic arrays used in magnetic

51. Coupled Oscillators

motors, particle accelerator undulators, free electron lasers (ubitrons) and magnetic levitation. For relativistic velocity electrons, the undular magnetostatic field is equivalent to a high intensity EM wave. For power transfer between the EM field and the electron to occur, the electron must have a velocity component transverse to the EM wave. Lateral acceleration of the electron in the undulator or wiggler field results in magneto-bremsstrahlung radiation with a quiver frequency proportional to the product of the Lorentz factor, wave number and velocity. Refer to Fig. 51-23.

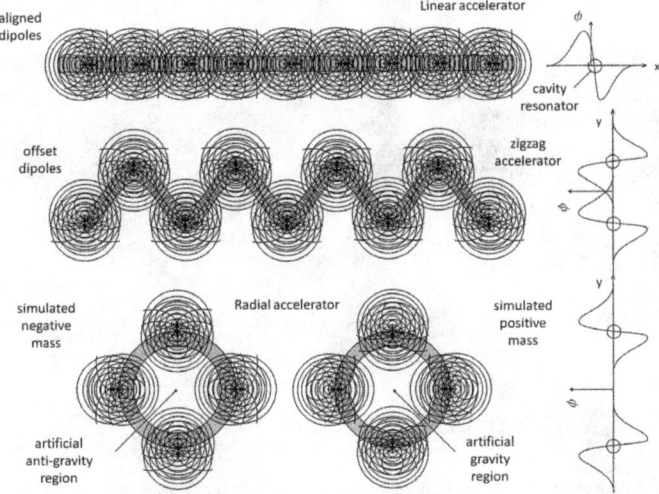

Fig. 51-23. Rhythmodynamic oscillator arrays of disparate frequencies can produce accelerative force field effects.such as collective electron exotic vacuum object excitation..

51.6 Standing wave levitation and propulsion

In recent years, demonstrations of experimental EM drives have been conducted with a goal to advance propulsion technology beyond simple expulsion of reaction mass. Such devices typically involve low power, high-Q magnetron-driven frusto-conical resonant cavities with dielectric resonator inserts operating at a single, relatively low microwave frequency providing barely measurable thrust as there is virtually no energy gradient to impart acceleration. Nor is there significant wavefunction dissonance interaction energy to convert to kinetic energy of motion as there is negligible phase or frequency difference. Net translation velocity is attributed to the difference in group velocities of opposing travelling waves or Lorentz force gradients. These modest attempts, thus far, suffer from lack of sufficient test confirmation to convince skeptics and lack theoretical development to provide credible performance prediction. Questions regarding conservation of momentum exchange with the external environment, thermal effects, nonuniform field polarization/dielectrophoretc effects, resonance shifts, EM flux leakage, radiation pressure imbalance, vacuum performance, etc remain to be answered to the satisfaction of critics. Development of EM wave-based propulsion potentially allows realization of levitation and flight without wings, stress or strain, fulfilling fanciful dreams of flying magic carpets of long ago or the enduring staple of science fiction that of futuristic gravitic or space warp

51. Coupled Oscillators

drives such as suggested in Figs 51-24 and -25. Known EM wave-based effects are depicted in Figs. 51-26. Theoretical concepts and technological means that may enable realization of such goals and aspirations are discussed in the following sections.

Fig. 51-24. Ancient flying carpet dreams reinterpreted in terms of a moving standing wave within a resonator. The waving magic carpet is a metaphor for coherent wavefronts of moving standing waves – a synthesized matter wave modulated by Lorentz Doppler beams. Velocity of a moving standing wave system composed of counter-propagating travelling waves varies with wave amplitude and/or phase velocity while acceleration varies as a difference in frequency. Mass is a result of and proportional to change in wave frequency.

Fig. 51-25. Futuristic contragravity vehicles with electromagnetic matter wave-based propulsion. Synthesized Lorentz-Doppler and gravito-magnetic Poynting vector effects offset local gravitational spectral energy density gradient.

51. Coupled Oscillators

Standing wave levitation of objects has been widely demonstrated in a variety of applications by multiple means including magnetic, electric, optical, aerodynamic and acoustic forces to counteract downward gravitational force. Such force-field effects do not nullify the Earth's gravitational field by out-of-phase EM wave cancellation, but rather oppose it. Magnetic levitation of small objects such as plastic spheres, liquid droplets, nuts and bolts, biological organisms, etc and applications such as magnetic bearings, contactless melting and maglev trains are well known examples. Electromagnetic current is typically controlled via proportional and derivative (PD) control. Levitation involving suspension of supercooled superconductors in a magnetic field by flux pinning due to the Meissner effect is likewise easily demonstrated. Acceleration or deceleration of beams of ions by changing the relative frequencies of two counterpropagating laser beams to create a moving standing wave has been achieved in a number of experiments including laser cooling demonstrations.

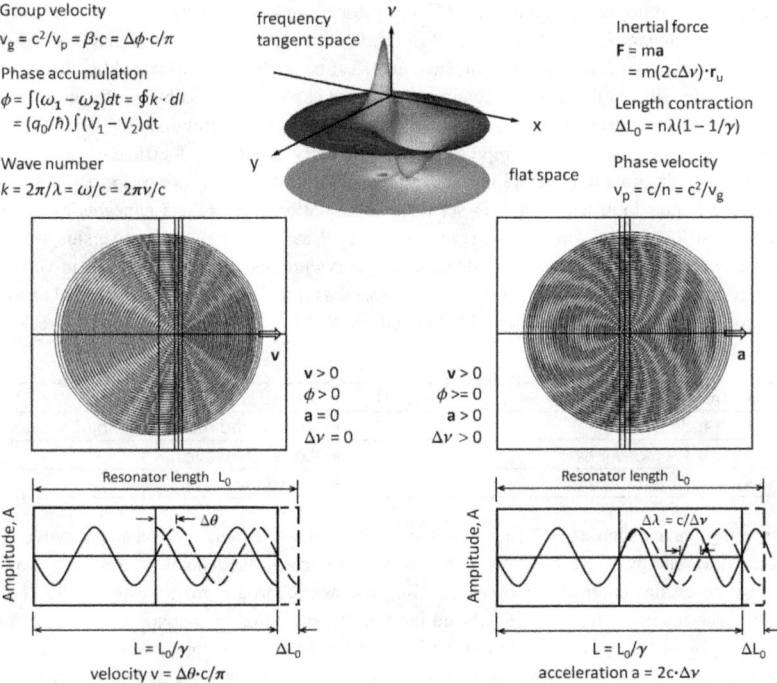

Fig. 51-26. Induced motion of a resonator with synthesized Lorentz-Doppler shifted modulated standing waves. Contracted moving standing waves are created by superposition of counterpropagating red- and blue-shifted travelling waves. Translation of the wave system shown lies in a direction along a line connecting the emitter array. Translation velocity is controlled by modulating phase differential of the counter-propagating traveling waves. Acceleration is controlled by the frequency difference between each pair of oscillators. Acceleration is in the direction of the frequency gradient. The wave system is propelled by the self-generated ponderomotive matter wave.

Simplicity is the ultimate sophistication. – Leonardo da Vinci

485

51. Coupled Oscillators

Although long a staple of science fiction, electromagnetic propulsion without repulsion of reaction mass has yet to be realized. Experiments on microwave resonant cavities thrusters known as EM drives to date have not conclusively demonstrated measurable thrust to the satisfaction of critics who cite apparent violation of conservation of momentum and test measurement errors as lack of proof of concept. Is there any hope of future development of an EM propulsion drive based on known physics? The physical mechanism for gravity and the origin of inertia are still subject to much debate. Consider, for example, a falling rock which moves without expulsion of reaction mass. Mass in motion constitutes a mass current with associated Doppler effect and internal de Broglie matter waves. Acceleration of gravity arises as a result of the EM spectral energy density gradient of the Earth's gravitational field. The associated gravitational frequency shift results in Lorentz-Doppler shifted contracted moving standing waves within the fermionic standing wave structures (i.e., elementary particles and composites) which make up matter. At non-relativistic velocities, the wavelength of de Broglie matter waves is minute. For example, a moving rock with a mass of 1 kg traveling at a velocity of 50 m/s has a de Broglie wavelength λ_{dB} (= h/mv) of 1.326E-35 m and a corresponding frequency ν_{dB} (= E/h = (h/mv)($\Delta\nu/\Delta\lambda$)) of 2.26E43 Hz. Can such an effect be artificially duplicated by electronic means? We will briefly review resonator theory and explore the possibility of synthesis of matter waves as a potential means to realize matter wave impulse propulsion.

Expulsion of reaction mass represents a particularly crude and inefficient method of induced motion of matter. Do we not understand sufficiently the physics of force fields and wave mechanics to at least begin to set forth some notional theoretical concepts as to the method for EM propulsion without reaction mass? What sort of energy conversion would be required? Energy is a measure of wavefront curvature and may be conveyed in waves. Energy may be stored as standing waves or released as traveling waves. Consider what sort of wave transformation is needed. As detailed Table 51-1, waves occur in any of several forms including:

• Traveling waves	• Partial standing waves
• Standing waves	• Contracted moving standing waves
• Transverse waves	• Coherent waves
• Longitudinal waves	• Soliton waves

Traveling waves such as EM radiation represent a release of energy. Matter consists of stored EM energy as resonant standing wave structures. Movement of standing wave resonators creates contracted moving standing waves (de Broglie matter waves). See Fig. 51-27. The inverse effect of self-induced motion of matter may potentially be realized, for example, by utilizing synthesized red- and blue-shifted Lorentz-Doppler waves to generate self-induced motion of a wave system without wheels, friction, or expulsion of reaction mass. Kinetic energy of motion is provided by direct conversion of electromagnetic energy in the amplified contracted moving standing wave formed within a phase-locked cavity resonator. Inverse effects are not without precedent as, for example, inverse Doppler effect, inverse Sagnac effect, inverse Faraday effect, inverse Compton effect, inverse spin Hall effect, inverse Cherenkov effect, inverse Raman effect, inverse Cotton-Mouton effect, inverse Barnett effect (Einstein de Haas effect), inverse piezo-electric effect, etc.

A rock has no detectable opinion about gravity. – Terry Prachett

Inventing is like sending a message in a bottle to the future. – Claude Shannon

To study music we must learn the rules. To create music we must forget them. – Nadia Boulanger

51. Coupled Oscillators

Table 51-1. Example wave types.

Description	Example	Remarks
traveling wave	photon wave train	Freely propagating wave energy with zero rest mass in vacuo. Wave velocity depends on the energy density of the wave medium.
standing wave	electron	Confined wave energy within a cavity resonator. An electron is a topologically bound, high energy photon. Exhibits rest mass with respect to an inertial reference frame.
transverse waves	electromagnetic waves	Electric & magnetic fields orthogonal to direction of propagation.
longitudinal waves	sound waves, scalar EM waves, plasma waves	Compression & rarefaction pressure pulsations. Tesla radiation.
partial standing waves	mufflers, organ pipes, Helmholtz resonators	A mixture of a standing wave & traveling wave. Impedance mismatch.
contracted moving standing waves	de Broglie matter waves, acoustic waves, transverse & longitudinal EM waves	Matter in motion with respect to an observer. Modulated standing wave. Lorentz contracted resonator.
coherent waves	lasers, masers	Frequency & phase-locked overlapping waves.
phase conjugate waves	optical phase conjugation	Spatial and temporal coherence.
soliton waves	water waves topologically confined in a waveguide; optical soliton waves	Nondispersive wave.

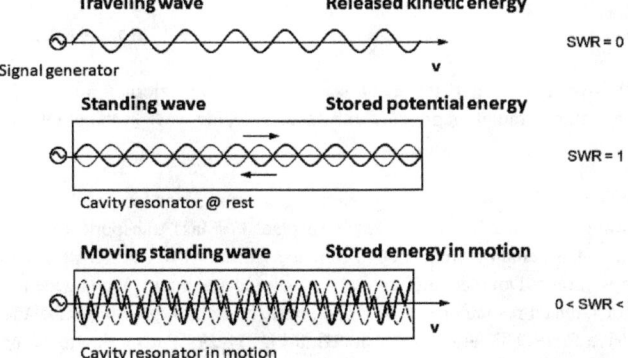

Fig. 51-27. Comparison of traveling waves, standing waves and moving standing waves. Motion of the driven quantum system alters the internal wave function. A phase-locked state has identical mean phase velocities. A Lorentz-Doppler frequency shift alters the accumulated phase of the matter wave function resulting in a change in resonator motion.

487

51. Coupled Oscillators

A major stumbling block over the 1-½ decades in development of EM microwave propulsion without expulsion of reaction mass is using a standing wave cavity resonator with single frequency excitation. The frequency source is a magnetron typically operating at a few GHz with power up to ~100 watts and measured max. thrust in vacuo on the order of μN[121]. Nonuniform field effects associated with asymmetric frusto-conical resonators are minimal due to very modest EM field density gradients and nonlinear polarization effects for the given low energy input conditions. Resonator walls which are free to move tend to settle in at node points minimizing dissonance. With a single frequency excitation (and associated harmonics), oppositely directed nodal displacement occurs only when changing frequency modes. Standing waves stay put with end nodes locked at boundary walls with no net thrust. This represents a consonance condition in which the number of contained waves are in whole integer numbers with harmonic intervals forming Pythagorean ratios. The resonator remains at rest unless acted upon by a force imbalance such as shown in Fig. 51-28. The matter wave frequency is proportional to acceleration.

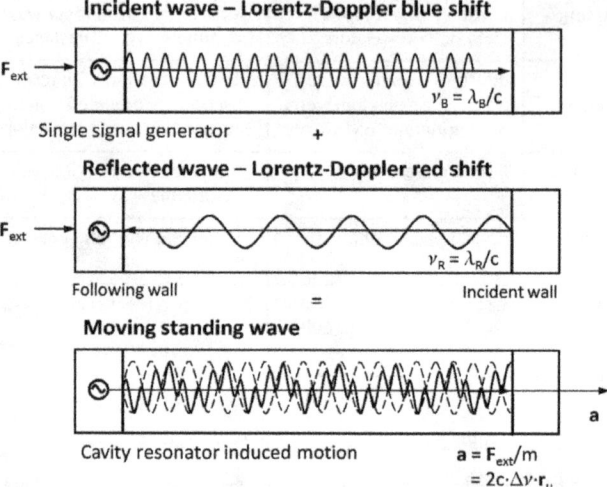

Fig. 51-28. Standing wave resonator set in motion by external force. The impulse generates an internal radiation pressure imbalance inducing wave system motion.

If, however, using two frequency signal generators, one for each traveling wave, standing wave nodal end points are easily displaced. Mass transport is thus effected by varying the phase and/or frequency difference of the two counter-propagating waves simulating a Lorentz Doppler shift. Voilá! - A moving standing wave is generated which is essential for induced resonator motion. The internal radiation pressure imbalance provides the propulsive force. This effect is illustrated in Fig. 51-29. For motion to occur, a state of dissonance is required at one or both of the boundary walls. Dissonance is marked by number of contained waves forming Fibonacci sequence of standing wave damping ratios. Phi ratios mark boundary conditions on propagation of harmonic energy flow.

We cannot conceive of matter with negative inertia or mass. – James Clerk Maxwell

51. Coupled Oscillators

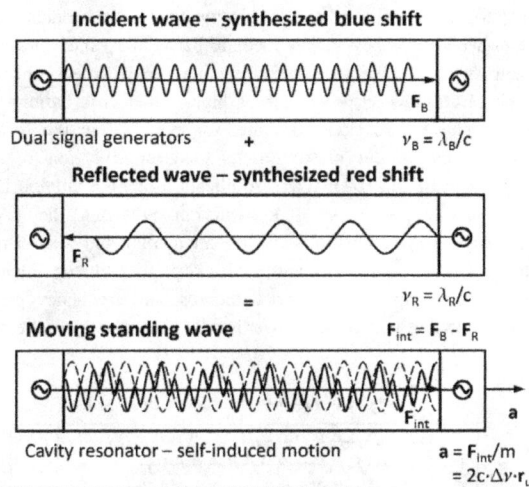

Fig. 51-29. Standing wave resonator set in motion by internal force. A ponderomotive force results from an internal radiation pressure imbalance of disparate source oscillators.

Using, say, a minimum of three signal generators to create a Shepard-Risset-glissando as shown in Fig. 51-30, many more photons are put to work as the resonator driven endplate is bombarded with a barrage of wavefronts increasing radiation pressure and resultant thrust. Acceleration is proportional to the frequency difference between paired frequency sweep bands. The forcing function inputs include phase and frequency difference, offset interval, sweep period and ramp rates to precisely control acceleration. Up chirp provides acceleration while down chirp provides deceleration. High internal radiation pressure is provided by high frequency standing wave modulation of paired Shepard tone frequencies with discrete phase and frequency differential swept over a wide frequency band.

Matter waves have been poorly understood – even de Broglie and Schrödinger had evident problems in representing ψ as a state vector in real or abstract coordinate space. Phase waves were thought to be 'unphysical' or some sort of 'pilot' waves guiding motion of matter or probability amplitudes in configuration space. A contrasting view is that matter waves are in fact physical and may be synthesized and electronically controlled. Observable characteristics of matter wave propulsion are summarized in Table 51-2. A multifrequency drive and associated harmonics increases dimensionality of parameter space allowing increased performance capability over a dual frequency source drive.

Given known wave mechanics and phase conjugation effects, further developmental opportunities may be possible such as illustrated in Figs. 51-31 & 32. The inverse effect of self-induced motion of matter may potentially be realized by utilizing synthesized red- and blue-shifted Lorentz Doppler waves in a phase conjugate four-way mixing process using parametrically amplified Lorentz Doppler pump beams to generate self-induced motion of a wave system without wheels, friction, or expulsion of reaction mass. Kinetic energy of motion is provided by direct conversion of electromagnetic energy in the amplified pump beams to the contracted moving standing wave formed from the signal wave and its phase conjugate wave within a phase-locked cavity resonator. A conceptual example of a standing wave levitation system employing a microwave high power, multiple frequency, phased antenna array and phase conjugate reflection is indicated in Fig. 51-33. As shown, an array of oscillators with frequencies v_1 and v_2 may focused to produce a holographic Bragg

51. Coupled Oscillators

interference pattern in a nonlinear polarizable medium. An incident wave from a third central oscillator with frequency v_3 intermediate between v_1 and v_2 results in a reflected phase conjugate wave. The nonlinear medium may consist of high third-order susceptibility photorefractive crystals such as high-K dielectric barium titanate ($BaTiO_3$) with a tetragonal phase Perovskite structure, for example, in the form of nanocrystal waveguides in a quantum dot matrix to augment amplification in short pulse non-degenerate four-wave mixing (NDFWM). Application of high voltage DC potential with electric field intensities of tens of kilovolts/cm polarizes the crystals increasing nonlinearity. Phase conjugation exhibits low loss attenuation and counter-propagating pump beams in four-way mixing can resonantly self-excite to build up intensity. Parametric amplification with a pump beam at twice the resonant frequency provides low noise amplification at significantly increased power levels. The strength of the phase conjugate

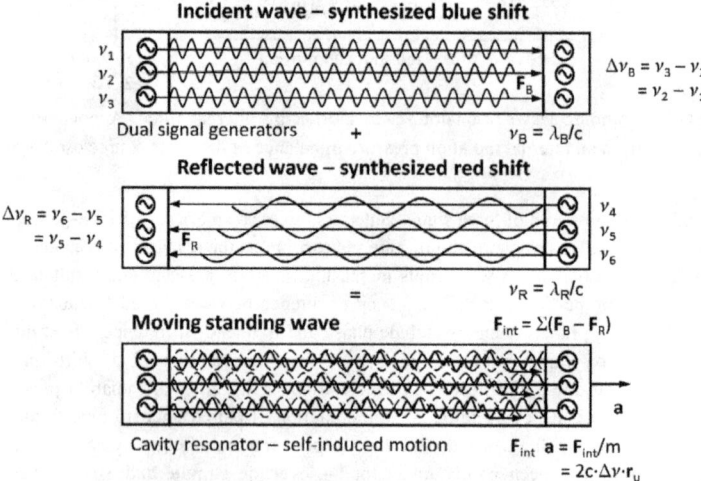

Fig. 51-30. Shepard-Risset-glissando phased matter waves allow increased thrust and control of rate-of-change of acceleration. Spatial and temporal coherence of matter waves allow uniform acceleration/deceleration, minimization of catastrophic jerk and avoidance of excessive inertial stresses/strains. Minimizing progressive acceleration prevents buckling.

Table 51-2. Matter wave propulsion characteristics	
• High velocity & acceleration without jerk or excessive inertial strain	• Extreme agility & near instantaneous vectoring
• No visible exhaust	• No sonic boom
• No wings or control surfaces	• No reaction mass expulsion
• High altitude operation	• Opearable in a vacuum
• Silent motion & hover	• Low thermal signature
• Submersible capability	• Apparent inertia neutralization

It may be possible to modify the [polarizable vacuum] refractive index by applying an intense superposition of fields within a single frequency mode to locally affect "g". – Ricardo C. Storti

51. Coupled Oscillators

beam increases nonlinearly with the pump beam power. Frequency modulation of a phase conjugate contracted moving standing wave produces a synthesized de Broglie wave with adjustable momentum which may be varied to produce either a repulsor or tractor beam depending on phase velocity direction.

Objects within the phase conjugate beam experience radiation pressure tending to induce motion towards the standing wave anti-nodes. Modulation of the phase difference between oscillators of different frequency imparts a velocity $\mathbf{v}$ (= $c/\pi \cdot \Delta\phi$) in the direction of the phase gradient $\nabla \phi$. A fourth variable frequency oscillator(s) of frequency v_4 located above the disc interacts with the lower frequency oscillators to produce an acceleration $\mathbf{a}$ (= $2c \cdot \Delta v \cdot \mathbf{r}_u = (2c^2/h)(mv/\Delta\lambda) \cdot \lambda_{dB} \cdot \mathbf{r}_u$) in direction of the ∇v frequency gradient. As shown, a set of pump beam oscillators of frequency v_1 in outer rings and a second set of oscillators of frequency v_2 in inner rings is disposed in a concentric arrangement within a phase conjugate circular disc forming a 2-dimensional planar phased antenna array. The relative phase of the oscillators may be adjusted to control the azimuthal direction from the boresight axis perpendicular to the disc and focus the pump beam as in conventional electronically scanned radar with digital beam forming.

The antenna elements array spacing should preferably be less than half-wavelength of the incident radiation to minimize radiation pattern gating lobes which add incoherently and element-to-element mutual coupling. The frequencies of the pump beams v_1 and v_2 are adjusted to vary the respective wavelengths λ_1, λ_2 to the desired velocity motion state (synthetic de Broglie wavelength λ_{dB}). A coaxial oscillator of intermediate frequency v_3 ($v_2 > v_3 > v_1$) is located at the center of the disc generating a focused beam downward towards the focal point which is reflected from the interference grating forming a phase conjugate standing wave. Modulating the pump beam frequencies v_1 and v_2 results in a contracted standing wave with group velocity v_g and phase velocity v_p that vary accordingly. The resultant Poynting energy flow is from the high frequency oscillator towards the low frequency oscillator with velocity $V_E = (v_1 - v_2)/(v_1 + v_2)$. Varying the phase of each dipole pair enables modification of the radiation pattern of each element set (antenna directive gain). With phase $\phi = 0$, the radiation is broadside to the disc plane.

In any field, find the strangest thing and then explore it. – John Archibald Wheeler

To change something, build a new model that makes the existing model obsolete.
– R. Buckminster Fuller

If you are worried that something in an experiment might be a potential source of error, take the step of deliberately exaggerating the error sources to determine just how bad they can get. This enables you to quantify the magnitude of the problem and determine a way to deal with it. – Henry Cavendish

Maxwell made use of quaternions in EM theory and mathematics beyond three-dimensional vectors, but scientific community of the time accepted only vector components and cutoff the "transcendal and fantastic elements". – Ron Croix

I could trust a fact and always cross-question an assertion. – Michael Faraday

In order to reach the Truth, it is necessary, once in one's life, to put everything in doubt...so far as possible. – Descartes

A good conjecture should fit on a T-shirt. A great conjecture should be outrageous.
– John Conway

The present is theirs; the future, for which I have really worked, is mine. – Nikola Tesla

51. Coupled Oscillators

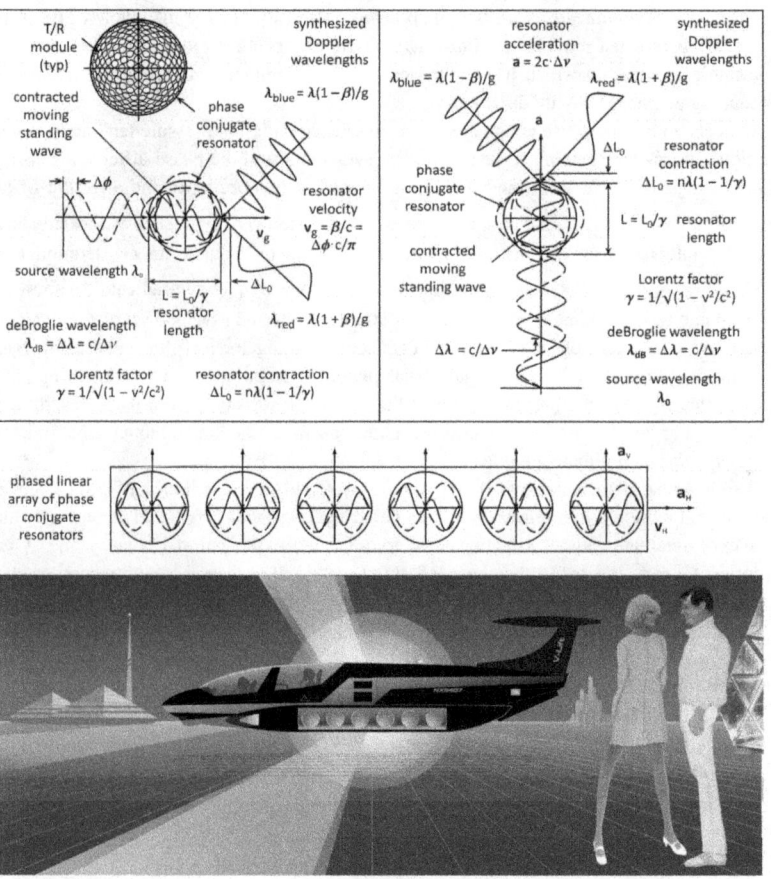

Fig. 51-31. Illustration of an antigrav vehicle concept. An array of phase–locked, phase conjugate standing wave resonators pumped by parametrically amplified, synthesized Lorentz Doppler-shifted beams may be used to generate self-induced motion in horizontal and/or vertical directions. Direction of motion may be rapidly changed by redirecting the vector orientation of the incident and phased array phase conjugate beams enabling levitation and high acceleration, darting, zigzag motion without expulsion of reaction mass. Modulation of $\Delta\phi$ phase overlap and rate-of-change of acceleration $\Delta a/\Delta t$ enables inertial damping reducing excessive strains. Kinetic energy of motion of the wave system is provided by direct conversion of electromagnetic energy in resonant amplified EM pump beams to modulate a synthesized matter wave within a phase conjugate resonator.

'Imaginary' universes are so much more beautiful than this stupidly contructed 'real' one.– G. H. Hardy

For there is much I must think upon... A life of endless adventure beckoning before me. – Silver Surfer

While the circumnavigation of the solar system seems farfetched, it may not be once the problem of effective anti-gravitational control is solved. – Donald A. Wollheim

51. Coupled Oscillators

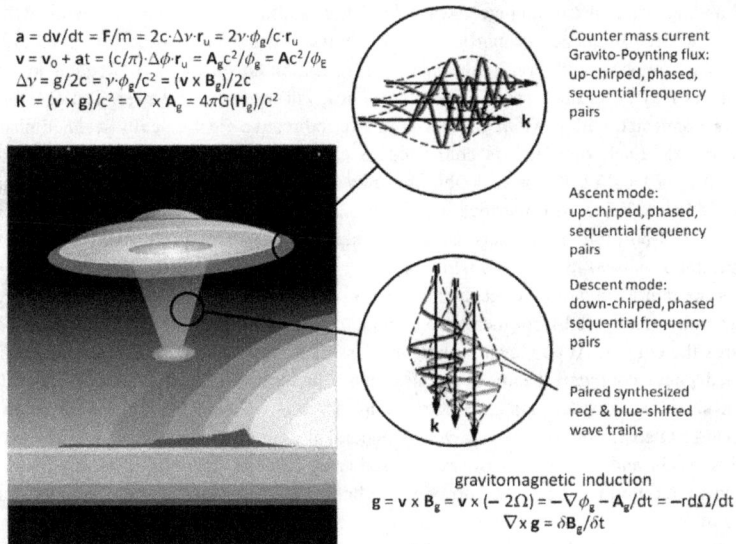

Fig. 51-32. Contragravity may be sustained by modification/inversion of the naturally occurring EM spectral density gradient of the gravitation frequency spectrum. The spectral energy density represents the energy density per frequency mode. The gravitational field spectral energy density $\rho_0(\omega)$ varies with the cube of the frequency ($= 2\hbar\omega^3/c^3$). EM field energy density U is a function of electric field strength **E** and magnetic flux density **B** ($= \frac{1}{2}(\epsilon_0\mathbf{E}^2 + \mathbf{B}^2/\mu_0)$). High intensity electric and magnetic fields augments nonlinear vacuum polarization and via the Faraday effect can create dark halos or dark zones by optical rotation of polarized incident atmospheric light and emitted radiant light. Extreme bandwidth allows frequency roaming above and below visible light. A phase conjugate, phased array beam antenna provides agile beam focusing and steering of synthesized matter waves. Phase conjugation provides time-reversed beam formation. Parametric amplification of phase conjugate reflected ground return signal allows intensification of resonant standing wave beam augmenting nonlinear spectral density profile gradient.

With phase $\phi = 90°$, the radiation pattern is asymmetrical, while a phase $\phi = 180°$ results in a symmetrical figure-8 pattern. Variable frequency oscillators of frequency ν_4 are located above the disc plane and interacts with oscillators of frequency ν_3 below to generate an upward or downward acceleration depending on frequency difference $\nu_4 - \nu_3$ according to the Ivanov relation: $a = 2c\cdot\Delta\nu\cdot\mathbf{r}_u = (2c^3/h)(m\nu/\Delta\lambda\cdot\nu_{dB})\cdot\mathbf{r}_u$. For 1g acceleration, for example, the frequency differential $\Delta\nu$ is 1.634E-08 Hz. The EM potential U_{em} is minimal when $\langle v_p^2 \rangle$ is minimal and $\langle p_p^2 \rangle$ is maximal at the points of radiation pressure anti-nodes. The relative phase of the pump beams (ϕ_1, ϕ_2) and/or PC beams (ϕ_3, ϕ_4) may be varied to adjust velocity. Reversing the direction of the phase velocity v_{ph} alters the direction momentum p ($= \rho_e\cdot\mathbf{v}/c^2$) and wave vector k ($= p/\hbar$) enabling tractor or repulsor beam operation. Simultaneous operation of opposed repulsor and tractor beams allows tilt edge suspension hover mode. The size of the standing wave system object is not limited by the wavelength. The standing wave system can be levitated at positions of multiples of the half wavelength above the phase conjugate wave focal point.

51. Coupled Oscillators

Standing wave/phase conjugate wave levitation requires high intensity beams at high power and high frequency proportional to the levitated mass on the order of ~1-25 kw/kg depending on size and configuration. For operation at microwave frequencies, the oscillators may be in the form of Gunn diodes or klystrons for high power. The oscillators may be connected with phase lag filters as in a retrodirective electronically scanned phased array radar. Each oscillator is connected to a phase shifter, mixer and amplifier and pumped by a local oscillator at double the frequency of the incident wave. The incoming signal is heterodyned with the local oscillator signal in the mixer to produce an outgoing wave. A 3-port circulator acts as a duplexer allowing high-power waves from the transmitter to pass to the antenna while blocking their way to the receiver and, conversely, passes the return signal to the receiver bypassing the transmitter. Scattered reflection from the ground is redirected from the phase conjugate disc with a portion that is reflected back towards the emitter. At an elevation height h less than the focal length f_0 of the downward focused phase conjugate beam, the return signal includes scattered ground return. The return signal is boosted in a parametric amplifier and mixed with the source beam to resonantly intensify the standing wave. A general system schematic diagram is illustrated in Figs. 51-34 and -35 of a microwave phased array antenna disc with phase conjugation. As shown, a retrodirective phased array of oscillators of frequencies v_1 and v_2 is located in a series of annular waveguides in a disc arrangement. The transmit/receive antenna array is fed in a conventional corporate architecture of local oscillators, phase shifters and amplifiers to produce an agile, steerable beam with adjustable focus. The control electronics includes a plurality of local oscillators operating at synthesized Doppler frequencies. Signal output is frequency and phase shifted in accordance with command input signals and sensed velocity and acceleration error signals. Output signal is parametrically amplified and summed with phase conjugate beam of reflected return increasing intensity.

A phase conjugate beam is generated by wave interaction of a central oscillator of frequency v_3 with oscillators of pump frequencies v_1 and v_2. The phase conjugate standing wave produced by the sum of the ingoing and outcoming signals is modulated by a synthesized Doppler signal to generate a contracted moving standing wave imparting momentum to the disc array. A contracted moving standing wave is induced by a simulated Doppler shift Δf_D. The phase conjugate standing wave produced by the sum of the ingoing and outcoming signals is modulated by a synthesized Doppler signal to generated a contracted moving standing wave imparting momentum to the disc array. The reflected phase conjugate beam is resonantly amplified by a parametric amplifier connected to a mixer diode at twice the resonant frequency to boost field intensity.

Altitude station-keeping may be accomplished by further mixing with the measured Doppler shift by means of a synchronizer circuit and Doppler processor. The oscillators may be in the form of varactor tuned Gunn diodes within resonant cavities with biased Schottky-effect mixer diodes. The antenna elements may be in the form of waveguide slots, microstrip arrays or equivalent. A 3-port circulator serves to decouple the transmit and receive signals from the transmitter and receiver. Phase shifters enable beam steering and focusing. Acceleration and velocity control may be accomplished by varying the frequency differential and phase differential between oscillators.

Concept illustrations of operating modes are depicted in Figs. 51-35 through -37. If the antenna array is moving upward with respect to the ground, the ground return signal will be Doppler redshifted with respect to the reference frequency f_0 with a frequency difference $-\Delta f\ (=2f_D \cdot v/c)$. Likewise, if the antenna array is moving downward, the return ground signal is blueshifted with a corresponding frequency difference $+\Delta f$. A synchronous detector may be used to determine the phase difference between the Doppler frequency f_D

51. Coupled Oscillators

and the reference frequency f_0 producing an output voltage V_{out} proportional to the amplitude and phase difference (= $A\cos\phi$). The received Doppler shifted frequency may be combined with the pump beam frequencies in a microwave four-wave mixer to produce a blueshifted phase conjugate beam with frequency $+\Delta f$ to automatically compensate for relative motion and maintain resonant amplification.

As illustrated, a tractor and repulsor phase beam array may be adapted for levitation and suspension modes. High power, high field intensity is required sufficient to generate the interference grating and develop sufficient acceleration for the given mass to counter gravitation. Objects within the standing wave beam tend to move from nodes to antinodes and move in unison with the phase velocity. If, in addition, the center beam is circularly polarized, objects will acquire orbital angular momentum as well simulating the effect of an Archimedes' screw. Reversal of Shepard tone pitch sequence switches tractor/repulsor beam modes. A dynamic spira mirabilis field may be generated by counter-rotating beams producing, in addition, a radial force effect.

By inducing phase unlocking between master and slave oscillators, it may be possible to reduce coupling strength by injection of a range of coherent quasi-monochromatic frequencies over a selected narrow band from a third oscillator (pump) in a phase-locked loop (PLL) to form a pulse comparable to the frequency differential $\Delta v = a/2c$ of a standing wave system restrained from free-fall. Adjustment of the tracking offset $\Delta \omega_t$ above or below the selected plateau loop center frequency ω_0 acts to trigger phase locking or unlocking. The loop tracking error may be controlled by comparison with a signal proportional measurement of wave system acceleration $\mathbf{a}$ (= $\mathbf{F}/m$ = $2c \cdot \Delta v \cdot \hat{\mathbf{r}}$ = $(2c^2/h) \cdot (mv/\Delta\lambda) \cdot v_{dB} \cdot \hat{\mathbf{r}}$ = $\mathbf{v} \times \mathbf{B}_g$). With a system of slave oscillators combined in a phased array antenna for adaptive phase control and beam steering, the individual antenna channels may be phase-locked to the resultant sum so that the individual signals add coherently.

The phase velocity of the wave associated with the particle is inversely proportional to the velocity of the particle itself; it is infinite in the proper system where the particle velocity is zero. – Louis de Broglie

Inertia is simply proportional to the total energy of the electromagnetic field. – Arthur Eddington

There can be no inertia relative to space, but only an inertia of masses relative to one another. – Albert Einstein

Problems worthy of attack prove their worth by fighting back. – Piet Hein

I think there is a moral to this story, namely that it is more important to have beauty in one's equations than to have them fit experiment. – P. A. M. Dirac

I do not believe on your perception of 'holes' even if its existence of the 'anti-electron' is proved. – Wolfgang Pauli

Nonsense will fall of its own weight by a sort of intellectual law of gravitation. – Cecilia Payne

Strange as it may seem, the strength of mathematics lies in the avoidance of unnecessary thought form already a system of wonderfully simplicity and economy. – Ernest Mach

The spectral density of black body radiation... represents something absolute, and since the search for absolutes has always appeared to me to be the highest form of research. I applied myself vigorously to its solution. I had to obtain a positive result, under any circumstances and at whatever cost. – Max Planck

Elegance and simplicity are the hallmarks of truth. – Geoffrey S. Diemer

We're always searching for new frontiers. We're drawn to the unknown. – Ridley Scott

51. Coupled Oscillators

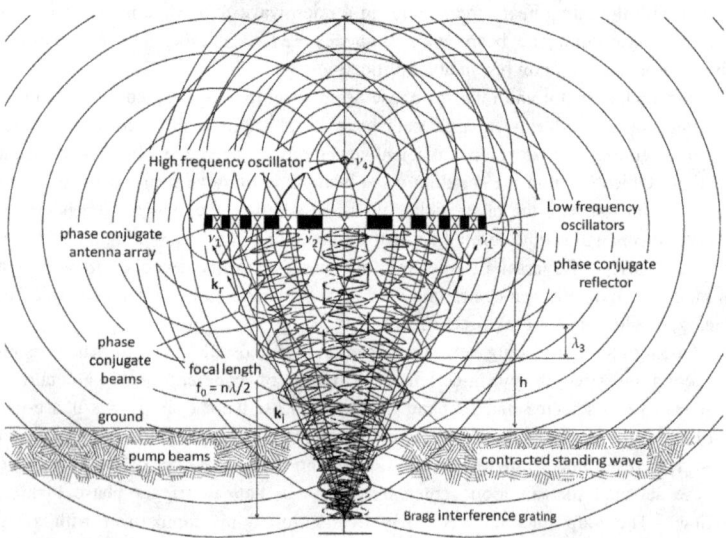

Fig. 51-33. Phase conjugate standing wave levitation with multi-phase, multi-frequency oscillator array in a planar disc arrangement. Reversal of generated contracted moving standing wave phase velocity enables either tractor or repulsor modes. A phase conjugate beam is reflected from the Bragg refraction grating generated by interference of overlapping beams. Modulation of the pump beams at synthesized Doppler shift frequencies generates a contracted moving standing wave. Parametric amplification of the beams increases beam intensity. Phase modulation allows chromatic or achromatic focusing. Levitation is accomplished by neutralization of the gravitational frequency differential $\Delta v = g/2c = v \cdot \phi_g/c^2$ which is a measure of the spectral energy density gradient. Kinetic energy of motion is imparted by EM energy input into the wave system.

It is better to do the right problem the wrong way than the wrong problem the right way. – Richard Hammimg

The story of civilization is, in a sense, the story of engineering – that long arduous struggle to make the forces of nature work for man's good. – L. Sprague de Camp

Man studied birds for centuries, trying to learn to make a machine to fly like them... his final success came when he broke away entirely and tried new methods. – John W. Campbell

When it does not exist, design it. – Sir Henry Royce

Fine if it works in practice, but does it work in theory? – Philip Bourke

If nature existed on endless levels, so might intelligence. – Richard Matheson

A lack of seriousness has led to all sorts of wonderful insights. – Kurt Vonnegut

Up above the world you fly, like a tea tray in the sky. – Lewis Carrol

How much easier it is to be critical than to be correct. – Benjamin Disraeli

51. Coupled Oscillators

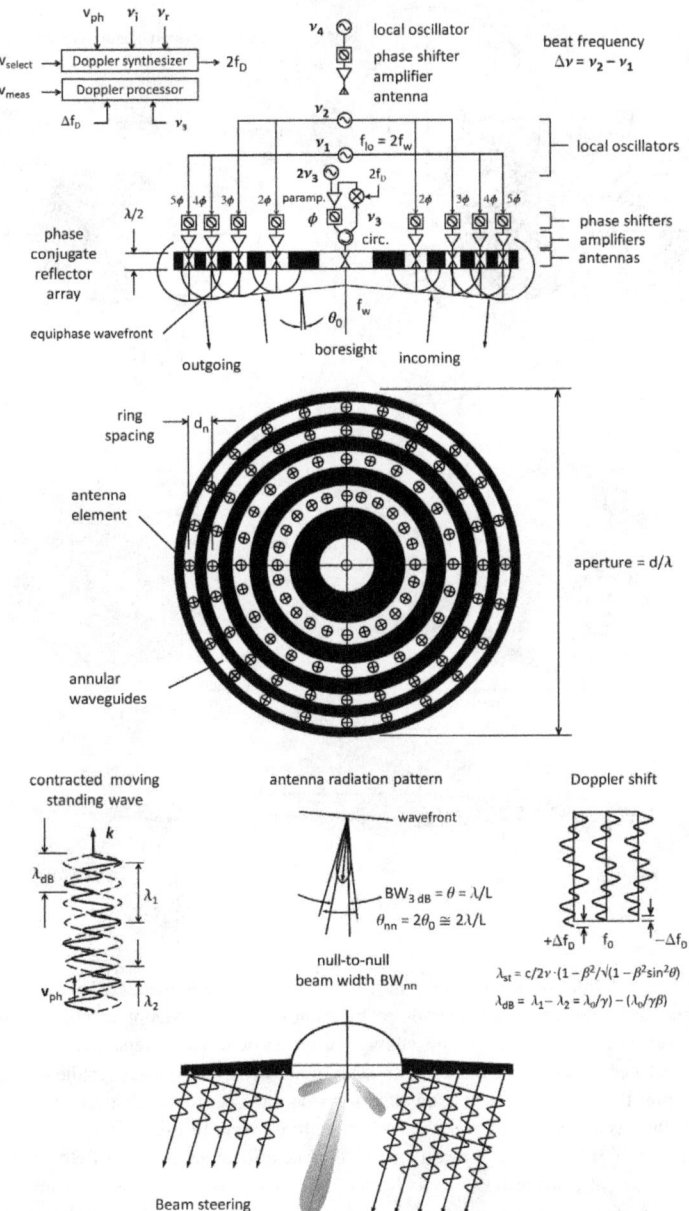

Fig. 51-34. Synthesized Doppler phased emitter array antenna disc with agile beam steering and focusing. Multiple microresonator arrays may be arranged, for example, in ganged, linear fashion, triangular, circular or spherical groupings.

51. Coupled Oscillators

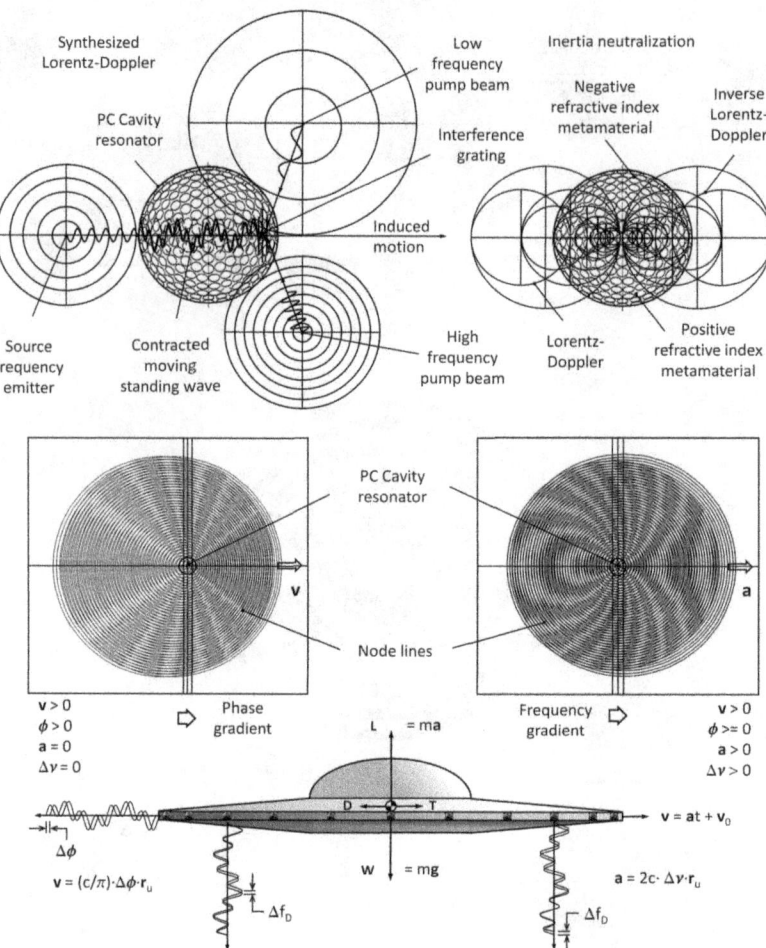

Fig. 51-35. A synthesized contracted moving standing wave is generated within a phase conjugate cavity resonator and modulated by amplified Lorentz-Doppler pump beams to provide self-induced motion of the coupled wave system via inverse Lorentz-Doppler effect. Induced acceleration direction is determined by the frequency gradient. Energy flow is from the high frequency oscillators towards the low frequency oscillators. Nodal line curvature is a measure of acceleration. A multiple array of oscillators may be used to produce a zone of inward or outward energy flow or a local acceleration field. Similarly, uniform motion at constant velocity is induced by a phase differerential. An interference grating is created in the pump beam overlap region producing a phase conjugate mirror enabling four-way mixing with the incident and reflected source beam. Very high velocities and accelerations are thus achievable with extreme degree of control. Inertial effects without stress or strain may be effected with an enveloping synthesized matter wave field for uniform acceleration on matter contained within thereby preventing adverse effects of progressive wave acceleration/deceleration by minimizing jerk ($\Delta a/\Delta t$) and jounce ($\Delta j/\Delta t$).

51. Coupled Oscillators

Tractor/Repulsor beam generated by phased array antenna

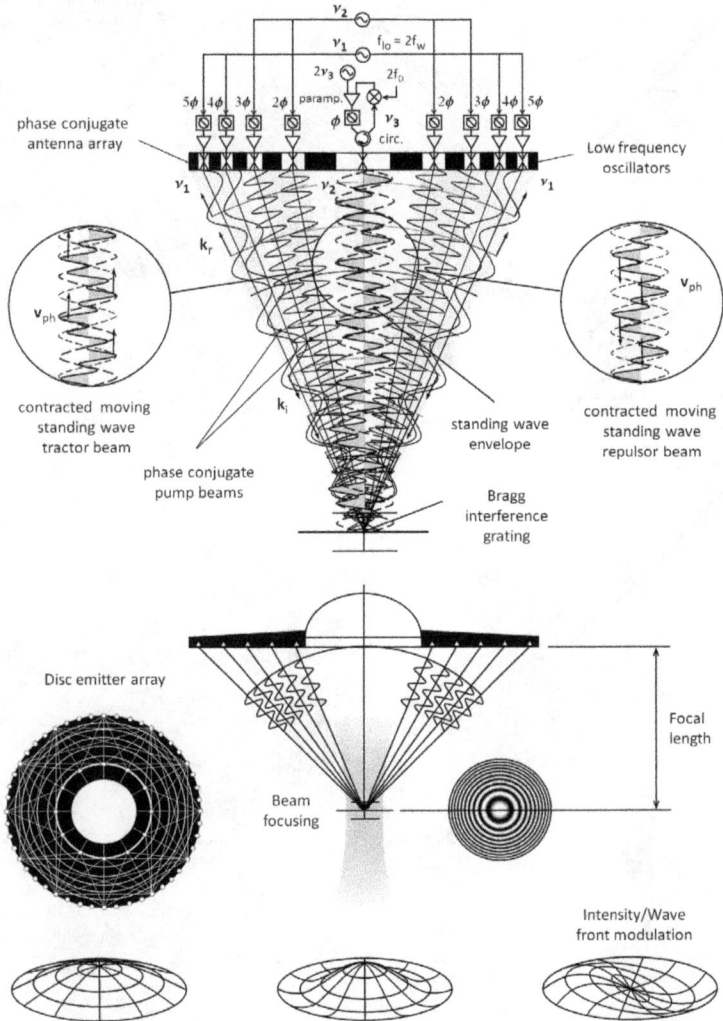

Fig. 51-36. Concept illustration of tractor and repulsor phase conjugate beam forming and focusing. Attraction or repulsion is determined by the direction of the contracted moving standing wave vectors which is controlled by the relative phase and frequency of the synthesized Doppler signals fed to the emitter array. The phased array allows control of the beam boresight direction, focus length, intensity and wave front surface contour. With cross-polarization, the phase front may be warped or the main lobe replaced by multiple smaller lobes.

The scientist describes what is; the engineer creates what never was. – Theodore von Kàrmàn

51. Coupled Oscillators

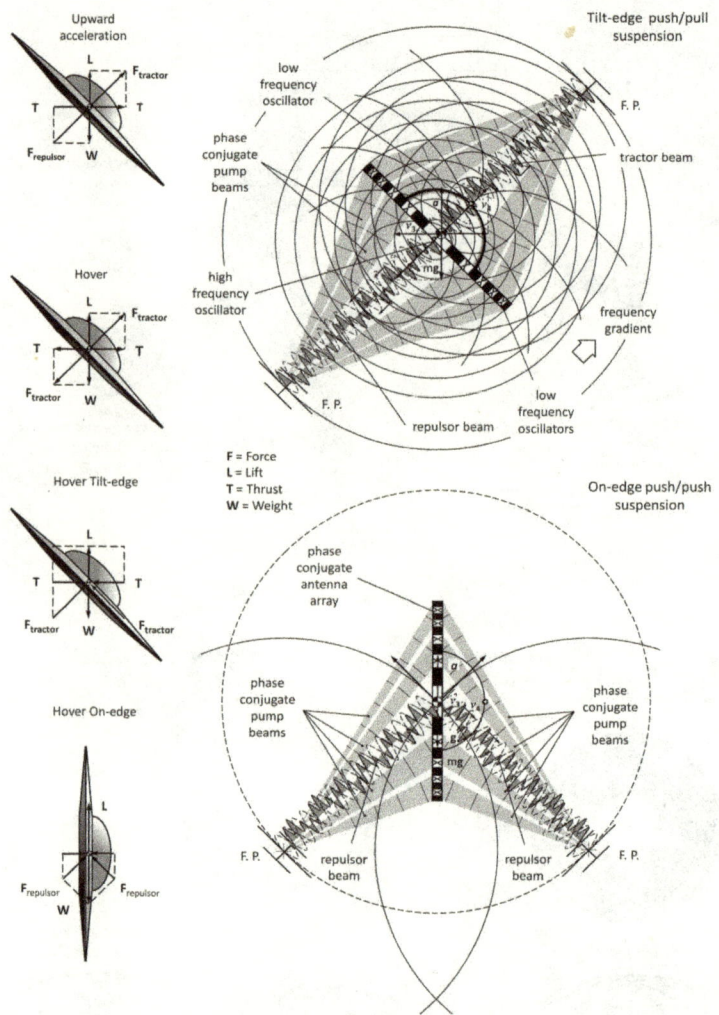

Fig. 51-37. Tilt-edge/on-edge suspension hover modes enabled with opposed, contracted moving standing waves in a tractor/repulsor, tractor/tractor or repulsor/repulsor mode configuration. Chromatic focusing results in a frequency gradient along the boresight augmenting tractor/repulsor beam effect.

Must we continue for ever to beat the air as though it were the enemy of flight? – Sir Victor Goddard

Everything is hard before it is easy. – Johann Wolfgang von Goethe

It was impressive. It was fast. It was maneuverable, and I'd really like to fly it. – David Fravor

500

51. Coupled Oscillators

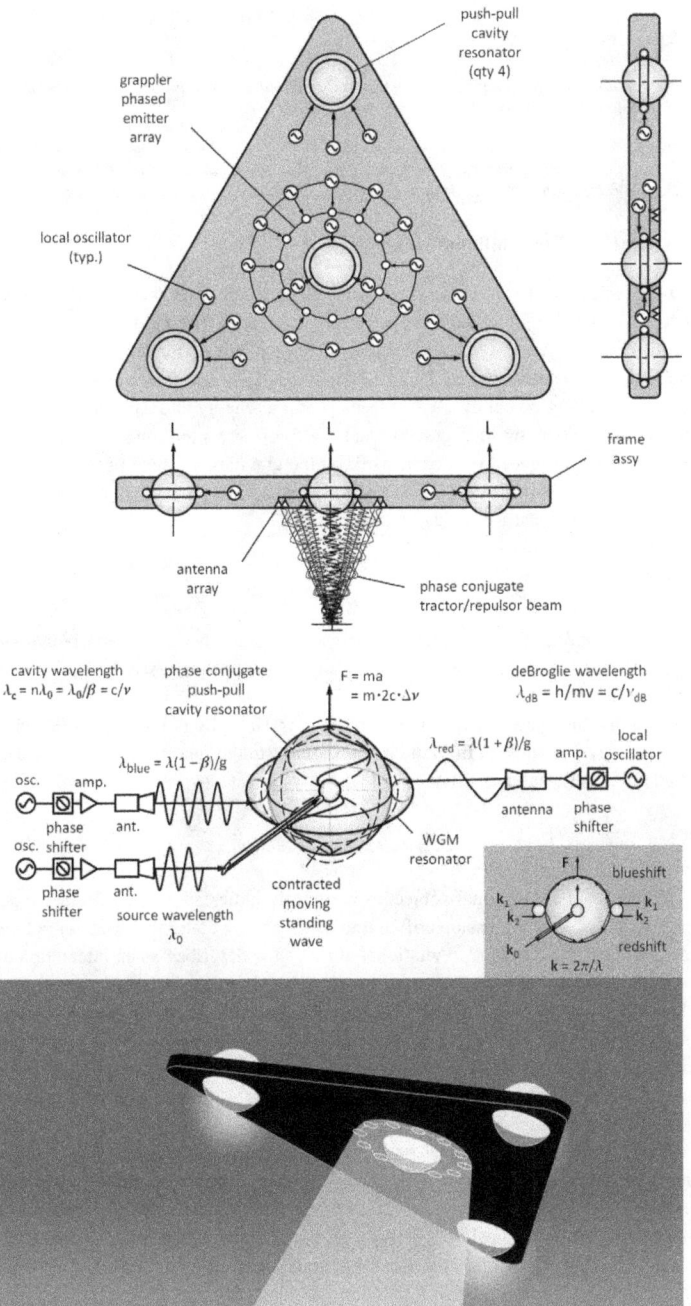

Fig. 51-38. Triangular planform flight vehicle concept with phase conjugate push-pull cavity resonators and phased array push-pull phase conjugate beam.

501

52. Anti-gravity

I have explained the phenomena of the heavens and of our sea by the force of gravity, but I have not yet assigned a cause to gravity. – Isaac Newton

There exists a profound similarity between Newton's law of universal gravity and Sir Humphrey Gilbert's law for the interaction of magnetic poles. And if one can shield electrical and magnetic forces, why can it not also be done with gravitational forces? – George Gamow

52.1 Alteration of gravitational potential

Gravity is described as a conservative force field, hence, work performed is independent of path taken between initial and final positions of an object during displacement within the central force field. Work W is a measure of energy and equals force x distance ($= F \cdot d$) or, equivalently, to power x time ($= P \cdot t$). Energy E equals mass x acceleration x distance ($= m \cdot a \cdot d$). A falling object of mass m loses potential energy and gains kinetic energy as a result of work done by the gravitational field of a central mass M on the object. Conversely, a rising object must exert work on the gravitational field of the central mass to increase potential energy. The change in potential energy ΔU equals the negative of work W done by the gravitational force $F(r)$ on the object.

$$\Delta U = U_f - U_i = -\int_{r_i}^{r_f} F(r)dr = -W \tag{52-1}$$

where $F(r) = -GMm/r^2$. The potential energy is negative because it is a bound state and energy is required to increase separation. Binding energy is released during contraction as radiated waves.

Substituting the Newtonian expression for $F(r)$ into Eqn 52-1 gives the change in potential energy U evaluated from the initial radius r_i to the final radius r_f as measured from the center of a spherical mass M

$$U_f - U_i = -GMm(\frac{1}{r_f} - \frac{1}{r_i}) \tag{52-2}$$

The gravitational field of a mass object as previously indicated is of electromagnetic origin generated by EM wave emissions of component mass oscillators with extreme bandwidth and negative energy density. Gravitational attraction is described as an interaction between coupled oscillators which results in an accelerative force as the mass objects attempt to synchronize in frequency reducing the overall energy level. The acceleration of gravity is a measure of the spectral energy density gradient. The gravitational Poynting vector is proportional to the rate of energy transport. Photons and gravitons respond to variation in EM energy density. Is gravity strictly an attractive force? Purported dark energy seems to suggest an accelerating universe with repulsive gravity effect. The energy density of gravitational and co-gravitational fields are negative with respect to the surrounding vacuum, hence, energy must be added to alter the energy gradient and gravitational acceleration. A negative energy density is dimensionally equivalent to negative pressure (compression) while a positive energy density is equivalent to a positive pressure (expansion). Is there any prospect that the gravitational field may be subject to electronic manipulation and control via EM radiation to alter the energy density and refractive properties of the local vacuum? Can a gravitational field be simulated by electronic means? We will briefly review some implications of an optical theory of gravity and consider an example model based on extrapolations of present day technology to assess plausibility.

52. Anti-gravity

Gravity may be described in terms of complex permittivities ϵ^* and refractive index $K_{PV}(r,\omega,M)$ with real and imaginary components. The Fourier transform representation of the gravitational field frequency spectrum may be described by superposition of odd number of harmonics with real and imaginary frequency components. The complex exponential form of the Fourier series $e^{i\omega t}$, unlike the – form, yields an EM spectrum with positive and negative angular frequencies $(+\omega, -\omega)$. Negative frequencies refer to the rotating complex phasor components $e^{i\omega t}$, $-e^{i\omega t}$ in which the real cos component lags the imaginary sin component [$(e^{i\omega t} = \cos(\omega t) + i \cdot \sin(\omega t)$]. See Fig. 52-1. If the imaginary parts of the signal have the same magnitude, with opposite rotation, the imaginary parts cancel out, leaving a real-valued signal with a positive temporal frequency in the real time domain.

Types of carrier wave modulation by a signal wave

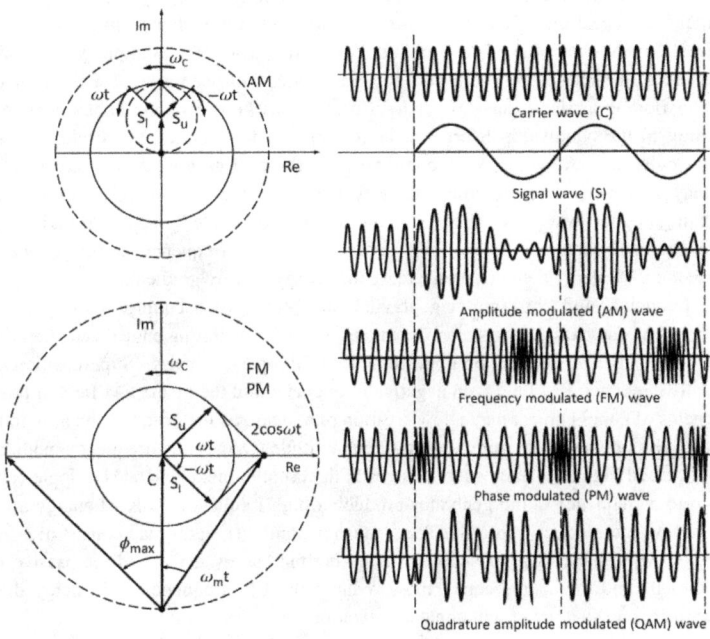

Fig. 52-1. Phasor diagram of amplitude, frequency and phase modulation of a carrier frequency illustrating positive and negative frequency components.

A positive frequency $+\omega = d\phi/dt$ indicates the phase ϕ accumulation increases with time while a negative frequency $-\omega = -d\phi/dt$ indicates the phase ϕ accumulation decreases with time. When the positive and negative frequency envelopes are in phase, one is a mirror image of the other. The negative frequency component $-\omega$ is equivalent to the complex conjugate of the positive frequency component ω^*. Thus, the effect of negative frequency EM emissions in a real-valued optical field may be analyzed in numerical simulations of a gravitational field and cavity mass resonator interactions. The negative gravitational energy of a mass object cancels the positive energy in matter. An imbalance of negative gravitational energy may be expected to impart a repulsive effect of neighboring matter.

52. Anti-gravity

Electronic duplication of the entire gravitational frequency spectrum of any mass object is beyond present day technology. However, it appears it may be possible to selectively modulate a portion of the low end of the gravitational EM spectrum to interfere with the resonant coupling of mass oscillators to decrease or increase the force of attraction ($F_g = m2c \cdot \Delta v \cdot r_u$) by altering the frequency gradient between the active and passive mass. As shown in Fig. 52-1, various types carrier wave modulation by a signal wave schemes may be applied such as amplitude, frequency and phase modulation or combinations of each. Amplitude modulation (AM) of a carrier frequency f_C by a lower modulating frequency f_m produces upper and lower sideband frequencies $f_c \pm f_m$. Frequency modulation (FM) and phase modulation (PM) also produce sideband frequencies but differ from AM modulation in that the phase of the sidebands is shifted by 90°. Quadrature amplitude modulation (QAM) consists of amplitude modulation (AM) and phase modulation (PM).

Mass represents an EM wave interference effect. Resonant out-of-phase EM waves may serve to interfere with mass oscillator frequency coupling, in effect, jamming a portion of the frequency spectrum. The EM wave emissions of a transmitter may be extended somewhat beyond the range of the oscillator carrier frequency, for example, by modulation with a series of very sharp square pulses. A truncated Fourier series of a finite function results in ringing artifacts due to overshoot oscillations known as the Gibbs phenomena. According to the bandwidth theorem, a large range of frequencies $\Delta \omega$ results in a short duration pulse Δt. A synthesized square wave pulse with abrupt corners creates a high frequency spectrum of ripple components which can serve as relatively low frequency modulating signal waves of a high frequency carrier wave generating higher order harmonics further augmented by four-wave mixing in a frequency comb generator to increase the EM energy density and decrease the energy density gradient.

The frequency and corresponding gravitational profile of a coupled mass pair of an active mass M and passive mass m is shown in Fig. 52-2 illustrating a localized alteration to equalize the potential gradient on opposite sides of the passive mass. Superposition of an artificially generated EM field with negative frequencies and the innate EM field of positive frequencies of a mass object may allow creation of a localized distortion of the gravitational potential with opposite curvature. Quadrature amplitude (QAM) plus frequency modulation (FM) of pulsed high frequency square waves with abrupt corners generates a ripple barrage as input to a frequency comb generator producing high frequency radiated energy altering the local EM energy density profile. The frequency comb generator may consist of a fractal sized array with one comb per octave. Increasing the energy density of the passive mass gravitational field facing the active mass reduces the EM frequency and energy density differential decreasing net gravitational acceleration.

Gravitational acceleration is equal to the negative of the gravitational potential gradient ($\mathbf{g} = -\nabla \phi$) and is proportional to the frequency gradient. The maximum EM frequency difference $v_1 - v_2$ equals $-GM/2cr^2$ for a given radius r. In a nonorbital interaction, the minimum gravitation potential occurs at a radius $r = R(m/M)$. For an orbiting mass pair, the minimum potential corresponds to the L1 Lagrange point. The passive mass Hill sphere radius is bounded by the L1 and L2 Lagrange points. The gravitational field energy density $u = -g^2/8\pi G = [(\epsilon_g/\Gamma)/2](\mathbf{E}_x^2 + \mathbf{E}_y^2 + \mathbf{E}_z^2) + [1/2(\mu_g/\Gamma)](\mathbf{B}_x^2 + \mathbf{B}_y^2 + \mathbf{B}_z^2) = [(\epsilon_g/\Gamma)/2]\mathbf{E}^2 + [1/(2\mu_0/\Gamma)]\mathbf{B}^2$ where the zero curvature permittivity and permeability of free space is $\epsilon_0 = \epsilon_g/\Gamma$ and $\mu_0 = \mu_g/\Gamma$, respectively. The energy density of the gravitational field u_V is negative, hence, energy must be added to the field to offset and counter gravitational attraction with an opposing gravito-Poynting vector $\mathbf{S}_{gi} = -\mathbf{S}_g$.

If gravity can be controlled artificially, then by necessity, it must be EM in nature. – Ricardo C. Storti

52. Anti-gravity

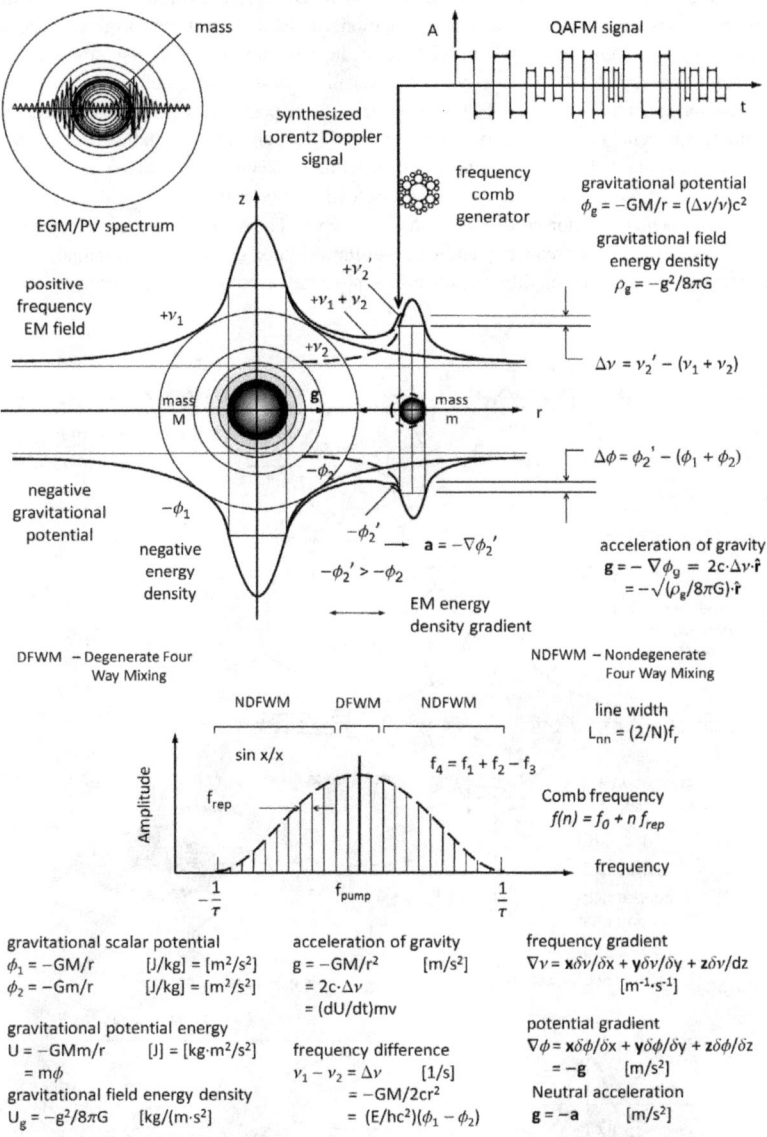

Fig. 52-2. Modified gravitational potential and frequency profile of a coupled mass pair with localized field gradient flat metric plateau on the nearside of the passive mass. Gravitational potential ϕ_g varies as c^2, i.e., $\phi_g = c^2/\Gamma^2 = (1/\epsilon_0\mu_0)/\Gamma^2 = v_p v_g/\Gamma^2 = (\Delta\nu/\nu)c^2$.

The [gravitational] equations correspond largely to those of electrodynamics, [...] up to the sign, and [...] up to a factor ½. – Albert Einstein

52. Anti-gravity

Negative refractive index metamaterial (NIM) allows creation of artificial negative mass and augmentation of gravitomagnetic effects. NI metamaterials exhibit inverse Lorentz-Doppler effects including inversion of gravitational redshift ($\Delta v/v$) over a given spectral bandwidth set by inductance and capacitance of the resonant elements. The gravitational scalar potential ϕ_g (= $-GM/r$ = $(\Delta v/v)c^2$). Negative mass ($-M$) results in negation of gravitational potential $-\phi_g$ (= $+GM/r$ = $-(v/\Delta v)/c^2$. Superposition of positive and negative metamaterials enables partial or complete nullification of gravitational potential. Gravitational potential ϕ_g may be expressed in terms of electrostatic scalar potential ϕ_g = $(A_g/A)\phi_E$ where A_g = gravitomagnetic vector potential (= $(v/c^2)\phi_g$ = $g \cdot v/c^2$ = $(\phi_g/\phi_E)A$) and A = electromagnetic vector potential (= $v\phi_E c^2$) = mv/e). The height of the frequency hill is reduced by the negative frequency addition and the depth of gravitational potential well is reduced accordingly. The modified gravitational potential resembles the sombrero function.

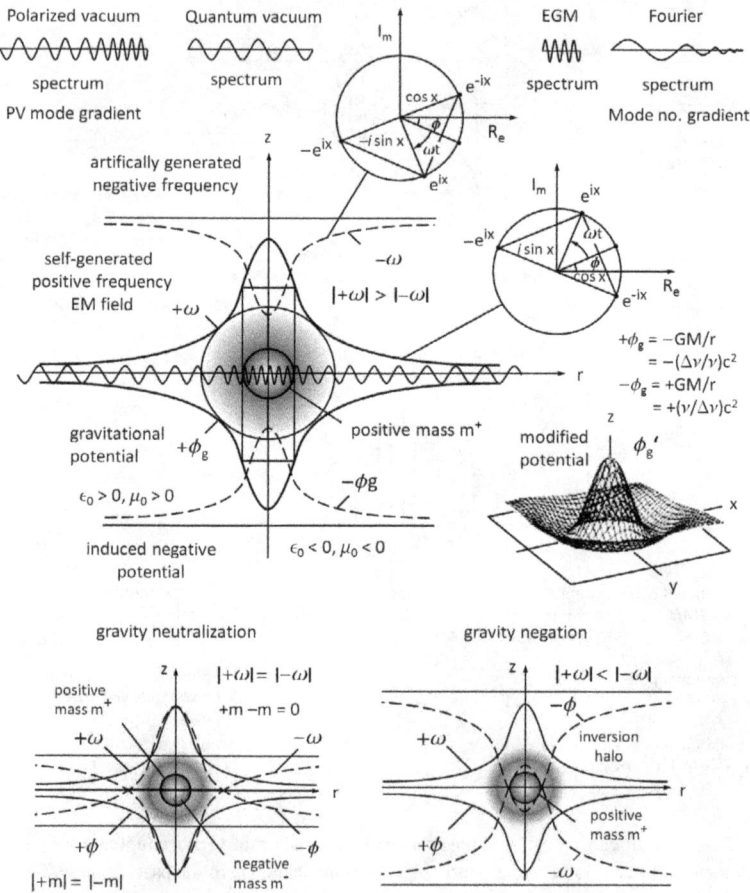

Fig. 52-3. Modified gravitational frequency and gravitational potential profile of spherical mass object by superposition of positive and negative refractive index metamaterials. An inverse gravitational potential with negative mass yields an anti-gravitational effect.

52. Anti-gravity

As shown in Fig 52-3, the reduction in amplitude decreases the acceleration of gravity and the escape velocity. Rest mass energy E ($= mc^2 = ne\Phi$) is proportional to the enclosed volume under the curve. The rest mass energy density $\rho = U(r,m) = -\nabla\cdot g/4\pi G = 3mc^2/4\pi r^3$. The gravitational field energy stored surrounding a solid spherical mass $U = -Gm^2/2r$. The gravitational energy density $u = -g^2/8\pi G$ where the acceleration of gravity $\mathbf{g} = -\nabla\phi \simeq -G(m/r^2)\mathbf{r}_u - G^2(m/2c^2r^3)\mathbf{r}_u$. For mass in motion $\mathbf{g} = -\nabla\phi - \mathbf{A}_g/dt = -\nabla\phi + \mathbf{g}_k$. The mass energy density of the gravitational field $\rho_{gm} = g^2/8\pi Gc^2 = G(m^2/8\pi c^2 r^4)$. The mass energy density of the gravitational field is negative and orders of magnitude less than the rest mass energy density $|\rho_g| \ll \rho$. The force of gravity is always attractive and cumulative in proportion to the number of constituent mass oscillators. A positive mass is an analog of a sink while a negative mass is an analog to a source. Negative mass corresponds to a positive scalar potential (negative frequency) and is equivalent to a negative permittivity difference $\Delta\epsilon$, i.e., $m^- = (Rc^2/4\pi G)\cdot(\Delta\epsilon/\epsilon_0)$. Negative mass acts as a diverging lens refracting light oppositely. An anti-gravity effect may be simulated with a surrounding medium of negative index of refraction meta-material with an index of refraction $n = \sqrt{(\epsilon(\omega)\cdot\mu(\omega))} < 0$ with simultaneous real parts of frequency dependent electric permittivity $\epsilon(\omega)$ and magnetic permeability $\mu(\omega)$. In such an arrangement, a grazing incident light beam or positive test mass will be deflected away from the central mass simulating repulsion.

An example of a rotating electromagnetic field with positive and negative frequencies in the form of a counter-rotating quadrupole spira mirabilis field with forward and backward swept flux is shown in Fig. 52-4. The field pattern, for example, may be in the form of an electric field intensity **E**, magnetic field strength **H**, magnetic vector potential **A**, Planck dipole density ρ_P wave or delta frequency $\Delta\omega$. The field pattern represents a combined source/sink and vortex flow and resembles that of two coaxial, counter-rotating magnetrons. Scissoring action of counter-rotating wavefronts tends to sweep charged particles away from nodal interference points. A positive frequency $+\omega$ is equivalent to positive energy ($\omega = E/n\hbar$) while a negative frequency $-\omega$ is equivalent to negative energy ($-\omega = -E/n\hbar$). Energy $E_n = ne\Phi$ where n = no. of charges, e = electric charge and Φ = potential (voltage). With backward swept flux arms resembling a radial outflow compressor, counter-rotation tends to pump charged particles outward reducing charge density. Conversely, with forward swept flux arms resembling a radial inflow turbine, counter-rotation tends to pump charged particles inward increasing charge density. A positive angular frequency ω is associated with positive energy density ($+\omega = E/n\hbar$) while a negative frequency $-\omega$ is associated with negative energy density ($-\omega = -E/n\hbar$) where E is rest mass energy ($= mc^2 = m_0/\sqrt{(1 - v^2/c^2)} = ne\Phi$) and $\hbar$ is the Dirac constant ($= E/\omega$). A torsional field may be created by a swept frequency gradient $d\omega/dt$ with radial and azimuthal frequency step change.

A whispering gallery mode (WGM) resonator may be configured to interconvert linear and rotational momentum forming in effect an inertial damper or inerter. In addition to mass, other sources of gravity in the stress-energy momentum tensor include kinetic energy, linear momentum, angular momentum and stress energy. Elementary particles contain considerable amounts of spin angular momentum. If such angular momentum could be converted, a significant amount of linear momentum would be produced as suggested by the relation $E = mc^2 = pc = Lc/L_p$ where p = linear momentum ($= mv = \rho_e\cdot v/c^2 = qA = S/c^2$), **L** = angular momentum ($= \mathbf{p}L_p$) and L_p = Planck length ($= \sqrt{(hG/2pc^3)}$).

As Sabbata and Sivaram[122] have shown, torsion is a geometric effect of spin in spacetime analogous to curvature as a geometric effect on spacetime. All EM wave interactions can be understood as originating in spin-curvature coupling. Under the influence of curvature and torsion, a closed path becomes open such that the tetrad undergoes both rotation and translation inducing a closure failure spin defect.

52. Anti-gravity

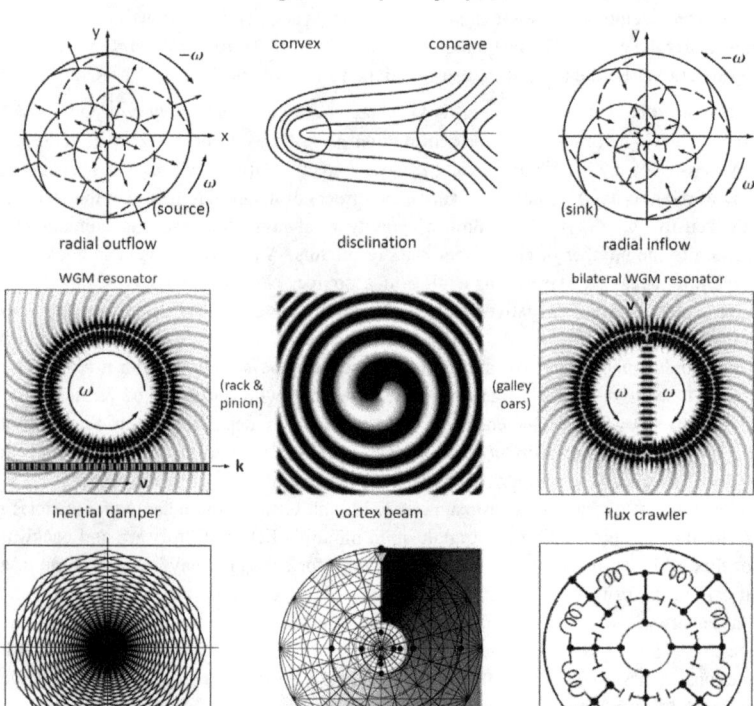

Fig. 52-4. A spira mirablis field may generated to produce a source or sink vortical flow pattern. A rotating EM field of a whispering gallery mode (WGM) metamaterial resonator may be coupled with an external wave guide to provide a means for conversion of linear momentum into rotational momentum to form an inertial damper or inerter. Bilaterally symmetric D-shaped WGM resonators may be combined to generate a flux crawler for conversion of EM wave rotary motion into resonator linear motion. Reactance is analogous to inertial mass. A positive energy source yields an increased heat signature whereas a negative energy sink results in a reduced heat signature subject to any frequency shift.

Although quite remarkable progress has been demonstrated at the lower frequencies, it is still possible to gain fame and glory by making a contribution at the higher frequencies. – Ben Munk

Scalar waves and longitudinally propagating standing waves are described by the wave equation. Every emitted wave contains both longitutinal and transverse parts! – Konstantin Meyl

He that removeth weight doeth as much advantage motion as he that addeth wings. – John Pym

Because EGM [Electro-Gravi-Magnetics] models a mass-object as existing in equilibrium with the QV [Quantum Vacuum], the local energy state of the vacuum may be considered to be equivalent to the mass-energy of the object it encapsulates. – Geoffrey S. Diemer

52. Anti-gravity

In Fig. 52-5 is a notional sketch of the gravitational potential well of a mass object in which the gravitational frequency spectrum is modified by superposition an external EM beam of asymmetric intensity profile with positive and negative frequency components. In this speculative concept, the complex index of refraction $K_{PV} = \sqrt{(\epsilon(\omega) \cdot \mu(\omega))}$ is modified by addition of a positive frequency to the real part and a negative frequency to the imaginary part increasing the degree of vacuum polarization. The extreme high frequency content of the gravitational field induces a nonlinear polarization of the vacuum. For small input electric fields, the polarization remains approximately linear; but with electric fields of large intensity, the polarization results in waveform distortion with increased harmonic components. Alteration of the local refractive index $K_{PV}(r,\omega,M)$ yields a non-uniform gravitational field gradient decreasing escape velocity towards the region of reduced EM field intensity. The energy density gradient results in an acceleration of mass in the direction of reduced potential. Energy flow is in the direction the gravito-Poynting vector.

Alteration of **E/B**, **E_g/B_g**, **A_g/A** ratios, matter wave beat frequency $\Delta \nu$ interference, internal wave system phase shift modulation, synthesized scalar ϕ and vector potential **A**, gravito-magnetic field augmentation via negative index metamaterials, etc may allow inertial mass reduction (m = E/c^2 = $E/(E/B)^2$ = $2L/(E^2 - c^2B^2)$ = $F/2c \cdot \Delta \nu \cdot \hat{r}$ = $E \cdot \epsilon_0 \mu_0$ = $E \cdot \epsilon_g \mu_g$). Negative index metamaterials enable creation of artificial negative mass (m = $\mp 2(T - V)/(\epsilon_0 E^2 - c^2 B^2/\mu_0)$. Gravitomagnetic effects can be dramatically increased by alteration of the gravitomagnetic permeability μ_g utilizing artifically created negative mass using negative refractive index metamaterials and creation of the gravitomagnetic induction equivalent of the Faraday disc dynamo. Electromagnetic induction expressed as **E** = F/e = **v** x **B** and $\nabla \times \mathbf{E} = \delta \mathbf{B}/\delta t$ may be analogously expressed in gravitoelectromagnetic induction as **g** = F/m = **v** x **B_g** and $\nabla \times \mathbf{g} = \delta \mathbf{B}_g/\delta t$, respectively.

As previously noted, the gravitational field of the Earth is characterized by EM frequencies concentrated well above the THz range of present day technology to synthesize. Pulse repetition of artificially generated sawtooth waves with abrupt corners enables synthesis of higher frequencies in the frequency domain with a sinc function profile. Pulse amplification of a portion of lower accessible frequencies with high frequency comb, prescaler or aliasing techniques to generate negative frequencies may potentially allow localized alteration of the gravitational frequency spectral profile such that the EM energy density gradient is inverted outward of the passive mass and increasing the synchronization frequency decreasing the force of attraction. Gravitational acceleration may be offset by Gravito-Poynting vector **S_{gi}** outflux countering the natural Gravito-Poynting vector **S_g** influx by which gravitational potential energy ϕ_g is converted to kinetic energy **A_g**.

A man may imagine things that are false, but he can only understand things that are true. – Isaac Newton

No great discovery was ever made without a bold guess. – Isaac Newton

Surely the force of gravitation and its probable relation to other forms of force may be attacked by experiment. Let us try to think of some possibilities. – Michael Faraday

Now what is the there analogous to magnetic force in the gravitational case? And if it has an analogue, what is there to correspond with electric current?. – Oliver Heaviside

Rather than application of an external force, the speed of an object may be influenced in other ways, such as by a phase shift from inside the system to change the mode of motion. In order to achieve anti-gravity, it is only necessary to equalize frequencies neutralizing the frequency gradient in accordance with the principles of Rhythmodynamics. – Yuri N. Ivanov

52. Anti-gravity

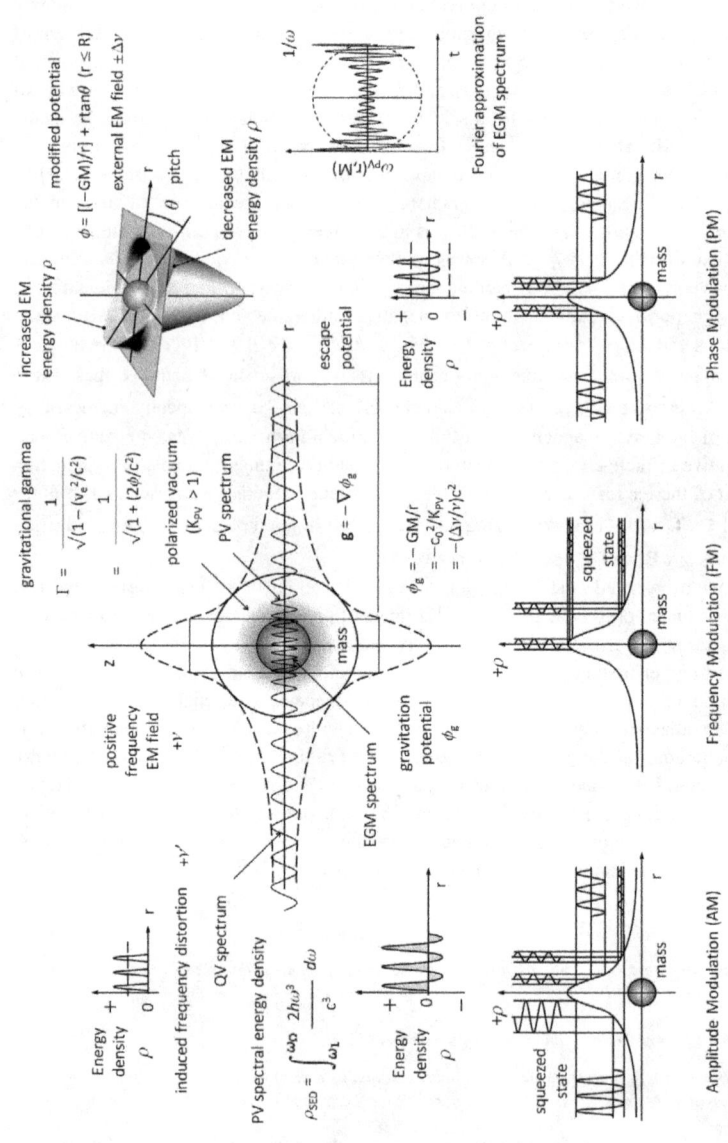

Fig. 52-5. Induced gravitational frequency distortion of a mass object by irradiation with modulated external electromagnetic beam. Asymmetric distortion of the polarized vacuum (PV) spectrum results in acceleration $\mathbf{g} = 2c \cdot \Delta\nu \cdot \hat{\mathbf{r}} = 2\nu \cdot \phi_g \cdot \hat{\mathbf{r}}/c$.

52. Anti-gravity

52.2. Spectral energy density modulation

I had already decided that to create anything of value you had to cover your own ground, observe, and make your own rules and decisions. – Syd Mead

Rest mass m as a function of energy according to the Einstein relation (= E/c^2) may be described in terms of the ratio of electric field intensity to the magnetic flux density in accordance with the Maxwell relation (= $E/(E/B)^2$. In the far field region, E/B = c for r > $\lambda/2\pi$, hence, mass varies as the square of the ratio of E/B. Mass reduction corresponds to minimizing the amplitude, divergence and curl of E and maximizing the magnitude and curl of B. Modulation of the relative phase ϕ of E and B likewise leads to alteration of wave impedance and velocity of light. As shown in Figs. 1-4 & -6, the ratio of $E/H = E/\mu_0 B$ is a measure of electromagnetic wave impedance Z_0 which varies as phase change with distance from the source generator. In the near-field region of the emitter, the E-field leads the H-field with a consequent increase in phase velocity ($v_p = 1/Z = H_\theta/E_\theta$). To effect a change in apparent inertial mass m_i (= E/c^2 = F/a = W/g = $E/(E/B)^2$ = $2L/(E^2 - c^2B^2)$ = $F/2c\cdot v\cdot\hat{r}$), how can the relative phase be modified? One potential option may be a cross-field antenna which allows independent control of the phasing of the E and H vectors unlike a conventional dipole Hertzian antenna. To offset the acceleration of gravity g, an opposing acceleration a (= $(F/E)\cdot(E/B)^2$) = $2v\cdot\phi_g\cdot\hat{r}/c$ may be generated (radiated spectral power blue-shift). In terms of frequency, the acceleration may be expressed as

$$a = (F/E)\cdot(E/B)^2 = (F/mc^2)\cdot(E/B)^2 = 2(E/B)\cdot\Delta\omega\cdot r_u \qquad (52\text{-}3)$$

where
- a = acceleration ($F/m = (c^2/\phi)E = ((2c^2/h)(mv/\Delta\lambda)/\lambda_{dB})\cdot\hat{r} = F/2c\cdot v\cdot\hat{r}$) [m/s^2]
- F = force (= $ma = m2c\cdot\Delta v\cdot r_u = dL/dx$) [N = kg(m/s^2)]
- E = energy (= $mc^2 = m/(\rho_e\cdot v/p) = T + V = (c^2/\phi)E$) [J = N·m = kg(m^2/s^2)]
- E = electric field strength (E/q^+) [volts/m]
- B = magnetic flux density (E/c) [T = Wb/m]
- c = speed of light (= $E/B = \omega/k = 1/\sqrt{(\epsilon_0\mu_0)} = \sqrt{(v_p v_g)} = c_0/\Gamma$) [m/s]
- $\Delta\omega$ = oscillator frequency difference (Shepard tone band offset) [Hz]

To concentrate energy toward the extreme upper frequency range, the antenna may, for example, be in the form of a high gain, asymmetrical, self-similar, multi-resonant, microstrip fractal element antenna (FEA) array fed with a mode-locked quadrature amplitude frequency modulated (QAFM) triangular or square wave pulse signal. The number of resonant modes of a fractal antenna increases with element iteration number N. The number of repeating antenna elements (N iteration order fractals) is ultimately limited in physical size to the scale of atomic dimensions. Nano-scale antenna elements such as 1D quantum wells, 2D quantum wires, or 3D quantum dots with characteristic dimensions on the order of < 0.1 μm may be formed, for example, from C_{60} buckyballs, nanotubes, graphene sheets, etc. A graphene repeating subunit is ~0.001 μm (= 1 nm) in size corresponding to an EM wavelength of 1.0 x 10^{-6} m and frequency of 2.99E14 Hz. In a transmitter (TX) mode, the antenna may be used to increase EM energy density in the beam direction or in a receiver (RX) mode absorb incident EM energy.

A notional sketch of a nonlinear fractal antenna (endfire) with rectangular cross-section waveguide is shown in Fig. 52-6. The scaling of the fractal elements is varied with frequency to enhance higher resonant frequency modes. Odd-numbered harmonics may be realized with odd-number of repeating elements. Multiple-element high gain fractal arrays facilitate near field forward gain. A nonlinear contraction of fractal antenna element length shifts the overall frequency response of the antenna array towards higher frequencies. Each

52. Anti-gravity

smaller fractal element than the driven element acts as a director increasing antenna directivity (gain). Spin Hall or magnon excitation facilitate subwavelength propagation increasing radiation pressure. Fractal elements larger than the driven element act as reflectors. Collectively, a large number of active dipoles results in a broadband, high gain antenna spaced in a periodic function of frequency and is similar to a conventional log-periodic dipole array. Independent control of the **E**- and **H**- field vectors allows variable phasing in the near-field reactive zone. The maximum number of N-iterations of patch elements required equals the number of octaves within the spectral bandwidth $\Delta\omega_{PV}$ for a given mass. Radiation resistance, gain loss and ohmic losses of a fractal antenna increases with frequency. Electrical resistance of wires at the nanoscale exhibit resistance in discrete multiples of the Klitzing constant R_K (= $h/e^2 \cong 25{,}812$ Ω). Ohmic resistance is typically large compared to the radiation resistance for small antennas and lowers antenna efficiency. Effects of electromigration become problematic at such scales although mitigated by increase in redundancy. The number of N-iterations of repeating fractal units is equal to the number of octaves of the operating bandwidth $\Delta\omega$.

To increase EM energy density of a directed antenna radiation pattern, in addition to conventional beam forming techniques, one method may be to employ a phase conjugate reflector behind the antenna array to suppress backward radiation by reflecting the energy normally lost in back lobes in the forward direction. This effect is illustrated in Fig. 52-7. Antenna directivity may be increased by increasing antenna diameter, increasing frequency, image reflection, synchrotron acceleration, illumination tapering or backside lobe suppression by amplified phase conjugation reflection. For a uniformly illuminated aperture, sidelobes can result in a mainlobe powerloss of ~1.5 dB (~25 percent). Pump beam amplification can increase the power of reflected backward lobes increasing forward gain and beam intensity. See Fig.52-8.

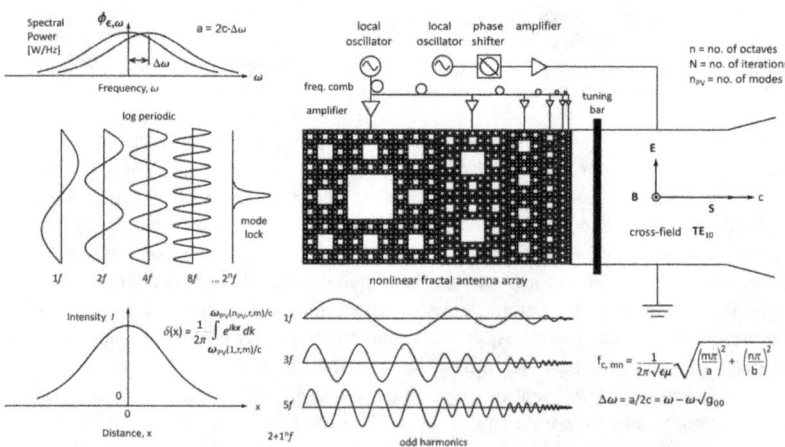

Fig. 52-6. Concept illustration of a N-iteration, ultrawideband, nonlinear fractal cross-field antenna (CFA) with AM/FM modulation. Fractal elements are microresonator arrays. Induced acceleration of matter is proportional to spectral power frequency shift.

The formation of a problem is often more essential than its solution. – A. Einstein and L. Infeld

52. Anti-gravity

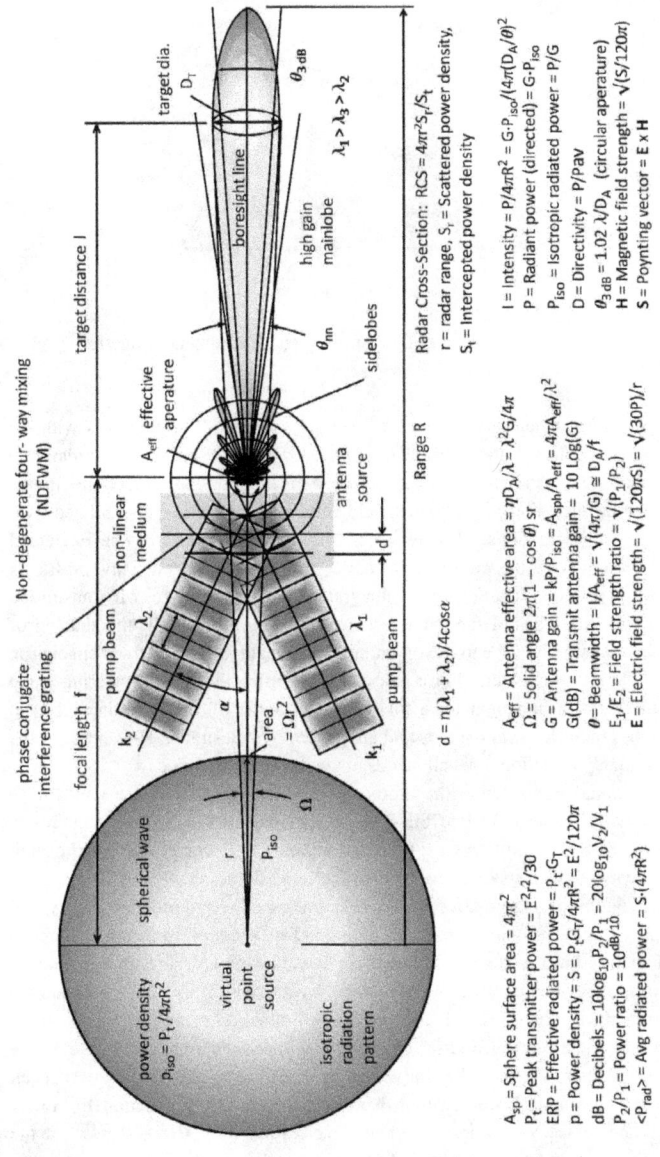

Fig. 52-7. Backward radiation suppression and forward antenna gain by amplified phase conjugate reflector. Negative refractive index metamaterials may inhibit radar tracking and velocity measurement.

52. Anti-gravity

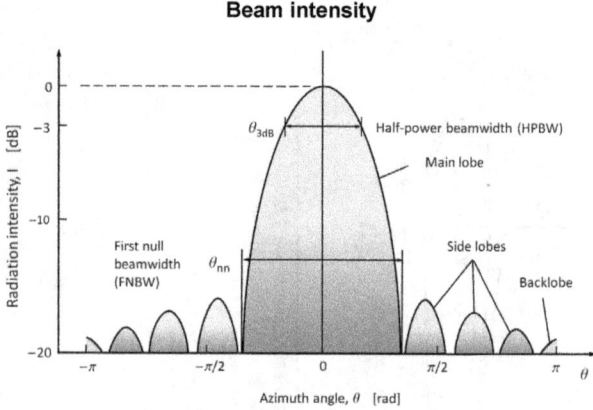

Fig. 52-8. Forward gain and beam intensity by amplified phase conjugate reflection.

The Earth's gravitational field is estimated to span a frequency ~77-80 octaves with the bulk of the energy concentrated above the PHz range. As a consequence, to simulate the Earth's gravitational field bandwidth would require a like number of fractal iterations. Smaller masses exhibit a smaller gravitational field bandwidth and may be simulated with correspondingly fewer interactions. Lower order low energy modes may be neglected allowing a very high waveguide cutoff frequency f_c. The radiated intensity profile is approximated by the delta function $\delta(x)$ integrated over the range of transmitted frequencies. Grating side lobes of the main antenna beam are produced if the spacing of radiating elements d is large relative to the operating wavelength λ. The element spacing in terms of Bragg angle is $d = n\lambda/\sin\theta_n$. For a phased planar array, the useable steering angle is ~±60° about the antenna boresight for a total included angle of 120°. Gating lobes in addition to side lobes radiate power off axis and are generally undesirable. However, for the given application, highly directional pencil beams or stealth is not required.

The peak gravitational frequency at the Earth's surface is estimated at $\cong$ 5.19-8.29E26 Hz equivalent to a wavelength of 3.61-5.76E-19 m. This length is smaller than a hydrogen atom for example (i.e., $\cong$ 0.06 nm = 6.0E-11 m). At extreme frequencies, the usual design of micro-sized arrays with dimensions smaller than the effective resonant wavelength (e.g., $\lambda/10$) allowing simulation of a large effective area becomes the inverse problem of physical dimensions $\gg \lambda$. Cavity resonators may be combined to produce sum and difference frequencies. At about 10 nm, corresponding de Broglie wavelength of a room-temperature electron, electrons may be trapped in standing waves within quantum-mechanical resonators. For practical structures, we are physically limited to structures > ~10 nm = 1.0E-8 m in size and must rely on Fourier or wavelet synthesis to generate such extreme gravitational frequencies. As will be shown later, smaller masses have lower peak gravitational frequencies and lower overall bandwidth requiring proportionally fewer octave doublings and lower N-iterations. For an acceleration of 1 g (= 9.8 m/s^2) at the Earth's surface, the associated frequency difference according to the Ivanov relation (= $2c\cdot\Delta\nu$) is 1.63E-08 Hz (equivalent to a wavelength 1.839 m) and is easily generated.

The EGM spectrum is a wave function representation of mass-energy obeying a Fourier distribution such that the number of modes decreases as energy density increases, – Geoffrey S. Diemer

52. Anti-gravity

A periodic square wave may be represented as harmonic multiples of a given fundamental frequency f_0. Square waves when expanded as a Fourier series yield an infinite number of odd harmonic terms in the frequency domain. See Fig. 52-9. Ripple is reduced by increasing the number of harmonic modes. Harmonics are increased by chirping the signal. To produce a square wave, the harmonics must be in phase such that positive or negative peaks occur the same time as the maximum. Since the duration of positive and negative excursions for each harmonic are equal, the amplitude of even harmonics cancel out. In generation of square waves, ringing occurs with the finite frequency response of real world devices due to inability to accommodate an abrupt rate-of-change in rise and fall. As described by Storti *et al*, the acceleration of gravity **g** may be represented as a Fourier harmonic series in the time domain. The summation of Fourier polynomials over the spectral bandwidth $\Delta\omega$ yields a square wave. The bandwidth $\Delta\omega$ is the frequency interval ($\omega_\Omega - \omega_{PV}(1,r,m)$) where ω_Ω is the harmonic cut-off frequency and $\omega_{PV}(1,r,m)$ is the fundamental frequency. The amount of energy [Watt·sec] transmitted in a square wave pulse is equal to the peak power [Watts] times the pulse width [sec]. The total amount of energy transmitted in a given period equals the average power times the pulse period. Average power P_{avg} may be increased in any of three ways: 1) by increasing the pulse repetition frequency f_r, 2) by increasing the pulse width τ, or 3) by increasing the peak power P. Power is the rate of flow of energy or work per unit time (P = E/t) whereas action equals energy times time (S = E·t). The amount of energy in a transmitted pulse is equal to the output power times the pulse duration. For a square wave, the energy per pulse is equal to the peak power times the pulse width. The duty cycle is a measure of the fraction of time τ/T during which power is transmitted. As an example, for a pulse repetition frequency f_r = 1 kHz with pulse duration $\tau = 1\ \mu\text{sec}$ and an interpulse period T = 1 msec, the peak power P = 1 MW while the average power P_{avg} = 1 kW. The energy per pulse is 1 J with a duty factor = 0.001. Refer to Fig. 52-10.

The Fourier transform defines the relationship of a signal in the time domain and its equivalent representation in the frequency domain. Modulation of a continuous carrier wave by a pulsed square wave produces frequency sidebands above and below the carrier frequency f_c. The number of spectral lines in each sideband increases as the pulse repetition frequency f_r is reduced. The spectrum becomes continuous as the pulse repetition frequency goes to zero. Superposition of a high intensity short duration EM signal of high pulse repetition frequency focused in the near field of the passive mass m opposing the active mass M increases the local EM energy density and, if sufficiently intense, equalizes the field gradient across the passive mass. The full EM frequency spectrum of the active mass need not be replicated if additional lower energy photons are produced to offset the energy deficit. With a high amplitude frequency pulse in excess of the corresponding gravitational frequency amplitude of the passive mass, a comparable gravitational field energy may conceivably be produced.

As shown by Storti *et al*, the local spectral energy density (SED) per frequency mode varies with the cube of the frequency $\rho_{SED}(\omega) = 2h\omega^3/c_0^3$. An energy density spectrum that is Lorentz invariant, according to stochastic electrodynamic theory, is one in which the spectral energy density ρ_{SED} varies with frequency ω^3.

The energy spectrum associated with matter; termed the "EGM spectrum" is a wavefunction representation of matter obeying a Fourier distribution such that the number of (harmonic) modes decreases as energy density increases. – Riccardo C. Storti

The EGM construct yields a quantized description of gravity. – Geoffrey S. Diemer

52. Anti-gravity

A zero-point energy of $E = h\nu/2$ is associated with each harmonic mode of a quantized EM radiation field. Summing the energy per mode over the frequency bandwidth corresponds to the classical zero-point fluctuations (ZPF) of SED theory. The change in energy density per odd frequency mode increases with frequency. The acceleration of gravity **g** is a measure of cumulative effect of change in energy density summed over the gravitational frequency bandwidth $\Delta\omega$.

With a narrow, high intensity, high frequency repetitive EM pulse of bandwidth $\Delta\omega$, the local gravitational field energy density U_ω may be correspondingly altered:

$$U_\omega(r,m) = \sum_{\omega_1}^{\omega_0} \rho_{SED}(\Delta\omega) + \sum_{\omega'_1}^{\omega'_0} \rho'_{SED}(\Delta\omega') \tag{52-4}$$

where ρ_{SED} is the natural spectral energy density and ρ_{SED}' is the artificially generated spectral energy density. With sufficient energy density, the field gradient as measured by the acceleration of gravity across the passive mass may thus be offset.

A spectrum of a train of rectangular pulses in the time domain may be represented in the frequency domain. The mathematical transformation of the signal from the time domain to the frequency domain is effected by means of the Fourier Transform. The inverse transformation of a signal spectrum from the frequency domain to the time domain is accomplished by means of the Inverse Fourier Transform.

Square wave pulse in time and frequency domain

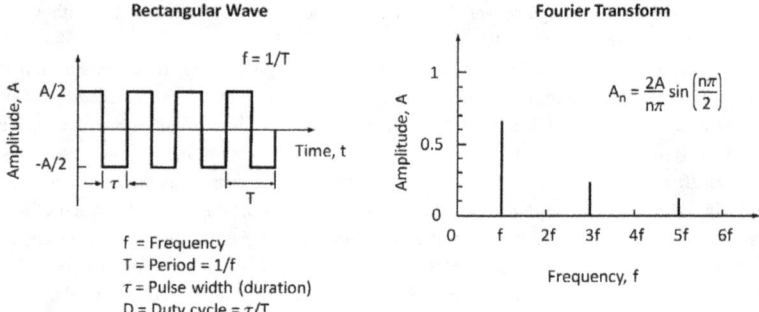

Fig. 52-9. Fourier transform illustrating a square wave signal pulse in the time domain represented in the frequency domain as a spectrum of equally spaced spectral lines consisting of the fundamental and harmonic frequencies. Transformation from the frequency domain to the time domain is called the inverse Fourier transform.

If we consider a finite energy flux E_f given by the integral of an energy signal f(t) as a function of time

$$E_f = \int_{-\infty}^{\infty} f^2(t)dt \tag{52-5}$$

and substitute the inverse Fourier transform of f(t) yields an expression represented as a function of frequency

52. Anti-gravity

Pulse energy

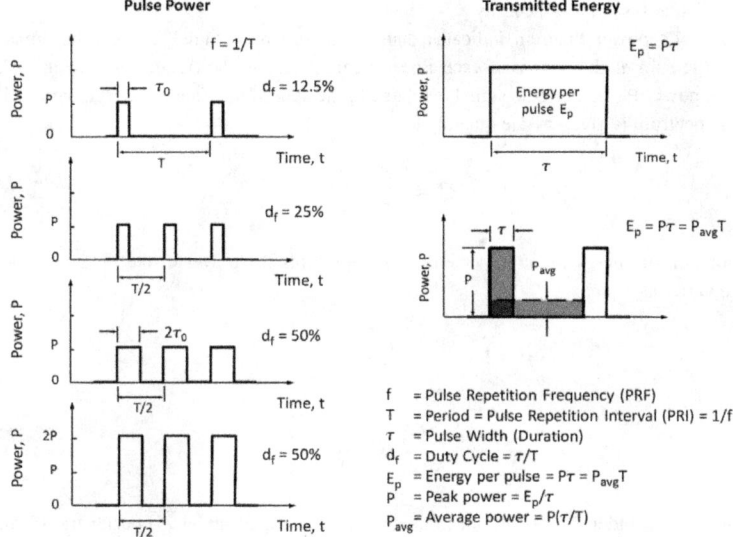

Fig. 52-10. The amount of energy in a transmitted pulse is equal to the output power times the pulse duration.

$$E_f = \frac{1}{2\pi} \int_{-\infty}^{\infty} F(\omega)F(-\omega)d\omega \qquad (52\text{-}6)$$

As $F(\omega) = F^*(-\omega)$ are conjugates for a real valued f(t), Parseval's theorem for a Fourier transform is

$$E_f = \frac{1}{2\pi} \int_{-\infty}^{\infty} |F(\omega)|^2 d\omega \qquad (52\text{-}7)$$

where $|F(\omega)|^2$ is the spectral energy density of the function.

The energy contained in a given frequency band $\omega_1 < \omega < \omega_2$ is given by

$$E_f(\omega_1, \omega_2) = \frac{1}{2\pi} \int_{\omega_1}^{\omega_2} |F(\omega)|^2 d\omega \qquad (52\text{-}8)$$

The increment of energy lying in a increment of bandwidth $\Delta\omega$ is

$$\Delta E_f = (1/2\pi) \cdot |F(\omega)|^2 \Delta\omega \qquad (52\text{-}9)$$

The distributed energy of a radiated signal from a directional antenna in a pencil beam is approximated by

$$E = K[\sin x/x] \qquad (52\text{-}10)$$

52. Anti-gravity

where $x = \pi(L/\lambda)\sin\theta$ in which E is field strength, L is length of antenna array, λ is signal wavelength and θ is the azimuth angle of the point of measurement from the boresight perpendicular bisector of the array.

Parseval's power theorem indicates that the energy of a signal in the time domain equals the sum of the square of each Fourier component in the frequency domain. The average power P of a periodic signal is related to the sum of the Fourier coefficients. The power spectrum is given by the quantity $|c_n|^2$.

$$P = \sum_{n=-\infty}^{\infty} c_n c_n^* = \sum_{n=-\infty}^{\infty} |c_n|^2 \qquad (52\text{-}11)$$

The amount of energy radiated per unit time represents the power of the radiated waves which varies as $K[\sin x/x]^2$

$$P = \int_0^\infty \frac{\sin^2(x)}{x^2} dx \qquad (52\text{-}12)$$

$$P = \frac{1}{2}\int_{-\infty}^\infty \frac{\sin^2(x)}{x^2} dx = \frac{\pi}{2} \qquad (52\text{-}13)$$

The total area under the power spectrum of a pulsed signal under a Fourier transform is equal to the total radiated signal energy. A rectangular pulse modulated sinusoidal carrier wave is illustrated in Fig. 52-11.

Belief gets in the way of learning. – Robert A. Heinlein

Mathematics is the tool specially suited for dealing with abstract concepts of any kind and there is no limit to its power in this field. A great deal of my work is playing with equations and seeing what they give. – P.A.M. Dirac

Just by studying mathematics we can hope to make a guess at the kind of mathematics that will come into the physics of the future... If someone can hit on the right lines along which to make this development, it may lead to a future advance in which people will first discover the equations and then, after examining them, gradually learn how to apply them... – P.A.M. Dirac

Mathematics is not only one of the most valuable inventions – or discoveries – of the human mind, but can have an aesthetic appeal equal to that of anything in art. Perhaps even more so, according to the poetess who proclaimed, "Euclid alone hath looked at beauty bare." – Arthur C. Clarke

Everything starts as somebody's daydream. – Larry Niven

Go confidently in the direction of your dreams. – Henry David Thoreau

Clarke's First Law: If an elderly but distinguished scientist says that something is possible, he is almost certainly right; but if he says that it is impossible, he is very probably wrong. – Arthur C. Clarke

There are three classes of people: those who see, those who see when are shown, those who do not see. – Leonardo Da Vinci

Thus any form of gravity control would also be a propulsion system. – Arthur C. Clarke

Come on, Rory! It isn't rocket science, it's just quantum physics! – Dr. Who

52. Anti-gravity

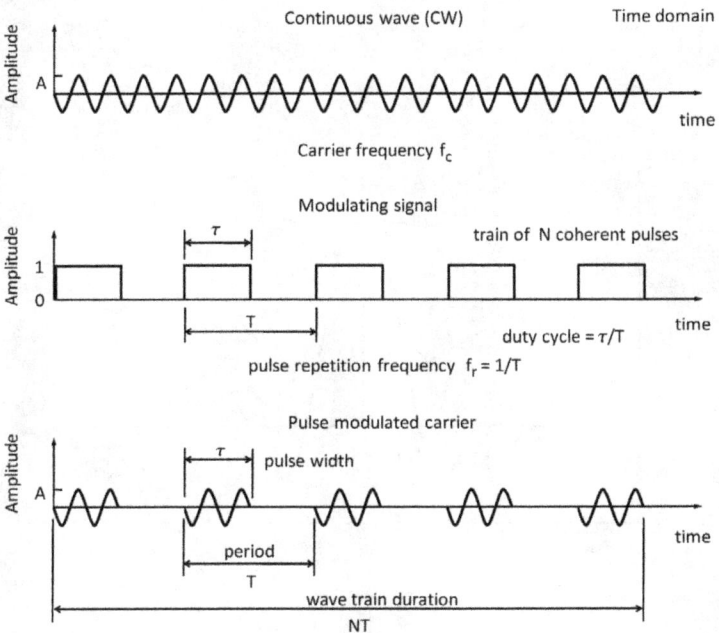

Fig. 52-11. Pulse modulated carrier formed by amplitude modulation of carrier wave by rectangular pulse train.

The spectral lines and envelope of a pulsed signal is illustrated in Fig. 52-12. Field strength of a pulse modulated signal varies as [sin x/x]. Power varies as [sin x/x]^2. The total energy of a power spectrum equals the total energy of the pulsed signal. The time and frequency domain representation of a single rectangular positive voltage pulse is illustrated in Figs. 52-13 and -14, respectively. The bandwidth may be determined from the pulse spectral energy density. The spectral energy density is obtained by squaring the magnitude of the Fourier transform spectrum which describes the pulse energy distribution as a function of frequency.

The spectral energy density of a single rectangular positive voltage pulse is shown in Fig. 52-15. The spectral energy density of a negative pulse is the same as a positive pulse. The averaged normalized energy density of a series of N pulses is equal to N times the energy density of a single pulse. The power of a series of pulses is equal to the averaged normalized energy density divided by the time required to transmit the series of pulses.

There is no space without aether, and no aether without space. – Sir Arthur S. Eddington

You pilot into an unknown future; facts are your single clue. – Robert A. Heinlein

We're busier than an electron in a mistuned waveguide here. – Bob Davis

52. Anti-gravity

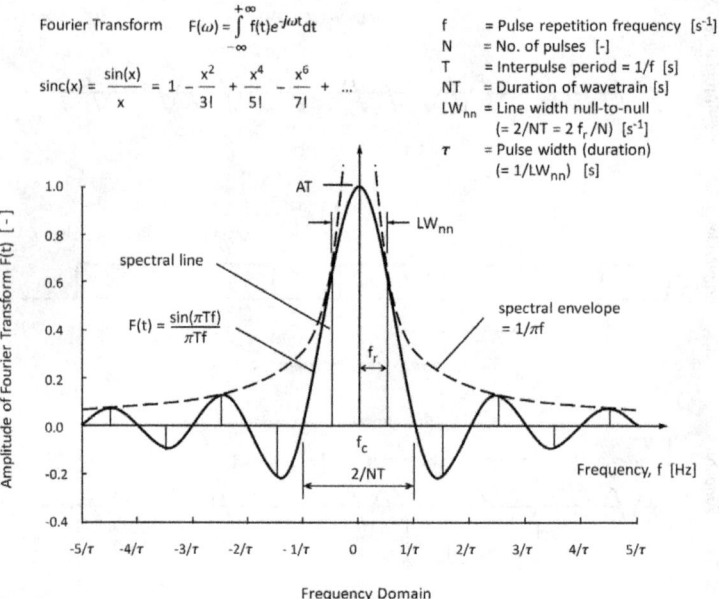

Fig. 52-12. Spectral lines for a square wave modulated signal.

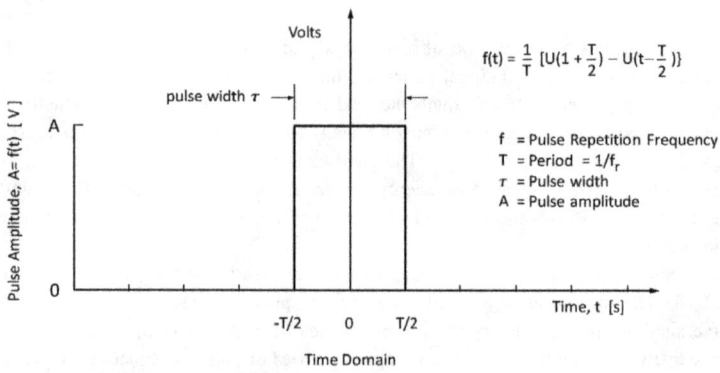

Fig. 52-13. Time domain representation of a single rectangular voltage pulse.

This wonderful elixir of light is the thing that actually connects the immaterial with the material...
– James Turrell

Old radar troops never die, they just 'Phase Array' – Robert L. Fate

52. Anti-gravity

Fourier representation of a single rectangular pulse

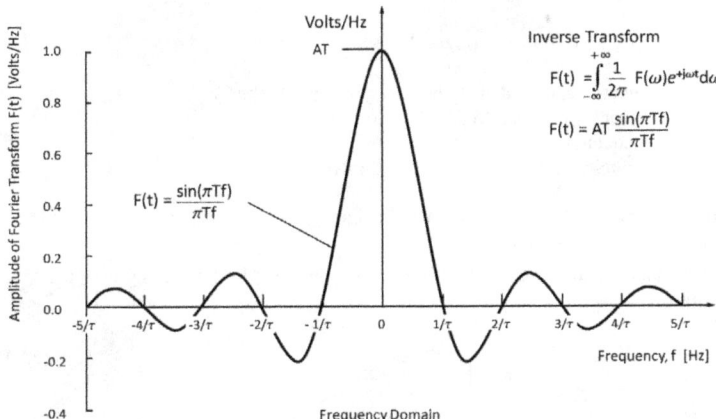

Fig. 52-14. Frequency domain representation of a single rectangular voltage pulse. The Fourier amplitude spectrum of a single rectangular pulse exhibits a sinc (= sin(x)/x) profile.

Normalized energy spectral density of a single rectangular pulse

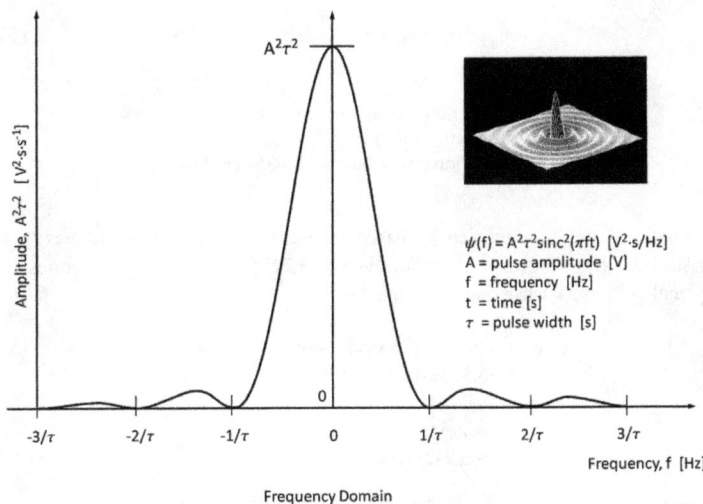

Fig. 52-15. Normalized energy spectral density of a single rectangular pulse.

The limits of the possible can only be defined by going beyond them into the impossible. – Arthur C. Clarke

Fortitudine Vincimus – "By endurance we conquer". – Ernest Schackleton

52. Anti-gravity

The gravitational field spectral energy density (SED) as previously noted varies with the cube of the frequency

$$\rho_0(\omega) = 2h\omega^3/c^3 \tag{52-14}$$

where
$\rho_0(\omega)$ = spectral energy density per frequency mode (kg/s)
h = Planck's constant (6.626093E-34 J··s)
ω = frequency (Hz)
c = velocity of light (2.99792458E8 m/s)

The spectral energy density represents the energy density per frequency mode with dimensions [= J/m³/s = kg/m·s = Pa/Hz].

The gravitational field energy density may be computed by integrating over the spectral bandwidth $\Delta\omega$

$$\rho_E = \int_{\omega_1}^{\omega_0} \frac{2h\omega^3}{c^3} d\omega \tag{52-15}$$

where
ρ_E = energy density (kg/m·s² = J/m³)

According to the above, the spectral bandwidth energy density of the Earth's gravitational field for the estimated frequency bandwidth $\Delta\omega$ at the Earth's surface is ~8.9987E47 J/m³ (= kg/m·s²). For comparison, the gravitational field energy density in terms of the acceleration of gravity is

$$\rho_g = u_g = -g^2/8\pi G = GM^2/(8\pi r^4) \tag{52-16}$$

where
u_g = gravitational field energy density (J/m³ = kg/m·s²)
g = acceleration of gravity (m/s²)
G = Newtonian gravitational constant = −6.68541E-11 (N·m²/kg²).

The gravitational field energy density based on the local acceleration of gravity at the Earth's surface (g = −9.8 m/s²) is estimated as ~5.7275E10 kg/m·s². The corresponding gravitational field mass density and field mass is

$$\rho_g/c^2 = -g^2/8\pi G/c^2 = GM^2/(8\pi r^4 c^2) \tag{52-17}$$
$$= -6.383\text{E-07} \quad [\text{kg/m}^3]$$

$$m_f = 8\pi^2 G \rho_0^2 r_0^5/(45c^2) \tag{52-18}$$
$$= 6.9329\text{E}14 \quad [\text{kg}]$$

For comparison, the estimated mass of the Earth M_E is ~5.977E24 kg.

The mass of the external field (r > r₀) is

$$m_f' = GM^2(1/r_0 - 1/r)/(2c^2) \tag{52-19}$$

The estimated spectral energy density of the Earth's gravitational field is illustrated in Fig. 52-16. For comparison a 10,000 kg vehicle mass object is shown in Fig. 52-17.

52. Anti-gravity

The spectral energy density corresponds to a scalar field and does not change under coordinate transformation. The gradient of a scalar field is a one-form (also known as a covariant vector). Acceleration is a contravariant vector lying on the direction of the one-form basis vector tangent to the parametized coordinate surface. Under a uniform acceleration a, the spectral energy density $\rho_0(\omega)$ in a moving reference frame undergoes a Lorentz transformation which may be expressed as

$$\rho(\omega)d\omega = [\omega^2/\pi^2 c^3][1 + (a/\omega c)^2][(\hbar\omega/2) + \hbar\omega/(\exp(2\pi c\omega/a) - 1)]d\omega \qquad (52\text{-}20)$$

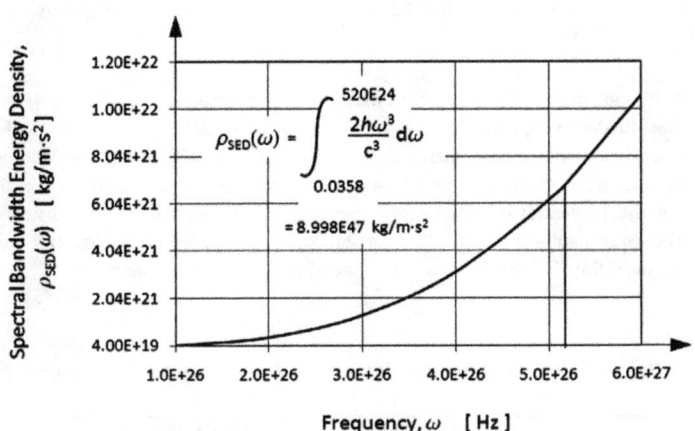

Fig. 52-16. Estimated spectral energy density of Earth's gravitational field. The area under the curve equals the energy density contained within the field for a given bandwidth $\Delta\omega$. The 1st 415 harmonic modes account for 99.99% of gravitational spectral energy.

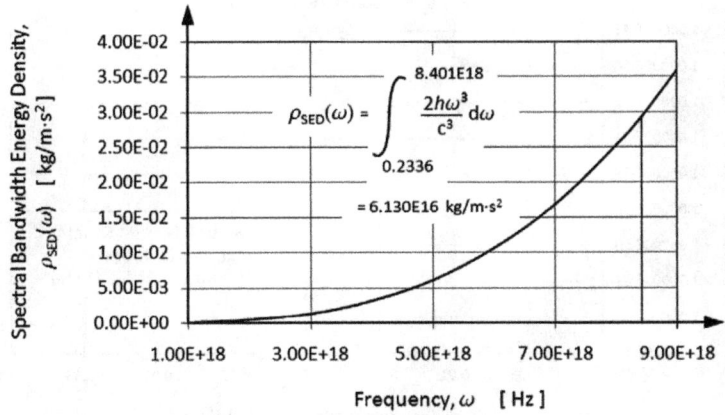

Fig. 52-17. Estimated spectral energy density of 10,000 kg test object. The spectral bandwidth is orders of magnitude less than that of Earth and contains significantly less energy density in comparison.

52. Anti-gravity

Storti[34] observes that the gravitational flux density of the PV spectrum may be approximated by a single-valued Poynting vector wavefunction $S_\omega\Omega(r,M)$

$$S_\omega\Omega(r,M) = 3Mc^3/(4\pi r^3) \quad (52\text{-}21)$$
$$= c \cdot U_m(r,M)$$

where
$S_\omega\Omega(r,M)$ = gravitational field Poynting vector (kg/s³ = W/m²)
M = mass (kg) (= $V_0/c^2 \int \eta(\nu)\rho_{SED}(\nu)d\nu$) where V_0 = vol., η = No. of modes
c = velocity of light (= $\nu\lambda = \sqrt{(v_g v_p)} = E/B = c_0/\Gamma$ = 2.99792458E08 m/s)
r = radius (m)
$U_m(r,M)$ = mass energy density of uniform solid sphere (= $3Mc^2/(4\pi r^3)$) (J/m³ = kg/(m·s²) = Pa)

For the Earth modeled as a solid spherical mass of uniform density, $S_\omega\Omega(1,r,M) \approx 1.48341E29$ kg/s³. The fundamental frequency $\omega_{PV}(1,r,M)$ is approximated as $\sqrt[3]{(M_E/S_\omega\Omega(r,M))}$ = 0.03429 Hz. Gravitational field energy density $U_\omega(r,M)$ of 10,000 kg test object compared to that of Earth as a function of fundamental mode frequency is illustrated in Fig. 52-18. Dimensionally, gravitational field energy density (T44 component of the Einstein field equation) is equivalent to pressure. The gravitational Poynting vector $S_\omega\Omega(r,M)$ represents the gravitational flux density which is a measure of power per unit area.

Gravitational field energy density vs. Fundamental frequency

Parameter	radius	mass	Acceleration of Gravity	Fundamental Frequency (1st Harmonic)	Gravitational Field Energy Density	Gravitational Poynting vector
	r	M	g	$\omega_{PV}(1,r,M)$	$U_\omega(r,M)$	$S_\omega\Omega(r,M)$
units	m	kg	m/s²	Hz	kg/(m·s²)	W/m²
vehicle	5	1.00E+04	2.67E-08	2.73663E-08	6.90293E-90	5.14117E+26
moon	1.74E+06	7.35E+22	1.62E+00	0.009391937	9.57609E-68	8.99725E+28
Earth	6.37718E+06	5.977E+24	9.81E+00	0.034416721	1.72681E-65	1.48105E+29

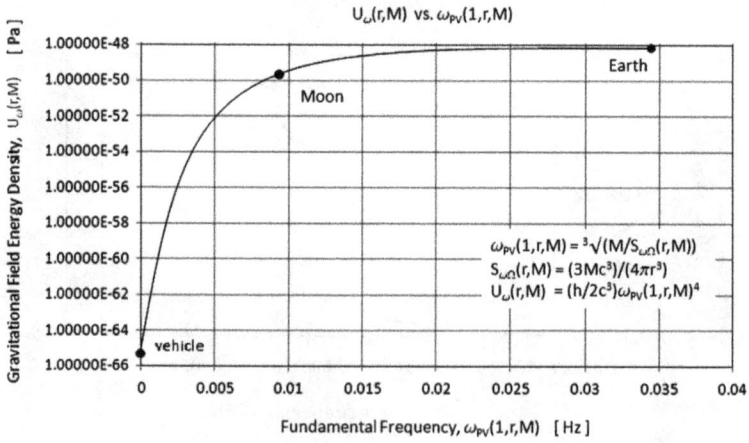

Fig. 52-18. Gravitational field energy density comparison of Earth, Moon and a 10,000 kg vehicle test mass object for fundamental mode spectral frequency.

52. Anti-gravity

The acceleration of gravity **g** is a measure of cumulative effect of change in energy density per odd frequency mode summed over the gravitational frequency bandwidth $\Delta\omega$. Modification of the naturally occurring spectral energy density ρ_{SED} profile enables a change in the local acceleration of gravity. For nullification, the required electromagnetic density is equal to the gravitational field energy density. For example, to offset the acceleration of gravity in a localized volume, a high energy density electromagnetic pulse is applied over a localized region with a bandwidth $\Delta\omega$ equal to the frequency shortfall between maximum transmitter frequency ω_{max} and the maximum gravitational frequency $\omega_\Omega(r,M)$ with spectral energy density ρ_{pulse} approximately equal to the corresponding spectral energy difference $\Delta\rho_{SED}$. The resulting spectral energy density profile is depicted in Fig. 52-19. A localized spectral energy density inversion generated by an onboard EM field generator to counter Earth's gravitational field for vehicle suspension is illustrated in Fig. 52-20. Induced acceleration $\mathbf{g}_i$ equals the induced electromagnetic field energy density gradient $-\nabla\rho_{em}$. For hover mode ($\mathbf{g}_i = \mathbf{g}$), EM field energy density $\rho_{em} = \rho_g = g^2/8\pi G$.

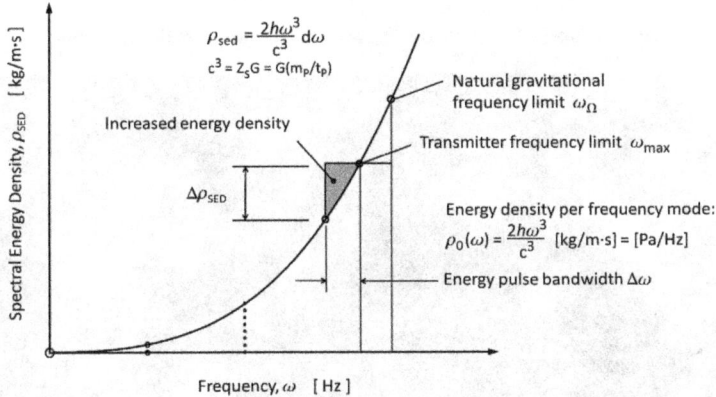

Fig. 52-19. High energy, short duration EM energy pulse with equivalent integral area matches gravitational energy density of frequencies generated by the gravitating mass above transmitter frequency range. Inability to generate extreme gravitational frequencies approaching ω_Ω is compensated by increased number of lower energy photons emitted.

A saturated signal (clipped sine wave) creates harmonics of the carrier frequency. Cross modulation of two or more saturated signals results in sidebands. Creation of harmonics and cross modulation products by signal saturation increases pulse modulation bandwidth. The null-to-null bandwidth $BW_{nn} = 2/\tau$ where τ = the pulse width. The narrower the pulse width, the wider the central spectral lobe. Energy pulse width $\Delta\omega$ required may be generated by cross-modulated pulses of short duration at multiple pulse repetition frequencies. Cancellation occurs when the electromagnetic field energy density ρ_{em} (= ½ ($\epsilon_0 \mathbf{E}^2$) + $\mathbf{B}^2/\mu_0$) is equal to the gravitational field energy density ρ_g (= $g^2/8\pi G$) [kg/m·s²] → $\mathbf{E} \simeq 9.07E8$ m/rad·s and $\mathbf{B} \simeq 0.02E-8$ rad⁻¹ equipartition for acceleration g = 9.8 m/s². Gravitational field energy density is proportional to gravitation potential $\rho_g = -(\rho_m/6)/\phi_g$.

52. Anti-gravity

Gravitational acceleration **g** is proportional to the net frequency difference (= $2c \cdot \Delta v \cdot \hat{r}$) which is a function of the gradient in EM energy density ρ_e (= $\frac{1}{2}(\epsilon_0 \mathbf{E}^2) + \mathbf{B}^2/\mu_0$) and, hence, may be subject to control. As shown in Fig 52-21, overlapping electromagnetic frequency bands generated over a selected frequency band analogous to a paired Shepard multiple frequencies tone sequence provides a continuous frequency differential. Acceleration g^μ (= $dV^\mu/d\tau$) is proportional to the frequency difference Δv (= $g/2c = v \cdot \phi_g/c^2 = -(GM/r) \cdot v/c^2$. The frequency sweep rate dv/dt corresponds to the rate of acceleration. Wave system propagation is in the direction of the frequency gradient. A rising/falling tone sequence corresponds to increasing acceleration/ deceleration. A constant carrier frequency and frequency differential (flat spectrum) results in a constant acceleration. Overlapping /interlacing of paired frequency bands prevents accelerative jumps lessening inertial strains. A frequency comb swept through multiple frequencies increases spectral energy density in proportion to the number of teeth or harmonics of the fundamental carrier frequency. The number of harmonics increases with each added frequency pair further increasing energy density. Generation of continuous Shepard tone sequences of rising or falling frequency difference over a selected EM bandwidth may allow reduction or augmentation of the local acceleration of gravity as suggested in Fig. 52-22.

Neutralization of gravity by EM energy density gradient inversion

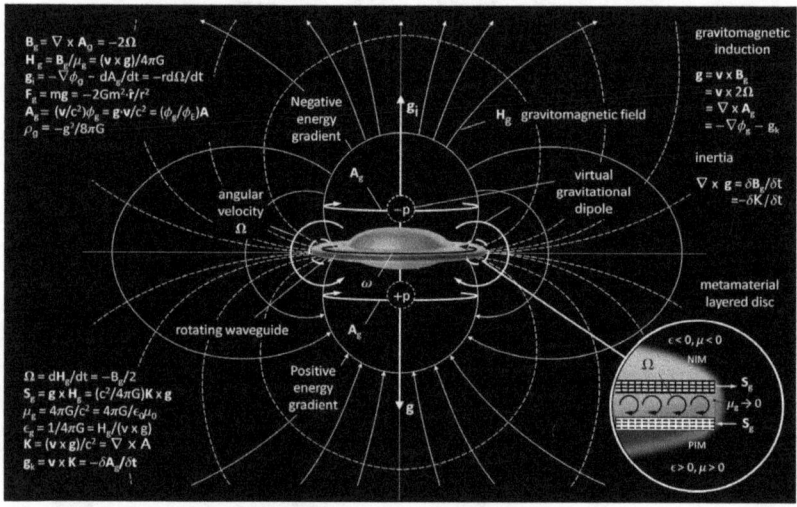

Fig. 52-20. Inversion of the local spectral energy density gradient offsets the local acceleration of gravity yielding a negative mass anti-gravity effect. For hover mode, generated EM field energy density ρ_{em} equals local gravitation field energy density ρ_g (=$-g^2/8\pi G$). Induced Gravito-Poynting efflux $\mathbf{S}_{gi}$ (= $-\mathbf{S}_g$) opposes the naturally occuring Gravito-Poynting influx $\mathbf{S}_g$ countering conversion of gravitation potential energy. The induced gravito-magnetic field $\mathbf{H}_{gi}$ (= $-\mathbf{H}_g$) alters the local EM field energy density ρ_{em}. The gravito-magnetic field is generated within a stacked disc of positive and negative metamaterials creating a simulated mass dipole with unidirectional thrust. Combination of positive and negative metamaterials, (PIM) and (NIM), respectively, allows alteration of gravitomagnetic permeability decreasing c^2 term greatly increasing gravitomagnetic effect. Upward radiation pressure P_{rad} (= $-(\rho_g/c^2)\phi_g$) acting over the antenna disc provides lift.

52. Anti-gravity

As indicated in Fig. 52-23, overlapping electromagnetic frequency bands generated over a selected frequency band analogous to a paired Shepard-Risset multiple frequency tone sequence provides a continuous frequency differential. Acceleration g is proportional to the frequency difference Δv. The frequency sweep rate dv/dt corresponds to the rate of change of acceleration. Wave system propagation is in the frequency is in the direction of the frequency gradient. A rising/falling tone sequence corresponds to increasing acceleration/deceleration. A constant carrier frequency and frequency differential (flat spectrum) results in a constant acceleration. Overlapping/interlacing of paired frequency bands prevents accelerative jumps lessening inertial strains. A frequency comb swept through multiple frequencies increases spectral energy density in proportion to the number of harmonics of the fundamental carrier frequency. The number of harmonics increases with each added frequency pair further increasing energy density. Generation of continuous Shepard tone sequences over a selected bandwidth may allow reduction or augmentation of the local acceleration of gravity. In a gravitational field, the EM energy density gradient results in a standing wave dissonance which impels a resonator to move in response to an imbalance of radiation pressure. Acceleration of gravity arises as a result of a frequency arrthymia of coupled mass oscillators attempting to synchronize. Inertial control (i.e., modulation of time rate of change in acceleration (jerk = $\Delta a/\Delta t$) may be accomplished by overlapping frequency sweeps over time analogous to a Shepard-Risset-glissando tone sequence and control the phase and frequency differential to ensure uniform acceleration by adjusting the starting and ending frequency sweep rate reducing inertial stresses and strains. This is akin to gently curved freeway on/off ramp transitions minimizing sudden accelerations.

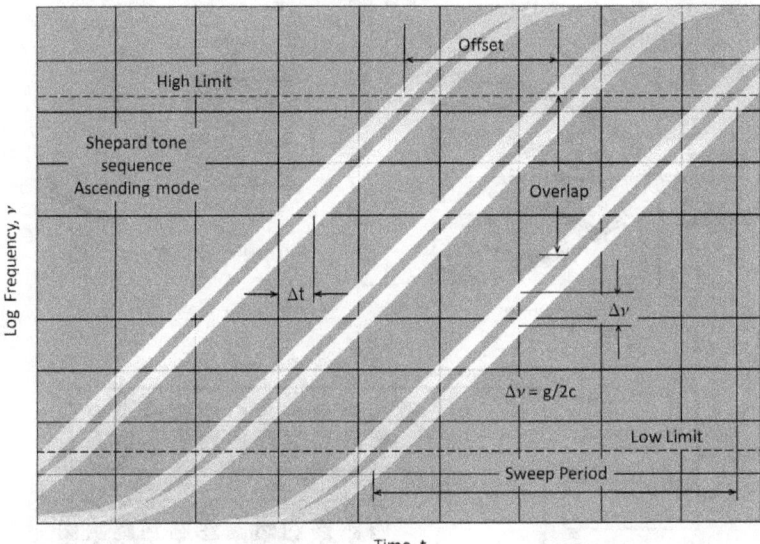

Fig. 52-21. Paired EM Shepard tone sequence of overlapping multi-band swept frequencies with a discrete frequency differential Δv. Acceleration is proportional to the frequency differential ($\Delta v = 2c\Delta v$). Modulation of $\Delta v/\Delta t$ ramp rate allows reduction of inertial stresses and strains by minimization of jerk, jolt, lurch, yank and jounce..

52. Anti-gravity

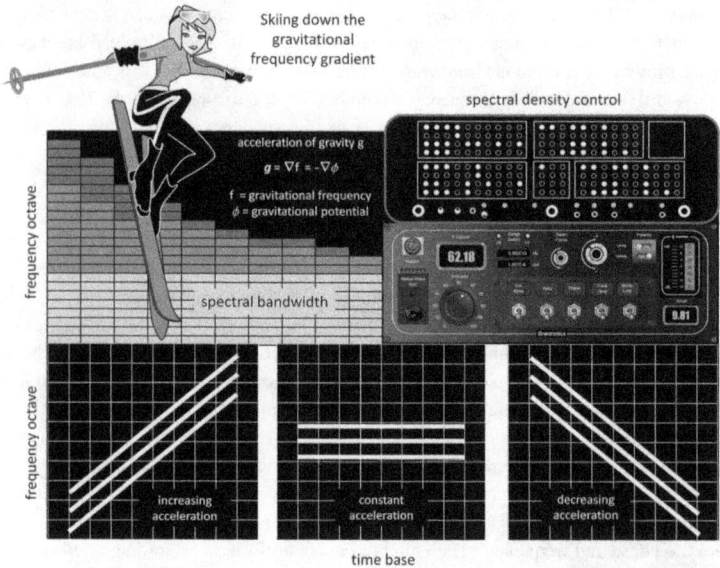

Fig. 52-22. The gravitational spectral energy density gradient $\nabla \rho_{\text{SED}}$ may be subject to alteration allowing control of acceleration by electronic control means. Such alteration gives rise to contracted moving standing waves inducing motion of matter.

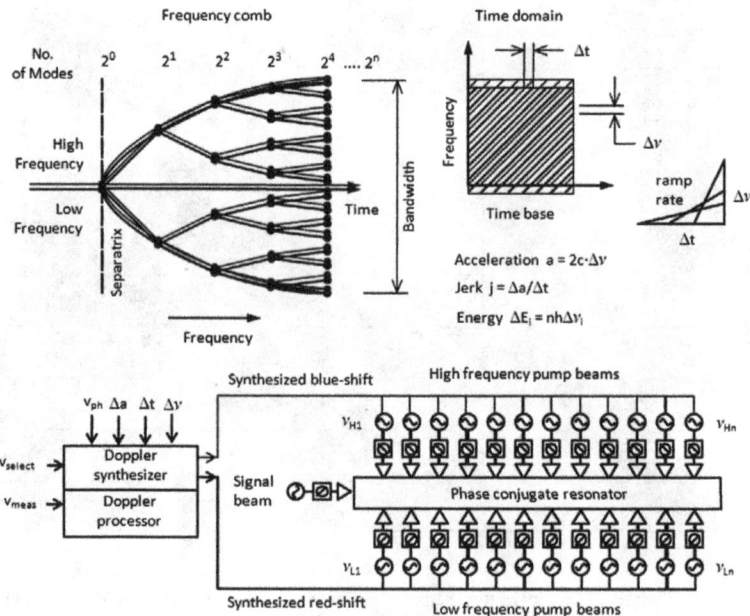

Fig. 52-23. Inertia control using acceleration ramp rate $\Delta v/\Delta t$ modulation. Inertial mass m_i is proportional to the Lagrangian interaction energy ($= 2L/(\mathbf{E}^2 - c^2\mathbf{B}^2) = E/(\mathbf{E}/\mathbf{B})^2 = E \cdot \epsilon_g \mu_g$).

52. Anti-gravity

An inverse Lorentz-Doppler effect may be produced using negative index metamaterials. Superposition of Lorentz and inverse Lorentz-Doppler effects may potentially allow cancellation of acceleration within the enclosed resonator volume as illustrated in Fig. 52-24.

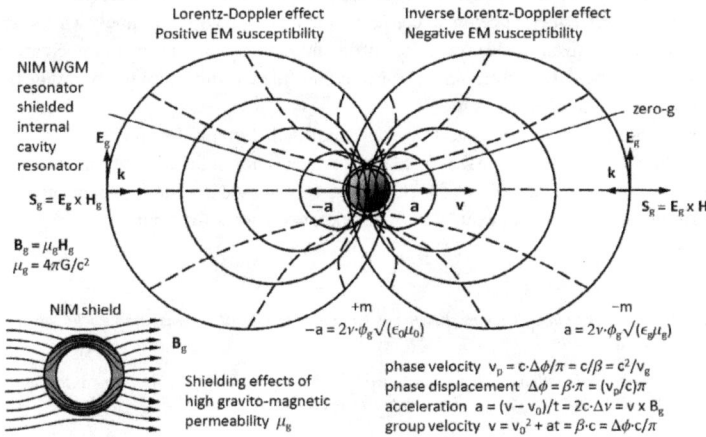

Fig. 52-24. Acceleration neutralization using inverse Lorentz-Doppler effect utilizing negative index metamaterial (NIM) resonator to offset Lorentz-Doppler effect. A side effect is possible loss of microwave Pulse Doppler Radar tracking during acceleration jumps.

Unlike naturally occuring materials, metamaterials composed of periodic resonant microstructures may be designed to react resonantly out-of-phase with external excitation. Standing wave interaction (resonant coupling) only occurs at discrete frequencies. Electromagnetic waves traveling through metamaterials with negative permitivitty and negative permeability have been demonstrated to exhibit an inverse Doppler effect. Reducing the local velocity of light in a negative index metamaterial (NIM) may result in alteration of antenna radiation impedance Z_0 (= $\mu_0 \cdot c$), spacetime impedance Z_s (= c^3/G), de Broglie matter waves λ_{dB} (= $h/\gamma m_0 \cdot (c/\pi) \cdot \Delta\phi$), etc. An inverse Lorentz-Doppler effect for a NIM resonator under positive acceleration generates red-shift waves in the direction of forward motion and blue-shifted waves propagating in the backward direction producing a ponderomotive force opposing motion. Combining a positive index material resonator with a negative index metamaterial whispering gallery mode resonator may potentially allow neutralization of internal acceleration as the forward acceleration is offset by backward acceleration. This concept is illustrated in Fig. 52-25. A mass in motion constitutes a mass current generating an induced gravito-magnetic field H_g (= B_g/μ_g) analogous to an electric current generating an induced magnetic field H (= B/μ_0).[6] The Lagrangian (interaction) energy of the EGM field is L_{EGM} = ½($E_g^2 - B_g^2$). Reducing the local velocity of c in a NIM increases the gravitomagnetic permeability μ_g (= $4\pi G/c^2$) and increasing gravitomagnetic field H_g.

In order to avoid dissipative losses such as formation of drag vortices or shock waves, etc, the surrounding media must be uniformly accelerated around the moving resonator. How might such an external influence zone be established? If the external emission is sufficient to ionize the surrounding air, such as ball lightning, we may create an external microwave

529

52. Anti-gravity

cavity resonator enveloping the internal standing wave resonator. To reduce energy requirements, drag, wake disturbance, thermal signature, sonic boom, deleterious biologic effects, etc, we might imagine a phase conjugate reflector formed by synthesized Lorentz-Doppler pump beams focused ahead of the resonator in direction of motion and bounce a signal beam off the wave interference zone to create external contracted moving standing waves acting on the surrounding medium. An omni-directional outward flux of matter waves in a surrounding spherical irradiation zone may serve to offset compressive forces during motion in air or submergence in a dense liquid medium without shock waves or wake turbulence. Synthesized matter waves of sufficient intensity may result in the creation of an gravitomagnetic field $\mathbf{H}_g$ acting on the surrounding media uniformly accelerating the fluid media around the object in motion.

Direct matter wave synthesis utilizing amplified Lorentz-Doppler beams and inverse Lorentz-Doppler effects appears as a plausible means for induced motion of matter, inertia control and contra gravity effect. Such matter wave-based technology may enable impulse drives for propulsion without expulsion of reaction mass, stress or strain.

Dreams mean everything. They're the stories we tell ourselves of what could be, who we could become. – Robert Ford

I'm a mathematical optimist. I deal only with positive integers. The hardest thing being a mathematiciam is that they always have problems. – Tendai Chiteware

A fact is a thought that is true. – Gottlob Frege

No intelligent idea can gain general acceptance unless some stupidity is mixed in with it. – Fernando Pessoa

We must strive to be more than we are. It does not matter that we will not reach our ultimate goal. The effort itself yields its own reward. – Gene Roddenberry

We can assume that when we go to the quantum theory, the lines of force become all discrete and separate from one another. We have so a model where the basic physical entity is a line of force. – P.A.M. Dirac

Your theory is crazy but not crazy enough to be true. – Niels Bohr.

Gravitation, ever since Newton, had remained isolated from the forces in nature; various attempts had been made to account for it, but without success. The immense unification effected by electromagnetism apparently left gravitation out of its scope. It seemed that nature had presented a challenge to the physicists which none of them were able to meet. – Hendrik Antoon Lorentz

All the standard equations of mathematical physics can be separated and solved in Kerr geometry. – S. Chandrasekar

The Janus model involves positive and negative masses. – Jean-Pierre Petit

Schwere masse ist triâ masse. (Gravitational mass is inertial mass). – Albert Einstein

If you are worried that something in an experiment might be a potential source of error, take the step of deliberately exaggerating the error sources to determine just how bad they can get. This enables you to quantify the magnitude of the problem and determine a way to deal with it. – Henry Cavendish

Maxwell made use of quaternions in EM theory and mathematics beyond three-dimensional vectors, but scientific community of the time accepted only vector components and cutoff the "transcendal and fantastic elements". – Ron Croix

The EGM spectrum is a harmonic description of mass-energy obeying a Fourier distribution. The local energy state of the vacuum may be considered equivalent to the mass-energy of the object it encapsulates. – Geoffrey S. Diemer

52. Anti-gravity

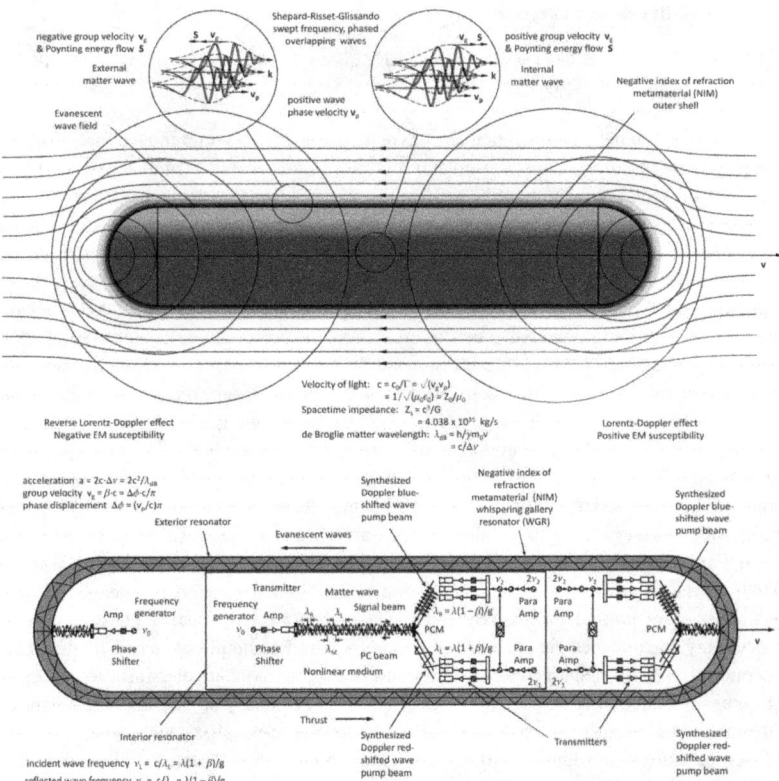

Fig. 52-25. Lorentz Doppler and Inverse Lorentz Doppler effects may potentially be combined using a cavity resonator and NIM Metamaterial whispering gallery resonator to provide internal acceleration neutralization. In this speculative concept, an induced matter wave in the NIM exterior shell generates a gravitomagnetic field $\mathbf{H}_g$ (= $(\mathbf{v} \times \mathbf{g})/4\pi G$) and associated circumferential co-gravitation field $\mathbf{K}$ (= $(\mathbf{v} \times \mathbf{g})/c^2$) in the surrounding media creating a counter-propagating mass current effect. This is analogous to the magnetic field $\mathbf{H}$ generated by a time variable electric current creating an electrokinetic $\mathbf{E}_k$ (= $d\mathbf{A}/dt$) field opposing a change in electric current. A time-varying gravitomagnetic field generates an induced acceleration of mass. Under acceleration a mass experiences an inertial force $\mathbf{F} = m_i(\mathbf{g}_i - \mathbf{a})$ where $\mathbf{g}_i$ is the induced gravitomagnetic acceleration. The c^2 term is suggestive of interconversion options (c^2 = E/m = $\rho_c \cdot \mathbf{v}/\mathbf{p}$ = V·q/m = $4\pi G/\mu_g$ = $\mathbf{g} \cdot \phi/\mathbf{E}$ = $\mathbf{v} \cdot \phi_g/\mathbf{A}_g$ = $\mathbf{v} \times \mathbf{g}/\mathbf{K}$).

What we observe is not nature itself but nature exposed to our method of questioning.
– Werner Heisenberg

Are not gross bodies and light convertible into one another, and may not bodies receive much of their activity from the particles of light which enter their composition. – Isaac Newton

Like Dirac's equation, Maxwell's wave equation for light has two solutions, the so-called "retarded solution" that describe a wave travelling forward in time and the "advanced solution" that describe a light wave travelling backward in time. – Nick Herbert

52. Anti-gravity

52.3 Speculative design exercise

We may learn to deprive large masses of their gravity, and give them absolute levity, for the sake of easy transportation. – Benjamin Franklin

Gravitational repulsion may perhaps involve some method to produce out-of-phase desynchronization of n-coupled oscillators of a passive mass m and active mass M composed of a total population of N oscillators. Electronically, we might imagine a super resonant LRC circuit with a suitable transmit/receive (T/R) antenna which senses the local EM field over a sufficiently wide bandwidth, parametrically amplify the detected signal, introduce a phase delay and emit a pulse modulated phase conjugate reflected wave with an amplified out-of-phase signal for inversion of the local EM energy density gradient. Alternatively, utilizing negative index of refraction metamaterials with negative electric permittivity ϵ and magnetic permeability μ may allow simulation of a repulsive negative mass as part of a gravitic drive composed of a coupled resonant dipole of positive and negative mass with net propulsive force to counter the local field gradient. See Fig. 52-26.

Extrapolating on concepts reviewed, it is tempting to speculate on various potential future means for electronic alteration of local gravitational energy density. Falling into temptation, we will briefly outline a simplified sample case study for a hypothetical model configuration as an exercise to get an idea of parameter and solution space. As an example, consider the EM energy requirements for levitation of a sphere of 10 m diameter and a 10,000 kg mass. Illustrated in Fig. 52-27 is a self-contained frequency generator set and transmitter array housed in a spherical volume with a circular disc of phase conjugate phased array antenna elements in a Saturn configuration. For simplicity, a non-rotating disc is supposed without negative mass, gyroscopic rotors, homopolar electrostatic charging, multi-phasic induction coils or external electrodes. An arbitrary station keeping height of 100 m above the Earth's surface is assumed with standard atmospheric ambient air pressure and temperature. A summary of analysis variables and estimated values for this hypothetical case is shown in Table 52-1.

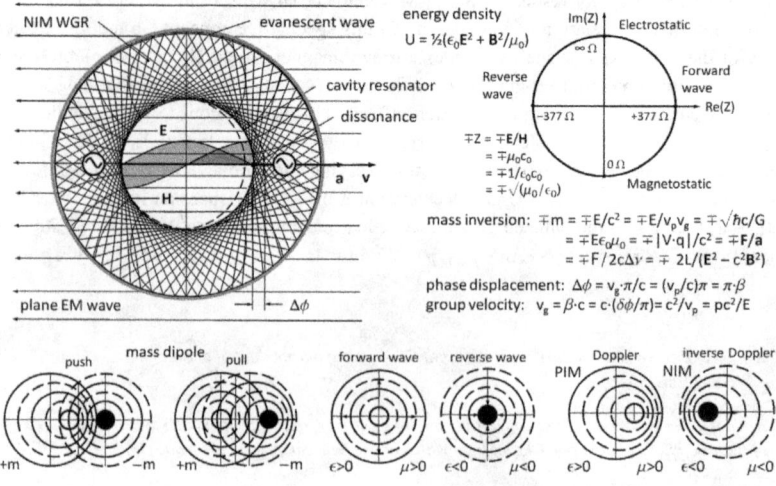

Fig. 52-26. Negative index metamaterials may allow artificially created negative mass.

52. Anti-gravity

Table 52-1. 10,000 kg mass – Saturn antenna configuration – Hover mode

Parameter	Symbol	Relation	Value	Units	Remarks
Earth – Active mass					
Mass	M_E	$= -\phi_g \cdot R/G = \hbar \mathbf{k}/\mathbf{v} = |V \cdot q|/c^2$ $= E/\mathbf{a} \cdot \mathbf{d} = \mathbf{F}/\mathbf{a} = E/(|\mathbf{E}/\mathbf{B}|)^2$ $= \mathbf{F}/2c\Delta v \cdot \hat{\mathbf{r}} = E \cdot \epsilon_0 \mu_0 = E \cdot \epsilon_g \mu_g$	5.977E24	kg	
Acceleration of Earth gravity (surface)	g	$= -GM/(R+h)^2 = -\phi/(R+h)$ $= W/m = -\nabla\phi_g - \delta \mathbf{A}_g/\delta t$ $= -GM/r^2 - G^2(M/2c^2 r^3)$ $= \phi/r = -2c(GM/cr^2) = E \cdot d/m$ $= -2cr\nabla v = (c^2/\phi)\mathbf{E}$ $= 2c\Delta v = \{2c^2/h\}(mv/\Delta\lambda)\lambda_{dB}$ $\nabla \times \mathbf{A}_g = v \times \mathbf{B}_g = v \times 2\Omega$	9.80	m/s^2 $= N/kg$	Direction opposite to unit vector $\mathbf{r}_u$.
Escape velocity	v_e	$= \sqrt{(2GM/r)}$	11,187	m/s	
Gravitational potential energy	U	$= -GMm/r$	-3.9928E17	$J = kg \cdot m^2/s^2$	Total potential.
Ionization threshold	TI	Measured	~12–16	eV	S.L. @ S.T.P
Gravitational potential energy difference	ΔU	$= mgh = Wh = m\Delta\phi$	-9.80E6	$J = kg \cdot m^2/s^2$	h = 100 m (h << R_E) uniform field approx.

cont

52. Anti-gravity

Table 52-1. 10,000 kg mass – Saturn antenna configuration – Hover mode (Cont)

Parameter	Symbol	Relation	Value	Units	Remarks
Earth (Cont)					
Gravitational potential	ϕ_g	$= -GM/r = U/m = c^2 \cdot \mathbf{A}_{ag} \cdot \mathbf{v}$ $= -c_0^2/K_{PV} = (\mathbf{A}_g/\mathbf{A})\phi_E$ $= (M/m)c^2 = (\Delta v/v)c^2$	-6.26E07	$J/kg = m^2/s^2$	Potential/unit mass
Gravitational potential difference	$\Delta\phi$	$= \phi_0 - \phi = -gh/(1 + h/R)$ $= c^2(v_2 - v_1)/v_1$	-980	$J/kg = m^2/s^2$	uniform field approx. @ Earth's surface
Gravitational field-energy density	ρ_g	$= -g^2/8\pi G = (\rho_m/6)\phi_g$ $= -GM/(8\pi r^4)$	-5.7368E10	$J/m^3 = kg/m \cdot s^2$	stress-energy tensor element [T_{00}]
Gravitational pressure	p_G	$= \rho_g/3$	-1.9122E10	$N/m^2 = kg/m \cdot s^2$ $= Pa$	@ Earth's surface. stress-energy tensor element [T_{11}, T_{22}, T_{33}]
Gravitational field mass density	ρ_{gm}	$= -g^2/8\pi Gc^2$ $= -GM^2/8\pi r^4 c^2$	6.3727E-07	kg/m^3	
Lorentz-Doppler beat frequency	Δv	$= |v_2 - v_1| = -GM/(2cr^2)$ $= g/2c = v \cdot \phi_g/c^2 = (\mathbf{v} \times \mathbf{B}_g)/2c$	1.64E-08	Hz	uniform field approx. @ Earth's surface

cont

52. Anti-gravity

Table 52-1. 10,000 kg mass – Saturn antenna configuration – Hover mode (Cont)

Parameter	Symbol	Relation	Value	Units	Remarks
Earth (Cont)					
Mass-energy density	$U_m(r,m)$	$= 3Mc^2/4\pi r^3$	4.9448E20	$J/m^3 = kg/m \cdot s^2$	Solid homogeneous sphere approx.
Mass-energy stored in gravitational field	$U_\omega(r,m)$	$= (h/2c^3) \cdot \omega_{PV}(1,r,M)^4$	1.4727E-14	$J/m^3 = kg/m \cdot s^2 = Pa$	
Fundamental PV frequency (Earth)	$\omega_{PV}(1,r,m)$	$= (1/r) \cdot \sqrt[3]{((2cGM)/(\pi r))} \cdot \sqrt{K_{PV}}$ $= \sqrt[3]{(M/S_{\omega\Omega}(r,M))}$	0.03584	Hz	
Avalanche breakdown (air)	E_a	Measured	~26–32	KV/cm	@ S.T.P.
Fundamental PV wavelength	$\lambda_{PV}(1,r,m)$	$= c/\omega_{PV}(1,r,M)$	5.765E-19	m	
Harmonic mode cut-off function	$\Omega(r,m)$	$= \sqrt[3]{108 \cdot U_m(r,M)/U_\omega(r,M) + 12 \cdot \sqrt{768 + 81 \cdot (U_m(r,M)/U_\omega(r,M))^2}}$	1.73968E29	–	
No. of harmonic modes	$n_\Omega(r,m)$	$= [\Omega(r,M)/12] - [4/\Omega(r,M)] + 1$	1.44974E28	–	
Harmonic cut-off frequency	$\omega_\Omega(r,m)$	$= n_\Omega(r,M) \cdot \omega_{PV}(1,r,M)$	520E24	Hz	

cont

52. Anti-gravity

Table 52-1. 10,000 kg mass – Saturn antenna configuration – Hover mode (Cont)

Parameter	Symbol	Relation	Value	Units	Remarks
Earth (Cont)					
Bandwidth	$\Delta\omega_{PV}$	$= \omega_\Omega(r,M) - \omega_{PV}(1,r,M)$	520E24	Hz	$\omega_1 = 0.03584$ Hz, $\omega_\Omega = 5.20$E26 Hz
Spectral bandwidth energy density (Earth)	$\rho_{SED}(\omega)$	$= \int_{\omega_1}^{\omega_\Omega} \frac{2\hbar\omega^3}{c^3} d\omega$	8.9987E47	$J/m^3 = kg/m \cdot s^2$	ω = freq. [Hz] h = Planck's const. = 6.626093E-34 J·s
Spectral energy density/ frequency mode	$\rho_0(\omega)$	$= 2\hbar\omega^3/c^3$	varies	$kg/m \cdot s = Pa/Hz$	
Gravitational energy/ change in odd mode	$U_\omega(n_{PV},r,M)$	$= (h/2c^3)\omega_{PV}(1,r,M)^4[(n_{PV}+2)^4 - n_{PV}^4]$ $= U_\omega(r,M) \cdot [(n_{PV}+2)^4 - n_{PV}^4]$	varies	$J/m^3 = kg/m \cdot s^2$ $= Pa$	$S_m(n_{PV},r,M)$
Gravitational Poynting vector	$S_m(n_{PV},r,M)$	$= c \cdot U_m(r,M) = (c^2/4\pi G) \times \mathbf{K}$ $= (3M^2)/(4\pi R^3) = \mathbf{g} \times \mathbf{H}_g$	1.48242E29	$kg/s^3 = W/m^2$	for $n_{PV} = 1$
Newtonian gravitational constant	G	$= c^2\Sigma(ri/mi) = c^3/Z_s = G_0/I^3$ $= c^3(t_P/m_P) = G_P/2\pi = \hbar c/m_P^2$ $= c^2\mathbf{K}/4\pi\mathbf{H}_g = \mathbf{F} \cdot \mathbf{r}^2/m_1 m_2 \cdot \mathbf{r}_u$ $= (\mathbf{v} \times \mathbf{g})\mu_g/4\pi\mathbf{B}_g = R_s c^2/2M$	6.62762E-11	$N \cdot m^2/kg^2$ $= m^3/kg \cdot sec^2$	

cont

Table 52-1. 10,000 kg mass – Saturn antenna configuration – Hover mode (Cont)

Parameter	Symbol	Relation	Value	Units	Remarks
Earth (Cont)					
Gravitational Poynting vector/ change in odd mode	$S_\omega(n_{PV},r,M)$	$= c \cdot U_\omega(n_{PV},r,M)$ $= c \cdot (c^2 h / 2c^3) \omega_{PV}(1,r,M)^4$	2.03103E-65	$kg \cdot m \cdot s^{-2}] = [Pa]$	for $n_{PV} = 1$
Gravitational time dilation	t	$= t_0 \sqrt{(1 - 2GM/rc^2)} = \Gamma t_0$ $= t_0/\sqrt{(1 - 2gr/c^2)}$ $= t_0/\sqrt{(1 + 2\phi/c^2)}$	1.000000000 69626	s	
Gravitational redshift	z	$= (\lambda_o - \lambda_e)/\lambda_e$ $= \nu_0/\nu_\infty - 1$ $\approx GM/r_0 c^2$	varies	Hz	wavelength λ_o = observed λ_e = emitted
Gravitational Gamma	Γ	$= dt/d\tau = 1/\sqrt{(1 - v_e^2/c^2)}$ $= 1/\sqrt{(1 - (2\phi/c^2))} = \sqrt{K_{PV}}$ (weak field) $= \sqrt[4]{K_{PV}}$ (strong field)	1.000000000 69626	-	
Wavelength ratio	λ/λ_0	$= (hc/\lambda_0)(\lambda/hc)$	varies	-	
Frequency ratio	ν/ν_0	$= 1 - GM/rc^2 = E_0/E$	varies	-	$\nu = \nu_0 [1 + gh/c^2]$

cont

52. Anti-gravity

Table 52-1. 10,000 kg mass – Saturn antenna configuration – Hover mode (Cont)

Parameter	Symbol	Relation	Value	Units	Remarks
Earth (Cont)					
Energy change ratio	$\Delta E/E$	$= gh/c^2$	2.5E-15	-	Pound-Rebka result for $h \ll R_e$.
Earth radius	R_E	$= \sqrt{(x^2 + y^2 + z^2)}$	6.37718E06	m	sphere
Schwarzschild radius	r_S	$= 2GM/c^2$	8.88517E-03	m	
Atmosphere					
Density of air	ρ	$= m/vol$	1.275	kg/m^3	@ S.L.
Air temperature	T_a	$15°C - 1.98°C/1000$ ft $\times$ PA	288.15	°K	$= 15$ °C (S.T.P.)
Air pressure	P_a	$= 1$ Atm	101.325	kPa	$= 760$ mm Hg (S.T.P.)
Dynamic viscosity	μ_g	$= v_g \rho_g$	1.8	$N \cdot s/m^2$	
Real gas constant	R	$= PVT/n$	287.04	$m^2/°K \cdot s^2$	
speed of sound	a_0	$= \sqrt{(dp/d\rho)}$	340.294	m/s	ISA

cont

52. Anti-gravity

Table 52-1. 10,000 kg mass – Saturn antenna configuration – Hover mode (Cont)

Parameter	Symbol	Relation	Value	Units	Remarks
Vehicle (De-energized) – Passive mass					
Object rest mass	m_o	$= \rho \cdot V = m/\gamma = E_o/c^2 = m/\Gamma$ $= \|V \cdot Q\|/c^2 = 4V\rho_e/3c^2$ $= F/a = E/a \cdot d = \hbar k / v = E/c^2$ $= E/(E/B)^2 = 2L/(E^2 - c^2 \mathbf{B}^2)$ $= F/2c\Delta v \cdot \hat{\mathbf{r}} = h \cdot v_{dB}/c \cdot v$ $= \phi_E q/c^2 = E/v_p v_g = E\epsilon_0 \mu_0$ $= 2(T-V)/(\epsilon_0 E^2 - c^2 \mathbf{B}^2/\mu_0)$	10,000.0	kg	Assumed
Object weight	W	$= mg$	98,000.0	N	
Sphere radius	a	$= r_{sphere}$	5.0	m	Assumed
Sphere volume	V_s	$= (4/3 \pi a^3)$	523.5	m^3	
Disc radius (outer)	R_o	$= 3a$	15.0	m	Assumed
Disc radius (inner)	R_i	$= a$	5.0	m	Assumed
Disc area	A_d	$= \pi(R_{do}^2 - R_{di}^2)$	6.28E02	m^2	
Disc height	D_h	$=$ const.	1.5	m	Assumed

cont

52. Anti-gravity

Table 52-1. 10,000 kg mass – Saturn antenna configuration – Hover mode (Cont)

Parameter	Symbol	Relation	Value	Units	Remarks
Vehicle (De-energized) – Passive mass					
Disc volume	V_d	$= A_d \cdot D_h$	9.42E02	m^3	
Total volume	V_t	$= V_s + V_d$	1.47E03	m^3	
Mass density	ρ	$= m/(V_s + V_d)$	6.82	kg/m^3	average
Rest energy	E	$= mc^2 = m(\mathbf{E}/\mathbf{B})^2 = m(\mathbf{v}_p/\mathbf{v}_g)^2$	8.9875E20	$J = N \cdot m = kg \cdot m^2/s^2$	
Mass-energy density	$U_m(r,m)$	$= E/V_t$	6.13E17	$J/m^3 = Pa = kg/m \cdot s^2$	
Mass-energy stored in gravitational field	$U_\omega(r,m)$	$= (h/2c^3) \cdot \omega_{pv}(1,r,m)^4$	3.6661E-62	$J/m^3 = Pa = kg/m \cdot s^2$	natural (de-energized state)
Gravitational energy	U_g	$= -1/8\pi \, G \int g_0^2 \, dv$	-1.1157E-04	$J = kg \cdot m^2/s$	sphere of radius a
Gravitational field energy density	ρ_g	$= -g_0^2/(8\pi G) = Gm^2/8\pi \, r^4$	-2.660E-08	$J/m^3 = Pa = kg/m \cdot s^2$	natural (de-energized state)
gravito-magnetic permeability	μ_g	$= 4\pi G/c^2 = 4\pi G \cdot \varepsilon_0 \mu_0 = \mathbf{B}_g/\mathbf{H}_g$	9.3287E-27	m/kg	

cont

52. Anti-gravity

Table 52-1. 10,000 kg mass – Saturn antenna configuration – Hover mode (Cont)

Parameter	Symbol	Relation	Value	Units	Remarks
Vehicle (De-energized) – Passive mass					
Gravitational field energy density	ρ_g	$= -g_0^2/(8\pi G) = Gm^2/8\pi r^4$	-2.660E-08	$J/m^3 = Pa$ $= kg/m \cdot s^2$	natural (de-energized state)
Gravitational pressure	p	$= \rho_g/3$	-8.866E-09	$N/m^2 = Pa$ $= kg/m \cdot s^2$	
Gravitational field mass density	ρ_{gm}	$= -g_0^2/8\pi Gc^2$ $= Gm^2/(8\pi c^2 r^4)$ for $r > a$	-2.959E-25	kg/m^3	
Gravitational potential	ϕ_g	$= -Gm/r = c^2 \mathbf{A}_{g}/\mathbf{v} = (\phi_g/\phi_E)\mathbf{A}$	-6.685E-08	$J/kg = m^2/s^2$	
Acceleration of gravity of object	$\mathbf{g}_{obj}$	$= -G(m/r^2)\mathbf{r}_u$ for $r > a$	6.69E-09	$m/s^2 = N/kg$	Acceleration measured at object surface $r = a$.
Delta frequency	$\Delta\nu$	$= -Gm/(2cr^2) = g_{obj}/2c$ $= (c/\Delta\lambda)\lambda_{dB}$	1.1150E-17	Hz	
Fundamental PV frequency	$\omega_{PV}(1,r,m)$	$= (1/r) \cdot \sqrt[3]{(2cGm)/(\pi r)} \cdot \sqrt{K_{PV}}$ $= \sqrt[3]{(M/S_{\omega\Omega}(r,M))}$	0.2336	Hz	$K_{PV}(r,m) \approx 0.9999...$

cont

52. Anti-gravity

Table 52-1. 10,000 kg mass – Saturn antenna configuration – Hover mode (Cont)

Parameter	Symbol	Relation	Value	Units	Remarks
Vehicle (De-energized) – Passive mass					
Fundamental PV wavelength	$\lambda_{PV}(1,r,m)$	$= c/\omega_{PV}(1,r,m)$	1.283E09	m	
Harmonic mode cut-off function	$\Omega(r,m)$	$= \sqrt[3]{108 \cdot U_m(r,m)/U_\omega(r,m)} + 12 \cdot \sqrt{768 + 81 \cdot (U_m(r,m)/U_\omega(r,m))^2}$	1.0812E27	-	
No. of harmonic modes	$n_\Omega(r,m)$	$= [\Omega(r,m)/12] - [4/\Omega(r,m)] + 1$	9.0105E25	-	
Harmonic cut-off frequency	$\omega_\Omega(r,m)$	$= n_\Omega(r,m) \cdot \omega_{PV}(1,r,m)$	2.1055E25	Hz	
Bandwidth	$\Delta\omega$	$\cong \omega_\Omega - \omega_{PV}(1,r,m)$	2.1055E25	Hz	
Energy density per frequency mode	$\rho_0(\omega)$	$= 2\hbar\omega^3/c^3$	varies	kg/m·s	
Spectral bandwidth energy density (passive mass)	$\rho_{SED}(\omega)$	$= \int_{\omega_1}^{\omega_\Omega} \frac{2\hbar\omega^3}{c^3} d\omega$	$= 6.130\text{E}16$	$J/m^3 = kg/m \cdot s^2$	$\omega_1 = 0.2336$ Hz, $\omega_\Omega = 8.401\text{E}18$ Hz

cont

52. Anti-gravity

Table 52-1. 10,000 kg mass – Saturn antenna configuration – Hover mode (Cont)

Parameter	Symbol	Relation	Value	Units	Remarks
Vehicle (De-energized) – Passive mass					
Blackbody spectral energy density	$\mu_\nu(T)$	$= (8\pi h\nu^3/c^3)[1/(e^{h\nu/kT} - 1)]$	varies	$J \cdot m^{-3} \cdot s^{-1}$	black-body radiation energy density per unit frequency interval
Spectral energy density of ZPF	$\rho_{ZPF}(\nu)$	$= 4\pi h\nu^3/c^3$	varies	$kg/m \cdot s$	uniform acceleration
Equiv. blackbody radiated power	P	$= \sigma T^4 A$ $= nh\nu$	6.127E5	W	Ambient temp T = 288 °K. σ = Stefan-Boltzman const
Radiated intensity	I	$= 2\pi h\nu^3 \Delta\nu/(c^2(e^{h\nu/kT} - 1)$	varies	W/m^2	Freq. range $\Delta\nu$
Blackbody spectral radiance	$B_\nu(\nu, T)$	$= (2h\nu^3/c^2)[1/(e^{h\nu/kT} - 1)$ $= I(\nu, T)$	varies	$W \cdot m^{-2} \cdot sr^{-1} \cdot s^{-1}$	Power per unit surface area per unit solid angle per unit frequency
ZPF energy density	$\rho(\nu, T_a)$	$= (8\pi\nu^2/c^3) \cdot [1 + [a/2\pi c\nu)^2][h\nu/2 + h\nu/(e^{h\nu/kT} - 1)]$	varies	$J/m^3 = m^2 kg/s^2$	a = acceleration
Degrees of freedom/unit vol.	DOF	$= (8\pi\nu^2/c^3)$	varies	m^{-3}	No. modes per unit freq. per unit vol.

cont

52. Anti-gravity

Table 52-1. 10,000 kg mass – Saturn antenna configuration – Hover mode (Cont)

Parameter	Symbol	Relation	Value	Units	Remarks
Vehicle (De-energized) – Passive mass					
Thermal energy/ mode	$KE_{avg}(n)$	$= h\nu_n/e^{h\nu/kT} - 1$	varies	J	Avg. energy per mode
Unruh temperature	T_a	$= \hbar a/(2\pi c k_B)$	4E-20	°K	a = acceleration k_B = Boltzman's const. = 1.381E-23 J/°K
de Broglie frequency	ν_{dB}	$= Pc/h = c/\lambda_{dB}$ $= (h/mv)(\Delta\nu/\Delta\lambda)$	0	Hz	
de Broglie wavelength	λ_{dB}	$= h/\gamma m_0 v = c/\nu_{dB} = h/p$ $= h/\gamma m_0(c/\pi)\Delta\phi$	0	m	velocity $v = 0$
Acceleration of gravity ratio	g_ω/g	$= g_\omega/g$	6.8265E-10	—	Referenced to Earth
Gravitational Poynting vector	$S_m(r,m)$	$= c \cdot U_m(r,m) = (3m^2)/(4\pi r^3)$	6.4324E25	kg/s^3 = W/m^2	Object (passive mass)
	$\mathbf{S_g}$	$\mathbf{g \times H_g} = (c \cdot \mathbf{B_g}) \times (\mathbf{B_g}/\mu_g)$ $= (c^2/4\pi G)\mathbf{K} \times \mathbf{g}$		kg·rad/s^3	
Gravitational Poynting vector per change in odd mode	$S_\omega(n_{PV},r,m)$	$= c \cdot U_\omega(n_{PV},r,m)$ $= c \cdot (h/2c^3)\omega_{PV}(1,r,m)^4$	1.0991E-53	kg/m·s^2 = [Pa]	

cont

52. Anti-gravity

Table 52-1. 10,000 kg mass – Saturn antenna configuration – Hover mode (Cont)

Parameter	Symbol	Relation	Value	Units	Remarks
Energized Vehicle					
Required lift	L_{reqd}	$= mg = W = p_{reqd} \cdot A_d$	9.80E04	$N = kg \cdot m/s^2$	
Required altitude	h	$=$ const.	100	m	Assumed
Required energy	E_{reqd}	$= \Delta U = mg\Delta h = W \cdot h = m \cdot a \cdot d$	9.80E6	J	$h \ll R_E$
Potential energy density	U_P	$= E_{reqd}/V$	6.313	$J/m^3 = m^2 kg/s^2$	V = object vol.
Required pressure	p_{reqd}	$= W/A_d$	155.9	N/m^2	
Required EM energy density	u_{reqd}	$= 3 \cdot p_{reqd}$	4.679E02	$J/m^3 = N/m^2$	
Oscillator low freq.	ν_1	$= \Delta\nu - \nu_2$	100.0E09	Hz	100 GHz (Assumed)
Oscillator high freq.	ν_2	$= \nu_1 + \Delta\nu$	$\cong$ 100.0E09	Hz	
Power absorption	P	$= P_r \times \{\nu_1^2/\nu_2^2\} \, [(\nu_2^2 - \nu_1^2)^2 + (\nu_1^2/\nu_2^2)]\}$	varies	Watts = J/s	P_r = Power absorption @ resonance $\nu_1 = \nu_2$

cont

52. Anti-gravity

Table 52-1. 10,000 kg mass – Saturn antenna configuration – Hover mode (Cont)

Parameter	Symbol	Relation	Value	Units	Remarks
Energized Vehicle					
Oscillator beat frequency	$\Delta \nu$	$= (\nu_2 - \nu_1) = g/2c = (\nu \cdot \phi_g)c^2$ $= (\mathbf{v} \times \mathbf{B}_g)/2c$	1.64E-08	Hz	$\Delta \nu$ req'd for 1 g (= 9.8 m/s²) acceleration.
Pump osc. low freq.	ν_3	$= \nu_1$	$\approx$ 100.0E09	Hz	PCM emitters
Pump osc. high freq.	ν_4	$= \nu_2$	$\approx$ 100.0E09	Hz	PCM emitters
Energy difference	ΔE	$= nh\nu_2 - nh\nu_1 = E_2 - E_1 = \Delta U$	9.80E6	$J = kg \cdot m^2/s^2$	
Gravitational redshift	$\delta \nu$	$= -(GM/2r^2) \cdot (c/g \cdot h)$	-1.50025E06	Hz	Referenced to surface.
Frequency ratio	$\Delta \nu / \nu$	$= gh/c^2 = GM/Rc^2 = \phi_g/c^2$	1.09039E06	—	h = 100 m
Vertical acceleration	**a**	$= 2c \cdot \Delta \nu \cdot \mathbf{r}_u = (\Delta \nu / \nu) \cdot (c^2/h)$ $= -\nabla \phi - d\mathbf{A}_g/dt$ $= c \cdot \mathbf{B}_g = 2c^2/h)(m\nu/\Delta \lambda) \cdot \lambda_{dB} \cdot \mathbf{r}_u$ $\mathbf{F}/m = (\mathbf{F}/E) \cdot (\mathbf{E}/\mathbf{B})^2$ $E/m \cdot d = c^2 \cdot \mathbf{A}/\nu = (dL/dx)/m$ $\nabla \times \mathbf{A}_g = \mathbf{B}_g c/\theta = \mathbf{v} \times \mathbf{B}_g$	9.80	m/s²	c = 2.99792458E08 m/s

cont

52. Anti-gravity

Table 52-1. 10,000 kg mass – Saturn antenna configuration – Hover mode (Cont)

Parameter	Symbol	Relation	Value	Units	Remarks
Energized Vehicle					
Radiation pressure	p_{rad}	$= L/A_d = (1+f)\cdot S/c$ $= \langle u \rangle/n = \rho_{rad}/3$ $= (1/3)aT^4$	1.559E02	$N/m^2 = kg/m\cdot s^2$ $= Pa$	f = Amt of incident beam reflected. n- dimensional space. $u \propto T^{n+1}$
Energy density	ρ_{rad}	$= 3\cdot p_{rad} = u = n\cdot p = c^2\cdot \mathbf{p/v}$ $= \frac{1}{2}(\epsilon_0 \mathbf{E}^2 + \mathbf{B}^2/\mu_0) = (c^2\cdot \mathbf{p})/\mathbf{v}$ $= \frac{1}{2}(\mathbf{E}\cdot\mathbf{D} + \mathbf{B}\cdot\mathbf{H})$ $= \frac{1}{2}(\epsilon_0 \mathbf{E}^2 + \mathbf{B}^2/\mu^0)$	4.159E02	J/m^3	Virial theorem $n = 3$ Phase conjugation $n = 1$
Pressure force (lift)	L	$\mathbf{F}_{grad} = m\mathbf{a}$ [mode 1] $F_{rad} = p_{rad} \times A_d$ [mode 2]	9.8E04	$N = kg\cdot m/s^2$	
Radiated power per unit time period	P_{rad}	$= \Delta E_p$ $= \Delta\nu h\nu/(e^{h\nu/\kappa T} - 1)$	9.80E06	$W = J/s =$ $kg\cdot m^2/s^3$	
Vertical velocity	$\mathbf{V}_v$	$= \mathbf{V}_{SW} = (c/\pi)\cdot\Delta\phi = c^2 \mathbf{p}/\rho_e$ $= \mathbf{v}^p/c^2 = \mathbf{K}\cdot c^2/\mathbf{g} = \mathbf{v}_o + \mathbf{a}t$	0	m/s	
Oscillator phase delta	$\Delta\phi$	$= \phi_2 - \phi_1 = \pi\,(v/c) = \pi\cdot\beta$	0	rad	Velocity $v = 0$

cont

52. Anti-gravity

Table 52-1. 10,000 kg mass – Saturn antenna configuration – Hover mode (Cont)

Parameter	Symbol	Relation	Value	Units	Remarks
Energized Vehicle					
Linear momentum	**p**	$= mv = m \cdot (c/\pi)\Delta\phi = F \cdot t$ $= \rho_e \cdot \mathbf{v}/c^2 = \mathbf{S}/c^2$	0	kg·m/s = J·s	
Power	P	$\Delta E/\Delta t = E_{reqd}/dt = (\mathbf{E} \times \mathbf{H}) \cdot \mathbf{S}$ $= (\mathbf{g} \times \mathbf{H_g}) \cdot \mathbf{S_g}$	9.80E6	J/s	1 sec ascent
Carrier frequency	f_c	$= c/\lambda_c$	100.0E09	Hz	100 GHz (Assumed)
Carrier wavelength	λ_c	$= c/f_c$	0.00299	m	
de Broglie frequency	ν_{dB}	$= pc/h = c/\lambda_{dB}$ $= (h/mv)(\Delta\nu/\Delta\lambda)$	0	Hz	momentum p = $\gamma mv = 0$
de Broglie wavelength	λ_{dB}	$= h/\gamma m_0 v = c/\nu_{dB} = h/p$ $= h/\gamma m_0(c/\pi)\Delta\phi$ $= \Delta\lambda/(c/h)(mv/\Delta\nu)$	0	m	
Electric field energy density	u_E	$= \frac{1}{2}(\epsilon_0 \mathbf{E}^2) = (\epsilon_{00}/2\Gamma)\mathbf{E}^2$	7.80E01	N/m² = J/m³	
Magnetic field energy density	u_B	$= \frac{1}{2}(\mathbf{B}^2/\mu_0) = (\Gamma/2\,\mu_{00})\mathbf{B}^2$	3.89E02	N/m² = J/m³	

cont

52. Anti-gravity

Table 52-1. 10,000 kg mass – Saturn antenna configuration – Hover mode (Cont)

Parameter	Symbol	Relation	Value	Units	Remarks
Energized Vehicle					
Total EM field energy density	u_t	$= \tfrac{1}{2}m(\epsilon_0 E^2 + B^2/\mu_0)$ $= u_E + u_B = U/V = \rho_e = c^2 \cdot \mathbf{p}/\mathbf{v}$	4.16E02	$N/m^2 = J/m^3$	
Electric field strength	E	$= c \cdot B = (\phi_E/\phi_B)B$	7.27E06	$V/m =$ $m/(s \cdot rad)$	$\mathbf{E} = \mathbf{B} \cdot c = \mathbf{B} \cdot \omega \cdot n/k$
Magnetic flux density	B	$= E/c = \tfrac{1}{2}m(\mathbf{B}^2)/\mu_0 = \mu_0 \mathbf{H}$	3.13E-02	$T = 1/rad$ $= W_b/m^2$	$\mathbf{B} = \mathbf{E}/c = \mathbf{E} \cdot \mathbf{k}/\omega\, n$
Power density (EM flux Poynting vector)	S	$= 1/\mu_0(\mathbf{E} \times \mathbf{B}) = \mathbf{E} \times \mathbf{H}$ $= \langle P_{rad} \rangle \cdot c = c \cdot B_{max}^2/2\,\mu_0$ $= E_{max} B_{max}/2\,\mu_0 = E_{max}^2/2\,\mu_0 c$ $= c \cdot B_{max}^2/2\,\mu_0 = \mathbf{v} \cdot \rho_e$ $= (\mathbf{c} \cdot \mathbf{B}) \times (\mathbf{B}/\mu)$	4.156E10	$W/m^2 = J/(s \cdot m^2)$ $= kg/s^3$	Power per unit area. EM wave intensity $I = \langle S \rangle$.
Equiv. blackbody temperature of emitter	T	$= \sqrt[4]{E/(\epsilon\sigma)} = \sqrt[4]{j^*/(\epsilon\sigma)}$ $= \sqrt[4]{L\pi/\sigma}$ $= \sqrt[4]{P/(A\epsilon\sigma)}$ $= P/k\Delta\nu$	varies	°K	ϵ = Emissivity ($0 < \epsilon < 1$), σ = Stephan-Boltzmann const. $= \pi^2 k_B^4/60\,\hbar^3 c^2 =$ 5.67037E-8 $J \cdot m^{-2} s^{-1} K^{-4}$

cont

52. Anti-gravity

Table 52-1. 10,000 kg mass – Saturn antenna configuration – Hover mode (Cont)

Parameter	Symbol	Relation	Value	Units	Remarks
Energized Vehicle					
Gravito-Poynting vector	$\mathbf{S}_{gi}$	$= (c^2/4\pi G) \times \mathbf{K} = \mathbf{g} \times \mathbf{H}_g$	0	kg/s^3	$\mathbf{S}_{gi} = -\mathbf{S}_g$; $v_H = 0$
Unruh temperature	T	$= \hbar a/(2\pi c k_B)$	4E-20	°K	$k_B = 1.381$E-23 J/°K; a = acceleration = 1g
Effective antenna aperture	A_e	$= \eta \cdot A = \eta(\pi R^2) = G\lambda^2/4\pi$	1.757E03	m^2	η = aperture efficiency. $\eta \cong 0.7$ assumed for planar array.
Isotropic ionization radius	R_{ion}	$= \sqrt[3]{(E_t/(4/3)\pi \cdot u_{bd})}$	38.7	m	ionization energy density ~ 39.8 N/m^2
Isotropic radiant energy density	u_{rad}	$= E_t/((4/3)\pi \, R^3$	2.34	J/m^3	@ Earth's surface.

Most of the important things in the world have been accomplished by people who have kept on trying when there seemed to be no hope at all. – Dale Carnegie

Phase displacement leads to the shift of potential holes (the nodes of the standing waves) relative to their initial position, and consequently to the position of the sources (oscillators). The sources come under the influence of the wave field, and naturally drift toward their potential holes. – Yuri N. Ivanov

I consider radiation to be a high species of vibration in the lines of force which are known to connect particles and also masses of matter together. – Michael Faraday

A body moves in a straight line through the tendency of its elements to their 'natural places'. – Aristotle

52. Anti-gravity

In the example illustrated, a relatively low carrier frequency of 1.0E11 Hz (= 100 GHz) coinciding with an atmospheric window is arbitrarily assumed to examine feasibility of microwave frequencies accessible with present technology in lieu of generally inaccessible higher gravitational frequencies that fall outside the range of common digital frequency synthesizers, harmonic amplifiers and nonlinear photonic crystals. Higher frequencies may be produced, for example, by optical oscillators using frequency combs and resonantly enhanced four-wave mixing in crystalline, nonlinear whispering gallery mode (WGM) resonators. Using microwave radiation pressure to generate ponderomotive force has previously been demonstrated in microwave and plasma supported levitation of objects in various forms on smaller objects in research projects for many decades. Microwaves, ultraviolet and x-rays present a number of significant thermal and ionizing radiation hazards and, in this case, the immediate vicinity and local neighborhood is irradiated at high intensity levels. Extreme gravitational frequencies (> PHz), however, would tend to reduce EM interference effects with conventional electronic devices and disruptive biologic effects as these are above the range of fundamental and low order harmonic atomic, interatomic and molecular resonance frequencies.

An energy density gradient is required for pressure to do work. For stationary hover, the required lift L equals the radiation pressure acting over the area of the disc (L = $p_{rad} \cdot A_d$) to equal the object weight W (= mg). For the given disc area A_d (= 628 m^2), without phase conjugation, the minimum energy density difference ρ_{rad} = 3·p_{rad} needed equals 4.67E02 N/m^2. Upward emitted EM radiation generates a downward acting radiation pressure on the source emitters. This radiation pressure may be offset with downward emitted radiation. If the upwardly directed radiation is in-phase, the waves constructively interfere increasing the intensity. Conversely, if the downwardly directed radiation is out-of-phase, the waves destructively interfere decreasing intensity. In the illustrated example, the emitted radiation is directed downward with a resultant upward thrust. The high frequency emitters are disposed below the low frequency emitters generating an upward frequency gradient ∇f and an upward acceleration $\mathbf{a}$ (= $2c \cdot \Delta v \cdot \mathbf{r}_u = (2c^3/h)(mv/\Delta\lambda \cdot v_{dB}) \cdot \mathbf{r}_u = 2(\mathbf{E}/\mathbf{B}) \cdot \Delta v \cdot \mathbf{r}_u$). A phase conjugate interference grating in a nonlinear medium formed by high amplitude pump beams (optical Kerr effect) allows amplification of backward reflected radiation from the dual planar emitter array below. The downward phase conjugate beam augments the energy density ρ_{rad} below the disc plane. The Poynting vector $\mathbf{S}$ (= $\mathbf{E} \times \mathbf{H} = \mathbf{p} \cdot c^2$) of reflected radiation provides an upward force $\mathbf{F}$ (= 2P/c) acting on the diffraction grating and associated pump beam emitters and support structure.

As shown, coaxial planar arrays of antenna elements driven at two different frequencies produce a beat frequency Δv differential corresponding to an upward 1 g acceleration. High amplitude counter-propagating pump beams acting on a nonlinear medium create a phase conjugate reflector allowing energy buildup in a nonlinear standing wave. Parametric amplification of the pump beam(s) in a non-degenerate four-way mixing process produce an amplified reflection of the upward probe beam (downward phase conjugate) augmenting the energy density below the disc. Vertical and horizontal velocity is controlled by phase modulation of the high and low frequency emitters.

For larger mass weight, as well as accelerations greater than 1 g, increased energy input is required. By increasing the energy excitation towards optical light frequencies, the resultant increase in acceleration would be accompanied by a visible ionized plasma glow below the emitter array immediately preceding the onset of acceleration. The dielectric strength of air is ~3 x 10^6 V/m which decreases with increasing frequency. The corresponding energy density for conventional breakdown u_b is $\cong$ 39.8 J/m^3 at standard atmospheric pressure. At energies greater than ~12 eV ionization effects are produced with visual corona due to electron/ion collisions and recombination of air molecules. The visible

52. Anti-gravity

color of the glow discharge may be modeled in terms of the irradiance of an equivalent blackbody radiation source to determine corresponding color temperature. Using Wein's law, the maximum power per unit wavelength of a blackbody emitter is $\lambda_m T = b = 2898E-3$ m °K where λ_m is the peak emission wavelength, T = absolute temperature and b = Wein's displacement constant. In terms of spectral flux per unit frequency, the peak optical emission frequency is given by $\nu_m = (\alpha/h)k_B T$ where $\alpha \cong 5.879E10$, h = Planck's constant, k_B = Boltzman's constant and T = absolute temperature. The total power P (= $\sigma T^4 A$) radiated by a blackbody obeys the Stefan-Boltzman law proportional to T^4 absolute temperature where σ = 5.67E-8 J/m²·K⁴. The total power density of a blackbody source is determined by integrating the spectral irradiance over all wavelengths given by H = σT^4 where σ = Stefan-Boltzman's constant. A plasma may be sustained by continuing absorption of microwave energy by air molecules (e.g., nitrogen and oxygen), water molecules and temporarily held in metastable states. With an EM radiation intensity sufficient to produce a plasma, the irradiated zone in the vicinity the antenna disc may serve as a resonant microwave cavity adjusting its radius to a multiple of the wavelength of the radiation such that the resonance condition is maintained. In equilibrium, the trapped EM energy plasma pressure (sum of the ion and electron pressures) is equal to the atmospheric pressure with glow aura due to slow energy release from nonionized metastable states.

A somewhat similar plasma effect occurs in the form of ball lightning (BL) which acts a resonant microwave cavity. Ball lightning is typically observed in cloud-to-ground lightning strikes and appears to be created by release of ionizing x-ray radiation in a lightning strike electron avalanche resulting in a standing wave EM resonant spherical cavity of plasma in which binding energy due to electrostatic separation of electric charges of ions and electrons is balanced by EM wave radiation pressure. One proposed mechanism is that carbon particulates and vaporized silicon from silicates in the soil reacts with oxygen releasing heat: $S_iO^2 + C \rightarrow Si + CO_2$. The energy of atmospheric ball lightning is estimated to vary from $10^2 - 10^9$ J with an average energy density on the order of ~10^5 J/m³. Hot, energetic electrons 'boiled' off the surface of the solid particle core ricochet within a reduced air pressure volume surrounding the core and are reflected off the surrounding ionized plasma sheath (whispering gallery mode). The emission of electrons leaves the particle core a net positive charge. Air pressure adjacent the core is reduced owing to the increased gas temperature. The radiation pressure p_{rad} is related to the internal energy density (= u/3). The ionized spherical plasma shell is maintained by the trapped standing wave microwave radiation. The plasma bubble decays typically within 1 – 5 sec once the internal radiation is dissipated. The color of the ionized plasma provides an indication of the ion temperature. With a ball radius R = ½m($v_x + v_y + v_z$)²/eE = ½m(eE/mω)² = (3/2)kT_e/eE and BL diameter (~0.01 – 0.2 m typ.) suggests trapped EM microwave radiation of frequencies ~100 MHz to 1 GHz with electron temperatures T_e on the order of 10^9 °K and ion temperatures T_{ion} of ~5,000 – 10,000 °K (yellow – white). See Fig 52-28. Ball lightning is generally spherical or pear-shaped floating plasmoid typically the size of a golf or tennis ball which may appear stationary, hovering above the ground moving slowly, move erratically, following powerlines/fences or dropping to the ground.

The most interesting aspect of the effect of the gravitational energy on gravitational fields is the possibility of the existence of mass distributions creating antigravitational fields in free space. The question arises therefore: can there exist a mass distribution producing an attractive field at all points within itself, but a repulsive field outside. – Oleg D. Jefimenko

It's not rocket surgery! – David Lee Roth

52. Anti-gravity

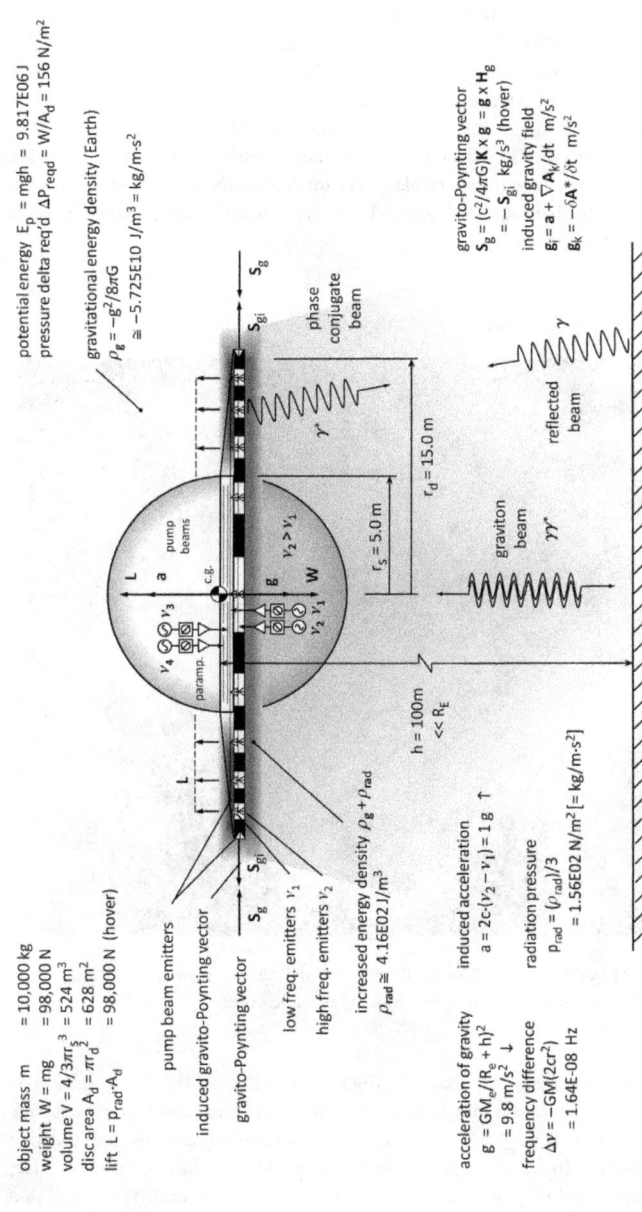

Fig. 52-27. EM energy density gradient creates radiation pressure differential generating a lift force acting over the antenna disc. For hover mode, acceleration $a = -g = 2c(v_2 - v_1) = 2c\Delta v = -2v\phi_g/c = 2\Delta v/\sqrt{(\varepsilon_g\mu_g)}$. The induced gravito-Poynting vector efflux S_{gi} cancels the natural gravito-Poynting vector influx S_g preventing conversion of gravitational potential energy into kinetic energy and zeroing Lorentz-Doppler frequency shift preventing fall. A phase conjugate reflector interferes with mass oscillator coupling. Negative refractive index metamaterials enables inverse Lorentz-Doppler effects, inertia neutralization and creation of artificial negative mass.

52. Anti-gravity

The objects may sometimes explode or split into smaller globes or simply fade away. Contact may prove fatal. Often appears white, blue, green, yellow, orange or red in color with fuzzy outline and occasionally glowing filamentary structures within. A frying or sizzling sound or acrid odor has been reported. Lightning ground strikes are thought to ionize silicate minerals in soil resulting in ionized silicon. Subsequent oxidation of silicon provides observed heat and glow. The observed glow is due to microwave emissions and is visible even in daylight. Ability to pass through walls and windows is consistent with microwaves or wave tunneling effect owing to partial overlap of wave functions broadening effective wave length. An irregular standing wave pattern may produce localized spikes in electric field intensity. The very high electric field gradient presumably enables restoration of atmospheric nitrogen or oxygen metastable states for temporary storage of microwave energy on the farside of the barrier allowing reconstitution. An attraction to metal objects is ascribed to conductor mirror imaging.

Ball lightning model

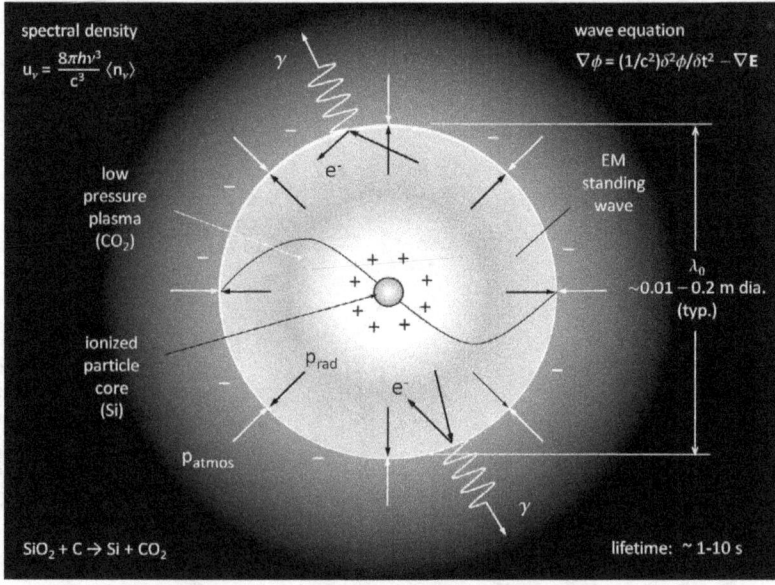

Fig. 52-28. Atmospheric ball lightning model consisting of plasma confined in a microwave resonant cavity surrounding a hot, ionized particle core.

Such copious EM emissions required for flight may be expected to produce EM interference, thermal and ionization effects and communication disruption in the immediate locale and to result in an enhanced EM cross-section facilitating object detection, tracking and collision avoidance. In the example illustrated, the EM signature varies with size, radiated power output, velocity and acceleration with distinctive beat frequency and vertical/horizontal polarization components depending on operating mode as well as atmospheric scintillation and non-ionic metastable photoluminescence effects. The pronounced EM radiation of the emitter array yields a prominent radar return signature.

For example, the radar cross-section of a perfectly reflecting sphere is equal to the physical cross-section (= $\pi a^2 \cdot v/c$) when the radius a $\gg \lambda$. The received power is augmented by the target emissions. As a radar target, the antenna gain is

$$G = (8\pi r^2/\lambda a)\sqrt{(P_r/P_t)} \qquad (52\text{-}22)$$

where
- r = distance from antenna to sphere (m)   r $\gg$ a
- a = radius of sphere (m)
- λ = wavelength (m)
- P_r = received power (W)
- P_t = transmitted power (W)

Painting a large target return is contraindicated for evasion/stealth purposes. Besides ECM jamming, a defensive close-in counter-measure, in such a case, is to increase position uncertainty with random zigzag jinking motion which may also facilitate perceptual saccade refreshing. Extreme phase, frequency agility coupled with prodigious power capability allows rapid vector changes in acceleration, velocity and direction to confuse, outrun or outmaneuver various material object potential threats. To reduce external radiated fields, it may be possible to produce a zero-amplitude energy state by superposition of out-of-phase standing waves such as illustrated in Fig. 52-29. As shown, wave phase reversal of π radians occurs at the antenna disc boundary without reflection. The wavelength illustrated is approximately twice the antenna disc diameter. Phase reversal occurs when changing phase passing through a resonance frequency with the bulk of the phase shift and impedance occurring within the width of the resonance. Alternatively, induced motion by synthesized Lorentz Doppler-shifted matter waves of very short wavelengths minimizes external radiation as the bulk of the wave energy is confined within a phase-locked, phase conjugate resonator cavity.

An alternate means for levitation and vertical acceleration in lieu of massive external EM emissions to generate an energy density to offset the gravitational energy density gradient is to confine the EM waves within a standing wave resonator. As previously noted, motion of matter may be induced by generation of synthesized Lorentz Doppler frequency shift of counter-propagating waves in a standing wave resonator. Parametric pumping of a phase conjugate reflector provides wave energy for conversion into kinetic energy of motion. Lorentz-Doppler frequency modulation of a standing wave creates a contracted moving standing wave generating an internal radiation pressure imbalance acting to impart motion to the cavity resonator. External EM emissions are limited to longitudinal and transverse waves generated by pulsation of the standing wave resonator and any internal EM radiation leakage. A conceptual schematic diagram is illustrated in Figs. 52-30 and -31.

An example geometry of a double-ended phase conjugate cavity resonator is illustrated in Figs. 52-32 and -33. The nonlinear phase conjugating medium, for example, may consist of a high K dielectric piezoelectric crystal or metamaterial of periodic nanostructure elements. The polarization nonlinearity may be enhanced by application of an external high voltage electric field. A phased array allows increased depth of field with multiple holographic Bragg gratings and adjustable number of modes per resonator length. Substantially greater reflectivity can be achieved with mode-locking by synchronous pumping with short pulses allowing compression of reflected pulses. The phased array emitters, for example, may be in the form of slotted waveguides or complementary microstrip slot array antennas. A vector potential **A**' and scalar potential ϕ' allows the possibility to generate **E**' and **B**' components to modify scattered diffraction fields.

Initially, the laser was called an invention looking for a job. – G. Harry Stine

52. Anti-gravity

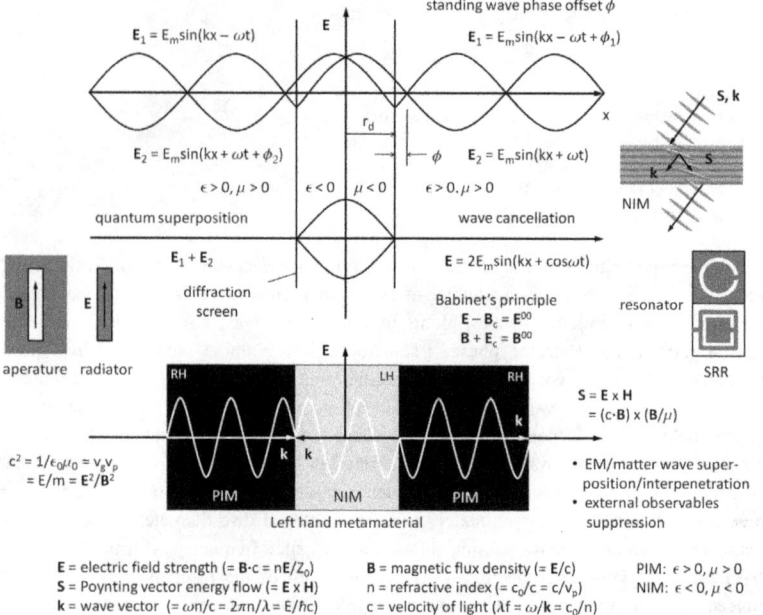

Fig. 52-29. External zero amplitude EM standing wave cancellation produced by phased offset. In the central region, wave superposition results in a monadnock. Negative index of refraction reverses wave propagation vector. Matter wave cancellation may allow matter interpenetration without displacement. According to Babinet's equivalence principle, diffraction fields from an aperature screen and its complement combine to reconstruct the incident plane wave ($\psi + \psi_c = \psi_0$). A waveguide slot is equivalent to a linear antenna.

Modulation of the simulated high intensity, parametrically amplified Lorentz Doppler blue- and red-shifted pump beams may be timed to augment or diminish inertial response to application of an external force by varying the internal energy density. Varying the boresight angle allows control of the mode-locked wave front group velocity of the moving contracted standing wave in either direction. Unidirectional response may be augmented by varying the dielectric constant gradient in the axial direction and/or varying the cross-sectional area to produce a nonuniform EM field gradient in the direction of motion.

Matter creation is possible because the energy of matter is actually cancelled by its gravitational potential energy. – John Preskill

The elevator was of the new sort that ran on gravitic propulsion. – Isaac Asimov (Foundation)

Far from being "purely qualitative", as some critics fondly imagine, catastrophe theory can be extremely quantitative. – René Thom

Elegance comes first; if the problem that is solved is physically interesting or important that's a bonus! – Freeman. J. Dyson

52. Anti-gravity

Vertical motion – Ascent/Descent

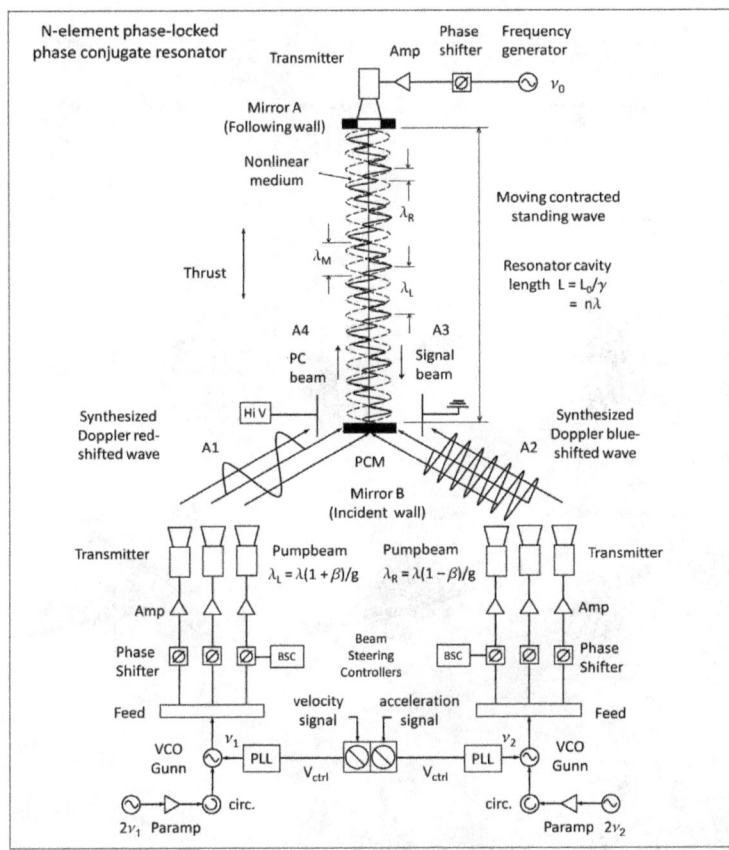

Fig. 52-30. Schematic of a broad band matter wave impulse drive using 4-way phase conjugate mixing. A contracted moving standing wave corresponds to a matter wave system in motion. Phase and frequency modulation allows velocity and acceleration control. Generation of partial standing waves corresponds to an intermediate state between travelling waves with no rest mass and standing waves with rest mass. Control of EM wave number and nodal width by modulation of frequency and reflection coefficient of the phase conjugate reflector, respectively, enables inertial mass modification by alteration of EM wave volumetric nodal density. Excessive acceleration-induced stress/strain may be countered by pulse width modulation of the continuous contracted moving standing wave. The interpulse nodal distance may be adusted down to molecular spacing intervals forming nodal pockets to lessen intermolecular bond stresses and strains. Pumping of the phase conjugate mirror by parametric amplification of simulated Lorentz Doppler red- and blue-shifted EM waves provides energy of motion.

Anything one man can imagine, other men can make real. – Jules Verne

It is the first duty of a hypothesis to be intelligible. – Thomas H. Huxley

52. Anti-gravity

Phased array beam operating modes

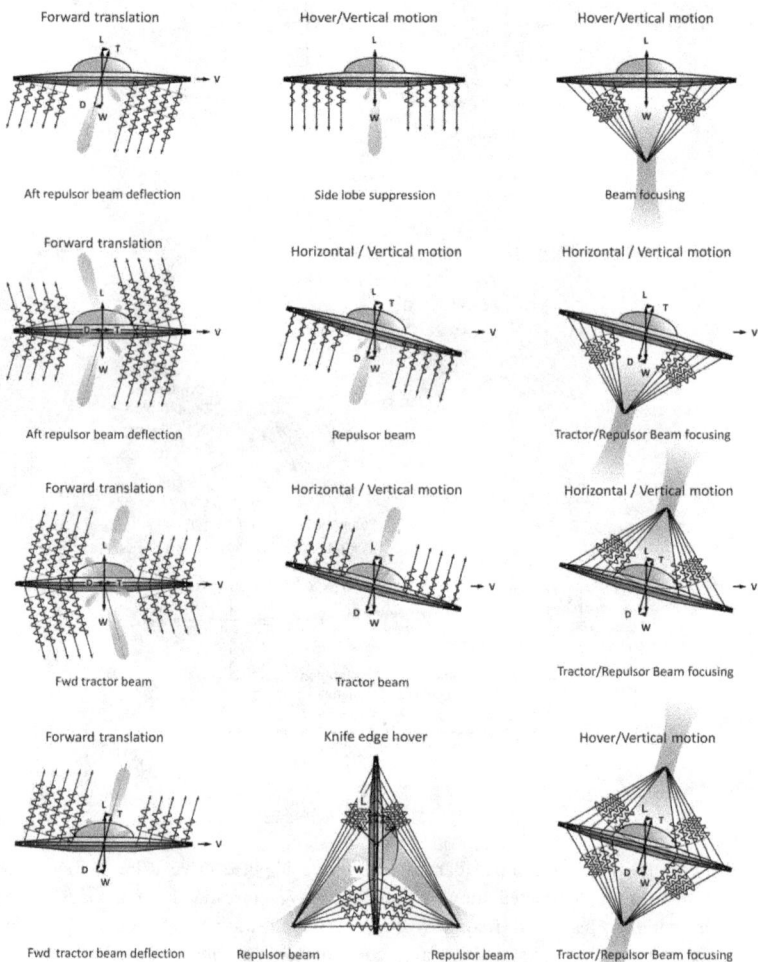

Fig. 52-31. Example phased array beam configurations and flight operating modes. Horizontal and vertical flight may be accomplished with either repulsor or tractor beams or combination thereof. Similarly, sustained hover including tilt-edge and edge-on modes may be achieved with a combination of tractor and repulsor beams, dual repulsor beams or dual tractor beams. Objects located within the beam focus will tend to be attracted/repelled toward the vehicle with the tractor/repulsor beam engaged. Relative movement is dependant on the ratio of the mass of object and the vehicle. Near the ground, phase conjugate beam is augmented by phase reflected return. With a focused beam close to the ground or water surface, the soil or water may be repelled or drawn upward depending on the direction of the beam wave vector. Spin angular momentum may be imparted by azimuthal phasing of the projected beam increasing magnetic energy density.

52. Anti-gravity

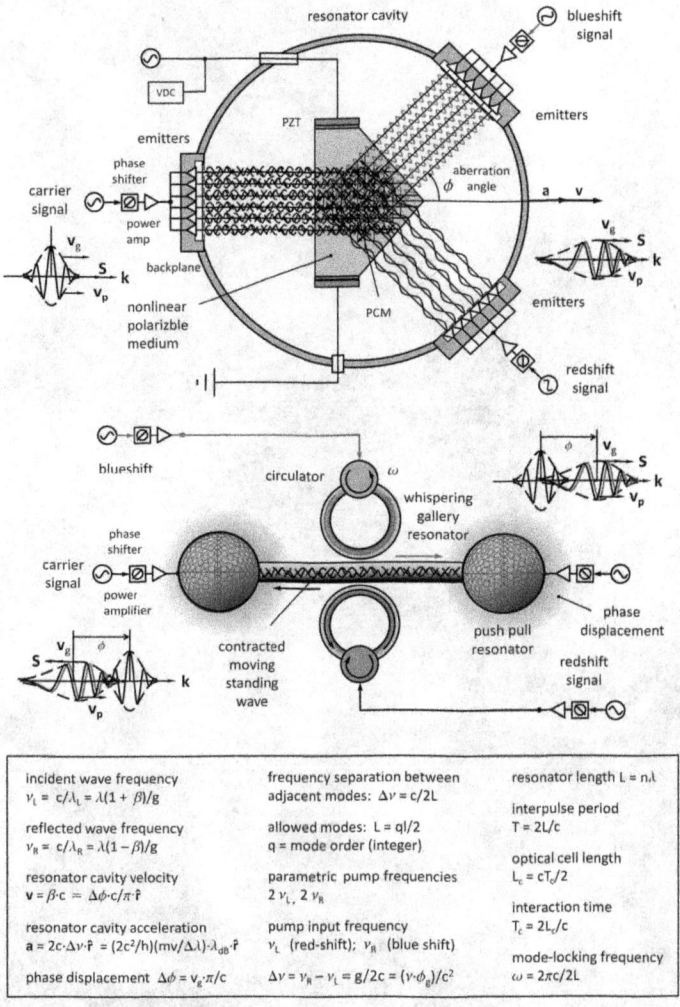

Fig. 52-32. Double-ended phase conjugate cavity resonator assembly with slotted emitter antenna arrays. A phase conjugate array may be constructed of a plurality of such resonator cells to provide a graded field intensity profile. Microwaves may be confined within a dielectric resonator by the abrupt change in permittivity and reflected by the large change in conductivity by the walls of a metallic cavity resonator. Contracted moving standing wave formation results in resonator motion. Phase conjugation is accomplished via four-wave mixing using photorefractive crystals, nonlinear Kerr media or metamaterials.

I especially placed the particle into the continuous wave and assumed that the propagation of the wave carried the particle with it, The psi wave is in a sense 'shows the way' to the traveling particle. – Louis de Broglie

52. Anti-gravity

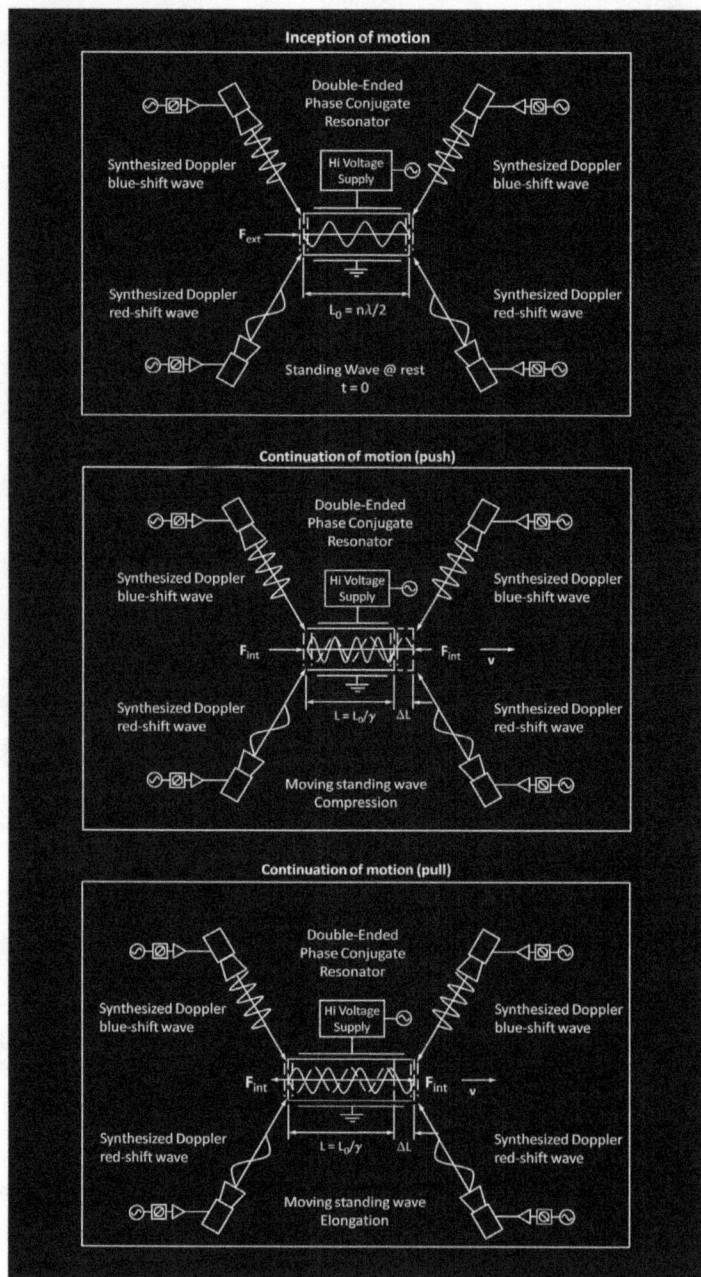

Fig. 52-33. Concept sketch of a double-ended phase conjugate phased-locked cavity resonator with four-way mixing generating induced contracted moving standing wave system motion via synthesized Lorentz-Doppler effect.. High voltage induces polarization in the phase conjugate reflector medium increasing nonlinearity of response.

52. Anti-gravity

Continuing our speculative example, let us consider the requirements for high speed horizontal motion. Assuming the same standing wave resonator configuration as before, how might translation at, say 225 m/s (~500 mph), be effected? A moving electromagnetic wave source produces Doppler and Lorentz Doppler shifted waves, and as previously noted, it should be possible to induce motion by parametric pumping of a phase-locked, phase conjugate resonator using synthesized Lorentz Doppler EM waves. A constant translation velocity v ($= \beta/c = c\cdot\Delta\phi/\pi = \mathbf{p}\cdot c^2/\rho_e$) of 225 m/s, according to the wave model illustrated, involves a relative phase difference $\Delta\phi$ ($= \beta\cdot\pi = \rho\cdot\pi/\gamma$) of ~2.357E-06 radians. A performance summary is shown in Table 52-2.

Using a phase conjugate reflector, the intercepted ground return is reflected back towards the surface forming a phase conjugate beam. The superposition of the upward directed photons (γ) and the downward directed phase conjugate photons (γ^*) forms in effect a graviton beam ($\gamma\gamma^*$) increasing local EM energy density. The phase conjugate mirror, for the given frequency range $\Delta\nu$, constitutes a perfectly reflecting surface which doubles the radiation pressure ($p_{rad} = 2S/c$) over that of a perfect absorber ($p_{rad} = S/c$) where S = the energy flux. In addition, as the radiation energy is not dispersed in a phase conjugate beam, the radiation pressure equals the energy density $p_{rad} = u_e/n$ where n = 1 rather than the dispersive case $p_{rad} = u_e/3$. Hence, the radiation flux requirements may be reduced beyond that shown. A phase conjugate reflector, additionally, by preventing absorption of incident radiation and heating of the levitated object provides a form of radiation shielding above the disc plane. Lastly, for higher acceleration, increased power may be provided by pumping the phase conjugate reflector to augment the reflected (phase conjugate) signal in which the pumping frequency is an oscillatory function of time such as described by the Mathieu equations regarding parametric resonance. Parametric excitation, in this instance, would involve mixing the weak reflected ground return signal ν_r with local oscillator signal tuned to the desired acceleration ($\Delta\nu = \mathbf{a}/(2c\cdot\mathbf{r}_u)$) as input to a parametric amplifier resulting in an amplified phase conjugate beam containing sum and difference frequencies. A concept sketch for a phased array for horizontal flight translation mode is shown in Fig. 52-31.

Planar array wave system horizontal velocity $\mathbf{v}_h$ is controlled by varying the relative phase of two counter-propagating moving standing waves ($= \Delta\phi\cdot c/\pi\cdot\mathbf{r}_u$). Motion is in direction of the phase change. Horizontal acceleration $\mathbf{a}_h$ is controlled by altering the frequency difference ($= 2c\cdot\Delta\nu\cdot\mathbf{r}_u = (2c^3/h)(mv/\Delta\lambda\cdot\nu_{dB})\cdot\mathbf{r}_u$). The phase conjugate reflector in a nonlinear medium is formed and modulated in nondegenerate four-wave mixing (NDFWM) by EM waves of simulated red- and blue-shifted Lorentz Doppler frequencies, ν_1 and ν_2, respectively, corresponding to the desired acceleration. Application of high voltage to the phase conjugate medium such as a high-K crystalline dielectric increases polarization nonlinearity. Parametric 2X amplification of the pump beams modulated by voltage controlled oscillator (VCO) signals effectively increases intensity of the standing wave. A plurality of resonators disposed in a circular array and selectively energized allows motion in any selected azimuthal direction. Self-induced motion is effected by modulation of synthesized matter wave(s) giving rise to ponderomotive force imbalance. Kinetic energy of motion is derived from the pump beam energy input.

EGM is a method permitting one to mathematically represent a mass-object in spectral form. – Riccardo C. Storti

I think there is nothing worse than inertia...the key is to keep flying. – Ridley Scott

The prospect of being wrong need not frighten us a bit. – Alan E. Nourse

52. Anti-gravity

Table 52-2. 10,000 kg mass – Saturn antenna configuration – Translation mode (Cont)

Parameter	Symbol	Relation	Value	Units	Remarks
Energized Vehicle					
Object rest mass	m	$= \rho \cdot V = E/c^2 = \mathbf{W/g} = E/(\mathbf{E/B})^2$ $= 2(T-V)/(\epsilon_0 \mathbf{E}^2 - c^2 \mathbf{B}^2/\mu_0)$	10,000.0	kg	Assumed
Object radius	a	$= r_{sphere}$	5.0	m	Assumed
Object volume	V	$= (4/3\, \pi a^3)$	523.6	m^3	sphere
Mass density	ρ	$= m/V$	19.1	kg/m^3	average
Rest energy	E_0	$= m_0 c^2$	8.987E20	J = N·m = kg·m^2/s^2	Potential energy
Potential energy	E_p	$= mgh$	9.80E06	J	
Kinetic energy	E_k	$= \frac{1}{2} mv^2 = E_t - E_0$	2.53E08	J	
Total energy	E_t	$= E_P + E_k = H$	2.63E08	J	
Velocity	**v**	$= c\Delta\phi/\pi = k_c \cdot \Delta\phi = (\mathbf{A} \cdot c^2)\phi_E$ $\mathbf{p} \cdot c^2/\rho_e = c^2/v_p\,(A_g \cdot c^2)/\phi_g$	225	m/s	Assumed. $k_c = c/\pi =$ 95426903.18 m/(s·rad)

cont

52. Anti-gravity

Table 52-2. 10,000 kg mass – Saturn antenna configuration – Translation mode (Cont)

Parameter	Symbol	Relation	Value	Units	Remarks
Energized Vehicle					
Mach no.	M	$= v/v_s$	0.66	–	$V_s = 343$ m/s @ 30 °C
Cruise altitude	h	Above MSL	100	m	Assumed
Gamma	γ	$= 1/\sqrt{(1-v^2/c^2)}$	1.000000000000028	–	
Phase difference vertical	$\Delta\phi_V$	$= \phi_2 - \phi_1$ $= \pi(v_V/c)$	0	rad	$V_H = 0$
Phase difference horizontal	$\Delta\phi_H$	$= \phi_4 - \phi_3$ $= \pi(v_H/c) = \pi\cdot\beta$	2.3578E-06	rad	
Velocity ratio	β	$= v_H/c = \Delta\phi/\pi = \sqrt{(1-g^2)}$	7.50519214E-07	–	$g = \sqrt{(1-\beta^2)}$
Wavelength ratio	R	$= \lambda_b/\lambda_r = (1+\beta)/(1-\beta)$	≈ 1.0	–	
Redshift	z	$= \sqrt{((1+\beta)/(1-\beta))} - 1$ $= \sqrt{(R)} - 1$	≈ 0	–	

cont

52. Anti-gravity

Table 52-2. 10,000 kg mass – Saturn antenna configuration – Translation mode (Cont)

Parameter	Symbol	Relation	Value	Units	Remarks
Energized Vehicle					
Carrier frequency	f_c	$= \nu_0 = (\nu_1 + \nu_2)/2$	1.0E11	Hz	100 GHz
Carrier wavelength	λ_c	$= c/f_c$	2.9979E-03	m	
Doppler blue shift frequency	ν_{DB}	$= c/\lambda_B = c/\lambda_c(1-\beta)$	1.00000020E11	Hz	
Doppler red shift frequency	ν_{DR}	$= c/\lambda_R = c/\lambda_c(1+\beta)$	9.9999979E10	Hz	
Doppler blue shift wavelength	λ_{DB}	$= \lambda_c(1-\beta)$	2.99792395E-03	m	
Doppler red shift wavelength	λ_{DR}	$= \lambda_c(1+\beta)$	2.99792520E-03	m	
Lorentz Doppler blue shift frequency	ν_{LDB}	$= c/\lambda_{LDB} = c/\lambda_0(1+\beta)/g$	1.00000209E11	Hz	
Lorentz Doppler red shift frequency	ν_{LDR}	$= c/\lambda_{LDR} = c/\lambda_0(1-\beta)/g$	9.9999792E10	Hz	

cont

52. Anti-gravity

Table 52-2. 10,000 kg mass – Saturn antenna configuration – Translation mode (Cont)

Parameter	Symbol	Relation	Value	Units	Remarks
Energized Vehicle					
Lorentz Doppler blue shift wavelength	λ_{LDB}	$= c/\nu_{LDB} = \lambda_0(1-\beta)/g$	2.9979240E-03	m	
Lorentz Doppler red shift wavelength	λ_{LDR}	$= c/\nu_{LDR} = \lambda_0(1+\beta)/g$	2.9979252E-03	m	
Oscillator low freq.	ν_1	$= \Delta\nu - \nu_2$	100.0E09 = 1.0E11	Hz	100 GHz (Assumed)
Oscillator high freq.	ν_2	$= \nu_1 + \Delta\nu$	$\approx$ 100.0E09	Hz	
Oscillator beat frequency diff.	$\Delta\nu$	$= (\nu_2 - \nu_1) = g/2c = \nu\phi_g/c^2$ $= (\mathbf{v} \times \mathbf{B}_g)/2c$	1.64E-08	Hz	$\Delta\nu$ req'd for 1 g
Pump beam high frequency	ν_3	$= \nu_0 = c/(\lambda_c/\gamma)$	1.00000E11	Hz	Horizontal mode. Para. amp. freq. = $2\nu_3$
Pump beam low frequency	ν_4	$= \nu_1 - \nu_3 = c/(\lambda_c/\gamma\beta)$	1.00000E11	Hz	Horizontal mode. Para. amp. freq. = $2\nu_4$
Pump beam short wavelength	λ_3	$= c/\nu_3 = \lambda_0(1-\beta)/g$	2.99792E-03	m	Horizontal mode.

cont

565

52. Anti-gravity

Table 52-2. 10,000 kg mass – Saturn antenna configuration – Translation mode (Cont)

Parameter	Symbol	Relation	Value	Units	Remarks
Energized Vehicle					
Pump beam long wavelength	λ_4	$= c/\nu_4 = \lambda_0(1+\beta)/g$	2.99793E-03	m	Horizontal mode.
Vertical acceleration	a_V	$= \mathbf{F}/m = 2c\cdot\Delta\nu\cdot\mathbf{r}_u = c^2\mathbf{A}/v$ $= (\mathbf{F}/\mathbf{E})\cdot(\mathbf{E}/\mathbf{B})^2 = c\cdot\mathbf{B}_g/\theta$	9.80	m/s^2	c = 2.9979245E08 m/s
Oscillator freq.diff.	$\Delta\nu_V$	$= a_V/2c = |\nu_2 - \nu_1|$	1.6344E-08	Hz	
Radiation pressure (vertical)	p_{emv}	$= L/A_d = \rho_{rad}/3 = (1+f)\cdot S_v/c$	1.3864E02	N/m^2 = kg/m·s^{-2}	f = Amt of incident beam reflected
Pressure force (lift)	L	$F_{grad} = ma$ [mode 1] $F_{rad} = p_{rad} \times A_d$ [mode 2]	9.8E04	N = kg·m/s^2	0° tilt
Radiation pressure (horizontal)	p_{emh}	$= F_T/(2\cdot R_d\cdot D_h)$	1.24E04	N/m^2 = kg/m·s^{-2}	0° tilt
Pressure force req'd (thrust)	F_T	$= F_D = P_k/v$	5.58E05	N = kg·m/s^2	[Mode 1]
Aerodynamic drag	F_D	$= \frac{1}{2}C_{D0}\rho v^2\cdot A$	5.58E05	N = kg·m/s^2	C_{D0} = 0.25 assumed.

cont

52. Anti-gravity

Table 52-2. 10,000 kg mass – Saturn antenna configuration – Translation mode (Cont)

Parameter	Symbol	Relation	Value	Units	Remarks
Energized Vehicle					
Thrust to weight	T/W	$= F_T/W$	5.96	–	Cruise
Linear momentum	**p**	$= \gamma m_0 \mathbf{v} = \gamma m_0 \cdot (c/\pi)\Delta\phi \cdot \mathbf{r}_u$ $= \mathbf{F} \cdot t = \rho_e \mathbf{v}/c^2 = \mathbf{S}/c^2$	2.25E06	kg·m/s² = J·s	v = 225 m/s
de Broglie frequency	v_{dB}	$= pc/h = \gamma m_0 c^2 /h = c/\lambda_{dB}$	1.017995114E48	Hz	
de Broglie wavelength	λ_{dB}	$= h/\gamma m_0 v = c/v_{dB} = h/p$ $= h/\gamma m_0 (c/\pi)\Delta\phi$	2.94493022E-40	m	v = 225 m/s
Impulse	**I**	$= \mathbf{F} \cdot t = m\Delta \mathbf{v} = \Delta \mathbf{p}$ $= m \cdot (c/\pi) \cdot (\Delta\phi_2 - \Delta\phi_1)$	0	N·t = kg·m/s	$\Delta v = 0$
Horizontal acceleration	a_H	$= (\mathbf{T} - \mathbf{D})/m = \mathbf{F}/m = c^2 \mathbf{A}/\mathbf{v}$ $= (V_2 - V_2)/(t_2 - t_1)$ $= (c/\pi)(\Delta\phi_2 - \Delta\phi_1)/\Delta t$	0	m/s²	Constant velocity $V_2 - V_1 = 0$
Oscillator freq.difference	Δv_H	$= a_H/2c$	0	Hz	Horizontal mode.
Radiated energy	E_{rad}	$= \Delta E_k + \Delta E_p$	2.63E08	J	

cont

52. Anti-gravity

Table 52-2. 10,000 kg mass – Saturn antenna configuration – Translation mode (Cont)

Parameter	Symbol	Relation	Value	Units	Remarks
Energized Vehicle					
Isotropic ionization radius	R_{ion}	$= \sqrt[3]{E_t/((4/3)\pi \cdot u_{bd})}$	116	m	ionization energy density ~ 39.8 N/m^2
Isotropic radient energy density	u_{rad}	$= E_t/(4/3\pi\, r^3)$	62.7	J/m^3 = N/m^2	@ r = 100 m
Equiv. blackbody temperature of emitter	T	$= \sqrt[4]{E/(\epsilon\sigma)} = \sqrt[4]{(j^*/(\epsilon\sigma)}$ $= \sqrt[4]{(I\pi/\sigma)}$ $= \sqrt[4]{(P/(A\epsilon\sigma)}$ $= P/k\Delta v$	varies	°K	ϵ = Emissivity (0 – 1), σ = Stephan-Boltzmann const. $= \pi^2 k_B^4/60\, \hbar^3 c^2) =$ 5.67037E-8 J·m^{-2} s^{-1} K^{-4}
Reynold's no.	R_e	$= \rho v L/\mu$	1.415E05	-	Charac. length L = r_s
Dynamic (absolute) air viscosity	μ	$= \rho \cdot \nu$	1.785E-05	kg/(m·s) = N·s/m^2 = Pa·s	
Kinematic viscosity	ν	$= \mu/\rho$	14.55E-06	m^2/s	
Gravito-Poynting vector	S_{gi}	$= (c^2/4\pi G) \times \mathbf{K} = \mathbf{g} \times \mathbf{H_g}$	2.576E13	rad·m/s^3	Radial efflux = $-S_g$
Air density	ρ	$= p/R \cdot T$	1.225	kg/m^3	R = Univ. gas const = 8.3144 J/(K·mol)

cont

52. Anti-gravity

Table 52-2. 10,000 kg mass – Saturn antenna configuration – Translation mode (Cont)

Parameter	Symbol	Relation	Value	Units	Remarks
Static air pressure	p	$= p_0 e^{\wedge}(-Mgh/kT)$	101,325	Pa	@ STP @ 15°C $k = R/N_A$ where k = Boltzmann's const, R = Univ. gas const, M = molecular wt. of air N_A = Avogadro's no.
Dynamic air pressure	q	$= \frac{1}{2}\rho v^2$	28,413	Pa	v = 225 m/s
Speed of sound	v_s	$= \sqrt{(K_s/\rho)}$	340	m/s	@STP @ 15°C
Total air temperature	T_t	$= T_0(1 + ((\gamma - 1)/2) \cdot M^2)$	313 °K	°K	T_s = 288.2 °K

Elevate your thinking and you'll levitate to greater levels. – Constance C. Friday

How can you have a wave without a wave equation? – Peter Debye

There is, however, another possible road through matter (interpenetration) – a tortutous and badly sign posted road, for it leads us to the fourth dimension. Another possibility is that, even if a fourth dimension or direction of space does not exist in nature, we may be able to create such an extension artificially. – Arthur C. Clarke.

One can even imagine the bulk movement of freight or raw materials along "gravity pipelines". – Arthur C. Clarke

Under the action of a gravitational field, changes occur within material bodies and they fall to earth. But what are these changes? And is it possible to influence them in any way to eliminate the tendency to fall and get the effect of nullifying their weight? – Yuri. Nikozevich Ivanov

52. Anti-gravity

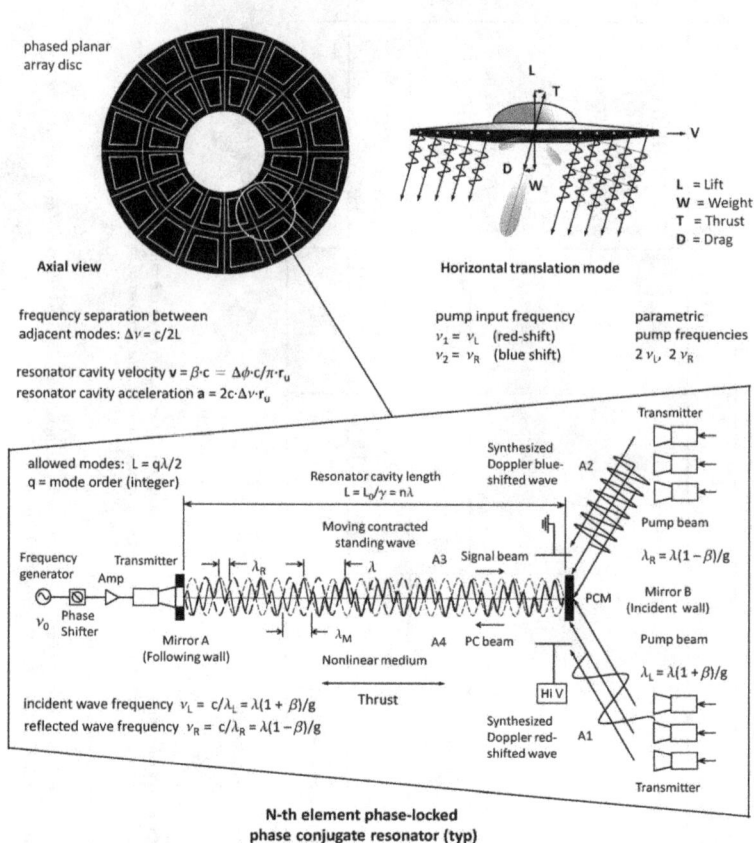

Fig. 52-34. Planar array wave system composed of n-element phase conjugate, phase-locked resonators. Horizontal velocity v_h is controlled by varying the relative phase of two counter-propagating moving standing waves. Motion is in direction of the phase change. Horizontal acceleration a_h is controlled by altering the frequency difference.

According to the equivalence principle, a gravitational field is locally indistinguishable from effects of an accelerated frame of reference and can be counteracted locally by transformation to an accelerated frame. As previously noted, inertial control may conceivably be achieved by modulating or simulating the gravitomagnetic effect. A mass in motion constitutes a mass current (i_m) generating an induced gravitomagnetic field ($\mathbf{H_g}$) analogous to an electric current (i) generating an induced magnetic field ($\mathbf{H}$). Conversely, a changing co-gravitational field creates a gravitational field just as changing magnetic field creates an electric field. This effect is illustrated in Figs. 52-35 through -37. A mass in motion at constant velocity creates a constant circumferential gravitomagnetic field. A mass under acceleration creates a time-varying circumferential gravitomagnetic field. Conversely, a time-varying circumferential gravitomagnetic field may be expected to

52. Anti-gravity

generate an induced acceleration of mass. The gravitomagnetic field $\mathbf{H}_g$ is analogous to the electromagnetic field $\mathbf{H}$ but is considerably weaker. For example, the ratio of mass current i_m to electrical current i varies as the ratio of electron mass m_e to electron charge e ($i_m/i = m_e/e = 5.685\text{E-}12$). The calculated gravitomagnetic permeability μ_g ($= 4\pi G/c^2 = 1/\epsilon_g\epsilon_0\mu_0 = 9.328\text{E-}27$ m/kg) in free space is extremely small such that detection of a gravitomagnetic field outside of a mass current i_m is exceedingly difficult to measure. The low value of the gravitomagnetic permeability results from the c^2 divisor ($= 1/\epsilon_0\mu_0 = v_p v_g = (\mathbf{E}/\mathbf{B})^2 = 1/\epsilon_g\mu_g = (\mathbf{v} \times \mathbf{g})/\mathbf{K} = (\mathbf{v} \times \mathbf{E})/\mathbf{B} = (\mathbf{v}/\mathbf{A}_g)\phi_g = (\mathbf{v} \times \phi_E)/\mathbf{A} = (\Delta\nu/\nu)\phi_g$). Negative index metamaterials allow local alteration of gravitomagnetic permeability. The physical mechanism for generation of a gravitomagnetic field $\mathbf{H}_g$ is theorized to involve the same mechanism for generation of a magnetic field $\mathbf{H}$ in terms of induced rotation of Planck dipoles of positive and negative Planck mass (m_P^+, m_P^-) of a quantum vacuum. For gravitomagnetic fields, the Planck dipole angular velocity is much lower with random phase rotation reducing the gravitomagnetic field strength compared to electromagnetic field strength.

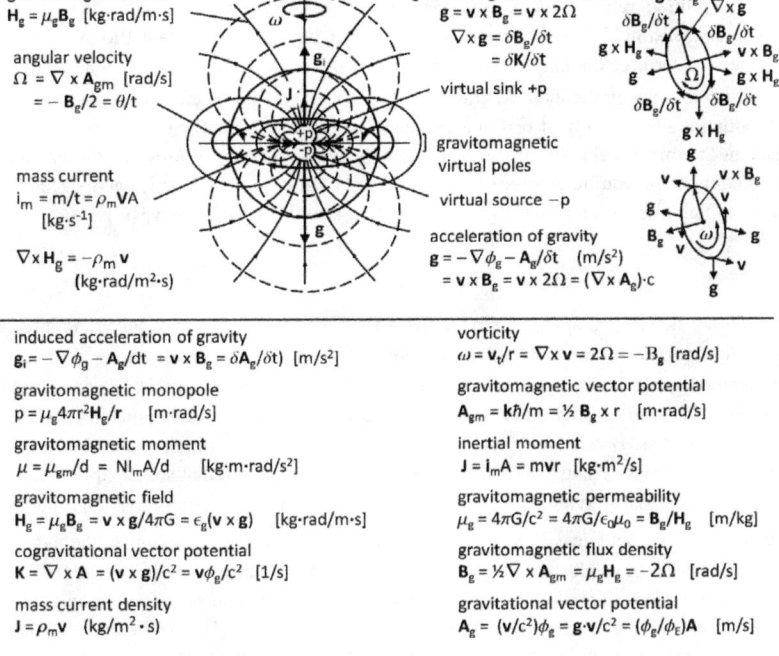

Fig. 52-35. Gravitomagnetic field (closed flux lines) of a steady-state mass current loop is analogous to a magnetic field of a steady-state electrical current in a superconducting loop with repulsion characteristics. Similarly, a gravitomagnetic mass current loop may be described in terms of a virtual positive and negative mass pole with an inertial moment analogous to a virtual north and south magnetic pole with a magnetic moment. Gravitomagnetic induction involving mass currents is analogous to Faraday magnetic induction associated with electric currents (e.g., unipolar generator, disc dynamo, etc.).

52. Anti-gravity

Motion induced gravitomagnetic field

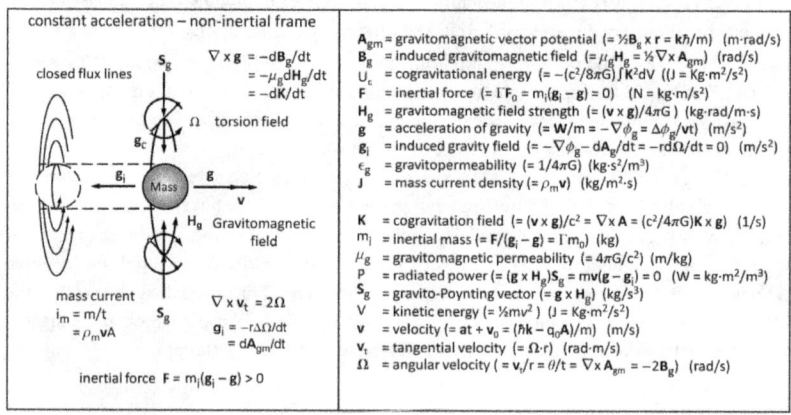

Fig. 52-36. Motion of mass constitutes a mass current (i_m) with an induced gravitomagnetic field ($\mathbf{B_g}$) analogous to an electric current (i) with an associated magnetic field (**B**). For mass in motion relative to an observer, the increase in inertial mass resides in the gravitomagnetic field and is proportional to the square of the change in EM field nodal density. Under acceleration, a mass experiences an inertial force $\mathbf{F} = m_i(\mathbf{g_i} - \mathbf{a})$ resisting acceleration in which the induced gravitomagnetic field $\mathbf{g_i}$ opposes the applied acceleration **a**. Falling matter initially at rest in a gravitational field radiates gravitational energy until reaching terminal velocity. In free-fall gravitational radiation ceases as the induced gravitational field equilibrates with the acceleration of gravity. The kinetic energy acquired by the falling mass is equal to the gravitational energy influx ($dU/dt = m|\mathbf{v}| \cdot |\mathbf{g}|$).

A falling slender rod of mass m in the Earth's gravitational field **g** experiences no inertial force in free-fall as the induced gravitational acceleration $\mathbf{g_i}$ equals the gravitational acceleration **g**. A falling mass with velocity **v** constitutes a mass current i_m generating a left-handed circumferential co-gravitational field $\mathbf{K_c}$ producing an analogous gravitational equivalent to the Barnett and Einstein-deHaas effects. The total gravitational field of the falling object (passive mass m) is the vector sum of the inward radial self-field gravitation $\mathbf{g_c}$ and the external gravitational acceleration of the Earth (active mass M). The induced gravito-magnetic field density $\mathbf{B_g}$ (= $\frac{1}{2}\nabla \times \mathbf{A_{gm}} = \mu_g \mathbf{H_g} = 2\mathbf{\Omega_k}$) is analogous to the induced magnetic field **B** of an electric current i. The gravikinetic field $\mathbf{g_k} = -d\mathbf{A^*}/dt = \mathbf{v} \times \mathbf{K}$ where $\mathbf{A^*}$ is the retarded cogravitation vector potential, **v** is velocity and **K** is the cogravitational field. The cogravitation field **K** acts only on moving masses. The cogravitation field is left-handed relative to the mass current J_m and has negative field energy.

The precession rate of the falling rod $\Omega = \frac{1}{2}(\mathbf{g} \times \mathbf{v})/c^2$ due to the Lense-Thirring "frame-dragging" effect is vanishingly small. The precession results in a tilting of light cones in the direction of motion. The antithetical case of a slender rod corresponds to a flattened disc. The axial gravitational field of a falling disc has the same general equation form as the that of a rotating disc due to similarity of mass currents. The Newtonian gravitational acceleration in the axial direction for a disc of density ρ and thickness t is $\mathbf{g} = 4G\rho t$.

52. Anti-gravity

Co-gravitational field of a free-falling slender rod

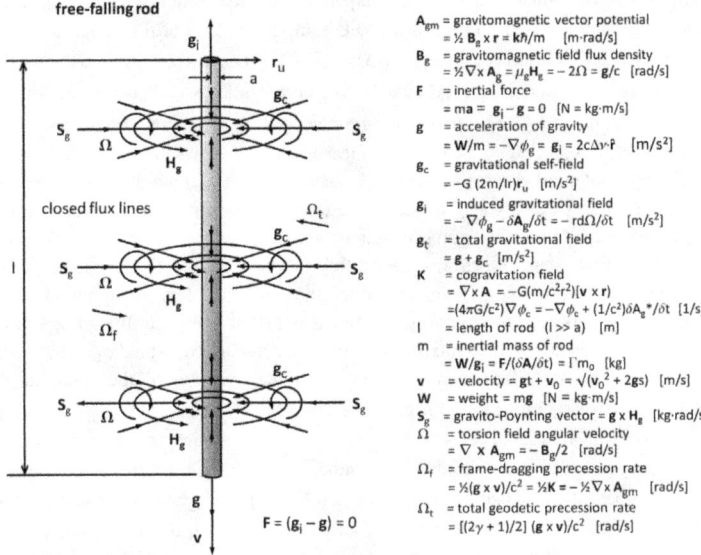

Fig. 52-37. A falling mass with velocity **v** constitutes a mass current i_m generating a circumferential gravitomagnetic field $\mathbf{B}_g$. The inertial force in free-fall is nullified as the induced gravitational acceleration $\mathbf{g}_i$ equals the gravitational acceleration $\mathbf{g}$. The de Broglie matter wave components undergo Lorentz-Doppler shift under acceleration and become phase-locked at zero-g. A synthesized right-handed cogravitational field would yield an anti-gravity effect by opposing the radial influx of gravitational potential energy via the gravitational Poynting vector $\mathbf{S}_g$ (= $c^2/4\pi G$)$\mathbf{K} \times \mathbf{g}$) with radial energy efflux $\mathbf{S}_{gi} = \mathbf{g}_i \times \mathbf{H}_{gi} = -\mathbf{S}_g$. The gravito-Poynting vector influx is proportional to the matter wave frequency shift.

In the Lorentz-invariant theory of gravitation, the gravitational four-potential includes both time-like scalar and space-like vector components which represented in covariant form is given by

$$A_\mu = (\phi_g/c, -\mathbf{A}) = (\phi_g/c, -A_x, -A_y, -A_z) \qquad (52\text{-}23)$$

where

A_μ = gravitational four-potential = $(\phi_g/c^2)v_i = (\phi_g/c\Gamma, \phi_g \mathbf{v}/c^2\Gamma)$ (m^2/s^2)

μ = index ranging from 0 to 3; $x_\mu = g_{\mu\nu} x^\nu = (-ct, x, y, z)$

ϕ_g = gravitational potential = $-Gm/r = (A_g/A)\phi_E = (\Delta\nu/\nu)c^2$ (J/kg = m^2/s^2)

V_i = covariant 4-velocity (= $\gamma(c,\mathbf{v})$) (m/s)

c = velocity of light (= $c_0/\Gamma = \sqrt{(1/\epsilon_0\mu_0)} = \sqrt{(1/\epsilon_g\mu_g)} \simeq 2.99792458\text{E}8$ m/s)

Floating has so much more gravitas. – Gail Carriger

We live in a minute island of known things. – Damon Knight

All imagination begins by analyzing nature. – Jacob Bronowski

52. Anti-gravity

For mass in motion, the induced gravitational field strength $g_i = -\nabla\phi_g - \delta A_g/\delta t = -\nabla\phi_g + g_k = -\nabla\phi_g - (v \times K)$ [m/s^2] while the accompanying torsion field $\Omega = \nabla \times A_{gm}$ [rad·s]. The induced gravitational acceleration is a function of the gradient of the scalar potential ϕ_g which changes due to motion ($\phi_g = \phi_{0g}\Gamma = (\Delta\nu/\nu)(E/B)^2 = (A_g/A)\phi_E = (A_g/A)(2E\phi_B/h)$) and the rate of change in time of the gravitation vector potential $\delta A_g/\delta t$ where gravitational gamma $\Gamma = 1/\sqrt{(1-v_e^2/c^2)}$. The inductive gravitational field is directly proportional to acceleration. The scalar potential term is a measure of the electromagnetic field density while the vector potential derivative term is a measure of the temporal rate of change of electromagnetic energy density and, hence, is suggestive of means of electronic control.

Although naturally occurring gravitomagnetic fields associated with mass motion are considerable weaker than electromagnetic fields associated with electrical current, conceivably it may be possible to boost the effect of induced acceleration by radial outward EM energy efflux opposing the radial gravitational potential energy influx. For example, we might imagine creation of a rotating circular electromagnetic standing wave with a rapidly varying circular polarization spin rate. Angular rotation of the standing wave corresponds to a mass current loop gravitomagnetic field H_g. Varying the circular polarization angular frequency Ω corresponds to a time-varying gravito-magnetic field dH_g/dt. Perhaps, it may be possible to synthesize the effect of mass motion while amplifying the magnitude to useful levels such that the required gravitomagnetic field self-energy is equivalent to the kinetic energy of motion ($E_g = E_k = \frac{1}{2} mv^2$). Suppose, the disc is made to revolve about the central mass generating radially outward directed EM waves creating a synthesized gravito-Poynting energy flow S_{ig} ($= -g_i \times H_g$) opposing the gravito-Poynting energy influx S_g ($= g \times H_g$). The gravitomagnetic field H_g may be increased by use of Negative Index Metamaterials (NIM) which allow dramatic increase in gravitomagnetic permeability μ_g ($= 4\pi G/c^2$) by reduction of local velocity of light.

Rather than attempting, for example, to rotate a large massive disc of negative mass at relativistic velocities, a technologically achievable option may be to generate an equivalent rotating EM field to counter the predominant gravitational mode-locking frequency. The resultant torsion (verdrehung) field forms the equivalent to a co-gravitational field **K** inducing an acceleration in the axial direction perpendicular to the disc plane. The corresponding co-gravitational field $K = (g \times v)/c^2 = (2c\Delta\nu \times v)/c^2 = (mv \cdot \lambda_{dB}/h\Delta\lambda) \times v$. The induced gravitational field is proportional to the difference in circular beam and pump beam frequencies. The direction of the induced acceleration g_i based on the Fleming left hand rule would point upward if spun clockwise or downward if spun counter-clockwise. Disc rotation may be induced via the inverse Sagnac effect. For vertical lift-off, the disc must be spun cw up to speed with a rate of change of velocity corresponding to the desired acceleration. In-flight, the gravitomagnetic induced acceleration may be used to augment acceleration to counter gravitational acceleration or conversely increase gravitational acceleration. As shown, the induced acceleration is concentrated in the central region of the disc. In vertical acceleration or hover mode, the induced g-field g_i ($= E \cdot c^2/\phi$) is opposed to the Earth's acceleration of gravity g ($= \nabla \times A_g) = B_g \cdot c/\theta = \theta/(V \cdot d) \cdot r_u = -\Omega_k \cdot c/2\theta$).

A performance summary for the given model configuration is given in Table 52-3 for a zero-g and 1-g acceleration. For a circular travelling EM wave of 5 m radius, the required frequency to offset the Earth's gravitational potential for a positive mass of 10,000 kg in the given model is on the order of 6.41E68 Hz which is well beyond technological reach. The required frequency becomes less if mass is reduced such as by negative mass offset. For a circulating electromagnetic standing wave confined in circular waveguide, an effective negative mass may conceivably be realized by pumping the beam in the direction of motion

52. Anti-gravity

in a whispering mode gallery with external energy input. Beam intensity may be enhanced in a synchrotron emission pattern that is sharply collimated. In-phase spin wave coupling results in phase accumulation while anti-phase coupling results in reduced phase accumulation with corresponding modulation of electrodynamic inertia. This results in an azimuthal variation in torsion and angular frequency. The variation in torsion is equivalent to a variation in effective mass, in particular, an increase in torsion corresponds to negative mass. An increase in frequency is equivalent to an increase in energy. The induced vortical flow may be augmented by amplitude squeezing of the EM wave. If an azimuthal variation in circular polarization is imparted to a series of coaxial circular standing waves or spira mirabilis spin field, a nonuniform field gradient is introduced producing a lateral force in the disc plane. Thus, both vertical and horizontal acceleration may be generated with a simulated gravito-magnetic torsion field. A concept sketch is shown in Fig. 52-38. Examples of notional spin wave augmentation effects are shown in Figs 52-39 and -40.

It is ironic that Einstein's most creative work, the general theory of relativity should boil down to conceptualized space as a medium when his original premise [in special relativity] was that no such medium existed. – Robert Lauglin

We fret about how to keep going the same old way when we should be casting around for another way that's better. – John Brunner

Don't shoot for the stars; we already know what's there. Shoot for the space in between because that's where the real mystery lies. – Vera Rubin

We arrive at the astonishing conclusion that dark matter is present with a much greater density than luminous matter. – Fritz Zwicky

The knowledge of motion inevitably leads to knowledge of nature. – Aristotle

Motion with respect to the universal ocean of aether eludes us. – Sir Arthur Eddington

The quantity of action is the product of mass of the bodies times their speed and the distance they travel. – Pierre Louis Maupertius

I admire Einstein's theory of gravity as a work of art. – Max Born

Scientists are the true driving force of civilization, – James Burke

Much of what we see in the universe... starts out as imaginary. Often you must imagine something before you can come to terms with it. – Clifford Simak

Nothing is more fearful than imagination without taste. – Johann Wolfgang von Goethe

Without our dreams, we wouldn't be where we are. – Wally Schirra

Of all the words of mice and men, the saddest are 'It might have been.' – Kurt Vonnegut

But I abhor the dull routine of existence. I crave mental exhaltation. – Sherlock Holmes

Knowing is not enough; we must apply. Willing is not enough; we must do. – Johann Wolfgang von Goethe

"So I need to lay down some ground rules." "Rules for use of the ground?" He's gazing out the window. "Am I still allowed to step on it?" – Arthur Conan Doyle

To let the brain work without sufficient material is like racing an engine. It racks itself to pieces. – Arthur Conan Doyle

I pressed down the mental accelerator. The old lemon throbbed fiercely. I got an idea. – P.G. Wodehouse

52. Anti-gravity

Induced gravitational acceleration due to EM spin wave

closed flux lines

L = lift

Hover mode
$S_{gi} = -S_g$

S_g S_{gi}

induced gravito-Poynting vector

W = weight

H_g gravitomagnetic field (angular momentum density)

induced acceleration
$g_i = -rd\Omega/dt = c \cdot B_g/\theta$
$= -\nabla\phi_g - \delta A_g/\delta t$

S_{gi} S_g gravito-Poynting vector

acceleration of gravity
$g = W/m_i$
$= -\nabla\phi_g = -Gm/r^2$
$= (\nabla \times A_g) \cdot c = K \cdot c$
$= 2c \cdot \Delta v \cdot r_u$
$= (2c^2)(mv/\Delta\lambda) \cdot \lambda_{dB} \cdot r_u$

modified gravitational potential
$\phi_{gm} = A_{gm}/(v/c^2) = -G[nh(v_1 - v_2)/c^2]/r$

$\phi_{gi}(r) = -(\phi_{gm} + \phi_{gM})[1 - (r/a)^2]e^{-\frac{1}{2}(r/a)^2}$

$\phi_g = -Gm/r$

radius r

$v = \Omega \cdot r = \theta \cdot r/t$

$\Omega = \theta/t = v/r$
$= -B_g/2$
$= \nabla \times A_{gm}$

$\delta H_g/\delta t = (\nabla \times g)/\mu_g$

$A_g = (v/c^2)\phi_g$
$= (\phi_g/\phi_E)A$

gravitomagnetic vector potential
$A_{gm} = k\hbar/m$

modified gravitational potential ϕ_{gm}

Mexican hat potential wavelet

gravitomagnetic field
$B_g = \mu_g H_g = \frac{1}{2}\nabla \times A_{gm}$
$= -2\Omega$

acceleration $a = F/m = (\delta A_g/\delta t)c^2/\phi_g = v \times B_g$ [m/s²]	gravitomagnetic field $H_g = (v \times g)/4\pi G = -2\Omega/\mu_g$
angular velocity $\Omega = \theta/t = v/r = -B_g/2$ [rad/s]	$= \epsilon_g(v \times g) = K \cdot c^2/4\pi G = B_g/\mu_g$ [kg·rad/m·s]
cogravitation field $K = (v \times g)/c^2 = \nabla \times A$ [s⁻¹]	induced gravitational acceleration $g_i = -rd\Omega/dt$
gravitational acceleration $g = W/m = -\nabla\phi_g = \Delta A_g/vt$	$= -\nabla\phi_g - \delta A_g/\delta t$ [m/s²]
$= 2c \cdot \Delta v \cdot \hat{r} = v \times 2\Omega$ [m/s²]	induced scalar potential $\phi_{gi} = A_i c^2/v_i$ [J/kg = m²/s²]
gravitational scalar potential $\phi_g = -Gm/r = (\Delta v/v)c^2$	inertial force $F = m(g_i - g)$ [N = kg·m/s²]
$= (A_g/A)\phi_E = g \cdot c/2v \cdot \hat{r}$ [J/kg = m²/s²]	inertial mass $m_i = F/(g_i - g) = \Gamma m_0 = m_0/\sqrt{(1-\beta^2)}$ [kg]
gravitomagnetic potential $A_{gm} = k\hbar/m$ [rad·m/s]	gravikinetic field $g_\kappa = -dA^*/dt = -(v \times K)$ [m/s²]
gravito-Poynting vector $S_g = g \times H_g$	gravitomagnetic vector potential $A_g = (v/c^2)\phi_g$
$= (c^2/4\pi G)K \times g$ [kg/s³]	$= (\phi_g/\phi_E)A$ [m/s]

Fig. 52-38. Modified gravitational potential energy due to effects of an induced gravitomagnetic field. The induced gravitational acceleration is proportional to the altered de Broglie matter wave frequency and may be modulated to counter or augment the Earth's gravitational field. An effective negative mass current loop is created by external energy input augmenting the gravito-Poynting vector of the circulating EM beam. The energy input is directly converted to the kinetic energy required for induced motion.

Victory belongs to the most persevering. – Andre Norton

52. Anti-gravity

Gravitic drive concept utilizing negative index metamaterials

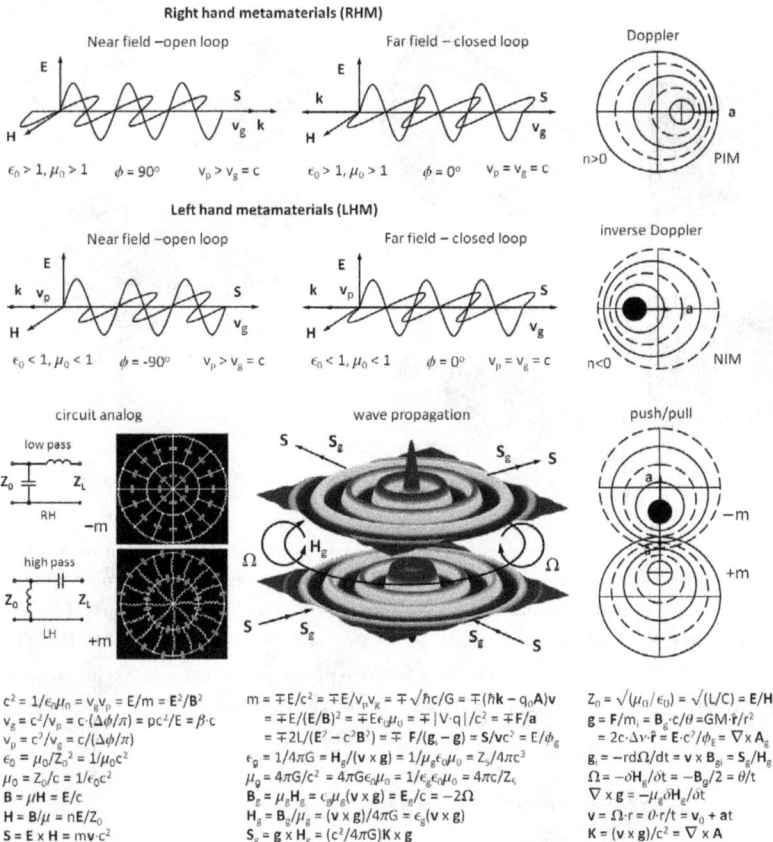

Fig. 52-39. Notional sketch of a mass dipole formed by coupling of a positive mass and negative mass which interact to produce unidirectional thrust. The positive mass is attracted to the negative mass while the negative mass is repelled by the positive mass. Effective negative mass is created by negative index of refraction metamaterials which exhibit inverse Lorentz-Doppler effect. Metamaterials are composed of repeating elements of artificial structures which may be resonant or nonresonant. In the example illustrated, the repeating elements consist of transmission line LC filters: High pass left hand (LH) filter corresponds to positive mass; Low pass right hand (RH) filter corresponds to negative mass. The filter elements are preferentially arranged as graded fractal arrays sized to simulate the spectral energy density gradient of a positive energy mass and negative energy gravitational field. Negative impedance converters and gyrators which convert impedance to negative and the inverse, respectively, are other potential element structures.

It has always been the dream of philosophers to have all of matter built up from one fundamental particle, so it is not altogether unsatisfactory that we have two in our theory. – P.A. M. Dirac

577

52. Anti-gravity

Gravitomagnetic propulsion concept using LH metamaterials

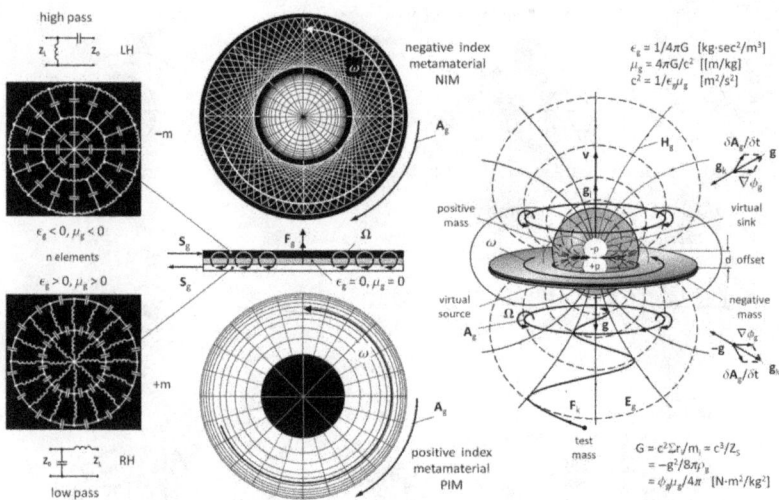

Fig. 52-40. Negative refractive index metamaterials (NIM) enables creation of artificial negative mass allowing enhanced gravitomagnetic effects and realization of a gravitic drive. For null acceleration, induced acceleration g_i ($= \mathbf{E} \cdot c^2/\phi = -r\delta\Omega/\delta t = (\mathbf{S}_g/K) \cdot (4\pi G/c^2)$ $= \mathbf{S}_{gi}/\mathbf{H}_{gi}$) opposes acceleration of gravity $\mathbf{g}$ ($= -\nabla\phi_g - \delta\mathbf{A}^*/\delta t = \nabla \times \mathbf{A}_g = Kc^2/v = \mathbf{B}_g \cdot c/\theta$). The torsion field Ω ($= \nabla \times \mathbf{A}_g = -2\mathbf{B}_g$) is a function of the gravitomagnetic vector potential $\mathbf{A}_g$ ($= \mathbf{v} \cdot \phi_g/c^2 = \mathbf{v} \cdot (\Delta v/v) = (\phi_g/\phi_E)\mathbf{A}$ where $\mathbf{A}$ is the magnetic vector porential.($= \epsilon_0\mu_0 \cdot \mathbf{v} \cdot \phi_E$).

Research is what I'm doing when I don't know what I'm doing. – Wernher von Braun

In a very real sense, we are shipwrecked passengers on a doomed planet. – Norbert Weiner

The rocket worked perfectly, but it fell on the wrong planet. – Wernher von Braun

Gravity pulls on every atom of an object simultaneously not just the ones on the outer surface. – Sherlock Holmes

There's something that doesn't make sense. Let's go and poke it with a stick. – Steven Moffat

If your experiment needs a statistician, you need a better experiment. – Lawrence Rutherford

When I can't do something, this always impels me to study it. – Theodore Sturgeon

If you can dream it, you can do it. – Walt Disney

Speculation is perfectly alright... If you test it, you've started a science. – Algis Budrys

What now, dear reader, shall we make of our telescope? Shall we make a Mercury's magic wand to cross the liquid aether with, and like Lucian, lead a colony to the uninhabited evening star, allured by the sweetness of the place? – Johannes Kepler

The lure of flying is the allure of beauty. – Amelia Earhart

I am and always will be the optimist. The hoper of far-flung hopes and the dreamer of impossible dreams. – Dr. Who

52. Anti-gravity

Table 52-3 Gravitomagnetic induced acceleration

Parameter	Symbol	Relation	Value	Units	Remarks
Acceleration of gravity	**g**	$= \mathbf{F}/m_i = \mathbf{E} \cdot c^2/\phi$ $= -\nabla \phi_g - d\mathbf{A}_g/dt$ $= -G(m/r^2)\mathbf{r}_u = \mathbf{K} \cdot c^2/\mathbf{v}$ $= 2c \cdot \Delta \nu \cdot \mathbf{r}_u = 2c^2 \cdot \Delta \lambda \cdot \mathbf{r}_u$ $= \mathbf{B}_g \cdot c/\theta = \nabla \times \mathbf{A}_g$ $= \mathbf{v} \times \mathbf{B}_g = \mathbf{v} \times 2\mathbf{\Omega}$	9.80	m/s²	@ Earth's surface
Induced acceleration	$\mathbf{g}_i$	$= -\mathbf{g} = 2c \cdot \Delta \nu \cdot \mathbf{r}_u$ $= \mathbf{E} \cdot c^2/\phi$ $= -r d\mathbf{\Omega}/dt = \mathbf{S}_{gi}/\mathbf{H}_{gi}$	−9.80	m/s²	Free-fall: **g** = **g**$_i$ Accel.: **g**$_i$ ≥ **g** Decel: **g**$_i$ ≤ **g**
Inertial mass	m_i	$= \mathbf{F}/\mathbf{a} = \hbar \mathbf{k}/\mathbf{v}$ $= \mathbf{E}/(\mathbf{E}/\mathbf{B})^2 = \mathbf{F}/2c\Delta \nu \cdot \hat{\mathbf{r}}$ $= 2L/(\mathbf{E}^2 - c^2\mathbf{B}^2)$	10,000	kg	Assumed
Mass current	i_m	$= m/t = \rho_m \mathbf{v} \mathbf{A}$	varies	kg/s	
Mass current density	$\mathbf{J}_m, \mathbf{J}_g$	$= \rho_m \mathbf{v}$	845.56	kg/m²·s	
Gravito-Poynting vector	$\mathbf{S}_g$	$= \mathbf{g} \times \mathbf{H}_g$ $= (c^2/4\pi G)\mathbf{K} \times \mathbf{g}$ $= -(c^2/4\pi G)\mathbf{E}_g \times 4\mathbf{B}_g$	230.19	kg·rad/s³	Free-fall: **S**$_g$ = 0 Accel.: **S**$_g$ > 0 Decel.: **S**$_g$ < 0

cont

52. Anti-gravity

Table 52-3 Gravitomagnetic induced acceleration (Cont)

Parameter	Symbol	Relation	Value	Units	Remarks
Potential energy	E_P, V	$= mgh$	9.80E06	$J = kg \cdot m^2/s^2$	For $h = 100$ m (hover)
Kinetic energy	E_k, T	$= \tfrac{1}{2} m\mathbf{v}^2$ $= (-\mu_g/2)\mathbf{H}_g^2)\text{Vol}$	9.80E06	$J = kg \cdot m^2/s^2$	For $v = 44.27$ m/s @ $t = 4.517$ sec. (ascent to 100 m)
Gravitomagnetic field strength Angular momentum density	$\mathbf{H}_g$	$= m\mathbf{v}r/\text{Vol} = i_m/2\pi r$ $= -(1/\mu_g)\mathbf{p}/4\pi r^2$ $= (c^2/4\pi G)\mathbf{K} = \mathbf{g} \cdot \mathbf{n}/Z_0$ $= (\mathbf{v} \times \mathbf{g})/4\pi G$ $= (c^2/4\pi G)\nabla \times \mathbf{A}$ $= \mathbf{B}_g/\mu_g = \epsilon_g(\mathbf{v} \times \mathbf{g})$	5.172E11	(kg·rad)/m·s	p = gravito-magnetic pole charge
Time rate of change of gravitomagnetic field strength	$\delta \mathbf{H}_g/\delta t$	$= (\nabla \times \mathbf{g})/\mu_g$	1.145E11	kg/m·s^2	Ascent to 100 m. $t = \sqrt{(100/0.5 \cdot 9.8)}$ $= 4.517$ sec
Gravitomagnetic flux density, Induced gravito-magnetic field, torsion field	$\mathbf{B}_g$	$= \mu_g \mathbf{H}_g = \tfrac{1}{2}\nabla \times \mathbf{A}_{gm}$ $= (G/2c^2)\mathbf{J}/r^3 = -2\boldsymbol{\Omega}$ $= (\mu_g/\mu)(m_e/e)\mathbf{B}$ $= \mathbf{g} \cdot \theta/c = (\mathbf{K}c^2)\mu_g/4\pi G$ $= -\nabla \times \mathbf{v} = \epsilon_g \mu_g(\mathbf{v} \times \mathbf{g})$ $= (4\pi G/c^2)\mathbf{H}_g$	4.825E-15	rad/s	Ascent to 100 m @ 1g

cont

52. Anti-gravity

Table 52-3 Gravitomagnetic induced acceleration (Cont)

Parameter	Symbol	Relation	Value	Units	Remarks
Cogravitational field	$\mathbf{K}$	$= \nabla \times \mathbf{A_g} = \mathbf{v}\phi_g/c^2$ $= (\mathbf{v} \times \mathbf{g})/c^2$ $= -Gm(\mathbf{v} \times \mathbf{r_u})/c^2 r^3$ $= (\mathbf{r} \times \mathbf{g})/cr$ $= -\nabla\phi_c + 1/c^2\, d\mathbf{A_g}/dt$ $= (2\Delta\nu) \times \mathbf{v}/c^2$ $= (mv/h\Delta\lambda)\cdot\lambda_{dB}) \times \mathbf{v}$ $= \mathbf{H_g}\cdot 4\pi G/c^2$	4.830E-15	s^{-1}	Ascent to 100 m @ 1g
Earth gravitational potential/unit mass	ϕ_{gE}	$= -GM_e/r = U/m$	-6.26E07	$m^2/s^2 = J/kg$	$M_e = 5.977E24$ kg
Mass object gravitational potential	ϕ_{go}	$= -Gm/r = -(\Delta\nu/\nu)c^2$	-1.334E-07	$m^2/s^2 = J/kg$	m = 10,000 kg r = 5.0 m
Induced gravitational potential	ϕ_{gi}	$= \oint \mathbf{A_g}\cdot d\mathbf{l}$	+6.26E07	$m^2/s^2 = J/kg$	Hover mode
Gravitational constant	G	$= 1/4\pi\,\epsilon_g = c^3/Z_S$	6.673E-11	$m^3/kg\cdot s^2$	$G = c^2\Sigma(r_i/m_i)$
Electric permittivity	ϵ_0	$= 1/\mu_0 c^2 = \epsilon_0 \Gamma$	1.725E8	$kg\cdot rad^2/m^3$	
Magnetic permeability	μ_0	$= \mathbf{B}/\mathbf{H} = Z_0/c = \mu_0\Gamma$	6.449E-26	$m\cdot s^2/(kg\cdot rad^2)$	

cont

52. Anti-gravity

Table 52-3 Gravitomagnetic induced acceleration (Cont)

Parameter	Symbol	Relation	Value	Units	Remarks
Gravitopermittivity	ϵ_g	$= 1/4\pi G = \mathbf{H}_g/(\mathbf{v} \times \mathbf{g})$ $= Z_s/4\pi c^3$	1.193E9	kg·s^{-2}/m^3	
Gravitopermeability	μ_g	$= 4\pi G/c^2 = \mathbf{B}_g/\mathbf{H}_g$ $= 4\pi G/\epsilon_g \mu_0 = 4\pi c^3/Z_s$	9.3287E-27	m/kg	
Gravikinetic field	$\mathbf{g}_k$	$= -\delta \mathbf{A}^*/\delta t$ $= \mathbf{v} \times \mathbf{K}$	-9.8	m/s^2	@ 1g
Electrokinetic field	$\mathbf{E}_k$	$= -\delta \mathbf{A}^*/\delta t$ $\sim -\delta \mathbf{A}/\delta t$	1.180E13	volts/m	Ascent to 100m @ 1g
Time rate of change of gravitomagnetic field	$\delta \mathbf{B}_g/\delta t$	$= \nabla \times \mathbf{g} = -\nabla \times \mathbf{g}_i$ $= \mu_g \delta \mathbf{H}_g/\delta t$ $= \delta/\delta t \, (\nabla \times \mathbf{A}_g)$ $= \delta/\delta t(-2\Omega)$	1.0683E-15	rad·s^{-2}	Curl of induced acceleration
Time rate of change of cogravitational field	$\delta \mathbf{K}/\delta t$	$= -\nabla \times \mathbf{g}_k$	1.0687E-15	s^{-2} rad/s^2	Ascent to 100 m. t = √(100/0.5*9.8) = 4.517 sec

cont

52. Anti-gravity

Table 52-3 Gravitomagnetic induced acceleration (Cont)

Parameter	Symbol	Relation	Value	Units	Remarks
Vorticity	ω	$= d\theta/dt = r \times \mathbf{v}/r^2$ $= \omega_0 + at = \mathbf{v}_t/r$	17.70	rad/s	Curl of tangential velocity
Torsion field angular velocity	Ω	$= (\nabla \times \mathbf{v}_t)/2 = -2\mathbf{B}_g$ $= \nabla \times \mathbf{A}_{gm}$	8.85	rad/s	1 g acceleration
Angular momentum	J, L	$= I\omega = rm\mathbf{v}_t = \mathbf{r} \times \mathbf{p}$	varies	kg·m²/s	$I_{disc} = \tfrac{1}{2}mR^2$ $I_{ring} = mR^2$
Angular momentum density	$\nabla \times \mathbf{H}_g$	$= -\rho\mathbf{v} - \epsilon_g \delta\mathbf{g}/\delta t$	varies	kg/s·m² rad·kg/s·m²	ρ = mass density $\delta g/\delta t$ = jerk
Divergence of gravitational field	$\nabla \cdot \mathbf{g}$	$= \nabla \cdot \mathbf{E}_g = -\rho_m/\epsilon_g$ $= -4\pi G \rho_m$	−1.6016E-08	s⁻²	
Gravitational field energy density	ρ_g	$= -g^2/8\pi G = GM^2/8\pi r^4$ $= \nabla \cdot \mathbf{g}/4\pi G$ $= -(\epsilon_g/2)g^2$	−5.7368E10	J/m³	
Rhythmodynamic mass	m_{RD}	$= mc/\pi$	9.542E11	(kg·m)/(s·rad)	

cont

52. Anti-gravity

Table 52-3 Gravitomagnetic induced acceleration (Cont)

Parameter	Symbol	Relation	Value	Units	Remarks
Gravitomagnetic energy density	ρ_{gm}, W_g	$= -(\mu_g/2)\mathbf{H_g}$ $= -(1/8\pi G)(\mathbf{g}^2 + c^2\mathbf{B_g}^2)$	−2.412E-15	J/m^3	
Velocity (vertical)	$\mathbf{v}$	$= \mathbf{a}t + \mathbf{v}_0 = (\hbar k - q_0\mathbf{A})/m$ $= \mathbf{A}/\epsilon_0\mu_0\phi_g = \mathbf{A}c^2/\phi_E$ $= \mathbf{A_g}c^2/\phi_g = \mathbf{p}\cdot c/\rho_e$ $= (\mathbf{A}q)/m = \mathbf{A_g}(\phi_E/\phi_g)$ $= K\cdot c^2/\mathbf{g} = c\cdot(\Delta\phi/\pi)\cdot\hat{\mathbf{r}}$	44.27	m/s	Ascent to 100m @ 1g
Mass density	ρ_m	$= m/\mathrm{Vol}$	19.1	kg/m^3	
Mass current density	$\mathbf{J_m}, \mathbf{J_g}$	$= \rho_m\mathbf{v}$	845.56	$kg/m^2\cdot s$	Ascent to 100m @ 1g
Magnetic vector potential Retarded vector potential	$\mathbf{A}$ $\mathbf{A^*}$	$\epsilon_0\mu_0\mathbf{v}\phi_E = m\mathbf{v}/q$ $\mathbf{v}\phi_E/c^2 = \mathbf{E}\cdot\mathbf{v}/c^2$ $= -\int \mathbf{E_k}\, dt + \mathrm{const}$ $= LI/l = \tfrac{1}{2}\mathbf{B}\times\mathbf{r} = \hbar\mathbf{k}/q$ $= \mathbf{v}\phi_E\epsilon_\mu\mu_g = \mathbf{A_g}(\phi_E/\phi_g)$	−5.334E13	m/s m/rad	Ascent to 100m @ 1g
Gravitomagnetic vector potential	$\mathbf{A_g}$ $\mathbf{A^*_g}$	$= \epsilon_g\mu_g\mathbf{v}\phi_g = (\phi_B/\phi_E)\mathbf{A}$ $= \mathbf{v}\phi_g/c^2 = \mathbf{g}\cdot\mathbf{v}/c^2$ $= \mathbf{v}\phi_g\epsilon_g\mu_g = \mathbf{v}\phi_g\epsilon_0\mu_0$	varies	m/rad	

cont

52. Anti-gravity

Table 52-3 Gravitomagnetic induced acceleration (Cont)

Parameter	Symbol	Relation	Value	Units	Remarks
Magnetic flux density	**B**	$= \mu_0 \mathbf{H} = \mathbf{E}/c$ $= \mu_0 I/2\pi r = (\mathbf{v} \times \mathbf{E})c^2$	varies	T (= Wb/m^2 = 1/rad)	
Magnetic field strength	**H**	$= \mathbf{B}/\mu_0 = (\mathbf{v} \times \mathbf{E})/\mu_0 c^2$ $= \mathbf{E} \cdot \mathbf{n}/Z_0 = I/2\pi r$	varies	Amp/m	
Curl of gravito-magnetic field	$\nabla \times \mathbf{B}_g$	$= -(4\pi G/c^2)\mathbf{J}_m +$ $(1/c^2)\delta \mathbf{E}_g/\delta t$	varies	rad·m^{-1}·s^{-1}	
Curl of cogravitational field	$\nabla \times \mathbf{K}$	$= -(4\pi G/c^2)\rho \mathbf{v}$ $= -(4\pi G/c^2)\mathbf{J}_m +$ $(1/c^2)\delta \mathbf{g}/\delta t$	varies	m^{-1}·s^{-1}	
Gravitomagnetic monopole charge	p	$= \mu_g(4\pi r^2) \cdot \mathbf{H}_g$	varies	m/s	
Gravitational force	$\mathbf{F}_g$	$= \mathbf{W} = m\mathbf{g} = -m\nabla \phi_g$ $= m2c\Delta v \cdot \hat{\mathbf{r}} = m(c^2 \cdot \mathbf{A})/\mathbf{v}$ $= m(2c^2/h)(mv/\Delta t)\lambda_{dB} \cdot \hat{\mathbf{r}}$ $= dL/dz$ $= -(GMm/r^2)\mathbf{r}_u -$ $((GMmv^2)/2c^2 r^2)\hat{\mathbf{r}}$	9.80E04	kg·m/s^2 = N	

cont

52. Anti-gravity

Table 52-3 Gravitomagnetic induced acceleration (Cont)

Parameter	Symbol	Relation	Value	Units	Remarks
Inertial force	F_i	$= m_i a = (\hbar k/v)a$ $= m_i(g_i - g)$	-9.80E04	$kg \cdot m/s^2 = N$	Ascent to 100m @ 1g
Cogravitational force, ala Kinemassic, Prorotational force	F_C	$= m(\mathbf{v} \times \mathbf{K})$	2.13E-09	$kg \cdot m/s^2 = N$	Ascent to 100m @ 1g
Total force	F_t	$= m(\mathbf{g} + \mathbf{v} \times \mathbf{K})$	9.80E04	$kg \cdot m/s^2 = N$	gravitational + cogravitational
gravitomagnetic dipole moment	μ_g	$= \tfrac{1}{2} \Sigma\, \mathbf{r} \times (m_i \mathbf{v}_i)$	varies	$kg \cdot m^2/s$	
Coriolis force	F_C	$= -m(\mathbf{v} \times 2\boldsymbol{\Omega})$ $= m(\mathbf{v} \times \mathbf{B}_g)$	varies	$kg \cdot m/s^2 = N$	
gravitoelectric field strength	E_g	$= -\nabla \phi_g - (1/c)\delta \mathbf{A}_g/\delta t$ $= -\nabla \phi_g - (\phi_g/c^2)\delta \mathbf{v}/\delta t$ $= \mathbf{H}_g \cdot Z_0/n$ $= c \cdot \mathbf{B}_g = \mathbf{g}$	varies	m/s^2	
de Broglie frequency	ν_{dB}	$= c/(h/mv) = \lambda_{dB}/c$ $= (h/mv)(\Delta\nu/\Delta\lambda)$	1.64E-08	$Hz = 1/s$	

cont

52. Anti-gravity

Table 52-3 Gravitomagnetic induced acceleration (Cont)

Parameter	Symbol	Relation	Value	Units	Remarks
gravitational wave velocity	c_g	$= \sqrt{(G/H)} = 1/\sqrt{(\epsilon_g \mu_g)}$ $= c = 1/\sqrt{(\epsilon_0 \mu_0)}$	2.99792E8	m/s	$H = 7.3\text{E-}28$ m/kg
Radiated power	P_g	$= dE/dt$ $= m(\mathbf{g} - \mathbf{g}_i)\mathbf{v}$ $= (\mathbf{g} \times \mathbf{H}_g) \cdot \mathbf{S}_g$	4.338E06	Watts	$P_{gin} = (\mathbf{g} \times \mathbf{H}_g) \cdot \mathbf{S}_g$ $P_{gout} = (\mathbf{g} \times \mathbf{H}_g) \cdot \mathbf{S}_g$
EM standing wave frequency	ν	$= (c^2 \cdot \phi_g \cdot \mathbf{r})/-Gh$	varies	Hz	
EM standing wave angular frequency	ω	$= 2\pi(c^2 \cdot \phi_g \cdot \mathbf{r})/-Gh$	varies	rad/s	
Phase shift	φ	$= \int(\omega_1 - \omega_2)dt = \pi\beta$ $= \oint \mathbf{k} \cdot d\mathbf{l} = (v_g \cdot \pi)/c$ $= (m/\hbar)\int \mathbf{A}_g \mu dx^\mu$	varies	rad	dX^μ = spacetime displacement (AB effect)

I firmly believe that before long man will acquire the ability to build an electromagnetic contra-gravity mechanism that works. Much of the same line of reasoning that enabled scientists to split up atomic structures also enable them to learn the nature of gravitational attraction and ways to counter it. – Grover Loening

The Euler-Mascheroni constant γ, first discovered by Euler in 1735, represents the limit of a harmonic sequence. – Riccardo. C. Storti

I have not yet lost a feeling of wonder, and delight, that this delicate motion should reside in all ordinary things around us, revealing itself only to those who looks for it. – Edward Mills Purcell

52. Anti-gravity

The circulating EM beam may be expected to produce a gravitational Faraday effect resulting in rotation of the plane of polarization of an incident EM wave. The rotation angle θ (= $V \cdot \mathbf{B} \cdot d$) in analogy to the optical Faraday effect is estimated as

$$\theta = V_g \cdot \mathbf{B}_g \cdot d = 2G\rho/\sqrt{\mathbf{B}c^4} \tag{52-24}$$

where

$V(\lambda)$	= Verdet constant (= $-\frac{1}{2}(e/m)(\lambda/2c^2)(dn/d\lambda)$ [rad/T·m = rad^2/m]
$\mathbf{B}_g$	= gravitomagnetic flux density (= $\frac{1}{2}\nabla \times \mathbf{A}_{gm} = g \cdot \theta/c = -2\Omega_k$) [rad/s]
d	= propagation distance [m]
G	= gravitational constant (= $c^3/(m_P t_P) = c^2 \Sigma(r_i/m_i) = c^3/Z_s$) [m^3/kg·s^2]
ρ_λ	= linear energy density [J/m = kg·m/s^2]
B	= magnetic flux density (= $\nabla \times \mathbf{A} = \mathbf{E}/c = \mu_0 \mathbf{H}) = \mu_0 I/2\pi r$) [1/rad]
c	= velocity of light ($c_0/\Gamma = \mathbf{E}/\mathbf{B} = \phi_E/\phi_B = \sqrt{v_g v_p} = v/\Delta\phi \cdot \pi$) [m/s]

Analogous to $\mathbf{E}/\mathbf{B} = c$, a gravitomagnetic relation is $\mathbf{g}/\mathbf{B}_g \cdot \theta = c$ where $\theta = s/r = \Omega_k \cdot t$. Incident reflected sunlight with a slight sky polarization may be expected to produce a darkened zone at the center of the beam ring against the background illumination.

A positive mass attracts other masses while a negative mass repels other masses. Consequently a coupled negative and positive mass will self-accelerate. This effect is illustrated in Fig. 52-41. The action resembles that of a predator and prey which are equally mobile. The motion of a gravitational dipole proceeds in a quantized oscillating fashion comparable to a single mass in motion. In effect, a single mass in motion corresponds to a half positive and half negative mass moving in tandem. Although negative mass has not been observed in isolation in nature, negative mass, as H. Bondi in 1957 noted, is not prohibited according to General Relativity[122]. Moreover, negative effective mass has been demonstrated in experiments with polaritons (photons combined with excitons), photons in fiber optic loops, nonlinear optic mesh lattices and electrons in plasma waves. In the photon model previously described, negative effective mass corresponds to decreased torsion in a reduced vacuum energy density such as occurs in the Casimir effect. An effective negative mass may be created confining the circular traveling wave in a suitable photonic crystal (PC) or metamaterial resonator wave guide enabling optical or electrical pumping. Excitation of a wave packet propagating around a circular waveguide may be utilized to produce a spatio-temporal dependent potential. To reduce bending losses, the wave guide may preferably be in the form of an annular Bragg resonator (ABR). Due to low group velocity and high quality Q factor, the achievable field intensity is greater than a conventional whispering mode circular waveguide employing total internal refection (TIR). A second-order annular Bragg resonator may be utilized to emit EM radiation nearly perpendicular to the disc plane surface to pump energy into a confined vortical spin wave and modulate circular polarization. The angular wave velocity, in such a case, is very low as the incident beam is nearly perpendicular to the Bragg layers. A pair of contra-rotating, co-axial, annular Bragg resonators may be employed to create a spin torsion field. A spira mirablis torsion field may be produced by annular Bragg resonators of spiral configuration. If phase conjugate reflectors are utilized, a graviton beam array may conceivably be generated to augment vorticity and create a axial magnetic dipole moment/optical dipole force. The spin field may be modulated with synthesized coherent de Broglie EM waves in a simulated matter wave circuit to produce a Sagnac or inverse Sagnac effect.

If you do not change direction, you may end up where you are heading. – Buddha

52. Anti-gravity

Gravitational dipole of positive and negative mass

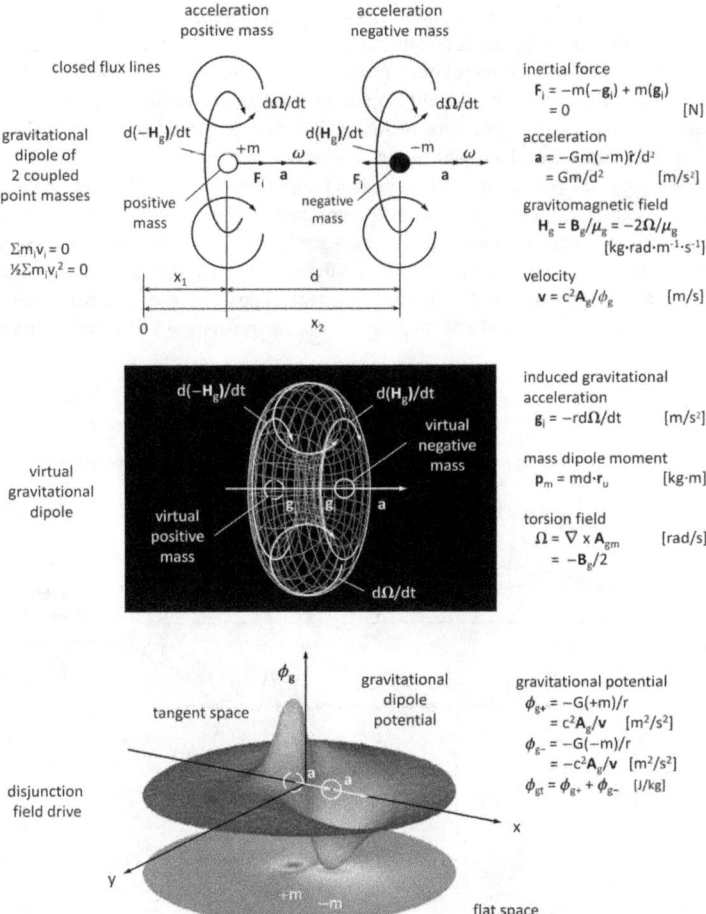

Fig. 52-41. A hypothetical gravitational dipole formed of a positive and negative mass may be expected to accelerate in the direction of the field gradient from the negative mass towards the positive mass. Likewise, a changing gravitomagnetic mass current loop may be expected to generate a net acceleration from the virtual negative mass pole towards the virtual positive mass pole. Two coupled, co-axial gravitomagnetic loops may similarly generate a net acceleration force in the direction of the inertial moment. Conservation of energy and momentum is maintained. Gravitational dipoles resemble active matter or self-propelling fluids consisting of particles or microtubules immersed in a fluid that can extract energy extracted from the surroundings into directed motion.

Nothing in either Newtonian or relativistic gravitation theory precludes the existence of negative mass, but it is an empirical fact that it has never been observed. – Joseph Weber

What it takes to produce gravitational repulsion is a negative pressure. – Alan Guth

52. Anti-gravity

As detailed by Bondi[122], Sabbata/Sivaram[123] and Evans[124], a coupled positive and negative mass may be expected to produce a gravitational dipole exhibiting a net acceleration in the direction of the dipole moment. A gravitomagnetic vortex ring provides a means of creating a virtual gravitational dipole. A changing gravitomagnetic field $(d(-\mathbf{H_g})/dt)$ current loop corresponding to a negative mass in proximity to a changing gravitomagnetic field $(d(\mathbf{H_g})/dt)$ current loop corresponding to a positive mass results in net acceleration of the coupled pair. An increased EM flux density is produced in the waist region of a vortical beam increases the local vacuum refractive index K_{PV} and the local gravitation potential ϕ_g. A virtual gravitational dipole of two coupled tori of unequal angular frequencies forms a resonator system resulting in a leapfrogging effect as momentum is exchanged with a net acceleration $\mathbf{a} = 2c \cdot \Delta\Omega/2\pi \cdot \mathbf{r_u}$ of the wave system in the direction of the frequency gradient. A pair of vortex rings topologically confined forms in effect a rotational gravitational dipole of a coupled positive and negative mass with opposing gravitational potential and frequency states corresponds to a disjunction drive.

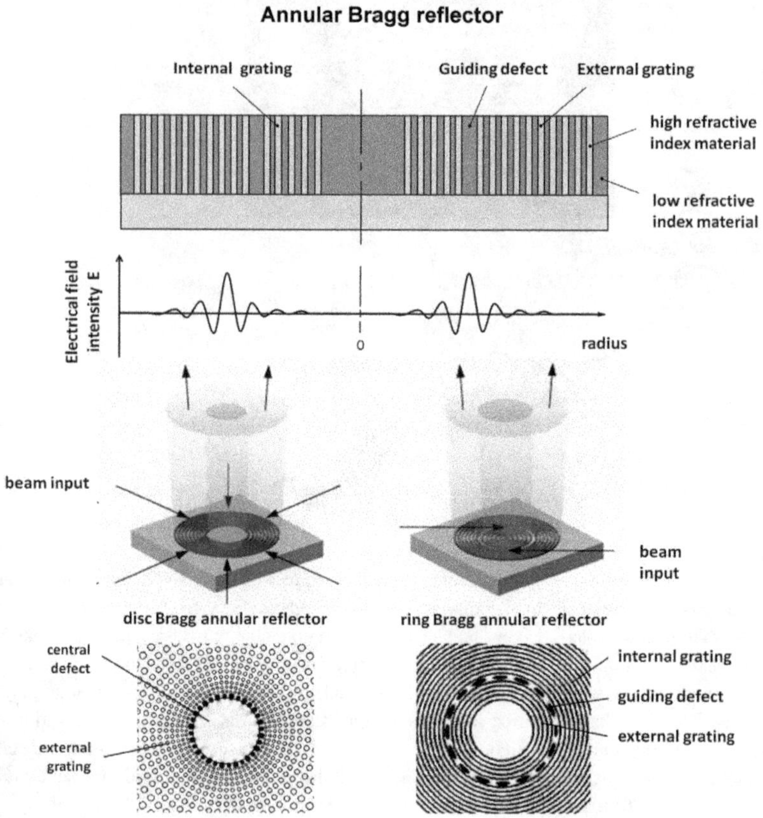

Fig. 52-42. Examples of representative annular Bragg reflector configurations for optical beam generation. The internal and external gratings may be suitably formed from positive and/or negative index of refraction metamaterials for inverse Lorentz-Doppler effects.

52. Anti-gravity

Illustrated in Fig. 52-42 are examples of disc and ring annular Bragg reflectors for optical beam formation. Coupled with a phase conjugate reflector, the optical beam and its phase conjugate beam may used to generate a graviton beam. A concept sketch of annular Bragg reflector for vortical gravitomagnetic beam generation is shown in Fig. 52-43. A synthesized magnetic and electric field may be generated by spatial and temporal variation of an effective vector potential induced by a vortical EM beam. A vortical beam allows creation of a ring vortex with toroidal and poloidal spin components altering the gravitational potential within the beam generating an induced gravitation acceleration g_i in direction of the optical axis. A phased array annular Bragg reflector generates a steerable focused EM beam. Phase may be modulated to adjust orbital angular frequency and shifted to higher harmonics for high frequency generation. A metamaterial lens is illustrated for vortical beam focusing. Phase conjugation enables spin 2 graviton generation allowing higher spin than that of spin 1 photons to increase vorticity. In addition, a phase conjugate reflector enables pumping with synthesized Lorentz Doppler frequencies to induce overall motion of the wave system.

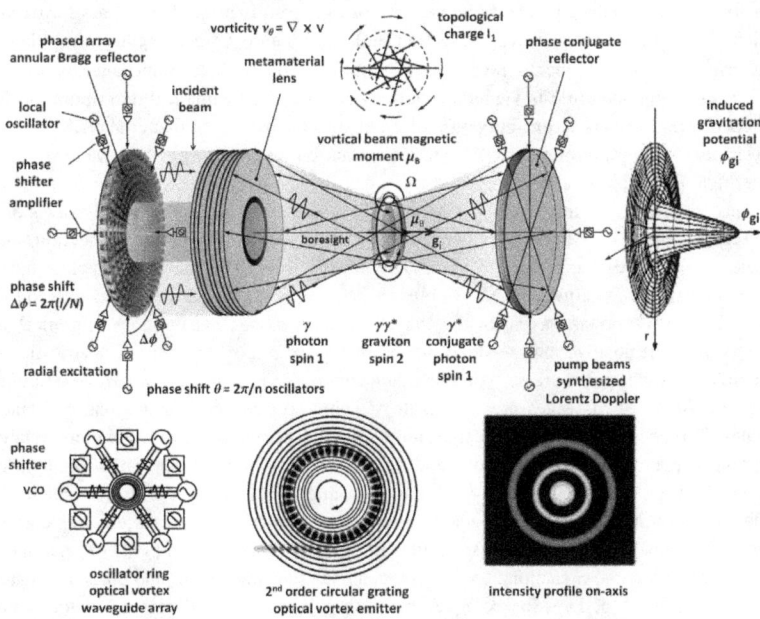

Fig. 52-43. Concept sketch of annular Bragg reflector for vortical gravitomagnetic beam generation. The graviton spin-2 tensor field $h_{\mu\nu}(=T_{\mu\nu}/j^2)$ is related to the source tensor $T_{\mu\nu}$ as an interaction between opposing photon currents $\pm e_\mu j^\mu$. Vorticity $v_\theta = \nabla \times v$.

Even spins lead to attractive forces... spin-2 theory leads to an interaction energy which has $\sqrt{(1-v^2/c^2)}$ in the denominator, in agreement with the experimental results on the gravitational effect of binding energies... Thus, the spin-0 theory is out, and we need spin-2 in order to have a theory in which attraction will be proportional to energy content. – Richard P. Feynman

52. Anti-gravity

52.4 Engineering the vacuum

I will no longer be a thing confined to limitations, cares or thoughts that bind me to an even smaller universe illusion, petty ways of matter's curse. I shall break loose from three dimensional being to fourth dimensional inner sounds and seeing and shed all binds and bounds and limitations to never more be tied to worldly stations. – Eyvind Earle

The primordial universe appears to have undergone a transition from an undifferentiated null state into a bimodal frequency decompostion of chaotic states of increasing modes and entropy. The overall energy constant is thought to have remained constant, but becoming more diffuse as the universe expands. Dark energy locked within the vacuum is thought to drive the expansion due to repulsive effect of negative energy. A task for future scientists and engineers is to elucidate the cause of the Big Bang and experimentally locally cohere the vacuum at will by resonant EM wave superposition. As previously noted, the physical vacuum is of very high energy density and exceedingly stiff. The relatively large velocity of light is a consequence of the large elastic restoring forces within the vacuum. Altering the vacuum energy density to produce force field effects or alter the velocity of light in vacuo, for example, may be realizable as evidenced by the Casmir effect in which the EM mode density is slightly reduced. The finite velocity of light results from Planck mass formation and resultant Planck impedance. Creation and annihilation of positive/negative Planck mass $\pm m_P$ appears energetically reversible with nullification of Planck impedance.

The spacetime manifold in General Relativity (GR) is a mathematical description of the effects of presumed spacetime curvature of gravity on the motion of matter and EM waves. Physically, one may reinterpret GR spacetime manifold as a metric of EM wave patterns rather than a metaphysical bending of space and time. The warp and weft of the "fabric" of the spacetime continuum or metric corresponds to the nodal distance of EM waves and wavefront direction. Curvature of the metric corresponds wavefront deflection and acceleration towards the region of increased energy density. The vacuum refractive index K_{PV} is a measure of energy density. Ripples in the spacetime continuum corresponds to EM waves and gravitational waves. In the Planck vacuum model, synchronous alignment of coupled pairs of positive and negative Planck masses results in creation of electric fields. Rotation of such pairs create vortical filaments of spin aligned dipoles constituting magnetic fields. EM waves are the result of spin wave density fluctuations of Planck dipoles. The observed continuing expansion of the universe appears to require continuous creation of Planck dipoles of positive and negative energy with no net continued input of energy. Perhaps a technological means may be found to resonantly excite the vacuum to simultaneous collect positive and negative energy and extract same in an alternating current circuit of concurrent out-of-phase opposing energy flows composed of nonlinear resistances, inductors, capacitors, memistors, metamaterial waveguides and phase conjugate reflectors arranged, perhaps for example, in a push-pull resonant tunneling oscillator circuit with voltage controlled density of metastable states for generation of soliton-like pulses. As the pulse propagates along a waveguide, the pulse frequency is increased raising energy level. A negative value of charge-dependant resistance or memresistance $M(q) = d\phi_m/dq$ corresponds to extracted energy with superconducting alternating concurrent $I(t) = I^+(t) + I^-(t)$ where q = electric charge and ϕ_m = magnetic flux linkage. In analogy with power consumption characteristic of a resistor, the generated power $P(t) = I(t)V(t) = I^2(t)M(q(t))$.

The existence of electromagnetic waves and gravitational deflection of same demonstrates the vacuum is polarizable and that we can resonantly interact with it. The electron represents a stable EM standing wave resonant structure in a polarizable vacuum.

52. Anti-gravity

In the Planck vacuum model described, parameters that may potentially be subject to manipulation and control for vacuum modification include that shown in Table 52-4.

	Table 52-4. Potential vacuum modifications
1)	Conversion and extraction of positive and negative energy from the vacuum in a nonlinear polarization process. For example, introducing a phase shift temporarily flattening the vacuum releasing high frequency content stored in the Planck quantum foam. Spontaneous reformation of positive/negative Planck charges and subsequent recombination may allow energy conversion on demand in analogy to an electric generator, fuel cell or battery. Induced precession of spinning Planck dipoles in a toroidal magnetic field such as pinched plasma filaments may conceivably be used to augment/suppress charge or generate electrical power with a voltage/current ratio equal to the vacuum radiation resistance.
2)	Alteration of Planck mass dipole volumetric density (scalar potential, scalar waves, strain, strain rate, etc). E.g., Squeezed vacuum positive and negative curvature and torsion states. Cancellation of Planck mass $\pm m_P \rightarrow$ Alteration of local speed of light and time rate of change.
3)	Orientation and synchronous alignment of Planck mass dipoles (electric field).
4)	Dipole spin and co-axial spin alignment of Planck mass dipoles (magnetic field).
5)	Orientation and spatial displacement of Planck mass dipoles (gravitic field).
6)	Polarization, displacement and directed motion of coupled positive and negative Planck masses. Negative RI metamaterials allow creation of negative mass enabling a gravimetric drive.
7)	Modulation of mass electromagnetic (EM) energy spectrum (effective bandwidth, modal density) to enable alteration or inversion of local gravitational field by microresonators driven by additive and subtractive frequency comb arrays.
8)	Alteration ot cancellation of inertial mass. Mass is a function of voltage and charge ($m_i = E/c^2 = |\text{Volts·charge}|/c^2 = \hbar k/v$). Electric charge ($q = mc^2/\text{Volts} = \hbar k/A$) is an effect of electron spin precession due to an imbalance of magnetostatic and electrostatic energy. Inertial mass is a measure of imbalance of momentum (mv) and electrodynamic momentum (qA) and may be subject to cancellation by alteration of vector potential A or magnetostatic/electrostatic ratio to close the precession torsion defect by reducing wave function dissonance (phase shift).
9)	Spin angular momentum conversion to linear momentum (spin drive / spin generator). Linear momentum p (= L/l_P = $\gamma mV + qA$) where L = angular momentum (= $I\omega$ = r x p) and l_P = Planck length (= $\sqrt{(\hbar G/c^3}$ = ct_P = 1.61619E-35 m). One unit of spin angular momentum $\hbar$ (= 1.0545E-34 m²·kg·s⁻¹) corresponds to 6.5 kg·m/s linear momentum. Conversion of a neutrino with spin ½ would yield half of this amount. Perhaps conversion may be accomplished by altering the local electrostatic/magnetostatic energy ratio or by using the Faraday optical rotation effect utilizing, say, a center-fed circulator of diameter λ/π or a system of ring resonators arranged around a central waveguide to transform a tangential disgyration (whirlpool) into a radial disgyration (hedgehog). Energy conversion $E = mc^2 = pc = L/cl_P$ = 22E-5 J. Unwinding of a spinwave in a transition process from one vacuum state at t = 0 to another at t = $t_0 + \Delta t$, resembles that of an instanton with periodicity in space and time.
10)	Spin glass-like vacuum defects. A frustrated lattice of Planck dipoles may admit possibility of generation of electric charges or magnetic charges at the ends of string-like defects. In analogy to exotic properties of ³He superfluid, a vacuum BEC superfluid may allow various textures (field line patterns), quantized vortices, vortical filaments, solitons, skyrmions, particle-like topological defects, disclinations, boojums (point defects at the end of a continuous vortex), hedgehogs, spin disgyrations, etc. Creation of defect states & modulation of quasi-particle motion.

But I like the idea of getting our energy from the subdimensions. – Rudy Rucker

52. Anti-gravity

The quantum vacuum consisting of a BEC superfluid of Planck dipoles manifests itself in coherent, correlated motion of positive and negative Planck masses without dissipation. A black hole represents a maximally rigid arrangement of Planck masses. A cluster is minimally rigid with at least $3n-6$ sphere-to-sphere contacts and at least 3 points of contact per sphere. A collection of spatially oriented Planck dipoles $\langle M_P^+|\rightarrow\rangle$ and $\langle M_P^-|\rightarrow\rangle$ would form an aggregation of gravitational dipoles with directed motion without inertia. Ability to locally increase or decrease the Planck mass dipole density would potentially enable EM warp drive propulsion by "surfing" the energy density gradient in effect. In the aforementioned spacetime fabric analogy, warp corresponds to EM nodal distances. The Planck mass density may be potentially altered by locally cohering the vacuum effectively annihilating Planck dipoles and decreasing vacuum impedance. Phase shift reversal may allow conversion of real mass back into imaginary mass. The technical means to induce such a phase shift at the Planck energy level is well beyond current technological reach. Alternatively, ability to locally align (polarize) Planck positive and negative mass dipoles in a given direction may yield a net gravitational translation of a wave system.

As previously noted, an electrostatic field of a charged sphere or point source may be simulated by a spherical capacitive network. Similarly, a magnetostatic field may be simulated by a spherical inductive network. These fields exhibit the same radial symmetry as that of a gravitational field. A spherical arrangement of inductive and capacitive L-network elements may be used to create an omni-directional EM antenna. Supposing we represent such an array of LC reactive elements forming a Riemann sphere and stereographically project a conformal mapping onto a circular disc in the Euclidean E^2 plane in a Saturn geometry with the sphere centered midplane. A circular array of such elements combined with metamaterial slotted waveguides may be used to generate a plane wave front forming in effect a phased array antenna with positive or negative phase velocities. Addition of tunnel diodes allows reflection amplification, while Gunn diodes enable microwave generation. Gyrotron high frequency, high power microwave amplifiers can provide THZ radiation at megawatt power levels. Metamaterial with negative ϵ and μ reverse energy flow opposite to the direction of motion of wave phase fronts. Such an arrangement with sufficiently intense spectral energy densities may allow synthesis of EM fields equivalent to gravitational fields of positive, negative and imaginary mass.

Three types of mass may be distinguished: 1) inertial mass ($F = ma$), 2) gravitational mass ($F = Gm_1m_2/r^2$) and 3) mass-energy ($E = mc^2$). Mass is equivalent to a localized concentration of photonic energy. Negative mass represents an absence of such energy below that of the surrounding vacuum. Negative mass with negative mass-energy requires energy input to push it away. Mass represents an obstruction to change in energy flow, and, in particular, condensation of matter slows the universe expansion and polarizes the local vacuum. As Storti et al[34] have theorized, a mass-object recompresses the local vacuum into fewer EM modes at higher frequencies. The more massive the object, the fewer modes in its PV spectrum, i.e., the number of frequency modes is inversely proportional to energy density. Alteration of the local EM spectral energy density gradient of a mass-object would enable increase or decrease in the acceleration of gravity. In the Electro-Gravi-Magnetic (EGM) construct, the acceleration of gravity represents a spectral energy density gradient in the local polarized vacuum spectrum. EGM models the spectral energy of a gravitational field equivalent to the mass-energy of the object generating the field, expressible in terms of a PV (Polarizable Vacuum) spectrum analogous to spacetime curvature within GR.

The mass-energy within an object is energetically equivalent to the gravitational field surrounding the object. – Riccardo C. Storti

52. Anti-gravity

Simple frequency model of a central mass and gravitational field

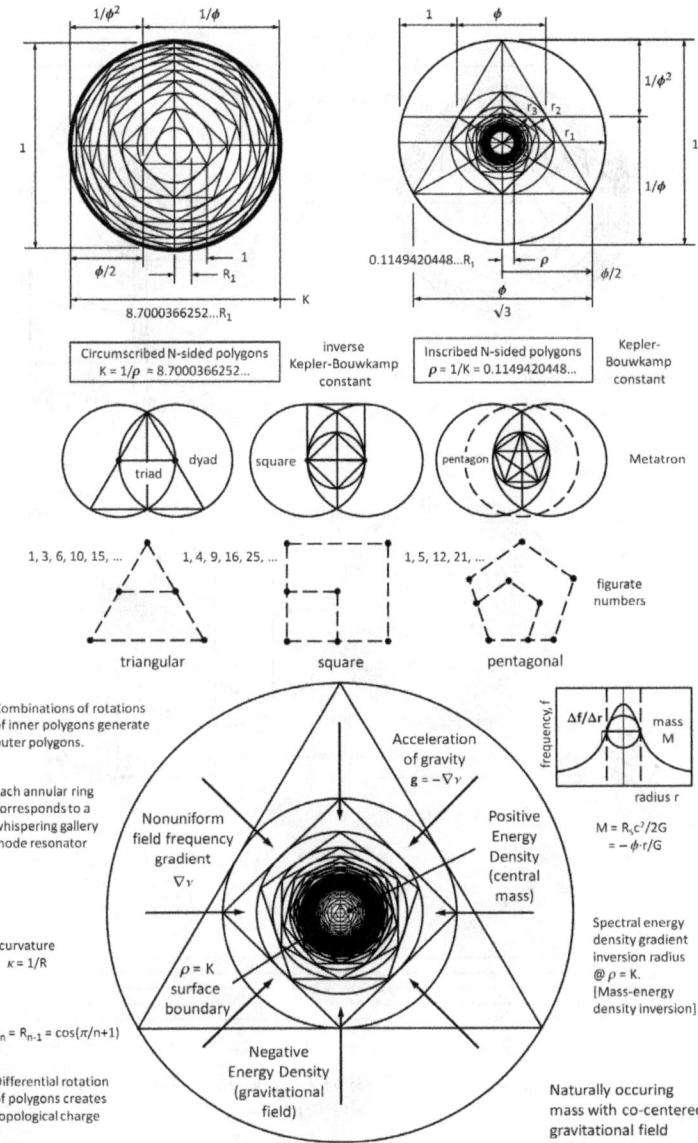

Fig. 52-44. N-scribed polygon representation of a naturally occurring positive energy mass object and a negative energy gravitational field with coincident incenter and circumcenter. No net translation occurs as the force field is symmetric. The symmetrical bound state resembles that of a stationary de-excited lattice exciton consisting of an electron and electron hole. Setting $K = \rho$ corresponds to a flat primordial vacuum state $[+1, -1 \to 0, 0]$ of zero curvature. The number of modes decreases as the energy density increases.

52. Anti-gravity

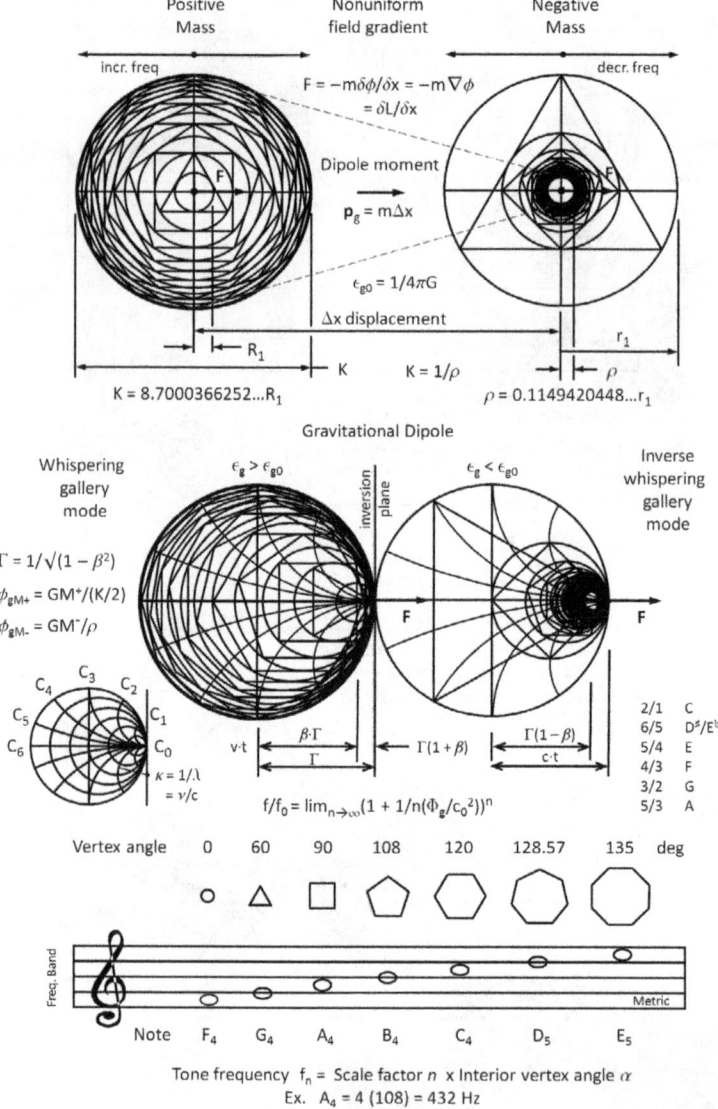

Fig. 52-45. N-scribed polygon representation of a gravitational dipole formed by displacement of positive and negative mass centers. Displacement creates a nonuniform field gradient and dipole moment. A oscillating ½-wave dipole mass resonator provides means for absorption or radiation of gravitational energy. Notes correspond to particle mass-energy harmonic cut-off frequencies (eg., Planck mass → Planck frequency). Planck mass m_P corresponds to precession shift of N-gons ($m_P = |V_P \cdot Q_P|/c^2 = E_P/c^2$) where V_P = Planck voltage, Q_P = Planck charge, E_P = Planck energy, c^2 = Energy per unit mass.

52. Anti-gravity

Frustrated vacuum lattice of Planck dipoles

[Figure: Frustrated vacuum lattice of Planck dipoles, with labels: negative Planck mass M_p^-, positive Planck mass M_p^+, vacuum energy $\langle 0|\hat{H}|0\rangle$, positive mass deficit, negative mass deficit, Planck mass dipole, symmetry-breaking → collective excitations]

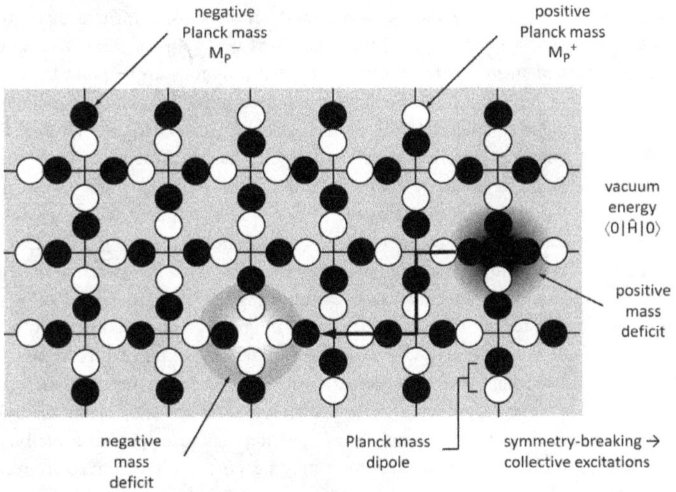

Fig. 52-46. Frustrated vacuum lattice of Planck dipoles yields localized regions of excess positive and negative mass which when coupled form a gravitational dipole. A Planck mass dipole of positive and negative mass ($1 = -1$) is equivalent to a gravitational dipole quasi-particle with net zero mass or $2 = 0$. This is comparable to a polaron condensed matter quasiparticle. Much of the vacuum energy is locked inside the Planck foam formed of such dipoles with net Planck mass cancellation. Gravitation is an effect of vacuum polarization.

Conversion of real mass into imaginary mass (phase rotation) may allow inertia neutralization such that helicity equals chirality. Chiral metamaterials with negative index of refraction enable reversal of energy flow and wave phase velocity direction. Alteration of vacuum impedance and dispersion characteristics may allow alteration of EM wave phase and group velocity. The group velocity is the velocity at which energy propagates. In the Scharnhorst-Casmir effect, reduction in EM mode density enables reduction in the local velocity of light. Reduction in vacuum impedance by alteration of the vacuum electric permittivity and magnetic permeability may enable increase in the local velocity of light. A gravitic displacement drive may perhaps be realized by locally altering Planck dipole density and polarization. Cleaving of the quasi-fluid lattice of the Planck vacuum increases the dipole spacing, reducing local energy density and vacuum impedance. Augmenting the nonlinear polarization of the vacuum by increasing the negative energy in a localized region by irradiation with an external EM pump beams with asymmetric curvature/torsion characteristics, the local velocity of light may conceivably be altered. Bunching up EM wavefronts ("scrunching up spacetime") to produce octave doubling (expanding vesica pisces) may allow local EM wave acceleration a ($= 2Ec/h$) and increased curvature enabling acceleration of an object in front riding a wave of increased frequency and energy density.

Truth is found if ever in the simplicity, never in the multiplicity and confusion of things. – Sir Isaac Newton

The theory is pretty, but is there also some truth in it? – Albert Einstein

52. Anti-gravity

Positive and negative mass Planck dipoles of characteristic dimensions on the order of the Planck length may visualized as regions of positive and negative curvature of the Planck vacuum with naturally occurring radius (wavelength) limits. Inscribed and circumscribed N-sided polyhedra yield a geometrical representation of gravitational field spectral energy densities of a spherical mass. Circumscribed N-sided polygons correspond to a positive mass while inscribed polygons correspond to a negative mass. Coincident incenters and circumcenters correspond to a naturally occurring mass and associated gravitational field. A coupled offset pair corresponds to a gravitational dipole. Displacement of the incenter and circumcenters creates a mass dipole moment. The n-line elements correspond to wave vectors. A positive energy mass is analogous to EM waves confined within a whispering gallery mode resonator. A negative energy gravitational field is analogous to EM waves confined externally to a central void in an inverse whispering gallery resonator such as may be constructed from a suitable metamaterial. Refer to Figs. 52-44, -45 and -46. The vertex density may be augmented and/or modulated with frequency combs coupled to a nested concentric configuration of N-polyhedra. Changes in gauge (scale) results in nonuniform field gradients with accompanying transformation forces. Bound states of Planck positive and negative mass in some respects resemble excitons (bound states of electrons and holes in lattice structures) which under external laser excitation may propagate as travelling wave packets. Wave vectoring of external radiation may be potentially achieved by means of concentric or offset Riemann sphere dielectric antennas using conformal optics methods.

How might n-gon patterns such as shown in Fig. 52-44 naturally arise? According to Ramsey theory which describes how order arises from chaos, within a large set of randomly dispersed points on a plane, there will always be found well-organized subsets. In particular, the subset of points may be connected to form a convex n-gon. Erdös and Szekeres in 1935 proposed that the minimum number of points needed to yield a convex polygon is $2^{(n-2)} + 1$ where n is the number of sides. The conjecture of order from chaos was dubbed the "happy ending" problem. Randomly connecting pairs of points confined within a spherical volume produces a random, chaotic distribution of line elements corresponding to wave vectors connecting EM oscillators that replicates the gravitational field potential of a spherical mass. Nodal density corresponds to spectral energy density.

According to the EGM modeling method developed by Storti et al[34], the quantum vacuum undergoes a recompression of spectral energy density in the vicinity of mass. The spectral frequency upper limit increases in proportion to mass while the number of modes decreases. The acceleration of gravity corresponds to the negative gradient of the spectral energy density. See Fig. 52-47. Polarization of the vacuum at the Planck scale corresponds to alignment of coupled pairs of positive and negative Planck mass. The quantum state of a Planck dipole can assume 0 or 1 quantum wave logic states or any state in between corresponding to a quantum bit of information or qbit. A spinning dipole can exist in a spin up state $|\uparrow\rangle$, spin down $|\downarrow\rangle$ or any superposition state $|\nearrow\rangle$. Qbit states may be represented on a Bloch sphere as a qbit vector. Multiple qbits can represent quantum entanglement.

A gravitational dipole corresponds to a moving standing wave system of two interacting EM oscillators. A standing wave system at rest with respect to an observer represents an equilibrium condition in which mass oscillators lie at potential hole minima. The forward and backward traveling waves are equal in frequency, i.e., no Doppler shift. If the wave system is subjected to an external or internal radiation pressure imbalance, the system as a whole undergoes a phase shift and is set in motion. Once disturbed from equilibrium, the wave system moves to restabilize at a new equilibrium position such that the source oscillators settle in time displaced potential hole minima. The wave system undergoes a Lorentz contraction and a Lorentz-Doppler shift generating a contracted moving standing wave (de Broglie matter wave). The velocity of the wave system v_g ($= \beta \cdot c = c \cdot \Delta\phi/\pi = $ at $+$

52. Anti-gravity

$v_0 = \sqrt{(v_0^2 + 2as)}$ is proportional to the phase difference. Acceleration of the wave system a $(= dv/dt = 2c \cdot \Delta v)$ is proportional to the frequence difference of the blue- and red-shifted counterpropagating waves. The difference frequency Δv is proportional to the de Broglie wave frequency v_{dB} $(= pc/h)$. Motion of mass oscillators constitute a mass current and generates a gravitomagnetic field $\mathbf{B}_g$ $(= \mu_g \mathbf{H}_g = \frac{1}{2} \nabla \times \mathbf{A}_g = \mathbf{g} \cdot \theta/c)$ analogous to a magnetic field $\mathbf{B}$ $(= \mu \mathbf{H})$ induced by an electric current. A gravitomagnetic field $\mathbf{A}_g$ $(= v/c^2)\phi_g$ is likewise generated as a result of motion analogous to the magnetic vector potential $\mathbf{A}$.

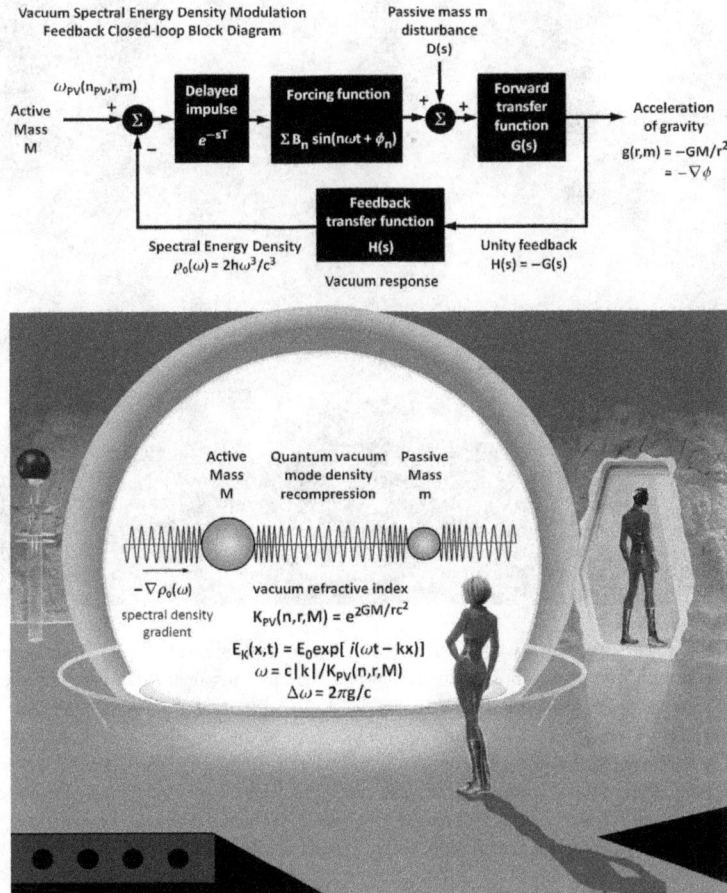

Fig. 52-47. The presence of mass induces a local recompression of the quantum vacuum spectral energy density (SED). This increases the EM frequency limit ω_Ω and reduces the number of modes. The increase in EM energy density corresponds to an increase the local polarized vacuum refractive index $K_{PV}(n,r,m)$. Change in frequency $\Delta\omega$ is proportional to g/c. The acceleration of gravity g is a measure of the EM spectral frequency gradient.

52. Anti-gravity

Acceleration of a wave system generates an gravitoelectric field $\mathbf{E}_g$ ($= -\delta\mathbf{A}_g/\delta t = (\mathbf{a}/c^2)\phi_g = \mathbf{v} \times \mathbf{K}$) analogous to the electrokinetic field $\mathbf{E}_k$ ($= -\delta\mathbf{A}/\delta t$) which acts to oppose mass motion. A moving mass generates a cogravitation force $\mathbf{F}_k = m(\mathbf{v} \times \mathbf{K})$ where $\mathbf{K}$ is cogravitation field ($= -GM(\mathbf{v} \times \hat{\mathbf{r}})/c^2 r^2 = (\mathbf{v} \times \mathbf{g})/c^2$) which acts only on moving masses. See Fig. 52-48.

Matter wave resonator

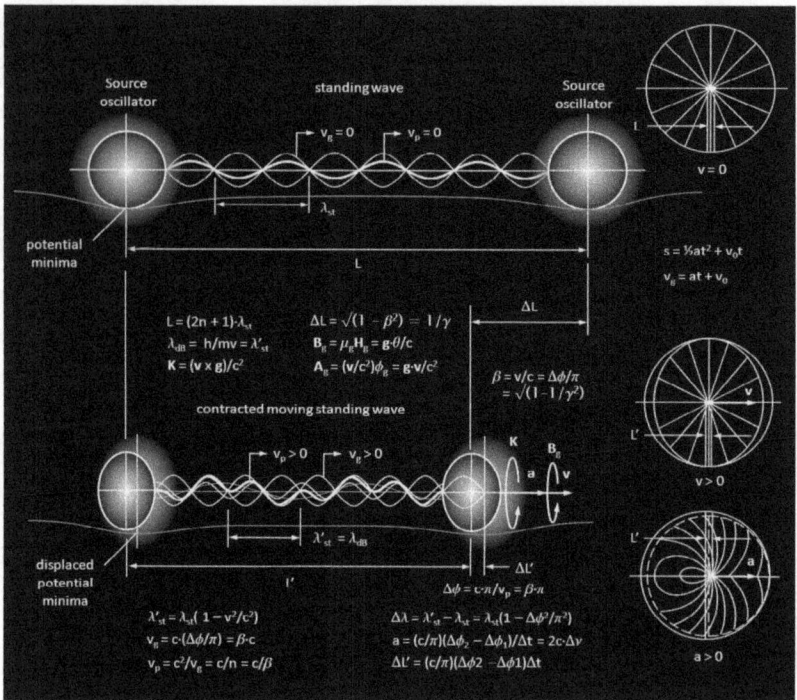

Fig. 52-48. Mass oscillator wave system consisting of two interacting source oscillators. Once disturbed from an equibrium, consonant rest state to a dissonant state, the nodal end point(s) undergo phase displacement towards the displaced potential minima to reestablish resonance. As a result of relative motion with respect to an observer inertial frame, the wave system undergoes an apparent Lorentz-Doppler frequency shift and a Lorentz contraction. Mass motion constitutes a mass current and generates a gravitomagnetic field $\mathbf{B}_g$ and a gravitomagnetic vector potential field $\mathbf{A}_g$. An accelerating mass generates, in addition, a vector cogravitational field $\mathbf{K}$ which acts on nearby matter opposing motion.

It seems probable that once the machine thinking method has started, it would not take long to outstrip our feeble powers. – Alan Turing

The mind is a machine that is constantly asking: What would I prefer? – George Saunders

Plasma seems to have the kinds of properties one would like for life…It has at least the potential for organizing itself in interesting ways. – Freeman Dyson

53. Gravitation Tonality

Arithmetic is pure number, geometry is number in space; music is number in time and astronomy is number in space and time. – Quadravium.

Opposition brings concord. Out of discord comes the fairest harmony. – Heraclitus

Music is the pleasure of the human mind experiences from counting without being aware it is counting – Gottfried Leibniz

Music would lose its charm were not dissonance interspersed at frequent intervals. – Max Heindel

The whole universe is based on rhythms. – John Hartford

Motion of a mass object through a gravitational field produces ultrahigh frequency tones as the mass moves through EM wave resonance points. The chord progression corresponds to the trajectory path through a tone net. An inverse effect may be demonstrated in motion of objects on a vibrating Chladni plate in which motion may be controlled by a melody of note progression. Musical pitch, ratios, and rate of change define harmony, melody, meter and rhythm invoking a sense of motion. Musical intervals are apparent in the EM spectrum of oscillation modes of standing waves. At nodal points, kinetic energy of material particles is minimum and at anti-nodes is maximum. Wave motion transfers energy from one point to another. Nodal displacement induces mass transport. Standing wave resonance results in localized concentration of matter density at nodal points. Resonance occurs at simple, whole number ratios requiring the least energy per oscillation period. Dissonant musical intervals such as tritones require more energy than octaves with thirds, fourths and fifths in between. The distance between tones may be represented by a Minkowski or taxicab metric on a circular timeline where the circular arc coordinates of sounded notes or chords X and Y are $x_1, x_2,...x_n$ and $y_1, y_2, ...y_n$, respectively is given by $d_s(X,Y) = (|x_1-y_1|^n + |x_2-y_2|^n +...+ |x^n-y^n|^n)^{1/n}$. A logarithmic frequency spiral as represented in conventional music theory is depicted in Figs. 53-1 and -2 illustrating harmonic resonance points.

An octave is defined as an interval of frequency such that $f_2 = 2f_1$ or $f_2/f_1 = 2$. The 2f or f/2 octave equivalency is described as circularity of pitch. The center of frequency of the octave band is defined as the geometric mean of the interval upper and lower frequencies: $f_0 = (f_1/f_2)^{1/2}$. Equal temperament divides the octave into twelve equal semitones described as the chromatic scale. The number twelve is of special significance in music and represents the sum of the three smallest integers $3 + 4 + 5 = 12$ and relates to the Pythagoras Theorem ($a^2 + b^2 = c^2$) for a 3:4:5 right triangle. In music theory, the first octave ranges from C_1 (32.7 Hz) to C_2 (65.4 Hz) in equal temperament of 12 semitones. Audible frequencies employed in music in an equal tempered scale of C encompass 16.351 Hz (C_0) to 16,744 Hz (C_{10}) or ~10 octaves. The EM frequency range of the Earth's gravitational field is estimated to extend up through 520E24 Hz (C_{77}) or ~77 octaves. For comparison, the electron Compton frequency is 1.2356×10^{20} Hz (C_{18}). The overall EM frequency spectrum extends up through C_{138} (2.913E43 Hz) which corresponds to the Planck frequency where the number of octaves = log(f2/f1)/log(2). Pythagorean tuning logarithmic spiral of perfect fifths (3:2) is mapped onto a 12-note equal temperament circle of fifths. In the chromatic scale of equal temperament, the octave is divided into 12 intervals of tempered half tones of frequency ratios $f^n = \sqrt[n]{2}$ where $f^{12} = 2$. Twelve major keys in ascending fifths and minor keys in descending fourths correspond to even (sine) and odd (cosine) frequencies. The frequency spiral corresponds to a golden mean pathway of harmonic cascade energy flow.

53. Gravitation Tonality

Equiangular or logarithmic spirals are of Fibonacci proportions. Adjusting the perfect fifths to a 12-note equal temperament closes the spiral into a circle.

Gravitational frequencies encompass the entire EM frequency spectrum with field energy concentrated at higher frequencies. In tonal music theory, gravitation refers to flow of energy from dissonant notes to consonant notes. Who has not felt the virtual gravitational pull of the meter of recurring beats of the rhythm of a song or the melodic inertia of tonal progression toward harmonic consonance? Stronger attraction occurs between tonic I and adjacent supertonic II and subtonic Vii notes than more distant dominant V and subdominant IV notes. Rhythm is associated with duration and grouping of a series of notes and rests. Temporal grouping of notes is described by meter. For example, a measure in 4/4 time indicates the song is organized into groups of four notes with a basic note length of one quarter of a whole note. Cadence refers to the beat, rate or measure of rhythmic movement as in melodic or harmonic chord progression. Melody refers to a succession of notes in time, tempo relates to the pace and vibrato a modulation of pitch. Syncopation may be defined in terms of hesitation and anticipation in the onset of notes with syncopated rhythm relating to displacement or offset of beats and accents. Two tones sounded at the same time is termed a harmonic interval. The frequency interval is expressed as a ratio of the frequencies or by a logarithm of this ratio. A musical fifth, for example, is expressed as 3:2. Glissando is the passage from a tone of one frequency to another (transition between notes) in a continuous glide through all the intervening frequencies. A related term, portamento, refers to a frequency sweep in quantized (stepped) changes – discrete glissando. A sense of gravity is conveyed by the low frequency notes such as a tuba or double bass.

Chords consist of a combination of three, four, five, six, or seven tones sounded simultaneously bearing a harmonic relation. Harmony is related to consonance and dissonance – A tonic interval ratio of 1:1 and an octave of 2:1 are consonant with increasing dissonance as the interval ratio decreases. Other consonant intervals formed by frequency ratio of small integers include the major third 5:4 and major sixth 8:5. A consonant harmonious blending of tones in an interval corresponds to the absence of beat frequencies between harmonics. Two notes separated by a semitone contains interfering harmonics and is described as dissonant. A barbershop seventh chord having frequency ratios of 4-5-6-7 has multiple of overlapping overtones with a minimum of dissonance. Resolution in tonal music is movement of a note or chord towards a tonal center from unstable dissonance to stable consonance. A harmonic progression sounds unresolved if tonally roving and unstable, and never settling down for long on a definite tonic. Symmetric chords sound unresolved or unstable whereas asymmetric chords are stable. Consonant chords are distributed asymmetrically about the tonal center. Major and minor three-note chords (triads) are resonant as the intervals are in a 3-4-5 proportion. Tritones are resolved by inward or outward chromatic progression towards the root chord. An end on a tonic chord (Rallentando) is stable or balanced, an end on a fourth or seventh is unstable. Tension and release may be effected by placement of intermediate beat notes. Descending pitch corresponds to falling under gravity. Tremolo is a repeated variation in loudness (amplitude) while holding pitch constant while vibrato is a repeated variation in pitch. A Shepard tone sequence of paired overlapping frequency bands of varying pitch and frequency may be constructed to simulate effects of constant or varying acceleration and deceleration. A sequential pair of Shepard tones separated by half an octave interval produces a tritone paradox that may appear to be either ascending or descending pitch.

Emma Peel: Do you know my wavelength? John Steed: I do indeed.

53. Gravitational Tonality

Extended chromatic equally-tempered scale

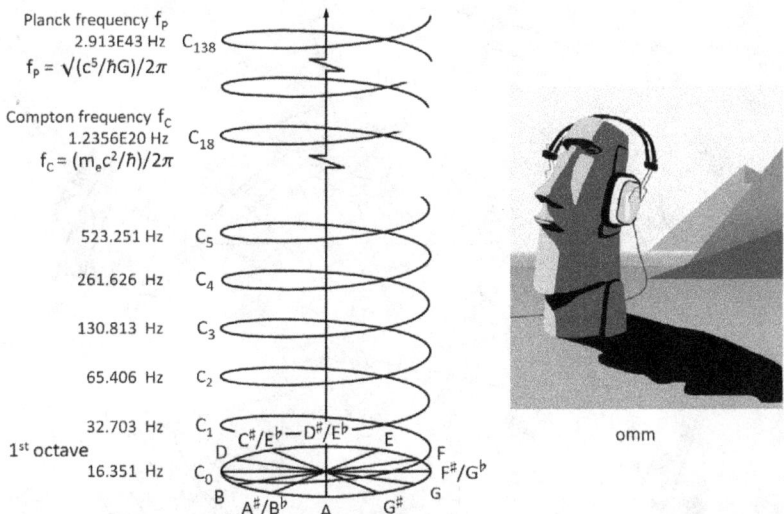

Planck frequency f_P
2.913E43 Hz C_{138}
$f_P = \sqrt{(c^5/\hbar G)}/2\pi$

Compton frequency f_C
1.2356E20 Hz C_{18}
$f_C = (m_e c^2/\hbar)/2\pi$

523.251 Hz C_5
261.626 Hz C_4
130.813 Hz C_3
65.406 Hz C_2
32.703 Hz C_1
1st octave
16.351 Hz C_0

omm

Semi-tone No.	Note	Musical Interval	FR	Frequency Ratio	Cents	Degrees
0	C	Unison	1	1.000000	0	0°
1	C♯ D♭	Minor second (Semitone)	f^1	1.059463	100	15°
2	D	Major second (Whole tone)	f^2	1.122462	200	30°
3	D♯ E♭	Minor third	f^3	1.189207	300	45°
4	E	Major third	f^4	1.259921	400	60°
5	F	Perfect fourth	f^5	1.334840	500	75°
6	F♯ G♭	Augmented fourth, Diminished fifth	f^6	1.414214	600	90°
7	G	Perfect fifth	f^7	1.498307	700	105°
8	G♯ A♭	Minor sixth	f^8	1.587401	800	120°
9	A	Major sixth	f^9	1.681793	900	135°
10	A♯ B♭	Minor seventh	f^{10}	1.781797	1,000	150°
11	B	Major seventh	f^{11}	1.887749	1,100	165°
12	C	Octave	f^{12}	2.000000	1,200	180°

Fig. 53-1. Logarithmic frequency spiral of EM frequency spectrum represented as a logarithmic spiral of notes. Frequencies of musical notes in a chromatic scale with a reference frequency of C_0 to C_{10} span 16 Hz to 16 kHz in 10 octaves. For comparison, the electron Compton frequency f_C of 1.24×10^{20} Hz corresponds to 18 octaves. The Planck frequency f_P of 2.91×10^{43} Hz, taken as the cutoff frequency of the vacuum, corresponds to 138 octaves. The Compton to Planck frequency interval f_C/f_P is $\sim 4.0029 \times 10^{24}$ semitones.

Mozart music is so pure and beautiful that I see it as a reflection of the inner beauty of the universe. – Albert Einstein

There are more riddles in a stone than in a philosopher's head. – Damon Knight

53. Gravitation Tonality

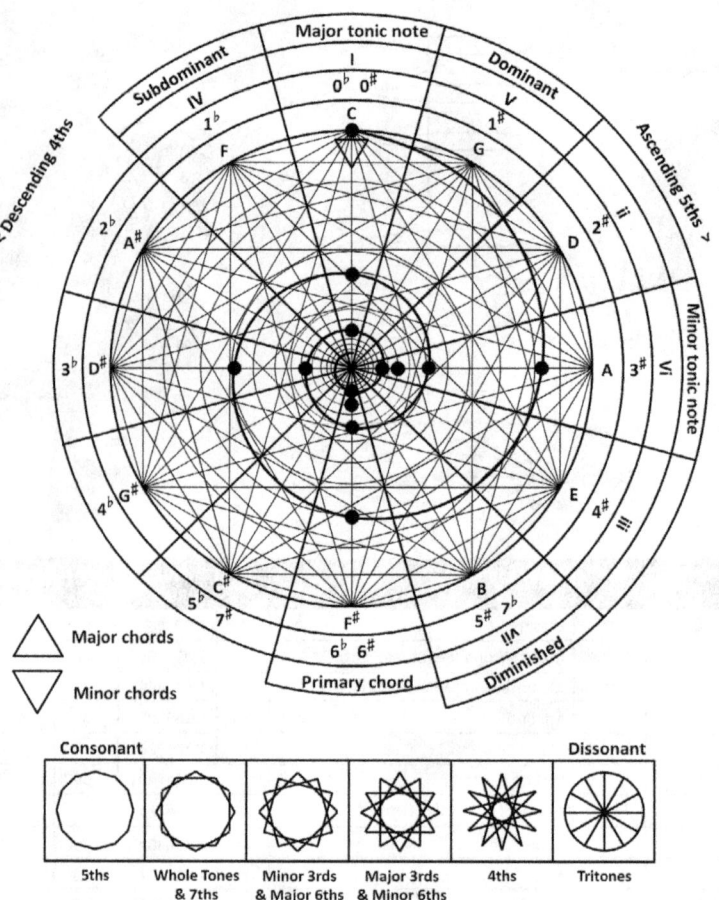

Fig. 53-2. Chromatic circle of fifths of 12-integer tone frequencies of an octave with logarithmic spiral superimposed illustrating harmonic resonance points. Twelve major keys in ascending fifths are shown on circle bezel with a base note neutral tonic C major key illustrating harmonic progression. A melody, say, in the key of C, refers to a central set of pitches or series of notes that return to the note C as the starting point or tonal center of the frequency progression. A unison interval 1:1 and an octave interval 2:1 are consonant. A consonant ratio of 3:2 defines the interval of a perfect fifth. A dissonant tritone interval corresponds to a frequency ratio of $\sqrt{2}:1$, an irrational number. Two notes can sound dissonant, when played together or in sequence, if superposition of two difference frequencies results in formation of interference dissonance. Energy flows in direction from dissonance to consonance increasing frequency synchronization and eurhythmia.

I discovered that the most interesting music of all was made by simply lining the loops in unison and letting them slowly shift out of phase with each other... – Steve Reich

53. Gravitational Tonality

Similar to release of tension in music theory, an object released to free fall in a gravitational field converts potential energy residing in the surrounding cogravitational field into kinetic energy reducing frequency discordance. An object restrained from falling corresponds to a state of tension or discordance. A state of tension corresponds to an asymmetry of spatial distribution of nodal points or Poynting vectors about the center-of-gravity. The internal oscillators within each coupled mass are partially phase-locked but not synchronized. As the masses are attracted together, the strength of coupling increases resulting in phase coherence. A mass oscillator descending in a gravitational field decreases in natural frequency whereas an ascending mass oscillator in a gravitational field increases in natural frequency in a rhythmic and melodic progression as a result of the difference in gravitational energy density. A gravitational field represented as a tone net of integer frequencies in a series of inscribed and circumscribed circles of fifths is illustrated in Fig 53-3. Mass is proportional to node number. Planck mass corresponds to 138 octaves.

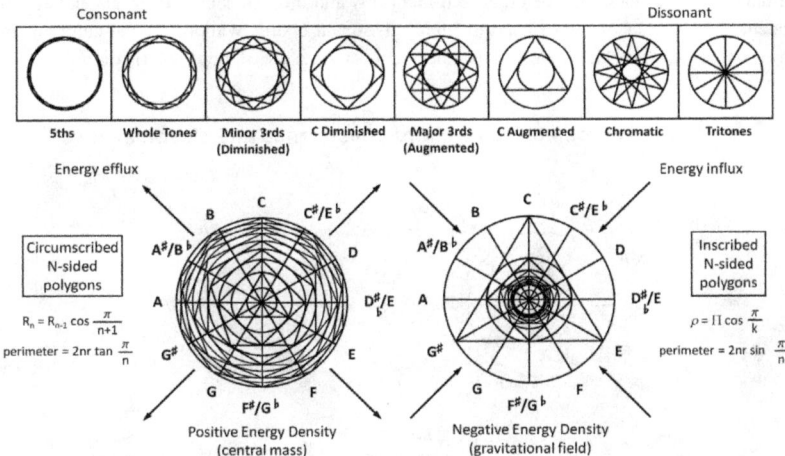

Fig. 53-3. Gravitational tone net analog of a central mass and surrounding gravitational field represented as nested n-sided polygons connecting integer tone frequencies. Division of inscribed and circumscribed tonal n-gons in tritones yields a 12-tone chromatic scale. Each circle corresponds to a 2:1 frequency octave of equal temperament division into 12 equal intervals. N-gon orientation is shown in the key note C. Tritones are resolved by inward or outward chromatic progression. Energy efflux for circumscribed N-sided polygons propagates outwardly from the tonal center whereas energy influx for inscribed N-sided polygons proceeds inwardly. Force field effects are associated with changes in scale. The vertices of an inscribed n-gon corresponds to the nth roots of $\omega = e^{\wedge}2\pi ri/n$ on a unit circle in the complex plane. Planck charge represents n-gon precession ($q_P = m_P\omega_p$).

Music is the arithmetic of sounds as optics is the geometry of light. – Claude Debussy

There was one man who was interested in the color of music, the connection between light and music and that was Einstein. – Léon Theremin

605

53. Gravitation Tonality

Frequency consonance and dissonance of chord sequences of a basic 12-tone equally tempered chromatic scale are illustrated in Fig. 53-4. As shown, consonant chords are asymmetric about the tonal center while dissonant chords are symmetric. A 3-tone triad is a chord consisting of three notes sounded simultaneously. Major and minor chords are commonly described as stable or resolved. Each triad contains three intervals with the difference in intervals and overtones may be described as tension. An asymmetrical chord such a C major or C minor chord shown is stable whereas a symmetrical C augmented or C diminished chord sounds unresolved or unstable. Major and minor chords form scalene triangles of differing angles and lengths of sides. An augmented chord divides the octave into three equal parts and forms an equilateral triangle as does a diminished triad symmetrically dividing the octave. A fully diminished chord divides the octave into four equal parts forming a square. In lyrical composition, symmetric chords have high tension. An asymmetrical triad and its inversions are consonant chords, symmetrical triads are dissonant. The symmetry group patterns of triadic transformations may be represented in triangular mesh tone nets as displayed on a tonnetz torus in Neo-Riemannian theory. Ordered systems of chords and intervals is known as functional tonality. The twelve notes of the chromatic scale may be organized in an row and ordered sets of intervals and atonal cadences or sequences constructed of highly dissonant chords without a tonal center as in twelve-tone serial music. Centroid displacement is a measure of tonal polarization.

12-tone chromatic scale chord consonance and dissonance

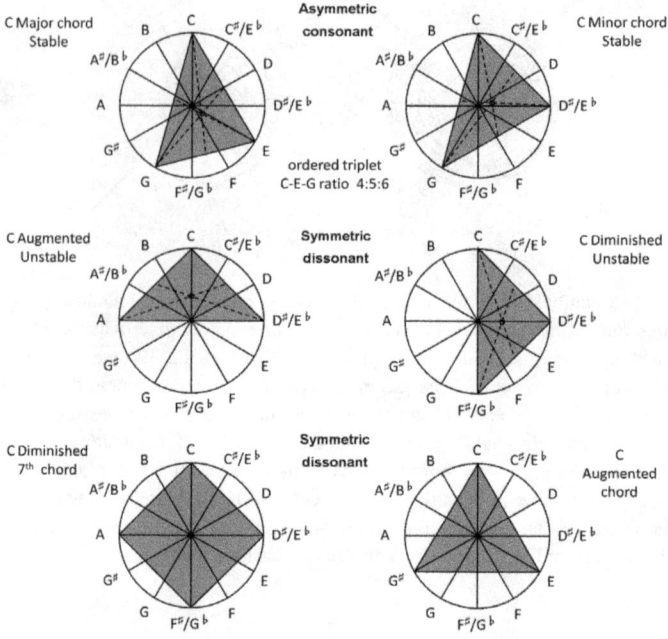

Fig. 53-4. Frequency consonance and dissonance of a 12-tone chromatic scale. The centroid of consonant chords is displaced from the tonal center while dissonant chords are mirror symmetric about refection plane(s) through the center of the chromatic circle.

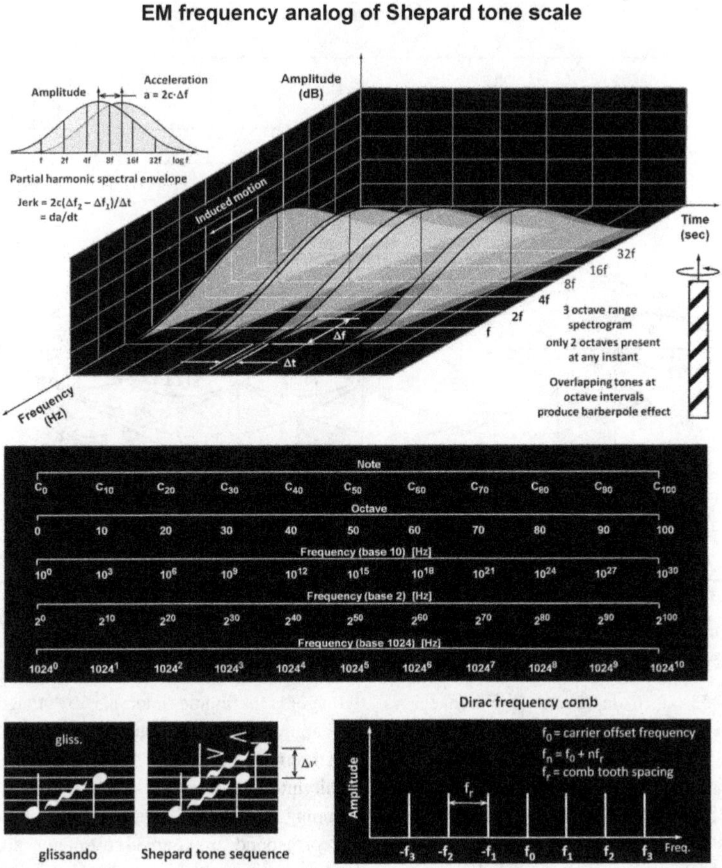

53. Gravitational Tonality

EM frequency analog of Shepard tone scale

Fig. 53-5. Shepard-Risset glissando of continuous glide of ascending pitch constructed of a phased sequence of a tritone interval of a pure tone and staggered, octave-spaced, Shepard tone harmonics with Gaussian spectral envelopes is shown. The frequency spectrum consists of a pure tone center frequency and partial harmonics separated by equal musical intervals. Odd-numbered tones produce a rising scale while even-numbered tones produce a descending scale. Skiing down the frequency hill is equivalent to climbing against gravitational potential well. Zero g (freefall) = zero frequency difference Δf. Continuous frequency surfing is equivalent to transposition of an ever falling pitch sequence of phased, interlaced harmonics in which a new tone is generated as another fades out. Acceleration varies with the frequency gradient. Jerk varies with change in acceleration. Musical meter corresponds to the gravitational frequency space metric.

Examples of Fibonacci and Pythagorean harmonic and geometric relationships are illustrated in Figs. 53-6 through -9. Phi ratios are a leitmotif involving wave resonance interactions defining damping constraints on standing wave resonances. The intervals illustrated combine ingredients of beauty and aesthetics: 1) Alternation of tension and release, 2) Realization of expectations, 3) Surprise at the unexpected, 4) Perception of unexpected relationships, 5) Patterns that fit together in a harmonious way.

607

53. Gravitation Tonality

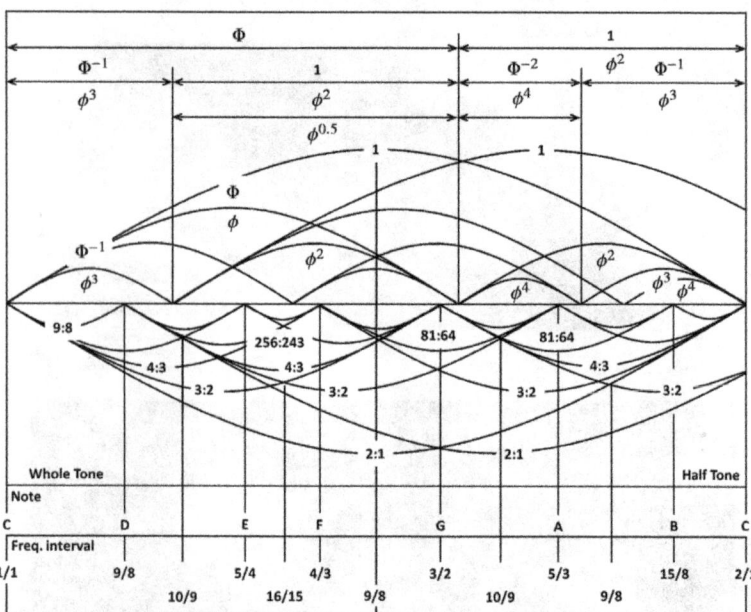

Fig. 53-6. Comparison of Fibonacci vs. Pythagoras harmonic intervals. Pythagoras musical octave (2:1) and fifth (3:2) intervals approximate the Fibonacci ϕ ratio (= $(1\pm\sqrt{5})/2$). Nodal points represent zones of constructive interference. The distribution of anti-nodes is non-Gaussian. Pythagorean harmonic intervals represent resonant conditions while Fibonacci intervals represent resonance damping ratios marking levels of support and resistance. The Pythagorean frequency ratios correspond to coupled oscillator stable synchronization states. One half sine intervals shown are equivalent to one half standing wavelengths. Motion involves alternating jumps from dissonance to consonance states.

A resonant condition of three standing waves in a Fibonacci ratio referred to as the "golden section" or "divine proportion" minimizes higher frequency harmonics thereby limiting energy losses. Thus, for a line of wavelength L = A + B, the wavelength proportions are given by A/B = B/(A + B). The geometric mean (= $\sqrt{(a \times b)}$) of both ratios is equal to ϕ = 0.618. Nonlinear effects result in harmonic distortion yielding higher order harmonics and energy dissipation. The mode interactions for nonlinear standing waves formed by counter-propagating travelling waves may be evaluated with the non-homogeneous Burgers' equation ($\delta u/\delta t + u\delta u/\delta x = 0$).

The juxtaposition of shapes and colors creates order out of chaos just as a conglomeration of noises arranged with taste, order, harmony and rhythm becomes a symphony. – Eyvind Earle

I'm interested in time, fame, death, beauty, truth, all those things. – Marianne Faithfull

There are wavelengths that people cannot see, there are sounds that people cannot hear, and maybe computers have thoughts that people cannot think. – Richard Hamming

53. Gravitational Tonality

Fibonacci, Pythagorean and binary sequences

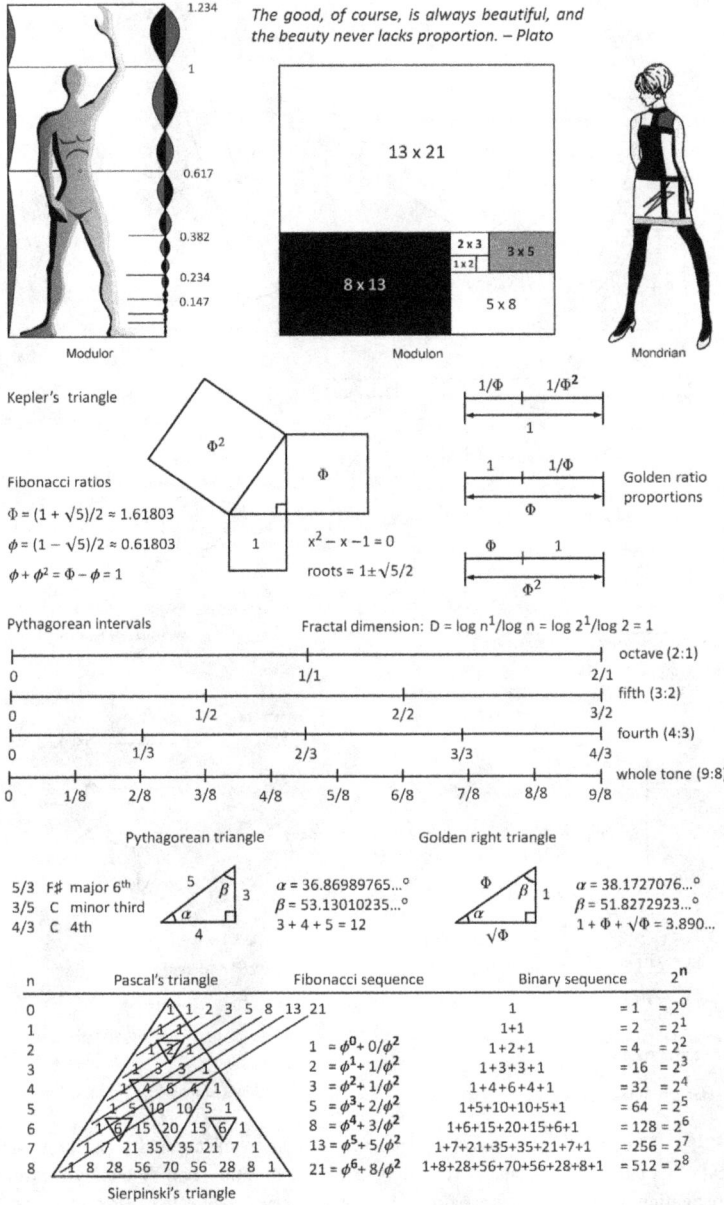

Fig. 53-7. Fibonacci and Pythagorean ratios are ubiquitous in nature. Standing wave damping correspond to Fibonacci ratios while resonant conditions correspond to Pythagorean whole number ratios. Illustrated are examples of various Pythagorean, Fibonacci, Pascal, fractal, binary sequences and geometric relationships.

53. Gravitation Tonality

Fibonacci Phi dissonance ratio examples

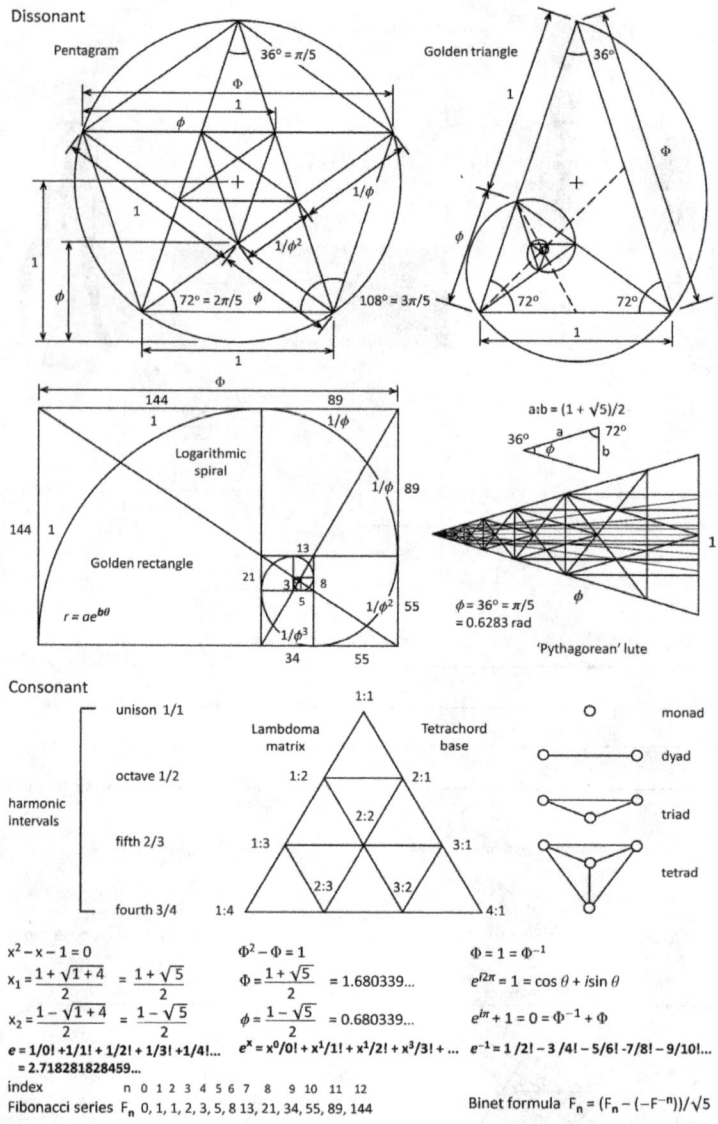

Fig. 53-8. The golden triangle of a pentagram, Pythagorean lute and logarithmic spiral illustrate Fibonacci Phi dissonance (damping) ratios. Phi ratios mark boundary conditions on propagation of harmonic energy flow. Harmonic intervals occur in consonant (resonant) whole number ratios as illustrated by the tetrachord diagram. The 4th represents an optimal tuning compromise between harmonic and melodic consonance.

In order to understand the universe you must know the language in which it is written and that language is mathematics. – Galileo Galilei

53. Gravitational Tonality

Rational and Irrational Winding Ratios

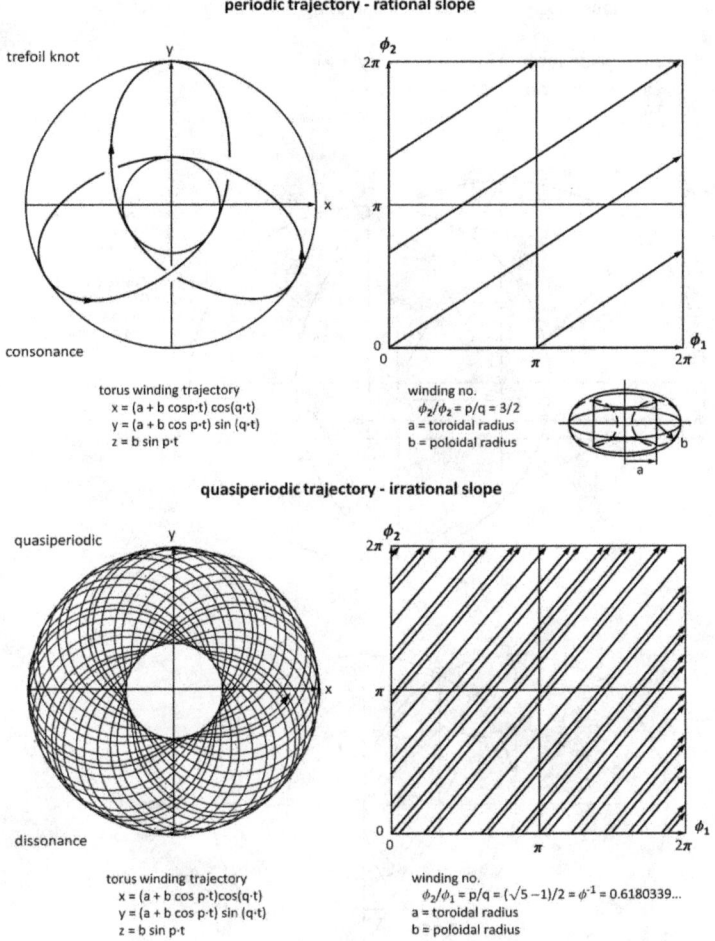

Fig. 53-9. Trajectory of traversal of a point on the surface of a torus is depicted for rational winding ratio. A rational winding ratio corresponding to a Pythagorean musical fifth (3:2) generates a trefoil knot. A rational value causes the winding to close after a finite number of orbital transits and represents a resonant condition. Trajectory of traversal of a point on the surface of a torus is depicted for a irrational winding ratio. An irrational winding number ratio corresponding to the golden ratio $\Phi_0 = 1.6180339887...$ generates a quasiperiodic orbit representing dissonance. An irrational value causes the winding to cover the torus densely but never quite closing. An irrational ϕ number ratio taken as a winding number of a periodic orbit of a dynamic system results in orbital stability. The more irrational the winding number (ratio of the resonant frequencies of two coupled oscillators) is, the more periodic orbit tends to stabilize as the frequencies lock preferentially onto small integer ratios creating a resonance resistant to destabilization.

611

53. Gravitation Tonality

An example of arithmetic, geometric and harmonic ratios of waves emitted from two interacting oscillators for arbitrary frequency ratio and dipole spacing is depicted in Fig. 53-10. In 3-dimensions of x, y and z, the corresponding ratios are given by

Arithmetic mean $AM = 1/3 \cdot (x + y + z)$
Geometric mean $GM = (x \cdot y \cdot z)^{1/3}$
Harmonic mean $HM = 3/(1/x + 1/y + 1/z)$

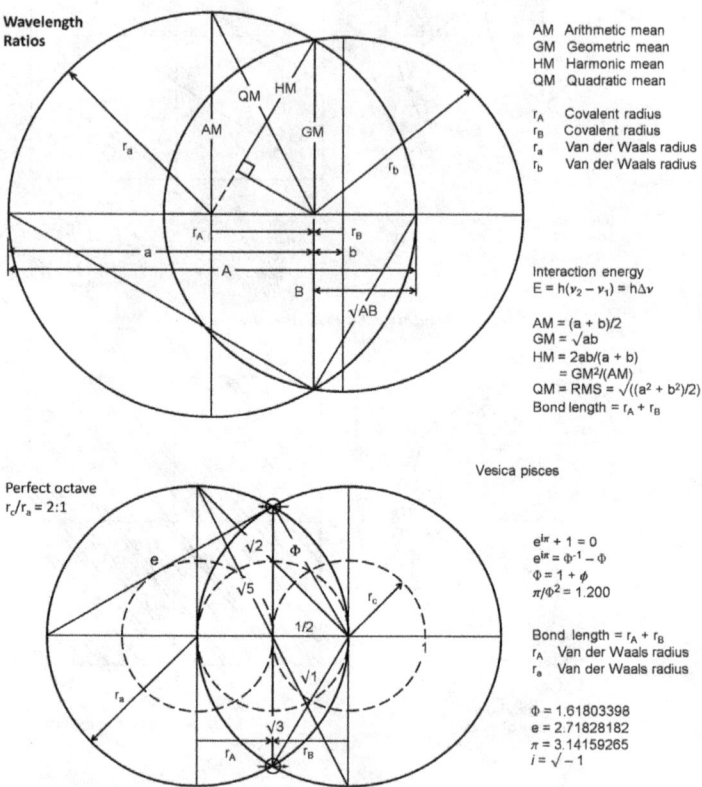

Fig. 53-10. Illustration of various wavelength ratios in 2D for two interacting oscillators of different and identical frequencies. The later corresponds to a vesica pisces. The interaction zone overlap corresponds to a zone of destructive interference of two wave functions of opposite phase. The interaction energy E is proportional to the frequency difference ($= \Delta \nu$). The 3D geometry case is descriptive of a covalent bond. Van der Waals radius is greater than the covalent radius resulting in weaker attractive forces than strength of covalent bonds and usually overcome random thermal motion at room temperature.

Less is more. – Ludwig Mies van der Roche

If less is more, zero is everything. – Bill Schweber

Less but better. – Dieter Rams

612

53. Gravitational Tonality

A 12 tone chromatic scale of seven octaves may be constructed from a sequence of expanding vesica pisces. Refer to Fig. 53-11. The diameter of each circle represents a fundamental wavelength. Each succeeding pair of overlapping dyads increases in diameter by a factor of 1.5. The musical interval between notes is a fourth when descending as each note is 4 letters ahead. When ascending, each note is 5 letters ahead constituting a fifth. The expanding circles form an included angle of 60°. The vertical dividing line of each dyad represents a nodal line. A simultaneous or sequential sounding of adjacent notes produces a harmonious, concordant tone. A similar transition pattern may occasionally be observed in wind-driven waves in calm water with a light breeze of ~1 m/sec in which bunched wave fronts undergoes dispersion shifting to a stretched out wave pattern with sharpened focus of increased curvature which rapidly accelerates and ultimately dissipates typically within 1-5 seconds over a distance of about 3-4 meters. An expanding vesica pisces suggests a model for universe expansion in terms of Planck mass pair creation.

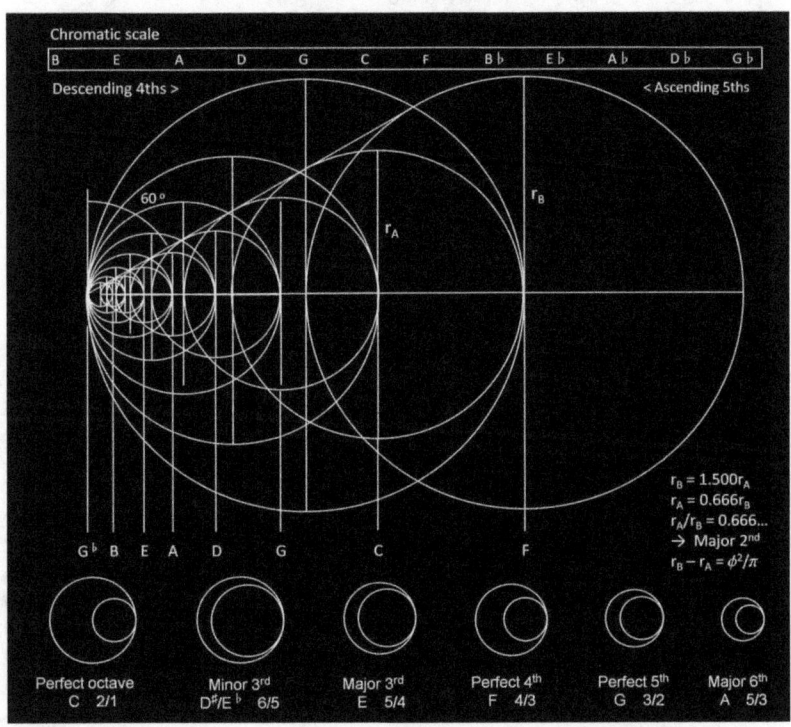

Fig. 53-11. The notes of the Pythagorean musical scale of whole tones may be constructed from a series of vesica pisces. Each pair of circles is equivalent to a pair of Villarceau circles produced by intersection of a torus and an oblique cutting plane. A 12-tone chromatic scale may be represented as a 12-component fibrillar winding on a tonnetz torus.

I call architecture frozen music. – Johann Wolfgang von Goethe

53. Gravitation Tonality

The frequency of a standing wave in the form of a tensioned string varies with string length, tension and mass. A contracted moving standing wave in an inertial frame undergoes Lorentz contraction with corresponding change in length, tension and mass. Proper length, mass and frequency likewise vary with gamma factor of Special Relativity γ (= $1/\sqrt{(1 - v^2/c^2)}$) and gravitational gamma Γ (= $1/\sqrt{(1 - v_e^2/c^2)}$). Potential energy corresponds to a scalar ϕ while kinetic energy corresponds to vector potential **A** of the electromagnetic 4-potential given by $A^\mu = (\phi/c, \mathbf{A})$. Potential energy represents energy of position or arrangement and may be considered as a form of hidden kinetic energy of string structural elements of component atoms. Fibonacci and Pythagorean harmonic ratios remain invariant in a transition between inertial frames. Refer to Fig. 53-12.

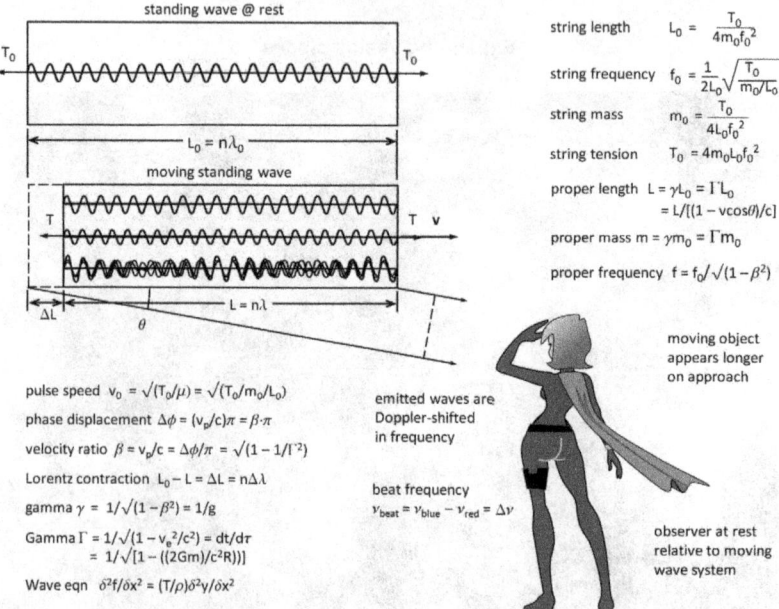

Fig. 53-12. Illustration of effect of motion on a standing wave system at constant velocity relative to a system at rest. For a standing wave generated by an oscillating tensioned string, the mass (m) of a string length (L), string tension (T) [energy/unit length] and frequency (f) vary with gamma factor γ. The rest mass of a tensioned string is a function of the tension force, string length and natural frequency. A spinning tensioned string with spin axis perpendicular to the direction of motion will appear to slow down at relativistic velocities. The Fibonacci and Pythagorean ratios remain invariant in a Lorentz transformation (each point in spacetime has the same value in all inertia frames).

There is geometry in the humming of strings, there is music in the spacing of the spheres. – Pythagoras

Since I was a kid, I always wanted to figure out how to make a bass line that was pendulum–like, gravity would control it and then you could make it play different notes. – Bjork

53. Gravitational Tonality

Musical keyboards typically reflect Pythagorean tuning in whole number ratios as illustrated in Fig. 53-13. A harmonic interval is created when two tones are sounded at the same. Differing frequencies overlap creates a beat frequency. When two tones are sounded in sequence, the result is a melodic interval. Three or more notes sounded simultaneously constitutes a chord. A fourth is an interval encompassing four staff positions while a fifth encompasses five staff positions. A tritone consists of three adjacent whole tones. In the chromatic scale there are twelve possible tritones, six of which are classified as augmented fourths and six as diminished fifths. Tritones tend toward resolution from dissonance to consonance. An augmented fourth resolves outward to a minor or major 6th while, in the inverse process, the diminished 5th resolves towards a major or minor third.

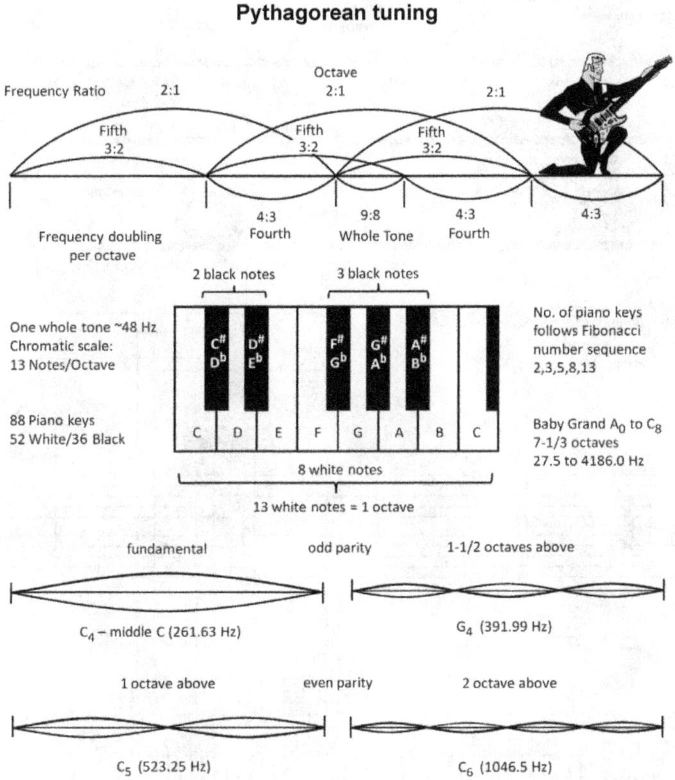

Pythagorean tuning

Fig. 53-13. Pythagorean standing wave tuning on a musical keyboard in a 12-tone chromatic scale. Harmonics occur in whole number multiples alternating in even and odd parity. Harmony produces stability. Pitch shifts are inevitable with pure harmonic ratios and in adaptive tuning sustained chord progression gravitates toward pure ratios.

> *It [theory of relativity] occurred to me by intuition, and music, was the driving force behind the intuition. My discovery was the result of musical perception. – A. Einstein*
>
> *I do know the effect that music still has on me. I am completely vulnerable to it. I'm seduced by it.*
> *– Debbie Harry*

53. Gravitation Tonality

A comparison of frequency ratios of common musical scale intervals is shown below in Fig. 53-14. The eight notes of the Pythagorean scale may be derived from a interval of four notes with ratio 4/3 and an interval of five notes with ratio 3/2 corresponding to a harmonic mean and arithmetic mean, respectively. A musical interval f_2/f_1 between the first note f_1 and second note f_2 is usually expressed in cents in musicology as a logarithmic ratio 1200 $\log(f_2/f_1)$. A comma pump circular chord progression is tempered by a syntonic comma (81:80) ratio. In Lydian chromatic theory common to modal jazz, tonal gravity decreases proceeding up the first seven tones of the circle of fifths, e.g., C, G, D, A, E, B, F♯.

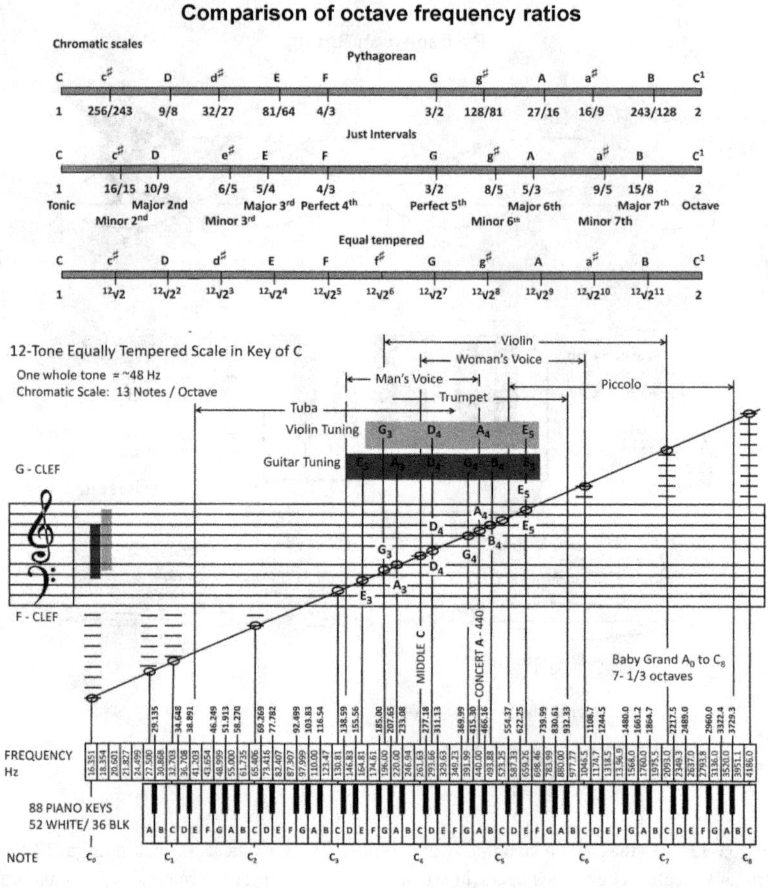

Fig. 53-14. A frequency ratio comparison of the Pythagorean, Just Interval and Equal Tempered scales for the 1-2 Octave (C → C^1).

It is proportion that beautifies everything, the whole universe consists of it and music is measured by it. – Orlando Gibbons

It must be possible to solve the task of controlling nature and yet simultaneously create a new freedom. – Ludwig Mies van der Roche

53. Gravitational Tonality

Geometrical relationships illustrating similarities to internal and external gravitational EM fields of a spherical mass are illustrated in Fig. 53-15. The lengths of each side of the polygon corresponds to an EM wavelength and vertexes to nodal points. The nodal density corresponds to EM energy density. The positive mass energy density of a spherical mass is approximated by circumscribed N-sided polygons. The negative energy density of a gravitational field of a spherical mass is approximated by inscribed N-sided polygons. Similarly, a positive Planck mass M_P^+ is analogous to a circumscribed N-sided polygon while a negative Planck mass M_P^- corresponds to an inscribed N-sided polygon.

Gravitational field density analog of a spherical mass

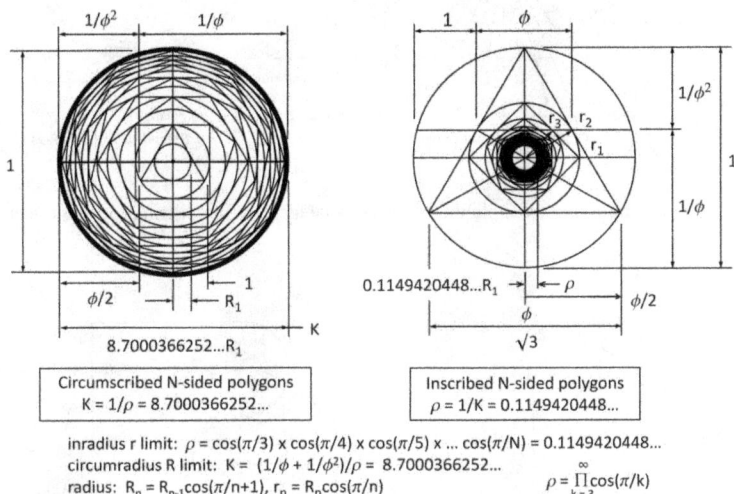

inradius r limit: $\rho = \cos(\pi/3) \times \cos(\pi/4) \times \cos(\pi/5) \times \ldots \cos(\pi/N) = 0.1149420448\ldots$
circumradius R limit: $K = (1/\phi + 1/\phi^2)/\rho = 8.7000366252\ldots$
radius: $R_n = R_{n-1}\cos(\pi/n+1)$, $r_n = R_n\cos(\pi/n)$

$$\rho = \prod_{k=3}^{\infty} \cos(\pi/k)$$

Fig. 53-15. N-sided polygon constructions exhibits vertex density similar to gravitational EM field density internal and external to a spherical mass. An infinitely nested set of circumscribed polygons in three-dimensional polyhedra approximates the internal gravitational potential of a finite spherical mass while an infinitely nested set of circular inscribed polygons approximates the external gravitational potential. Each demonstrate a naturally occurring radius limit. In plane geometry, the Kepler-Bouwkamp or n-polygon inscribing constant $\rho = 0.1149420448\ldots$ where the inradius r of the incircle is related to the circumradius R of the circumcircle ($r = R\cos(\pi/n)$). The geometric mean of the two curvatures equals unity ($= \sqrt{(K \cdot \rho)} = 1$). At the surface, the distribution and orientation of line segments approach consonance while, at the center, the pattern of line segments is increasingly dissonant. The length of each line segment corresponds to an EM wavelength or wave vector amplitude. Radii spacing corresponds to the frequency gradient and proportional to acceleration of gravity. The resulting structure is equivalent to a tone net of frequencies. A consonant, coherent state is associated with reduced temperature whereas a discordant state is associated with increased temperature and energy dissipation. The Planck length l_P ($= \sqrt{\hbar G/c^3}$) may represent a quantum scale limit of tone set frequencies (138 octaves) corresponding to a minimum torsion defect size in space time.

Number rules the universe – Motto of the Pythagoreans

53. Gravitation Tonality

Other examples of Fibonacci Phi ratios for circular and spherical geometries is illustrated in Figs. 53-16 and 53-17.

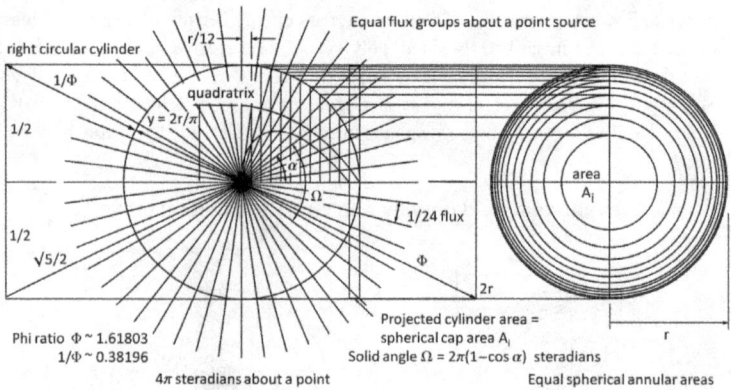

Fig. 53-16. Illustration of example Phi ratios associated with flux distribution about a point source.

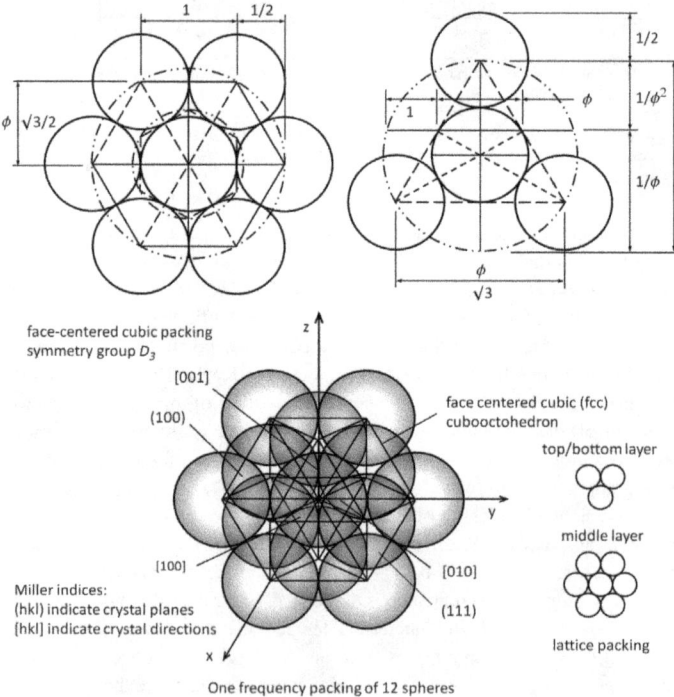

Fig. 53-17. Examples of Phi damping ratios apparent in clustered spheres.

53. Gravitational Tonality

Electromagnetic wave ray path trajectories in a dense mass may be represented as a form of quantum chaos. In Figure 53-18a, a randomized pattern of chord lengths simulates the field intensity within a mass object which increases at a surface boundary. A series of concentric shells of graded index of refraction is shown Fig. 53-18b illustrating effect of ray curvature external to a mass object. Similar examples in the form of a GRIN lens is shown in Fig. 53-19 and as whispering galleries in Fig. 53-20.

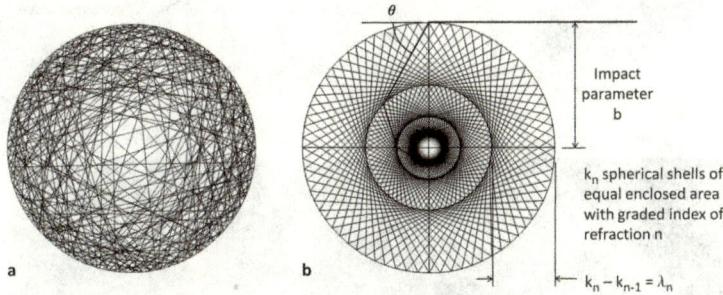

Fig. 53-18. Quantum chaos representation of ray trajectories formed by randomly connecting of two points on a circle confined within a spherical volume is shown in (a). The random, chaotic distribution illustrates random chord lengths representing a random distribution of wavelengths. The randomized pattern is scale, rotation and translation invariant and results in maximum field intensity at the reflecting surface boundary. A series of concentric shells of graded index of refraction external to a central mass is shown in (b) illustrating increasing energy density with decreasing radius.

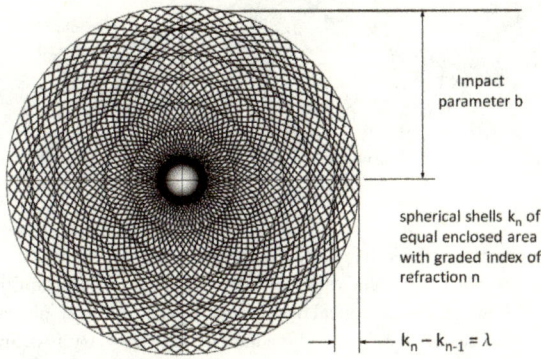

Fig. 53-19. Concentric shells (constant thickness) of graded refractive index. Light rays traversing a stratified layer of optical density follows a least time path described by the equation of Brachitochrone: $y[1 - (dy/dx)^2]$ = constant with velocity $v = \sqrt{(2gy)}$ where $x = a[\beta - \sin(\beta)]$, $y = a[1 - \cos(\beta)]$ and g = acceleration of gravity.

To be ignorant of motion is to be ignorant of nature. – Aristotle

53. Gravitation Tonality

Whispering galleries

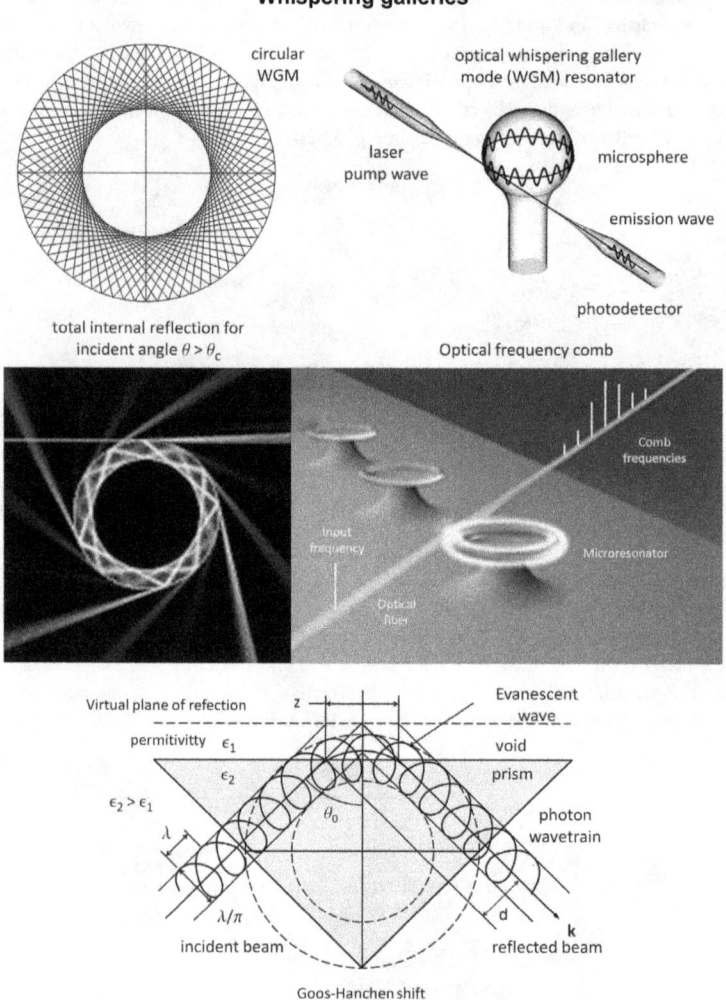

Fig. 53-20. Concentric shells correspond to whispering galleries common to reflection of EM waves from Fresnel zone boundaries, reflection of radio waves from the ionosphere, and reflection of light in micro optical resonators. The Fresnel or plasmoid reflection boundaries are opaque to the external radiation of wavelength equivalent to the characteristic dimensions of the zone boundary and provide a measure of radiation shielding to the interior. Optical resonance is created by localizing coherent light within a micro- or nanoscale structure such as a sphere or toroid ring resonator so that it interferes constructively and generates strong spatial confinement with long duration. Internal reflection results in a Goos-Hänchen shift associated with the evanescent wave propagation along the boundary surface. Coupled circulators allow control of transparency and opacity.

Fiat lux – Genesis 1.3

53. Gravitational Tonality

A gravitational field has negative energy and may be represented in terms of hyperbolic space with negative curvature. A triangle of constant Gaussian curvature K on a hyperbolic surface has an area $\Delta = [(\alpha + \beta + \gamma) - \pi]/K$ where α, β and γ are interior angles in which $\alpha + \beta + \gamma < 180°$. The EM gravimetric nodal density is similar to hyperbolic tessellations of n-gons and k-gons in quasi-periodic tilings on a Lobachesky sphere. A conformal representation of a Poincaré disc in 2D is illustrated in Fig. 53-21 with an isometric hyperbolic triangular tiling which resembles that of a gravitational spectral density gradient.

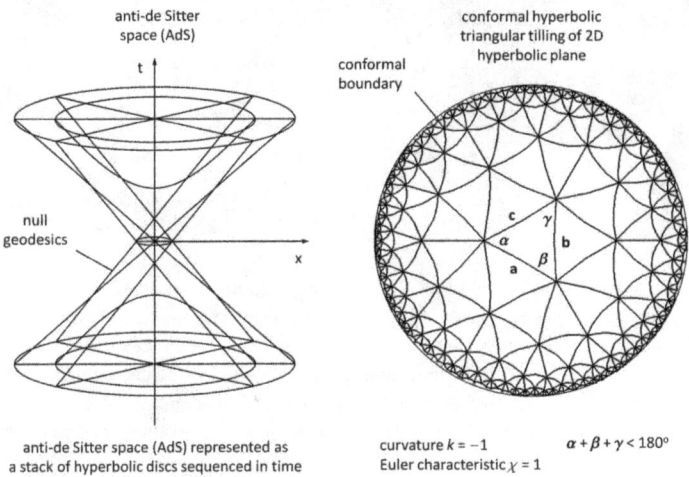

Fig. 53-21. Poincaré disc with hyperbolic triangular tiling. The sum of the triangle interior angles is $< \pi$ rad. Number of frequency modes varies as number of nodes.

An example of a wave interference nodal pattern is the Spira mirabilis field which exhibits graded density variation similar to the gravitational field of a mass object. See Fig. 53-22. Acceleration is in the direction of the frequency gradient.

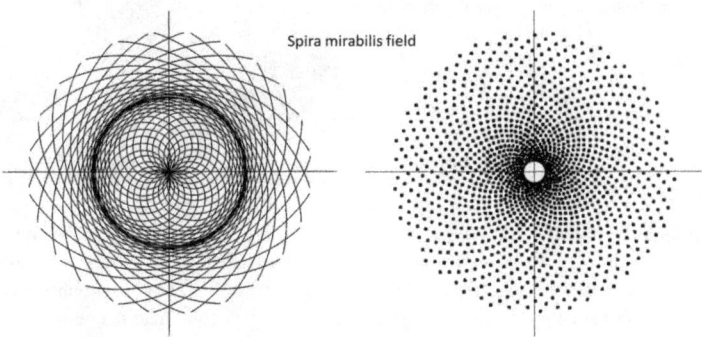

Fig. 53-22. Spira mirabilis field – Phi double spiral. Nodal pattern represents scattering centers of wave interference anti-nodes.

621

53. Gravitation Tonality

Examples of synthesized quadrapole curvature wave field constructions are shown in Figs. 53-23 and -24.

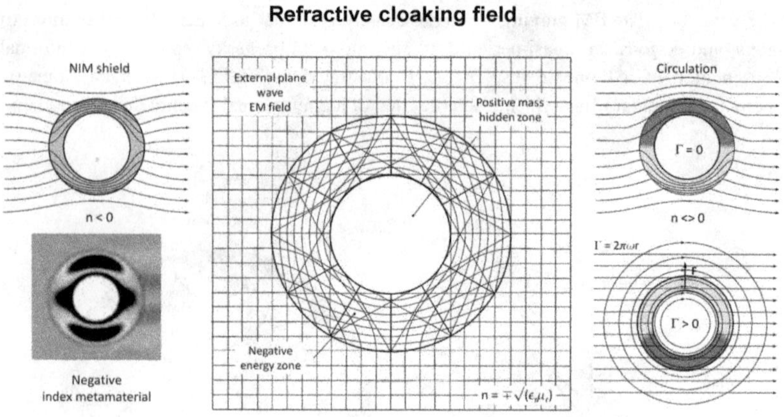

Refractive cloaking field

Fig. 53-23. Geometrical model of an EM cloaking field in which energy flow is directed around an object. A positive mass object is enveloped in a medium of negative refractive index deflecting incident radiation away from the central mass unlike an Einstein ring. A negative index metamaterial NIM may enable analogous EGM effects including Lorentz-Doppler inversion, Cerenkov inversion, artificial negative mass, inertia and mass cancellation, enhanced electrogravitomagnetic vector potential and anti-gravitic effects.

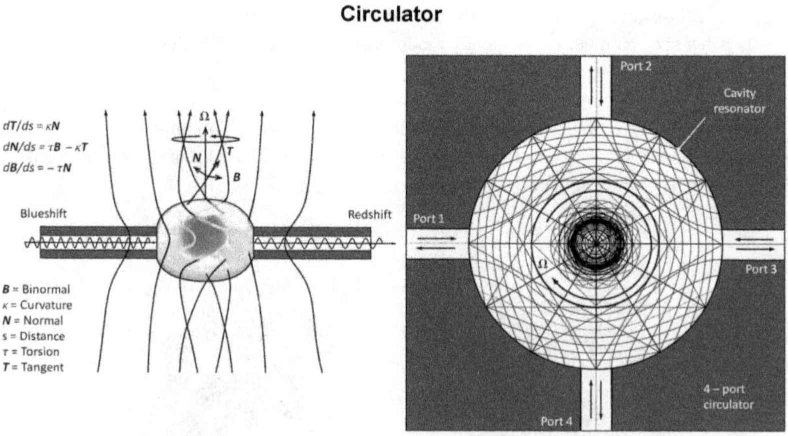

Circulator

Fig. 53-24. Differential rotation of a co-axial positive and negative EM energy field imparts circulation of external energy flow similar to that of a duplexer or gyrator. Circulation enables feedback elimination and unidirectional wave propagation as well as nonlinear rotational Doppler effects. Longer wavelength rejection than the period of Bragg gratings at the intermediate ports results in chirped output pulses. In a 4D spacetime analogy, circulation at constant acceleration corresponds to closed time-like curves. A standing wave resonator with variable transparency may be formed by coupled circulators.

54. Visualization of Dimensional Relationships

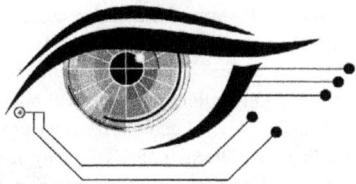

The beginning of wisdom is the definition of terms. – Socrates

Not only do we measure the movement by the time, but also time by the movement, because they define each other. – Aristotle

The laws of nature are written in the language of mathematics...the symbols are triangles, circles and other geometric figures, without which means it is humanly impossible to comprehend a single word - Galileo Galilei

The description of right lines and circles, upon which geometry is founded belongs to mechanics. Geometry does not teach us to draw these lines, but requires them to be drawn. – Isaac Newton

The mathematical rules of the universe are visible to men in the form of beauty. – John Michell

If you cannot describe it simply, you do not understand it well enough. – Michel de Montaigne

54.1 Ontological structure

An emphasis has been placed in this book on geometrical and dimensional descriptions of electromagnetic and gravitational quantum fields. A variety of geometrical relationships between physical quantities and fundamental interactions in physics are apparent such as in group theory with the Lie E^8 group being a prominent example. Dimensions for various physical quantities and MKS derived units are detailed in Table 54-1 and fundamental definitions and relationships are briefly summarized. Scientific International SI units (rationalized MKS) are used throughout unless otherwise noted. The term "rationalized" refers to the inclusion of the factor 4π into equations involving point quantities and thus 2π involving line quantities and disappears in uniform fields as well as Maxwell's equations. Use of 'geometrized' units where fundamental constants $G = c = 1$ and non-dimensionalized Planck units where $k = \hbar = 1$ is, in general, avoided due to quantitative misrepresentation, obfuscation of physical meaning, dimensional error check, and labor of reconversion for practical use.

An example visual summary of an organizational structure is an illustration of a hypercube ordering of physical quantities based on four dimensions of length, time, mass and angle represented as matrix elements $[L^a\ T^b\ Q^c\ M^d]$ including derived dimensions for mechanical and electromagnetic quantities such as momentum, angular momentum, charge, action, energy, etc together with physical constants c, G and h is depicted. The hypercube is illustrated as an extension of a Bronstein-Penrose cube revealing the interrelationships of quantum mechanics, classical mechanics, relativity theory, quantum field theory and quantum gravity.

All mathematics is a tale about groups. – Henri Poincaré

Logic will get you from A to B. Imagination will take you everywhere. – Albert Einstein

54. Visualization of Dimensional Relationships

54.2. Dimensional conversions

A tabulation of dimensions of various physical quantities and meter-kilogram-second (MKS) SI units is shown in Table 54-1. Derived dimensions based on conversion of electric charge to mechanical equivalents are included. For comparison, Planck unit conversions are shown in Table 5-1. Parameters are arranged in groupings: Fundamental (basic), geometrical, kinematic, mechanics, gravitation, mathematical operators, electromagnetism, radiometric and photo-metric (optics). Use of geometrized or natural units taking $c = G = \hbar = k_B = 1$ is avoided as tending to conceal underlying physical meaning and hinders dimensional analysis. For example, electrical charge e becomes a dimensionless number ($= 1/\sqrt{137}$). Likewise, CGS system of units is avoided in which the use of electrostatic units (ESU) or magnetostatic units (EMU) give rise to different dimensions for the same quantities revealing the error in setting ϵ_0 and μ_0 as dimensionless when both are related to c by Maxwell's equation $c = 1/\sqrt{(\epsilon_0\mu_0)}$.

A distinction may be made as to a fundamental dimension and degree-of-freedom. The dimension of a space represents the number of linearly independent vectors comprising the basis set. In the MKS system, electric charge is taken as fundamental and given its own dimension measured in Coulombs. As illustrated in Table 54-1, electric charge [Q] may be described purely in terms mass, rotation and time as [$M\Theta T^{-1}$]. Electric charge represents a spin precession effect with mechanical dimensions the same as spin angular momentum. In the Kaluza-Klein theory, electromagnetism was posited to be a result of an unseen, rolled-up, compactified fifth dimension. More recently, one version of superstring theory maintains the universe is not 4-dimensional spacetime, but is 10-dimensional with 6 dimensions compacted and hidden from view. Other such theories describe a universe of 26 dimensions of which 22 are likewise unseen. It remains to be proven whether such extra dimensions exist or are simply degrees-of-freedom within a 4D spacetime.

Mathematical objects possess symmetries defined as transformations of the object that preserve its shape. All such symmetries form groups. The invariants of groups of transformations are describable in terms of geometry. The study of such groups constitutes group theory. Mathematical objects parametized by n-real numbers may be treated as a point of n-dimensional space such as Riemann or Bloch spheres. In vertiginous use, selected properties of mathematical objects and relationships depending on classification type may be represented in affine or projective space, topological space, configuration space, rotation space, vector space, phase space, tangent space, probability space, Hilbert space, etc. Manifolds are higher dimensional analogs of a smooth curve or surface. Intersections of manifolds, connectivity and transversality are characterized by their dimensions. Local symmetry transformations are associated with force fields (gauge symmetry) to restore global symmetry. Phasor rotation, if locally symmetric, is gauge invariant. Symmetry-breaking is associated with a change in state. Symmetries and interrelationships of various physical theories including Gaussian (G), Newtonian Mechanics (NM), Rhythmodynamics (RD), Special Relativity (SR), General Relativity (GR), Classical Electrodynamics (CE), Quantum Electrodynamics (QED), Quantum Mechanics (QM), Quantum Gravity (QG), Relativistic Quantum Gravity (RQG), Electro-Gravi-Magnetics (EGM) may all be represented in subspace domain mappings in higher dimensional spaces. Geometrical representations and relationships of dimensions of physical quantities are illustrated as projections of higher dimensional abstract spaces in the following sections.

Thinking is the best way to travel. – Moody Blues

54. Visualization of Dimensional Relationships

Table 54-1. Dimensions of Physical Quantities

Parameter	Symbol	Dimensions	Derived Dimensions	MKS Units	MKS Derived Units*	Remarks
Basic SI quantities						
Length	l, x	L		meter (m)		$l = c/t = l_0/|\sqrt{K}| = l_0/\Gamma$
Mass	m	M		kilogram (kg)		$m = \mathbf{F}/\mathbf{a} = q\mathbf{E}/\mathbf{a} = A\mathbf{e}/\mathbf{v}$ $= q/\omega = m_0/\Gamma = E/c^2 = kh/v$ $= m_0\sqrt{K_{PV}} = |V\cdot q|/c^2$ $= E/(\mathbf{E}/\mathbf{B})^2 = R_S c^2/2G$ $= \mathbf{F}/2c\cdot\Delta v \cdot \mathbf{r}_u = E/\phi_g$; $m_P = \sqrt{\hbar c/G}$, $m_W = q\phi/3c^2$; $\Delta m = \hbar/\Delta\phi_m$
Time	t	T		second (s)		$t = c/l = t_0/|\sqrt{K}| = t_0/\Gamma$
Current	i, I	$T^{-1}Q$	$\Theta M T^{-2}$	Ampere (A) = C/s	[rad·kg/sec^2]	Charge/Time $i = q/t = e\omega/2\pi$, $I = l_0/\Gamma$
Temperature	T	T°		Kelvin degree (K°)		$T = E/k_B$; Energy/Entropy $\Delta t \Delta T = \hbar/k_B$; $T = pV/R$ $= mv^2/3 k_{KB}$; $\Delta t = q/\dot{m}c_p$
Mole, Amount of substance	N, n	mol		mol M = (kg·mol^{-1})		

cont

54. Visualization of Dimensional Relationships

Table 54-1. Dimensions of Physical Quantities (Cont)

Parameter	Symbol	Dimensions	Derived Dimensions	MKS Units	MKS Derived Units*	Remarks
Basic SI quantities						
Luminous intensity (density), Luminosity	I_V	cd		Candela (cd) = lm/sr, J		Luminous Flux/ Solid Angle
Geometrical quantities						
Planar rotation angle, Phase angle, Angular displacement, Angle deficit, Twist angle	θ, ϕ	Θ		radian (θ) = $kg^0 m^0 s^0 A^0$		Arc/Radius $\theta = s/r$ 1 rev = 2π rad = 360°
Arc length	s	ΘL		m·radians		$s = r\theta$
Solid angle, Spherical angle	Ω	–		Steradian (sr) = m^2/m^2		Surface area/radius^2 $\Omega = A/r^2$ sr = A/r^2 Unit sphere = 4π sr
Winding number, Number of turns, Topological charge	N, F	–		No. of turns (N) $N\cdot\theta$ = N·radians		Winding no. F = $2\pi N$ = topological charge = $2\pi Q$ = Flux Torus winding no. = p/q

cont

54. Visualization of Dimensional Relationships

Table 54-1. Dimensions of Physical Quantities (Cont)

Parameter	Symbol	Dimensions	Derived Dimensions	MKS Units	MKS Derived Units*	Remarks
Geometrical quantities (Cont)						
Curvature (Disclination defect)	κ, R	L^{-1}		1/meter		Intrinsic: $K = k_1 k_2$ $= -1/\rho_1\rho_2$ (Gaussian) Extrinsic: $K = k_1 + k_2 = 1/\rho_1 + 1/\rho_2$ Helix: $k = u/(c^2+u^2)$ $\kappa = (\mathbf{v}_x\mathbf{a}_y - \mathbf{v}_y\mathbf{a}_x)/\sqrt{(v_x^2+v_y^2)^3}$
Torsion (Dislocation defect)	τ, Q	$\Theta\, L^{-1}$		Θ/meter		Helix: $\tau = 2\pi c/(c^2+u^2)$ $\tau = 1/\sigma$ $\Delta Q \cdot \Delta R = l_P^{-3}$
Kinematics						
Velocity	**v**	LT^{-1}		meters/s (m/s)		Displacement/Time $v = l/t = ds/dt = c/\pi \cdot \Delta\phi$ $= V_0^2 + \mathbf{at} = c^2 \cdot \mathbf{A}/\phi_g$ $= V_0/K = \Gamma v_g = \beta \cdot c = p \cdot c^2/\rho_e$ $= \phi_E/\phi_B,\ \mathbf{v} = \hbar\mathbf{k}/m;\ v_g = c^2/v_p$ $= d\omega/d\mathbf{k};\ v_p = c^2/v_g = \omega/\mathbf{k}$

cont

54. Visualization of Dimensional Relationships

Table 54-1. Dimensions of Physical Quantities (Cont)

Parameter	Symbol	Dimensions	Derived Dimensions	MKS Units	MKS Derived Units*	Remarks
Kinematics (Cont)						
Areal velocity Areal speed	dA/dt	$L^2 \Theta T^{-1}$		$meter^2 \cdot rad/sec$		Area/Time $A = r^2 (d\theta/dt)/2 = L^2 m$ $dA/dt = r \times v/2 \, dA/dt$
Acceleration, linear	**a**	LT^{-2}		$meters/s^2$		ΔVelocity/Time $\mathbf{a} = d\mathbf{v}/dt = dx^2/dt^2 = \mathbf{F}/m$ $= 2c \cdot \Delta v \cdot \hat{\mathbf{f}} = \mathbf{E} \cdot c^2/\phi$ $= (2c^3/\hbar)(m\mathbf{v}/\Delta\lambda \cdot v_{dB})$
Angular frequency	ω	ΘT^{-1}		radians/s		$\omega = d\theta/dt = 2\pi f$
Angular velocity	ω	ΘT^{-1}		radians/s		Angular Displacement /Time $\omega = d\theta/dt = E/\hbar = \omega_0/\Gamma$ $\omega(k) = \hbar k^2/2m$
Angular acceleration	α	$\Theta^2 T^{-2}$		$radians^2/s^2$		ΔAngular velocity/Time $\alpha = d\omega/dt = d^2\theta/dt^2$
Vorticity	ζ	ΘT^{-1}		radians/s		$\zeta = \nabla \times \mathbf{v}_t$

cont

54. Visualization of Dimensional Relationships

Table 54-1. Dimensions of Physical Quantities (Cont)

Parameter	Symbol	Dimensions	Derived Dimensions	MKS Units	MKS Derived Units*	Remarks
Kinematics (Cont)						
Linear momentum, Impulse	p	MLT^{-1}		kg·m/sec = N·s		Mass x Velocity $\mathbf{p} = m\mathbf{V} = \mathbf{F}\cdot t = \mathbf{S}/c^2 = \beta E/c$; $p = mc \cdot \Delta\phi/\pi = \rho_e \cdot V/c^2$ $\mathbf{p} = m\mathbf{v} + q\mathbf{A} - mT\delta S/\delta \mathbf{p}$
Angular Momentum, Angular Impulse	L	$\Theta ML^2 T^{-1}$		rad·kg·m²/sec = rad·J·s		Moment of Inertia x Angular Velocity $\mathbf{L} = I\omega = \mathbf{r} \times \mathbf{p} = \mathbf{r} \times m\mathbf{v}$
Yank	Y	MLT^{-3}		N/sec = kg·m/s³		ΔForce/ΔTime
Jerk, Jolt, Lurch	j, y	LT^{-3}		m/sec³		ΔAcceleration/ΔTime
Jounce, Snap	s	LT^{-4}		m/sec⁴		ΔJerk/ΔTime
Mechanics						
Entropy	S	$ML^2 T^{-2}$		J/°K		$S = K_B \ln(W)$. W = # of states, K_B = Boltzman's const.
Enthalpy	H	$L^2 T^{-2}$		J/kg		$H = E + PV$; $\Delta h = c_p \Delta T$

cont

54. Visualization of Dimensional Relationships

Table 54-1. Dimensions of Physical Quantities (Cont)

Parameter	Symbol	Dimensions	Derived Dimensions	MKS Units	MKS Derived Units*	Remarks
Mechanics (Cont)						
Energy, total	E	ML^2T^{-2}		Joule (J) = N·m $= kg·m^2·s^{-2}$		$E = K.E. + P.E. = T + V = H$ $E = \pm\sqrt{[(pc)^2 + (mc^2)^2]}$
Energy, kinetic	K, T	ML^2T^{-2}		Joule (J) = N·m $= kg·m^2 s^{-2}$		Translation: $T = \frac{1}{2} mv^2$ $T = mc^2(\Delta\phi)^2/2\pi^2 = m·a·d$ Rotation: $T = \frac{1}{2} I\omega^2$
Energy, potential	U, V	ML^2T^{-2}		Joule (J) = N·m = Volt·Coul		$E = mc^2 = |V·q| = E_0/T$ $E = mgh = E_0/\sqrt{k}|$
Energy, Hamiltonian	E, H	ML^2T^{-2}		Joule (J) = N·m $= kg·m^2/sec^2$		Force · Distance, Power · Time $H = T + V = \frac{1}{2}m(\mathbf{E}^2 + c^2\mathbf{B}^2)$
Energy flux (Power)	J, ϕ, P	ML^2T^{-3}		W = J/s		Energy/Time $P_0 = \Gamma^2 P = K_{PV}$; $P = \sigma E^2$
Energy density	U, W, ρ_E	$ML^{-1}T^{-2}$		J/m^3 = Pa $= m^{-1}·kg·s^{-2}$		Energy/Volume $U_0 = \Gamma U = \sqrt{k}U$ $U = \frac{1}{2}(\epsilon_0 E^2 + \mathbf{B}^2/\mu_0)$

cont

54. Visualization of Dimensional Relationships

Table 54-1. Dimensions of Physical Quantities (Cont)

Parameter	Symbol	Dimensions	Derived Dimensions	MKS Units	MKS Derived Units*	Remarks
Mechanics (Cont)						
Energy, Lagrangian,	E, L	ML^2T^{-2}		Joule (J) = N·m = kg·m²/sec²		Force · Distance, Power · Time $L = T - V = m(E^2 - c^2B^2)$
Gravitomagnetic field strength	H_g	$M\Theta L^{-1}T^{-1}$		(kg·rad)/(m·s)		$= B_g/\mu_g = (c^2/4\pi G)K$ $= (v \times g)/4\pi G = g \cdot n/Z_0$ $= (c^2/4\pi G)\nabla \times A_g$
Gravitomagnetic flux density	B_g	ΘT^{-1}		rad/s		$= \mu_g H_g = \tfrac{1}{2}\nabla \times A_g = g \cdot \theta/c$ $= (G/2c^2)J/r^3 = -2\Omega$ $= \theta/\nabla d = (v \times g)\mu_g/4\pi G$ $= (Kc^2)\mu_g/4\pi G$
Cogravitation field	K	T^{-1}		1/s		$= \nabla \times A = (v \times g)/c^2$
Gravitomagnetic force	F_{gm}, F_c	MLT^{-2}		N = kg·(m/s²)		$= m(v \times K)$
Magnetic force	F_m	MLT^{-2}		N = kg·(m/s²)		$= q^+(v \times B)$
Gravito-Poynting vector	S_g	MT^{-3}		kg/s³		$= (c^2/4\pi G) \times K = g \times H_g$

cont

54. Visualization of Dimensional Relationships

Table 54-1. Dimensions of Physical Quantities (Cont)

Parameter	Symbol	Dimensions	Derived Dimensions	MKS Units	MKS Derived Units*	Remarks
Mechanics (Cont)						
Energy flux density	$S, \langle I \rangle, \phi$	MT^{-3}		$W \cdot m^{-2} = J \cdot m^{-2} \cdot s^{-1}$		Energy/(Area × Time)
Force (mechanical)	**F**	MLT^{-2}		Newton (N) $= kg \cdot (m/s^2)$		Mass · Acceleration Energy/Length $\mathbf{F} = m\mathbf{a}$, $\mathbf{F} = q\mathbf{E}$, $F = E/s$ $\mathbf{F} = E/r = \mathbf{F}_0/T = 2mc \cdot \Delta \mathbf{v} \cdot \hat{\mathbf{r}}$ $\mathbf{F} = -(\delta V/\delta x)m = -\nabla \phi \cdot m$ $= \delta L/\delta x = \delta \mathbf{p}/\delta t$
Centrifugal force	**F**	MLT^{-2}		$N = kg \cdot (m/s^2)$		$\mathbf{F} = m\mathbf{v}^2/r = mc^2(\Delta\phi)2/\pi^2 r$
Coriolis force	**F**	MLT^{-2}		$N = kg \cdot (m/s^2)$		$\mathbf{F} = m(\mathbf{v} \times 2\mathbf{\Omega})$; $\mathbf{\Omega} = \nabla \times \mathbf{v}$
Impulse	**I**	MLT^{-1}		$N \cdot s = kg \cdot m/s$		Force · Time, Mass · ΔVelocity, $\mathbf{I} = m\Delta \mathbf{v} = \mathbf{F} \cdot t = \Delta \mathbf{p}$
Torque, Couple, Moment of force	τ	$\Theta ML^2 T^{-2}$		$N \cdot m = m^2 \cdot kg \cdot s^{-2}$		Force · Moment Arm $\tau = \mathbf{r} \times \mathbf{F} = mr^2 \boldsymbol{\alpha} = l\alpha$ $= |\mathbf{r}| \cdot |\mathbf{F}| \sin\theta$

cont

54. Visualization of Dimensional Relationships

Table 54-1. Dimensions of Physical Quantities (Cont)

Parameter	Symbol	Dimensions	Derived Dimensions	MKS Units	MKS Derived Units*	Remarks
Mechanics (Cont)						
Spring constant	k	MT^{-2}		$N/m = kg \cdot s^{-2}$		Force/Length F/L
Stiffness	k	MT^{-2}		$N/m = kg \cdot s^{-2}$		$k = F/\delta$, $c^4/8\pi G$ (spacetime)
Pressure, Stress	p	$ML^{-1}T^{-2}$		$N/m^2 = kg/ms^2$ $= Pa$		Force/Area $p = F/A$, $P = P_0/T = nRt/V$
Molar energy	U	$ML^2T^{-2}N^{-1}$		$J/mol =$ $m^2 \cdot kg \cdot s^{-2} \cdot mol^{-1}$		Energy/Quantity
Work	W	ML^2T^{-2}		Joule (J) = N·m $= kg \cdot m^2/sec^2$		Force · Distance $W = F \cdot d = P \cdot t$ $= V \cdot I \cdot t = \tau \theta$
Power, Energy flux	P	ML^2T^{-3}		Watt = J/s $= kg \cdot m^2 \cdot s^{-3}$ $= V \cdot A$		ΔEnergy/ΔTime, Work/Time $P = W/t$, $P = Fv = \tau\omega = \sigma E^2$
Action	S	L^2MT^{-1}		$J \cdot s = m^2 kg \cdot s^{-1}$		Energy · Time $S = K\Delta t = mv\Delta x = \int L dt$

cont

54. Visualization of Dimensional Relationships

Table 54-1. Dimensions of Physical Quantities (Cont)

Parameter	Symbol	Dimensions	Derived Dimensions	MKS Units	MKS Derived Units*	Remarks
Mechanics (Cont)						
Inertial mass, Weber mass	m, m_i, m_W	M		kg		$m_i = \mathbf{F}/\mathbf{a} = \hbar \mathbf{k}/\mathbf{v} = q\phi_E/3c^2$
Gravitational mass	m, M	M		kg		$m = E/\phi = E/c^2 = \mathbf{F}_g \cdot \mathbf{r}^2/GM \cdot \hat{\mathbf{r}}$
Angular moment of inertia, Mass moment of inertia, Rotational Inertia	I	ML^2		$kg \cdot m^2$		Mass · Radius of Gyration^2 $I = L/\omega = kmr^2 = \tau/\alpha,$ $= \Sigma m_i r_i^2 = I_p = I_{cm} + md^2$
Radius of gyration	r_g, k	L		m		$r_g = \sqrt{(I/m)}, k = \sqrt{(I/A)}$
Circulation	Γ, C	L^2T^{-1}		$J \cdot s \cdot kg^{-1}$	irrotational flow. circulation = angular momentum	Vorticity · Area, Velocity/Loop Length, Length^2/Time = Velocity · Length $C = 2\Omega\pi R^2$
Energy/mass	c^2	L^2T^{-2}		m^2/s^2 J/kg		$c^2 = E/m = Gm_P/l_P = \epsilon_q\mu_0^{-1}$ $= V \cdot q/m = 2GM/R_S = v_P v_g$ $= \mathbf{g} \cdot \phi/\mathbf{E} = \mathbf{v} \cdot \phi_g/\mathbf{A} = \mu\epsilon/LC$ $= \mathbf{v} \times \mathbf{g}/\mathbf{K} = 4\pi G/\mu_g$

cont

Table 54-1. Dimensions of Physical Quantities (Cont)

Parameter	Symbol	Dimensions	Derived Dimensions	MKS Units	MKS Derived Units*	Remarks
Mechanics (Cont)						
Vorticity	Ω, ζ, ϕ_m	ΘT^{-1}		rad/sec s^{-1} = Hz		Curl of velocity field $\Omega = \nabla \times \mathbf{v} = 2\omega$
Wave number	k	L^{-1}		m^{-1}		$k = 2\pi/\lambda = 2\pi f/c = |\mathbf{k}|$ $= n\omega/c = E/\hbar c$ $\mathbf{k} = k_x \mathbf{i} + k_y \mathbf{j} + k_z \mathbf{k}$ $= m\mathbf{v}/\hbar = n\omega/c$
Wavelength	λ	L		meter (m)		$\lambda = v/f = 2\pi/k = f\omega/k$, $\lambda_{st} = c/2\nu$
Standing wavelength	λ_{st}	L		meter (m)		$\lambda_{st} = c/2\nu = (c/2\nu) \cdot$ $(1-\beta^2)/\sqrt{(1-\beta^2 \sin^2\theta)}$
Frequency	f, ν	T^{-1}		Hertz (Hz) = cps = s^{-1}		1/Time Period $f = 1/T = v/\lambda = \omega/2\pi$
Angular frequency	ω	ΘT^{-1}		radians/s		$\omega = 2\pi f = ck$
Period	T	T		second (s)		$T = 1/f$

cont

54. Visualization of Dimensional Relationships

Table 54-1. Dimensions of Physical Quantities (Cont)

Parameter	Symbol	Dimensions	Derived Dimensions	MKS Units	MKS Derived Units*	Remarks
Gravitation						
Acceleration of gravity, Gravitational field strength	$\mathbf{g}$	LT^{-1}		$N/kg = m/s^2$		Force/Mass $\mathbf{g} = \mathbf{W}/m = \mathbf{g}_0 / |K^{3/2}|$ $= \mathbf{g}_0 \Gamma^2 = -\nabla \phi_g - d\mathbf{A}_g/dt$ $= GM/r^2 = 2c \cdot \Delta v \cdot \mathbf{r}_u$ $= (2c^2/h)(mv/\Delta\lambda)\cdot\mathbf{r}_u$ $= \nabla \times \mathbf{A}_g = \mathbf{B}_g c/\theta$
Gravitational potential (scalar)	ϕ_g, U	$L^2 T^{-2}$		Joules/kg $= m^2 \cdot s^{-2}$		Energy/Mass $\phi_g = -GM/r = \mathbf{A}_g c^2/v$
Gravitational constant	G $G_e^{(2)}$ G_N G_P	$M^{-1}L^3T^{-2}$	$QVLM^{-2}$ $ITVLM^{-2}$	$N \cdot m^2/kg^2$ $= m^3/s^2 \cdot kg$	[coulomb·volt· meters/kg^2] [(Coulomb·volt ·m/kg^2)]	Force x Length^2/Mass^2 $G = c^4/F_P = \hbar c/m_P^2$ $G = F_g r^2/m_e^2 = G_0 \epsilon_{12}^2/\epsilon_0$ $G_0 = \Gamma^3 G_g = K^{3/2} G_g$ $G_P = l_P^3/(m_P t_P^2) = t_P c^3/m_P$ $= c^3/Z_5 = c^2\Sigma(r_i/m_i)$ $= 6.67262E-11 \, m^3/kg \cdot s^2$
Gravitational coupling constant	α_G	—		Dimensionless		$\alpha_G = Gm_e^2/\hbar c = (t_P \omega_C)^2$ $\cong 1.7518E-45.$

cont

54. Visualization of Dimensional Relationships

Table 54-1. Dimensions of Physical Quantities (Cont)

Parameter	Symbol	Dimensions	Derived Dimensions	MKS Units	MKS Derived Units*	Remarks
Gravitation (Cont)						
Gravitational gamma	Γ	—		—		$\Gamma = dt/d\tau = \sqrt{K_{PV}} = n$ $= 1/\sqrt{(c/c_0)} = 1/\sqrt{g_{00}}$ $= 1/\sqrt{(1 - v_e^2/c^2)}$ $= 1/\sqrt{(1 + (2\phi/c^2))}$
Vacuum refractive index	K_{PV}, K_{DEP}	—		—		$K_{PV} = \Gamma^2 = (dt/d\tau)^2$ $= 1/(1+(2\phi/c^2)) = 1/g_{00}$ $= g_{11} = g_{22} = g_{33} = c_0^2/\phi$ $= n^2 = \epsilon\mu = e^{\wedge}2GM/rc^2$ $= e^{\wedge}r_S/r$. $K_{DEP} = 1/\sqrt{(1 - 2GM/rc^2)}$
Gravitational field energy density	ρ_g, U_v	$ML^{-1}T^{-2}$		$J/m^3 = kg/m \cdot s^2$ $= Pa$		$\rho_g = -g^2/8\pi G = 3\rho$, $U_v = \frac{1}{2}(\phi_x^2 + \phi_y^2 + \phi_z^2)$, $\kappa = \rho_g/\rho_0$
Gravitomagnetic permeability	μ_g	LM^{-1}		m/kg		$= 4\pi G/c^2 = \mathbf{B}_g/\mathbf{H}_g = 4\pi c/Z_s$ $= 9.32877E-27$
Spacetime impedance	Z_s	MT^{-1}		kg/s		$= m_P/t_P = c^3/G = \pi c^3/\phi_g\mu_g$

cont

54. Visualization of Dimensional Relationships

Table 54-1. Dimensions of Physical Quantities (Cont)

Parameter	Symbol	Dimensions	Derived Dimensions	MKS Units	MKS Derived Units*	Remarks
Gravitation (Cont)						
Gravitational field mass density	ρ_{gm}	ML^{-3}		kg/m^3		$\rho_g = U_v c^2 = -\mathbf{g}^2/8\pi Gc^2$ $= Gm^2/8\pi c^2 r^4$
Mass density	ρ	ML^{-3}		kg/m^3		$\rho = mass/vol. = \Gamma \rho_0$
Mass flux, Mass flow rate	J_m	MT^{-1}		kg/sec		$J_m = \Delta m/\Delta t$, $dm/dt = \mathbf{B}q/\omega$
Mass current density	J_g	LMT^{-2}		$m \cdot kg/sec^2$		$J_g = \rho_g v_p$
Operators						
Length derivative	d/dx	L^{-1}		m^{-1}		Referenced to distance
Time derivative	d/dt	T^{-1}		s^{-1}		Referenced to time
Nabla, del operator, differential operator, gradient of a function, vector derivative	∇	L^{-1}		m^{-1}		$\nabla = \delta/\delta x + \delta/\delta y + \delta/\delta z$ $\nabla\phi = [\delta\phi/\delta x_i]$ (scalar) $\nabla\mathbf{g} = [\delta g_i/\delta g_j]$ (vector) $\nabla F = \nabla \cdot F + \nabla \wedge F$

cont

54. Visualization of Dimensional Relationships

Table 54-1. Dimensions of Physical Quantities (Cont)

Parameter	Symbol	Dimensions	Derived Dimensions	MKS Units	MKS Derived Units*	Remarks
Operators (Cont)						
Laplacian	∇^2	L^{-2}		m^{-2}		Referenced to distance Gradient of divergence $= \nabla \cdot \nabla$ $= \delta^2/\delta x^2 + \delta^2/\delta y^2 + \delta^2/\delta z^2$
Electromagnetism						
Electric charge Electric flux	q, e ψ	$\dfrac{Q}{T^{-1}I}$	ΘMT^{-1}	Coulomb (C) = A·s = J/V StatC = kg	[kg·rad/sec]	Current*Time $q = \mathbf{F/E} = \hbar \mathbf{k/A}$ $q = V \cdot C = 6.242 \times 10^{18}$ e $C \approx 4.43 \times 10^9$ kg·rad/s $e \approx -1.602 \times 10^{-19}$ C $e = q = E/\phi = E/V$ $\approx -7.07 \times 10^{-10}$ kg·rad/s $= m(\omega_c + \omega_p) \simeq m\omega_c$ $= \psi = \mathbf{D} \cdot \mathbf{A} = \epsilon \mathbf{E} \cdot \mathbf{A}$
Capacitance, Electric capacity	C	$\Theta M^{-1} L^{-2} T^2 Q^2$ $\Theta M^{-1} L^{-2} T^{-4} I^2$	$\Theta M L^{-2}$	Farad (F) = C/V = s/Ω = $kg^{-1} \cdot m^{-2} \cdot s^4 \cdot A^2$	$(rad^2 \cdot kg)/m^2$	Charge/ΔPotential $C = q/\Delta V = LI^2/V^2$ $= W/(½ V^2)$

cont

54. Visualization of Dimensional Relationships

Table 54-1. Dimensions of Physical Quantities (Cont)

Parameter	Symbol	Dimensions	Derived Dimensions	MKS Units	MKS Derived Units*	Remarks
Electromagnetism (Cont)						
Charge volume density	ρ_V	QL^{-3} $L^{-3}TI$	$\Theta MT^{-1}L^{-3}$	Coul/meter^3 (C/m^3) = m$^{-3} \cdot$s$\cdot$A	kg·rad/(sec·m^3)	Charge/Volume
Charge surface density	ρ_S, σ	QL^{-2} $L^{-2}TI$	$\Theta MT^{-1}L^{-2}$	C/m^2	kg·rad/(sec·m^2)	Charge/Area
Charge linear density	ρ_L, λ	QL^{-1} $L^{-1}TI$	$\Theta ML^{-1}T^{-1}$	C/m	kg·rad/(sec·m)	Charge/Length
Electric dipole moment	$\mathbf{p}, \mathbf{p}_e$	LQ LTI	ΘLMT^{-1}	Coul·meter (C·m) A·s·m	kg·rad·m/sec	Charge · Distance, Charge · Translational Displacement $\mathbf{p = qd}$
Electric quadrapole moment	Q_{ij}, eQ	QL^2 L^2IT	$\Theta MT^{-1}L^2$	C·m^2	kg·rad·m^2/sec	Electric dipole · Distance, Electric charge · Distance^2
Perveance	p	$Q^{-1}TL^{-5/2}M^{-3/2}$	$\Theta^2 MT^{-1}L^{-5/2}$	A/V$^{3/2}$	kg·rad^2/s·m$^{5/2}$	Beam current /Voltage$^{3/2}$ $P = I/V^{3/2}$

cont

54. Visualization of Dimensional Relationships

Table 54-1. Dimensions of Physical Quantities (Cont)

Parameter	Symbol	Dimensions	Derived Dimensions	MKS Units	MKS Derived Units*	Remarks
Electromagnetism (Cont)						
Electric potential, Voltage, Electric scalar potential	V, ϕ, ϕ_E	$ML^2T^{-2}Q^{-1}$ $ML^2T^{-3}I^{-1}$	$\Theta^{-1}L^2T^{-1}$	volt (V) = J/Coul = W/A = C·F^{-1} = kg·m^2·s^{-3}·A^{-1}	m^2/(sec·rad)	Power/Current, Energy/Charge, Work/Charge $V = W/q$ = Joul/Coul $V = q/(4\pi\epsilon_0)\, r = q/C$ $= IZ = V_0/T$ $= p/(4\pi\epsilon_0)\, r^2$ $\phi = E/e$
Electromotive force, (No load potential), Electric potential difference, Electromotance	$\mathcal{E}$, EMF	$ML^2T^{-2}Q^{-1}$ $ML^2T^{-3}I^{-1}$	$\Theta^{-1}L^2T^{-1}$	volt (V) = J/Coul = J·A^{-1}·s^{-1} = W/A = kg·m^2·s^{-3}·A^{-1}	m^2/(sec·rad)	ΔPotential open circuit $\mathcal{E} = E/q = W/q = V + IR$ $\mathcal{E} = -N \times d\phi/dt$ $= -d\phi_m/dt$ $\mathcal{E} = -NA(dB/dt) = \int E \cdot dl$
Electric flux	Φ_E, ψ	$ML^3T^{-2}Q^{-1}$ $ML^3T^{-3}I^{-1}$	$\Theta^{-1}L^3T^{-1}$	volt·meter = N·m^{2}C^{-1} = kg·m^3·s^{-3}·A^{-1}	m^3/(sec·rad)	Q lines of flux about a point. $\psi = 4\pi r^2$ $Q/4\pi\epsilon r^2 = Q$ $\Phi_E = E \cdot A = EA\cos\theta$

cont

54. Visualization of Dimensional Relationships

Table 54-1. Dimensions of Physical Quantities (Cont)

Parameter	Symbol	Dimensions	Derived Dimensions	MKS Units	MKS Derived Units*	Remarks
Electromagnetism (Cont)						
Electric flux density, Electric displacement, Induction	D	$L^{-2}Q$ $L^{-2}TI$	$\Theta L^{-2}MT^{-1}$	Coul/meter^2 (C/m^2) $m^{-2} \cdot s \cdot A$	(kg·rad)/ (m^2·sec)	Charge/Area $D = \varepsilon E + P = Q/A = \varepsilon_r \varepsilon_0 E$ $= \varepsilon_0 E(1 + \chi_e)$
Electric displacement, Electric polarization, polarity	P	$L^{-2}Q$ $L^{-2}TI$	$\Theta L^{-2}MT^{-1}$	Coul/meter^2 (C/m^2) $= m^{-2} \cdot s \cdot A$	(kg·rad)/ (m^2·sec)	Charge/Area
Electric field strength, Electric field intensity, Electric potential gradient (negative)	E	$MLT^{-2}Q^{-1}$ $MLT^{-3}I^{-1}$	$\Theta^{-1}LT^{-1}$	volts/meter = N/Coul = Joules/Coul·m $= m \cdot kg \cdot s^{-3} \cdot A^{-1}$	m/(s·rad)	Voltage/Distance, Force/Charge, Energy/Charge·Length, ΔPotential/Distance $E = V/d$, $E = F/q$, $E = B \cdot c$ $E = -\nabla \varphi - \delta A/\delta t$ $E = -\nabla \varphi + E_k$ $E = -\Delta V/\Delta x = H \cdot Z_0/n$ $E = q/4\pi \varepsilon_0 r^2$, $E = J/\sigma$ $\nabla \cdot E = \delta \beta/\delta t$, $\nabla \times E = -\delta B/\delta t$

cont

54. Visualization of Dimensional Relationships

Table 54-1. Dimensions of Physical Quantities (Cont)

Parameter	Symbol	Dimensions	Derived Dimensions	MKS Units	MKS Derived Units*	Remarks
Electromagnetism (Cont)						
Electric field gradient	dE/dx, V_i, E_β, $q_{\alpha\beta}$	$MT^{-2}Q^{-1}$ $MT^{-3}I^{-1}$	$\Theta^{-1}T^{-1}$	Volts/meter^2 $= J \cdot Cm^{-2}$ $= kg \cdot s^{-3} \cdot A^{-1}$	$(sec \cdot rad)^{-1}$	ΔPotential/Area, ΔE Field Strength/Distance
Coulomb force	$\mathbf{F}_e$	MLT^{-2}		N $= kg \cdot s^{-2}$ (m/sec^2)		Source · Field Strength $F_e = k_C q/r^2$, $\mathbf{F}_e = q^+\mathbf{E}$
Current	$I, \mathbf{I}$	QT^{-1} I	ΘMT^{-2}	Amp (A) = C/s	kg·rad/sec^2	Charge/Time $i = dq/dt = (\hbar k - qA)n \cdot q/m_e$ 1 C/s = 4.43 kg·rad/sec^2 $i = (V \cdot C)/t = i_0/T = E/\phi_b$
Current density	$\mathbf{j}, \mathbf{J}$	$L^{-2}T^{-1}Q$	$\Theta L^{-2}MT^{-2}$	amp/meter^2 $= (A/m^2)$	$(kg \cdot rad)/(m^2 \cdot s^2)$	Current/Area $\mathbf{J} = \epsilon \delta \mathbf{E}/\delta t = \delta \mathbf{D}/\delta t = \sigma \mathbf{E}$ $= \nabla \times \mathbf{H} - \delta \mathbf{D}/\delta t = \nabla^2 \mathbf{A}/\mu$
Scalar magnetic field	β	Θ^{-1}		1/rad		$\beta = (1/c^2)\mathbf{v} \cdot \mathbf{E}$ $\beta = -\nabla \cdot \mathbf{A} - (1/c^2)\delta V/\delta t$ $\beta_c = (\mu/4\pi)(q_e \mathbf{v} \cdot \hat{\mathbf{r}}/r^2$

cont

54. Visualization of Dimensional Relationships

Table 54-1. Dimensions of Physical Quantities (Cont)

Parameter	Symbol	Dimensions	Derived Dimensions	MKS Units	MKS Derived Units*	Remarks
Electromagnetism (Cont)						
Inductance	L	ΘML^2Q^{-2} $\Theta ML^2T^{-2}I^{-2}$	$\Theta^{-2}M^{-1}L^2T^2$	Henry (H) = V·s/A = Ω·s = Wb/A = kg·m²·s⁻²·A⁻²	$(m^2 \cdot s^2)/(kg \cdot rad^2)$	Voltage/(Current/Time), ΔPotential/ΔCurrent/Δt, Magnetic Flux/ Current $L = V/(dI/t) = N\phi/I = \Lambda/I$ $= \mu_0 N^2 A/l = \mu_0\mu N^2 A/l_c$ $= CV^2/I^2$
Induction constant	l_c	Dimensionless		$l_c = 37$		$l_c = S_2/S_1 \cdot \Phi^2 = 37 \cong 1/e$.
Reluctance	$\mathcal{R}$	$M^{-1}L^{-2}Q^2$	$\Theta^2ML^{-2}T^{-2}$	$s^2 \cdot A^2/(m^2 \cdot kg)$ = A/Wb	$kg \cdot rad^2/(m^2 \cdot s)$	1/Permeance
Charge-to-mass ratio	g	QM^{-1}	ΘT^{-1}	$= e/m_e = 2V\sqrt{B_r^2 r^2}$ $= \mathbf{v}/r \cdot \mathbf{B} = \mathbf{E}/r \cdot \mathbf{B}^2$	rad/sec	= 1.75820088E11 C/kg [C/kg]= [rad/sec]
Weber's inertial mass	m_W	M		kg		$= q\phi/3c^2 = qV/3c^2$
Electric field mass	m	M		kg		$= 2\mathcal{E}_E/c^2;\ \mathcal{E}_E = \tfrac{1}{2}mc^2$ $\mathcal{E}_B + \mathcal{E}_\beta = \mathbf{v}^2/c^2 \mathcal{E}_E$

cont

54. Visualization of Dimensional Relationships

Table 54-1. Dimensions of Physical Quantities (Cont)

Parameter	Symbol	Dimensions	Derived Dimensions	MKS Units	MKS Derived Units*	Remarks
Electromagnetism (Cont)						
Electrostatic energy, Electric potential energy	E_U, U_E	ML^2T^{-2}		Joules = N·m = kg·m^2/sec^2		Charge · Charge/Dist. $U = k_e Qq/r$, $U_E(r) = q\Phi(r)$, $U_E = \frac{1}{2}CV^2$
Electrostatic field energy density (volumetric)	u_E	$ML^{-1}T^{-2}$		Joules/m^3 = m^{-1}·kg·s^{-2} = N·m^{-2} = Pa		Energy/Unit Vol. $u_E = (\epsilon_0/2)\mathbf{E}^2$
EM field total energy density (volumetric)	U, η	$ML^{-1}T^{-2}$		Joules/m^3 = m^{-1}·kg·s^{-2} = N·m^{-2} = Pa		Vacuum: $U = \frac{1}{2}(\epsilon_0 \mathbf{E}^2 + \mathbf{B}^2/\mu_0)$ $U = -3P = \rho c^2$ Linear, nondispersive media: $U = \frac{1}{2}(\mathbf{E}\cdot\mathbf{D} + \mathbf{H}\cdot\mathbf{B})$ Energy momentum density = $\mathbf{E}^2 + \mathbf{B}^2 + \mathbf{E}\times\mathbf{B}$
EM radiation pressure	P, η	$ML^{-1}T^{-2}$		Joules/m^2 = kg/m·s^2 = N/m^2 = Pa		$P = F/A = -U/V = -3\rho$ $P_{rad} = \langle S \rangle /c = \Phi_E/c$ EM waves: $P = \eta/2$ $= \frac{1}{2}\epsilon_0 E_0^2 = \langle u \rangle$

cont

645

54. Visualization of Dimensional Relationships

Table 54-1. Dimensions of Physical Quantities (Cont)

Parameter	Symbol	Dimensions	Derived Dimensions	MKS Units	MKS Derived Units*	Remarks
Electromagnetism (Cont)						
Potential momentum	p	$MT^{-1}\Theta^{-2}$			m/s·rad^2	= **E** x **B** (non-reactive momentum available for extraction)
Radiative mass shift	Δm	$L^2 T^{-2}$		m^2/s^2 (=J/kg)		$\Delta m = e^2 \langle \mathbf{A}^2 \rangle / 2m_0$
Pole Strength Pole strength, magnetic	p	LQT^{-1} LI $ML^3T^{-1}Q^{-1}$	$\Theta L M T^{-2}$	Amp·meter (A·m) = C·m/s = T·m^2	m·kg/sec^2	Work/Current $p = W/I$ $p = F/\mu_0 H$
Magnetic charge	g, q_m b	$L^2\Theta^{-1}$ ML^{-1}		m^2/rad kg/m		$g = h/e$; $g = n/2(hc/e)$ where $n = 0, \pm1, \pm2,...$ $b = -e/c = -2.359E-18$ kg/m
Magnetic moment, Orbital magnetic moment	$m, \mu,$ $\mathcal{M}_L$	$L^2T^{-1}Q$ $L^2 I$	$\Theta L T^{-2} M$	Amp·meter^2 (A·m^2) N·m/T = J·T^{-1}	m^2·kg·rad/sec^2	Magnetic Pole Strength · Distance, Current · Area $\mu = I \cdot A$, $m = n i \pi r^2$, $\mu = p l$, $\mu = NIS$

cont

Table 54-1. Dimensions of Physical Quantities (Cont)

Parameter	Symbol	Dimensions	Derived Dimensions	MKS Units	MKS Derived Units*	Remarks
Electromagnetism (Cont)						
Magnetic moment (loop)	μ, j, m, p_m	$L^2T^{-1}Q$ L^2I $ML3T^{-1}Q^{-1}$	$\Theta L^{-2}T^{-2}M$	Wb/m Amp·meter^2 J·T^{-1}	m^2·kg·rad /sec^2	Charge Rotational Displacement, Current x Area $M = ml, \mu = I \cdot A$
Magnetic moment (dipole)						
Magnetic torque	τ	$L^{-2}T^{-2}Q^2$	$\Theta L^{-2}T^{-4}M^2$	(Amp/meter)^2	(kg^2·rad^2)/ (m^2·s^4)	$\tau = M \times H$
Magnetization by volume, Moment per unit volume, Magnetic dipole moment per unit volume	M	$L^{-1}T^{-1}Q$ $L^{-1}I$	$\Theta M L^{-1}T^{-2}$	Amp-turn/meter (A·N·m^{-1}) $= C \cdot s^{-1} m^{-1}$ $= JT^{-1}m^{-3}$ $= kg \cdot m^2 A^{-2} s^{-2}$	(kg·rad)/(m·s^2)	Magnetic Moment/ Volume $M = (N/V)m$ $M = m/V$ $M = \chi_v H$
Magnetization by mass	Ω, σ, M	$L^2M^{-1}QT^{-1}$ $L^2M^{-1}I$	ΘL^2T^{-2}	A·m kg^{-1}, Wb·m/kg	rad·m^3·sec^{-2}	Magnetic Moment/Mass M/m

cont

54. Visualization of Dimensional Relationships

Table 54-1. Dimensions of Physical Quantities (Cont)

Parameter	Symbol	Dimensions	Derived Dimensions	MKS Units	MKS Derived Units*	Remarks
Electromagnetism (Cont)						
Magnetic polarization, Magnetization, Intensity of Magnetization	J, I	$\Theta^{-1}MT^{-2}Q^{-1}$ $\Theta^{-1}MT^{-2}I^{-1}$	$\Theta^{-1}T^{-1}$	$T, Wb/m^2$	$1/rad \cdot sec$	$J = \mu_0 M$
Magnetic field strength, Magnetic intensity, Magnetizing force, Applied field	H	$L^{-1}T^{-1}Q$ $L^{-1}I$	$\Theta L^{-1}T^{-2}M$	Amp-turn/meter $= C \cdot s^{-1} \cdot m^{-1}$ $= N/Wb$ $= kg \cdot m^2 A^{-2} s^{-2}$	$(kg \cdot rad)/(m \cdot s^2)$	Current/Distance Charge/(Time·Length) $H = B/\mu = nE/Z_0$ $= (v \times E)/\mu_0 c^2 = (1/\mu_0)\nabla \times \mathbf{A}$ $= -\nabla \psi_m = -\text{grad } \mathcal{F}$ $H = N \times I/lc = i/2\pi r = DV$ $\mathbf{H}_m = \mathbf{B}_m/\mu \simeq \mathbf{B} \cdot \mathbf{v}/c$ $H = \sqrt{(\mathbf{H}_0^2 + \mathbf{H}_m^2)}$
Magnetic flux, Magnetic pole strength, Magnetic charge, Magnetic scalar potential	$\Phi_B, \Phi_M,$ $\Phi_m(r),$ q_m N	$ML^2T^{-1}Q^{-1}$ $ML^2T^{-2}I^{-1}$	$\Theta^{-1}L^2$	Weber (Wb) = volt·sec = J/A $= J \cdot s/C = T \cdot m^2$ $= (kg \cdot m^2) \cdot (s^{-2} A^{-1})$	m^2/rad	Magnetic Field · Area, ΔPotential · Time, ΔCurrent/Δt, $\Phi_M = h/2e = \hbar\varphi/q_0$ $\Phi_M = B \cdot A = BA\cos\theta,$ $\Phi = F_m/R_m, \phi_B = LI = \int Vdt$ $\phi_B = \oint \mathbf{A} \cdot d\mathbf{l} = E/I = h\phi_E/2E$

cont

54. Visualization of Dimensional Relationships

Table 54-1. Dimensions of Physical Quantities (Cont)

Parameter	Symbol	Dimensions	Derived Dimensions	MKS Units	MKS Derived Units*	Remarks
Electromagnetism (Cont)						
Magnetic flux density, Magnetic induction, Magnetic field (strength), Displacement density	B	MT^1Q^{-1} $MT^{-2}I^{-1}$	Θ^{-1}	Tesla (T) = Webers/meter^2 = V·s/m^2 = N/(A·m) = kg·s^{-2}·A^{-1} = J/A·m^2	1/rad	Magnetic Flux/Area Energy·Time/(Charge·Length^2) $B = \mu H = -\mu \nabla \psi_m = \nabla \times A$ $B = \mu_0(H + M) = \mu_r\mu_0 H =$ $\mu_0 H(1 + \chi_m)$, $B_m = B \cdot v/c$ $B = v/c^2 \times E = -\mu_0 \nabla \Phi_m$ $B = \sqrt{(B_0^2 + B_m^2)} = E/c$ $B = \mu_0 I/2\pi r = (v \times E)/c^2$ $\nabla \cdot B = 0$ $\nabla \times B + \nabla \beta = (1/c^2)\delta E/\delta t$
Magnetostatic energy, Magnetic potential energy	$E_m, U_B,$ $E_{p,m}$	ML^2T^{-2}		$J = N \cdot m =$ kg·m^2/sec^2		Magnetic Moment · Magnetic Field $U_B = -M \cdot B = \frac{1}{2}Li^2$ $E_m = -m \cdot B \cos\theta$
Magnetic field energy density (volumetric)	$\eta_B, u_B,$ u_m	$MT^{-2}L^{-1}$		$J/m^3 = m^{-1}\cdot kg\cdot s^{-2}$		Energy/Vol $u_B = B^2/2\mu = \frac{1}{2}HB$

cont

54. Visualization of Dimensional Relationships

Table 54-1. Dimensions of Physical Quantities (Cont)

Parameter	Symbol	Dimensions	Derived Dimensions	MKS Units	MKS Derived Units*	Remarks
Electromagnetism (Cont)						
Magnetostatic pressure	P_m	$ML^{-1}T^{-2}$		N/m^2 $= kg/m \cdot s^{-2} = Pa$		P_m = Force/Area = Energy/Vol = $-U_m/3$ $P_m = B^2/2\mu_0$
Magnetic permeability, Relative permeability, Absolute permeability, Inductivity	μ, μ_r, μ_0	MLQ^{-2} $MLT^{-2}I^{-2}$	$\Theta^{-2}M^{-1}LT^2$	Henrys/meter $H/m = Wb/(A \cdot m)$ $= Wb^2/J \cdot m$ $= m \cdot kg \cdot s^{-2} \cdot A^{-2}$ $= N/A^2$	$(m \cdot s^2)/(kg \cdot rad^2)$	Magnetic Flux Density/ Magnetic Field Strength μ = energy·(time)^2/ (charge)^2·length $\mu = B/H = \mu_0 K = \mu_0 \Gamma$ $\mu_0 = Z_0/c = 1/\epsilon_0 c^2$ $= 1.25664 \times 10^{-6}$ m·kg/C^2 $\mu_0 = 6.4498 \times 10^{-26}$ m·s^2/(kg·rad^2)
Magnetic permeance	$\mathcal{P}$, Λ	L^2MQ^{-2}	$\Theta^{-2}M^{-1}L^2T^2$	$(m^2 \cdot kg)/(s^2 \cdot A^2)$	$(m^2 \cdot s^2)/(kg \cdot rad^2)$	$\mathcal{P} = 1/\mathcal{R} = \phi/NI = \mu A/l$
Magnetic reluctance	$\mathcal{R}$, R_m	$M^{-1}L^{-2}Q^2$ $M^{-1}L^{-2}T^2I^2$	$\Theta^{-2}ML^{-2}T^{-2}$	inverse Henry $(H^{-1}) \cdot m \cdot H^{-1}$ Amp·turn/Weber $s^2 \cdot A^2/(m^2 \cdot kg)$	$(kg \cdot rad^2)/(m^2 \cdot s^2)$	Circuit length/ permeability x area $l_m / \mu \cdot A_m$ $\mathcal{R} = \mathcal{F}/\Phi_B$

cont

54. Visualization of Dimensional Relationships

Table 54-1. Dimensions of Physical Quantities (Cont)

Parameter	Symbol	Dimensions	Derived Dimensions	MKS Units	MKS Derived Units*	Remarks
Electromagnetism (Cont)						
Magnetomotive force, Magnetic potential delta	$\mathcal{F}, F_m$, MMF	$Q\Theta \over TI$	ΘMT^{-2}	Ampere-turns $(A \cdot N)$ = J/Wb	$kg \cdot rad/sec^2$	Current × No. of Turns $\mathcal{F} = NI = \Phi_B \mathcal{R} = \int H_t dl$
Magnetic susceptibility, bulk (volume)	χ_v, χ_b, κ	$L^{-2}T^{-1}Q$	$\Theta ML^{-2}T^{-2}$	H/m = Wb/(A·m)	$[kg \cdot rad/m^2 \cdot s^2]$	$\chi_v = $ **M/H**
Magnetic susceptibility, Magnetic mass	χ_ρ, κ_ρ, χ_m	$L^3 M^{-1}$	$m^3 kg^{-1}$ = m^3/mol = H·m²/kg			$\chi_\rho = $ Vol/mol
Magnetic susceptibility, molar	χ_M, κ_{mol}	$L^3 mol^{-1}$		$m^3 mol^{-1} = m^3$/mol = H·m²/mol		$\chi_M = \mu_r - 1$
Electric permittivity, Relative permittivity, Absolute permittivity Capacitivity	ϵ, ϵ_r, ϵ_0	$M^{-1}L^{-3}T^2Q^2$ $M^{-1}L^{-3}T^{-4}I^2$	$\Theta^2 ML^{-3}$	farads/meter (F/m) = $C^2/n \cdot m^2$ = $m^{-3} \cdot kg^{-1} \cdot s^4 \cdot A^2$ = $C \cdot V^{-1} \cdot m^{-1}$	$kg \cdot rad^2/m^3$	Electric Flux Density / Electric Field Strength $\epsilon_0 = 1/\mu_0 c^2, \epsilon = \epsilon_0 \Gamma$ $\epsilon_0 = 8.854 \times 10^{-12} C^2/n \cdot m^2$ = $1/(4\pi k_c)$ $\epsilon_0 = 1.725 \times 10^8 kg \cdot rad^2/m^3$

cont

54. Visualization of Dimensional Relationships

Table 54-1. Dimensions of Physical Quantities (Cont)

Parameter	Symbol	Dimensions	Derived Dimensions	MKS Units	MKS Derived Units*	Remarks
Electromagnetism (Cont)						
Resistance	R	$ML^2T^{-1}Q^{-2}$ $ML^2T^{-3}I^{-2}$	$\Theta^{-2}M^{-1}L^2T$	Ohm $(\Omega) = V/A =$ $m^2 \cdot kg \cdot s^{-3} \cdot A^{-2}$	$(m^2 \cdot s)/(kg \cdot rad^2)$	Voltage/Current $R = V/I = \rho \cdot L/A$
Resistivity	ρ	$ML^3T^{-1}Q^{-2}$ $ML^3T^{-3}I^{-1}$	$\Theta^{-2}M^{-1}L^3T$	Ohm·meter $(\Omega \cdot m)$ $= m^3 \cdot kg \cdot s^{-3} \cdot A^{-2}$	$(m^2 \cdot s)/(kg \cdot rad^2)$	Resistance · Length/Area $\rho = RL/A$
Impedance	Z	$ML^2T^{-1}Q^{-2}$ $ML^2T^{-3}I^{-2}$	$\Theta^{-2}M^{-1}L^2T$	Ohm $(\Omega) =$ Volt/Amp	$(m^2 \cdot s)/(kg \cdot rad^2)$	$Z = V/I = R + jX$ $Z = \sqrt{R^2 + (X_l - X_c)^2}$ $Z = \sqrt{(L/C)} = \sqrt{(\mu_0/\epsilon_0)}$
Wave Impedance Transverse Electric	Z_{TE}	$ML^2T^{-1}Q^{-2}$ $ML^2T^{-3}I^{-2}$	$\Theta^{-2}M^{-1}L^2T$	Ohm $(\Omega) =$ Volt/Amp	$(m^2 \cdot s)/(kg \cdot rad^2)$	$Z = \mu\omega/ck = k_0/k\sqrt{(\mu/\epsilon)}$ $k_0 = (1/c)\omega\sqrt{(\mu\epsilon)}$ $Z_{TE} = Z_0/\sqrt{(1 - v_c/v)^2}$
Wave Impedance Transverse Magnetic	Z_{TM}	$ML^2T^{-1}Q^{-2}$ $ML^2T^{-3}I^{-2}$	$\Theta^{-2}M^{-1}L^2T$	Ohm $(\Omega) =$ Volt/Amp	$(m^2 \cdot s)/$ $(kg \cdot rad^2)$	$Z = ck/\epsilon\omega = k/k_0\sqrt{(\mu/\epsilon)}$ $k_0 = (1/c)\omega\sqrt{(\mu\epsilon)}$ $Z_{TM} = Z_0\sqrt{(1 - v_c/v)^2}$
Wave guide cutoff frequency	ν_c	T^{-1}		Hz		$\nu_c = c\sqrt{(m/2a)^2 + (n/2h)^2}$ m, n = mode indices a = height, h = width

cont

54. Visualization of Dimensional Relationships

Table 54-1. Dimensions of Physical Quantities (Cont)

Parameter	Symbol	Dimensions	Derived Dimensions	MKS Units	MKS Derived Units*	Remarks
Electromagnetism (Cont)						
Impedance of free space, Radiation impedance, Intrinsic impedance	Z_0	$ML^2T^{-1}Q^{-2}$ $ML^2T^{-3}I^{-2}$	$\Theta^{-2}M^{-1}L^2T$	$kg \cdot m^2/s \cdot Coul$	$(m^2 \cdot s)/(kg \cdot rad^2)$	$Z_0 = \sqrt{(\mu_0/\epsilon_0)} = \mu_0 c = 1/\epsilon_0 c$ $= n \cdot |E|/|H|$ $= 4\pi \cdot 10^{-7} \, H^{-1} \cdot c$ $= 1.92 \times 10^{17} \, m^2 \cdot s/kg \cdot rad^2$ $= 376.73031 \, \Omega$
Spacetimeimpedance	Z_S	MT^{-1}		$= kg/s$		$Z_S = c^3/G = m_P/\omega_P = F_P/c$ $= 4.0302E35 \, kg/s$
Reactance, Inductive	X_L	$ML^2T^{-1}Q^{-2}$ $ML^2T^{-3}I^{-2}$	$\Theta^{-2}M^{-1}L^2T$	Ohm (Ω)	$(m^2 \cdot s)/(kg \cdot rad^2)$	[Ang. Freq. / Inductance]/ $X_L = 2\pi \cdot f \cdot L$
Reactance, Capacitance	X_C	$ML^2T^{-1}Q^{-2}$ $ML^2T^{-3}I^{-2}$	$\Theta^{-2}M^{-1}L^2T$	Ohm (Ω)	$(m^2 \cdot s)/(kg \cdot rad^2)$	1/[Angular Freq. /Capacitance]/ $X_C = \frac{1}{2}\pi \cdot f \cdot C$
Admittance	Y	$M^{-1}L^{-2}TQ^2$ $M^{-1}L^{-2}T^3I^3$	$\Theta^2ML^{-2}T^{-1}$	Siemens (S), mhos ($1/\Omega$) = A/V $= kg^{-1} \cdot m^{-2} \cdot s^3 \cdot A^2$	$(kg \cdot rad^2)/$ $(m^2 \cdot sec)$	1/Impedance $Y = 1/Z = G + jB$

cont

54. Visualization of Dimensional Relationships

Table 54-1. Dimensions of Physical Quantities (Cont)

Parameter	Symbol	Dimensions	Derived Dimensions	MKS Units	MKS Derived Units*	Remarks
Electromagnetism (Cont)						
Conductance	S, G	$M^{-1}L^{-2}TQ^2$ $M^{-1}L^{-2}T^3I^2$	$\Theta^2ML^{-2}T^{-1}$	Siemens (S) $= 1/\Omega$ = mhos = A/V $= m^{-2} \cdot kg^{-1} \cdot s^3 \cdot A^2$	$(kg \cdot rad^2)$ $/(m^2 \cdot sec)$	Current/Voltage $G = 1/R = I/V$
Conductivity	σ	$M^{-1}L^{-3}TQ^2$ $M^{-1}L^{-3}T^3I$	$\Theta^2ML^{-3}T^{-1}$	$(ohm \cdot m)^{-1} = 1/\rho =$ Siemens/m (S/m) $= m^{-3} \cdot kg \cdot s^3 \cdot A^2$	$(kg \cdot rad^2)$ $/(m^3 \cdot sec)$	1/Resistivity $\sigma = 1/\rho$ $\sigma = \mathbf{J}/\mathbf{E} = \mathbf{P}/\mathbf{E}^2$
Electric Power, Apparent Power, Real Power	P	ML^2T^{-3} $ML^2T^{-3}I^{-1}$		$V \cdot A$ = Watt (W) $= J \cdot s$ $= kg \cdot m^2/s^3$		Work/Time Energy/Unit Time $P = V \times I = I^2R = dE/dt$
Reactive Power, True Power	VAR, P	ML^2T^{-3}		$kg \cdot m^2/s^3$		$P = VI \cos\theta,$ θ = power factor
Work function	W	ML^2T^{-2}		$J = kg \cdot m^2 \cdot s^{-2}$		$W = -e\phi - E_v$
Energy flux	**S**	MT^{-3}		$J/m^2 \cdot s = kg \cdot s^{-3}$		$\mathbf{S} = c^2\mathbf{g} = c^2\mathbf{p} = \mathbf{v} \cdot \rho_e = \mathbf{v} \cdot \mathbf{u}$

cont

54. Visualization of Dimensional Relationships

Table 54-1. Dimensions of Physical Quantities (Cont)

Parameter	Symbol	Dimensions	Derived Dimensions	MKS Units	MKS Derived Units*	Remarks
Electromagnetism (Cont)						
Magnetic potential, Magnetic scalar potential, Magnetomotance	ψ_m, ϕ_m, Ω	QT^{-1} I	$\Theta Q T^{-1}$	$A = J/Wb = C/s$	$(kg \cdot rad)/(sec^2)$	Magnetic Field Strength · Distance $\psi_m = \frac{1}{2} \mathbf{r}_1 \times \mathbf{H}$ $\psi_m = \mathbf{m} \cdot \mathbf{r}/4\pi r^3$ $= m \cdot \cos\theta / 4\pi r^2$ $\phi_m = h\phi_E/2E = I\,\Omega/4\pi$
Magnetic vector potential, Electrodynamic potential	**A**	$MLT^{-1}Q^{-1}$ $MLT^{-3}I^{-1}$	$L\Theta^{-1}$	Joule/Amp·m $=$ volt·sec/m $=$ Wb/m $=$ T·m $=$ N/A $=$ kg·m·s^{-2}·A^{-1} $=$ kg·m/Coul·sec $=$ (kg·m/sec)/Coul	m/rad	Magnetic Flux Density · Distance, Magnetic Field Strength/Unit Charge, Force/Unit Current, Momentum/Unit Charge $\mathbf{A} = (\mu_0 i / 4\pi)\oint d\mathbf{l}/r = A' - \phi_m$ $\mathbf{A} = \epsilon_0 \mu_0 \mathbf{v}\phi = \mathbf{v}\phi/c^2 = L\mathbf{I}/l$ $\mathbf{A} = \frac{1}{2} \mathbf{B} \times \mathbf{r}; \mathbf{A} = (A, jV/c)$ $\mathbf{A} = \mathbf{p}/q - \hbar \mathbf{k}/q = e/\hbar \mathbf{k}$ $\mathbf{A} = \mu_0/4\pi = Q\mathbf{v}/4\pi\epsilon_0 c^2$
Retarded magnetic vector potential	**A***					
Electrokinetic field	$\mathbf{E}_k$	$MLT^{-2}Q^{-1}$	$LT^{-1}\Theta^{-1}$	$=$ volt·sec/m/sec $=$ volts/m	m/(rad·s)	$\mathbf{E}_k = -d\mathbf{A}^*/dt$

cont

54. Visualization of Dimensional Relationships

Table 54-1. Dimensions of Physical Quantities (Cont)

Parameter	Symbol	Dimensions	Derived Dimensions	MKS Units	MKS Derived Units*	Remarks
Electromagnetism (Cont)						
Lorentz force	F_L	MLT^{-2}		$N = kg \cdot m/sec^2$		$F_L = q^+(v \times B)$ RH rule
Electrodynamic (potential) momentum	p_{el}, Q	MLT^{-1}		$= kg \cdot m \cdot s^{-2}$		Charge-Vector Potential $p_{el} = q_0 A$
Total momentum, canonical momentum	p	MLT^{-1}		$= kg \cdot m \cdot s^{-2}$		$p = \hbar k/v = p_{el} + p_{mv}$ $= q_0 A + mv; p = \hbar k + eA$
Total energy	E, E_t, H	ML^2T^{-2}		$Joule (J) = N \cdot m$ $= kg \cdot m^2/sec^2$		$E = \hbar\omega + e\phi = H = T + V$ $= P.E. + K.E.$ $= \pm\sqrt{((pc)^2 + (mc^2)^2)}$ $= m(E^2 + c^2B^2)$
Poynting vector	S	MT^{-3}		$W/m^2 = J \cdot s^{-1} \cdot m^{-2}$ $= kg \cdot s^{-3}$		Electric Field Strength x Magnetic Field Strength $S = (1/\mu_0) E \times B = E \times H$
Magnetic field gradient	$\delta B/\delta x$	$ML^{-1}T^{-1}Q^{-1}$ $ML^{-1}T^{-2}I^{-1}$	$\Theta^{-1}L^{-1}$	$T \cdot m^{-1}$ $= kg \cdot m^{-1} \cdot s^{-2} \cdot A^{-1}$	$1/rad \cdot m$	ΔMagnetic Flux Density/Distance
Momentum density	p, g	$ML^{-1}T^{-1}$		$kg/s \cdot m^2$		$g = (1/c^2) E \times H = \epsilon_0 E \times B$

cont

54. Visualization of Dimensional Relationships

Table 54-1. Dimensions of Physical Quantities (Cont)

Parameter	Symbol	Dimensions	Derived Dimensions	MKS Units	MKS Derived Units*	Remarks
Electromagnetism (Cont)						
Momentum flux	f	$ML^{-2}T^{-1}$		$kg \cdot m^{-2} s^{-1}$		$f = I \cdot g = (v/c^2)S$
Magnetic quadrupole moment	Q	$L^3T^{-1}Q$ L^3	$\Theta L^3 T^{-2} M$	$m \cdot J \cdot T^{-1}$ $m^3 \cdot A$	$m^3 \cdot rad \cdot kg/sec^2$	Magnetic Dipole · Distance
Gyromagnetic ratio, Magnetogyric ratio	γ	QM^{-1} $M^{-1}TI$	ΘT^{-1}	$rad \cdot s^{-1} \cdot T^{-1}$ $= Hz \cdot T^{-1}$ $= C \cdot kg^{-1}$ $= kg^{-1} \cdot s \cdot A$	rad/sec	Magnetic Moment / Angular Momentum $\gamma = 2\pi f/B = q/2m$
Radiometric						
Radiant energy	Q_e	ML^2T^{-2}			$J \ (= W \cdot s)$	EM radiation energy
Radiant energy density	U, w_e	$ML^{-1}T^{-2}$		$J/m^3 \ (= W \cdot s/m^2)$	Radiant Energy /Unit Vol.	Energy Current Density
Flux	ϕ	ML^2T^{-3}		J/s $= W$		Energy flow rate $\phi = dQ/dt; Q_e = $ radiant, $Q_v = $ luminous, $Q_q = \gamma$

cont

54. Visualization of Dimensional Relationships

Table 54-1. Dimensions of Physical Quantities (Cont)

Parameter	Symbol	Dimensions	Derived Dimensions	MKS Units	MKS Derived Units*	Remarks
Radiometric (Cont)						
Irradiance, Illuminance, Radiant flux density	I_i, E, E_e, E_v	MT^{-3}		$W/m^2 = kg \cdot s^{-3}$ (= dose rate)		Radiant Energy /(Area x Time). Power Incident/Unit Area $I = \frac{1}{2} E_0 H_0 = n/2Z_0 \|E_0\|^2$
Radiance, Luminance	L, R	$MT^{-3}\Theta^{-1}$		$W/(m^2 \times steradian) = (W/m^2/sr)$		Energy flux density/solid angle. Irradiance/Unit Solid Angle
Radiance	L_e, $L_{e\Omega}$	WT^{-3}		$W \cdot sr^{-1} \cdot m^{-2} = kg \cdot s^{-3} \cdot sr^{-1}$		Radiant Power / Area/ Solid Angle
Radiant Intensity, Luminous Intensity	I_e, $I_{e\Omega}$, I_v J	ML^2T^3		$W/steradian$ = (W/sr)		Energy flux/solid angle. Power Radiated/Unit Solid Angle $I = d\phi/d\omega$
Fluence, Radiant energy fluence, Radiant exposure	Ψ, Q, H_e	MT^{-2}		$J/m^2 = kg/s^2$ (= dose)		Radiant energy/Area

cont

54. Visualization of Dimensional Relationships

Table 54-1. Dimensions of Physical Quantities (Cont)

Parameter	Symbol	Dimensions	Derived Dimensions	MKS Units	MKS Derived Units*	Remarks
Radiometric (Cont)						
Radiant flux, Radiant power, Luminosity	F, ϕ, ϕ_e P	ML^2T^{-3}		Watts (W) $= J/s$ $= m^2 \cdot kg/s^3$		Radiated energy/Unit Time, Radiant Energy Flow Rate, Radiated Pwr $\phi = dQ/dt$
Spectral energy density	μ_v, υ_s $\rho_0(\omega)$ ρ_{SED}	$ML^{-1}T^{-1}$		$J/m^3/s$ $= kg/m \cdot s$ $= Pa/Hz$		Energy density/freq mode. Spectral energy/Unit Vol., [Spectral power/unit area]/flow speed $\mu_v = 1/c \int I \nu d\Omega$ $\rho_{SED} = \int (2\hbar\omega^3/c^3)d\omega$
Radiant incidance	E_e, H	ML^{-2}		W/m^2		Flux/Unit Area $E = df/dA$
Flux density	ϕ_v, ϕ_λ	MT^{-2}		$W \cdot m^{-2} \cdot s^{-1}$		Flux/Unit Area/Unit Time $\phi = dQ/dA/dt$
Spectral irradiance, Radiant incidence, Irradance	$E_{e,\lambda}$	$ML^{-1}T^{-3}$		W/m^3		Irradiance of a surface/wavelength

cont

54. Visualization of Dimensional Relationships

Table 54-1. Dimensions of Physical Quantities (Cont)

Parameter	Symbol	Dimensions	Derived Dimensions	MKS Units	MKS Derived Units*	Remarks
Radiometric (Cont)						
Spectral irradiance (cont)	$E_{e,\nu}$	MT^{-2}		$W \cdot m^{-2} \cdot Hz^{-1}$		Irradiance of a surface/Unit freq.
Spectral radiance	$L_{e\lambda}, L_{e,\Omega,\lambda}$	$ML^{-1}T^{-3}$		$W \cdot sr^{-1} \cdot m^{-3}$		Power/Steriadian/Vol
	$L_{e\nu}, L_{e,\Omega,\nu}$	MT^{-2}		$W \cdot sr^{-1} \cdot m^{-2} \cdot Hz^{-1}$		Power/Sterradian/Area/Freq
Spectral power, spectral flux	$\phi_{e,\lambda}$	MLT^{-3}		$W \cdot m^{-1}$		Radiant Power /Wavelength
	$\phi_{e,\nu}$	ML^2T^{-2}		W/Hz		Radiant flux/unit freq.
Spectral intensity	$I_{e\lambda}, I_{e\Omega\lambda}$	MLT^{-3}		$W \cdot sr^{-1} \cdot m^{-1}$		Radiant Intensity/Wavelength
	$I_{e\nu}, I_{e\Omega\nu}$	ML^2T^{-2}		$W \cdot sr^{-1} \cdot Hz^{-1}$		Radiant Intensity /Unit Freq.
Radiosity	J_e	MT^{-3}		W/m^2		Radiant flux

cont

54. Visualization of Dimensional Relationships

Table 54-1. Dimensions of Physical Quantities (Cont)

Parameter	Symbol	Dimensions	Derived Dimensions	MKS Units	MKS Derived Units*	Remarks
Photometric (Optics)						
Luminous energy	Q_v	TJ		lumen·sec (lm·s) = cd·sr·s = talbots		Luminous Flux x Time, Luminosity x Time
Luminous energy density	U_v	$L^{-2}TJ$		lumen·sec/m² (lm·s)		Luminous Energy/Area
Luminous energy density (volumetric)	ω_v	$L^{-3}TJ$		lm·s·m⁻³		Luminous Energy/Vol
Luminous flux, Luminous power	$F_v, \phi_v, \Phi, \Phi_v$	ML^2T^{-3} J		lumen (lm) = cd·m² = cd·sr (Ref 555 nm)		Energy/Unit Time, Radiated Power, Luminosity/Solid Angle
Irradiance, Illuminance, Luminous incidence	E, E_v	$L^{-2}J$ cd·sr·m⁻²		lux (lx) = lm·m⁻² = cd·sr·m⁻²		Photometric flux /Unit Area, Visible flux density, Luminous Flux/Area $I = \frac{1}{2} E_0 H_0 = (n/2Z_0)\|Z_0\|^2$ $E_v = I_v/d^2$

cont

54. Visualization of Dimensional Relationships

Table 54-1. Dimensions of Physical Quantities (Cont)

Parameter	Symbol	Dimensions	Derived Dimensions	MKS Units	MKS Derived Units*	Remarks
Photometric (Optics) (Cont)						
Radiance, Luminance, Sterance	L_v	$kg \cdot s^{-3} \cdot sr^{-1}$		lumens/m² /steradians (lm/m²/sr)		Illuminance /Unit Solid Angle $L = dI/dA$
Luminosity, Radiant Intensity, Luminous Intensity	I, I_v	cd J		candela (cd) = lumens/ sr (lm/sr)		Visible Power Radiated /Unit Solid Angle, Luminous Flux /Solid Angle; $I_v = E_v \cdot d^2$
Luminance, Brightness	L_v	$L^{-2} J$		cd/m² = nits		Luminosity/Area, Candles/Area

Logic and truth are two very different things, but they often look the same to the mind that's performing the logic. – Theodore Sturgeon

We can assume that when we go to the quantum theory.... we have so a model where the basic physical entity is a line of force. – P.A.M. Dirac

Group theory... will never be of any use in physics. – Sir James Hopwood Jeans

There is, however, another possible road through matter [interpenetration], - a tortuous and badly signposted road, for it leads us into the fourth dimension. – A.C. Clarke

To say that in the presence of large bodies space becomes curved is equivalent to stating that something can act on nothing. I, for one, refuse to subscribe to such a view. – Nikola Tesla

54. Visualization of Dimensional Relationships

54.3 Graphical representations

The distribution of energy follows definite paths which may be studied by geometric construction.
– Samuel Coleman

Numbers are the sources of form and energy in the world. – Theon of Smyrna

The chessboard is the world, the pieces are the phenomena, and the rules of the game are what are called the laws of nature. The player on the other side is hidden from us. – Thomas Henry Huxley

Vision is the art of seeing what is invisible to others. – Johnathan Swift

The examples of physical quantities illustrated in Table 54-1 are shown loosely grouped by category, however, the overall ontological organization is not readily apparent. What sort of organizational structure may we peruse to categorize and depict relationships between physical quantities and concepts to decode the nature of the universe? Mechanical dimensions may be depicted in a directed graph form using dimensions of length L, mass M, Time T and plane angle θ such as illustrated in Fig 54-1. With four dimensions, the derived dimensions may be represented as matrix elements $[L^a\ T^b\ \Theta^c\ M^d]$ which may be projected on an n-dimensional hypercube such as shown in Fig. 54-2 with each vertex corresponding to a physical quantity. An illustration of a matrix array and a hypercube of an ordered topology of regular occurring cells depicting relationships of example physical quantities in terms of the derived dimensions is shown in Fig. 54-3. Translation along an edge on the base manifold corresponds to a multiplication or division of a vertex parameter by a matrix element (e.g., multiplication of Momentum P [MLT^{-1}] by a Length l [L] equals Action S [ML^2T^{-1}]; division of Action S [ML^2T^{-1}] by an angle θ [Θ] equals Angular Momentum L [$\Theta^{-1}ML^2T^{-1}$]). Similarly, translation along a line in the tangent space corresponds to multiplication or division by a physical constant, c, G or h depending on the axis direction (e.g., multiplication by the Gravitational constant G converts an inertial frame in SR to an accelerated frame in GR; division by Planck's constant h converts an expression in QFT to SR. The incorporation of the velocity of light c converts an expression in Newtonian mechanics to SR. Increasing the number of dimensions by multiplying an expression in QM with 3 degrees of freedom (DOF) by n allows conversion to QFT with nDOF. The Euler number χ = v – e + f where v = no. of vertices, e = no. of edges and f = no. of faces. In the hypercube, for the base manifold, χ = 16 – 32 + 20 = 4 and for the tangent manifold χ = 12 – 24 + 8 = –4. According to the Buckingham Π theorem, an equation involving "n" variables may be equivalently rewritten as an equation of "n–m" dimensionless parameters where "m" is the number of fundamental dimensions. For example, equations involving electric charge Q as a separate dimension may be expressed in equivalently mechanical dimensions ΘMT^{-1} where Θ represents spin precession rate.

In mathematics, there are only four elementary arithmetical operations, i.e., addition, subtraction, multiplication and division with addition being the inverse of subtraction and multiplication being the inverse of division. Operations may be combined to yield arithmetic, geometric or power series. Nested ratios of the Fibonacci recursive series result in a convergent limit known as the golden ratio, Phi: ϕ = 1 + 1/[1 + 1/[1 + 1/[1 + ...]]] = (1 ± √5)/2 = 0.6180339... Successive operations of nested fractions give rise to the base of natural logarithms: e = 1/0! + 1/1! + 1/2! + 1/3! + ...1/n! = 2.718281828.... which may be interrelated to the irrational number π to yield Euler's identity $e^{i\pi}$ = –1 = Φ^{-1} – Φ = cos π + i sin π. Similarly, $e^{i2\pi}$ = 1 = Φ – ϕ = 1.6180339...–0.6180339...

54. Visualization of Dimensional Relationships

Fundamental dimensions are described by whole integer numbers. Fractal (Hausdorf) dimensions have non-integer values. The fractal dimension is a measure of complexity. For example, the Cantor set has dimension log2/log3 ≅ 0.6309 and the triadic Koch curve has dimension log4/log3 ≅ 1.2618. Fractal curves are characterized by self-symmetry under magnification and have a Hausdorf dimension greater than their topological dimension. The Euclidean or topological dimension D_T is 0 for sets of points, 1 for sets of lines, 2 for sets of surfaces, and 3 for sets of volumes. Such curves may be generated with polygons as initiators and displaced segmented edges recursively iterated as generators.

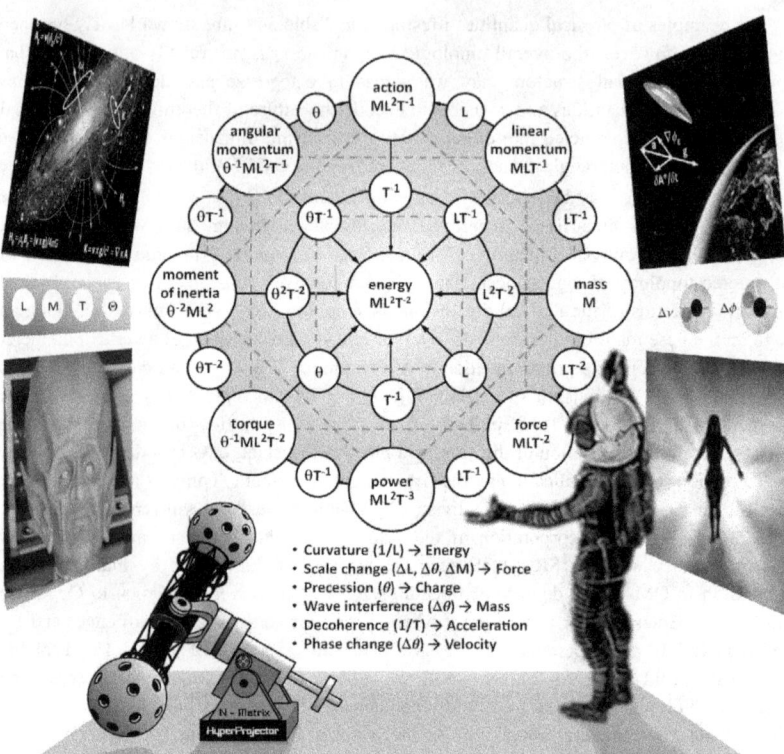

Fig. 54-1. Dimensions and conversions of mechanical quantities expressed in basic dimensions of length L, mass M, time T and rotation angle θ. The dimensions shown in the smaller circles are conversion multipliers to obtain quantities shown in the larger circles. The arrows indicate the order of multiplication. For example, to convert mass with dimension [M] (kg) to energy with dimension [ML^2T^{-2}] (Joules), the multiplier has dimensions of [L^2T^{-2}] (m²/s²) which corresponds to potential energy/unit mass (J/kg).

Dimension regulated the general scale of the work, so that the parts may all tell and be effective.
— *Vitruvius*

54. Visualization of Dimensional Relationships

Dimensional matrix represented on a 4-D hypercube

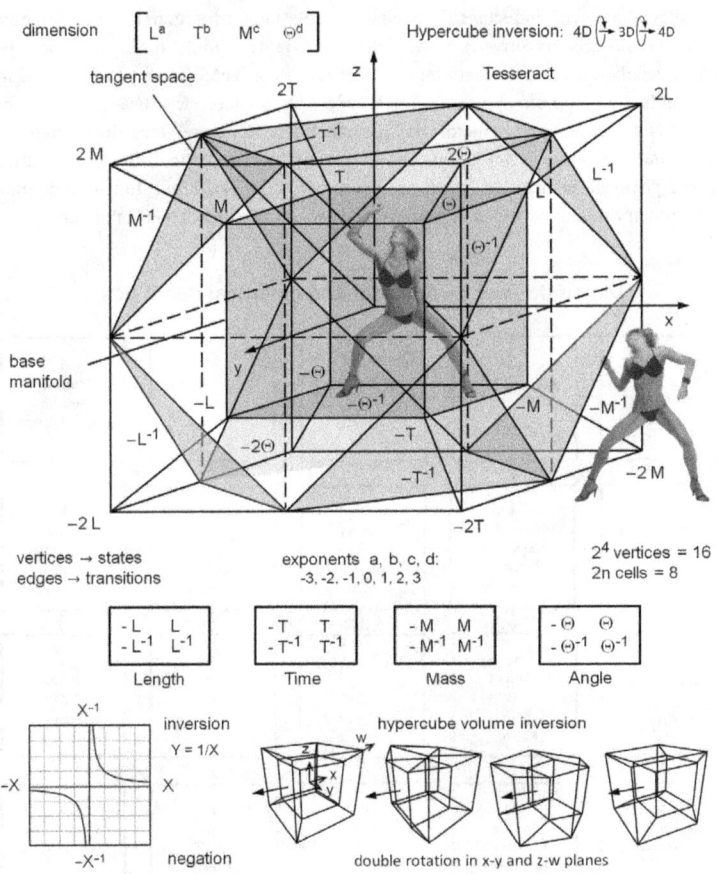

Fig. 54-2. A 4-D hypercube illustrating a base manifold of matrix elements $[L^a\ T^b\ \Theta^c\ M^d]$ and a tangent manifold of reciprocal space for a, b, c, d = −1 and 1. Mechanical quantities are identified with vertices with energy corresponding to the axis origin. Translation along edges in a positive direction corresponds to multiplication or division in a negative direction by the respective fundamental dimensional quantities. Inversion involves unfolding such that the initial internal trivector volume envelops the external enclosed volume. Revolution in 3-D of the inverse function y = 1/x yields a pseudospherical figure with infinite surface area, but finite volume known as Toricelli's trumpet (aka Gabriel's horn). The converse of Toricelli's hyperbolic solid is a surface of revolution with finite surface area, but an infinite volume. A 4-D hypercube network resembles a maximally stiff 3-D octonion Fano cube cellar structure composed of triangular plate elements (triads).

Imaginary time is a new dimension, at right angles to ordinary, real time. – Stephen Hawking

Great design is making something memorable and meaningful. – Dieter Rams

54. Visualization of Dimensional Relationships

Opposing vertices in the hypercube represent duals. Examples of duality include positive/negative curvature, inside/outside, source/sink, plus/minus topological charge, resistance/conductance, inductance/capacitance, rotational/irrotational flow, top/bottom, right/left, front/back, positive/negative electric charge, north/south magnetic poles, coherence/decoherence, potential/kinetic energy, positive/negative torsion, positive/negative poles, compression/tension, convergence/divergence, etc. Duality, a pairing of opposites, is a fundamental principle in mathematics and nature. This dual symmetry of complementary variables is ubiquitous and a defining characteristic of the known universe. Conserved physical quantities of matter or energy are connected with four-fold continuous global symmetry transformations in space or time according to Noether's theorem.

Dimensions of mechanical and electrical quantities

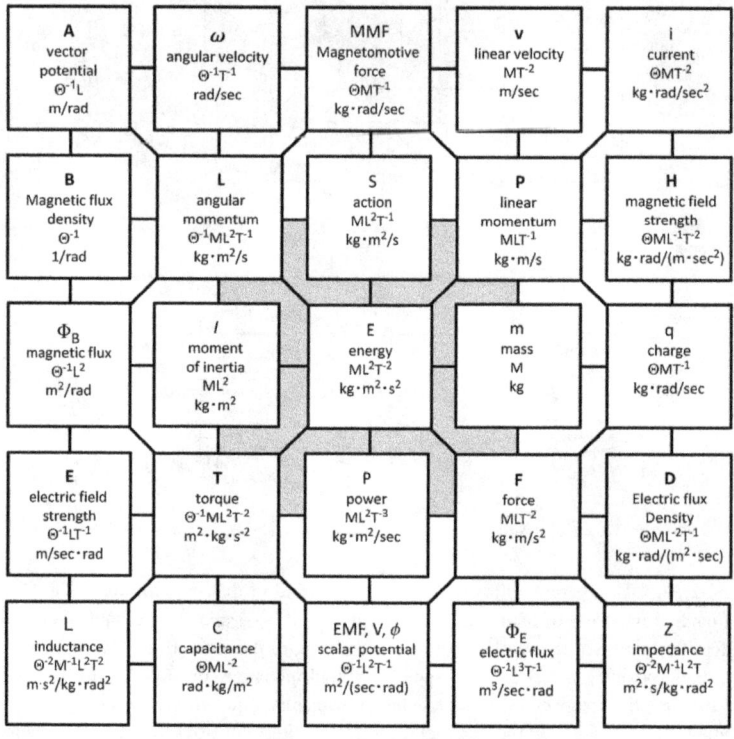

Fig. 54-3. An example 5 x 5 matrix formed from a 4-dimensional array illustrating conversion relationships of various physical quantities including electric charge in terms of derived [$L^a\ T^b\ \Theta^c\ M^d$] MKS dimensions. Other patterning of unfolded cutting planes may be generated from the 8-fold symmetry planes by box pleating methods such as in technical origami or sequenced rotations as in a Rubik's cube to reveal group symmetries of physical objects. For example, mass = |volts·charge| = |volts·volts·capacitance| = force/acceleration.

Mon royaume est de la dimension de l'univers... On m'appelle la Science. – G. Flaubert

54. Visualization of Dimensional Relationships

The matrices formed from a 4-dimensional array resemble the Brillouin zones formed by a square reciprocal lattice with reciprocal lattice vectors. As shown in Fig. 54-4, energy corresponds to the first Brillouin zone denoted by the darkened square. The four points defining a square whose sides touch the corners of the inscribed square indicate the second Brillouin zone. Similarly, the eight points describing a square whose sides touch the second square mark the third Brillouin zone consisting of eight triangular elements. The patterning is suggestive of an organizing method of dimensional elements of a 4-D hypercube.

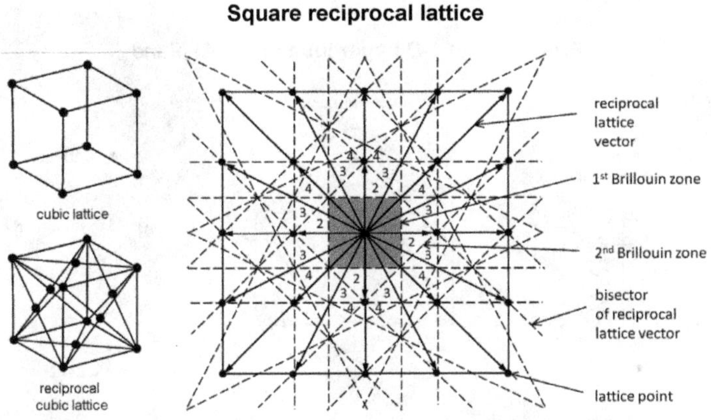

Square reciprocal lattice

Fig. 54-4. Illustration of a square reciprocal lattice with reciprocal lattice vectors.

A 4-D hypercube (tesseract) projected on a 2-D plane forms a hexahedron also known as a Metatron cube. The geometry as shown consists of 13 equal circles (12 circles surrounding 1 central circle) connected by lines from the center of each circle with a total of 13 vertices. Projections of the 5 Platonic solids (cube, icosahedron, tetrahedron, octahedron, dodecahedron) are encoded within connected points or nodes. Dividing the vertices according to the golden ratio Φ forms various triangles and hexagons. Various symmetry groups may be connected via Fano planes. In the Fano plane (7,3,2) design, each design contains 3 points. An example application is the Hamming (7,4) code to correct single bit binary errors with 3 parity bits and 7 total bits which is essentially equivalent to the Fano plane structure. The Hamming distance is the shortest length connecting any two bit strings. As another example, a triangular symmetry group in the form of a directed graph provides a representation of a multiplication of basis elements. See Fig. 54-5. Each facet corresponds to a geometric algebra bivector. The base manifold cube is defined by the interior grouping of circles. Higher dimension polyhedral structures represent allowed space of conformal field theories (CFT) with vertices that correspond to critical junction phase transitions. The corners may be described by state vector correlation functions describing conformal transitions such as particle collision outcome probabilities.

Algebra is generous; she often gives more than is asked of her. – Jean D'Alembert

All the modern higher mathematics is based on a calculus of operations... – Mary Boole

Communicative algebra is a lot like topology, only backwards. – John Baez

54. Visualization of Dimensional Relationships

A Metatron cube diagram provides an example of a Boolean logic lattice of system attributes. An example is a color lattice describing logical relations between primary colors (red, green, blue) used in additive color mixtures and process colors (yellow, magenta, cyan) used in subtractive color processes. The lattice diagram illustrates all of the possible between attributes using Boolean logic AND, OR, NOT operators. Refer to Fig. 54-6. A similar diagram illustrating relation of particle electric charge symmetry is shown in Fig. 54-7. An illustration of relationship of various electrical quantities represented as graph manifolds and geometric algebra expressions is depicted in Fig. 54-8.

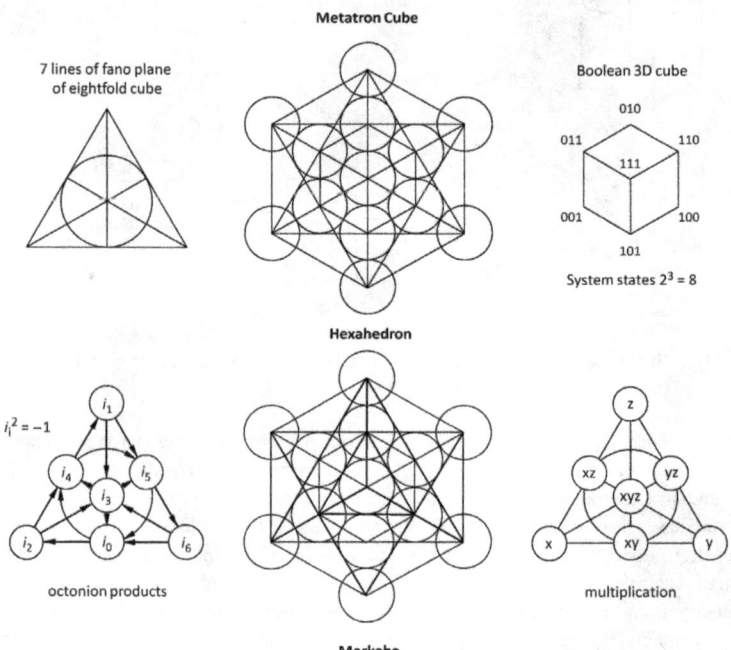

Fig. 54-5. An alternate representation of a 4-dimensional array is that of a hexahedron (Metatron's cube) containing various 2-dimensional images of Platonic solids which includes an embedded hypercube. A hexahedron corresponds to a 2D representation of a simple cubic Bravais crystalline lattice with equal sides (a = b = c) and equal interior angles ($\alpha = \beta = \gamma = \pi/2$). There are 14 different Bravais types forming 7 lattice arrangements used to categorize atomic crystalline structures. Miller indices (*hkl*) indicate crystal planes while [*hkl*] indicate crystal directions. Boolean algebraic logic may be represented as arrangements of squares or cubes. For two variables, the possible combinations of IF, AND, NOT operations results in 16 Boolean squares. Similarly, for three variables, the combinatorial space corresponds to 256 Boolean cubes which reduce to 14 basic arrangements once rotations and reflections are taken into account.

Matter's lattical waves are spaced at intervals corresponding to the frets on a harp or guitar, with analogous sequence of overtone from each fundamental. – Armstrong Gerhard Christian

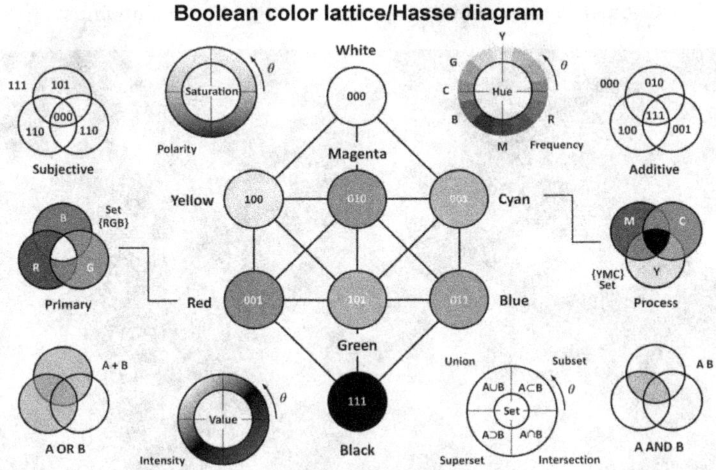

54. Visualization of Dimensional Relationships

Boolean color lattice/Hasse diagram

Fig. 54-6. Boolean color attribute logic lattice describes logical AND/OR operations and relations between primary light colors (RGB) and process reflection colors (YMC). Operations remain invariant in discontinuous optical illusions as in a Necker cube.

Stacked tetrahedrons

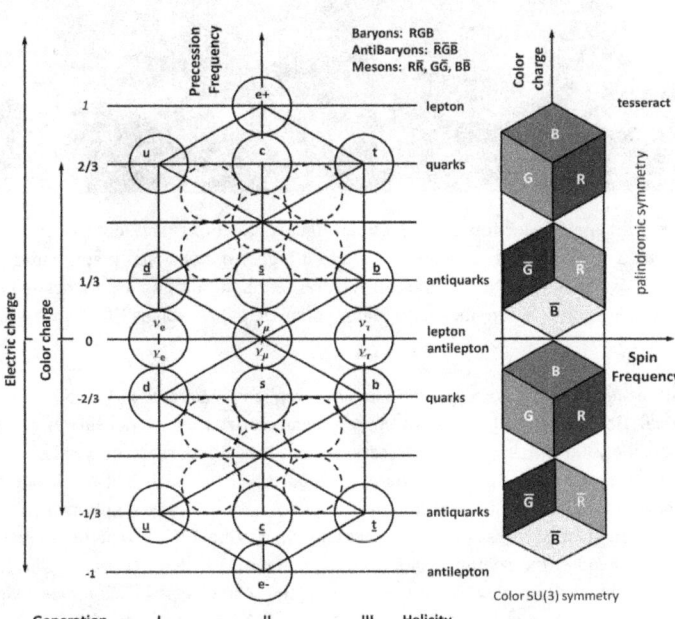

Fig. 54-7. The arrangement of fractional and integral electric charges of leptons/anti-leptons and quarks/anti-quarks shown as vertices of four tetrahedrons in a stacked reflected cube arrangement corresponding to a spin rotation and spin precession network.

669

54. Visualization of Dimensional Relationships

Geometric representation of electromagnetic relationships

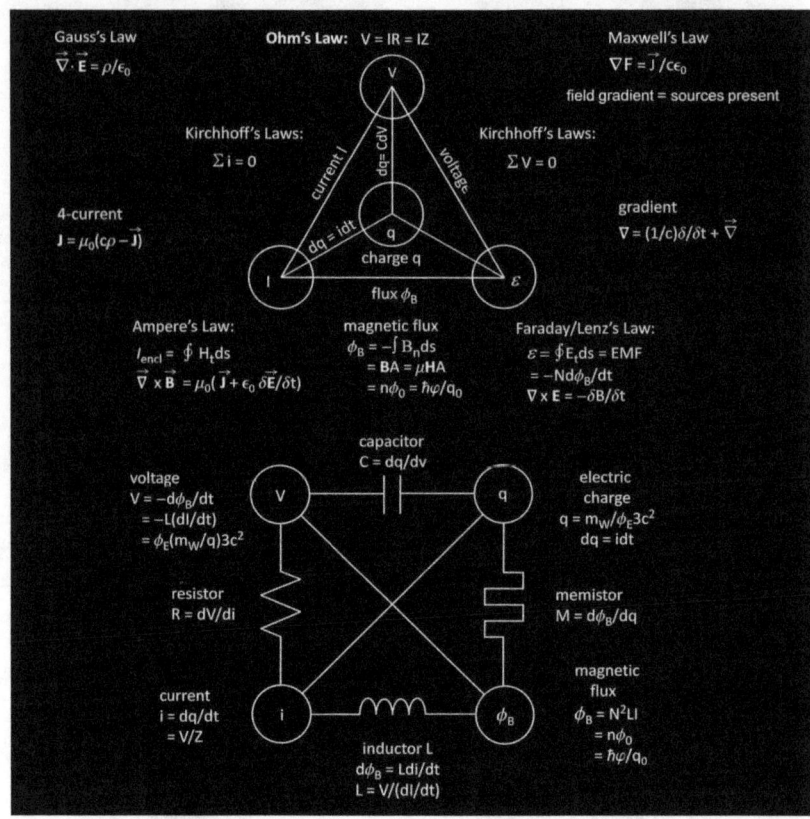

Fig. 54-8. Example electromagnetic circuit theory relations and components. Such diagrams are suggestive of a rearrangement into a higher dimensional graph with addition of Gauss's law, Maxwell's laws, Poisson relation, etc and additional parameters such as energy, mass, velocity, action, magnetic vector potential, etc as well as EGM analogs.

A 16-dimensional array corresponding to a hypercube is illustrated in Fig. 54-9. A three-dimensional Boolean cube has $2^3 = 8$ corners characterized by 3 coordinates whereas a four-dimension Boolean hypercube (tesseract) has $2^4 = 16$ corners or vertices. A 4D quaternion inversion $z \mapsto 1/z$ turns the hypercube inside out such that the inside proper volume becomes larger than the outside coordinate volume. Increase in volume corresponds to a decrease in quantum pressure and momentum. Squeezing a wave function in real space increases its spread in momentum space. A wave function ψ (= $\mathbf{E} + i\mathbf{H}$) may be inverted in a transform from the complex z-plane to the w-plane where $w = z^{-1}$. Complex mappings corresponding to Lorentz transformations in 4D spacetime are Möbius transformations.

The act of choosing a representation for a problem involves the specification of a space where the search for solution can take place. – Anon.

54. Visualization of Dimensional Relationships

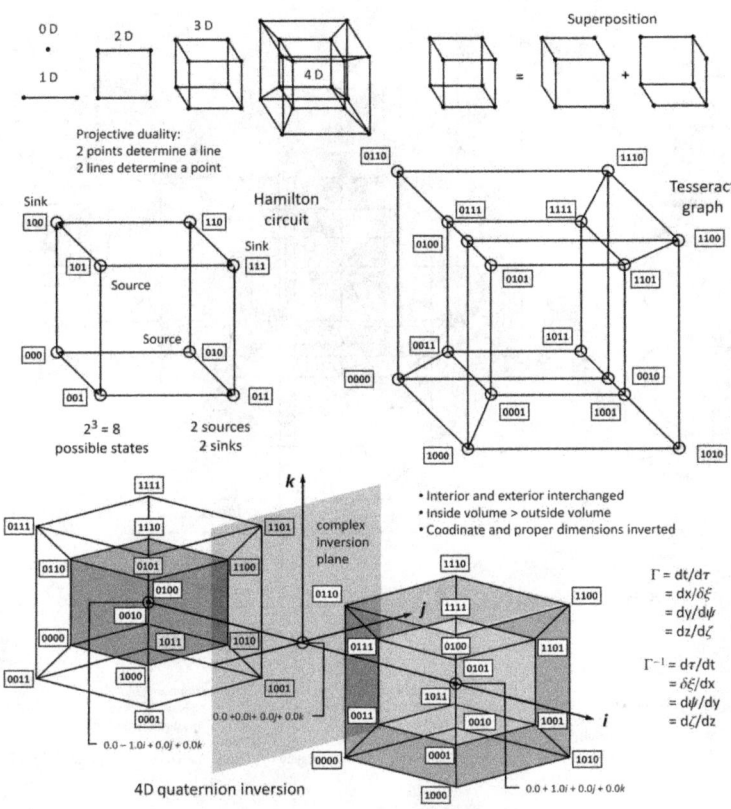

Fig. 54-9. Boolean four-dimensional hypercube tesseract network and inversion lattice. Graph nodes represent mathematical operations while graph edges represent multidimensional data arrays communicated as tensor flow interchanges between nodes.

Identifying symmetry planes of a hypercube with physical constants provides a geometrical representation of various physical theories in a Bronstein Penrose cube with extension into higher n-dimensional space. See Fig. 54-10. As shown the base manifold unit cell exhibits an 8-fold symmetry about each orthogonal axis. Lines or edges describe connections between item vertices. Translation along an orthogonal axis corresponds to multiplication or division depending on direction by the respective physical constant α, c, G, k_C, h, q and Φ_0. Categorization of physical theories, e.g., SR, GR, NM, QM, QED, QFT, EGM, QM, etc, is represented by which of the constants c, G, or h is incorporated in the theory. The addition of c adds relativity, addition of G adds gravity, addition of h adds energy quantization. Addition of the fine structure constant α and Coulomb's constant k_C adds electromagnetism. Increasing the number of dimensions from 4D to 5D in spacetime as in the Kaluza-Klein theory by addition of electric charge q, a spin precession effect, extends geometrical representation of physical theories to include incorporation of Maxwell's electromagnetism and expressible via geometric algebra using bivectors..

671

54. Visualization of Dimensional Relationships

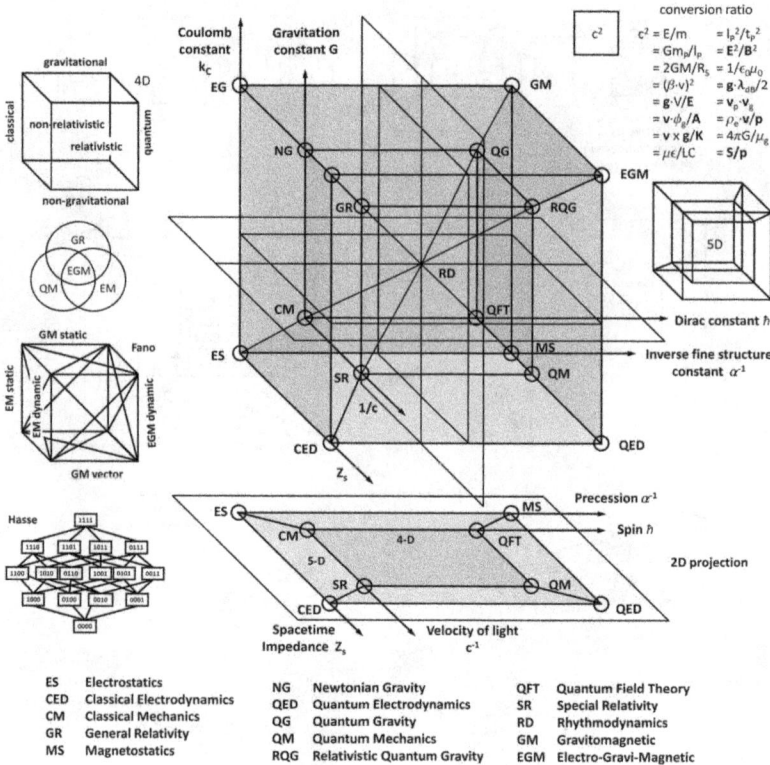

Fig. 54-10. Illustration of a Bronstein Penrose cube represented as a hypercube in higher dimensional space encompassing electromagnetism which depicts symmetries and interrelationships of Newtonian Mechanics (NM), Newtonian Gravity (NG), Special Relativity (SR), General Relativity (GR), Electrostatics (ES), Magnetostatics (MS), Classical Electrodynamics (CED), Rhythmodynamics (RD), Quantum Electrodynamics (QED), Quantum Mechanics (QM), Quantum Field Theory (QFT), Quantum Gravity (QG) and Electro-Gravi-Magnetics (EGM). Orthogonal axes correspond to physical constants velocity of light c, gravitation constant G, Dirac constant $\hbar$, Coulomb's constant k_C and fine structure constant α. Addition of electric charge (spin precession) adds a fifth dimension.

After the relativistic quantum theory is created, the task will be to develop the next part of our scheme, that is to unify quantum theory (constant h), special relativity (constant c) and the theory of gravitation (constant G) into a single theory. – Matvei P. Bronstein

We can move in the other three [dimensions]: Up, down, forward, sideways, backward. But when it comes to time, we are all prisoners. – H. G. Wells (The Time Machine)

All mathematical truths are relative, conditional. – Charles Proteus Steinmetz

When in a dilemma, introduce novelty. – Stewart Brand

54. Visualization of Dimensional Relationships

54.4 Creation of the universe

It was a place without a single feature of the space-time matrix that he knew. It was a place where nothing yet had happened – an utter emptiness. There was neither light nor dark: there was nothing here but emptiness. – Clifford D.Simak

I don't believe in God but I'm very interested in Her. – Arthur C. Clarke

One man's theology is another man's belly life. – Robert Heinlein

We are the miracle of force and matter making itself over into imagination and will. Incredible...The Universe has shouted itself alive. – Ray Bradbury

There was no 'before' the beginning of the universe, because once upon a time there was no time. – John Barrow

In the beginning the universe was created. This has made a lot of people very angry and been widely regarded as a bad move. – Douglas Adams

Listen: there's a hell of a good universe next door: let's go. – E.E. Cummimgs

The universe at large and on the quantum scale is characterized by duals – a pairing of opposites in the ultimate zero sum game. Thought, language and perception without dualism is unthinkable. The present degenerate state of duality provides a clue as the preexisting state of the universe. This duality may be represented as a pairing of sets resembling winding of a Cornu spiral of opposite curvature. The creation of the universe is compable to the deterministic generation of surreal numbers of the form $x = \{L|R\}$ in a set sequence $S_0 = \{0\}$, $S_1 = \{-1 < 0 < 1\}$, ... The sequence of paired sets X_L and X_R is started by taking both to be the empty set $\{\phi\}$ where, $0: = \{\phi|\phi\}$. Unlike Nero's fiddling due to indifference to outcomes or the opposite case of Buridan's logic paradox of reasoned preference determinism, vacuum instability may be a case of the inevitability principle – Anything that is not forbidden is compulsory. The curious dualism in life and nature is abstractly symbolized in Fig. 54-11. This duality forms an essential organizing principle of dimensional and physical quantities and may suggest a causal origin of the universe. The creation of the universe from a void may be described in terms of the generation of surreal numbers where every number corresponds to the sum of two sets of previously created numbers. The state of nothingness or vacuum state ϕ is represented as the null set $\{\ :\ \}$.

Interaction energy corresponds to union of sets, e.g. $\{-1\} \cap \{0\} \cap \{1\}$. The evolutionary sequence in iterative steps may be represented in relational steps of sets $\{\ L: R\}$ as

Step No.	Set	Sequence						
0	$\{\ :\ \}$				0			
1	$\{0:\ \}$		-1				1	
2	$\{0|1\}$	-2		-1/2		1/2		2
ω	$\{0,1,2,3...|\phi\}$	...	...	...	...	...	...	...
$\omega+1$	$\{\omega|\ \}$	...	...	...	...	...	...	...

Cardinality is a measure of the size of a set $N = \{0, 1, 2, 3, 4, 5,...n\}$. The transfinite sequence of cardinals numbers includes: $0, 1, 2, 3, ..., n,...; \aleph_0, \aleph_1,...\aleph_2,...\aleph_\alpha,...$ The real, countably infinite or transfinite numbers is represented as ω. The number sequence includes paired sets of opposite polarity R_n, $-R_n$ and inverse sets R_n^{-1}, $-R_n^{-1}$. At the end of aleph-null $\aleph_0$ steps (smallest transfinite cardinal numbers) corresponding to the countably rational numbers, a "big bang" occurs with an explosion of reals. At $\omega+1$, one gets infinity $+1$ and infinity -1 which corresponds to an uncountable ordinal number set.

54. Visualization of Dimensional Relationships

A surrealistic world in a universe of dualism

Fig. 54-11. The universe exhibits a fundamental dualism of opposite pairings in all aspects of life and nature even in a surrealistic world of thought and imagination. This dualism may be represented as successive paired sets of surreal numbers in the form x = {L|R}.

I began to lose heart. The appalling desert of darkness and barren fire, the huge emptiness so sparsely pricked with scintillations, the colossal futility of the whole universe, hideously oppressed me. – William Olaf Stapledon

Inner space is so much more interesting, because outer space is so empty. – Theodore Sturgeon

But what a universe, anyhow! No use blaming human-beings for what they were. Everything was made so that it had to torture something else... Here and there some speck of a planet dominated by some half-awake intelligence like humanity. – William Olaf Stapledon

Is it credible that our world should have two futures? ... Two entirely distinct futures lie before mankind, one dark, one bright; one the defeat of all man's hopes, the betrayal of all his ideals, the other their hard-won triumph. – William Olaf Stapledon

So do flux and reflux – the rhythm of change – alternate and persist in everything under the sky. – Thomas Hardy

When you reach the stars, boy, yes, and live there forever, all the tears will go, and Death himself will die. – Ray Bradbury

You may have the universe if I may have Italy. – Giuseppe Verdi

It may be that the old astrologers had the truth exactly reversed, when they believed that the stars controlled the destiny of men. The time may come when men control the destinies of stars. – Arthur C. Clarke

Is it not impossible that we shall, in the course of our travels, meet other intelligent creatures far more worthy than man to rule the universe? – A. E. van Vogt

54. Visualization of Dimensional Relationships

The creation of the universe may be modelled in a geometric metaphor in terms of catastrophe theory begining with a flat, featureless void corresponding to the empty set { | }. A fold in a catastrophe manifold produces a cusp. The trajectory path from the singularity event may diverge smoothly or exhibit a sudden catastrophic change in state jumping from a fold curve to another in a Cornu spiral. The projection of the generation of surreals exhibits a cusp geometry illustrating the sudden expansion of infinite and infinitesimal ordered subsets of number pairings of equivalence classes. The bifurcation set of surreal numbers includes an infinite subset of real numbers represented as $\{L_n|R_n\}$ and dyadic rationals including transfinite numbers and infintitesimals. Refer to Fig. 54-12.

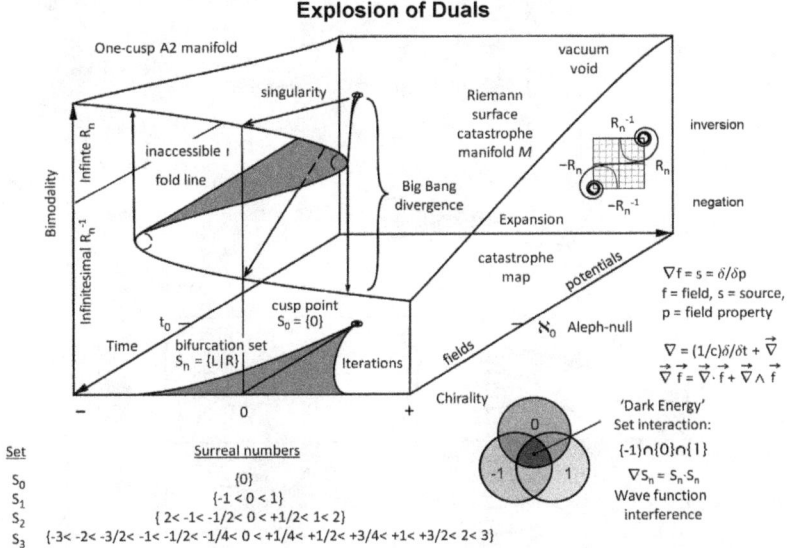

Fig. 54-12. Castastrophe representation of vacuum state { | } instability with decomposition into an explosion of surreal infinite and infinitesimal dyadic pairings. The cusp catastrophe consists of two folds linked at a topological singularity at the point of divergence. At the end of Aleph-null iterations, the universe explodes into being in a Big Bang from a ordered, coherent state of low mode numbers and high energy per mode into an unordered, incoherent state of high mode numbers and low energy per mode. A closed universe has zero net energy. The negative gravitational energy equals the positive mass-energy ($\Sigma mc^2 = 0$). Dark energy driving expansion represents set interaction interference.

Evolution of the universe may be represented as a catastrophe machine of 2^n frequency modes in (u, x, t) state space. As illustrated in Fig. 54-13, the Big Bang singularity corresponds to a degenerate critical point on a catastrophe manifold M subject to a phase shift perturbation (Devil's pitchfork rotation) resulting in symmetry-breaking into two stable equilibrium points of opposite sign. The catastrophe manifold corresponds to a potential surface of u evolving in time t in *n* iterative steps over mode expansion range x. State points within the catastrophe cusp, corresponding to the bifurcation set, represent a region of instability where sudden discontinuous state transitions can occur. The Lagrangian provides energy for mode expansion. At time t = 0 marking the Big Bang

54. Visualization of Dimensional Relationships

singularity event, the vacuum degenerates from a single critical point S_0 into a bimodality of paired opposites in a succession of iterative states S_n. Energy density is proportional to the square of the field $\nabla f = f \cdot f$. Entropy varies with the volume of state phase space as $k_B \log V_{ph}$ where k_B = Boltzman's constant (= $E/T = \hbar/\Delta t \Delta T$).

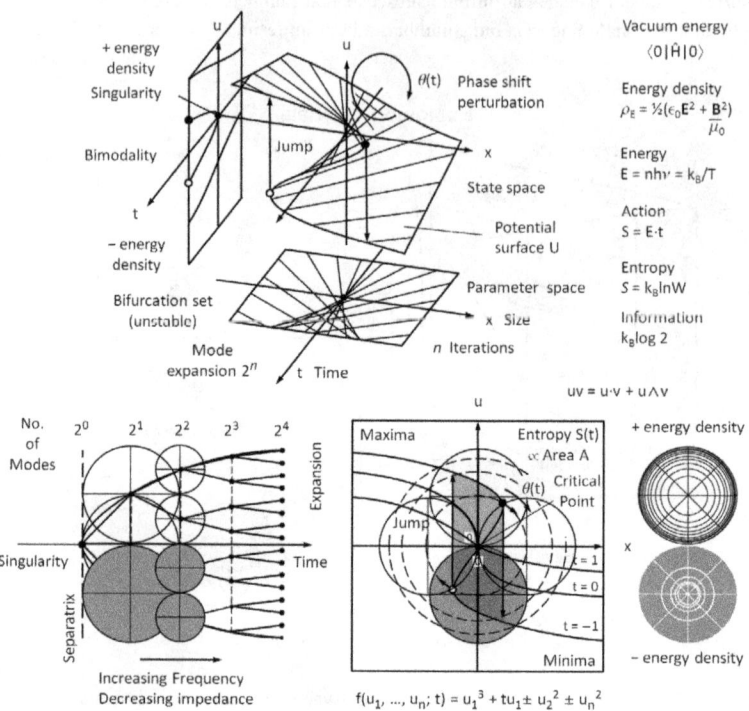

Fig. 54-13. Expansion of the universe modeled as bimodal binary progression of mode/frequency states from an initial unstable (degenerate) vacuum singularity. As the universe evolves with time, the number of modal states and information bits equivalent to the number of paired sets increases while the entropy corresponding to the area enclosed by the bifurcation set in state space increases. Mode expansion is in the direction of the interaction energy gradient. Complexity arises as pairings couple in groups of characteristic sizes such as formation of photons and electrons from Planck mass dipoles, protons and neutrons from aggregates of electrons, atoms of coupled groupings of electrons, protons and neutrons, molecules, planets, stars, galaxies, clusters, superclusters, etc.

Even in the realm of things which do not claim actuality, and do not even claim possibility, there exist beyond dispute sets which are infinite. – Bernard Bolzano

We all have our time machines, don't we. Those that take us back are memories....And those that carry forward, are dreams. – H.G. Wells

What's past is prologue. – William Shakespeare

54. Visualization of Dimensional Relationships

Increasing modal density of frequencies is graphically portrayed in Fig. 54-14 as nested Riemann spheres with a zero point centered on the complex tangent plane. Each nested sphere represents a discrete scale mass grouping. A nested Apollonian sphere packing resembles a three dimensional Smith chart of the complex s-parameter plot of normalized electrical impedance $z_t = Z_T/Z_0 = \sqrt{[(R/Z_0)^2 + (X/Z_0)^2]}$ where Z_0 equals characteristic impedance of the transmission line and Z_T equals the terminating load impedance. A degenerate singularity corresponds to an open circuit. In analogy to the Smith chart, the real and complex axes may be represented in terms of normalized Planck impedance Z_P (= $1/4\pi\epsilon_0 c$ = 29.9724558 Ω). Mass represents an impedance to electromagnetic wave energy flow. For a given mass impedance Z_m, the normalized impedance is $z_m = Z_m/Z_P$. For comparison, the vacuum impedance for electromagnetic waves $Z_0 = \sqrt{\epsilon_0\mu_0} = \mu_0 c = 120\pi$ Ω = 376.99 $\Omega \cong 4\pi Z_P$ which is a measure of antenna radiation resistance. Electromagnetic waves are characteristically much larger in wavelength than the Planck wavelength ($l_P = \sqrt{\hbar G/c^3} = ct_P = 1.616199 \times 10^{-35}$ m) and, hence, see Planck masses as point obstructions. The impedance of a black hole as a perfect absorber of radiation is a perfect match to that of free space and is equal to 120π Ω. The Big Bang singularity in analogy to the 3D Smith chart corresponds to a perfect reflector or white hole and represents a perfect impedance mismatch. Energy flow is in the direction of decreasing impedance.

Planck vacuum impedance analogy

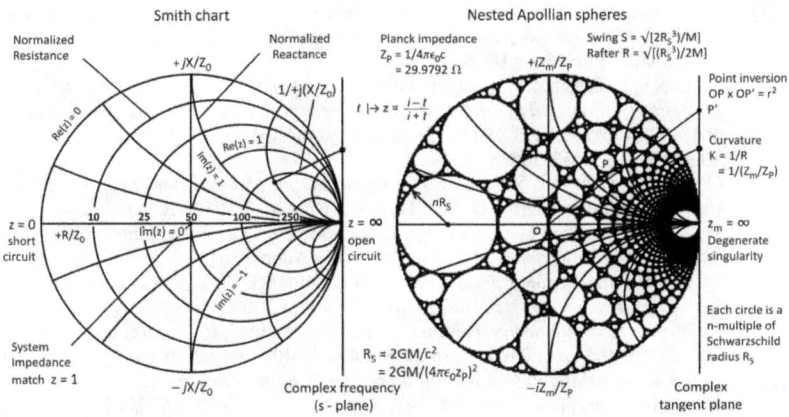

Fig. 54-14. Hierarchical arrangement of infinity of *n*-multiples of the Schwarzschild radius R_S as an illustration of expansion of the universe. The expansion represented as nested Riemann spheres of complex Planck impedance is analogous to a 3D Smith chart of complex electrical impedance. A mapping of points on the sphere onto the complex plane corresponds to inversion of the sphere by stereographic projection. Any affine plane together with a single point at infinity forms a Möbius plane or inversion plane. The nested configuration is an example of Möbius transformations (automorphisms) on to the hyperbolic plane D. Like "suds in the kitchen sink" mass accretion and contraction result in formation of bubble voids similar to that apparent in the universe at large.

Geometry existed before the creation. – Plato

It's all in there, implicit in those equations. – Isaac Asimov [Foundation]

References

[1] Geoffrey Hunter, Marian Kowalski and Camil Alexandres-cu, "The Bohr Model of the Photon", Proceedings of SPIE, Vol 5866: *The Nature of Light: What is a Photon?* (August 2005). SPIE, Bellingham, WA http://spie.org/SPIEDigitalLibrary.

[2] Daniele Funaro, "From Photons to Atoms The Electromagnetic Nature of Matter", arXiv.1026.3110v1. [physics.gen-ph], 7 May 2012, Department of Mathematics, University of Modena and Reggio Emilia.

[3] Daniele Funaro, "A Full Review of the Theory of Electromagnetism", Cornell University Library, arXiv.org>physics>arXiv:physics/0505068, (May 2005).

[4] Daniele Funaro, "Electromagnetism and the Structure of Matter", World Scientific, Singapore, 2008.

[5] M. W. Evans, *The Spinning and Curving of Spacetime: The Electromagnetic and Gravitational Fields In The Evans Unified Field Theory Paper*, M.W. Evans, http://www.aias.us/documents/uft/a15thpaper.

[6] M.W. Evans, *Generally Covariant Unified Theory, The Geometrization of Physics*, Springer,van der Merwe Series.

[7] Friedwardt Winterberg, *The Planck Aether Hypothesis: An Attempt for a Finitistic Non-Archimedean Theory of Elementary*, C.F. Gauss Academy of Science Press (January 1, 2002).

[8] Paramahamsa Tewari, "On the Space-Vortex Structure of the Electron", *What is the Electron?* /ed. V. Simulik, C. Roy Keys Inc. (Montreal: Apeiron 2005).

[9] V. L. Dyatlov, *Polarization Model of the Inhomogeneous Physical Vacuum*, Series: Problems of the Inhomogeneous Physical Vacuum, Novosibirsk Institute of Mathematics Publishing House 1998.

[10] M. Urban, F. Couchot, X. Sarazin and A. Djannati-Atai, *The quantum vacuum as the origin of the speed of light*, Cornell University Library, arXiv.org [physics.gen-ph]1302.6165v1 (21 Feb 2013); Euopean Physical Journal D, DOI 10.1140/epjd/e2013-30578-7.

[11] Gerd Leuchs and Luis Sanchex-Soto, *A sum rule for charged elementary particles*, European Physical Journal D, DOI 10.1140/epjd/e2013-30577-B.

[12] W. C. Daywitt, The Planck Vacuum, *Progress in Physics*, Vol 1, January 2009, *Galilean Electrodynamics*, Vol. 21, No. 4, July/August 2010.

[13] W. C. Daywitt, Magnetic and Faraday Fields as Planck Vacuum Responses, *Galilean Electrodynamics*, Vol 21, No. 5, Mar/Apr 2009.

[14] P. Tuisku, T. Pernu, and A. Annila, "In the light of time", *Proceedings of the Royal Society* Vol 465 Issue 2104, April 2009. DOI. 10.1098/rspa.2008.0494].

[15] Ernest J. Sternglass, *Before the Big Bang The Origins of the Universe and the Nature of Matter*, Four Wall and Eight Windows, New York, NY, 2001.

[16] Oleg D. Jefimenko, *Electromagnetic Retardation and Theory of Relativity*, 2nd ed. (Electret Scientific Company, Star City WV, 2000).

[17] Oleg D. Jefimenko, *Causality, Electromagnetic Induction and Gravitation*, 2nd ed. (Electret Scientific Company, Star City WV, 2004).

[18] Ivor Catt, Displacement Current, ivor catt.co.uk/x31192.htm (1978), *Wireless World* (Dec 1978) p 51/2

[19] Robert L. Oldershaw, Towards A Resolution Of The Vacuum Energy Density Crisis, Arxiv.org/pdf/ 0901.3381[physics-gen-ph], (Jan 2009)

[20] John A.Macken, "The Universe is Only SpaceTime," Website: www.onlyspacetime.com 2010, 2011.

[21] Carver A. Mead, *Collective Electrodynamics: Quantum Foundations of Electromagnetism*, (MIT Press, Cambridge MA, 2002).

[22] Jaroslav G. Klyushin, A Short Comment on Dimensionality, *Galilean Electrodynamics*, Vol. 23, Special Issues No. 1, p. 2, (Spr 2012).

cont

References

[23] Jaroslav G. Klyushin, Mechanical Dimensions for Electrodynamic Quantities, *Galilean Electrodynamics*, Vol. 11, Special Issues No. 5, p. 90 & 96, (2000).

[24] Jaroslav G. Klyushin, Fundamental Problems in Electrodynamics and Gravidynamics, Updated English Language edition, *Galilean Electrodynamics*, 2009.

[25] Morton E. Spears, An Electrostatic Solution for the Gravity Force and the Value of G, *Galilean Electrodynamics* Vol. 21, No. 2, 23-32, Mar/Apr 2010.

[26] W. Michaelis, H. Haars, and R. Augustin, "A new precise determination of Newton's gravitational constant", Metrologica Vol 32, No. 4,267, (1995).

[27] H. A. Wilson, "An Electromagnetic Theory of Gravitation", *Physical Review*, Vol XVII, No. 1, *Phys. Rev.* 17 54-59, (1921).

[28] R. H. Dicke, "Gravitation Without a Principle of Equivalence", *Rev. Mod. Physics*, **29**, 363-376 (1957).

[29] Puthoff, H. E, "Polarizable Vacuum (PV) Approach to General Relativity", *Foundations of Physics* **32**, 927-943 (2002).

[30] Puthoff, H. E, and Ibison, Michael, Polarizable Vacuum "Metric Engineering" Approach to GR-Type Effects, *Proc. Gravitational Wave Conf.*, MITRE Corp., McLean, VA (2003).

[31] Joseph G. Depp, Polarizable Vacuum and the Schwarzschild Solution, *Foundations of Physics*, April 14,2005, PACS numbers 03.03.+p, 03.50.De,04.20.Cv,04.25.Dm,04.40.Nr,04.50.+h quantum.site.nfoservers.com/Joe/PV.pdf. www.scribd.com/doc/17130549/Polarizable-Vacuum-PV-and-the-Schwarzschild-Solution.

[32] Krogh, Kris, "Gravitation without Curved Space-Time", NeuroScience Research Institute, U. of California, Santa Barbara, arXiv:astro-ph/9910325 v23 4 (July 2006).

[33] Todd J. Desiato, Riccardo C. Storti, *Electrogravimagnetics: Practical Modeling Methods of the Polarizable Vacuum, Physics Essays*, Vol 19, Issue 1, pp. 151-158, (March 2006). ISSN 0836-1398 http://dx.doi.org/10.4006/1.3025775.

[34] G.S. Diemer, R.C. Storti, *Quinta Essentia A Practical Guide to Space-Time Engineering Part 1-4*, Delta Group Engineering (dgE), 2nd ed., 2009, Melbourne, Australia, ISBVN 978-1-4092-0534-9, www.deltagroupengineering.com.

[35] Albert Einstein, "The Foundation of the General Theory of Relativity", Translated from "Die Gundlage der allegemeinen Relativitätstheorie", *Annalen der Physik*, 49, 1916. Reproduced in *The Principle of Relativity*, H.A. Lorentz, A. Einstein, H. Minkowski and H. Weyl, Dover, New York (1952).

[36] Yuri N. Ivanov, *Rhythmodynamics*, 2nd ed., (M:IAC Energia Publishing House, Moscow, Ru, 2007, ISBN 978-5-98420-018-9) Websites: www.rhythmodynamics.com, http://mirit.ru/rd_2007en.htm.

[37] Yuri N. Ivanov, "Compression of Standing Waves, Rhythmodynamics and the Third Condition of Rest", *Proceedings of the 1996 International Scientific Conference on New Ideas in Natural Sciences*, Anatoly Smirnov, ed. , Alexander Frolov, ed., PIK Publishing Co., Moscow RIA (1996), Website: www.worldsci.org.

[38] Gabriel LaFreniere, "Matter is made of waves", Website: web.archive.org/web/ 20110711095644/ http:glafreniere.com/matter/.htm/Alpha Transformations (2011).

[39] R. C. Jennison and A.J. Drinkwater, "An Approach to the Understanding of Inertia from the Physics of the Experimental Method", J. Phys Math Gen 10 (1977).

[40] R. C. Jennison, "Relativistic phase-locked cavities as particle models", *J. Phys. A: Math Gen* Vol 11, No. 8 (1978).

[41] R. C. Jennison, "The formation of charge from a traveling electromagnetic wave by reduction of the effective velocity of light to zero", J. Physics A: Math Gen 15(2) 405-408 (1982).

cont

References

[42] Alexandre A. Martins and Mario J. Pinheiro, "The Connection Between Inertial Forces and the Vector Potential", arix.org/physics/0611/0611167.
[43] Bernhard Haisch, Alfonso Rueda, and H.E. Puthoff, "Inertia as a Zero-Point Field Lorentz Force", *Physical Review A*, Vol 49, No. 2, Feb. 1994.
[44] A. H. Compton, "The Size and Shape of the Electron", *Phys. Rev.* **14** (1), 20-43 (1919).
[45] H. Stanley Allen, The Case for a Ring Electron, *Proceedings of the Physical Society of London*, V**31**, N1, pp 49-68 (Dec 1918).
[46] Llewellyn H. Thomas, "Motion of the spinning electron", *Nature*, vol. **117**, 514, (1926).
[47] R. C. Jennison, "A New Classical Model of the Electron", *Physics Letters A*, Vol 141, No. 8.9, 377-382 (20 Nov. 1989).
[48] David L. Bergman and J. Paul Wesley, "Spinning Charged Ring Model of Electron Yielding Anomalous Magnetic Moment", *Galilean Electrodynamics*, Vol. 1:5, p63-67 (Sept/Oct 1990), Arlington, MA 02476.
[49] David.L. Bergman, *"Theory of Forces"*, Physical Interpretations of Relativity Theory, London, Sept 1998, Preprint: *Proceedings of Common Sense Science*. (http:www.cormedia.com/css).
[50] David Hestenes, "The Zitterbewegung Interpretation of Quantum Mechanics," *Foundations of Physics,* 20, 1213 (1990).
[51] J. G. Williamson and M.B. van der Mark, "Is the electron a photon with toroidal topology?", *Annales de la Fondation Louis de Broglie*, Vol 22, No. 2, 133 (1997).
[52] Ph.M. Kanarev, "Model of the Electron", *APEIRON*, Vol 7, No. 3-4, p. 184-192, July-Oct, 2000.
[53] Ph.M. Kanarev, "The Electron", *The Bases of The Theory of Nanotechnologies*, Ch. 3, (Krasnodar 2007) 515 pg. (In Russian) http://www.new-physics.com/.
[54] Vladimir B. Ginzburg, "The Unification of Strong, Gravitational & Electric Forces", Helicola Press, Pittsburgh, PA 15238 (2003).
[55] R. L. Carroll, "The Toroidal Electron," *Galilean Electrodynamics, Volume 2, pp 94-97 (1991).*
[56] Robert J. Heaston, "Quantum Gravity and the Structure of the Electron", Proceedings of the NPA, Vol 2, 12th *Natural Philosophy Alliance Conference*, Storrs, CT, pp 58-62, (2005). Galilean-Electrodynamics, Arlington, MA 02476.
[57] Charles W. Lucas, Jr., "A Classical Electromagnetic Theory of Everything", 12th NPA, Spinning Charged Ring Fiber, 23-27 (May 2005).
[58] Martin Rivas, "The Spinning Electron", *What is the Electron*? /ed. V. Simulik, C. Roy Keys Inc. Montreal: Apeiron (2005), 59.
[59] Richard Gauthier, "The Electron is a Charged Photon with the de Broglie Wavelength", www.Academia.edu/ 10527918 (2015). DOI: 10.13140/2.1.2775.1681.
[60] Jaroslav Klyuschin, "Electricity, Gravity, Heat – Another Look, academia.edu, Saint-Petersburg, Russia:International Scientists' Club, 2015. ISBN 978-5-9906744-1-7.
[61] J. Schwinger, "On Quantum-Electrodynamics and the Magnetic Moment of the Electron", *Phy. Rev.* 73(4) 416-417 (1948).
[62] O. F. Shilling, "A Phenomenological Model for the Electromagnetic Origin of Mass in Particles, and Its Quantitative Application to the Electron, the Muon, the Proton, and the Neutron", *Journal of Modern Physics*, 2013, 4, 1189-1193 (http://www.scirp.org/journal/jmp).
[63] W. C. Daywitt, "The Planck Vacuum", *Progress in Physics*, Vol 1, January 2009.
[64] W. C. Daywitt, "The Compton Radius, the de Broglie Radius, the Planck Constant, and the Bohr Orbits", *Progress in Physics*, Vol 2, April 2011, 32-33.
[65] André Michaud, "From Classical to Relativistic Mechanics via Maxwell", *International Journal of Engineering Research and Development*, Vol 6, Issue 4,(March 2013), pp 01-10.

cont

References

[66] S. Ghosh, A. Choudhury and J.K. Sarma, "Radius of electron, magnetic moment and helical motion of the charge of the electron", *Apeiron*, Vol. 19, No. 3, July 2012.

[67] Y. Nambu, "An empirical mass spectrum of elementary particles", *Prog. Theo. Phys*, 1952, v.7, 595.

[68] M. H. MacGregor, "The Power of α, Electron Elementary Particle Generation with a-Quantized Lifetimes and Masses", World Scientific, Singapore, 2007.

[69] Colin Greenshields, Robert Stamps and Sonja Franke-Arnold, "Vacuum Faraday effect for electrons", *New J. Phys.* **14** 103040 (2012), IOP Publishing and Deutsche Physikalische Gesellschaft.

[70] Wilheim Weber, "Determinations of Electrodynamic Measure: Particularly in Respect to the Connection of the Fundamental Laws of Electricity with the Law of Gravitation", *Werke*, Bd 4. pp 479-525, Berlin, Springer (1894). English translation in *21st Century Archive*.

[71] A. K. T. Assis, "Gravitation as a Fourth Order Effect", *Advanced Electromagnetism – Foundations, Theory and Applications*, T.W. Barrett and D. Grimes (eds.), World Scientific, Singapore, pp. 314-331 (1995).

[72] A. Miller, *Deciphering the Cosmic Number*, W.W. Norton, New York, 2009.

[73] R. Feynman, *QED The Strange Theory of Light and Matter*, Princeton University Press, New Jersey, 1985.

[74] R. Penrose, *The Road to Reality*, A. Knopf, New York, 2004.

[75] P. A. Davies and R. C. Jennison, "Experiments involving mirror transponders in rotating frames, *J. Physics A: Math Gen.*, Vol **8**, No.9, 1390-1397 (1975).

[76] D. G. Ashworth, "A New and Simple Deduction of the Thomas Precession", *Il Nuovo Cimento*, Vol. 40 B, N. 1, (11 Luglio 1977).

[77] J. J. Thomson, "A Treatise on the Motion of Vortex Rings", MacMillan and C, London, 1883.

[78] G. I. Shipov, "Physical vacuum theory", NT-Center, Moscow, 1993, p.362.

[79] G. I. Shipov, "Theoretical and experimental research of inertial mass of a four-dimensional gyroscope", p1-38, www.shipov.com.

[80] Konstantin Meyl, *Scalar Waves*, INDEL GmbH,Verlagsabteilung, Cillingen-Schwenninggen 1996-2003. ISBN 3-9802 542-4-0.

[81] V. G. Shulgin, "Gravitation Results from Gradient of Ether Density", Proceedings of the International Congress on New Ideas in Natural Sciences, 1996, pp. 415-420.

[82] Eric Laithwaite, *Propulsion without Wheels*, The English Universities Press Ltd, St. Paul's House, Wawick Lane, London, E.C.4, (1966).

[83] M. W. Evans, "Gravitational Poynting theorem: interaction of gravitation and electromagnetism", *Journal of Foundations of Physics and Chemistry*, vol. 1 (4), 2011, p. 433-440.

[84] Donald G. Saari, *Collisions, Rings, and Other Newtonian N-Body Problems*, American Mathematical Society, Providence, RI, 2005.

[85] James T. Dwyer, "Inappropriate Application of Kelper's Empirical Laws of Planetary Motion Created the Perceived Galaxy Rotation Problem", fqxi.org/data/essay/Dwyer_FQXi_2012 (8/23/2012).

[86] Bjerknes, V., *Fields of force; supplementary lectures, applications to meteorology; a course of lectures in mathematical physics*, Cornell Library, Historical Math Monographs (Dec 1 to 23, 1905).

[87] George William Von Tunzelmann, "A Treatise on Electrical Theory and the Problem of the Universe", Charles Griffin and Company, Philadelphia, p. 434, 1910.

[88] A. H. Leahy, "On the pulsations of spheres in an elastic medium", *Transactions of the Cambridge Philosophical Society*, Cambridge University Press, Vol XIV, p. 45-62.(1884).

cont

References

[89] G. T. Boutorin, "To the Question about Quantum-Mechanical Nature of Gravitation," VINITI, Moscow, 1987, dep. N 5135-B87, p.49.
[90] B. P. Bershadsky, "Structured discretization of the basic types of composite relations of types of matter," VINITI, Moscow 1990, dep. N40-B90, p.11, in Russian.
[91] A. E. Akimov, "Frequency Spectrum of physical Fields in general interpretation",VINITI, Moscow, 1990, dep.K2826-B90, p.6, in Russian]Mekhed'kin.
[92] Robert L. Oldershaw, "Discrete Scale Relativity", Astrophysics and Space Science. 311(4), 431-433. DOI: 10.107/s10509-007-9557-x (arXiv:physics/0701132v3), Dec 2007.
[93] John Barrow and Frank Tipler, *The Anthropic Cosmological Principle*, Clarendon Press, Oxford, 1986.
[94] John D. Barrow, *"Outer space: A matter of gravity"*, plus.maths.org *+plus magazine*, (March 2006). University of Cambridge.
[95] Ofer Lahav and Lucy Calder, "Dark Energy: back to Newton?", *Astronomy & Geophysics* A&G Vol 49, February 2008, p1.13-1.18 homepages.ucl.ac.uk arxiv.org//0712.2196.pdf.
[96] Ya. B Zel'dovich, "Cosmological Constant and Elementary Particles", JETP Letters **6**,883 (1967).
[97] Ya. B Zel'dovich, "The Cosmological Constant and the Theory of Elementary Particles", Soviet Physics, Uspekhi **11**, 381(1968).
[98] Ben Kristoffen, "A New View of the Universe", Cornell University Library, arxiv.org/9612010v1, (Dec. 1996)
[99] Golden G. Nyambuya, "Four Poisson-Laplace Theory of Gravitation", *Mon. Not R. Astron. Soc.* 000, 1-5 (2012) vixra.org.
[100] A. Trautman, *Lectures on Relativity*, Prentice-Hall, Englewood Cliffs, NJ, 230 (1965).
[101] Todd J. Desiato, Riccardo C. Storti, "Electrogravimagnetics: Practical Modeling Methods of the Polarizable Vacuum", *Physics Essays*, Vol 19, Issue 1, pp. 151-158, March 2006. ISSN 0836-1398http://dx.doi.org/10.4006/1.3025775.
[102] Ignazio Ciufolini and John Archibald Wheeler, *Gravitation and Inertia*, Princeton Series in Physics, Princeton University Press, 1995. ISBN 0-691-07428-3.
[103] George Green, "Researches on the Vibrations of Pendulums in Fluid Media", Royal Society of Edinburgh Transactions (1836), p 315-324.
[104] P. A. M. Dirac,"The Cosmological Constants", Nature **165**, 199 (1937).
[105] A. D. Sakharov, "Vacuum quantum fluctuations in curved space and the theory of gravitation", *Soviet Physics* -Doklady Akad. Nauk S.S.S.R. 177, Vol 12, (11) 1040, p. 70-71 (1968).
[106] David G. Blair, The detection of gravitational waves, Cambridge University Press, p.45 (1991).
[107] Kevin Cahill, "The Gravitational Constant", *Gravity Research Foundation* essay (1983). GRF, Gloucester, MA.
[108] I. W. Roxburgh, "The Cosmological Mystery-The Relationship between Microphysics and Cosmology", *Encyclopedia of Ignorance*, Pergamon Press, 1977.
[109] John Hunter, www.gravity.uk.com.
[110] Michal Krizek and Lawrence Somer, "Antigravity-Its Manifestations and Origin", *International Journal of Astronomy and Astrophysics*, Vol.3, No.3(2013).
[111] Jean-Pierre Petit, "An Interpretation of Cosmological Model with Variable Light Velocity", *Modern Physics Letters A*, Vol. 3, No. 16, p. 1527 (Nov 1988).
[112] Pierre Midy and Jean-Pierre Petit, "Scale Invariant Cosmology", *International Journal of Modern Physics D*, 8; 271-280 (June 1989).
[113] O. F. Mossotti, "On the Forces Which Regulate the Internal Constitution of Bodies", R. Taylor, ed. *Scientific Memoirs*, Vol 1. pp. 448-469 (1843).

cont

References

[114] M. Fierz and W. Pauli, "Relativistic Wave Equations for Particles of Arbitrary Spin in an Electromagnetic Field", *Proceedings Royal Society London*, **A173**, 211-232 (1939).

[115] Theodor Kaluza, "Zum Unitätsproblem in der Physik",*Sitzungsber. Preuss. Akad. Wiss*, Berlin (Math Phys), p 966-972 (1921).

[116] Edmund T. Whittaker, *A History of the Theories of Aether & Electricity*, (2nd edition), vol 1   London: Nelson, ISBN 0-486-26126-3   Dover Publications (1989).

[117] Edmund T. Whittaker, "On an Expression of the Electromagnetic Field due to Electrons by means of two Scalar Potential Functions", *Proceedings of the London Mathematical Society*, Vol 1, 1904, p. 367-372.

[118] A. V. Frolov, "The Concept of Gravitation", *Proceedings of the International Congress on New Ideas in Natural Sciences*, 1996, pp. 481-490 .

[119] R. V. Pound, G.A Rebka, "Gravitational Red-Shift in Nuclear Resonance", *Physical Review Letters* **3** (9)(Nov 1, 1959), p. 439-441.

[120] R. Adler, "A study of the locking phenomena in oscillators", *Proceedings of the I.R.E. and Waves and Electrons*, 34:351-357, June 1946.

[121] H. White, P. March *et al*, "Measurement of Impulsive Thrust from a Closed Radio-Frequency Cavity in a Vacuum", *Journal of Propulsion and Power*, 33(4): 830-841. doi: 10.2514/ 1.B361120, (Nov. 2016)

[122] H. Bondi, "Negative Mass in General Relativity", *Revs. Modern Phys.* Vol 29, No. 3, 423 (1957).

[123] V. de Sabbata, C. Sivaran, *Spin and Torsion in Gravitation*, World Scientific, Singapore (1994), ISBN 981-02-1766-8.

[124] R. Evans, *Greenglow and the Search for Gravity Control*, Troubador Publishing Ltd, Leicestershire, UK, (2015), ISBN 978-1784620-233.

misc. videos

https://www.youtube.com/watch?v=zgYDEhZY2mQ [Origin of Fine Structure Constant – Quantum Wave Mechanics]
https://www.youtube.com/watch?v=jq-vkaYVOB4&list=PLjJB6KTdUYEI1-vsh3rWfs1XTvg3eJIDf. [A New Kind of Propulsion in Space – Proof of Principle – Neila-Tech]
https://www.rhythmodynamics.com/Gabriel_LaFreniere/matter.htm [Matter is Made of Waves]
https://www.keelynet.com/spider/b-104e.htm [Standing Wave Compression]
https://www.youtube.com/watch?v=MShclPy4Kvc [Shepard-Risset glissando reaktor visualization – ScrollingMusic]
https://www.youtube.com/watch?v=vt0f0dMojr8 [Shepard-Risset Tone Generator in Reaktor – Dig-it]
https://www.youtube.com/watch?v=MnKu4rTZyHk [Todd Desiato: An Engineering Model of Quantum Gravity – WarpDrive Tech]
https://www/youtube.com/watch?v=VwNq2Lx4UUU [EGM- Electro-Gravi-Magnetics – Practical Engineering Principles of the Polarizable Vacuum]
https://www.youtube.com/watch?v=0tJfqMYHaQwn [Professor Eric Laithwaite: The Circle of Magnetism - 1968 – Imperial College London]
https://www.youtube.com/c/JPPETITofficciel [Jean Pierre Petit modele janus]
https://www.youtube.com/watch?v=GpBLq7qJx9A [Phase Frequency Propulsion [Rhythmodynamics]]
https://www.youtube.com/watch?v=vXuUrF2iNYU [Phase-frequency method of moving in space/Boat test in the pool: Academician Ivanov Yu. N. – G.W.]
https://www.youtube.com/watch?v=mrruwLqy8ho [Gravitation, Antigravitation]

References

https://www.youtue.com/watch?v=LeU53-ceu-0 [Rhythmodynamics]
https://serpmedia.org/scigen/e4.4b.html [Wave interference simulator]
https://vk.com/video-92223300-456240339?list=6aac2a9b180447ace7 [The first experience with an inteferometer in a diving aircraft – Ivanov –Rhythmodynamics-Global Wave]
https://youtube.com/watch?v=DZUHyN_NCaQ [A history of the debate about the substance of the Universe by Jeff Yee]
https://youtube.com/watch?v=n20x4ddpxqw [Friedwart Winterberg- Can Quantum Mechanics be Derived from General Relativity?]
https://www.youtube.com/watch?v=LuMlrHlAf2Q [Syd Mead – Visual Futurism – LV-426nightops]

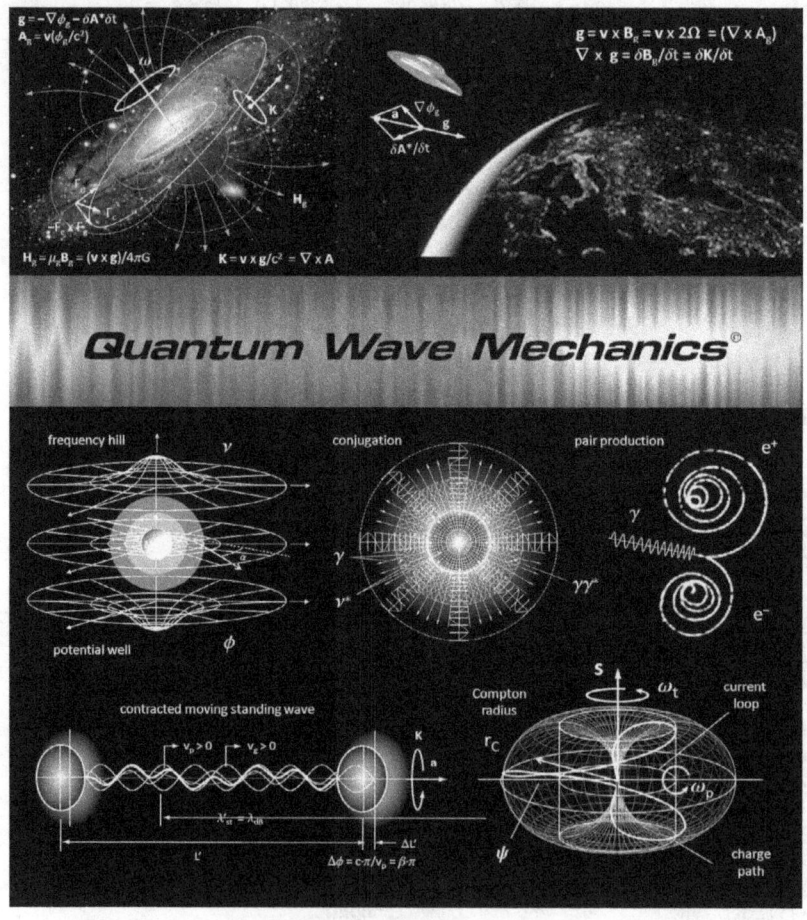

Quad erat demonstrandum

Index

A

acceleration
  centrifugal   289, 369, 425, 632
  centripetal   227
  of geodesics   80, 434-5
  of gravity   xi, 82, 328, 335, 365, 367, 369, 372-6, 381, 419-20, 424, 434, 436, 445-6, 448, 451-3, 456, 462, 470-1, 486, 502, 507, 511, 515-6, 522, 525-7, 541, 544, 572, 574, 579, 594, 598-9, 617, 636
action   166, 239, 243, 305-8, 330-2, 378, 405, 416, 429, 432, 450, 477, 515, 633
active mass   93-4, 363, 369, 371, 462, 504, 509, 515, 532-3, 572
Adler equation   457
aether   ix, 13-4, 25, 34, 132, 138, 141, 240, 319, 329, 419-20, 519
affine
  connection   18, 50, 52, 79
  isometries   40
  manifold   18
  parameter   435
  plane   677
  space   624
  torsion   52
  transformations   50
Aharonov-Bohm effect   25, 126, 269
Airy
  diffraction   288, 291-2
  disc   65, 288-9
Akimov, A.   357, 450
Allen, H. S.   148
amplituhedron   440
angular momentum   ix, x, 1, 15-8, 29, 34, 37, 39, 42, 47, 51, 132, 149, 151, 166, 205-7, 220, 228, 239, 243, 246-7, 250, 255-7, 267, 278, 284, 360, 378, 380, 410, 495, 507, 558, 583, 593, 623, 629, 657, 663
anholonomy   249-50, 252-3
annihilation, electron/positron   ix, x, 22, 124, 148-9, 208-11, 220, 222-3, 270, 289, 300
anomalous
  dispersion   9, 11,84, 108, 122, 302, 357, 451

gyromagnetic ratio   205-7, 240, 657
  magnetic moment   205, 207, 240
antenna   5, 6, 71, 88, 91, 95-9, 116, 122, 156, 159, 235-6, 357, 426, 430, 493-5, 509, 511- 2, 514, 518, 550, 552-3, 555, 594, 598 ,677
anti-gravity   71, 379, 502, 507, 526, 573
Archimedean spiral   152, 235-9, 380
arrhythmia   328, 448, 463, 471
Ashworth, D. G.   231
Assis, A.   227, 334

B

Babinet   236, 556
ball lightning   529, 552, 554
barium titanate   60, 321, 490
Barnett effect   xi, 15, 207, 487, 572
Barrow, J.   361, 363
baryons   275, 288, 290
basis vector   1, 52, 315, 523, 624, 667
beat frequency   86, 120, 369, 474, 478-9, 546, 551, 554-5, 602
beat frequency oscillator (BFO)   478
Beltrami flow   23, 156
Bergman, D.   148, 160-3, 230
binary sequence   609
binding energy   34, 84, 141, 284, 375, 502, 552
Birkeland current   23
Bjerknes effect   349-50, 352
blackbody radiation   45, 432-3, 451, 543, 549-50, 552, 568
black hole   18-9, 21, 71, 79, 80-1, 133, 234, 334, 357, 359, 372-3, 379, 403, 594, 677
Bloch sphere   95, 100, 598, 624
blueshift   xi, 83-4, 107-9, 116-7, 124, 128, 130, 166, 212, 485-6, 489, 511, 529, 556-7, 561
Bohr
  magneton   151, 205-7, 301
  radius   39, 164, 170, 228, 279
Boltzman   552\
Boscovitch, R.   349
Bose-Einstein
  condensate (BEC)   ix, 13-4, 17, 34, 47, 133, 147, 247, 317, 353
  statistics   246-7
boson   ix, xiii, 1, 13-4, 34, 40, 87, 149, 207, 210, 246-7, 273-4, 277, 288, 304, 405, 407-8, 413-4
Boutorin, G.   357, 450
Bragg
  angle   514
  diffraction   68-9, 111-2, 440-3

685

Index

Bragg (cont)
  interference 60, 432, 490
  law 68
  mirror 83-4
  reflection 69
  reflector 60, 111, 432, 496
  resonator 588, 590-1
Bravais lattice 69, 70, 440, 668
Breit-Wheeler
  decay 149
  pair production 209
bremsstrahlung 149, 299, 483
Brillouin 111, 113, 429-30, 667
Bronstein-Penrose hypercube 623, 671-2
Brown, Thomas Townsend 318

C

Cahill, K. 379
Calder & Lahov 361
Cantor set 664
Carroll, R. 148
Casimir effect 13, 40, 329, 426, 432, 588
Catt, I. 25
Cavendish 378
Cherenkov radiation xi, 8, 149, 299, 487
charge,
  definition and units of 239, 639
  color 247-8, 256-7, 275-6, 288-9, 295, 298
  electric ix-xii, 1, 15, 16, 32, 34, 37-9, 41, 92, 151-3, 159, 166, 225, 230, 239, 239, 240, 242, 246, 249, 269, 271-3, 335
  fractional 270-4, 290
  gravitational 331, 337
  hypercharge 276-7, 290, 298
  isotopic 276-7
  path trajectory 1, 148, 151, 153-4, 205
  Planck 7, 21, 57, 202, 384, 389, 391, 394
  radius 152, 205, 239, 279-83
  screening 13, 289, 294
  topological x, 16-7, 32, 37, 39, 41, 52, 220, 240, 249, 258, 269-72, 289, 626, 666
  vortex 16, 34, 246
chiral 10, 15, 65, 117, 152-3, 227, 239, 247, 271, 302, 304, 408, 412, 416, 526, 597
chromatic equally-tempered scale 601-6, 613, 615-6
Clausius-Mossotti factor 319-20
Clausius Vivial theorem 231
clock xi, 78, 88, 101, 108, 122, 328, 349, 391, 445, 447, 457

cogravitation, (See also gravito-electromagnetism)
  field 329, 334, 338, 340, 368, 424, 572, 578, 581-2, 585-6, 605, 631
  vector potential 340, 367
color charge 247-8, 256-7, 275-6, 288-9, 295, 298
commutator 250, 252
complex
  conjugate 124, 260, 312, 413, 497
  index of refraction 57-8, 509
  numbers 20, 111, 141, 258-64
  phasor 19-20, 88, 90, 111, 114, 124-6, 129, 210-11, 217, 263-4, 497, 503
Compton, A.H. 148
  frequency xii, 45, 49, 98, 110, 151, 153-7, 166, 210, 218, 230-1, 235-6, 238, 240, 334, 357, 388-9, 432, 451, 601, 603
  wavelength xii, 14, 39, 83-4, 98, 102, 110, 149, 151-2, 156, 212, 215, 217-8, 220, 236, 238, 287, 478
  radius x, 15, 19, 149, 151-2, 154-5, 160, 228, 231-2, 235-8, 278, 288, 292, 360, 388-9, 391, 393-4
  scattering 149, 209
conservation
  baryon no. 295
  charge 208, 263
  energy 17, 20, 50, 226, 253, 329, 435, 589
  momentum 17-8, 38, 50, 334, 408, 483, 589
consonance 426, 602, 604, 606, 610, 615, 617
contracted moving standing wave xi, 86, 88, 90, 98, 102, 104, 118-9, 128, 130, 166, 210, 212-3, 215-7, 465, 485-9, 490, 493-4, 498-9, 500, 528, 530, 555, 557, 560, 614
contragravity 493
Cooper pairs 160, 246, 318
covariance 26, 40, 52, 93-4, 225, 315-6, 334, 368, 408
Coulomb
  constant 44, 225, 228, 231, 245, 335, 341, 624, 639, 671-2
  dimensional conversion 37-8, 41, 162, 240, 242, 335, 624, 639
  force 38, 76, 225-7, 236-8, 288-9, 290, 294, 333, 335, 441, 643
  gauge 27
Coulomb's law, 225, 383
  generalized 226
Cornu spiral 135, 675
Cosmological constant 14, 132, 135, 316, 361, 368, 402

686

Index

coupled oscillators xii, 128, 328, 349-50, 375, 411, 457-62, 465, 471, 479, 502, 532, 608, 611
covariant theory of gravity (CTG) 334
cubic roots 270, 276
current
  electric x, 15, 21, 25, 27-8, 38, 87, 92, 96, 98, 126, 148, 151-3, 155, 158-60, 205, 225, 230, 263, 441, 625, 643
  mass 14-5, 32, 37, 92, 135, 270, 278, 303, 334, 337, 346, 424, 437, 529, 531, 570-4, 579, 589, 638
curvature
  extrinsic 146, 309, 440, 687
  Gaussian 52, 309, 418, 434-5, 440, 621, 627
  geodesic 15, 17, 50, 435
  intrinsic 50, 146, 250, 309, 418, 627
  mean 1, 309, 617
  principal 309
  radius of 51, 146, 595, 627
  Ricci 276, 316-7, 401, 406, 434-6, 440
  Riemann 276, 316, 402-3, 405-6, 434-5
  scalar 1, 3, 4, 43, 82, 146, 220, 244, 401, 406, 434, 627
  spiral 236
  wavefront xi, xiv, xv, 17, 18, 60, 76, 79-82, 308, 409, 432, 592
  wavefunction 15, 19, 164, 243, 308, 369
  Weyl 276, 316, 434-6, 440
cycloid 231
cyclotron
  frequency 195
  radius 169

D

Darboux frame 50
dark energy (quintessence) xiii, 14, 19, 21, 23, 132, 141, 317, 361-2, 380, 399-400, 416, 441, 496, 592, 675
dark matter 14, 21, 34, 300, 346-8, 362, 400, 597
Davies, P.A. 231
Davies-Unruh effect 13, 121, 472
Daywitt, W.C. 15, 379, 402
de Broglie
  contracted 212, 216
  frequency 87, 110-1, 366, 534, 544, 567
  matter wave xi, 84, 129, 166, 210, 218, 306, 353, 365, 529
  radius 169
  synthesized 113, 124, 126, 129-30, 141, 486-7, 588

phase velocity 141, 366
wavelength 84, 87, 102, 110, 141, 212, 216-8, 366, 514, 544, 548, 569
deceleration 84, 124, 140, 149, 162, 484, 498, 526-7, 603
deflection of light xi, xii, 76-9, 80-1, 101, 328, 441, 532, 593
Delbrück effect 13, 415
Depp, J. 76
Desiato, T. 369
de Sitter,
  anti de Sitter 250
  cosmology 379
  precession 231
  space 250
Devil's pitchfork rotation 675
Devil's staircase 457
Dicke, R. 76, 328, 441
dielectric constant xii, 15, 53, 56-7, 59, 76, 78, 94, 101, 321, 328-9, 371, 426, 431-2, 556
dielectrophoresis 318-9, 483
dielectrophoretic force 318-22
Diemer, G. 369, 416, 429, 452
differential forms
  0-forms 52
  1-forms 52
  2-forms 52
  3-forms 52
diffraction 65, 68-72, 11-2, 160, 269, 288, 291-2, 440-3, 551
diffractive lens 288, 426
dimensions,
  conversion of 37, 41-4, 242-5, 624-62, 664
dipole,
  antenna 5, 6, 91, 95-7, 156, 236, 487, 511-6
  gravitational 142-3, 588-9, 590, 594, 596-8
  moment 144, 206, 290, 299, 318-23, 586, 588, 590, 640, 637
  oscillator 1, 95, 478-9, 482-3
  Planck ix, xiii, 1, 13-21, 23, 29, 31-7, 132-3, 144, 151, 155, 159, 276, 289, 294, 303, 421, 597
  rotating 132-7, 145, 151, 155, 158, 295, 388, 592
  vortices 13, 132-7, 158, 235, 300, 405, 592
  waves 141, 156, 159, 166, 246, 362
Dirac, P.A.M.,
  constant 30, 35, 37, 95, 137, 201, 276, 353, 371
  large number hypothesis 378
  neutrino 299
  sea 13-4, 379

Index

spinors 135, 267
statistics 247
dissonance 350, 360, 426, 527, 602, 604, 606, 610-1, 615
Doppler
  effect 93, 104
  frequency shift xi, 101-2, 104, 110, 116, 375, 449, 471, 474, 555
  inverse effect xi, 116-7, 122, 128
  simulated/synthesized 113, 118-21, 128, 130, 479, 485-93, 530, 555, 560-1, 591
  wavelength 102, 108-9, 118, 215-16
dreibein (triad) 1, 98, 405
Drinkwater, A. x, 83, 375
Dröscher, W. and Häuser, L. 400
du Châtelet, E. 17
Dwyer, J. 348
dyad 19, 20-3, 134, 146, 315, 613, 675
Dyatlov, V. 15

E

Eaton lens 70, 74
Einstein, A.,
  Bose-Einstein condensate (BEC) ix, 13-4, 34, 47, 133, 147, 317, 353
  Bose-Einstein statistics 246-7
  constant 203, 234, 380, 401-2, 406
  cosmological constant 14, 132, 135, 316, 361, 368, 402
  Einstein-de Haas effect xi, 15
  energy-mass relationship 17, 39, 79, 83, 110, 133, 234, 306, 353, 374, 379, 419, 511, 594
  equivalence principle 365, 374, 570
  field equation 316, 361, 380, 401-2, 404
  general relativity (GR) xiii, xiv, 9, 76, 78, 166, 316, 328, 361, 372, 378, 401, 441, 445, 588, 614, 624, 671-2
  gravitational lens 71, 75-6, 79
  prinicple of covariance 40, 315, 332, 368, 408
  special relativity (SR) 94, 98, 260, 332, 372, 614, 624, 672
  stress-energy tensor 93, 98, 260, 332, 372, 614, 624, 672
electric charge, ix, x, xii, 1, 15-7, 32, 25-7, 34, 37-9, 41, 53, 56, 81, 92, 95, 98, 135, 137, 148-55, 159, 162, 166, 225, 230, 239, 239, 240, 242, 246, 249, 258, 269-74, 278, 295, 305, 337,393, 404, 624, 639
  dimensions and units of 239, 639
  elementary particles 170, 248, 279, 283, 295-7, 301, 669

electric dipole moments (EDM) 53, 144, 289-90, 299, 318, 320-1, 640
electric field intensity ix, 14, 17, 26, 29, 35, 38, 51, 53, 56, 76, 79, 95, 144, 320, 441, 507, 511, 554 ,642
electric flux lines 99, 100, 144, 155, 240, 343, 641-2
electric permittivity xii, 33, 56, 59, 76, 78, 92, 117, 160, 225, 240, 305, 328, 343, 382-4, 396, 398, 441, 503, 505, 507, 581, 597, 651
electrodynamic force
  Ampere 226
  Assis 227
  Grassman 226
  Klyuschin 226-7
  Lucas 227
  Neuman 226
  Weber 226, 334
  Whittaker 226
electro-gravi-magnetic (EGM) 452, 594, 672
electromagnetic spectrum 45-9, 121, 357, 451
electromagnetic vector potential xi, xiii, 14-7, 25-8, 32, 34, 37-8, 43, 78-9, 82-3, 95, 98, 124, 126, 159, 162, 239-40, 244, 259, 260, 266, 305, 334, 340, 353, 365-6, 405, 431, 507, 574, 584, 614, 655
electromagnetic energy density 223, 304, 319, 431, 502, 504, 509, 511-2, 515, 526, 551, 553, 561, 574, 617, 630
electromotive force (EMF) 13, 17, 158-9, 226, 228, 259, 288, 298-9, 333, 346, 394, 413, 415, 641
electrostatic field energy 230, 645
electron,
  capacitance 27, 160, 162, 165, 230, 639
  characteristics 148, 151-3, 168-204
  charge ix, x, 1, 15, 16, 32, 34, 37-9, 41, 92, 151-3, 159, 166, 225, 230, 239, 239, 240, 242, 246, 249, 269, 271-3, 335, 507, 552, 571, 624, 639, 670
  charge-to-mass ratio xii, 166, 206, 230, 239, 240, 644
  Compton frequency xii, 45, 49, 98, 110, 151, 156-7, 166, 168, 210, 218, 231, 235-7, 392, 415, 432, 451, 601, 603
  Compton radius x, 15, 19, 149, 151-2, 154-5, 160, 228, 231-2, 235-8, 278, 288, 292, 360, 388-9, 391, 393-4
  electrostatic field energy 230, 645

688

Index

electron (cont)
  energy storage   135, 162, 312
  g-factor   149, 205-6, 230
  gyromagnetic factor   205-6
  gyromagnetic ratio   206-7, 657
  inductance   27, 87, 92, 160, 162, 165, 230, 374, 664, 653, 666, 670
  magnetostatic field energy   x, xii, 43, 154, 159, 160, 163, 167, 230-1, 244, 261, 391, 396, 483, 593-4, 624, 649
  magnetic moment   15, 149, 205-7, 223-4, 240, 278, 280, 284, 300, 303, 322, 571, 646
  mass   ix-xi, 13-6, 21, 34, 38, 41, 82-4, 87, 89, 92-3, 98, 116, 140, 142-3, 151, 154, 223, 235, 238, 242, 247, 253, 273, 279, 374, 393, 624, 671
  mass deficit   141, 234-5
  orbital magnetic moment   206, 646
  point charge (source)   38, 95, 155, 367, 385
  potential well   168
  precession   x-xiii, 16, 32, 34, 38-9, 41, 135, 137, 144,150-1, 153-4, 158, 168, 205, 207, 223, 227, 229-32, 234, 239-40, 245-7, 252-5, 272, 289-90, 299, 391, 393, 593, 624, 663, 669, 671-2
  radius
    classical   149, 152, 164, 169, 189, 208, 228
      charge   152, 188, 205, 239
      Compton   x, 15, 39, 149, 151-2, 154-5, 160, 164, 169, 228, 232, 237-8, 360
        cyclotron   169
        de Broglie   169
        electromagnetic   188-9, 228
        mass   152, 188, 205, 232, 359
        poloidal   152, 188
        spindle   152
        toroidal   149, 152, 220
      Zitterbewegung   151-2, 164, 198, 231
      ring   189
      Schwarzschild   193
      sphere   149, 189
  spin magnetic moment   149, 205-7, 223, 278, 636
  spin density wave   159, 166, 246
  wavelength
    Compton   xii, 14, 39, 83-4, 98, 102, 110, 149, 151-2, 156, 212, 215, 217-8, 220, 236, 238, 287, 478

de Broglie   84, 87, 102, 110, 141, 212, 216, 218, 366, 514, 544, 548, 567
electromagnetic field tensors   404
electron impedance   228
electron-positron pair production   ix, 13, 50, 148-9, 208-11, 218, 220-1, 223
electron-positron annihilation   ix, 22, 124, 148-9, 208-11, 220, 222-3, 270, 289, 300
electron volts (eV)   16, 45, 83, 149, 159, 165-6, 208, 220, 279, 280-3, 296, 299, 300-4, 552
elementary particles   248
energy
  dark   xiii, 14, 19, 21, 23, 132, 141, 143, 317, 361-2, 379-80, 400, 416, 441, 496, 502, 592, 597, 675
  electrostatic   43, 98, 159, 160, 167, 230-1, 239
  kinetic   17, 25, 37, 79, 83, 91, 94, 117, 129, 132, 159, 162, 164, 166, 210, 219, 231, 234, 253, 289-90, 296, 300, 305, 307, 329, 364, 482, 496, 502, 507, 509, 561-2, 572, 574, 576, 601, 605
  magnetostatic   x, xii, 43, 154, 159, 160, 163, 167, 230-1, 244, 261, 391, 396, 483, 593-4, 624, 649
  potential   x, 17-8, 26, 39, 79, 84, 90-1, 129, 142, 149, 210, 226, 231, 260, 289-90, 307-8, 318-9, 329, 346, 353, 364, 375, 470, 472, 479, 482, 502, 509, 573, 594, 596, 605, 614,
  Planck   ix, 18, 29, 79, 132-3, 136, 202, 299, 389, 594
  vs. momentum   89
entangled pair   21, 95, 100, 156, 429
entropy   18-21, 29, 44, 134, 245, 592, 625, 629, 676, 629
epicycloid (nephroid)   271
epitrochoid (limacon)   153, 205
equivalence of gravitational and inertial
  mass   143, 365, 374, 570
etalon   65, 116
Euler's
  angles   265
  characteristic   52, 663
  identity   203, 663
  number   52, 663
  spiral   2
Evans, M.W.
  Evans-Cartan field equation   401, 406
  gravity and electromagnetism   26, 333, 335-344, 368, 404, 406
  tetrad   1-3, 26, 402, 406

689

Index

F

Fabry-Pérot interferometer 65-6
false vacuum 16, 136, 317, 409, 416
Fano
   planes 667-8
Faraday. D.
   field lines 17, 37, 144
   induction 25, 334
   inverse effect xi
   law 342
   rotation effect x, 220, 22, 224, 488, 556, 588
Fermi-Pasta-Ulam 35
Feynman, R. 229
Feynman diagrams 208, 408
Fibonacci
   damping ratios 48, 426, 607-10, 618
   Fibonacci (cont)
      harmonic intervals 602, 604, 607-8, 615
   sequence 604, 606, 609, 612, 615
Fierz, M. and Pauli, W. 404
fine structure constant xii, 7, 15, 18, 154, 205-6, 228-31, 236, 246-7, 334, 380, 394, 416, 671-2
   inverse xii, 154, 166, 228, 232, 240, 247
force
   acceleration xii, 15, 79, 83-4, 374, 482, 496, 594, 632
   buoyant 436
   Casmir 13, 40, 329, 426, 432, 578
   centifugal 137, 255, 425, 632
   cogravitation 341, 424, 437
   compressive 167
   Coriolis 575
   Coulomb 76, 225-7, 238, 240, 288-9, 441
   dielectrophoretic 27, 318-21
   dynamic gravitation 341
   electrokinetic 25-6, 82, 98, 124, 437
   electromagnetic 121, 228, 288, 298, 298-9, 333, 346, 394
   electromagnetic Magnus 323
   electromotive (E.M.F.) 17, 158-9, 259
   electrophoretic 298
   electrostatic 225, 236, 333, 335-6, 383-4, 394
   electroweak 247, 298-9, 406
   gravikinetic drag 438
   gravitomagnetic 341, 424, 437
   gravity ix, x, xii, 15, 38, 76, 78-9, 80, 82, 94, 98, 133, 141-2, 261, 298-9, 319, 330-6, 341, 346, 349, 351, 361, 363-4, 372, 374, 378-9, 383, 385, 388-91, 394-8, 413, 415, 424-5, 432, 438, 447-8, 470, 484, 502-3, 507, 511
   inertial 82, 84, 142, 167, 374, 380, 572-3
   lift 323, 325, 537, 541, 543, 556
   Lorentz 27, 82, 121, 207, 224, 323, 341, 346
   magnetic 341
   magnetophoretic 322
   maximum repulsion 228, 278, 391
   magnetomotive (M.M.F) 158-9, 649
   non-uniform 318
   Planck 298, 379, 391, 394, 402
   ponderomotive 87-8, 119, 124, 323, 551, 489, 529, 551, 561
   pressure 167, 352, 537, 556
   radiation 87-8, 119, 124, 323, 541
   strong nuclear 275, 288-90, 298-9, 406
   tension 167
   tidal 276, 380, 403, 435-7
   tension 167
   Van der Waals 321, 378
   Weber electrodynamic 226-7, 334
four-wave mixing (FWM) 55, 111-3, 117-8, 128, 440, 485-6, 489, 498, 551, 561
   degenerate (DFWM) 113
   nondegenerate (NDFWM) 113, 128, 485, 561
flat spacetime 33, 52, 77, 93, 162, 166, 250, 372, 404, 408, 442
   Euclidean xi, xii, 76, 78, 101, 250, 276, 307, 315, 371, 382, 402, 406, 417, 419, 595
Fleming left-hand rule 325, 574
Fleming, I. 229
flux
   cogravitational 340
   crawler 508
   electromagnetic radiation xii, 5, 17, 76, 88, 95, 97-9, 100, 144-5, 155, 157-8, 161
   electric 99, 100, 144, 155, 240, 343, 507-8, 641
   energy 14, 122, 329, 402, 426, 516, 561, 630, 633
   gravitational 101, 332-3, 339, 343, 368, 424, 432-3, 524
   luminous 659-61
   magnetic 16-7, 51, 76, 79, 92, 158, 160, 220, 224, 240, 322-4, 339-40, 441, 488, 511, 592, 670
   mass 16, 332-4, 424
   neutrino 299, 300
   radiant 659
   spectral 659-60
   tube 288
Frenet-Serret 1, 3, 50-3

Index

Fresnel
  coefficient 117
  zones 70, 72, 76, 95, 97, 288, 292, 329, 426-7, 429-33, 436-9, 440, 442-4, 620
frequency
  arrhythmia 328, 448, 463, 471
  Betatron 195
  chirp 4, 9, 122, 357, 515, 622
  comb 498-9, 504, 509, 526-7, 551, 593, 598
  Compton xii, 45, 49, 98, 110, 151, 153-7, 166, 210, 218, 230-1, 235-6, 238, 240, 334, 357, 388-9, 432, 451, 601, 603
  cut-off 47, 49, 91, 135, 219, 300, 378, 450, 452, 455, 474, 515, 535-6
  de Broglie 86-7, 90, 110-1, 161, 212, 366, 514, 529, 548, 586, 588
  definition and units of 4, 33, 45-7, 244, 603, 635
  difference 102-4, 108-9, 112-3, 115, 329, 349, 369, 375-7, 447, 462-3, 470-2, 487, 489, 498, 504,511, 514, 526-7, 561, 570
  dilation x-xii, 14, 76, 78, 83, 93-4, 101, 108, 232, 328, 371-2, 419, 447, 537
  EM spectrum 45-6, 49, 123, 357-8, 450, 503-4, 515-6, 532, 593
  gravitation 357-8, 503, 532, 601
  Larmor precession 195
  Planck 13, 45-7, 57, 133, 147, 235-6, 357, 413, 601, 603
  resonant 35, 57-8, 92, 120, 462, 489, 504, 511, 611
  Zitterbewegung 149, 151, 166, 205, 207, 210, 218, 220, 267
Frolov, A. 416
Fulling-Davies, Unruh effect 13, 121, 472
Funaro, D. 1

G

g-factor 149, 205-6, 230
Gabor
  lens 70
  zone plate 473
Gaussian
  beam 65, 67, 256
  curvature 52, 309, 418, 434-5, 440, 621, 627
  envelope 607
  filter 310
Gauss's law 342, 670
Gauss's law of gravity 368
gauge symmetry 27, 250, 259, 298, 411, 624

Gauthier, R 148
geodesic
  relative acceleration 15, 17, 80, 434-5
geodetic (de Sitter) precession 231
Gerber, P. 334
Ghosh, S. et al 148
Ginzburg, V. 148
glissando 489-90, 507, 602, 607
gluons 288, 292-4
gradient refractive index (GRIN) lens 70-1, 73-4, 426, 619
gravitation tonality 591
gravitational constant 18, 21, 34, 38, 43, 47, 82, 133, 225, 244, 316, 333, 335, 341, 359-60, 363-4, 369, 371, 378-86, 399, 401, 450, 522, 588, 636, 663
gravitational redshift 371-2, 380-2, 445-6, 470
gravitational energy 300, 343, 368, 400, 403, 419-20, 452, 498, 504, 572, 596, 675
gravitational field ix, xi-xii, 16, 34, 71, 76, 79-82, 160, 166, 261, 303-4, 329, 333-4, 338, 342, 361, 364-5, 367-8, 372, 374-8, 389-90, 401, 404, 413-4, 419-24, 432-3, 437-8, 445, 471, 601, 605, 621
  energy density 343, 502, 504-5, 507, 516, 522, 524-5, 555, 637
  mass density 82, 337, 365, 368, 420, 422-3, 436, 522, 534, 638
  nonlinear 419-23
gravitational flux 101, 332-3, 339, 343, 368, 424, 432-3, 524
gravitational force ix, xiii, 15, 38, 76, 78-9, 80, 82, 94, 98, 133, 141-2, 261, 298-9, 319, 330-6, 341, 346, 349, 351, 361, 363-4, 372, 374, 378-9, 383, 385, 388-91, 94-8, 413, 415, 424-5, 432, 438, 447-8, 470, 484, 496, 498, 502, 585
gravitational frequency
  differential xiii, 369, 375-7, 419, 433, 447, 452-3, 505, 514, 527-8, 561
  Fourier spectral analysis 450-456
  redshift 380, 445-6
  spectrum 357-8, 450-6, 504, 602
gravitational gamma 16, 33, 82, 160, 372-3, 381-2, 389, 413, 432, 446, 450, 537, 574, 637
gravitational lens 71, 75-6, 79
gravitational magnitude 372-3, 381-2, 388-9, 445
gravitational mass xii, 79, 132, 142-3, 253, 319, 329, 357, 363, 374, 378, 434, 443, 453, 594

691

Index

gravitational mass density   365, 368, 420-2, 638
gravitational parameter   330-1, 385
gravitation permeability   359, 380, 385, 401-2, 406
gravitational potential   76-7, 80, 261, 305, 309, 331, 336, 353, 363-7, 369-72, 375, 377, 401, 420, 434, 446-7, 450, 470, 478, 496, 498-502, 504-6, 523-4, 573, 617, 636
  normalized   364-5
gravitational Poynting vector   526, 544, 579
gravitational time dilation   xi, 14, 76, 78, 371-2, 419, 447, 537
gravitational vector potential   367, 424, 507, 574, 578, 582, 584, 591, 614, 631, 635
gravitational waves   332, 357-8, 414-5, 509, 592
gravitoelectromagnetism (GEM)   329, 334, 338, 340, 368, 424, 572-3, 581-2 (See also cogravitation)
gravity
  and electromagnetism   26, 333, 335-344, 368, 404, 406
  cogravitation   233, 329, 334, 338, 340, 368, 424, 572-3, 581-2
  force   xii, 15, 38, 76, 78-9, 80, 82, 94, 98, 133, 141-2, 261, 298-9, 319, 330-6, 341, 346, 349, 351, 361, 363-4, 372, 374, 378-9, 383, 385, 388-91, 394-8, 413, 415, 424-5, 432, 438, 447-8, 470, 484, 496, 498, 585, 594
  frequency spectrum   357-8, 603
  frequency differential   329, 349, 369, 375-7, 447, 462-3, 470-2, 487, 489, 498, 504, 514, 526-7
  Gauss's law   342, 368, 660
  Newton's law   330, 361, 378, 388
  optical theory of   xii, 71, 497
  variation with time   378-80
gravitons   ix, xii, 76, 203, 357, 399, 406, 408, 410, 413-8, 429-30, 432, 442, 502, 553
gravitomagnetic   (See also cogravitation)   160, 233, 237-9, 329, 334, 338, 340, 368, 424, 571-3, 581-2
Green, G.   374
group velocity   2, 4, 7-9, 11, 33, 35, 57, 65, 84, 87, 95, 109, 117, 119-20, 341, 416, 467, 487, 491, 556-7, 588, 597
gyromagnetic factor   205-6
gyromagnetic spin ratio   206-7, 230, 657
gyroscope inertial mass,   253

H

Haisch, B, Rueda, A., Puthoff, H.   124
Hall resistane   200, 246
Hamiltonian   90, 133, 305-8, 630
Hamiltonian circuit   671
Hamming code   671
Hasse diagram   671
Heaston, R.   148
Heaviside, O.
  EM field equations   15, 266
  gravitation   334, 367
Hegel dialectic   135
Heim, B.
  electric charge   400
  electron mass   400
  gravitational constant   399
  density action tensor   405
Heisenberg, W.
  fine structure constant   229
  uncertainty principle   250, 307
helicity   1, 96, 149, 152, 247, 259, 263, 269, 273-4, 299-300, 302, 413-4
Helmholtz, H.
  resonator   91
  vacuum   160
  vortex rings   299, 302
Hestenes, D.   148, 193, 198, 232
Hicks, W.M.   350
Higgs, P.
  field   x, 88
  mechanism   276
  potential   416
holonomy   252, 408
Holtzmüller, G.   334
Hooke, R.   330, 349
Hopf link   ix, xii, 50, 52, 148-50, 153, 155, 159, 223
Hubble
  constant   18, 300, 379-80
  redshift   227, 397
Hughes, W.   396
Humpty Dumpty   126
Hunter, G.   1
Hunter, J.   379
Huygens-Fresnel diffraction   68
Huygens
  clock synchronization   457
  wavelets   235
hypercharge   276-7, 290, 298
hypocycloid   271-3

I

imaginary number   2, 19, 258, 260, 663, 668

Index

impact
  cross-section 300
  line 411-2
  parameter 75, 80
impedance
  acoustic cavity oscillator 92
  black hole 667
  complex 135, 139, 263, 312
  antenna 97, 236
  cyclotron 199
  dimensions 652
  electromagnetic cavity oscillator 92
  electromagnetic oscillator 92
  electron 7, 220, 228, 236
  filter 311
  mechanical 92
  mismatch 210, 351, 354, 677
  Planck 8, 33, 667
  radiation 5, 13, 236
  ratio 7-8, 44, 245, 263, 677
  Smith chart 24, 135, 139-40
  spacetime 235, 330, 378
  triangle 140
  vacuum (free space) 5, 33, 82, 96, 220, 228, 241, 378, 400, 594, 597, 637, 653
  wave 5, 7-8, 652
  wave guide 8
index of refraction ix, 8, 14, 29, 33, 35, 51, 57-60, 71, 75-6, 82, 84, 87, 117, 133, 159-161, 372, 446, 507, 532, 574
  complex 57-8, 509
  graded 71, 117, 619
  negative 9, 71, 84, 116-7, 267, 507, 509, 533, 566, 597
  vacuum ix, 14, 29, 33, 35, 51, 58, 71, 75-6, 82, 133, 160, 372, 446, 591
inductance 27, 87, 91-2, 140, 159, 160, 162, 165, 230, 374, 644, 653, 666, 670
inerter 507-8
inertia x-xii, 14, 16, 38, 79, 82-4, 87-8, 98, 116-7, 121-2, 124, 126, 140-1, 142-3, 167, 232, 253, 329, 334, 374-5, 378, 380, 435, 492, 507-8, 511, 526-8, 530-1, 556-7, 570-3, 575, 579, 586, 594, 597, 602, 614, 634, 644
  gyroscopic 253
  origin of 83
  melodic 592
  moment 37, 571, 589, 629
  permeability 343
inertial damper 121, 507-8
inertial reference frame 39, 40, 78, 93, 98, 101, 103, 120-1, 162, 231-2, 234, 328, 332, 372, 463, 477, 614, 663

inertial mass xii, 16, 38, 83, 87, 98, 121, 124, 126, 132, 141-3, 232, 253, 329, 374, 378, 511, 557, 572, 579, 594, 634, 644
interference
  grating 113, 487, 489, 492, 551
  harmonic 352, 415
  oscillator 350, 353, 440, 443, 464-5, 473-7
  Fresnel zone xiv, 70, 288, 292, 329, 426-7, 429-33, 440, 442-4, 620
  interference nodes xii, 60, 69, 76, 288, 432-3, 440-1, 443, 471, 478, 601, 608, 621
  nodal lines 465, 477
  standing wave 350-1, 428, 463
  wave/beam 353, 430, 440-1, 443, 490
  wave patterns xii, 65, 68, 70, 98, 111-3, 120, 158, 288-9, 293-4, 332, 413-4, 426-7, 432, 440, 443, 463, 473, 474-9, 485, 621
  wave phase 350, 355-6, 556, 594
  zone 65, 289
inverse Barnett effect 486
inverse beta decay 296
inverse Cherenkov effect 486
inverse Doppler effect xi, 116-7, 122, 128, 529, 577
inverse Faraday effect 486
inverse fine structure constant xii, 154, 166, 228, 232, 240, 247
inverse Kepler-Boukamp 146, 595-6, 617
inverse Sagnac effect xi, 120, 486, 574, 588
inverse square law 332, 349, 353, 396-7
inversion 54, 134, 152, 493, 525-6, 532, 593, 606, 665, 670, 677
isospin 277
isotopic charge 277
Ivanov, Y.
  standing wave transformations xi, 98, 101, 104-110
  gravitation 80, 328-9, 375, 447, 463, 470-2
Ivanenko process 414

J

Jefimenko, O.
  co-gravitation 329, 338-340, 367
  electrokinetic field 25
  gravitational model 34, 329, 334, 338-344, 367-8, 374, 404, 420, 422-3
  induced electrokinetic force 25
  retarded potentials 98, 363
Jennison, R.C. x, 83-4, 116, 148, 231, 375
Jones vector 10

693

Index

K

Kaluza-Klein   16, 166, 239, 250, 404-5, 624, 671
Kanarev, Ph. M.   148
Kerr effect   9, 35, 55, 111, 113, 142, 551
Kepler-Bouwkamp constant   146, 617
Kepler's laws of motion   38, 330, 345-8, 374
Kepler's triangle   609
kinetic energy   17, 25, 37, 79, 83, 90, 91, 94, 117, 129, 132, 159, 162, 164, 166, 210, 219, 231, 234, 253, 289-90, 296, 300, 305-8, 329, 362, 364, 482, 496, 502, 507, 509, 555, 561, 572, 574, 576, 580, 601, 605, 614, 666
kink soliton   1, 14, 29, 31, 269
Klyuschin, J.   190, 226-7
Korteweg-de Vries equation   2, 35
Kristoffen, B.   366
Krizek, M. and Somer, L.   379
Krogh, K.   76, 79, 328, 364-6, 368, 446-7
Kuramoto model   457, 459

L

Lagrange (libration) points   348, 425, 498-9
Lagrangian   83, 90, 129, 143, 305-6, 343, 361, 377-8, 528-9, 631, 676
LaFreniere, G.
    gluon fields   288
    neutron model   290
    standing wave transformation   xi, 101-2, 104-110
    wave model   78, 83, 93, 101, 104, 328
Lahav, O. and Calder, L.   361
Lamb shift   13
Lambdoma matrix   610
Larmor precession   207, 231, 233
Landé g-factor   206
Laplace, P.
    equation   261, 309, 368
    force law   361
    gravitational potential   361, 363, 365 372
    operator   310, 368
    transforms   311-4
    equilibrium surfaces   310-1
Laplacian
    operator   219, 258-60, 353, 365, 368-9, 428
    wavefront curvature   309
Leahy, A.H.   350
Leibnitz, G.   17, 134
leptons   239, 247-9, 258, 277, 288
Lense-Thirring precession   233, 334, 572
Leuchs, G.   15

Levi-Civita,
    connection   52, 276
    effect   x, 220, 222, 224
levitation   319, 322, 324-7, 383-5, 483-5, 493, 500, 526, 532, 551, 553, 555, 557-8, 570-1, 576
Lie group   40, 260, 267, 271, 298, 613
Liénard-Wiechart potential functions   95
light,
    confinement   1, 35-6, 58, 83, 620
    curvature   x, 1-4, 43, 50-2, 76, 79, 80, 82
    deflection of   xi-xii, 76-9, 80-1, 101, 328, 441, 532, 592-3
    diffraction   65-70, 72, 111-2, 160, 291, 440-3
    reflection   35, 60-5, 68-70, 72, 83-4, 112-2, 413, 415, 432, 442-4, 512, 514, 551, 620
    speed of   ix-xi, 1-2, 5, 9, 11, 13, 19, 29, 33, 35, 47, 76, 78-9, 93, 95-6, 101-2, 105, 108-9, 128, 215, 218-9, 224, 226, 228, 237, 239, 289, 316, 328-9, 369, 371-2, 378-9, 391, 407, 422, 432, 441, 445-7, 463, 511, 515, 573, 577, 588, 593, 672
    torsion   1-4, 15, 17, 34, 38, 43, 50-1, 223, 226, 244, 404, 406, 416-8, 627
    wave   1-16, 29, 45-9, 54, 57, 60-75, 250, 362
Lissajous   126, 129, 263, 449, 467, 473-4
loop quantum gravity (LQG)   255, 267, 310, 407-8, 410
longitudinal mass   98
longitudinal wave   xii, 5-7, 160, 276, 353, 555
Lorentz, H.A.   98, 396
Lorentz boost   127, 308
Lorentz contraction   xi- xii, 17, 78, 83, 87-8, 90, 93, 101-3, 105, 108, 160-1, 210-1, 215, 217, 239, 278, 328, 349, 372, 424, 432, 437-8, 449, 479, 481, 577, 598, 600, 614
Lorentz Doppler shift   xi, 83, 88, 93, 101, 104, 108-10, 119, 128, 166, 215-7, 349, 449, 488, 529-32, 545-7, 551, 555-6, 561, 577, 590, 591
Lorentz factor   87, 93, 102-3, 105-6, 108, 128, 160, 163-4, 166, 211, 215, 230, 232, 278, 372, 385, 468-9, 483
Lorentz force   82, 121, 207, 224, 323, 341, 346, 656
Lorentz frame   121, 162, 332, 435
Lorentz group   250, 267
Lorentz-Fitzgerald transformations   xi, 39-40, 93-4, 98, 102, 104-10, 141, 225, 239, 260, 305, 614

Index

Lorentz invariance 13, 121, 315, 334, 515, 573
Lorentz-invariant theory of gravity (LITG) 334
Lorentz tensor 26, 401
Lorenz gauge 27, 260
Lucas, C. 148, 227, 361, 397-8
Lunar Laser Ranging tests 379
Luneberg lens 70, 73, 426

M

MacGregor, M. 148, 201, 283
Macken, J. 34, 83-4, 148, 158, 169, 183-4, 187, 195, 236, 372, 381-2, 388-9, 391, 394, 402, 415
magnetic charge 158, 646
magnetic field strength (intensity) 2-3, 6-8, 10, 38, 82, 145, 322, 338-9, 507, 585, 648
magnetic flux lines 16-7, 51, 76, 79, 92, 158, 160, 220, 224, 240, 322-4, 339-40, 441, 488, 511, 549, 585, 588, 592, 648-9
magnetic flux density 3, 6, 17, 51, 76, 92, 220, 224, 322-3, 339, 441, 488, 511, 549, 585, 588, 649
magnetic flux quantum (fluxoid) 16, 191
magnetic moment 15, 145, 149, 151-3, 159, 205-7, 223-4, 240, 278, 280, 284, 300, 303, 322, 571, 646
magnetic permeability 8, 28, 31, 33-4, 59, 76, 78, 92, 117, 160, 240, 319, 322, 328, 343, 441, 507, 529, 581, 650
magnetomotive force (MMF) 158-9, 649
magnetophoresis 319, 322
magnetostatic field xi, 158, 167, 261, 352, 396, 483, 594
magnetostatic field energy x, xii, 43, 154, 159, 160, 163, 167, 230-1, 244, 261, 391, 396, 483, 594, 649
magnetostatic pressure 160, 167, 650
manifold
    affine 18
    base 1, 80, 250, 402, 405-6, 663, 665, 667, 671
    Calibi-Yau 407
    catastrophe 675
    differentiable 52
    Riemann 254, 276, 316, 333, 435
    spacetime 16, 19, 50, 77, 93, 250, 254, 401, 429, 592
    supergravity 408
    tangent space 1, 76-7, 250, 369, 371, 406, 663, 665
    torus 148, 223
Martins, A. and Pinheiro, M. 98

mass
    active 93-4, 369, 371, 462, 498, 504, 509, 515, 532-3, 572
    complex 139, 141
    deficit angle 131, 164, 235
    energy 34, 93, 126, 133, 143, 149, 165-6, 219, 234, 279-84, 301, 303-4, 353, 378, 391, 452, 472,507
    gravitational xiii, 79, 132, 142-3, 253, 319, 329, 357, 363, 374, 378, 434, 443, 453, 594
    inertial xii, 16, 38, 83, 87, 98, 121, 124, 126, 132, 141-3, 232, 253, 329, 374, 378, 511, 557, 572, 579, 594, 634, 644
    passive 369, 462, 498, 504, 509, 515-16, 532, 572
    positive 15, 23-4, 29, 34, 132, 140, 142, 146, 419, 507, 574, 588-90, 596-8, 617, 622, 675
    reactive 94
    negative ix, 14-5, 19-20, 24, 29, 34, 128, 132-3, 135-6, 143-4, 146, 419, 421, 507, 526, 532, 571, 574-6, 588-90, 594, 596, 597-8
    scaling 359
master oscillator 349, 457, 462
matter wave xi, 70-1, 78, 84, 88, 93, 101, 109, 113, 116, 120-1, 124, 128-9, 130, 141, 160, 166, 210, 305-6, 328, 349, 457, 479, 484, 529-30, 555, 557, 561, 588
Maxwell, J.C.
    displacement current 25, 27
    equations 15, 20, 25, 33, 76, 226, 260, 266, 404-5, 511, 623-4
    gravitation 333, 348
Mead, C. A. 38, 83, 126
Merkaba 668
Meissner effect 264, 322, 484
metamaterial 4, 5, 8, 11, 73, 84, 111, 116-7, 123, 128, 131, 159, 267, 478, 529, 531, 555, 588, 591-4, 597-8
Metatron cube 667-8
metric
    coefficients 315, 381
    EM wave interference 18, 414, 419, 427, 592
    Minkowski 250, 315, 402, 408
    Riemann 276, 405-6, 434
    spacetime xii, 160, 276, 401-4, 406, 447
    tensor 79, 315-6, 328, 364, 372, 399, 401-5, 408, 419, 427, 434
Meyl, K. 299-300
Michaud, A. 173
Midy, P. 380

695

Index

Mikheyev-Smironov-Wolfenstein (MSW) effect   300
Minkowski spacetime   40, 76, 93, 121, 127, 250, 310, 315, 371, 402, 408, 414, 472, 601
Möbius   141, 677
mode
  cut-off   452, 455, 535-6, 542
  density   121, 592, 597
  electromagnetic   7, 24, 35, 60, 91-2, 453
  even-odd   452-3, 462, 503, 511
  locking   328, 391, 457, 462, 555-6, 574
  whispering gallery   507-8, 529, 531, 551-2, 574, 588, 598, 619-20
modulation
  amplitude (AM)   111, 115, 497-8, 504, 519, 600
  anharmonic   462
  carrier   115
  cross   525
  envelope   8, 115, 118, 119, 508
  frequency (FM)   115, 462, 497, 504, 511, 515, 522, 555, 557
  harmonic   518
  phase (PM)   9, 55, 115, 122, 487, 497-8, 504, 551
  pitch   602, 607, 615
  polarization   599
  refractive index   113
  quadrature (QAM)   497-8
  spectral energy density   xiii, 511
  standing wave   84, 87, 111, 113, 118, 119, 486, 490, 555-6
Möller scattering   209
Moiré
  interference fringe pattern   70, 332, 427-8
  wavefront   426-8
momentum
  angular   ix, x, 1, 15-8, 29, 34, 37, 39, 42, 47, 50, 132, 149, 151, 166, 205-7, 220, 228, 239, 243, 246-7, 250, 255-7, 267, 284, 360, 378, 380, 410, 489, 507, 558, 583, 593, 623, 629, 663
  canonical   26, 240, 656
  energy-momentum   8, 29, 221, 316, 401-2, 414, 507
  flux   317, 402
  linear   17, 28, 39, 132, 142, 223, 253, 255, 299, 300, 302, 353, 414, 507-8, 548, 593, 629
monad   19, 20, 22, 146, 610
monadnock   546
Mössbauer effect   446, 463, 470-1
Mossotti, O.   319-20, 396

muon (mu-meson)   49, 278, 282-3, 285, 288
muonium   283

N

Nambu, Y.   202
N-body gravitation   347-8
n-gon   229, 271, 387, 596, 598, 605, 621
negative curvature   xv, 1, 19-20, 52, 146-7, 353, 420, 434, 598, 621, 666
negative frequency   48, 482, 497-8, 502-9
negative energy   13, 16, 20-1,34, 48, 111, 122, 132, 146, 242, 260, 267, 353, 379, 416, 419-20, 496, 502, 506-8, 526, 532, 592-3, 595, 597-8, 617, 621
negative index metamaterial (NIM)   123, 128, 131, 526, 529, 532, 574, 577, 622
negative mass   ix, 14-5, 19-20, 24, 29, 34, 128, 132-3, 135-6, 143-4, 146, 353, 362, 419, 421, 507, 526, 532, 571, 574-7, 588-9, 590, 594, 596-8
nephroid (epicycloid)   271
neutrino
  absorption   296-7
  anti   296, 299-300
  electron   296-7, 299-301, 303
  flux   299-300
  muon   299-301, 303
  properties of   301, 303
  spin flip   304
  tau   299, 301, 303
neutron   22, 290, 243, 295, 296-7, 380, 676
Newton's laws
  gravitation   142, 227, 330-4, 361
  first law   x, 110, 162
  second law   83, 110, 124, 142, 374
  third law   38, 110, 253
Noether, E.   39, 666
Nyambuya, G.   368

O

octonions   266, 665, 668, 672, 697
Oldershaw, R.   34, 360
Ortho-positronium   279
oscillator,
  acoustic   91-2
  arrays   482
  cavity   84, 91-2, 487-92, 501
  compromise frequency   460-2, 470
  coupling   360, 457-9, 460, 463, 471, 498, 605
  electromagnetic   91-2, 487-92, 501
  electromagnetic cavity   91-2

Index

oscillator (cont)
  interference   xii, 350-1, 352-3, 355-6, 409, 413-5, 426-8, 442, 462-7, 473-9
  mass   124, 141, 329, 415, 442, 444, 450-1, 459, 461-3, 470, 496, 605
  master   349, 457, 462
  mechanical   92
  phase difference   128, 329, 350, 457, 460-1, 465, 467-9, 473, 487, 489
  phase-lock   84, 470, 459
  slave   349, 457, 462, 489-90
  synchronization   457-60, 462, 474, 522

P

Pachner move   440
pair production and annihilation   ix- x, 148, 208, 220
pair production,
  cross-section   208
  wave phasor diagram   210-1
parametric amplifier   113, 121, 126, 128, 486, 488-90, 532, 551, 561
parametric pumping   113, 117, 122, 555, 561
Para-positronium   279
partial standing waves   88, 210, 214, 352, 354, 486-7, 557
particle types   248
Pascal's triangle   599
passive mass   369, 462, 498, 504, 509, 515-6, 532, 572
Pauli, W.
  exclusion principle   247
  fine structure   229
  graviton   404
  spin matrices   267
Penrose, R
  Bronstein-Penrose cube   623, 672
  fine structure constant   229
  twistor theory   261
period
  coasting   x
  fundamental   454
  orbital   278, 331, 345, 347, 374, 611
  pulsation   104, 349, 352, 447, 517, 519, 619
  rotation   155, 158
  storage   5, 162
  dimension   635
permeability, magnetic   8, 28, 31, 33-4, 59, 76, 78, 92, 117, 160, 240, 319, 322, 328, 343, 441, 505, 507, 581, 597, 650
permittivity, electric   xii, 33, 56, 59, 76, 78, 92, 117, 160, 225, 240, 305, 328, 343, 382-4, 396, 398, 441, 505, 507, 581, 597, 651
Petit, J.P.   143, 380
phase
  accumulation   329, 391
  alignment   17, 19, 29, 133, 144, 151, 246, 385, 391, 465
  angle   88, 90-7, 114, 125, 129, 210, 251, 263, 626
  coherence   458, 605
  conjugation   111-3, 117, 124-5, 417
  conjugate mirror (PCM)   60-1, 64-5, 67, 86, 111, 116, 118-120
  conjugate phased array   122, 485-93, 501, 532, 555, 558, 561, 591, 594
  conjugate phasor   114, 124-5, 263-4
  conjugate wave (PCW)   111, 113
  displacement   84, 87, 323, 375, 448, 463, 473
  E & H relationship   7, 97-7
  geometric   18, 249-252
  interference   350, 352, 355-6
  lag   7, 96-7, 321, 448, 487
  lock   22, 118, 166, 246, 289, 555, 561, 570, 605
  locking boundary   232
  modulation (PM)   35, 115
  phase-locked resonator   83-88, 93, 98, 111, 116, 118-120, 122, 126-7, 131, 375, 570
  quadrature   5, 22, 84, 91, 97, 289, 293, 467
  rotation   114, 125, 259, 289, 414, 416, 426
  shift   117, 126, 128-9, 329, 365, 427, 448-9, 463, 465-9, 555, 587, 593-4, 675
  space   307-8, 624, 676
  synchronization   351, 448, 458-9
  velocity   2, 4, 7-10, 33, 35, 57, 65, 78, 82-4, 87, 95, 103, 108, 117, 141, 328, 413, 487
phasors   19-20, 88, 90, 111, 114, 124-6, 129, 210-1, 217, 263-4, 496
Phi ratios   48, 228-9, 360, 607, 610, 618, 663
photon   ix, x, xii, 1-3, 12-14, 16, 18, 29, 32, 35, 37-8, 50-1, 58, 60, 76, 78-9, 83-4, 87-8, 98, 111, 148-52, 166, 208, 210-11, 220-1, 222-224, 237, 263, 299, 304, 323, 357, 400, 407, 409, 413-18, 429-30, 432, 442, 445-7, 470, 502, 515, 525, 561, 589, 591, 599, 600
physical (quantum) vacuum   ix-xiv, 1, 5, 7, 9, 13-16, 18-35, 37, 45-7,49, 51, 56-8, 71, 76, 78-83, 94, 96-7, 101, 111, 121,

697

Index

physical quantum vacuum (cont)
  132-36, 141, 146-7, 155, 158, 160,
  166, 168, 210, 224, 235, 237, 241,
  276, 289, 299, 317, 319, 323, 328-9,
  353, 357-65, 369, 371-4, 378-9, 381-
  5, 389, 391-3, 396, 398, 402, 404-09,
  411, 413, 416, 419-23, 429-32, 435-7,
  441-2, 446, 452-3, 462, 502, 509, 571,
  592-9, 600, 603, 637, 653, 673,675-7
pilot wave 94, 217
pion (pi meson) 278, 280-1, 284, 286,
  290, 300
Planck
  aether ix, 13-4, 25, 34, 132, 138, 141,
    319, 329, 419-20
  constant, 2, 16-8, 37, 39, 42, 44, 47,
    79-80, 83, 87, 110, 141, 151, 162,
    201, 208, 215, 228, 244-5, 247,
    260, 307, 380, 388, 400, 416, 450,
    623, 663, 672
  reduced constant 2, 18, 21, 37, 42, 47,
    201, 228, 239, 243, 388, 400
  charge 30, 137, 170, 202, 245, 605
  dipole(s) ix, 1, 14, 16-9, 21, 23, 29, 32,
    34-7, 39, 57, 121, 132-3, 135, 137,
    141, 144-7, 151, 155, 158-9, 166,
    202, 235, 246, 276, 289, 294, 299-
    304, 362, 385, 391, 419-21,
    507,571, 592-4, 597-8, 600, 676
  energy ix, 18, 29, 42, 79, 132-3, 147,
    299, 353, 380, 594
  force 13, 42, 147, 201, 298, 379, 391,
    402
  frequency 13, 45-7, 49, 57, 133, 147,
    201, 235-6, 357, 413, 432, 601, 603
  impedance 8, 33, 44, 147, 677
  length 3, 13-4, 30, 34, 132-3, 146-7,
    201, 235, 300, 380, 389, 391, 507,
    593, 598, 617
  mass xiii, 7, 13-21, 29, 30-2, 35, 37,
    47, 98, 132-3, 135-8, 144, 146-7,
    155, 235, 303, 317, 353, 362, 379,
    385, 407, 419, 424, 450-1, 571,
    592-4,597, 617, 677
  time 22, 41, 132, 147, 334, 379-80,
    389
plane waves 1, 5, 7, 35, 71, 293, 473, 594
polarizability 13, 54, 56, 76, 318, 321,
  441
polarization,
  nonlinear 320, 389, 440, 462, 485,
    509, 593, 597
polaritons 117, 141, 207, 411, 588
polarons 276
poloidal (anapole) magnetic field 158,
  205, 240
poloidal radius 152, 188

poloidal spin x, 240, 299
Poincaré,
  disc 621
  gravitation 334, 348
  group 40, 298
  sphere 1, 9, 12, 250, 252
  transformations 39
positronium 279, 288
positrons ix-x, 19, 21, 148, 211, 221-22,
  247, 249, 278, 288-9, 290, 295-6,
  298-9, 357
potential energy x, 17-8, 26, 39, 79, 84,
  90-1, 129, 142, 149, 210, 226,
  231, 260, 289-90, 307, 308, 318-9
  329, 346, 364, 375, 470, 472, 479,
  482, 496, 502, 509, 533, 573-4, 576,
  580, 605, 614, 645, 649
Pound, Rebka, Snider experiment 446,
  470-1
power
  apparent (real) 654
  average 515, 518
  blackbody 552
  circulating 158, 184, 236
  definition and units of, 110, 158, 519,
    633
  electrical 158-9
  frequencies 57
  law 347
  luminous 661
  peak 517
  Planck 202
  radiant 659
  radiated 236, 518, 547,552, 554, 587,
    659
  reactive (true) 654
  spectrum 518-9
  loss 59
  series 20, 312, 320, 378, 663
  spectral 660
Poynting vector 5, 8, 14, 29, 65, 79, 82,
  94, 96-7, 117, 145, 329, 344, 426,
  428, 432, 456, 487, 496, 502, 505,
  509, 524, 551, 575-6, 579, 605, 631,
  656
precession
  charge-to-mass xii, 207
  de Sitter (geodetic) 231, 195
  electric charge 16, 34, 41, 150, 152-4,
    158, 168, 207, 223 230, 234, 238-9,
    246-7, 272,289, 290, 393,405, 593,
    624, 669, 671-2
  electron spin 16, 32, 141, 150, 152-4,
    158, 168, 193, 207, 223, 230, 238-
    9, 246-7, 272, 283, 289-90, 393,
    405, 593, 624, 669, 671-2

Index

precession (cont)
  fine structure constant   xii, 154, 204, 207, 246-7, 393
  gravitomagnetic   195, 231
  gyromagnetic spin   154, 179, 207, 247
  gyroscopic   253, 255
  Lamor   195, 207, 231
  Lense-Thirring   233, 334, 561
  orbital   133, 140-1, 164, 168, 193, 242, 252, 324
  perihelion advance   227, 334
  quadrapole   310
  retrograde   391, 231-2, 391
  spin   239-40, 246-7, 289, 299, 302-3
  Thomas-Wigner   154, 194, 207, 231-2, 252
proper acceleration   127, 472
proper frequency   111, 614
proper length   107, 109, 234, 371-3, 381, 445-6, 614,625
proper Lorentz transformations   40
proper mass   16, 381, 614, 625
proper time   16, 18, 93, 107, 109, 127, 160, 166, 371-3, 381, 391, 445-7, 614, 625
proton   288, 290, 292, 295-7, 299, 384
pump beams   69, 112-3, 116, 124, 126, 128, 486-7, 490, 530, 551, 556, 561, 597
pump wave   x, 113, 117-21, 130
push-pull cavity   xi, 85, 89, 119, 124, 501
Puthoff, H.   76, 78, 121, 133, 224, 328, 378, 441
Pythagoras theorem   591, 599
Pythagorean harmonic intervals   134, 349, 426, 601, 607-11, 613-4

Q

quantum chromodynamics theory (QCD)   275, 288, 295
quantum electrodynamics (QED)   166, 208-9, 260, 407, 624, 672
  diagrams   209
quantum gravity   255, 267, 310, 407-12, 427-8, 430-1, 433, 440, 623-4, 672
quantum field   ix-x, 15, 76, 400, 407-8, 411, 414, 623-4, 672
quark
  anti-   18, 270-1, 275, 288, 295, 298, 669
  charge   239, 247, 256-7, 270-2, 274, 277, 289-90, 295-7, 407, 669
  color   256-7, 275, 277, 288-9, 295, 298
  confinement   288
  decay   273
  generation   272-4
  flavor   273, 295
  mass   272-3, 277, 290, 295
  spin   277, 289, 293-4
quaternions   265-7, 671
quintessence   xv, 34, 361, 399-400

R

radiation
  blackbody   45, 432-3, 451, 542
  Cherenkov   8, 149, 299
  dipole   ix, 88, 91, 95-7, 99-100
  force   86-7, 98, 119, 389, 489-90, 526, 529, 551, 553, 561
  impedance   13
  pressure   8, 82-4, 110, 116, 119-120, 124, 126, 323, 329, 375, 436, 483, 487, 541, 550-51, 645-6
  resistance   5, 97, 239, 512, 677
  synchrotron   149
  zero-point   121, 123
Raman scattering   111, 443-4
rapidity   xii, 93, 127-8
Rayleigh criterion   291
Rayleigh-Wood zone plate   70
redshift
  cosmological (stellar)   361, 382
  Doppler   108-9, 215-6, 489
  gravitational   371-2, 380-2, 445-6, 470
redshift
  Hubble   227
  Tift   397
reflection coefficient   126, 352
refractive index   ix, x, xii, 4, 9, 16-7, 35, 54, 56, 65, 67-8, 70-1, 73-4, 76, 78-9, 82, 94, 113, 117, 131, 133, 20, 224, 256, 328-9, 357, 359, 369, 372-3, 382, 407, 413, 420-1, 426, 430-3, 441, 451, 456, 497, 503, 509, 590, 592, 599, 619, 622, 637
resonance   9, 16, 35, 57, 59, 65, 117, 122, 160, 166, 218, 228, 235, 237, 288, 300, 321, 347, 379, 389, 391, 393, 408-9, 413-5, 426, 482, 551-2, 555, 559, 561,601-2, 604, 607-8, 611, 620
resonator   x-xii, 8, 16, 35, 65, 71, 83-4, 87-9, 91-2, 95, 98, 103, 110-1, 113, 116-22, 124, 126-8, 131-2, 141, 147, 166, 210, 213, 329, 352, 465-7, 479, 481, 484-6, 495, 497-8, 501, 504, 507-8, 512, 514, 527-9, 529-31, 551, 555, 559-61, 570, 588, 590, 593, 596, 598, 620
rest mass   x, 16, 21, 29, 40, 78-9, 83-9, 116,140-1, 148-9, 154, 166, 220,223, 234, 240, 248, 254, 296, 364, 366, 375, 379, 381,393, 400, 419, 450, 487, 511,557, 614

Index

rest mass energy 126, 149, 166, 219, 234, 391, 450, 507
ribbon, twist xii, 32, 148, 150, 256-7, 275, 289
Ricci curvature 276, 316-7, 401, 406, 434-6, 440
Riemann curvature 276, 316, 402-3, 406, 434-5
Riemann manifold (metric) 254, 276, 316, 400, 435, 440
relativity
  discrete (DSR) 251
  general (GR) xii-xiii, 9, 76, 78, 166, 316, 328, 361, 372, 378, 401, 441, 445, 588, 623-1, 671-2
  special (SR) 93, 98, 162, 166, 260, 332, 372, 614, 624, 671-2
Rindler frame acceleration 93, 121, 470
Rivas, M. 148
rotating reference frame 38, 233-4, 437
Roxburgh, I. 379

S

Saari, D. 348
saccade 544
Sagnac effect, 120
  inverse xi, 120, 486, 574, 588
Sakharov 133, 378
Sanchez-Soto, L. 15
scalar
  curvature 82, 317-8, 402, 407, 435
  potential field 14-6, 18, 25-8, 43, 76, 78-9, 82, 95, 98, 126, 245, 267, 306, 310, 320, 334-5, 337, 341, 354, 364-9, 405-6, 414, 421, 432-3, 512, 574-6, 578, 580-1, 590, 614, 617, 636
Scharnhorst effect 13
Schrödinger wave equation 37, 353
Schwarzschild radius 18, 80-2, 133, 146, 148, 193, 235, 332, 360, 381, 454, 677
Schwinger
  critical field intensity 150, 223, 324
  correction 161, 231
Shepard tones 511, 526-7, 603, 607
Shilling, O. 186, 189
Shipov, G. 254
Shulgin, V. 320
sine-Gordon 2
skyrmions 271, 583
slave oscillator 349, 457, 462, 489-90
Smith chart 135, 139-41, 677
soliton(s) ix, 1-2, 9, 13-5, 29, 31-2, 35-6, 58, 84, 98, 257, 270, 290, 412-3, 486-7, 592-3
Sommerfeld 229

Soret zone plate 70, 72
spacetime
  impedance 235, 330, 378
  interval 16, 39-40, 83, 116, 124, 167, 251, 316, 403, 441, 448
  metric xii, 18, 51-2, 79, 161, 251, 255, 277, 308, 316-7, 329, 365, 373, 381-2, 290, 400, 402-7, 409, 415, 420, 422, 428, 435-6, 448, 505, 592, 601, 607
  stiffness 378
Spears, M. 38, 375, 384-6
special relativity (SR) 94, 98, 260, 332, 372, 614, 624, 672
spectral energy density xii, xiii, 24, 48, 121, 488, 496, 523, 525-8, 532, 536, 594, 598-9, 659
speed of light ix-xi, 1-11, 14-6, 19, 29, 33-35, 38, 47, 51, 76, 78-9, 82, 93, 95-6, 101, 128, 133, 146, 149, 160, 212, 215, 218-9, 224, 226, 228, 239, 289, 299, 302, 316, 328-9, 364, 369, 371-2, 375, 378-9, 388, 391, 394, 398, 401, 407, 422, 432, 441, 445-8, 462-3, 467, 470, 511, 522, 524, 529, 573, 588, 592-3, 597, 663, 672
spherical rotations 251
spin
  of bosons 1, 2, 4, 149, 210, 277, 413
  connections 590
  dimensions of 1, 18, 31, 37, 39, 47, 388, 629
  glass 15
  Hall resistance 200, 246
  of electron 149, 151-6, 168, 171-2, 277
  of graviton 277, 404, 408, 413, 415-8
  of leptons 277
  of neutrinos 277, 299, 301-4
  of photon 1-4, 223, 277
  of quarks 256, 274-5, 277
  poloidal xii, xii, 149, 151-3, 155, 240
  networks 15, 122, 255-6, 261, 408, 410-12, 598-9, 600
  toplogical 15, 52
  toroidal x, 149-159, 168, 171-3, 205, 240, 299, 389
  precession xi-xii, 16, 34, 154, 205, 207, 227, 230-32, 235, 239-40, 246-7, 253, 255, 334, 347, 361, 391, 400, 405, 437, 572, 593, 596, 605, 624, 663, 669, 671-2
spin angular momentum ix-x, 1, 5-6, 18, 29, 34, 37, 39, 42, 47, 51, 132, 149, 151, 166, 205-7, 220, 228, 239, 243, 246-7, 250, 255-6, 267, 284, 360, 378, 410, 507, 558, 593 619, 629

Index

spin wave ix, xiii, 2, 19, 21, 23, 31, 35-6, 147, 151, 156, 159, 238, 246-7, 317, 362, 389, 391, 431, 575-8, 588, 592
spinors x, 13, 84, 135, 148, 153, 267-8, 410
spira mirabilis field 2, 19, 21, 23, 31, 35-6, 147, 151, 156, 159, 238, 246-7, 317, 362, 389, 391, 432, 495, 507-8, 575, 588, 621
standing wave
  boundary effects 351-2
  moving contracted xi, 86, 90, 93-4, 102, 111, 118-9, 128, 130, 166, 210, 212-3, 466, 485, 487-93, 557, 570, 614
  interactions x-xii, 83, 90, 306
  interference 98, 288, 350-1, 428, 463
  levitation 483-5
  mode-locking 328
  partial 88, 210, 214, 352, 354, 547
  ratio (SWR) 102, 105, 117, 315, 352
  resonator 16, 65, 83-4, 86, 88, 90, 103, 110, 118-20, 17-8, 131, 141, 147, 166, 210, 213, 479
  system acceleration 86, 470, 475-6
  transformations xiii, 90, 101-2, 104, 108-10
Sternglass, E. 18-9, 22, 278, 284-7, 290, 299, 359, 380
stiffness 34, 122, 126, 146
Storti, R. 76, 300, 369, 416, 429, 441, 451-6, 462, 515, 524, 594
strain amplitude (resonance ratio) 218, 234-5, 347, 389, 391
strain xiii, 56, 126, 234, 239, 261, 315-6, 389, 391, 426, 484, 492, 498, 526-7, 530, 557, 593
stress xii, 17, 29, 126, 239, 261, 315-7, 401-4, 406, 422, 429, 435, 472, 484, 492, 498, 507, 527, 530, 557, 633
string theory 148, 166, 250, 407-8, 410, 414, 624
strong force 275, 288, 290, 298-9, 406
supersymmetry,
  theory 210
  transformations 210
surreal 258, 675
susceptibility 35, 53-5, 111, 322, 485, 651
symmetry
  Abelian 250
  action-reaction 142, 477
  breaking xiv, 20, 35, 122, 132, 142, 162, 239, 259, 264, 298, 409, 411, 416, 465, 477, 624, 675
  charge (C) 249, 290, 668, 670
  charge parity (CP) 290, 299

chiral 10, 15, 117, 227, 247, 271, 302, 304, 412, 526, 597
conformal scale 360
dual 573-5, 666
gauge 27, 250, 259, 624
global 624
groups 250, 260, 263, 295, 298, 380, 666-7, 676
inversion 54
local 258, 263
non-Abelian 250
parity (P) 290
space-time 39-40
transformations 39, 210, 247, 258-9, 268, 463, 624, 666
synthesized Lorentz-Doppler 485-6, 488-9, 491-4, 496-9, 530, 555, 560-1, 591
synthesized matter wave xi, 113, 116, 120-1, 124-5, 484, 492-3, 498, 530, 561

T

Taylor series expansion 53, 111
tangent space (manifold) 1, 50, 52, 76-7, 250-2, 369, 371, 382, 406, 430-3, 505-6, 510, 624, 663, 665
tangent vector 1, 17, 50, 52, 250, 316, 412, 435
tauon (t-meson) 49, 278, 287-8
tensors
  contravariant 26, 315-6
  covariant 26, 52, 315, 523, 573
  curvature 52, 80, 276, 316, 402-3, 434-5
  Kronecker 312, 315
  metric 79, 276, 315-6, 328, 364, 399, 401, 427, 447
  Ricci 276, 316-7, 401, 406, 434-6, 440
  Riemann 80, 276, 316, 401-3, 405-6, 434-5
  Weyl 276, 316, 434-6, 440
Tesla, N. 134, 239, 349, 649
tesseract graph 667, 671
tetrachord base 457, 609
tetrad 1-3, 146, 368, 401-2, 406
Tewari, P. 14, 148
Thomas, L.H. 231
Thomas-Wigner precession 154, 194, 207, 232-3, 252
Thomson, J.J. 234
Thomson, W. (Lord Kelvin) 349
Tipler, F. 365
Tisserand, F. 334
topological charge x, xii, 16-7, 32, 37, 39, 41, 52, 220, 239-40, 242, 246,

701

Index

topological charge (cont) 249, 256, 258, 269-72, 274, 276, 289, 626, 666
toroidal spin   x, 149-156, 158, 168, 171-2, 205, 240, 299, 389
torsion   x, 1-4, 13, 15, 17, 24, 29, 34, 38, 43, 50-1, 52, 220, 223, 226, 239, 244, 250, 273, 276, 339, 342, 391, 401, 404-6, 416-8, 482, 507-8, 574-5, 578, 580, 588, 593, 597, 617, 627, 666
torus
   curvature   52
   Euler characteristic   52, 199
   horn   50, 52, 150
   manifold   148
   of revolution   xi, 50, 150-3, 155, 223
   ring   148, 150, 299
   spindle   52, 150, 189
   tonnetz   606, 613
   winding trajectory   611
tractor/repulsor beam   8, 486-7, 489-90, 493-4, 558
transverse electric (TE)   8, 60, 91, 652
transverse electromagnetic (TEM)   7-8, 60, 91
transverse magnetic (TM)   8, 60, 91, 300, 652
transverse waves   103, 160, 226, 555
Trautman, A.   368
Tuisku, P.   18
Tusci couple   171
twist ribbon   32, 150, 256-7, 289

U

units and definitions of,   33-4, 37-8, 41-44, 169, 215-6, 239-245, 625-662
unit circle   19-20, 114-5, 125, 259, 264, 269-71
unit speed curve   1, 3
unit sphere   20, 140
Unruh temperature   121-3, 472
Urban, M.   16

V

vacuum
   energy   xi, xiii, 14, 34, 101, 133, 316, 319, 362, 379, 389, 402, 414, 588, 592, 597
   impedance   13, 33, 44, 82, 96-7, 135, 220, 228, 241, 245, 594, 597, 653, 677
   nonlinear polarization   320, 329, 389, 392-3, 419, 421, 423, 462, 509, 593, 597

index of refraction   x, xi, 4, 14, 16-7, 29, 33, 35, 51, 57-8, 71, 75-6, 78-9, 81-2, 94, 132, 160, 220, 224, 328-9, 359, 369, 372-3, 381, 407, 413, 420-1, 426, 429, 431-3, 441,444, 446, 451, 456, 503, 509, 590, 592, 599, 637
vector potential   ix, xi, xiii, 4, 14-7, 25-6, 27-8, 32, 34-5, 37-8, 43, 58, 71, 76-9, 81-3, 94-5, 98, 124, 126, 159, 162, 239, 240, 244, 259-60, 266, 305, 329, 333-4, 340, 353, 365-9, 372-3, 381, 391, 405, 407, 424, 431-3, 456, 503, 507, 509, 526, 574, 578, 582, 584, 590-2, 599, 614, 631, 655, 670
velocity
   escape   160, 166, 331, 371-2, 507, 513, 533
   group   2, 4, 7-9, 11, 33, 35, 37, 57, 65, 82, 84, 87, 95, 103, 105, 108-9, 329, 341, 416, 448, 466-7, 483, 487, 491, 556, 577, 588, 597
   phase   2, 4, 7-9, 11, 33, 35, 57, 78, 82-4, 87, 95-6, 103, 105, 108-9, 117, 141, 328, 413, 449, 466-7, 470, 484-7, 489-93, 511, 577, 594, 597
   ratio   xi, 44, 82, 87, 102,108, 119,128,160,163-5, 215, 245, 449, 468-9, 563
   staircase   85, 127
verdet constant   224, 578, 580, 588
verdrehung   564
vortex   14, 16-7, 132, 135-6, 138, 145, 221, 246, 256, 270, 299, 300, 302-4, 379, 507, 590-1, 593
vorticity   132, 246, 583, 588, 591, 635

W

W boson   248, 296-7
wave
   contracted moving standing   xi, 86, 88, 90, 98, 102, 104, 118-9, 128, 130, 166, 210, 212-3, 215-7, 465, 485-6, 489-91, 494, 496, 498-9, 500, 528, 530, 555, 557, 560, 614
   envelope   8-9, 35, 38, 86, 105, 119, 210, 213, 352, 407, 607
   equation   2, 20, 89, 161, 260, 309-10, 334, 344, 353, 365, 368, 413-4
   group velocity   2, 4,7-9, 11, 33, 35, 37, 57, 65, 82, 84, 87, 95, 103, 105, 108-9, 329, 341, 416, 448, 466-7, 483, 487, 556, 588, 597
   interference   x, xii, 60, 68, 70, 76, 111, 238, 289, 293-4, 350-1, 353, 355-6,

Index

wave interference (cont) 413, 427-8, 430, 432, 440-1, 443, 463-4, 470-1, 473, 477, 479, 621
longitudinal xi, 5-7, 160, 276, 353, 487, 555
matter xi, 70, 78, 84, 88, 93, 101, 109, 113, 116, 120-1, 124, 128-9, 141, 160, 166, 210, 305-6, 328, 349, 457, 479, 498, 529, 530-1, 555, 557, 561, 588
node displacement 84, 465
number 1-6, 10, 39, 31-2, 33, 37, 39, 47, 61-64, 67-70, 81-2, 86-7, 96, 98, 218-9, 378, 635
packet 1-2, 38, 93, 588, 598
phase velocity 2, 4, 7-9, 11, 33, 35, 57, 78, 82-4, 87, 95-6, 103, 105, 108-9, 117, 141, 413, 449, 466-7, 470, 484, 486-7, 489-90, 511, 594, 597
phase difference xi, 7, 27, 84, 128, 250, 252, 256, 289, 300, 329, 350, 427, 457, 460-1, 465-9, 473, 485, 487, 489, 498, 561
phasor 19-20, 88, 90, 111, 114, 124-6, 129, 210-1, 217, 263-4, 496, 624
standing x-xiii, 16, 21, 23, 83, 88, 90, 217, 498-9, 500, 514, 527-8, 530, 551-2, 554, 555-7,560-1,570, 573, 575, 587, 593, 601, 607-9, 614-5, 635
transverse xi, 1, 5, 7-10, 60, 88, 91, 95-8, 102-3, 108-9, 160, 226, 276, 555, 652
travelling x, xii, 1-4, 18-9, 21, 83, 88-90, 124, 126, 140-1, 150, 210, 223, 237, 323, 325-7, 351, 362, 433, 467, 479, 482-8, 529, 557, 574, 598, 608
vector 1-6, 10, 31-3, 37, 39, 47, 61-64, 67-70, 81-2, 86-7, 96, 98, 162, 218-9, 378, 415, 426, 493, 558, 598, 617, 635
wavefront x, xii-xiii, 1-2, 5, 17-8, 60-4, 68, 73, 76, 78-82, 84, 91, 95, 100, 102-3, 111, 122, 156-7, 238, 269, 289, 323, 409, 412, 426-7, 429-30, 432, 441, 488, 507, 592, 597
wave function 15, 17, 25, 29, 89, 126, 133, 160, 164, 260, 267, 270, 276, 307, 309, 344, 413, 462-3, 554, 612
wavelength
Compton xii, 14, 39, 83-4, 98, 110, 149, 151-2, 156, 169, 212, 215, 218, 220, 223-4, 236, 238, 288, 478
de Broglie 84, 87, 102,110, 141, 212, 216-8, 487, 514, 544, 548, 569
Doppler 102, 108-9, 118, 215-16, 564

Weber, W. 226, 396
constant 29, 226
force 226, 334
inertial mass 16, 634
Weber's law 226-7
units 340, 648
weight 29, 349, 369, 374, 436, 551
Wesley, J. 148
Weyl,
curvature tensor 276, 316, 434, 435-6, 440
gauge transformation 259
tidal deformation 435, 436-7
Wick rotation 250, 254, 260
Williamson, J. and van der Mark, M. 148
Wilson, H. 76, 328, 441
Winterberg, F. ix, 14, 16, 25, 34, 132, 135, 138, 148, 299, 353, 379, 419-20
whirl number xii, 154, 166, 205, 240
Whittaker, E. T.,
force between moving charges 226-7
scalar potential 413
scalar wave 429
winding number 16, 220, 270, 289, 295, 611, 613, 626

X

X-ray 45-6, 68, 111, 357, 443, 451, 551-2

Y

Yank 629

Z

Z boson 248
Zeeman interactions 224
Zel'dovich, Y. 323, 361
zero curvature vacuum 1, 4, 16, 19, 51, 78-9, 83, 133, 141, 210, 328, 381-2, 404, 420, 435, 446, 499
zero-point
energy 13, 20, 29, 34, 121, 132, 378, 516
field (ZPF) 13, 121, 133, 319
fluctuation 166, 429, 516
radiation spectrum 121, 123, 452
Zitterbewegung
frequency 149, 151, 166, 205, 210, 218, 220
motion 207, 267, 378
radius 151-3, 198, 231
Zollner, J. 396
zoom-orbit whirl 39, 154, 166
zone plate 70, 72, 288, 473
Zou Yan 134

NOTES